ORGANELLE HEREDITY

Organelle Heredity

Nicholas W. Gillham, Ph.D.

Professor of Zoology
Department of Zoology
Duke University
Durham, North Carolina

Raven Press ▪ New York

Raven Press, 1140 Avenue of the Americas, New York, New York 10036

Made in the United States of America

Library of Congress Cataloging in Publication Data

Gillham, Nicholas W
  Organelle heredity.

  Includes bibliographical references and index.
    1. Cytoplasmic inheritance. 2. Mitochondria.
3. Chloroplasts. 4. Cytogenetics. I. Title.
[DNLM: 1. Genetics. 2. Chloroplasts.
3. Mitochondria. QH452 G4790]
QH452.G54   575.2′1   75–43195
ISBN 0–89004–102–4

Preface

Within the past decade, the study of the genetics and biogenesis of chloroplasts and mitochondria has been transformed into a mature scientific endeavor. What began as black-box science has become decipherable. During that time I have had the good fortune to be engaged in research in this area. A major reason for the emergence of organelle heredity as a respectable field of scientific inquiry has been the discovery that chloroplasts and mitochondria contain DNA of their own. This discovery has given credence to the notion that cytoplasmically inherited genes affecting the phenotypes of these organelles may actually be located within them.

Although this book deals at length with organelle DNA, with the processes of transcription and translation within chloroplasts and mitochondria, and with the biogenesis of these organelles, it is basically a book about genetics. This emphasis reflects my own prejudice toward genetic solutions of biologic problems, as well as my belief that genetic arguments tend to be more difficult than most for the non-geneticist to appreciate. It is my opinion that the techniques and tools of the trade of organelle genetics have now been honed to the point that they have become highly effective and are unlikely to undergo major evolution in the future. Despite the seeming novelty of the field, the underlying genetic concepts are not really very different from those applied in transmission genetics generally. Organelle genes are part of the hereditary repertory of virtually all eukaryotic cells. Like other genes, they mutate, and in appropriate circumstances they recombine and can be mapped. These are subjects we will consider in detail. We will also consider the question why organelle genes are there in the first place: What do they do?

N. W. Gillham

Contents

Acknowledgments

The most rewarding period of my scientific life began 10 years ago when I undertook a collaboration with my good friend and colleague John Boynton that continues to flourish today. This collaboration has benefited this book both directly and indirectly in ways too numerous to mention. During this period Elizabeth Harris joined our group as a research associate, and her counsel, wisdom, and thoughtful experimental design have been appreciated by all. One of the things that has meant the most to all three of us has been our good fortune in having stimulating and productive postdoctoral fellows and graduate students in our laboratory. They have made important contributions to this book in terms of their results and their ideas. They have had the good sense to disagree with us frequently, and spirited debate has ensued. Speaking for myself, I know that more than one cherished hypothesis has crashed in flames. They include Paul Bolen, Jean Forster, David Grant, Bob Lee, Andy Wang, and Ed Wurtz, who have been or presently are postdoctoral fellows in our laboratory. As students we have been fortunate to have Mike Adams, Nancy Alexander, Mary Conde, Barb Sears, Hurley Shepherd, Karen Swift, Andy Wiseman, and most recently Alan Myers, all of whom elected to do graduate work with us. We have also been blessed with a wonderful group of hardworking research assistants, including Jean Chabot, Barbara Burkholder, Sue Fox, Connie Grabowy, Peter Joyner, and Wanda Wang. Their experimental contributions have been considerable, and to them we also owe the general feeling of well-being and good humor in our laboratory.

I owe a special debt of thanks to my good friend David Luck, who made it possible for my wife and me to spend a sabbatical at Rockefeller University in 1974–1975. There I found the peace and quiet to write much of this book, and I learned a great deal from being in David's laboratory and observing the very high caliber of work done there, as well as in the other laboratories in the Department of Cell Biology at Rockefeller. There I also had the opportunity to renew my friendship with Nam-Hai Chua; we had innumerable stimulating conversations concerning the sites of synthesis of thylakoid membrane polypeptides, how proteins get into organelles, and other topics in organelle biogenesis and genetics. We even did some experiments together that I greatly enjoyed. Also during that time I got to know Alan Lambowitz well. He taught me a great deal about *Neurospora,* the *poky* mutant in particular, and he was kind enough to read and criticize Chapter 8. Since then we have kept in touch and even collaborated on some *Chlamydomonas* work.

During the past year I have been writing to friends and colleagues all around the world asking for photographs of their work to be included in this book. They have all been most gracious and helpful in this respect, frequently providing me with unpublished photographs and interesting preprints. It gives me particular pleasure to thank Nancy Alexander, Charles Arntzen, Guiseppe Attardi, Bill Birky, Günter Blobel, Laurie Bogorad, Piet Borst, Ron Butow, Kwen-Sheng Chiang, Nam-Hai Chua, Frederick Crane, Jeff Davidson, Igor Dawid, Bernard Dujon, Jean Forster, Govindjee, Reinhold Herrmann, Rudolf Hagemann, Klaus Kowallik, R. Knoth, Alan Lambowitz, Tony Linnane, David Luck, Harvey Lyman, Maureen Hanson, George Palade, Phil Perlman, Murray Rabinowitz, Efraim

Racker, Irwin Rubenstein, Ruth Sager, Jeff Schatz, Jerry Schiff, Larry Simpson, Jon Singer, Piotr Slonimski, Karen Swift, Krishna Tewari, Achim Trebst, Ernst van Bruggen, Sam Wildman, and M. Wrischer for their help. Special thanks also go to John Boynton for his excellent micrographs of *Chlamydomonas* and to Sue Fox for printing a number of the illustrations.

I must also mention three people who have given me important aid, either directly or indirectly, with this book. My own investigations of chloroplast heredity in *Chlamydomonas* and those of my colleagues were stimulated originally by the work of Ruth Sager. Although we have often disagreed, I have always found her conclusions provocative and interesting. It is a pleasure to acknowledge this intellectual debt. Over the past few years Lynn Margulis has often provided me with timely words of encouragement when I needed them most. As this book became longer and longer, she commented with disgust that it was turning into a "tome" rather than the short introduction it was intended to be. I must also thank Jim Ebert, who has been interested in the progress of this book and was kind enough to put me in touch with Raven Press, whose personnel have been most helpful in expediting publication of the kind of book I really wished to do. In this regard it is a pleasure to acknowledge Berta Steiner Rosenberg whose expert editorial guidance has been so important in producing this book, and also Lorraine Moseley who prepared all of the original line drawings.

It gives me the greatest pleasure to acknowledge the help and support of my wife Carol. Although she is not a scientist, she has had to deal with the manic–depressive phases of my behavior during the long gestation period of *Organelle Heredity*. She has done so with grace, humor, and the greatest patience.

Finally, I am indebted to the Institute of General Medical Sciences of the National Institutes of Health for a Research Career Development Award that freed me from many academic responsibilities and gave me time to write, and I am indebted to my colleagues in the botany and zoology departments at Duke University for providing me with the salutary environment in which to write.

Introduction

Some 70 years have passed since Correns and Baur discovered that mutations affecting plastid phenotypes in higher plants frequently exhibit nonmendelian inheritance; yet only in the past 10 years has the study of chloroplast genetics (and, more recently, mitochondrial genetics) emerged from the realm of interesting phenomenology and become a respectable discipline. In the meantime the fields of bacterial genetics and viral genetics have advanced to high levels of sophistication. Perhaps it was necessary for the fields of bacterial and viral genetics to develop before organelle genetics could emerge as a legitimate field of investigation, for there are many similarities between the genomes of viruses, prokaryotes, and chloroplasts and mitochondria. These similarities were first given definitive treatment in the serial theory of endosymbiosis of Margulis (1–3), which was reiterated or modified by other students of the evolution of chloroplasts and mitochondria (4–7). In any event, the study of organelle genetics and biogenesis is now in high gear, as is reflected in the symposia, review articles, and original papers so numerous as to bewilder those uninitiated in the field. From this plethora of material we can now begin to define the rules of organelle genetics and the broad outline of the interaction of organelle and nuclear genomes in the biogenesis of chloroplasts and mitochondria. The aim of this book is to give the reader a working knowledge of what we presently understand about organelle genetics and biogenesis and what we would like to learn in the future. In this respect, it is hoped that *Organelle Heredity* will be regarded as a worthy successor to Ruth Sager's excellent treatise *Cytoplasmic Genes and Organelles* (22) which was published 6 years ago and as complementary to Paul Grun's thoughtful book *Cytoplasmic Genetics and Evolution* (21).

Organelle Heredity is divided into three sections. The first three chapters deal with structural and functional aspects of chloroplasts and mitochondria and their genetic apparatus. Chapter 1 presents a general discussion of the structure and function of the two organelles. Chapters 2 and 3 contain more detailed considerations of organelle DNA and the transcriptional and translational apparatus of chloroplasts and mitochondria. In regard to these chapters, several points should be stressed. First, organelle DNAs, with the possible exception of the chloroplast genome of the giant unicellular alga *Acetabularia,* have thus far proved to be of rather limited coding capacity with respect to the number of functions carried out by a chloroplast or mitochondrion. That is, most of the factors that define what these organelles are and what they do are determined by the nucleus. This may account for the fact that most recent work has focused on what the organelle genome does for the organelle rather than on what the nuclear genome does. By identifying those functions that are under the control of the limited number of genes present in the mitochondrion or the chloroplast, one can, by elimination, assign all other functions to the nucleus. Second, organelle genomes are highly redundant on a per cell basis. Redundancy can be achieved either by having many organelles per cell with a few DNA molecules each or by having a few organelles with many DNA molecules per organelle. Obviously, various combinations and permutations of these two extreme states are possible. Third, mitochondrial and chloroplast DNAs universally code for the RNAs of the ribosomes found in these organelles;

very likely they also code most, if not all, of the unique transfer RNAs. Fourth, we understand very little about how those chloroplast and mitochondrial proteins coded by nuclear genes whose messages are translated in the cytoplasm enter their target organelles. It seems likely that several mechanisms are employed, but at the moment we have only a primitive idea of what these are. Fifth, although there are undeniable similarities among the genomes and protein-synthesizing systems of chloroplasts, mitochondria, and prokaryotes, there are also striking differences. Chloroplast ribosomes, for example, are so similar to prokaryotic ribosomes that subunits can be exchanged between them, and the hybrid ribosomes will function in an *in vitro* protein-synthesizing system; but animal mitochondrial ribosomes are very different from all other ribosomes in that they are protein-rich and contain distinctive ribosomal RNAs. Sixth, from the limited evidence available, one cannot distinguish any tendency toward reduced organelle genome size as a function of evolutionary position. For example, mitochondrial DNA of most (possibly all) animals, except for the Protozoa, consists of 5-μm circles. The same is true of the green alga *Chlamydomonas.* Among fungi, mitochondrial DNAs of the yeasts alone range from 6 to 25 μm. In higher plants the circle size is 30 μm. In ciliate protozoans, but in no other organisms thus far described, mitochondrial DNA is a rod of approximately 15 μm. Chloroplast DNA seems uniformly to consist of 40 to 60-μm circles in higher plants and *Chlamydomonas,* but in *Acetabularia* the chloroplast genome could be greater than 800 μm in length. Three reasons may be offered to explain why no clear evolutionary trends can be discerned in the sizes and coding capacities of different organelle DNAs. First, organelle DNAs have not been characterized sufficiently in appropriate lower eukaryotes. Second, all organelle DNAs must code for a minimum number

of functions that cannot be reduced; this bottom line is realized in the 5-μm circular mitochondrial DNA and the 40 to 50-μm chloroplast DNA circle. Third, by the time this irreducible minimum has been reached, organelle DNAs may have undergone modification through acquisition of spacers, new regulatory functions, etc. For example, half of the mitochondrial DNA of bakers' yeast, which is a 25-μm circle, seems to consist of spacers. In the same vein, it is not clear why *Chlamydomonas* mitochondrial DNA is small and why the mitochondrial DNA of higher plants is the largest known. Do the latter DNAs also contain extensive spacers?

The second major section of *Organelle Heredity* concerns organelle genetics and consists of 15 chapters. Chapters 4–7 are devoted to the mitochondrial genome of bakers' yeast, and Chapters 12–15 are devoted to the chloroplast genome of *Chlamydomonas;* these appear to be the two model systems most desirable for the study of organelle genetics. Obviously, two systems are not nearly sufficient to reveal all that there is to know about organelle genetics in general; thus other chapters on mitochondrial heredity deal with *Neurospora* (Chapter 8), other fungi (Chapter 9), *Paramecium* (Chapter 10), and mammalian systems (Chapter 11). Chapters 17 and 18 deal with plastid genetics in higher plants. Chapter 18 is devoted to the *Nicotiana* hybrid system developed by Wildman and his colleagues, which provided the first concrete evidence concerning localization of structural genes for specific chloroplast proteins. Chapter 17 summarizes what has been learned from the study of chloroplast heredity in other higher plants. The reader may regard it as presumptuous to attempt to summarize in one chapter the vast literature on plastid genetics gathered over 70 years, but there are two reasons for this. First, Kirk and Tilney-Bassett (8) have done an excellent job of reviewing this classic literature in their book *The Plastids,* which is about to come out in a new edition. Second, genetic analysis of chloroplast genes (in the sense that it is carried out by use of various mutants with defined phenotypes that can be mapped by recombination analysis) cannot be performed in higher plants, although it is far along in *Chlamydomonas.* Chapter 16 is devoted to *Euglena.* This may seem bizarre, since this flagellate is without plastid genetics per se, but there is a fascinating literature on bleaching that relates to the discrepancy between the number of genetically competent copies of an organelle genome per cell and the number of physical copies of organelle genomes per cell. This is a theme we will return to again and again in this book.

The major points to be highlighted in this second section are the following: First, organelle genes may be transmitted either biparentally or uniparentally, depending on the organism and the organelle. When chloroplast or mitochondrial genes are transmitted biparentally, they rapidly segregate somatically, unlike nuclear genes. Second, the mechanisms that determine uniparental inheritance of organelle genes appear to vary. They include exclusion of organelles from one parent and elimination of the organelle DNA contributed by one of the parents by mechanisms that may involve preferential destruction and replication. Third, mapping of mitochondrial genes in yeast is proceeding apace with the aid of an armamentarium of recently devised techniques, some of which are unique to this system. Mapping of chloroplast genes is also continuing in *Chlamydomonas,* but at a somewhat slower rate. Several of the mapping methods used are common to both systems, but the powerful method of deletion mapping can presently be used only with the yeast mitochondrial genome. It is probably fair to say that mapping methods in both of these organelle genetic systems have undergone their initial shakedown periods and should be reasonably well standardized for future use. A major aim of the discussion of organelle genetics is to explain how these different mapping methods work. Fourth, mapping of organelle genes by genetic means is now being supplemented by physical mapping techniques that promise to be even more important in years to come. For example, several mitochondrial and chloroplast genomes have now been mapped using restriction endonucleases, and the stable RNAs of the organelle protein-synthesizing system (i.e., rRNAs and tRNAs) have been localized to specific points on the maps. Now these fragments are being cloned, and messenger RNAs isolated from chloroplasts and mitochondria are being hybridized to them. These include messengers whose products are known. This molecular approach is likely to prove particularly useful for animal mitochondrial genomes and the chloroplast genomes of higher plants, where genetic techniques, if they exist at all, are still rudimentary. Fifth, somatic segregation of organelle genomes has thus far precluded development of a test of gene function similar to the cis-trans test used so universally in other genetic systems. Therefore, gene definitions in organelle genetics must be based on tests of recombination, which situation is less than satisfactory. Sixth, a major mystery of organelle heredity, as mentioned previously, concerns the discrepancy between the numbers of genetic copies and physical copies of organelle genomes. In general, there seem to be few genetic copies, but there are many physical copies. Despite the existence of interesting and provocative hypotheses to explain this paradox, we have no physical evidence that explains the discrepancy. Fortunately, this problem has in no way proved an insurmountable obstacle, or even a significant obstacle to the mapping of organelle genes.

The third section of this book deals with the biogenesis of chloroplasts and mitochondria. The approach is to ask what the organelle does for itself. By definition, this eliminates most of the components that comprise chloroplasts and mitochondria, since they are coded by nuclear genes whose messages are translated in the cytoplasm. For this reason, Chapters 19 and 20 are devoted largely to the inner membranes of mitochondria and chloroplasts in which electron transport and ATP synthesis take place. We are now certain that a principal function of organelle genomes and protein-synthesizing systems is to make some of the components of these membranes. In addition, the chloroplast genome and protein-synthesizing system are involved in making part of the major soluble protein of the chloroplast, ribulose bisphosphate carboxylase or fraction I protein. Chapter 21 deals with the complex and poorly understood interplay between the organelle and nuclear genomes and the cytoplasmic and organelle protein-synthesizing systems in the biogenesis of the organelle protein-synthesizing systems themselves.

The major points to be made in these chapters are the following: First, cooperation is the governing principle in every case where the organelle genetic apparatus is involved. That is, synthesis of the mitochondrial inner membrane, the thylakoid membranes of the chloroplast, the organelle protein-synthesizing systems, and even the enzyme ribulose bisphosphate carboxylase involves participation of two genomes and two protein-synthesizing systems. Second, we have virtually no idea of how this cooperation is regulated. Third, although we know a great deal about which proteins are made via organelle protein synthesis, particularly in the case of the mitochondrion, we are just now beginning to determine which structural genes for organelle proteins are localized in the chloroplast and mitochondrial genomes.

In concluding this introduction, it should be noted that in 1976 major symposia were held at Bari (9), Munich (10), and Strasbourg (11), each of which dealt with aspects of organelle genetics and biogenesis. The proceedings of all three symposia have now been published. The first two volumes arrived as this book was being completed,

and they have been used extensively in bringing *Organelle Heredity* up to date, as the references at the end of each chapter indicate. Unfortunately, the third volume became available too late to be of use herein. In general, this book attempts to cover the pertinent topical material through the end of 1976, but in certain instances where the literature seemed to be of particular interest, the coverage is extended well into 1977. An attempt has been made to avoid including references to unpublished work, but this has not been entirely successful. Finally, two topics not discussed in this book should be noted. First, no consideration is given to theories of the origin of mitochondria and chloroplasts, as the purpose of the book is not to speculate on this topic, but rather to define our current body of knowledge about organelle heredity. In any event, theories of the evolution of chloroplasts and mitochondria have been discussed and reviewed extensively in the literature; for a discussion, the interested reader is referred to a recent *Symposium of the Society for Experimental Biology* on symbiosis (12), as well as other sources (1–7). The second topic not discussed here is cytoplasmic male sterility in maize. As this book was being completed, evidence was building up that this phenomenon may well involve mitochondrial genetic changes (13–20). However, for the sake of the sanity of both author and publisher it seemed inadvisable to delay publication of this book any longer to include yet one more chapter.

REFERENCES

1. Sagan, L. (1967): On the origin of mitosing cells. *J. Theor. Biol.,* 14:225–274.
2. Margulis, L. (1970): *Origin of Eukaryotic Cells.* Yale University Press, New Haven, Conn.
3. Margulis, L. (1975): Symbiotic theory of the origin of eukaryotic organelles: Criteria for proof. *Symp. Soc. Exp. Biol.,* 29:21–38.
4. Raff, R. A., and Mahler, H. R. (1972): The nonsymbiotic origin of mitochondria. *Science,* 177:575–582.
5. Raff, R. A., and Mahler, H. R. (1975): The

symbiont that never was: An inquiry into the evolutionary origin of the mitochondrion. *Symp. Soc. Exp. Biol.,* 29:41–92.

6. Bogorad, L., Davidson, J. N., Hanson, M. R., and Mets, L. J. (1975): Genes for proteins of chloroplast ribosomes and the evolution of eukaryotic genomes. In: *Molecular Biology of Nucleocytoplasmic Relationships,* Chapter 3. Elsevier, Amsterdam.

7. Taylor, F. J. R. (1974): Implications and extensions of the serial endosymbiosis theory of the origin of eukaryotes. *Taxon,* 23:229–258.

8. Kirk, J. T. O., and Tilney-Bassett, R. A. E. (1967): *The Plastids.* W. H. Freeman, San Francisco.

9. Saccone, C., and Kroon, A. M. (editors) (1976): *The Genetic Function of Mitochondrial DNA.* North Holland, Amsterdam.

10. Bücher, T., Neupert, W., Sebald, W., and Werner, S. (editors) (1976): *Genetics and Biogenesis of Chloroplasts and Mitochondria.* North Holland, Amsterdam.

11. Bogorad, L., and Weil, J. H. (1976): *Nucleic Acids and Protein Synthesis in Plants.* Plenum Press, New York.

12. Jennings, D. H., and Lee, D. L. (editors) (1975): Symbiosis. *Symp. Soc. Exp. Biol.,* 29:1.

13. Duvick, D. N. (1965): Cytoplasmic pollen sterility in corn. *Adv. Genet.,* 13:1–56.

14. Miller, R. J., and Koeppe, D. E. (1971): Southern corn leaf blight: Susceptible and resistant mitochondria. *Science,* 173:67–69.

15. Gengenbach, B., Koeppe, D., and Miller, R. (1973): A comparison of mitochondria isolated from male-sterile and nonsterile cytoplasm etiolated corn seedlings. *Physiol. Plant,* 29:103–107.

16. Flavell, R. B. (1975): Inhibition of electron transport in maize mitochondria by *Helminthosporium maydis,* race T, pathotoxin. *Physiol. Plant Path.,* 6:107–116.

17. Laughnan, J. R., and Gabay, S. J. (1975): An episomal basis for instability of S male sterility in maize and some implications for plant breeding. In: *Genetics and Biogenesis of Mitochondria and Chloroplasts,* edited by C. W. Birky, Jr., P. S. Perlman, and T. J. Byers, Chapter 10. Ohio State University Press, Columbus.

18. Levings, C. S., III, and Pring, D. R. (1976): Restriction endonuclease analysis of mitochondrial DNA from normal and Texas cytoplasmic male-sterile maize. *Science,* 193:158–160.

19. Barratt, D. H. P., and Peterson, P. A. (1977): Mitochondrial banding pattern difference in electrofocused polyacrylamide gels between male-sterile and nonsterile cytoplasm of maize. *Maydica,* 22:1–8.

20. Warmke, H. E., and Lee, S.-L. J. (1977): Mitochondrial degeneration in Texas cytoplasmic male-sterile corn anthers. *J. Hered.,* 68: 213–222.

21. Grun, P. 1976. *Cytoplasmic Genetics and Evolution.* Columbia University Press, New York.

22. Sager, R. (1972): *Cytoplasmic Genes and Organelles.* Academic Press, New York.

Chapter 1
Structure and Function of Chloroplasts and Mitochondria

One of the major features that distinguish eukaryote cells from the cells of prokaryotic organisms such as bacteria and blue-green algae is that they are highly compartmentalized into organelles. Of the many types of organelles found within the eukaryote cell, the chloroplast and the mitochondrion are distinctive in that they have high degrees of complexity and they have many similarities in structure and function (Table 1–1). Both chloroplasts and mitochondria are surrounded by double-membrane envelopes. The inner membrane is an infolded, complicated network in which structure and function are closely coordinated.[1] The inner membrane of each organelle contains components such as cytochromes and quinones that are required for electron transport, and attached to the inner membrane of each organelle are coupling factors for electron-transport-coupled phosphorylation. Within the double-membrane envelope surrounding each organelle is a matrix containing soluble enzymes, prominent among which are enzymes involved in different aspects of carbon metabolism. Chloroplasts and mitochondria are also unique among organelles in that they contain deoxyribonucleic acid (DNA) and protein-synthesizing systems.

The DNA of these organelles is distinctive with respect to the organelles themselves and also with respect to the nucleus (see Chapter 2). The protein-synthesizing systems of chloroplasts and mitochondria bear little resemblance to the protein-synthesizing systems of the eukaryote cytoplasm (see Chapter 3). What does organelle DNA do? Why do chloroplasts and mitochondria contain protein-synthesizing systems? Can one associate specific genes with organelle DNA? How do these genes segregate and recombine? These are some of the questions that this book will explore.

Consideration of chloroplast and mitochondrial heredity should be prefaced by a more detailed explanation of the organelles themselves. We have mentioned some similarities between chloroplasts and mitochondria, but they also have obvious differences. In nature, the two organelles complement one another in a fundamental sense. In the chloroplast, energy derived from light is used for oxidation of water and production of molecular oxygen. The electrons derived from the splitting of water are used via the photosynthetic electron transport chain to drive photosynthetic phosphorylation, and molecular carbon dioxide is ultimately re-

TABLE 1–1. *Some similarities between chloroplasts and mitochondria*

1. Both organelles are surrounded by double-membrane envelopes.
2. Inner membrane is highly infolded to form cristae in mitochondria and thylakoids in chloroplasts.
3. Inner membrane contains cytochromes, quinones, and other compounds required for electron transport.
4. Coupling factors involved in phosphorylation are attached to the inner membrane.
5. Double-membrane envelope surrounds a matrix containing soluble enzymes.
6. Both organelles contain DNA distinct from nuclear DNA.
7. Both organelles contain enzymes required for replication and transcription of the DNA contained within.
8. Both organelles contain protein-synthesizing systems distinct from the protein-synthesizing system of the cytoplasm.

[1] See footnote on page 38.

1

duced by the protons and electrons derived from the splitting of water and is converted into carbohydrate by the soluble enzymes of the chloroplast matrix or *stroma*. The mitochondrion, in contrast, catalyzes the aerobic oxidation of reduced carbon compounds via the soluble enzymes of the tricarboxylic acid cycle that are found in the mitochondrial matrix. The electrons produced by the oxidation of reduced carbon compounds flow via the respiratory electron transport chain and drive oxidative phosphorylation. The electrons and protons obtained from oxidation of reduced carbon compounds reduce molecular oxygen to water, and carbon dioxide is released as an oxidation product of the tricarboxylic acid cycle. In summary, the chloroplast reduces carbon dioxide and oxidizes water, and the mitochondrion oxidizes reduced carbon compounds and reduces molecular oxygen to water. Both processes generate energy in the form of adenosine triphosphate (ATP) via coupled electron transport chains.

MEMBRANES IN GENERAL

Since so much of chloroplast and mitochondrial function is intimately bound up with their membranes, a few general remarks about membranes are in order at the outset. Models of membrane structure vary from investigator to investigator and from year to year, depending on the combination of methods used for study and the assumptions made about what these methods are accomplishing. Most membranes contain about 40% lipid and 60% protein, but there are distinct variations, depending on the membrane in question (Tables 1–2 and 1–3). The lipid fraction includes both *polar* or *amphipathic* species (Fig. 1–1) and neutral lipids. Polar phospholipids are the predominant lipids in microsomal and mitochondrial membranes, whereas neutral glycolipids are abundant in the chloroplast inner (thylakoid) membrane (Tables 1–2 and 1–3). Various proteins are associated with membranes, and these proteins may be divided into two general categories (1,2). The first category is comprised of the *extrinsic* or peripheral proteins; these are only weakly bound to their respective membranes, and they are usually water-soluble (Table 1–4). The second category is comprised of the *intrinsic* or integral proteins; this includes 70 to 80% of all membrane proteins (2). Intrinsic proteins are hydrophobic and usually are associated with lipids (Table 1–4). The intrinsic proteins include transport proteins, membrane-associated enzymes, and hormone receptors. Almost all membranes are freely permeable to water and neutral lipophilic molecules, but they are much less permeable to polar mole-

TABLE 1–2. *Comparison of mitochondrial and microsomal membranes from guinea pig liver*

Component	Mitochondrial membrane		Microsomes
	Inner	Outer	
Phospholipid (mg/mg protein)	0.301	0.878	0.385
Cholesterol (mg/mg protein)	0.005	0.031	0.030
Phospholipid composition (expressed as % of total phospholipid)			
Phosphatidyl choline	44.5	55.2	62.8
Phosphatidyl ethanolamine	27.7	25.3	18.3
Phosphatidyl inositol	4.2	13.5	13.4
Cardiolipin	21.5	3.2	0.5
Others	2.2	2.5	5.6

Based on the work of Parsons et al. (22) and Parsons and Yano (23) as summarized by Ernster and Kuylentsierna (20).

TABLE 1–3. *Lipid composition of chloroplasts*

Component	Percentage of total
Total protein	48.0
Total lipid	52.0
Phospholipids	4.7
Glycolipids	23.1
Digalactosyl diglyceride	7.0
Monogalactosyl diglyceride	14.0
Sulfolipid	2.1
Quinones	1.7
Carotenoids	1.4
Chlorophylls a and b	10.8
Unidentified	10.0

Adapted from Zelitch (101) after Park and Biggins (88).

cules and are only slightly permeable to small ions such as Na^+ and Cl^-.

The first important hypothesis of membrane structure was that of Danielli and Davson (3); it proposed that the plasma membrane was a sandwich structure with a bimolecular layer of lipid at the center and protein films adsorbed to the inside and outside surfaces. The results of early attempts to measure the thicknesses of plasma membranes, as well as other membranes, by physical methods were compatible with this model. Development of the thin-sectioning technique for electron microscopy first made examination of membrane ultrastructure possible. It was found that in cross section, after fixation with osmium tetroxide, the plasma membrane appeared as two parallel dark (osmiophilic) lines sepa-

rated by a light (osmiophobic) region. Was each line a membrane? Or was the whole thing a membrane? In 1959 Robertson (4) made the important suggestion that the three lines together constituted a single membrane or *unit membrane* (Fig. 1 2). He concluded that the dark layers represented the polar ends of the phospholipid molecules, together with associated proteins, and that the light layer represented the nonpolar portions of the lipid bilayer. He thus extended the Danielli-Davson model to account for the electron microscopic observations. Strong support for the unit membrane concept came from the studies of Stoeckenius, who worked with artificial membranes (5,6). Although the exact dimensions of unit membranes vary somewhat, depending on the fixative employed, typical values of about 20 Å for each of the dark layers and 35 Å for the light layer are obtained (7).

In recent years it has become clear that the unit membrane concept does not provide an adequate description of any membrane. The principal problem is that during fixation for electron microscopy, globular proteins associated with membranes become denatured and subsequently uncoil, giving a spurious impression of uniformity of structure. Realization of this point came about principally because of the invention by Mühlethaler and colleagues (8) of the *freeze etching* technique for studying ultrastructure. The technique involves freezing a

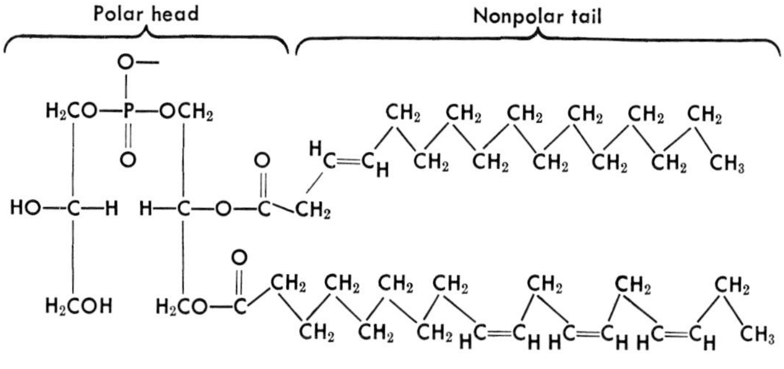

FIG. 1–1. Diagram of a polar lipid: phosphatidyl glycerol.

TABLE 1–4. *Criteria for distinguishing extrinsic and intrinsic proteins*

Property	Extrinsic protein	Intrinsic protein
Requirements for dissociation from membrane	Mild treatments sufficient: high ionic strength, metal ion chelating agents	Hydrophobic bond-breaking agents required: detergents, organic solvents, chaotropic agents
Association with lipids when solubilized	Usually soluble free of lipids	Usually associated with lipids when solubilized
Solubility after dissociation from membrane	Soluble and molecularly dispersed in neutral aqueous buffers	Usually insoluble or aggregated in neutral aqueous buffers

Adapted from Singer (2).

biologic structure, fracturing it, and then examining a carbon platinum replica of the fracture plane in the electron microscope. In a variation on this technique called *deep etching,* some of the water is sublimed off before the specimen is examined; in this way other surfaces previously covered with ice are exposed. Freeze etching of chloroplast and mitochondrial membranes clearly reveals the presence of globular membrane proteins of various sizes and kinds.

Unfortunately, study of membranes by the freeze etching method was plagued for some time by problems of interpretation. The major difficulty concerned interpretation of the fracture plane. Mühlethaler and associates (9) took the view that the fracture plane would be formed at the interface between the membrane and the aqueous environment. Branton and Park argued that in a frozen system the hydrophobic region within the membrane would provide a plane of weakness along which fracture would oc-

cur (10). In a liquid aqueous system, such a split would be unlikely to take place, because exposure of the hydrophobic interior of the membrane to the polar molecules would result in a marked increase in free energy. However, in a frozen system, exposure to water molecules does not occur, and polar bonding between the hydrophilic outer surface of the membrane and the surrounding aqueous environment will still exist, so that fracture at the interface between membrane and medium is rendered unlikely. The important point is that Mühlethaler's hypothesis assumed that the fracture plane would reveal the structure of the *outside* of the membrane, whereas the hypothesis of Branton and Park assumed that the fracture plane would reveal the structure of the *inside* of the membrane. Evidence summarized by Kirk (11) and Park and Sane (10) generally favors the Branton-Park interpretation. Mühlethaler himself (12) also came around to this view

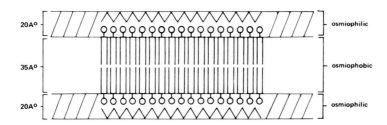

FIG. 1–2. Unit membrane model of Robertson. The membrane is depicted as a three-layered structure 75 Å thick consisting of two dense (osmiophilic) layers 20 Å thick bounding a light (osmiophobic) central zone 35 Å thick. The two osmiophilic outer layers contain proteins and the polar heads of phospholipids, whereas the central zone contains the nonpolar lipid tails of the phospholipid molecules.

because of a technique he devised that allows observation of complementary faces in the fractured region.

For a long time the designations A, B, C, and D were used to describe the faces seen in freeze etching, and these were interpreted in different ways by different workers. Recently, many scientists using the freeze etching technique have proposed a uniform nomenclature for the different surfaces they see (13). According to this nomenclature, the membrane half that is closest to the cytoplasm, nucleoplasm, chloroplast stroma, or mitochondrial matrix is designated P for protoplasmic, whereas the half that is closest to the extracellular space, exoplasmic space, or endoplasmic space (as the case may be) is designated E. The hydrophilic surfaces of these membrane halves are designated PS and ES, respectively, and the faces exposed by membrane fracture are called PF and EF.

Results of fractionation of membranes and their components by a variety of biochemical and mechanical methods, together with ultrastructural observations, support the view that membranes are very diverse. At the moment it seems most reasonable to view the structure of chloroplast and mitochondrial membranes in terms of the generalized lipid-protein mosaic model championed by Singer (1,2). In this model the membrane is considered to be a lipid bilayer in which the intrinsic proteins are embedded and to which the extrinsic proteins are attached (Fig. 1–3). The polar ends of the amphipathic lipid molecules are oriented toward the outside of the membrane, and the hydrophobic tails are oriented inward. Similarly, the intrinsic proteins are usually amphipathic molecules in which the hydrophobic portions are buried inside the membrane and the hydrophilic regions are exposed at the membrane surface.

The close integration of structure and function in the chloroplast and mitochondrial inner membranes has important genetic consequences. As we shall see, mutations affecting these organelles frequently produce a constellation of *pleiotropic effects* on organelle structure and function (see Chapters 19 and 20). The net result is that

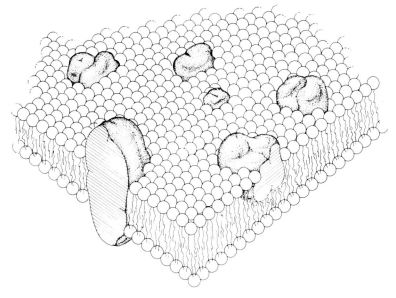

FIG. 1–3. Lipid–globular protein mosaic model with a lipid matrix (the fluid mosaic model); schematic three-dimensional and cross-sectional views. The solid bodies with stippled surfaces represent globular integral proteins; at long range they appear to be randomly distributed in the plane of the membrane; at short range, some may form specific aggregates, as shown. From Singer and Nicolson, ref. 1, with permission. Copyright 1972 by the American Association for the Advancement of Science.

it is often difficult to pin down the primary defect caused by the mutation.

MITOCHONDRIA

Mitochondria vary considerably in size and shape from one cell type to another. Within a given type of cell, mitochondria can undergo dramatic changes in shape and volume, depending on the physiologic state of the cell. For example, in the green alga *Chlamydomonas reinhardtii* the mitochondria appear small and ellipsoid when the cells are grown on a mixture of acetate and carbon dioxide as the carbon source, but they become much larger and more elongated when only carbon dioxide is supplied as the carbon source (Fig. 1–4). Interestingly enough, the total volumes occupied by mitochondria, as calculated from electron micrographs, are approximately the same in both kinds of cells; only the size, the shape, and possibly the number of mitochondria appear to vary.

With tacit understanding of the obvious limitations of the term, we can characterize an "average" mitochondrion, such as that obtained from a rat liver cell, as a structure 3 μm long and 0.5 to 1.0 μm in diameter (Fig. 1–5). The outer membrane of the double envelope is smooth and continuous, without obvious foldings or protuberances, and it has a diameter of 50 to 70 Å. The width of the inner membrane varies (14) from 75 to 100 Å; the inner membrane is also continuous, but it is highly convoluted into folds that frequently appear almost as transverse septa. These folds were named *cristae mitochondriales* or *cristae* by Palade (15), who published the first high-resolution electron micrographs of mitochondria. At that time they were interpreted to indi-

cate that the mitochondrion was surrounded by a single membrane that formed a number of infoldings or cristae. A year later Sjöstrand (16) discovered that the mitochondrion was surrounded by a double membrane. Sjöstrand interpreted Palade's cristae as separate units or septa distinct from the inner membrane. The existence of the double membrane was soon confirmed by Palade (17), who still maintained that the cristae were infoldings of the inner membrane. Subsequently, Sjöstrand's student Andersson-Cedergren (18) confirmed by careful analysis of serial sections that the cristae did have continuities with the inner membrane. What this controversy illuminates is the complexity of the cristae and the differences in morphology of cristae in different mitochondria. In the liver mitochondria used by Palade, the cristae do look like infoldings of the inner membrane, but in the kidney mitochondria studied by Sjöstrand, actual continuity between the inner membrane and a crista is only rarely seen.

The lumen between the inner and outer membranes is called the *intermembrane space,* and the lumen contained within the inner membrane is referred to as the *matrix* (Fig. 1–5). By means of observations made on isolated mitochondria exposed to different experimental conditions, much has been learned in recent years about the structural relationships of the two membranes. Hackenbrock (19) applied the terms *orthodox* and *condensed* to describe reversible states of isolated mitochondria that can be induced by altering their metabolic states (Fig. 1–6). These configurational changes primarily involve changes in the conformation of the inner membrane, with a corresponding shift in the ratio of

$\longrightarrow$

FIG. 1–4. Mitochondria in median sections of the green alga *Chlamydomonas reinhardtii.* $\times$37,000. A and B: Mitochondria from cells grown photosynthetically with carbon dioxide as sole carbon source (phototrophic growth). C: Mitochondria from cells grown in the light, with both carbon dioxide and acetate as carbon sources (mixotrophic growth). Note that mitochondria tend to be much larger in phototrophic cells than in mixotrophic cells. Key: Ch, cristae; CP, chloroplast; M, mitochondrion; N, nucleus; NM, nuclear membrane; NP, nuclear pore. (Courtesy of Dr. J. E. Boynton.)

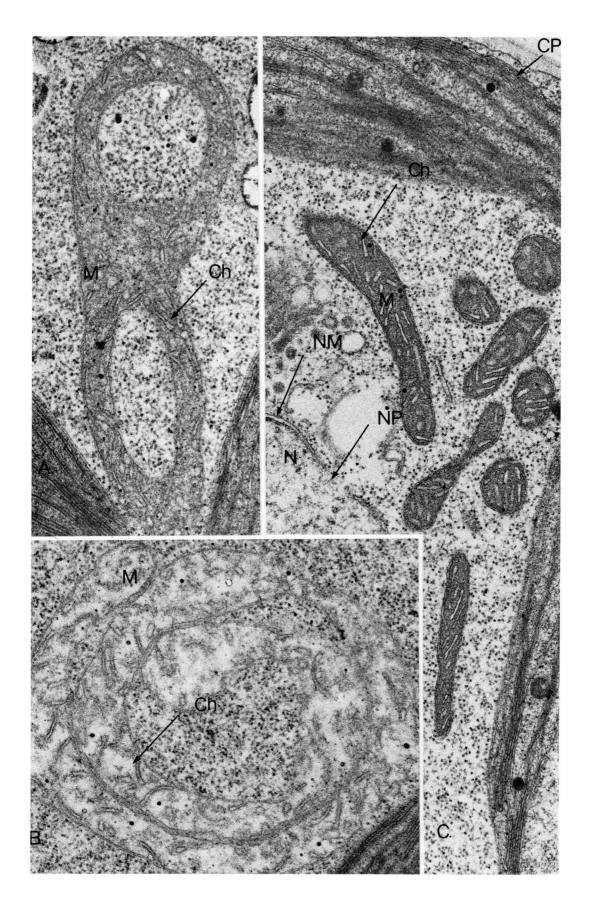

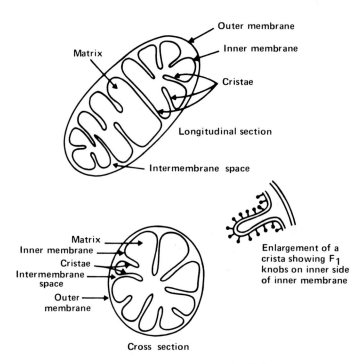

FIG. 1–5. Diagrammatic sketches of mitochondrial structure.

intermembrane space to matrix space. In mitochondria that are in the condensed state, frequent points of attachment can be seen between the outer and inner membranes. Exposure of mitochondria to a hypotonic medium, to an isotonic medium containing readily penetrating cations and anions, etc., leads to a swelling of the mitochondria, and subsequent exposure to a hypertonic medium or to ATP, Mg^{2+}, Mn^{2+}, etc., causes a contraction (20). A characteristic of swollen mitochondria is a distended and often ruptured outer membrane (Fig. 1–6). The inner membrane unfolds,

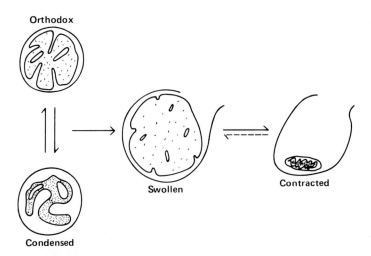

FIG. 1–6. Different conformational states of mitochondria. See text for discussion. (Adapted from Ernster and Kuylenstierna, ref. 20.)

but it retains its structural continuity. Contraction involves a refolding or reaggregation of the inner membrane without any restoration of the outer membrane, which tends to become detached from the inner membrane (Fig. 1–6). These studies with isolated mitochondria reveal important structural differences between the inner and outer membranes, and the differential responses of the two membranes to environmental stress have been of great importance in devising methods for separating the membranes (20). Separation of the inner and outer mitochondrial membranes has revealed that structural differences are accompanied by important functional differences. We will now turn to the functional differences among the two membranes, the intermembrane space, and the matrix.

MITOCHONDRIAL OUTER MEMBRANE

The outer and inner membranes differ markedly, not only in structure and osmotic properties but also in permeability and chemical composition, as well as in terms of the enzymes with which they are associated (Tables 1–2 and 1–5). The outer membrane is freely permeable to a variety of substances (both charged and uncharged) with molecular weights up to 10,000 d, whereas the inner membrane has very limited permeability to most substances, except uncharged molecules or molecules with molecular weights not greater than 100 to 150 d (20). The majority of substances that do pass through the inner membrane rely on specific transport systems (21).

As shown by Parsons and associates (22), the outer membrane is less dense (1.13 g/cc) than the inner membrane (1.21 g/cc); this difference, of course, aids in separation of the two membranes by density-gradient centrifugation. On a protein basis, the outer membrane contains two to three times as much phospholipid as the

TABLE 1–5. *Localization of enzymes in liver mitochondria*

Outer membrane
Rotenone-insensitive NADH–cytochrome c reductase (NADH–cytochrome b_5 reductase, cytochrome b_5)
Monoamine oxidase
Kynuronine hydroxylase
ATP-dependent fatty acyl-CoA synthetase
Glycerol phosphate acyl transferase
Lysophosphatidate acyl transferase
Lysolecithin acyl transferase
Choline phosphotransferase
Phosphatidate phosphatase
Phospholipase A_{II}
Nucleoside diphosphokinase
Fatty acid elongation systems
Xylitol dehydrogenase (NAD-specific?)

Inner membrane
Respiratory chain (cytochromes b, c_1, c, a, and a_3; succinate dehydrogenase; succinate–cytochrome c reductase; succinate oxidase; choline–cytochrome c reductase; rotenone-sensitive NADH–cytochrome c reductase; cytochrome c oxidase; respiratory chain-linked phosphorylation)
β-Hydroxybutyrate dehydrogenase
Ferrochelatase
α-Aminolevulinic acid synthetase?
Carnitine palmityl transferase
Fatty acid oxidation system?
Xylitol dehydrogenase (NADP-specific?)

Intermembrane space
Adenylate kinase
Nucleoside diphosphokinase
Nucleoside monophosphokinase
Xylitol dehydrogenase (NAD-specific?)

Matrix
Malate dehydrogenase
Isocitric dehydrogenase (NADP-specific)
Glutamate dehydrogenase
α-Ketoglutarate dehydrogenase (lipoyl dehydrogenase)
Citrate synthetase
Aconitase
Fumarase
Pyruvate carboxylase
Phosphopyruvate carboxylase
Aspartate aminotransferase
Ornithine carbamoyl transferase
Fatty acyl-CoA synthetase(s)
Fatty acid oxidation systems? (β-hydroxybutryl–CoA dehydrogenase)
Xylitol dehydrogenase (NADP-specific?)

Adapted from Ernster and Kuylenstierna (20).

inner membrane, which accounts for the density difference (23). Not only do the inner and outer membranes differ in their total amounts of lipids and proteins, but also the lipid and protein compositions of the two membranes distinguish them (Table 1–2). Cardiolipin, a major phospholipid component of the inner membrane, is present in only trace amounts in the outer membrane. Phosphatidyl inositol is an important constituent of the outer membrane, but it is much less conspicuous in the inner membrane. The neutral lipid cholesterol is six times more concentrated in the outer membrane than in the inner membrane on a protein basis (23). Schnaitman (24) compared the proteins present in the inner and outer membranes of rat liver by sodium dodecyl sulfate (SDS) polyacrylamide gel electrophoresis. This technique, which allows estimation of the apparent molecular weights of different proteins, revealed that the inner membrane contained 23 proteins and the outer membrane 12 proteins, but only 1 protein from the inner membrane and 1 protein from the outer membrane had similar molecular weights. It is interesting to note that while the inner and outer membranes are very dissimilar in protein and lipid content, there are notable similarities in both protein and lipid composition between the outer membrane and the membranes of the endoplasmic reticulum, although these membranes are by no means identical (Table 1–2).

The enzymes associated with the outer membranes are a heterogeneous group in comparison to the tightly integrated structural and functional respiratory chain assembly that comprises so much of the inner membrane. For example, monoamine oxidase is a characteristic extrinsic enzyme of the outer membrane, but the function of this enzyme in mitochondria is unknown (20). The NADH–cytochrome c reductase system, which is known to be insensitive to classic inhibitors of respiration (such as antimycin A and rotenone) and responsive to externally added NADH, is localized in the outer membrane. There are similarities between this system and the NADH–cytochrome b_5 reductase system of the microsomes (20). Numerous enzymes involved in phospholipid metabolism are associated with the outer membrane, which suggests that the outer membrane may play a role

in this process (Table 1–5). But, in summary, it is safe to say that we know much less about the diverse functions of the outer membrane than we do about the very specific functions of the inner membrane.

MITOCHONDRIAL INTERMEMBRANE SPACE

Characterization of the enzymes present in the intermembrane space presents serious difficulties. In order to isolate this soluble phase of the mitochondrion, it is necessary to remove the outer membrane of the mitochondrion without extensive rupture of the inner membrane. The membranes are then separated by centrifugation to yield heavy, light, and soluble phases (25). Ideally, the heavy fraction should include inner membranes and enclosed matrix, whereas the light fraction should be composed of outer membranes. The soluble fraction should contain whatever was enclosed in the intermembrane space. In practice, there are difficulties with such a fractionation procedure. Rupture of the inner membrane will lead to leakage of matrix components into the soluble fraction, as will loss of loosely bound components from the inner and outer membranes. However, if the fractionation is reasonably clean, there should be marked enrichment in those components that are restricted in part, if not entirely, to the intermembrane space. If a component is greatly enriched in the soluble phase, with respect to the heavy and light fractions, it is probably localized in the intermembrane space; but a component that is mildly enriched or not enriched at all may be a contaminant from another fraction. Given these limitations, it is not surprising that only a few enzyme activities have been attributed unambiguously to the intermembrane space (Table 1–5). There may be many other enzymes in this space, but they cannot be attributed with certainty to the intermembrane space if they also occur in reasonably large amounts in the other fractions.

MITOCHONDRIAL INNER MEMBRANE

The inner mitochondrial membrane is a highly specialized structure within which the respiratory assembly is located (Fig. 1–7). It is appropriate to pay more attention to the arrangement and function of the inner membrane than to the rest of the mitochondrion because its construction requires the participation of the mitochondrial genome and the mitochondrial protein-synthesizing system. It appears that the polypeptides of the outer membrane, as well as the enzymes of the mitochondrial matrix and intermembrane space, are coded largely, if not entirely, by nuclear genes whose messages are translated on cytoplasmic ribosomes.

The electron transport chain of the mitochondrial inner membrane can be isolated as four high-molecular-weight enzyme complexes (Fig. 1–7 and Table 1–6). Complex I catalyzes the oxidation of NADH by ubiquinone (coenzyme Q); it contains the flavoprotein electron acceptor NADH dehydrogenase (F_{P_D}), which accepts electrons from NADH in the mitochondrial matrix and donates them to ubiquinone. Complex II catalyzes the oxidation of succinate by ubiquinone. This complex contains the flavoprotein succinic dehydrogenase (F_{P_S}), which accepts electrons from succinate and transfers them to ubiquinone. Thus, ubiquinone communicates between complexes I and II and the rest of the electron transport chain. Complex III catalyzes the oxidation of reduced ubiquinone by cytochrome c and includes cytochromes b and c_1. Complex III is connected to complex IV by cytochrome c. Complex IV catalyzes the oxidation of reduced cytochrome c by molecular oxygen. The responsible enzyme is cytochrome oxidase, a complex of seven polypeptides (Table 1–6). Associated with this enzyme activity are cytochromes a + a_3. Complexes I, II, and III also have non-heme-iron proteins associated with them that probably play a role in electron transport (Table 1–6).

Electron transport via complex I is blocked by rotenone, and electron transport via complex III is blocked by antimycin

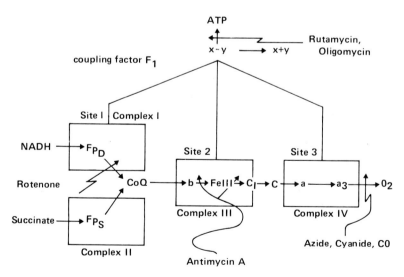

FIG. 1–7. Diagram of respiratory chain showing the three phosphorylation sites and the four complexes. The first phosphorylation site is located between NADH and cytochrome b (*site 1*); the second is between cytochromes b and c (*site 2*); the third is between cytochrome c and oxygen (*site 3*). Coupling factor F_1 catalyzes formation of a high-energy intermediate X ~ Y, which is used in formation of ATP from ADP and inorganic phosphate. The four complexes of the electron transfer chain are connected by coenzyme Q and cytochrome c. The components of the four complexes are listed in Table 1–6. The scheme for electron transport and phosphorylation shown is derived principally from Racker (28).

TABLE 1–6. *Intrinsic proteins of mitochondrial inner membrane.*

Complex	Designation	Components	Molecular weight of associated polypeptides (d)	Source
I	NADH–coenzyme Q reductase	NADH dehydrogenase I	42,000	Beef heart
		Non-heme-iron Ia	27,000	Beef heart
		Non-heme-iron Ib	16,000	Beef heart
		Polypeptides with unknown functions	74,000	Beef heart
			37,000	Beef heart
			23,000	Beef heart
II	Succinate–coenzyme Q reductase	Succinate dehydrogenase	69,000	Beef heart
		Non-heme-iron	27,000	Beef heart
		Polypeptide with unknown function	12,000	Beef heart
III	Coenzyme Q– cytochrome c reductase	Cytochrome b	26,000	Beef heart
			30,000	Neurospora
		Cytochrome c_1	37,000	Beef heart
			31,000	Yeast
		Non-heme-iron	30,000	Beef heart
		Polypeptide with unknown function	12,000	Beef heart
IV	Cytochrome oxidase	Cytochromes a + a_3	42,000	Yeast
		Specific functions not assigned to	34,500	Yeast
		these polypeptides, although one	23,000	Yeast
		probably binds heme, and all are	14,000	Yeast
		required for cytochrome oxidase	12,500	Yeast
		activity	12,500	Yeast
			4,500	Yeast

Data summarized from reviews by Crane and colleagues (14,31) for beef heart and by Schatz and Mason (102), Ross et al. (103), and Weiss and Ziganke (104) for yeast and *Neurospora*.

A (Fig. 1–7). Complexes I and III plus ubiquinone constitute an NADH–cytochrome c reductase system that can be distinguished from the NADH–cytochrome c reductase system of the outer membrane through the use of these inhibitors, since the outer membrane activity is insensitive to both of them (20). The activity of cytochrome oxidase is blocked by the inhibitors cyanide, azide, and carbon monoxide (Fig. 1–7). In many plants and fungi, respiration is only partially sensitive to cyanide (see 26, 27 for reviews). It can also be shown that this cyanide-resistant respiration is insensitive to inhibitors, such as antimycin A, which act between b- and c-type cytochromes (i.e., in the region of complex III). Hence, it appears that organisms with cyanide-resistant respiration have branched electron transport chains, with the branch point at or near the flavoprotein-containing dehydrogenase complexes (Fig. 1–8). One branch of the chain consists of the cytochrome-mediated electron transport chain; the second branch employs an "alternate oxidase" that is insensitive to inhibitors such as cyanide and antimycin A. The nature of alternate oxidase remains to be established. However, it may involve an iron-containing protein or proteins, as this pathway can be blocked specifically by iron-complexing agents such as the hydroxamic acids. The cyanide-resistant pathway lacks at least two and possibly all three of the phosphorylation sites associated with the cyanide-sensitive cytochrome-mediated electron transport pathway (Fig. 1–7).

The coupling of oxidative energy to the production of ATP depends on coupling factors residing in the inner membrane. Carefully prepared mitochondria are *tightly* coupled. They exhibit little adenosine triphosphatase (ATPase) activity, and they require added adenosine diphosphate (ADP) and inorganic phosphate (P_i) for respiration (28). In the presence of un-

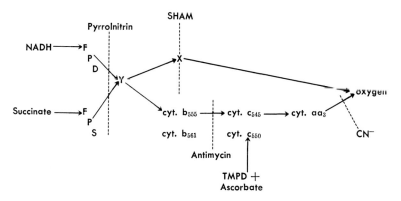

FIG. 1–8. Branched electron transport system. Y designates the branch point, and X designates the cyanide-insensitive oxidase. F_{P_D} and F_{P_S} are the flavoproteins. The sites of inhibition for antimycin and cyanide in the respiratory electron transport chain, as well as the site of inhibition of salicyl hydroxamic acid (SHAM) in the cyanide-insensitive chain, are shown. (From Lambowitz et al., ref. 106, with permission.)

coupling agents, tightly coupled systems become *uncoupled*. Respiration takes place in the absence of ADP and P_i, and added ATP is hydrolyzed. Submitochondrial particles (SMP) prepared for the purpose of isolating coupling factors are generally *loosely coupled*. In such loosely coupled systems the electron transport chain and ATPase are independently active as they are in an uncoupled system.

Three coupling sites are recognized in the respiratory electron transport chain (Fig. 1–7). Coupling between the electron transport chain and ATP formation is achieved by an enzyme complex that is usually referred to as the oligomycin-sensitive ATPase complex (OS-ATPase) (29,30), but sometimes it is called the rutamycin-sensitive ATPase complex (see Chapter 19). The distinction means nothing in functional terms; it simply indicates the inhibitor being used. Both oligomycin and rutamycin act to inhibit oxidative phosphorylation. The OS-ATPase consists of three parts: coupling factor 1 or F_1, the oligomycin-sensitivity-conferring protein (OSCP), and the membrane factor. F_1 is an extrinsic globular complex of five to six polypeptides (Table 1–7) with an aggregate molecular weight of around 360,000 d. Electron microscopic studies have revealed that F_1 is attached by a stalk to the inner surface of the inner membrane (Fig. 1–9). It has been claimed that OSCP, another extrinsic protein, is the connecting stalk. F_1 appears to catalyze the terminal oligomycin- and rutamycin-sensitive step in oxidative phosphorylation that results in ATP formation. When F_1 is bound to the inner membrane as a part of the OS-ATPase, its ATPase activity is quite low. However, the ATPase activity of the solubilized enzyme is considerable. In both instances the

TABLE 1–7. Subunit structures of OS-ATPase of yeast mitochondria and of chloroplast CF_1

Component	Subunit	Molecular weight
F_1	1	58,500
	2	54,000
	3	38,500
	4	31.000
	8a	12,000
OSCP	7	18,500
Membrane factor	5	29,000
	6	22,000
	8b	12,000
	9	7,500
CF_1	α	59,000
	β	56,000
	γ	37,000
	δ	17,500
	ϵ	13,000

Data from Panet and Sanadi (29) and Tzagoloff et al. (30).

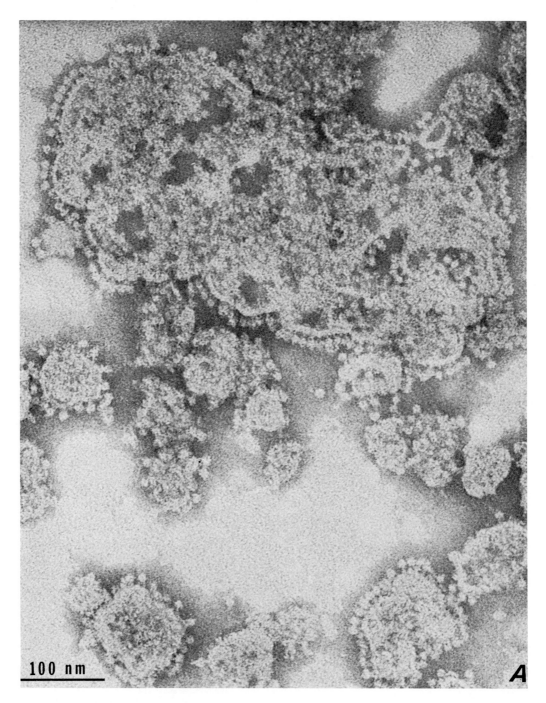

FIG. 1–9A. Electron micrograph of coupling factor F_1 attached to submitochondrial particles. $\times$234,000. (Courtesy of Dr. E. Racker.)

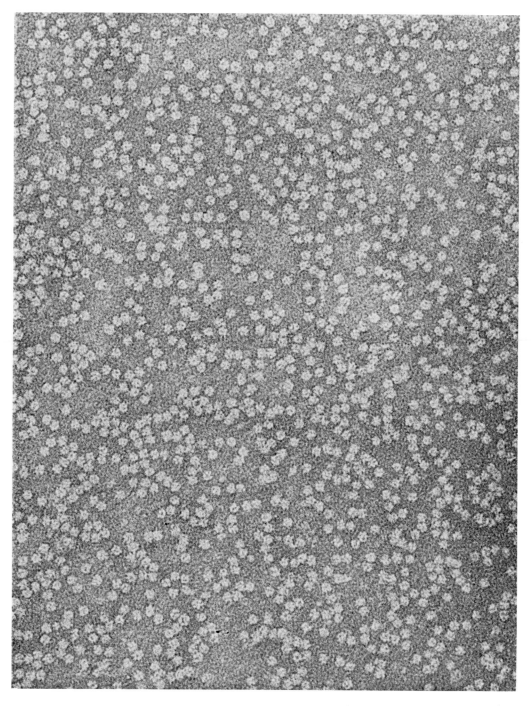

FIG. 1–9B. Electron micrograph of isolated F_1 molecules. $\times 234{,}000$. (Courtesy of Dr. E. Racker.)

ATPase activity is ordinarily used to assay the factor. In the membrane-bound state the ATPase activity is cold-stable and oligomycin-sensitive, but the activity of isolated F_1 is cold-labile and oligomycin-resistant. Such altered properties of an enzyme that depend on its association with a membrane have been termed *allotropy* by Racker (28). Clearly, they provide a convenient assay for establishing the state of F_1 *in vivo* under various conditions.

The OSCP does not confer oligomycin sensitivity directly on F_1 by itself; instead, it mediates the association of F_1 with the membrane factor to form the OS-ATPase. The membrane factor is a complex of four very hydrophobic intrinsic polypeptides of the inner membrane (Table 1–7) to which F_1 is attached by its stalk, the OSCP.

Studies on the topography of the inner membrane (31) have revealed a highly asymmetric arrangement of the constituents. Biochemical localization experiments have been greatly aided by the availability of wrong-side-out membrane vesicles called electron transport particles, which may be contrasted with right-side-out vesicles or with intact mitochondria in which proper membrane orientation is preserved (Fig. 1–10). An obvious indicator of orientation is F_1. In an electron transport particle the globular F_1 complexes can be seen by electron microscopy to be on the outside of the vesicle, whereas in a right-side-out vesicle or in an intact mitochondrion these complexes will be seen to face the inside or matrix side. Cytochrome c, an extrinsic membrane protein, has also been used to establish the orientation of inner membrane vesicles prepared by various methods. There is considerable evidence (31) that cytochrome c is on the outside of the inner membrane. Particles that are right-side-out can bind cytochrome c to the outside of the membrane. Addition of cytochrome c to these particles will stimulate electron transport, and this can be assayed as an NADH oxidase activity. Addition of cytochrome c

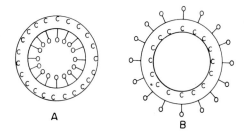

FIG. 1–10. Diagrammatic representation of mitochondrial membrane orientations. A: Properly oriented mitochondrial particle. B: Electron transport particle showing inverted orientation. The letter c indicates cytochrome c. (Adapted from Harmon et al. ref. 31, with permission.)

to electron transport particles does not stimulate NADH oxidase activity because the cytochrome c binding sites are on the inside of these vesicles. Therefore this assay can be used to establish the orientations of inner membranes in vesicles prepared by different methods. By use of this assay it can be shown that electron transport particles can be prepared with rather high efficiency when alkali is used.

By comparing the reactivities of various components of the electron transport chain in electron transport particles and mitochondria, it becomes possible to determine whether they are on the inside or the outside of the inner membrane. Consider the two flavoprotein electron acceptors NADH dehydrogenase and succinic dehydrogenase. Logically one would expect these dehydrogenases to be on the matrix side of the inner membrane, where they can accept electrons from their respective substrates in the matrix. The inner membrane is relatively impermeable to NADH and succinate. This means that oxidation of these substrates

TABLE 1–8. *Oxidation of NADH and succinate by O_2 in mitochondria and electron transport particles*

Component	NADH[a]	Succinate[a]
Electron transport particles	5.29	2.82
Mitochondria	2.08	1.20

[a] Data expressed as micromoles of substrate oxidized per milligram of protein per minute.
Adapted from Harmon et al. (31).

TABLE 1–9. *Locations of inner mitochondrial membrane components*

Component	Location	Nature of evidence
Cytochrome c	Outside	Good
Cytochrome a	Outside	Good
NADH dehydrogenase	Inside	Good
Succinate dehydrogenase	Inside	Good
F_1	Inside	Good
Cytochrome c_1	Inside	Tentative
Non-heme-iron III	Inside	Tentative
Cytochrome a_3	Inside	Tentative
Non-heme-iron Ib	Inside	Suggested

Adapted from Harmon et al. (31).

should proceed efficiently if they are added to electron transport particles, but not if they are added to a suspension of intact mitochondria. As shown in Table 1–8, the expected result is observed, which indicates that the two dehydrogenases are on the inside of the inner membrane. Similar experiments in which other constituents of the electron transport chain are assayed lead to the conclusion that the inner membrane is highly asymmetric (Table 1–9). This conclusion is also supported by freeze fracture studies that indicate that there is a denser concentration of particles on the matrix side of the inner membrane than on the side facing the intermembrane space.

Crane has arrived at the model shown in Fig. 1–11 by assembling the available biochemical and ultrastructural data; these depict a highly asymmetric arrangement of components. It is of particular significance that this membrane model is consistent with the chemiosmotic hypothesis of Mitchell (32), according to which the oxidation of substrates via the electron transport chain develops a proton motive force or "proton pump" across the membrane such that protons are translocated from this substrate side to the outside of the membrane.

Whereas the outer membrane is freely permeable to a wide range of low-molecular-weight substances, the inner membrane is permeable only to water, various small neutral molecules such as urea and glycerol, and short-chain fatty acids (33). The inner membrane is not permeable to the cations Na^+, K^+, and Mg^{2+}, to the anions Cl^-, Br^-, and No_3^-, or to sugars such as sucrose and

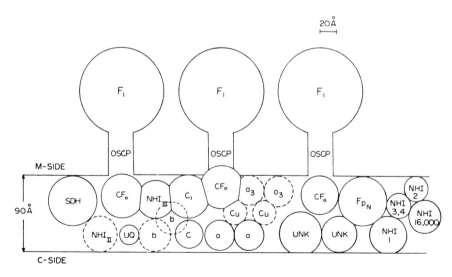

FIG. 1–11. Anisotropic binary arrangement of protein subunits in mitochondrial cristae. Subunit diameter is proportional to molecular weight. Dashed lines indicate uncertain locations. Key: NHI, non-hemi-iron protein; SDH, succinate dehydrogenase; F_1, coupling factor F_1; OSCP, oligomycin-sensitivity-conferring protein; UQ, coenzyme Q; F_{P_N}, NADH dehydrogenase; CF_0, protein fraction necessary for oligomycin-sensitive phosphorylation; $a + a_3 + Cu$, cytochrome–oxidase complex; b, c, and c_1, cytochromes b, c, and c_1, respectively. M-side is matrix side of inner membrane; C-side is side facing intermembrane space. (Reprinted from Harmon et al., ref. 31, with permission.)

TABLE 1–10. *Specific transport systems in mitochondrial membranes*

Substrate transported	Inhibitors
P_i, arsenate	p-Chloromercuribenzoate
Succinate or malate	2-Butylmalonate
Citrate, *iso*-citrate, *cis*-aconitate	2-Butylmalonate
α-Ketoglutarate	2-Butylmalonate
L-glutamate	2-Aminoadipate, 4-hydroxyglutamate
L-aspartate	
ADP, ATP	Atractyloside
α-Glycerophosphate	

Data from Pressman (21).

most amino acids. It is also impermeable to NAD⁺, NADP⁺, NADH, and NADPH, to nucleosides 5′-monophosphate, 5′-diphosphate and 5′-triphosphate, and to coenzyme A and its esters. The matrix contains an internal pool of these coenzymes and nucleosides whose physical distinctness from the cytoplasmic pool is maintained by the inner membrane. As we shall see, the apparent impermeability of the inner membrane raises some major problems, for it is abundantly clear (see Chapters 2 and 3) that the mitochondrial genome has very limited coding potential and that numerous proteins of the mitochondrial matrix and the inner membrane are synthesized in the cytoplasm rather than in the mitochondrial matrix.

Despite its general impermeability to most substances, the inner membrane does contain permeases or carriers specific for certain substances (Table 1–10). In rat liver mitochondria, carriers have been identified for ADP and ATP, for phosphate, and for certain intermediates of the tricarboxylic acid cycle.

MITOCHONDRIAL MATRIX

Although there is some disagreement, most workers believe that the enzymes of the tricarboxylic acid cycle are localized in the matrix (Table 1–5), with the exception of succinic dehydrogenase, which is an inner membrane component, as already discussed. Enzymes involved in related processes, such as glutamate oxidation, transamination, fatty acid oxidation, etc., are also found in the matrix (Table 1–5), as are mitochondrial DNA and the mitochondrial protein-synthesizing system (see Chapters 2 and 3).

CHLOROPLASTS

Although the variety of morphologically distinct chloroplast types found in algae and higher plants is bewildering, most fully developed chloroplasts possess a set of common structural features (Fig. 1–12). Like the mitochondrion, the chloroplast is surrounded by a double-membrane envelope. The inner membrane invaginates to form the chloroplast lamellae. The photosynthetic apparatus is located in lamellar membranes that are arranged in closed flattened vesicles called *disks* or *thylakoids* (Fig. 1–12). In higher plants the thylakoids are stacked in structures called *grana*. Individual grana are connected by thylakoids protruding beyond the ends of the grana called *stroma lamellae*. These lamellae take their name from the fact that they extend into the

→

FIG. 1–12. Chloroplasts of a wild-type tomato plant. ×50,000. Thylakoids are stacked up to form grana, and grana are connected by stroma lamellae. Key: G, grana; OG, osmiophilic granule; S, starch; SL, stroma lamella. (Courtesy of Dr. J. E. Boynton.)

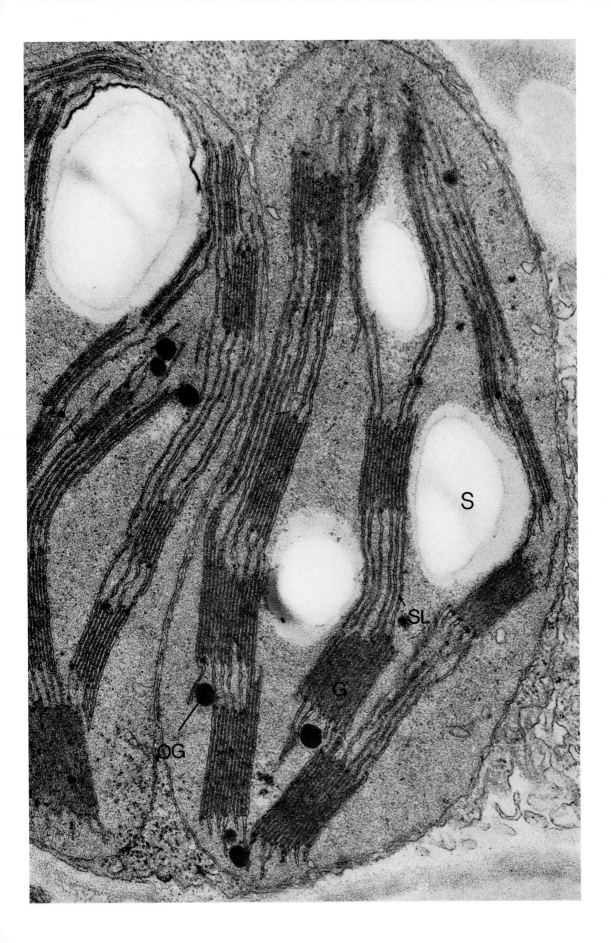

soluble matrix or stroma of the chloroplast. In algae the thylakoids lie free in the stroma and tend not to be stacked into grana, although they often appear in pairs or in groups of three or more (Fig. 1–13). The most common type of chloroplast structure among the algae is one in which the whole length of the chloroplast is traversed by sheets that appear in cross section as bands consisting of three closely appressed thylakoids. A starch-containing structure, the *pyrenoid,* is a characteristic feature of many algal chloroplasts (Fig. 1–13). An excellent review of plastid morphology can be found in the work of Kirk and Tilney-Bassett (34).

In 1885 Schimper classified the plastids into three groups based on their color. The *chloroplasts* were green, the *leucoplasts* were white, and the *chromoplasts* were yellow. Since that time a number of other plastid types have been described (Table 1–11). According to Kirk and Tilney-Bassett (34), the term proplastid is most correctly used to describe the small, ameboid, colorless or pale green plastid that occurs in the dividing meristematic cells of the shoots and roots of higher plants. Proplastids have relatively little internal structure, with the exception of a few apparently isolated vesicles and thylakoids and occasional invaginations from the inner membrane. In plants growing in the light, proplastids probably differentiate into chloroplasts by a process involving invagination of the inner chloroplast membrane to form lamellae. The process of invagination continues until there are many thylakoids lying in the stroma. During this process the attachment of the thylakoids to the inner membrane can be clearly seen in appropriate material (35).[1] As differentiation proceeds further, the number of thylakoids in each stack increases, until eventually the grana typical of the mature chloroplast are formed (Fig. 1–12). The whole process is accompanied by the synthesis of various chloroplast components, including chlorophyll and carotenoids.

If seeds are germinated and the young plants are grown in the dark, a different series of events takes place. The seedlings appear etiolated and possess no chlorophyll. The plastids present in such plants are also commonly termed proplastids (36) or dark-grown proplastids (37,38), but Kirk and Tilney-Bassett (34) have suggested that they be called etioplasts to distinguish them from the proplastids of light-grown plants. Although this usage has not found widespread acceptance, it seems most appropriate in view of the profound differences in structure observed in etioplasts and proplastids. In the dark, the vesicles present in the proplastid do not form lamellae; instead, they appear to bud off and aggregate to form a crystalline lattice called the *prolamellar body* (Fig. 1–14). The prolamellar body accumulates protochlorophyll(ide), a precursor of chlorophyll a. Thus the resulting etioplasts are distinctly different in structure from both proplastids and chloroplasts.

Exposure of etiolated plants to the light results in remarkable synchronous series of transformations that cause the etioplasts to differentiate into chloroplasts within 24 to 48 hr of illumination, depending on the age of the plant. Based on their studies with barley, von Wettstein and colleagues (37, 38) have distinguished three steps in the process. In the first step the protochlorophyll(ide) pigment, with an *in vivo* absorption maximum at 650 nm, is photoconverted to chlorophyll(ide), with an *in vivo* absorption maximum at 684 nm, following which the paracrystalline structure of the prola-

[1] See footnote on page 38.

$\longrightarrow$

FIG. 1–13. Wild-type cell of the unicellular green alga *Chlamydomonas reinhardtii.* A: Median section of whole cell. ✕15,900. B: Anterior end of cell showing the two flagella. ✕18,000. C: Portion of chloroplast showing the eyespot. ✕44,000. Key: CL, chloroplast lamella; CP, chloroplast; EY, eyespot; FL, flagella; M, mitochondrion; N, nucleus; NM, nuclear envelope; NP, nuclear pore; NU, nucleolus; P, pyrenoid; S, starch grain; V, vacuole. (Courtesy of Dr. J. E. Boynton.)

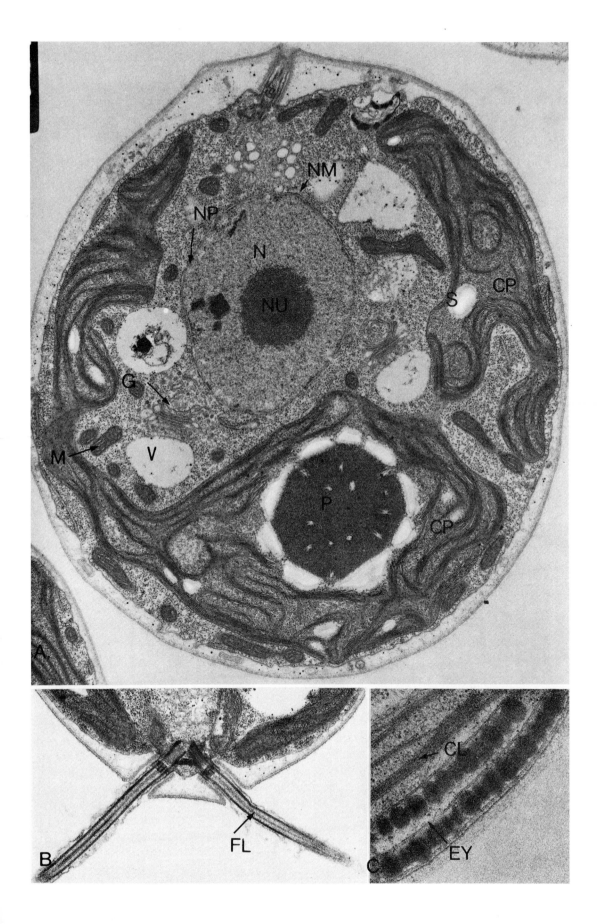

TABLE 1–11. *Classification of plastid types*

Plastid type	Characteristics
Chloroplast	Typical mature green plastid
Chromoplast	Yellow plastid containing carotenoid pigment globules
Leucoplasts (this group contains the following subtypes)	White plastids
Amyloplast	Starch-containing plastid
Elaioplast	Oil-containing plastid
Etioplast	Plastid of a dark-grown plant
Proplastid	Small, undifferentiated plastid of meristem in shoot or root
Proteinoplast	Plastid containing protein crystals

Data from Mühlethaler (36) and Kirk and Tilney-Bassett (34).

mellar body is lost, and the material is dispersed as primary lamellar layers. Whether or not these events are temporally separated depends on the age of the etiolated seedling and probably on other physiologic parameters as well. The protochlorophyll is attached to a lamellar protein called protochlorophyllide holochrome, which plays an essential role in the photoreduction of protochlorophyll to chlorophyll *in vitro* as well as *in vivo* (39). During dispersal of the prolamellar body material into primary layers, the photochemically formed chlorophyll with an *in vivo* absorption maximum at 684 nm is converted to the 672-nm form found in normal chloroplasts, as originally shown by Shibata (40). In the final light-dependent step, the formation of new thylakoids and their aggregation into grana occur in strict correlation with chlorophyll synthesis, resulting in an *in vivo* absorption maximum at 678 nm. This involves a series of correlated temperature-dependent enzymatic reactions. The photochemical activity of the chloroplast and its capacity for light-driven electron transport and oxygen evolution develop simultaneously with the formation of grana structure. Protein and lipid synthesis is also required. This is the slowest of the three steps.

Among algae, chloroplasts are usually derived from preexisting chloroplasts without the intervention of a morphogenetic sequence, except in certain multicellular forms where proplastids are found (41). However, in the two algae *Chlamydomonas* and *Euglena,* which will be discussed extensively in other chapters, chloroplast dedifferentiation can be induced by growing the cells in the dark. This dedifferentiation occurs naturally in wild-type *Euglena gracilis* (see Chapter 16) and in a very interesting mutant of *C. reinhardtii* called the yellow mutant (see Chapter 20). These dedifferentiated chloroplasts bear many resemblances to the dark-grown proplastids or etioplasts of higher plants. They accumulate protochlorophyll, and they contain structures similar to the prolamellar bodies of higher plant etioplasts (42,43). On exposure to light, protochlorophyll is converted to chlorophyll a, and the prolamellar-body-like structure disappears; synthesis of chlorophyll and membranes begins, and photosynthetic capacity is recovered. In *Euglena* the dedifferentiated plastid of dark-grown cells has generally been referred to as a proplastid (43).

The classic Calvin-cycle pathway of photosynthetic CO_2 fixation involves carboxylation of the pentose sugar ribulose-1,5-bisphosphate (RuBP) by the enzyme ribulose-1,5-bisphosphate carboxylase (RuBPCase) to form two molecules of the 3-carbon compound 3-phosphoglyceric acid (PGA). Certain plants, notably tropical grasses, such as sugar cane, fix CO_2 into 4-carbon compounds such as malic acid and aspartic acid. Such plants are often referred to as C^4 plants to distinguish them from C^3 or Calvin-cycle plants. Plants with the C^4 pathway of CO_2 fixation normally have specialized anatomy in which the vein of the leaf is surrounded by a bundle sheath consisting of a specialized layer of cells. Chloroplasts are most abundant in the cells of the bundle sheath and in the first layer of mesophyll cells immediately adjacent. The plastids of the two cell types are dimorphic. The chloroplasts of the bundle

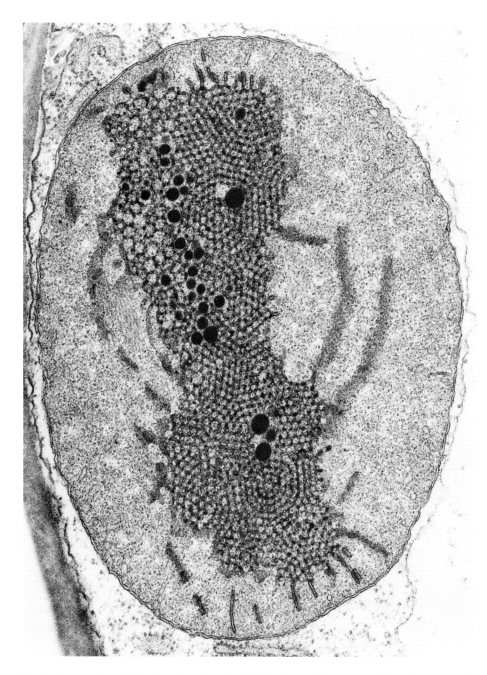

FIG. 1–14. Section through an etioplast from the primary leaf of a barley seedling grown in darkness for 7 days at 23°C and 80% relative humidity. The paracrystalline prolamellar body contains tubules with both wide and narrow spacing. Fixation: glutaraldehyde–osmium tetroxide. Staining: uranyl acetate–lead citrate. ×40,000. μm = 4.0 cm. (Courtesy of Dr. J. E. Boynton.)

sheath cells are nearly devoid of grana, whereas normal grana stacks connected by unpaired stroma lamellae are seen in the chloroplasts of the mesophyll cells. Fixation of CO_2 appears to occur principally by the C^3 pathway in the bundle sheath and by the C^4 pathway in the mesophyll.

CHLOROPLAST LAMELLAE

Like the mitochondrial inner membrane, the chloroplast lamellae depend for their biogenesis on chloroplast and cytoplasmic protein synthesis, and it is likely that the protein components of these lamellae are coded in both the nuclear and the chloroplast genomes (see Chapter 20). Thus it is appropriate to pay particular attention to the structure of these lamellae in this chapter.

Early electron microscopic studies of thin sections of chloroplasts confirmed and extended the general information previously obtained with the light microscope (44,45). The internal membrane system of the chloroplast was seen to consist of numerous flattened sacs in two general size ranges. The smaller size, less than 1 μm in diameter, occurred in stacks in which adjacent sacs were appressed to form the grana. The larger sacs interconnected the grana. This model was developed by Menke (46–48) as a general structure for chloroplast lamellae in higher plants; in 1962 he coined the name thylakoid to describe the flattened sacs (49). In Menke's terminology, the stroma lamellae were termed large thylakoids, and the grana lamellae were called small thylakoids.

Since sections give only a two-dimensional picture of a three-dimensional structure, different interpretations of the arrangement of chloroplast membranes were bound to arise. A simple three-dimensional model proposed by von Wettstein and colleagues (50) supposed that the lamellae of the stroma extended out as flat sheets going through the greater part of the chloroplast

and intersecting a number of grana. Weier and associates (51) set forth a different model in which it was imagined that the stroma lamellae were not flat sheets but rather were flattened tubules connecting the grana thylakoids together in a fretwork. The flattened tubules connecting the grana (i.e., stroma lamellae) were called *frets*. Heslop-Harrison (52) proposed a somewhat similar model based on studies of sections of hemp chloroplasts. In his model the fretwork consisted of more completely flattened tubes with changing diameters and without interconnections in the vertical plane, so that a single cross section produced a model very similar to the one proposed by Menke in 1960. An important ingredient of this model was that the entire lamellar system, including grana thylakoids and stroma lamellae, was assumed to constitute a single enormously complex membrane separate and distinct from the stroma. This general idea was pursued further by Paolillo and colleagues (53–57) and by Wehrmeyer (58, 59), who found that a single stroma lamella could exist in a spiral arrangement around a granum, being connected with individual small thylakoids. Paolillo (53) found in five different plant species that the helical pattern of a stroma lamella ascending a granum is always right-handed. In Paolillo's model each compartment of a granum can contact eight different stroma lamellae, which in turn can connect all other stacks of grana. The major distinguishing feature of all recent models of the chloroplast lamellar network is that they suggest that the entire system is continuous in much the same way that the mitochondrial cristae are continuous with the mitochondrial inner membrane. If this view is correct, chloroplasts, like mitochondria, should contain two membrane-bounded compartments. The innermost compartment, the stroma, would be bounded by the inner membrane, and its invaginations would be represented by the stroma lamellae and grana thylakoids. The outer compartment, which would be equiva-

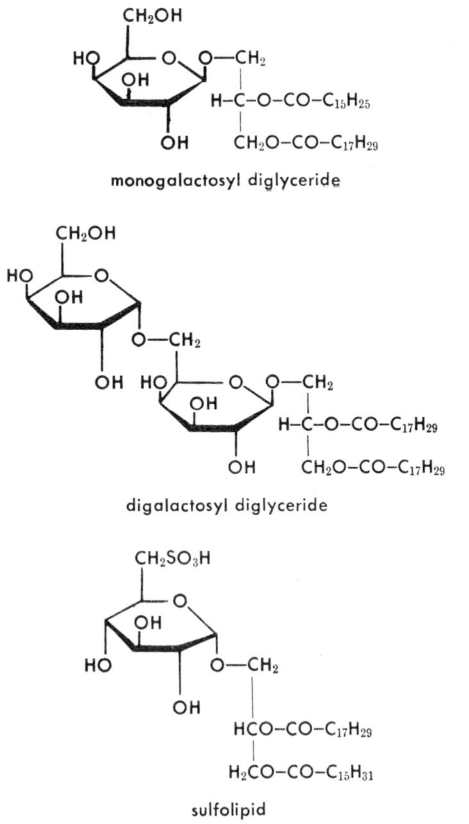

FIG. 1–15. Structures of the two galactolipids and the sulfolipid that are the predominant polar lipids of chloroplast lamellae.

lent to the intermembrane space of the mitochondrion, would be bounded by the inner and outer membranes of the chloroplast.

Chloroplast membranes are about half lipid and half protein (60–62), and the lipid composition of these membranes differs markedly from that of the mitochondrial and microsomal membranes (Table 1–3). While phospholipids are the predominant polar lipids in mitochondrial and microsomal membranes, glycolipids are the principal polar lipids in chloroplast lamellae (Table 1–3). Two glycolipids, the monogalactosyl and digalactosyl diglycerides, account for most of the glycolipid present in chloroplast lamellae, and a sulfolipid unique to plants makes up the rest (Table 1–3 and Fig. 1–15). The most characteristic aspect

of the galactolipids is their remarkably high α-linolenate $(18:3)$[1] content (Table 1–12). Mitochondrial and microsomal membranes are very poor in linolenic acid (63) (Table 1–12). A direct relationship between α-linolenate content and oxygen-producing capability was recognized by Erwin and Bloch (64). Except in rare cases where polyunsaturated fatty acids are not formed, linolenate production and photosynthetic capacity develop simultaneously. Benson (65) believed that either (a) galactolipids containing α-linolenate are involved in oxygen production or are required for proper conformation of the oxygen-producing system or (b) oxygen is a requirement for oxidative desaturation of linolenic acid $(18:2)$, and its production facilitates synthesis of α-linolenate. Although the former possibility seems more likely, the experimental data described by Benson (65) do not appear to rule out the latter.

The components of the photosynthetic electron transport system (PETS) are closely associated with the chloroplast lamellar system. The most commonly accepted formulation of the PETS is the series formulation or Z scheme originally proposed by Hill and Bendall (66), for which there has been much experimental verification in recent years. There have been numerous reviews (10,67–75) pertaining to various aspects of the process.

The scheme of Hill and Bendall supposes that there are two photosystems (PS I and PS II) linked in series by the PETS (Fig. 1–16). PS I responds to far red light (> 680–700 nm), and PS II responds to shorter wavelengths of light. Each of the photosystems contains light-harvesting photosynthetic units consisting of several hundred chlorophyll molecules. Each light quantum absorbed by any chlorophyll molecule in the photosynthetic unit must be transferred through the complex until it reaches the specialized chlorophyll a mole-

---

[1] Carbon:double bond ratio.

TABLE 1–12. *Percentage fatty acid composition of certain lipids found in rat liver mitochondria and spinach leaves*

Fatty Acid	Carbons: double bond	Phosphatidyl choline fraction		Phosphatidyl ethanolamine fraction		Rat liver cardiolipin fraction	Spinach monogalactosyl glyceride fraction
		Rat liver	Spinach	Rat liver	Spinach		
Palmitic	16:0	17	21	17	27	4	10
Hexatrienoic	16:3	—	—	—	—	—	25
Stearic	18:0	21	1	25	2	—	—
Oleic	18:1	13	16	8	7	12	—
Linoleic	18:2	20	32	10	42	79	—
Linolenic	18:3	—	30	—	22	—	65
Arachidonic	20:4	17	—	21	—	—	—
Docosahexanoic	22:6	6	—	12	—	—	—

Data from Chapman and Leslie (63) and Benson (65).

cules called energy traps or reaction centers that constitute less than 1% of the total chlorophyll. Once the excitation energy reaches the trap, the primary photoact facilitates transfer of a hydrogen atom (or electron) from a donor molecule to an acceptor molecule. In green plants and green algae, which are the only photosynthetic organisms we discuss in detail in this book, the photosynthetic units associated with the two photosystems differ in several ways. PS I contains a long-wavelength form of chlorophyll a (a_{690}), with the reaction center chlorophyll being in a form that has an even longer wavelength, called P_{700}. The photosynthetic unit of PS II contains the

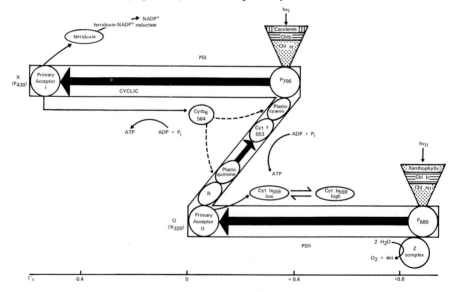

FIG. 1–16. The Z scheme for electron flow in photosynthesis. The two bold horizontal arrows represent the two light reactions; all others represent dark reactions. Flow of electrons from H_2O to $NADP^+$ is designated as noncyclic electron flow; flow from the primary acceptor of PS I to the intersystem intermediates (plastoquinone or plastocyanin) is designated as cyclic. Key: E'_0, oxidation–reduction potential at pH 7 in volts; Z complex, electron donor system for PS II; P_{680}, proposed energy trap of PS II; Chl a_{II}, bulk chlorophyll a of PS II; Chl b, chlorophyll b; Chl a_I, bulk chlorophyll a of PS I; Q (X320), electron acceptor for PS II (Q stands for the quencher of Chl a_{II} fluorescence); Cyt b_{559}, cytochrome b_{559}, with low and high referring to low- and high-potential forms; Cyt f_{553}, cytochrome f_{553}; Cyt b_6564, cytochrome b_6564; P_{700}, energy trap and reaction center of PS I; X (P430), PS I electron acceptor; R, quinone; ADP, P_i, and ATP, adenosine diphosphate, inorganic phosphate, and adenosine triphosphate, respectively; NADP, nicotinamide adenine dinucleotide phosphate. (Adapted from Govindjee and Govindjee, ref. 105, with permission.)

bulk, if not all, of a shorter-wavelength form of chlorophyll a (a_{670}), plus accessory pigments, including most, if not all, of the chlorophyll b, which absorb at wavelengths less than 670 nm. The reaction center chlorophyll of PS II is a special form of chlorophyll a with an absorption maximum between 680 and 690 nm.

Two basic types of light-induced electron flow patterns, termed noncyclic and cyclic, are associated with the two photosystems. In noncyclic electron transport, the flow of electrons is initiated by the absorption of light quanta by PS II. Absorption of excitation energy by PS II results in the production of a strong oxidant and a weak reductant. This photoreaction results in oxidation of water, with a standard potential (E'_0) of 0.8 volts (V), and reduction of Q, the quencher of fluorescence, also called component X320 ($E'_0 \cong 0V$) (Fig. 1–16). The electron donor for PS II is not water itself, but the complex Z, which contains manganese. From Q the electrons travel down the main part of the PETS to PS I. The components include two quinones (plastoquinone and R), a c-type cytochrome (cytochrome f), and the copper-containing protein plastocyanin. A b-type cytochrome (b_{559}) is also associated with PS II (75).

Absorption of excitation energy by PS I results in the production of a weak oxidant and a strong reductant. Electrons are transferred from plastocyanin ($E'_0 = +0.37V$) to the primary acceptor of PS I, X ($E'_0 \cong -0.6V$). X may or may not be equivalent with the ferridoxin-reducing substance (FRS) ($E'_0 = -0.55V$) that donates electrons to the iron-sulfur protein ferridoxin. Ferridoxin is oxidized by ferridoxin-NADP reductase, a flavoprotein, which reduces NADP. During transport of electrons between PS II and PS I, one or possibly two ATP molecules are formed per pair of electrons at a site prior to cytochrome f (68) via *noncyclic photophosphorylation*.

Cyclic electron transport is an electron transfer process that is associated with PS I (Fig. 1–16). It differs from noncyclic electron transport in that the high-energy electrons derived from excited P_{700} molecules are not transported to NADP, but are, instead, transported back to electron-deficient P_{700} molecules in the ground state. The members of the cycle include X as well as a photosynthetic cytochrome called cytochrome $b_6 564$. During the electron transfer process, ATP is formed via *cyclic photophosphorylation*.

ATP formation in photosynthesis is catalyzed by a coupling factor (CF_1) that is bound to the stromal surface of the thylakoid membrane and is generally similar in structure to the mitochondrial coupling factor F_1 (29,76,77). Purified CF_1 is a sphere roughly 90 Å in diameter with a molecular weight of 325,000 d. CF_1 is made up of five nonidentical polypeptide subunits designated α, β, γ, δ, and ϵ with molecular weights ranging from 59,000 to 13,000 d (Table 1–7). CF_1 comes off the thylakoid membrane with very mild treatment (e.g., EDTA washing), which indicates that it is not covalently bound to the membrane. CF_1, like F_1, has ATPase activity, but the two factors differ in that special treatments are required to activate the CF_1 ATPase activity. Illumination of chloroplasts in the presence of sulfhydryl reagents such as dithiothreitol activates Mg^{2+}-dependent ATPase activity of the membrane-bound enzyme. Soluble CF_1, when heated or treated with trypsin, also possesses ATPase activity, but this activity is dependent on Ca^{2+}. As in the case of F_1, the ATPase activity of solubilized CF_1 is cold-labile, whereas the activity of this enzyme in the bound state is cold-stable. It is interesting that CF_1, unlike F_1, is resistant to the energy transfer inhibitor oligomycin. Thus this drug can be used to block ATP formation preferentially in the mitochondrion without affecting the chloroplast.

The structure and organization of the thylakoid membrane components have been subjects of much interest in recent years, and they have been completely reviewed by several authors (73,78–81). In a book such as this, we can only sketch a bare outline of what is known. Like the mitochondrial inner membrane, chloroplast thylakoids and stroma lamellae are highly asymmetric membranes with which both extrinsic and intrinsic proteins are associated. The external surface of the thylakoid membrane is in contact with the stromal matrix. The main extrinsic protein of this surface is CF_1, although molecules of RuBPCase, most of which is in the stroma, are also seen along the surface (*vide infra*). Anderson (80), in her excellent review, also classified the chloroplast proteins of the PETS as probable extrinsic proteins; however, she pointed out that satisfactory classification is virtually impossible at this juncture, since with the exception of ferridoxin, these proteins remain to be characterized in detail. Opinion about the arrangement of electron carriers in the thylakoid membranes has been greatly influenced by Mitchell's chemiosmotic hypothesis of energy conservation (32), as in the case of the mitochondrial inner membrane. Trebst (74) has summarized the evidence for asymmetric arrangement of electron transport carriers across the thylakoid membrane and for vectorial proton flow. Through the use of immunologic and other techniques it has been shown that the electron acceptors of PS I (ferridoxin, ferridoxin–NADP reductase, and ferridoxin reducing substance) are located at the outer thylakoid surface. In contrast, the electron donors of PS I (cytochrome f and plastocyanin) appear to be localized at the inner surface. Most of the components of PS II have not been characterized in detail biochemically; thus evidence for their localization is still indirect. However, available data suggest that the electron donors of PS II are located toward the internal surface of the thylakoid membrane, whereas the acceptors are toward the outside, as in the case of PS I (Fig. 1–17).

The intrinsic proteins of the thylakoid membrane have been studied with the aid of detergents such as SDS. As shown by Thornber and others (79), SDS solubilization of chloroplast membranes followed by SDS gel electrophoresis yields three chlorophyll-containing bands: chlorophyll protein complex I (CP I), which has associated with it PS I activity; chlorophyll complex II (CP II), which is enriched with respect to PS II; a pigment-lipid-SDS complex (free pigment), which runs at the gel front. Thornber and associates (82,83) found that CP I has a high chlorophyll a content (chlorophyll a: chlorophyll b = 12:1) and accounts for 28% of the membrane protein, whereas CP II, which is the major chlorophyll-protein complex, accounts for 50% of the total membrane protein and has a high chlorophyll b content (chlorophyll a: chlorophyll b = 1:1). The CP I complex has a molecular weight of 110,000 d and contains 14 chlorophyll a molecules as well as P_{700}, the reaction center chlorophyll of PS I. The polypeptide or polypeptides associated with this complex have high apparent molecular weights (50,000–70,000 d). CP II has a molecular weight of 32,000 d and has associated with it one chlorophyll a and one chlorophyll b molecule plus a polypeptide or polypeptides of around 25,000 d apparent molecular weight. CP II has lost the reaction center chlorophyll of PS II. It is important to note that the amounts of the two complexes and thus the amount of associated protein can vary greatly with the degree of membrane stacking (*vide infra*).

While the proteins associated with the two chlorophyll-protein complexes can account for a significant fraction of the protein present in chloroplast lamellae (i.e., in fully developed chloroplasts containing stacked grana), it is clear from SDS gel electrophoresis of total membrane polypeptides that thylakoid membranes contain a great many other polypeptides as well. Perhaps the gel system with highest resolution is the SDS-gradient gel system devel-

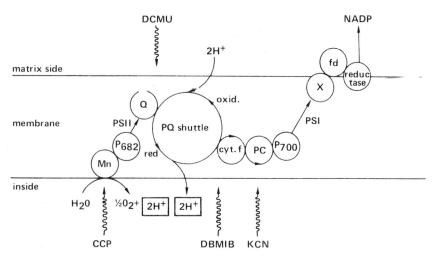

FIG. 1–17. Proposed vectorial distribution of photosynthetic electron transport components across thylakoid membrane. Key: CCP, carbonylcyanide-m-chlorophenylhydrazone; DBMIB, dibromothymoquinone, the plastoquinone antagonist; DCMU, dichlorophenyldimethylurea; and the thylakoid components: P_{700} and P_{682} are the reaction centers and X and Q are the primary electron acceptors of photosystems I and II; fd, ferridoxin; PC, plastocyanin; PQ, plastoquinone. (From Trebst, ref. 74, with permission.)

oped by Chua and Bennoun (84). In this system some 33 thylakoid membrane polypeptides can be separated; they range in apparent molecular weight from less than 10,000 to 68,000 d (Fig. 1–18). These polypeptides are presumed to be largely intrinsic polypeptides, and they include the polypeptides associated with CP I (polypeptide 2) and CP II (polypeptide 11). The functions of the remaining polypeptides, with a few exceptions (see Chapter 20), remain largely unknown.

The organization of the chloroplast thylakoids and stroma lamellae has also been studied using electron microscopic techniques such as negative staining and freeze fracture (for reviews see 81,85). The structural picture that emerges is one of asymmetry within membranes and marked organizational differences between stacked and unstacked regions (Figs. 1–19 and 1–20). Like other biologic membranes, the thylakoid membranes can be described according to the nomenclature for freeze etching electron microscopy discussed previously. The P side faces the stroma, and the E side faces the endoplasmic space of the thylakoid. The surfaces and fracture faces

are designated as indicated earlier; the subscripts S and U indicate whether the membrane is in a stacked or unstacked region. Thus, PF_U means the protoplasmic fracture face in an unstacked region.

The PS_U surface can be seen by negative staining to be covered with particles; Howell and Moudrianakis (86,87) showed that these particles are of two kinds (Fig. 1–19). First, there are 120-Å cuboidal particles that are released from the thylakoids by water washing. These particles have high specific activity of RuBPCase and thus appear to be fraction I protein molecules (see p. 34). The second type of particle, which is released by EDTA, is a five- or six-sided polygon with a diameter of 100 Å. These are CF_1 particles. On the endoplasmic surface (ES_S) of the thylakoids in regions of stacking, large particles can be seen in shadowed preparations. These are the quantasomes of Park and his colleagues (61,62, 88–90), to which we will return shortly. Freeze fracture studies reveal two other sets of faces. Fracture of the exterior, unstacked side of a thylakoid membrane or a stroma lamella reveals fracture faces PF_U and EF_U. The PF_U face contains two size classes of particles—about 40% with an average size of 80 Å and about 60% with an average size of 118 Å. The EF_U face is relatively smooth, with widely dispersed (but larger) particles embedded in it. In the stacked regions where two thylakoids are closely appressed together, two other fracture faces are seen. The PF_S face is characterized by the presence of many small, tightly packed particles with a uniform size of 80 Å. The EF_S face is studded with particles ranging in size from less than 60 Å to 200 Å, with maxima at 115 Å and 155 Å.

In summary, with the variations noted above, the freeze fracture studies reveal that relatively large particles are associated with the EF_S and EF_U faces, and smaller particles are found in the PF_S and PF_U faces. Particle densities are much higher in the stacked membrane faces than in the unstacked membrane faces. Branton and Park (91) suggested that the large particles, together with the matrix material of the membranes associated with them, were identical to the quantasomes described earlier by Park and colleagues that were thought to be repeating units in the membrane containing about 230 chlorophylls and being equivalent to the photosynthetic unit. The large freeze etching particle with matrix material removed was referred to as the quantasome core.

To investigate the two particle classes in a functional sense, it was necessary to dissociate the stroma lamellae from the grana and then fractionate the grana themselves. These aims have been achieved using combinations of mechanical and detergent fractionations of chloroplast membranes. Sane and associates (92) disrupted spinach chloroplast lamellae in the French pressure cell and separated fractions having PS I and PS II activity by differential centrifugation. The results indicate that the stroma lamellae contain mainly PS I activity and small freeze fracture particles, whereas the grana membranes, which have both PS I and PS II activity, contain both small and large freeze

fracture particles. Digitonin treatment of membranes also yields two main membrane fractions. The rapidly sedimenting fraction contains both PS I and PS II activity, whereas the slowly sedimenting fraction contains only PS I activity. The former fraction is derived from stacked membranes, and the latter fraction is derived from unstacked membranes. Thornber and associates (82) demonstrated that the digitonin PS I fraction was enriched for CP I, whereas the PS II enriched fraction also contained a greater amount of CP II. These kinds of results, together with abundant other findings from regreening plants and algae, studies of the dimorphic plastids of C^4 plants, and experiments with mutants, all of which have been competently reviewed (80,81, 85), suggest that the small particles contain CP I and the large particles contain CP II. However, studies of chloroplast development indicate that PS II activity may precede the appearance of CP II, and the same is true with respect to PS I activity and CP I. To explain these findings, Anderson (80) postulated that there are two levels of organization of the photosystems in green plants and algae. First, there are the initial or primary states of PS I and PS II. The organization of the two photosystems is at a less complex level in these states, and these primary photosystems are encountered in unstacked lamellae, such as those found in plastids during the initial stages of regreening. The most characteristic feature of these primary photosystems is the absence of CP II. Anderson suggested that absence of this major complex means that the membranes are unstacked and there are no large freeze fracture particles. In addition, although these membranes must possess an active reaction center for PS I, they probably lack the integrated assembly of CP I that is characteristic of the mature PS I. As the chloroplast membranes differentiate, the characteristic areas of stacked grana and unstacked stroma lamellae become apparent. The secondary photosystems are com-

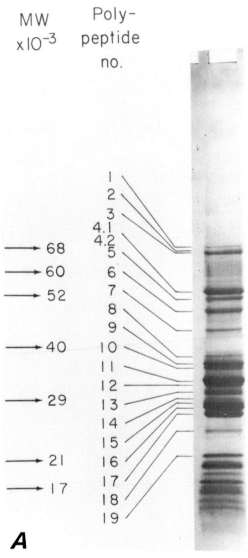

FIG. 1–18A. SDS electrophoretogram of thylakoid membrane polypeptides from wild-type *Chlamydomonas reinhardtii* equivalent to 25 μg of chlorophyll. Sample was heated to 100° for 1 min. Membrane polypeptides were stained with coomassie blue. The relationships between electrophoretic mobilities and molecular weights were established with the following markers: bovine serum albumin (68,000 d), catalase (60,000 d), α-amylase (52,000 d), creatine kinase (40,000 d), carbonic anhydrase (29,000 d), soybean trypsin inhibitor (21,000 d), and myoglobin (17,000 d). (From Chua and Bennoun, ref. 84, with permission.)

FIG. 1–18B. Densitometric tracing of thylakoid membrane polypeptides prepared from wild-type cells of *Chlamydomonas reinhardtii*. (From Chua and Bennoun, ref. 84, with permission.)

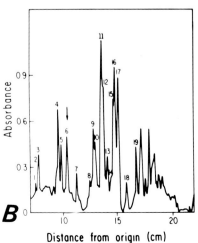

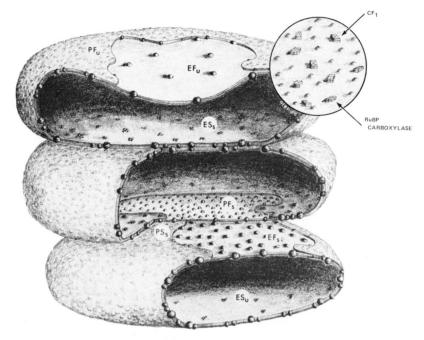

FIG. 1–19. Diagrammatic representation of membrane structure of three thylakoids in a grana stack. Within the partition regions (where two thylakoids are appressed) the "large" freeze fracture particles are densely distributed on the EF_S face. They are much less dense on the unstacked EF_U face. The "small" particles seen on the PF_S and PF_U faces are rather uniformly distributed throughout the membrane. The CF_1 portion of the coupling factor and RuBP carboxylase are localized on the protoplasmic surface (PS), as shown in the inset. For further discussion, particularly with respect to nomenclature, see text. (From Arntzen, ref. 81, with permission.)

pleted during this stage, and CP I and CP II become fully developed. The ultimate stage may be the assembly of CP II, following which the membranes stack. It is then and only then that PS II becomes restricted to grana membranes.

This dynamic view of chloroplast membrane organization has a great deal to recommend it, since it appears that there has been a certain static quality in our thinking about chloroplast membrane structure in recent years. A principal point at issue concerns the numbers of intrinsic membrane proteins and their functions. Largely through the work of Thornber and colleagues (79) it has been established that specific intrinsic polypeptides are associated with CP I and CP II. On the other hand, SDS gel electrophoresis of membranes washed with EDTA to remove CF_1 suggests that the total number of membrane polypeptides is large (Fig. 1–18)

although quantitatively the CP I and CP II polypeptides represent a major fraction of the total. Clear genetic evidence for association of several of these other polypeptides with photosystem function will be discussed in Chapter 20. The point is that for a complete understanding of thylakoid membrane organization it will eventually be necessary to establish the functions of these other intrinsic proteins and their topographic locations and relationships to the two chlorophyll-protein complexes and to the photosynthetic electron transfer chain.

CHLOROPLAST STROMA

The mobile phase of the chloroplast, the stroma, contains the soluble enzymes involved in CO_2 fixation. Most plants possess the classic C^3 or Calvin-cycle pathway in which the first product of photosynthetic CO_2 fixation is the 3-carbon compound

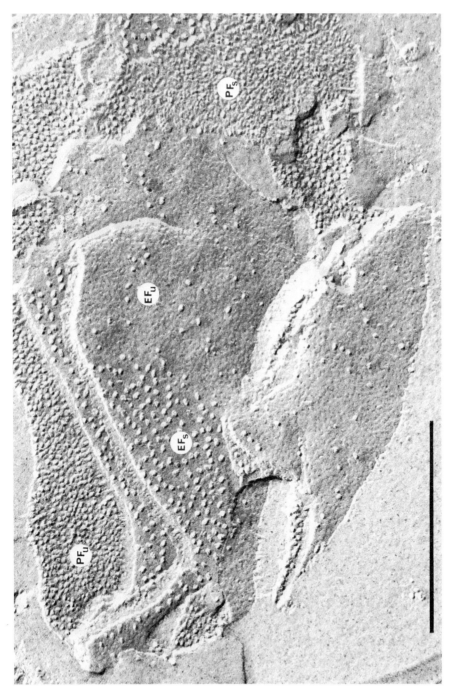

FIG. 1–20. Particulate substructure of pea (*Pisum sativum*) chloroplast membranes as revealed by freeze fracture technique. In upper left are portions of two appressed thylakoids that are part of a grana stack. The end membrane of this stack has been cross-fractured to reveal a PF_U face. A cross-fracture w thin the stack shows "large" particles on the EF_S face. The semicircular edge of a partition (the region where the grana membranes fuse) can be identified in l≥ft central portion as an area of transition from relatively densely packed EF_S particles of the grana stack to widely spaced EF_U particles of the stroma lamellae. EF_U particles are also observed on the end membrane of the circular thylakoid, which is partially visible in lower central portion. PF_S particles can be seen within a cross-fracture grana stack on right-hand side. Bar = 0.5 μm. (From Arntzen, ref. 81, with permission.)

PGA (93). The principal protein component of the stroma in such plants is a protein that was discovered by Wildman and Bonner (94) in 1947; they called it fraction I protein (for reviews see 95–97). This protein comprises more than 50% of the soluble protein in leaves of C^3 plants, and it appears to be identical with the CO_2-fixing enzyme RuBPCase. As this protein will be of particular concern to us in Chapters 18 and 20, some description of it is warranted here. The molecular weight of fraction I protein is about 560,000 d, and the particles themselves have diameters of 100 to 200 Å in negatively stained preparations. Fraction I protein is an aggregate of two polypeptide subunits, one with a molecular weight of approximately 55,000 d and one with a molecular weight of approximately 12,000 d. Each molecule of fraction I protein is currently believed to contain eight large and eight small subunits. It has been known for a long time that the primary function of fraction I protein is to catalyze the first CO_2 fixation step in photosynthesis, using RuBP and CO_2 as substrates to produce two molecules of PGA. Recently, fraction I protein has also been shown to catalyze the oxygenation of RuBP to form one molecule of PGA and one molecule of the 2-carbon compound phosphoglycollic acid. This is the first intermediate in the photorespiration pathway (for reviews see 93,98) which is the process by which light stimulates CO_2 production in green plants. The balance between photosynthetic CO_2 fixation and photorespiration is determined by the relative concentrations of oxygen and CO_2 in a photosynthesizing tissue. High oxygen concentrations stimulate photorespiration, and high CO_2 concentrations stimulate photosynthetic CO_2 fixation. Cells or tissues engaged in photorespiration can accumulate large amounts of glycollate.

As mentioned previously, some plants possess a C^4 pathway of photosynthesis; this was first studied in detail by Hatch and Slack (99,100). In such plants, CO_2 is fixed initially into malic and aspartic acids, which are 4-carbon compounds. The initial step in the pathway involves carboxylation of phosphoenolpyruvic acid (PEP) by CO_2 to form the 4-carbon compound oxaloacetic acid under the aegis of the enzyme PEP carboxylase. Oxaloacetic acid is then converted to malic acid, which reacts with the 5-carbon compound RuBP to yield a 6-carbon intermediate of the Calvin cycle and pyruvate, which is then converted to PEP. Plants having the C^4 pathway possess a specialized leaf anatomy, as previously described, in which the vein of the leaf is surrounded by a bundle sheath consisting of a specialized layer of cells. Chloroplasts are most abundant in cells of the bundle sheath and in the first layer of mesophyll cells immediately adjacent. The bundle sheath cells have a predominantly C^3 pathway of carbon metabolism, whereas the mesophyll cells fix CO_2 via the C^4 pathway.

In concluding this chapter, it is important to emphasize that only a sketchy account of chloroplast and mitochondrial structure and function is possible in a book of this kind. Those features of chloroplasts and mitochondria that seem clearly to involve the functioning of the genetic and protein-synthesizing apparatuses of the two organelles have been emphasized. Specifically, these are the electron transport membranes of both organelles and their constituents and the chloroplast enzyme RuBPCase. At the moment, with one or two exceptions (see Chapter 20), it does not appear that synthesis of the bounding outer membranes of the two organelles or the soluble enzymes of the organelle matrix (with the notable exception of RuBPCase) require the participation of either organelle genes or protein synthesis.

REFERENCES

1. Singer, S. J., and Nicolson, G. L. (1972): The fluid mosaic model of the structure of cell membranes. *Science,* 175:720–731.
2. Singer, S. J. (1974): The molecular organiza-

tion of membranes. *Annu. Rev. Biochem.,* 43:805–833.

3. Danielli, J. F., and Davson, H. (1935): A contribution to the theory of permeability of thin films. *J. Cell Comp. Physiol.,* 5:495–508.

4. Robertson, J. D. (1959): The ultrastructure of cell membranes and their derivatives. *Biochem. Soc. Symp.,* 16:3–43.

5. Stoeckenius, W. (1959): An electron microscope study of myelin figures. *J. Biophys. Biochem. Cytol.,* 5:491–500.

6. Stoeckenius, W. (1962): The molecular structure of lipid-water systems and cell membrane models studied with the electron microscope. In: *The Interpretation of Ultrastructure, Vol. 1, Symposium of the International Society of Cell Biology,* edited by R. J. C. Harris, pp. 349–367. Academic Press, New York.

7. DuPraw, E. J. (1968): *Cell and Molecular Biology.* Academic Press, New York.

8. Moor, H., Mühlethaler, K., Waldner, H., and Frey-Wyssling, A. (1961): A new freezing ultramicrotome. *J. Biophys. Biochem. Cytol.,* 10:1–13.

9. Mühlethaler, K. (1966): The ultrastructure of the plastid lamellae. In: *Biochemistry of Chloroplasts, Vol. 1,* edited by T. W. Goodwin, pp. 49–64. Academic Press, New York.

10. Park, R. B., and Sane, P. V. (1971): Distribution of function and structure in chloroplast lamellae. *Annu. Rev. Plant Physiol.,* 22:395–430.

11. Kirk, J. T. O. (1971): Chloroplast structure and biogenesis. *Annu. Rev. Biochem.,* 40: 161–196.

12. Wehrli, E., Mühlethaler, K., and Moor, H. (1970): Membrane structure as seen with a double replica method for freeze fracturing. *Exp. Cell Res.,* 59:336–339.

13. Branton, D., Bullivant, S., Gilula, N. B., Karnovsky, M. J., Moor, H., Mühlethaler, K., Northcote, D. H., Packer, L., Satir, B., Satir, P., Speth, V., Staehlin, L. A., Steere, R. L., and Weinstein, R. S. (1975): Freeze-etching nomenclature. *Science,* 190:54–56.

14. Hare, J. F., and Crane, F. L. (1974): Proteins of the mitochondrial cristae. *Subcell. Biochem.,* 3:1–25.

15. Palade, G. (1952): The fine structure of mitochondria. *Anat. Rec.,* 114:427–451.

16. Sjöstrand, F. S. (1953): A new ultrastructural element of the membranes in mitochondria and some cytoplasmic membranes. *J. Ultrastruct. Res.,* 9:340–361.

17. Palade, G. (1953): An electron microscope study of the mitochondrial structure. *J. Histochem. Cytochem.,* 1:188–211.

18. Andersson-Cedergren, E. (1959): Ultrastructure of motor endplate and sarcoplasmic components of mouse skeletal muscle fiber as revealed by three-dimensional reconstruction from serial sections. *J. Ultrastruct. Res. (Suppl. 1),* 1–191.

19. Hackenbrock, C. R. (1966): Ultrastructural bases for metabolically linked mechanical activity in mitochondria. I. Reversible ultrastructural changes with a change in metabolic steady state in isolated liver mitochondria. *J. Cell Biol.,* 30:269–297.

20. Ernster, L., and Kuylenstierna, B. (1970): Outer membranes of mitochondria. In: *Membranes of Mitochondria and Chloroplasts,* edited by E. Racker, pp. 172–212. Van Nostrand Reinhold, New York.

21. Pressman, B. C. (1970): Energy-linked transport in mitochondria. In: *Membranes of Mitochondria and Chloroplasts,* edited by E. Racker, pp. 213–250. Van Nostrand Reinhold, New York.

22. Parsons, D. F., Williams, G. R., and Chance, B. (1966): Characteristics of isolated and purified preparations of the outer and inner membranes of mitochondria. *Ann. N.Y. Acad. Sci.,* 137:643–666.

23. Parsons, D. F., and Yano, Y. (1967): The cholesterol content of the outer and inner membranes of guinea-pig liver mitochondria. *Biochim. Biophys. Acta,* 135:362–364.

24. Schnaitman, C. A. (1969): Comparison of rat liver mitochondrial and microsomal membranes. *Proc. Natl. Acad. Sci. U.S.A.,* 63: 412–419.

25. Sottocasa, G. L., Kuylenstierna, B., Ernster, L., and Bergstrand, A. (1967): Separation and some enzymatic properties of the inner and outer membranes of rat liver mitochondria. *Methods Enzymol.,* 10:448–463.

26. Ikuma, H. (1972): Electron transport in plant respiration. *Annu. Rev. Plant Physiol.,* 23:419–436.

27. Henry, M.-F., and Nyns, E. J. (1975): Cyanide-insensitive respiration. An alternative mitochondrial pathway. *Subcell. Biochem.,* 4:1–65.

28. Racker, E. (1970): Function and structure of the inner membrane of mitochondria and chloroplasts. In: *Membranes of Mitochondria and Chloroplasts,* edited by E. Racker, pp. 127–171. Van Nostrand Reinhold, New York.

29. Panet, R., and Sanadi, D. R. (1976): Soluble and membrane ATPases of mitochondria, chloroplasts and bacteria: Molecular structure, enzymatic properties, and functions. *Current Topics in Membranes and Transport,* 8:99–160.

30. Tzagoloff, A., Rubin, M. S., and Sierra, M. F. (1973): Biosynthesis of mitochondrial enzymes. *Biochim. Biophys. Acta,* 301:71–104.

31. Harmon, H. J., Hall, J. D., and Crane, F. L. (1974): Structure of mitochondrial cristae membranes. *Biochim. Biophys. Acta,* 344: 119–155.

32. Mitchell, P. (1966): Chemiosmotic coupling in oxidative and photosynthetic phosphorylation. *Biol. Rev.,* 41:445–502.

33. Lehninger, A. L. (1970): *Biochemistry: The Molecular Basis of Cell Structure and Function.* Worth Publishers, New York.

34. Kirk, J. T. O., and Tilney-Bassett, R. A. E. (1967): *The Plastids: Their Chemistry, Structure, Growth and Inheritance.* W. H. Freeman, San Francisco.

35. Menke, W. (1964): Feinbau und Entwicklung der Plastiden. *Ber. Dtsch. Bot. Ges.,* 77:340–354.

36. Mühlethaler, K. (1971): The ultrastructure of plastids. In: *Structure and Function of Chloroplasts,* edited by M. Gibbs, pp. 7–34. Springer-Verlag, New York.

37. von Wettstein, D. (1967): Chloroplast structure and genetics. In: *Harvesting the Sun: Photosynthesis in Plant Life,* edited by A. San Pietro, F. A. Greer, and T. J. Army, pp. 153–190. Academic Press, New York.

38. Henningsen, K. W., and Boynton, J. E. (1969): Macromolecular physiology of plastids. VIII. The effect of a brief illumination on plastids of dark-grown barley leaves. *J. Cell Sci.,* 5:757–793.

39. Smith, J. H. C. (1960): Protochlorophyll transformations. In: *Comparative Biochemistry of Photoreactive Systems,* edited by M. B. Allen, pp. 257–278. Academic Press, New York.

40. Shibata, K. (1957): Spectroscopic studies on chlorophyll formation in intact leaves. *J. Biochem. (Tokyo),* 44:147–173.

41. Bouck, G. B. (1962): Chromatophore development, pits, and other fine structure in the red alga *Lomentaria baileyana* (Harv.) Farlow. *J. Cell Biol.,* 12:553–569.

42. Friedberg, I., Goldberg, I., and Ohad, I. (1971): A prolamellar bodylike structure in *Chlamydomonas reinhardi. J. Cell Biol.,* 50: 268–275.

43. Schiff, J. A. (1970): Developmental interactions among cellular compartments in *Euglena. Symp. Soc. Exp. Biol.,* 24:277–302.

44. Hodge, A. J., McLean, J. E., and Mercer, F. V. (1955): Ultrastructure of the lamellae and grana in the chloroplasts of *Zea mays* L. *J. Biophys. Biochem. Cytol.,* 1:605–614.

45. Steinmann, E., and Sjöstrand, F. S. (1955): The ultrastructure of chloroplasts. *Exp. Cell Res.,* 8:15–23.

46. Menke, W. (1960): Weitere Untersuchungen zur Entwicklung der Plastiden von *Oenothera hookeri. Z. Naturforsch. [B],* 15:479–482.

47. Menke, W. (1960): Einige Beobachtungen zur Entwicklungsgeschichte der Plastiden von *Elodea canadensis. Z. Naturforsch. [B],* 15: 800–804.

48. Menke, W. (1960): Das allgemeine Bauprinzip des Lamellarsystems der Chloroplasten. *Experientia,* 16:537–538.

49. Menke, W. (1962): Structure and chemistry of plastids. *Annu. Rev. Plant Physiol.,* 22: 45–74.

50. Eriksson, G., Kahn, A., Walles, B., and von Wettstein, D. (1961): Zur makromolekularen Physiologie der Chloroplasten 3. *Ber. Dtsch. Bot. Ges.,* 74:221–232.

51. Weier, T. E., Stocking, C. R., Thomson, W. W., and Dreuer, H. (1963): The grana as structural units in chloroplasts of mesophyll of *Nicotiana rustica* and *Phaseolus vulgaris. J. Ultrastruct. Res.,* 8:122–143.

52. Heslop-Harrison, J. (1962): Evanescent and persistent modifications of chloroplast ultrastructure induced by an unnatural pyrimidine. *Planta,* 58:237–256.

53. Paolillo, D. J. (1970): The three-dimensional arrangement of intergranal lamellae in chloroplasts. *J. Cell Sci.,* 6:243–255.

54. Paolillo, D. J., and Falk, R. H. (1966): The ultrastructure of grana in mesophyll plastids of *Zea mays. Am. J. Bot.,* 53:173–180.

55. Paolillo, D. J., and Reighard, J. A. (1967): On the relationship between mature structure and ontogeny in the grana of chloroplasts. *Can. J. Bot.,* 45:773–782.

56. Paolillo, D. J., Falk, R. H., and Reighard, J. A. (1967): Effect of chemical fixation on the fretwork of chloroplasts. *Trans. Am. Microsc. Soc.,* 86:225–232.

57. Paolillo, D. J., Mackay, N. C., and Graffius, J. R. (1969): The structure of the grana of flowering plants. *Am. J. Bot.,* 56:344–347.

58. Wehrmeyer, W. (1964): Zur Klärung der strukturellen Variabilität der Chloroplastengrana der Spinats in profil und aufsicht. *Planta,* 62:272–293.

59. Wehrmeyer, W. (1964): Über Membranbildungsprozesse im Chloroplasten. II. Mitteilung zur Enstehung der Grana durch Membranüberschiebung. *Planta,* 63:13–30.

60. Lichtenthaler, H. K., and Park, R. B. (1963): Chemical composition of chloroplast lamellae from spinach. *Nature,* 198:1070–1072.

61. Park, R. B. (1963): Advances in photosynthesis. *J. Chem. Educ.,* 39:424–429.

62. Park, R. B., and Pon, N. G. (1961): Correlation of structure with function in *Spinacea oleracea* chloroplasts. *J. Mol. Biol.,* 3: 1–10.

63. Chapman, D., and Leslie, R. B. (1970): Structure and function of phospholipids in membranes. In: *Membranes of Mitochondria and Chloroplasts,* edited by E. Racker, pp. 91–126. Van Nostrand Reinhold, New York.

64. Erwin, J., and Bloch, K. (1963): Polyunsaturated fatty acids in some photosynthetic microorganisms. *Biochem. Zeit.,* 338:496–511.

65. Benson, A. A. (1971): Lipids of chloroplasts. In: *Structure and Function of Chloroplasts,* edited by M. Gibbs, pp. 129–148. Springer-Verlag, New York.

66. Hill, R., and Bendall, F. (1960): Function of the two cytochrome components in chloroplasts: A working hypothesis. *Nature,* 186: 136–137.

67. Arnon, D. I. (1971): The light reactions of photosynthesis. *Proc. Natl. Acad. Sci. U.S.A.,* 68:2883–2892.

68. Avron, M. (1971): Biochemistry of photophosphorylation. In: *Structure and Function of Chloroplasts,* edited by M. Gibbs, pp. 149–167. Springer-Verlag, New York.

69. Avron, M. (1975): The electron transport chain in chloroplasts. In: *Bioenergetics of Photosynthesis,* edited by Govindjee, pp. 374–384. Academic Press, New York.

70. Bishop, N. I. (1971): Photosynthesis: The electron transport system of green plants. *Annu. Rev. Biochem.,* 40:197–226.

71. Cheniae, G. M. (1970): Photosystem II and O_2 evolution. *Annu. Rev. Plant Physiol.,* 21: 467–498.

72. Levine, R. P. (1969): Analysis of photosynthesis using mutant strains of algae and higher plants. *Annu. Rev. Plant Physiol.* 20: 523–540.

73. Levine, R. P. (1974): Mutant studies on photosynthetic electron transport. In: *Algal Genetics and Biochemistry,* edited by W. D. P. Stewart, Chapter 14. Blackwell, Oxford.

74. Trebst, A. (1974): Energy conservation in photosynthetic electron transport of chloroplasts. *Annu. Rev. Plant Physiol.,* 25:423–458.

75. Horton, P. (1976): Organization and function of chloroplast photosystems. *Int. J. Biochem.,* 7:597–605.

76. Penefsky, H. S. (1974): Mitochondrial and chloroplast ATPases. In: *The Enzymes, Vol. 10,* edited by P. D. Boyer, pp. 375–394. Academic Press, New York.

77. Jagendorf, A. T. (1975): ·Mechanism of photophosphorylation. In: *Bioenergetics of Photosynthesis,* edited by Govindjee, pp. 413–492. Academic Press, New York.

78. Boardman, N. K. (1970): Physical separation of the photosynthetic photochemical systems. *Annu. Rev. Plant Physiol.,* 21:115–140.

79. Thornber, J. P. (1975): Chlorophyll-proteins: Light-harvesting and reaction center components of plants. *Annu. Rev. Plant Physiol.,* 26:127–158.

80. Anderson, J. M. (1975): The molecular organization of chloroplast thylakoids. *Biochim. Biophys. Acta,* 416:191–235.

81. Arntzen, C. J. (1978): Dynamic structural features of chloroplast lamellae. *Current Topics in Bioenergetics,* 7:(*in press*).

82. Thornber, J. P., Gregory, R. P. F., Smith, C. A., and Bailey, J. L. (1967): Studies on the nature of chloroplast lamella. I. Preparation and some properties of two chlorophyll-protein complexes. *Biochemistry,* 6:391–396.

83. Thornber, J. P., Stewart, J. C., Hatton, M. W. C., and Leggett-Bailey, J. (1967): Studies on the nature of chloroplast lamellae. II. Chemical composition and further physi-

cal properties of the two chlorophyll-protein complexes. *Biochemistry,* 6:2006–2014.

84. Chua, N. H., and Bennoun, P. (1975): Thylakoid membrane polypeptides of *Chlamydomonas reinhardtii:* Wildtype and mutant strains deficient in photosystem II reaction center. *Proc. Natl. Acad. Sci. U.S.A.,* 72: 2175–2179.

85. Arntzen, C. J., and Briantais, J.-M. (1975): Chloroplast structure and function. In: *Bioenergetics of Photosynthesis,* edited by Govindjee, pp. 52–113. Academic Press, New York.

86. Howell, S. H., and Moudrianakis, E. N. (1967): Function of the "quantasome" in photosynthesis: Structure and properties of the membrane-bound particle active in the dark reactions of photophosphorylation. *Proc. Natl. Acad. Sci. U.S.A.,* 58:1261–1268.

87. Howell, S. H., and Moudrianakis, E. N. (1967): Hill reaction site in chloroplast membranes: Non-participation of the quantasome particle in photoreduction. *J. Mol. Biol.,* 27:323–333.

88. Park, R. B., and Biggins, J. (1964): Quantasome: Size and composition. *Science,* 144: 1009–1011.

89. Park, R. B., and Pfeifhofer, A. O. (1968): The continued presence of quantasomes in ethylenediaminetetraacetate-washed chloroplast lamellae. *Proc. Natl. Acad. Sci. U.S.A.,* 60:337–343.

90. Park, R. B., and Pon, N. G. (1963): Chemical composition and the substructure of lamellae isolated from *Spinacia oleracea* chloroplasts. *J. Mol. Biol.,* 6:105–114.

91. Branton, D., and Park, R. B. (1967): Subunits in chloroplast lamellae. *J. Ultrastruct. Res.,* 19:282–303.

92. Sane, P. V., Goodchild, D. J., and Park, R. B. (1970): Characterization of chloroplast photosystems 1 and 2 separated by a non-detergent method. *Biochim. Biophys. Acta,* 216:162–178.

93. Whittingham, C. P. (1974): *The Mechanism of Photosynthesis.* American Elsevier, New York.

94. Wildman, S. G., and Bonner, J. (1947): The proteins of green beans. I. Isolation, enzymatic properties and auxin content of spinach cytoplasmic ribosomes. *Arch. Biochem.,* 14: 381–413.

95. Kawashima, N., and Wildman, S. G. (1970): Fraction I protein. *Annu. Rev. Plant Physiol.,* 21:325–358.

96. Wildman, S. G., Chen, K., Gray, J. C., Kung, S.-D., Kwanyuen, P., and Sakano, K. (1975): Evolution of ferridoxin and fraction I protein in *Nicotiana.* In: *Genetics and Biogenesis of Mitochondria and Chloroplasts,* edited by C. W. Birky, Jr., P. S. Perlman, and T. J. Byers, pp. 309–329. Ohio State University Press, Columbus.

97. Kung, S.-D. (1976): Tobacco fraction I protein: A unique genetic marker. *Science,* 191: 429–434.

98. Tolbert, N. E. (1971): Microbodies, peroxisomes and glyoxysomes. *Annu. Rev. Plant Physiol.,* 22:45–74.

99. Hatch, M. D., and Slack, C. R. (1966): Photosynthesis by sugarcane leaves. A new carboxylation reaction and the pathway of sugar formation. *Biochem. J.,* 101:103–111.

100. Hatch, M. D., and Slack, C. R. (1970): Photosynthetic CO_2-fixation pathways. *Annu. Rev. Plant Physiol.,* 21:141–162.

101. Zelitch, I. (1971): *Photosynthesis, Photorespiration, and Plant Productivity.* Academic Press, New York.

102. Schatz, G., and Mason, T. L. (1974): The biosynthesis of mitochondrial proteins. *Annu. Rev. Biochem.,* 43:51–87.

103. Ross, E., and Schatz, G. (1976): Cytochrome c_1 of Baker's yeast. II. Synthesis on cytoplasmic ribosomes and influence of oxygen and heme on accumulation of the apoprotein. *J. Biol. Chem.,* 251:1997–2004.

104. Weiss, H., and Ziganke, B. (1974): Biogenesis of cytochrome *b* in *Neurospora crassa.* In: *The Biogenesis of Mitochondria,* edited by A. M. Kroon and C. Saccone, pp. 491–500. Academic Press, New York.

105. Govindjee, E., and Govindjee, R. (1975): Introduction to photosynthesis. In: *Bioenergetics of Photosynthesis,* edited by Govindjee, pp. 1–50. Academic Press, New York.

106. Lambowitz, A. M., Slayman, C. W., Slayman, C. L., and Bonner, W. D., Jr. (1972): The electron transport components of wild type and *poky* strains of *Neurospora crassa. J. Biol. Chem.,* 247:1536–1545.

107. Chua, N.-H., Matlin, K., and Bennoun, P. (1975): A chlorophyll protein complex lacking in photosystem I mutants of *Chlamydomonas reinhardtii. J. Cell Biol.,* 67:361–377.

---

[1] Recent evidence indicates that in mature chloroplasts the thylakoid membrane may not be continuous with the inner chloroplast envelope despite the fact that this membrane undergoes frequent infoldings during ontogeny (cf. Gunning, B. E. S., and Steer, M. W. (1975): *Ultrastructure and the Biology of Plant Cells.* Edward Arnold, London.

Chapter 2
Genetic Material of Chloroplasts and Mitochondria

In 1951 Chiba (1) suggested that chloroplasts found in the club moss *Selaginella* and in the leaves of two flowering plants contained DNA. This conclusion was based on the observation that these chloroplasts contained areas that became colored with DNA-specific Feulgen stain. If the leaves were treated with hot trichloroacetic acid, which hydrolyzes nucleic acids, the staining could not be detected. Shortly thereafter, reports that isolated chloroplasts contained DNA began to appear (see Kirk and Tilney-Bassett, 2, for discussion). More precise results were obtained in 1962 by Ris and Plaut (3), who found that the chloroplast of the alga *Chlamydomonas* contained one or more bodies that became stained both by the Feulgen reaction and by acridine orange. Fluorescence microscopy revealed that the acridine-stained bodies emitted the yellow green fluorescence characteristic of DNA rather than the orange fluorescence that signifies the presence of RNA. These light microscopic studies were combined with electron microscopic observations. Ris and Plaut observed filaments in the DNA-containing regions that resembled those of bacterial nuclei and that, like DNA, could be stained by uranyl ion binding. Pretreatment of the cells with DNAse hydrolyzed the DNA, so that the Feulgen- and acridine-staining bodies disappeared and the uranyl-binding filaments could no longer be visualized in the electron microscope. Shortly thereafter, Nass and Nass (4,5) demonstrated that the uranyl-ion-binding fibrils they and other workers had observed in mitochondria were not seen if the mitochondria were first treated with DNAse.

Although these studies provided convincing evidence that chloroplasts and mitochondria contained DNA, they did not establish whether or not this DNA was different from the DNA of the nucleus. The first demonstrations that this was indeed the case came in 1963 for chloroplast DNA (clDNA). Kirk (6,7) reported that by chemical analysis the clDNA of the broad bean (*Vicia faba*) was found to have a guanine-cytosine (G + C) content of 37.4%, whereas the nuclear DNA (nDNA) of the same plant had a G + C content of 39.4%. That same year Chun and associates (8), using the technique of buoyant density centrifugation in cesium chloride, found that chloroplast preparations from spinach (*Spinacia oleracea*) and beet (*Beta vulgaris*) contained two kinds of DNA, β and γ, of 46% and 60% G + C content, respectively. They also detected a minor DNA component in the algae *Chlorella pyrenoidosa* and *Chlamydomonas reinhardtii,* but did not attempt to establish its cellular location. However, later that year Sager and Ishida (9) showed that in *C. reinhardtii* the minor DNA component, which constituted 6% of the whole-cell DNA, represented 40% of the DNA in chloroplast preparations from the alga, which suggested that it originated in the chloroplast. In another study, published in 1963, Leff and associates (10) reported that the flagellate *Euglena gracilis* had a satellite with a buoyant density of 1.688 g/cc, which represented 4% of the whole-cell DNA, and that this satellite was not found in an aplastidic mutant induced with ultraviolet light. Thus by the end of 1963 there was no doubt that clDNA was quite distinct from nDNA.

Demonstration that mitochondrial DNA (mtDNA) is different in base composition from nDNA came in 1964, when Luck and Reich (11) showed that the buoyant density of mtDNA from the mold *Neurospora crassa* was not the same as the buoyant density of its nDNA. In 1965 Rabinowitz and associates (12) showed that the same is true of mtDNA from chick heart and liver. Subsequently, clDNA and mtDNA from a wide variety of sources were characterized (see 13-21 for reviews). These DNAs are always double-stranded, and unlike the DNA of the eukaryote nucleus, they are not complexed with histones or other basic proteins.

BASE COMPOSITION OF ORGANELLE DNA

Direct analysis of the base compositions of clDNA and mtDNA reveals approximately molar equivalences of adenine (A) with thymine (T), and G with C, and significant quantities of unusual bases have not been reported. Data compiled from many species fail to show any predictable relationships among the G + C contents of nDNA, clDNA, and mtDNA (Fig. 2–1). Similarities in base composition, such as those seen for mammalian mtDNA and nDNA, are fortuitous.

Correct identification of clDNA from higher plants has posed certain problems, and these have been discussed by Kirk (19). Shortly after Chun et al. (8) reported that the clDNA of higher plants was more dense (1.705 g/cc for β-DNA and 1.719 g/cc for

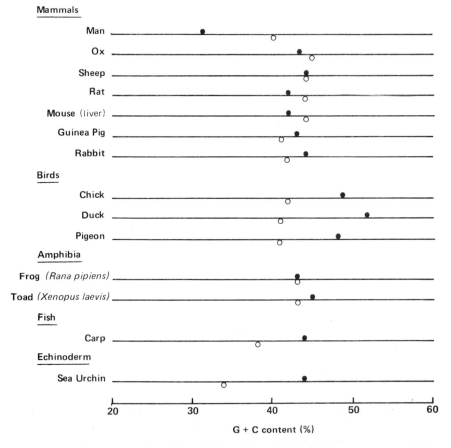

FIG. 2–1. Percentage G + C content of nuclear DNA (nDNA) and organelle DNA (clDNA and mtDNA) of different animals and plants: clDNA (*triangles*); mtDNA (*filled circles*); nDNA (*open circles*). (Data from Borst and Kroon, ref. 14, and Kirk, ref. 19.)

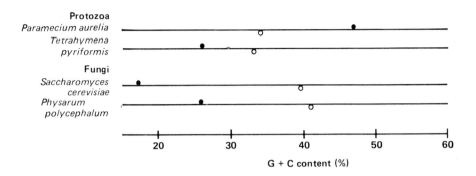

mt-DNA and n-DNA in Protozoa and fungi

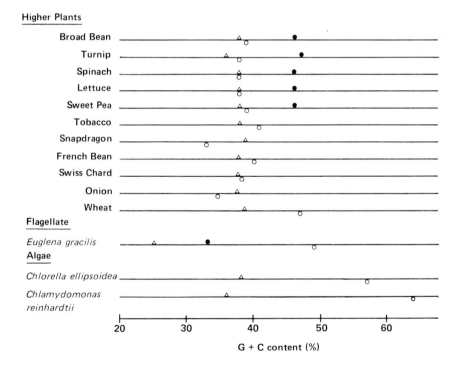

cl-DNA, mt-DNA, and n-DNA from higher plants, flagellates and algae

γ-DNA) than nDNA (1.695 g/cc), other authors began to publish similar findings. In each case the putative clDNA was found to have a buoyant density in the range of 1.700 to 1.707 g/cc, and the nDNA was less dense (1.694 to 1.697 g/cc). In almost every case it was also reported that the chlo-roplasts were contaminated with DNA of buoyant density similar to that of nDNA. In 1967 the picture began to change following a series of papers in which DNAs with buoyant densities close to those of nDNAs were reported from chloroplasts. The denser bands were no longer found.

Kirk (19) argued that the later reports were substantially correct and the earlier ones wrong. He concluded that clDNAs from higher plants have buoyant densities in the region of 1.679 g/cc and G + C contents of 37 to 38%. The earlier reports of dense satellites associated with chloroplasts were explained by bacterial or mitochondrial contamination of the isolated chloroplasts. This seems a plausible explanation, in view of the fact that higher-plant mtDNAs have buoyant densities in the range of 1.706 to 1.707 g/cc. Kirk also pointed out that plant nDNA is relatively rich in 5-methylcytosine, which is lacking from clDNA, and that these methylated bases lower the density of the DNA. Thus instances in which the buoyant densities of clDNA and nDNA appear very close do not necessarily entail close similarity in base compositions. Whether or not exceptions to Kirk's 38% G + C rule are to be found in higher plants remains to be seen, but as he pointed out, we should not be surprised if they occur.

Not only do organelle DNAs differ in average base composition, but also they can be distinguished by the method of nearest neighbor analysis (22), which permits determination of the frequencies of the 16 possible pairs of dinucleotides present in a single strand of DNA. The four doublets ending in G were compared in mtDNAs and nDNAs of the slime mold *Physarum polycephalum,* and definite differences were observed between the two (23). However, these differences could in part be artifactual, because the dinucleotide frequencies in the RNA product synthesized *in vitro* with RNA polymerase were examined rather than the DNA itself. If copying of the DNA were not complete, the dinucleotide frequencies measured would not provide an accurate estimate of the dinucleotide frequencies present in the total DNA. Sager (24) compared dinucleotide frequencies for all 16 possible dinucleotides in clDNA and

nDNA of *C. reinhardtii* and found them widely divergent.

FORM AND SIZE OF ORGANELLE DNA

Depending on the source and the conditions used for isolation of organelle DNA, one may find linear molecules of varying sizes, as well as circular molecules of different sizes and conformations. These can be visualized by electron microscopy (Figs. 2–2 and 2–3) and their sizes measured directly, or their sizes can be determined by sedimentation analysis in the ultracentrifuge. Because organelle DNA molecules are long and very thin, they often break during isolation. Consequently, the first organelle DNAs to be characterized satisfactorily were the circular mtDNAs of vertebrates, which are relatively small (approximately 4.5–6.0 μm) in comparison to the mtDNAs of certain fungi and higher plants (20–30 μm), and clDNA (40–60 μm), all of which are easily broken during preparation for electron microscopy.

Before discussing organelle DNAs in different organisms, it is important to distinguish further between the different forms of DNA that have been extracted from organelles. These are of three types: linear molecules, open circular molecules, and twisted supercoiled molecules. The first type needs no discussion. Open circular molecules are those in which there is a single-strand break in the covalent backbone of one of the strands of the duplex, but not in its complement (Figs. 2–2 and 2–4). Heat or alkali treatment will separate the complementary strands of both linear and open circular duplexes; in the first instance two linear single strands are produced; in the second instance one linear single strand and one circular single strand are produced (Fig. 2–4). The supercoiled molecules are double-strand, covalently linked, closed circular molecules with no breaks in

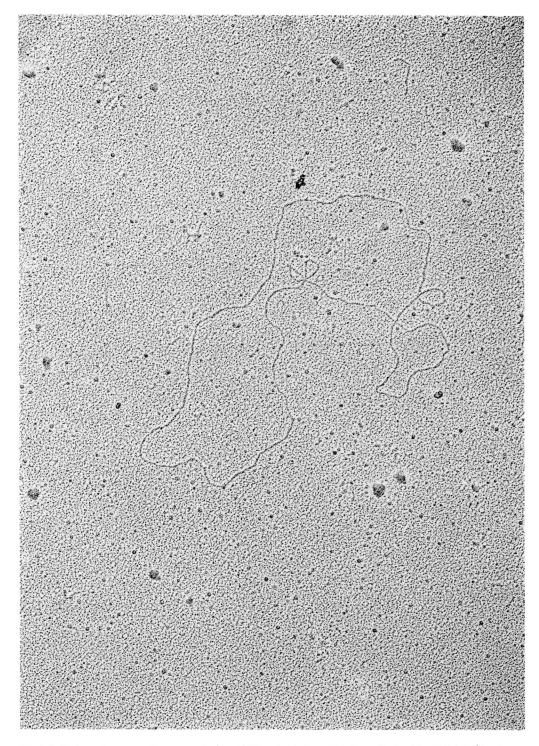

FIG. 2–2. Electron micrograph of an open circular mtDNA molecule from chick liver mitochondria. ✕68,300. (Courtesy of Dr. E. F. J. Van Bruggen.)

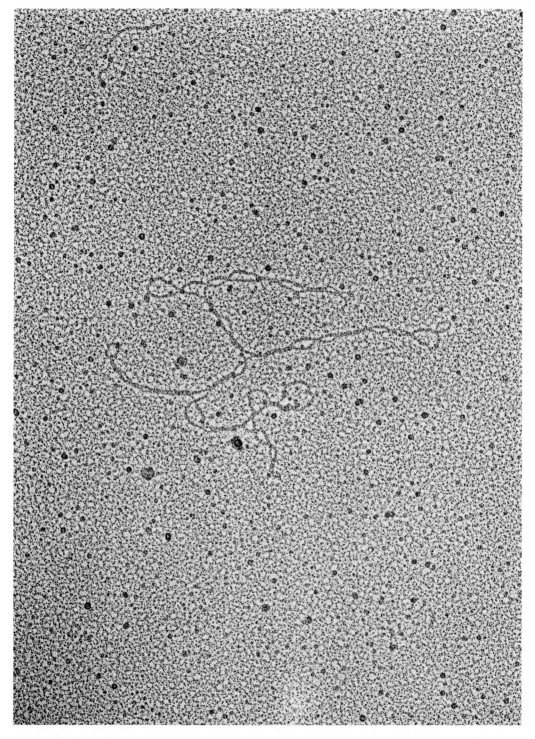

FIG. 2–3. Electron micrograph of a twisted circular mtDNA molecule from chick liver mitochondria. ×90,700. (From Borst and Kroon, ref. 14, with permission.)

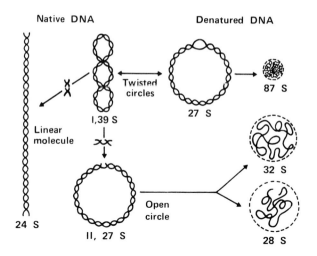

Native DNA   Denatured DNA

Twisted circles   87 S

1,39 S   27 S

Linear molecule

24 S   Open circle   32 S

II, 27 S   28 S

FIG. 2–4. Diagram of different forms of chick liver mtDNA. The denatured forms are those observed after complete denaturation by heating in the presence of formaldehyde. Numbers followed by S are sedimentation velocities of the molecular types in the ultra-centrifuge. (From Borst and Kroon, ref. 14, with permission.)

either strand (Figs. 2–3 and 2–4). These molecules are superhelically twisted on themselves because they contain more than the standard number of base pairs appropriate to their duplex length and consequently possess extra compensating negative turns (25). Supercoils are thought to originate during closure of the duplex, when part of the DNA is not base-paired but is complexed with protein. Removal of protein during isolation of the DNA, as well as changes in ionic conditions and temperature (26), can leave the circles underwound; as a result, superhelical turns are introduced in the purified molecules. Supercoiling does not arise in open circular molecules; in these molecules one strand is broken, and the broken and closed strands are free to rotate around each other.

The supercoiled closed duplex molecules have the characteristic of returning rapidly to their natural configuration after denaturation with heat or alkali. This phenomenon, called snap-back, is caused by the fact that even when denatured the two single-stranded circular molecules are interlocked and are maintained in close proximity so that they can renature rapidly.

Based on studies with the double-stranded circular DNA of polyoma virus and studies with mammalian mtDNA, Vinograd and colleagues (27–31) have developed general methods for separation and characterization of the closed and open circular forms of DNA. The open and supercoiled circles can be separated from each other and from the linear molecules because they sediment in different manners in neutral and alkaline sucrose gradients (Fig. 2–4). The supercoiled configuration also leads to an increase in buoyant density; thus the supercoiled and open circular forms can be separated by buoyant-density centrifugation in cesium chloride. The separation can be increased by use of the phenanthridine dyes ethidium bromide and propidium iodide, which intercalate between DNA base pairs. The closed circular form of DNA binds less dye than the open circular form at dye saturation. Since the dye is considerably less dense than DNA, the closed circular molecules (which bind less dye) will become heavier than the open circular or linear molecules; thus the effect of the dye is to increase the separation of the closed circular forms from the other forms in cesium chloride gradients. The different DNA bands in the gradients can then be detected optically by illuminating the centrifuge tubes with near-ultraviolet light (approximately 365 nm), since both dyes fluoresce under these conditions.

The available information on the size of organelle DNA is summarized in Tables 2–1 and 2–2. The mtDNA of the verte-

TABLE 2–1. *Sizes and molecular weights of mitochondrial DNAs from different organisms*[a]

| Source | Length (μm) | Molecular weight (× 10⁷ d) by following methods | | | Reference |
		Length	Kinetic complexity	Restriction fragments	
Human HeLa cells		1.03	—	—	137
Ox	5.1–5.3	1.05–1.1*	—	—	14
Sheep	5.4	1.1*	—	—	14
Rat	4.9–5.4	1.0–1.1*	1.0	0.96	14,86,138
Mouse	4.7–5.2	1.0–1.1*	—	—	14
Guinea pig	5.6	1.15	1.1	—	138
Chick	5.1–5.4	1.05–1.1*	1.0	—	32
Duck	5.1	1.05*	—	—	14
Frog (*Rana pipiens*)	5.56	1.15*	—	—	14
Toad (*Xenopus laevis*)	5.62	1.2	—	1.1	85,162
Carp	5.4	1.1*	—	—	14
Sea urchin (*L. pictus*)	4.45	0.9*	—	—	14
Housefly	5.2	1.1*	—	—	14
Drosophila	5.3	1.2	—	1.2	85,139
Roundworm (*Ascaris lumbricoides*)	4.79	1.0*	—		140
Tapeworm (*Hymenolepis diminuta*)	4.76	0.99*	—	—	140
Echiuroid worm (*Urechis caupo*)	5.85	1.21*	—	—	141
Tetrahymena	15.0	3.0	3.5	—	16,142
Paramecium	13.8	2.9*	3.0	2.9–3.0	35,142,143
Acanthamoeba	12.8	2.8	—	—	40
Yeast					
Schizosaccharomyces pombe	6.04	1.25	—	—	46
Torulopsis glabrata	6.0	1.28	1.5	—	47
Hansenula wingei	8.22	1.73	—	—	46
Candida parapsilosis	11.14	2.31	—	—	46
Kluyveromyces lactis	11.44	2.40	—	—	46
Saccharomyces cerevisiae	21–25	4.6–4.9	5.2	4.72–5.05	42,43,63,64
Saccharomyces carlsbergensis	25.6	4.9	5.0	4.52	42,64,144,145
Aspergillus	10	2.0	2.3	—	48,90
Neurospora	19–20	4.0	6.6[b]	4.0	44,45,89–91,146
Chlamydomonas	4.6	∼1.0*	1.6	—	54,55
Pea	29.9	7.0	7.4	—	52,53
Lettuce	30.0	7.0	7.4	—	53
Spinach	30.0	7.0	7.4	—	53
Beans	—	—	7.4	—	53

[a] All mitochondrial DNAs thus far characterized are circular, except those from *Tetrahymena* and *Paramecium*, which are linear. Molecular weights have been computed by three methods: length measurement, kinetic complexity, and summed molecular weights of restriction endonuclease fragments (see text for details). Internal standards of known length and molecular weights often were not included in the early length measurements. Molecular weights calculated from these measurements are marked with asterisks, and the masses per unit length of these DNAs have been calculated using T7 (2.07×10^6 d/μm) as a standard. Molecular weights calculated in this way are subject to some error (usually <10%).

[b] It seems likely that this estimate for kinetic complexity (146) is in error (15).

brates has been characterized in depth (14, 15,32); these DNAs are circular and relatively homogenous in size, ranging from 4.7 to 5.9 μm in length. Both open and closed circles have been isolated (Figs. 2–2 and 2–3). More complex forms of mtDNA are also found in vertebrates (15,33), in unfertilized sea urchin eggs (34), and in the trypanosome kinetoplast (see pp. 70–71). These complex forms can be catenated oligomers consisting of two or more interlocked circular duplexes connected to each other like links in a chain, or they can be circular duplexes with lengths that are multiples of the 5-μm monomer length (Fig. 2–5). Originally the circular dimers were

TABLE 2–2. *Sizes and molecular weights of chloroplast DNAs from different organisms*[a]

| Source | Length (μm) | Molecular weight ($\times 10^7$ d) by following methods | | | Reference |
		Length	Kinetic complexity	Restriction fragments	
Chlamydomonas	62.0	13.4	9.9–19.4	13	58,65–69
Euglena	40	8.2[b]	9.0[c]	—	21,50,148–150
Acetabularia	40–200	8–40	100–148	—	20,70,71,97
Pea	39	8.8–8.9	8.7–8.9	—	38,56
Lettuce	43	9.6–9.7	9.1–9.3	—	38
Spinach	45.7	9.4–9.5	8.8–11	9.5–9.6	37,38,92,147
Beans	—	—	8.9–9.1	—	38
Beets	44.9	9.3[b]	1.0–1.5	—	37,147
Antirrhinum	45.9	9.5[b]	1.0–1.5	—	37,147
Oenothera	45.2	9.4[b]	1.0–1.5	—	37,147
Oats	38	8.6	8.4–8.6	—	38
Corn	38	8.5	8.2–8.4	—	38

[a] All chloroplast DNAs thus far characterized rigorously have been circular. Molecular weight measurements were computed using the three methods described in Table 2–1. It should be noted that for higher plants there are small inconsistencies in molecular weight measurement among investigators that usually depend on the standards used. For higher plants, only more recent estimates of kinetic complexity (37,38,56,147) have been used. The large variation in molecular weight estimated by kinetic complexity for *Chlamydomonas* results in part because some measurements assumed the molecular weight of the T4 standard to be approximately 20×10^7 d, giving a kinetic complexity of 19.4×10^7 d for *Chlamydomonas* chloroplast DNA (65,66). The more recent estimate of 9.9×10^7 d assumes the molecular weight of T4 DNA to be 10.6×10^7 d (56).

[b] Indicates molecular weight calculations made by the author as described in Table 2–1.

[c] Adjusted for low G + C content (21).

thought to be restricted in mammals to cancer cells, which suggested that they might in some way be related to carcinogenesis (36); but they have since been found in normal animal tissues and cell lines, and there is no reason to believe that these circular dimers are indicative of any special abnormality of the mitochondrial genome of tumor cells

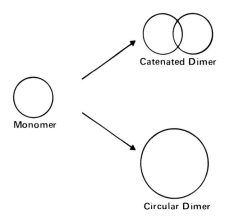

FIG. 2–5. Diagrams of a circular dimer and a catenated dimer.

(15). Circular dimers and catenates have also been reported in higher-plant clDNA (37,38).

The mtDNA of other animals, with the exception of the ciliate protozoans, appears to be very similar in size and form to that of the vertebrates (Table 2–1). Circle sizes range from 4.45 μm in a sea urchin to 5.85 μm in the echiuroid worm *Urechis caupo*. Another characteristic of animal mtDNA that is of considerable practical importance is that the two complementary strands have different buoyant densities; thus if they are not topologically linked together in the closed circular form, they can be separated on alkaline cesium chloride gradients. The two strands are referred to as the heavy (H) strand and the light (L) strand because the H strand has a higher equilibrium density in CsCl than the L strand. Not all mtDNAs show clean strand separation in CsCl, and in some cases alternative methods must be employed to improve strand separation. Most commonly synthetic polyribonucleotides are used which bind preferentially to one or the other of the two DNA strands. The resulting differential increase in buoyant density of the two strands is sufficient for the formation of two, usually well resolved, bands in neutral cesium chloride gradients. Thus the H strand of chicken mtDNA binds poly-cytidylic acid (poly-C), whereas the L strand binds polyuridylic acid (poly-U), polyinosinic acid (poly-I), and a copolymer composed of inosinic and guanylic acids (poly-I,G) preferentially (15).

The mtDNA of ciliated protozoans is linear, in contrast with all other organelle DNAs thus far characterized (Table 2–1). This was first demonstrated convincingly for *Tetrahymena* by Suyama and Miura (39) and later was confirmed by others (16). More recently, the mtDNA of *Paramecium* has also been shown to be linear (35). These are also the largest animal mtDNAs known, with lengths of 14 μm and 15 μm for *Paramecium* and *Tetrahymena*, respectively (Table 2–1). It is interesting that another protozoan, *Acanthamoeba,* contains mtDNA that is circular and has a length of 13 μm (40). Because of the central position of the yeast *Saccharomyces cerevisiae* in the investigation of mitochondrial heredity (see Chapter 4), many investigators have attempted to determine the size of its mtDNA. There were many early conflicting reports of linear and circular mtDNA molecules of varying lengths isolated from *S. cerevisiae* (18,32,41). In 1970 Borst and colleagues (42) succeeded in demonstrating the presence of circular molecules of mtDNA in osmotically lysed mitochondrial preparations from both *S. cerevisiae* and the related species *S. carlsbergensis* (Fig. 2–6). They reported measurements of 23 supercoiled circular mtDNA molecules, as well as one open circle. These circles formed a homogeneous population with an average length of 25 μm. Many different lengths (up to 27 μm) of linear molecules were also seen, and they were assumed to be degraded circular molecules. Later Petes and associates (43)

confirmed these observations, but their results indicated a contour length of 21 μm. Circular DNA molecules in the 20-μm range have also been isolated from the mitochondria of *N. crassa* (44,45).

It is important to note that these large mtDNA molecules are by no means necessarily typical of fungal mitochondria. Clark-Walker and colleagues (46,47) have isolated mtDNA from five other species of yeast (Table 2–1); all these mtDNAs are circular, but they range in size from 6 μm (*Schizosaccharomyces, Torulopsis*) to 11 μm (*Candida, Kluyveromyces*). In *Aspergillus* the mtDNA is circular and has a contour length of 10 μm (48). In short, the mtDNA of *Saccharomyces* and *Neurospora* may really be atypical. Although it seems likely that the heterogeneous-length linear molecules observed in preparations of fungal mtDNA are degradation products derived from intact circles, it is not possible to prove this by electron microscopic examination; satisfactory demonstration of this point has been achieved only by application of other techniques (e.g., analysis of mtDNA kinetic complexity, restriction endonuclease mapping, denaturation mapping) that will be discussed later.

We still know relatively little about the size and form of mtDNA in green plants. In one of the earliest papers on the subject, Wolstenholme and Gross (49) reported isolation of linear pieces of DNA from mitochondria of the red bean (*Phaseolus vulgaris*) ranging in length from 1 to 62 μm, with a mean of 19.5 μm. In *Euglena* Manning et al. (50) reported mtDNA lengths of 1 to 19 μm. Renaturation kinetics suggest a size around 20 μm (p. 440). Wong and Wildman (51) found lengths of 1 to 18 μm in mitochondrial preparations from tobacco (*Nicotiana tabacum*) leaves, with a mean of 14 μm when very small pieces (< 5 μm) were excluded. Quite obviously, all these studies were plagued with problems of mtDNA degradation. Finally, Kolodner and Tewari (52) succeeded in isolating intact mtDNA

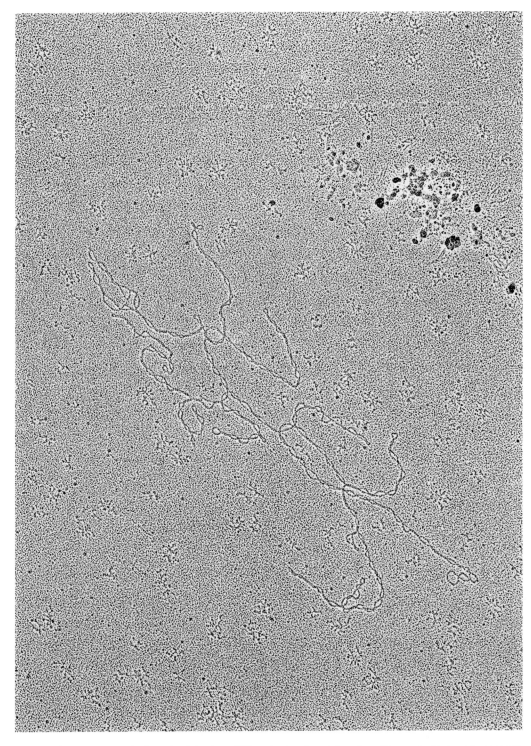

FIG. 2–6. Electron micrograph of a twisted circular mtDNA molecule released from *S. carlsbergensis* mitochondria. ×46,800. (From Hollenberg et al., ref. 154, with permission.)

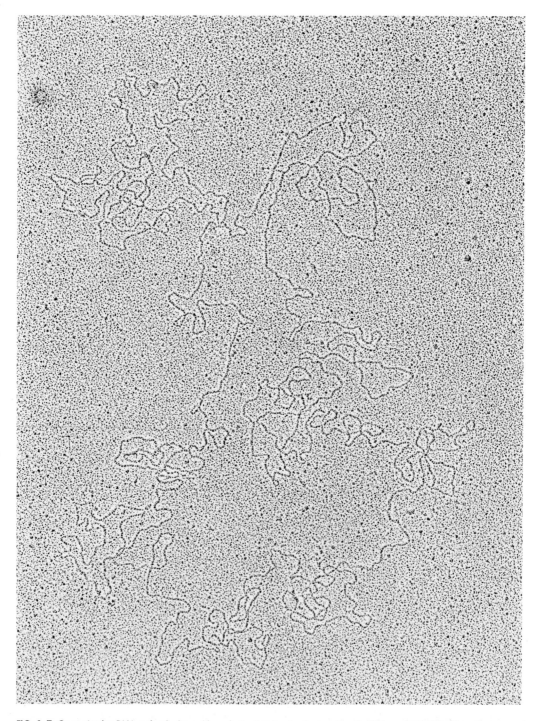

FIG. 2–7. Open circular DNA molecule from chloroplasts of *Spinacia*. Contour length 44.7 μm. ×48,000. (From Herrmann et al., ref. 37, with permission.)

molecules from the mitochondria of pea leaves of several other higher plants (53). These molecules were circular and had a mean contour length of 30 μm, making them the largest mtDNA molecules known (Table 2–1). Results to be discussed in a subsequent section on molecular weight and heterogeneity of organelle DNA suggest that the information content of pea mtDNA and the mtDNA of other higher plants is correspondingly larger.

In contrast, in *C. reinhardtii,* the only green alga whose mtDNA has been examined in any detail, it appears that the mtDNA consists of circles with contour lengths of 4.1 to 5.1 μm (54,55). This finding is supported by other studies to be discussed in the next section. In summary, it appears that mtDNA is rather constant in size (approximately 4.5–6 μm) in most animals and is circular. In ciliate protozoans, mtDNA is linear and about three times this size. In fungi, mtDNA is again circular, but it varies in size among different species. The scanty information available on plant mtDNA suggests that the size and form of mtDNA in *Chlamydomonas* are similar to those in animal mtDNA, but in higher plants one finds the largest mtDNA molecules known.

Intact clDNA was first isolated from *E. gracilis* by Manning and associates (50), who showed that this DNA consisted of circles 40 μm long. Subsequently, Kolodner and Tewari (56) reported that clDNA from pea leaves was circular and had a contour length of 39 μm. Manning and associates (57) and Herrmann and associates (37) reported contour lengths of 44 to 46 μm for clDNA from several different higher plants (Fig. 2–7 and Table 2–2). The two most recent studies by Herrmann and associates (37) and Kolodner and Tewari (38) reported the isolation of circular molecules in high yield as a fraction of the total clDNA, and in both cases significant portions of these molecules were found to be in the supercoiled form. Both groups reported

the presence of circular dimers in their clDNA preparations, and Kolodner and Tewari also observed catenated dimers, as mentioned previously. By making careful comparisons of size measurements using øX174 as an internal standard, Kolodner and Tewari were able to show that despite overall similarity in size among clDNA molecules from different higher plant species, there were small but significant size differences among these species. Circular clDNA molecules in the size range of 40 to 46 μm have now been observed in some 10 species of higher plants, including both dicotyledons and monocotyledons. Behn and Herrmann (58) reported that the clDNA of the green alga *C. reinhardtii* is a 62-μm circle, making it even larger than the clDNA of higher plants. Perhaps most extraordinary of all are the reports concerning the clDNA of the giant single-cell alga *Acetabularia* (20). Chloroplasts of *Acetabularia* lysed by osmotic shock release large masses of DNA. Although the DNA released in this way is too tangled to measure, it appears that the amount released is the same as is released from a typical bacterium. Molecules measuring up to 200 μm have been seen, although most appear to be smaller than 80 μm. Renaturation experiments that will be described in the next section also imply a much more complex chloroplast genome for *Acetabularia* than for the other plants thus far studied. Clearly, this is a situation that merits much further investigation.

MOLECULAR WEIGHT AND HETEROGENEITY OF ORGANELLE DNA

Molecular weights of organelle DNA molecules can be obtained by sedimentation analysis in an ultracentrifuge or by length measurements made on electron micrographs of organelle DNA preparations. The first method is most useful for smaller molecules, such as animal mtDNA, where breakage can be minimized. The second method is more generally applicable because gentle methods of organelle lysis that do not rely on purification of the DNA can sometimes be employed and because presumably intact molecules can first be identified and then measured. If one knows the mass per unit length of DNA, the molecular weight of the molecule can be calculated. Although this sounds simple, the method is subject to errors of two kinds. First, the method by which the DNA is prepared can influence the length. In discussing this problem, Kolodner and Tewari (56) pointed out that contour lengths varying from 1.64 to 2.26 μm have been reported for the circular DNA of bacteriophage ϕX174, which has been the subject of intensive study. This problem can be overcome to a considerable degree by mounting a standard DNA along with the DNA whose molecular weight is to be calculated. The molecular weight of the unknown DNA can then be estimated using the ratio of the lengths of the two DNAs and the molecular weight of the standard, which has been accurately determined. Second, the mass per unit length of standard DNAs also varies to some extent. Kolodner and Tewari (38) discussed this problem in relation to their studies on clDNA. They pointed out that the masses per unit length for ϕX174 and T7 DNAs (whose molecular weights are known) are slightly different. Errors up to 6.5% in molecular weight determination can be introduced, depending on the masses per unit length of the unknown DNA and the standard DNA. Most authors have taken the first (but not necessarily the second) source of error into account. Molecular weights for organelle DNAs obtained by electron microscopic measurement are given in Tables 2–1 and 2–2. It will be seen that molecular weights of mtDNAs vary widely, depending on source, as would be expected from their size variations (Table 2–1); in comparison, clDNAs are reasonably uniform in molecular weight (Table 2–2).

The molecular weight of an "intact" organelle DNA molecule, as determined by sedimentation analysis or electron microscopy, is not by itself a satisfactory guide to the potential genetic content of DNA in that organelle. There are two reasons for this: First, particularly in the case of the larger circular organelle DNA molecules, one usually obtains linear molecules the same length as the intact circles or shorter than the intact circles. Although these are almost certainly the result of breakage, the possibility remains that a given organelle may contain more than one species of DNA. Second, even if a population of organelle DNA molecules is very homogeneous in form and size, this does not mean that they are all identical genetically. Fortunately, there are two independent methods for determining whether the DNA molecules obtained from a given organelle constitute a homogeneous or a heterogeneous set in terms of base composition. The first method relies on the use of renaturation kinetics; the second relies on restriction enzymes that cleave double-stranded DNA at specific base sequences. A short description of each method is in order before we proceed.

Renaturation kinetic analysis was developed by Britten and Kohne (59) and Wetmur and Davidson (60). The method takes advantage of the fact that if DNA is denatured into its complementary single strands, either by heating or by alkali denaturation, the complementary strands can renature to form duplex DNA when incubated under appropriate conditions. Renaturation takes place because complementary strands of DNA "find" each other in solution, and reformation of H bonds occurs between complementary bases on the two strands. Renaturation rates can be measured by several different methods (59,61). One of the simplest and most widely used of these methods takes advantage of the fact that single-stranded DNA absorbs more ultraviolet light than double-stranded DNA. When duplex DNA is heated, there is a characteristic increase in absorbance at 260 nm (hyperchromic shift) at temperatures where the H bonds are "melted," and the highly ordered double-stranded duplex is converted to the disordered, randomly coiled, single-stranded state (Fig. 2–8). If the DNA is then cooled, the absorbance at 260 nm will decrease as renaturation of complementary strands to form duplex molecules occurs (Fig. 2–9). If the DNA renatured consists of a relatively homogeneous population of molecules (e.g., clDNA), renaturation will proceed relatively rapidly, and the decrease in absorbance will be of similar magnitude as the increase observed during denaturation (Fig. 2–9). On the other hand, if molecules are very heterogeneous in base composition (e.g., nDNA), renaturation will proceed slowly, and the absorbance of the "renatured" population of DNA molecules will be higher than that of the original undenatured population (because of imprecise pairing of nonhomologous molecules, single strands that have not found their complements, etc.) (Fig. 2–9).

The quantitative aspects of renaturation have been considered in detail by Britten and Kohne (59) and Wetmur and Davidson (60). Renaturation is a two-step proc-

ess that begins with a slow nucleation event in which short sequences of matching bases form pairs, followed by a rapid "zippering" of the neighboring sequences to form duplex. The rate of the overall process is governed by the second-order kinetics of the nucleation reaction. When DNA is sheared to appropriately small sizes and other conditions are held constant (i.e., total DNA concentration, salt concentration, and temperature), the rate of renaturation is directly proportional to the concentration of complementary sequences. The graphic method developed by Wetmur and Davidson (60) for determining kinetic complexities of DNAs is illustrated in Fig. 2–10. In this method the rate constant k_2 is determined from the rate of strand reassociation as a function of time. The rate of strand reassociation can be determined directly from the change in ultraviolet absorbance of the DNA, which will decrease as strands reassociate and assume the native duplex configuration. As shown in Fig. 2–10, guinea pig mtDNA renatures more rapidly than yeast mtDNA, which indicates that it has a lower kinetic complexity and therefore a smaller genetic content. A different kind of plot, called a Cot plot, was developed by Britten and Kohne (59). In a Cot plot the fraction of DNA reassociated is plotted against the logarithm of the product of DNA concentration (Co) in moles of nucleotides per liter and time (t) in seconds. Such a plot gives a second-order curve (Fig. 2–11) in which the middle two-thirds closely approximates a straight line. For DNAs that are comprised of unique sequences, the value for Cot at the time when the DNA is half reassociated ($Cot_{1/2}$) is directly proportional to the genome size in nucleotide pairs. Thus the molecular weight of an unknown DNA can be calculated using the ratio between $Cot_{1/2}$ of the unknown DNA and $Cot_{1/2}$ of a standard DNA with a similar Cot value whose molecular weight has been accurately determined. Figure 2–11 shows Cot plots for clDNA and nDNA from tobacco.

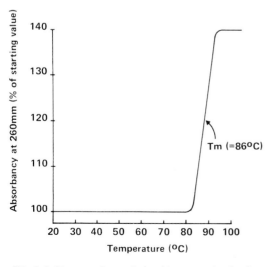

FIG. 2–8. Diagram of a typical melting curve showing T_m.

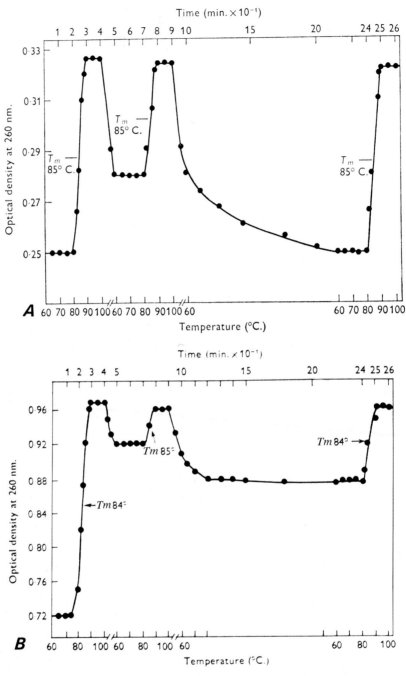

FIG. 2–9. Denaturation–renaturation curves for clDNA (A) and nDNA (B) of tobacco repeated over two cycles. (From Tewari and Wildman, ref. 159, with permission.)

The clDNA has a $Cot_{1/2}$ that is considerably lower than that for *Escherichia coli* ($Cot_{1/2}$ = 8), which indicates that clDNA is about 20-fold less complex than the *E. coli* genome. Nuclear DNA of tobacco, like other eukaryotic DNAs, contains many thousands of unique sequences and so renatures very slowly.

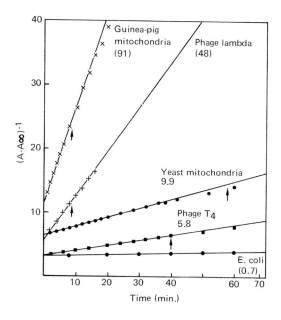

FIG. 2–10. Second-order rate plot of renaturation of two mitochondrial DNA preparations and three reference DNA preparations. Data are presented according to Wetmur and Davidson (60). In this plot the slope of the line is dependent only on the renaturation constant, not on the DNA concentration used. $(A - A_\infty)^{-1}$ is the reciprocal absolute value of the absorbance at 260 nm of the DNA at time $= t$ minus the absorbance of the DNA at time $=$ infinity. Values between brackets are second-order renaturation constants calculated by a procedure slightly modified from that of Wetmur and Davidson (60). Arrows indicate points at which 50% renaturation was reached. (From Borst, ref. 32, with permission.)

The methods devised by Britten and Kohne and Wetmur and Davidson have proved to be extremely useful in estimating the concentrations of unique sequences of DNA in preparations from many organisms. Proof of the validity of these measurements of the genomes of *E. coli* and several different phage DNAs comes from their close agreement with sizes estimated by other means, such as sedimentation analysis and electron microscopic measurement. Since it appears that *E. coli* and phage DNA molecules consist of unique (as opposed to repeated) sequences, and since accurate molecular weights are available for these DNAs, most researchers compare their results for organelle DNA with phage or E. *coli* DNA as standard (Figs. 2–10 and 2–11).

Restriction endonucleases are enzymes that recognize specific nucleotide sequences in DNA and cleave both strands of the duplex (62). All the restriction enzymes thus far characterized are prokaryotic in origin.

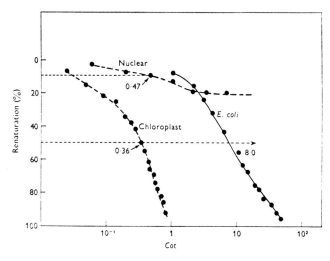

FIG. 2–11. Rates of reassociation of clDNA and nDNA from tobacco compared to the rate of reassociation of *E. coli* DNA. Cot is a parameter that measures the rate of reassociation and is expressed as moles of nucleotide reassociated $\times$ seconds/liter. DNA with a low Cot is composed of few unique sequences and reassociates rapidly, whereas DNA with a high Cot contains many unique sequences and reassociates slowly. (From Tewari and Wildman, ref. 159, with permission.)

Different restriction enzymes cleave in different nucleotide sequences. For example, the enzyme EcoRI (*E. coli* restriction endonuclease 1) cleaves the sequence

$$5' \ G {\downarrow} A \ A \ T \ T \ C$$
$$3' \ C \ T \ T \ A \ A {\uparrow} G$$

whereas the enzyme Hind III (*Haemophilus influenzae* serotype d restriction endonuclease III) cleaves the sequence

$$5' \ G \ G {\downarrow} C \ C$$
$$3' \ C \ C {\uparrow} G \ G$$

After a given DNA is cleaved into fragments by a restriction enzyme, the fragments can be separated on agarose-polyacrylamide gels, and the molecular weights of the fragments can be determined by comparison with appropriate standards. If the molecular weights of the fragments are summed, they should equal the molecular weight of the intact molecule, provided that each fragment has a unique molecular weight and the molecular population is homogeneous.

If the DNA molecules of a given organelle form a homogeneous collection in terms of base composition, the molecular weights of these molecules should be similar, regardless of the method used to establish molecular weight: electron microscopy, sedimentation analysis, renaturation kinetics, or restriction fragments.

The molecular weights of different organelle genomes determined by at least two independent methods are summarized in Tables 2–1 and 2–2. It is clear from these data that both mtDNA and clDNA, regardless of source, constitute unique collections of molecules homogeneous in base composition and that the molecular weight of an intact molecule determined by electron microscopic measurement is representative of the whole collection. Thus it seems that the linear pieces of DNA frequently seen in or-

ganelle DNA preparations are, with the notable exceptions of the mtDNA molecules of *Paramecium* and *Tetrahymena,* products of degradation. In addition to unit-length molecules, circular dimers are also found both in clDNA and mtDNA preparations. What are not found, except in the case of kinetoplast DNA (*vide infra*), are molecules that are heterogeneous in base composition. In short, the entire genome in both the mitochondrion and the chloroplast is localized in a single DNA molecule or organelle "chromosome."

A few specific comments are in order with respect to the data summarized in Tables 2–1 and 2–2 concerning the mtDNA of *S. cerevisiae* and the clDNA of *C. reinhardtii* and *Acetabularia.* The former two organisms have played a central role in the study of organelle genetics. The latter organism may have the most complex organelle genome thus far described. Length measurements and measurements of kinetic complexity are in agreement that the mitochondrial genome of *S. cerevisiae* is a circle of 21 to 25 μm with a molecular weight of about 5.0×10^7 d (42,43,63). Sanders and associates (64) have recently completed a restriction fragment analysis and a restriction enzyme map (*vide infra*) of the mitochondrial genome of *S. cerevisiae,* with results that are in close agreement with results from the other two methods. Thus the molecular weight and conformation of this mitochondrial genome are now firmly established. The situation with respect to the chloroplast genome of *C. reinhardtii* is not quite so clear, but it is rapidly improving. Bastia and associates (65) and Wells and Sager (66) originally reported kinetic complexities equivalent to 2×10^8 d, which would mean clDNA molecules from this alga should be about twice the size of clDNA from higher plants. In their experiments *E. coli* and T4 DNAs were used as controls with a value of 2.5×10^9 d being taken as the molecular weight of the *E. coli* chromosome. Using this value a kinetic complexity of 1.85×10^8 d was

TABLE 2–3. *Theoretical coding capacities of organelle DNAs for different proteins*[a]

| DNA | Source | rRNA | | Molecular weight of DNA available to code for proteins ($\times 10^7$ d) | Number of 20,000-d proteins that can be coded by this DNA | Reference |
		Number of genes/ molecule	Molecular weight ($\times 10^6$ d)			
Mitochondrial	HeLa cells	1.0	0.36 + 0.56	0.75	23	16
	Xenopus laevis	1.0	0.30 + 0.56	0.83	25	16,85
	Drosophila	1.0	0.28 + 0.50	0.94	29	85
	Tetrahymena	1.3	0.47 + 0.91	2.54	76	16
	Neurospora	1.0	0.6 + 1.28	3.52	106	151,152
	Saccharomyces	1.0	0.6 + 1.20	4.54	138	64,153
Chloroplast	Chlamydomonas	~2	0.56 + 1.1	12.24	371	154,155
	Euglena	2	0.55 + 1.1	8.24	250	156
	Spinacia	2	0.7 + 1.34	8.59	260	21,92

[a] Protein coding capacity is estimated by assuming that an average protein has a molecular weight of 20,000 d and is coded by 3.3×10^5 d of double-strand DNA. DNA available to code proteins is determined by subtracting the amount known to code for organelle RNA and by assuming that organelle DNA codes for a complete set of 20 tRNAs or 1×10^6 d of double-strand DNA (see Chapter 3). It is also assumed that none of the DNA has a spacer function or structural function. Molecular weights of organelle DNA are taken from Tables 2–1 and 2–2. References give information on rRNA.

obtained for T4 DNA. However, Kolodner and Tewari (56) pointed out that a value of 1.06×10^8 d is probably a more accurate value for the molecular weight of T4 DNA; if this value is used, the kinetic complexity of *C. reinhardtii* clDNA reduces to 9.9×10^7 d. More recently, Howell and Walker (67), using *E. coli* as a standard, have reported a kinetic complexity for clDNA in substantial agreement with the values reported earlier. Behn and Herrmann (58) reported that the clDNA genome of *C. reinhardtii* is a 62-μm circle. If we take 2.16×10^6 d/μm as the mass per unit length of DNA under their conditions of spreading (40), this will give a molecular weight of 13.4×10^7 d. Restriction fragment analysis indicates a genome size of 12×10^7 d to 13×10^7 d (68,69); thus the best estimate of the molecular weight of the chloroplast genome would now seem to be in the range of 10^8 d to 1.3×10^8 d.

Green and colleagues (70,71) reported the kinetic complexity of clDNA from *Acetabularia cliftoni* and *A. mediterranea* to be 1.48×10^9 d. It will be recalled that pieces of clDNA up to 200 μm (4.14×10^8 d) have also been seen in lysates of *Acetabu-laria* chloroplasts (71). As Green has pointed out in a recent review (71), the fact that *Acetabularia* can grow and differentiate in the absence of a nucleus suggests that the chloroplasts of the *Acetabularia* species may have more autonomy than those of other organisms.

INFORMATION CONTENT OF ORGANELLE DNA

Once the molecular size of an organelle genome is established, crude estimates of the number of proteins that can be coded by that genome can be made, provided that two assumptions are made and a correction is included to account for the fact that organelle genomes code for organelle ribosomal RNA (rRNA) and very likely a complete set of transfer RNAs (tRNAs). The assumptions are that each gene is represented only once and that there are no spacer DNA sequences that are not transcribed *in vivo*. The latter assumption, in particular, needs to be tested critically and is almost certainly incorrect in the case of *Saccharomyces* mtDNA.

Let us suppose that an average gene

codes for a protein having a molecular weight of 20,000 d. If we assume the molecular weight of an average amino acid to be 120 d, such a protein will contain 167 amino acids. Three nucleotide pairs specify one amino acid; thus the number of nucleotide pairs necessary to specify a protein of 20,000 d molecular weight is about 500. The molecular weight of an average nucleotide pair is 660 d; thus the molecular weight of DNA necessary to specify a protein of molecular weight 20,000 d is 3.3×10^5 d (660 d $\times 500$). Therefore a DNA molecule of 10^6 d could specify the sequence for 3 proteins of 20,000 d, and a DNA molecule of 10^8 d could code for 300 proteins. From this estimate it will be seen that animal mtDNA can code for 20 to 30 proteins, whereas that of *Saccharomyces* could code for five times that number (Table 2–3) although it almost certainly does not since 50% of its mtDNA may be spacers (Chapter 6). Chloroplast DNA may code for over 200 proteins. *Acetabularia* clDNA is, of course, out of line with this estimate; the coding capacity of this DNA will be on the order of the *E. coli* chromosome ($> 3,000$ proteins).

ORGANELLE DNA ORGANIZATION

Several methods are used for determining the number of DNA molecules per organelle. The simplest of these yields the average number of molecules per organelle, but it tells nothing about the variation in numbers of DNA molecules per organelle. In this method the average amount of DNA per organelle is determined, and its molecular weight is calculated. This number is then divided by the molecular weight of one of the DNA molecules contained within the organelle, which yields the number of DNA molecules of that molecular weight per organelle. For example, mouse L cells contain 9×10^{-17} g of mtDNA per mitochondrion, and the molecular weight of this DNA is

5.4×10^7 d (9×10^{-17} g/particle $\times 6.03 \times 10^{23}$ particles/mole). Since a mtDNA molecule from an L cell has a molecular weight of 1×10^7 d, the average L-cell mitochondrion contains 5.4 molecules of this molecular weight. A serious problem with this method, particularly in the case of mitochondria, is that it is sometimes hard to determine exactly where one organelle begins and another ends.

Calculations of the average numbers of DNA molecules per mitochondrion and chloroplast for several organisms are presented in Table 2–4. The data summarized in this table indicate that the average number of mtDNA molecules per organelle is of the order of three to seven, whereas in the case of clDNA, the numbers of molecules per organelle are severalfold higher, with the exception of *Acetabularia*, where there will be only about two of the giant ($> 10^9$ d) chloroplast genomes per organelle (71).

In one of the earliest studies on the organization of organelle DNA, Ris and Plaut (3) observed that chloroplasts in different *Chlamydomonas* species had one or more DNA-containing bodies adjacent to the pyrenoid. The precise number was not established in these experiments because serial sections were not done. In an elegant study, Gibbs (76) examined the structure of the chloroplast nucleoid in the chrysophyte alga *Ochromonas danica*. Cells were sectioned serially for electron microscopy, and DNA localization to the nucleoid was established through electron microscope autoradiography following labeling of the DNA with tritiated thymidine. The chloroplast nucleoid of this alga forms a continuous cord or ring encircling the rim of the chloroplast. In a separate study, Gibbs and Poole (77) presented evidence that a light-grown *Ochromonas* cell contains a minimum of 10 and probably 20 separate DNA molecules. Although it remains to be shown that these molecules are identical genetically, this conclusion seems almost inescapable in light of the data presented in Table 2–2. Hence

TABLE 2–4. Average number of DNA molecules per organelle[a]

DNA	Source	Molecular weight ($\times 10^7$ d)		DNA molecules organelle	Reference
		Single DNA molecule	Total DNA organelle		
Mitochondrial	Mouse (L cells)	1.05	5.4	5	14
	Rat (liver)	1.0	7.5	7.5	14
	Chick (liver)	1.0	4.2	4.2	14
	Tetrahymena	3.0	22.3	7.4	14
	Neurospora	4.0	13.0	3.2	146
	Saccharomyces	5.0	6–24	1–4	153,157
	Bean	7.0	60.3	8.6	14
Chloroplast	Chlamydomonas	13	1,040	80	112[b]
	Euglena	9	920–2,700	92–270	150,161[c]
	Pea	8.8	∼300	34	56
	Lettuce	9.1	200	22	21

[a] Molecular weights of organelle DNA molecules were taken from Tables 2–1 and 2–2. Molecular weights of DNA per organelle were calculated as described in text. References contain information on total DNA per organelle.

[b] Calculated for vegetative cells. Gametes would contain half as much c/lDNA per plastid (see Table 12–3).

[c] See discussion of c/lDNA amounts in Chapter 16 also.

it appears that each nucleoid probably contains a rather large number of chloroplast genomes. Gibbs (76) also reviewed the literature on nucleoid organization in algae and higher plants. It appears that the peripheral ring-shaped nucleoid is characteristic of five closely related classes of algae (Raphidophyceae, Chrysophyceae, Bacillariophyceae, Xanthophyceae, Phaeophyceae). In other algae it appears that clDNA is scattered more randomly throughout the chloroplast stroma. The three-dimensional structure of these scattered DNA areas has been studied in only one dinoflagellate (*Prorocentrum micans*) by Kowallik and Haberkorn (78). In this alga it appears that each of the cell's two large multilobate chloroplasts contains between 80 and 100 nucleoids. In summary, it appears that the numbers of nucleoids per algal chloroplast are rather variable, as is the nucleoid organization. In *Ochromonas* and its relatives there appears to be only one nucleoid, with many clDNA molecules. In other algae, nucleoid number and organization seem to be more variable. One last algal study should be mentioned. Using four independent methods, Woodcock and Bogorad (79) showed that in *A. mediterranea* only

20 to 35% of the plastids contained detectable amounts of DNA, and in these the clDNA content per plastid was variable.

In higher plants the most complete studies of clDNA organization have been done by Herrmann and colleagues on *Beta vulgaris* using serial sections and light microscopic autoradiography (80–84). Autoradiography revealed a variable number of centers per plastid that contained labeled DNA (82,83). The number was correlated with plastid size such that small plastids (4 μm) had labeled centers that varied from 1 or 2 up to 7 or 8, and large plastids (8 μm) contained anywhere from 3 or 4 up to 17 or 18 centers. These data are summarized in Fig. 2–12. Variability is also confirmed by parallel studies in which serial sections of plastids were prepared (83,84). It is clear from the sectioning data that the number of DNA-containing nucleoids per organelle is related to size and that an individual plastid can contain anywhere from 1 nucleoid in very small plastids to 10 or 12 nucleoids in large plastids (83,84). Estimates of clDNA content made from nucleoid volumes or from relative lengths of DNA per nucleoid suggest that clDNA content per nucleoid must also be variable. In a recent review, Nass

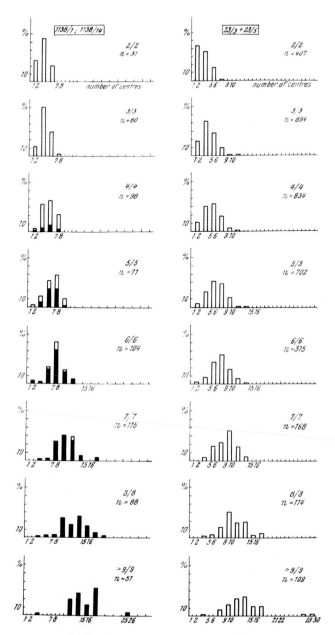

FIG. 2–12. Number of centers per plastid labeled with tritiated thymidine (%) related to chloroplast size (2/2, 3/3, etc. indicate diameters of plastids in micrometers, and n indicates number of chloroplasts that could be evaluated) in *B. vulgaris*. Euploid (small plastid) and trisomic (large plastid) plants are distinguished by no shading and dark shading, respectively. Left side of figure shows one set of progeny plants; right side shows a second set. (From Herrmann and Kowallik, ref. 83, with permission.)

(75) pointed out that the DNA-containing nucleoids of mitochondria are most conspicuous and tend to have the highest DNA content in embryonic cells, plant cells, unicellular organisms, and tissue culture cells,

but they are difficult to see in the dense, cristae-rich mitochondria of highly differentiated tissues.

After reviewing the foregoing papers, one is left with the profound impression that much more information is needed about the variability in amount of organelle DNA per organelle and per nucleoid. There is also a critical need for information on how nucleoids and the DNA molecules they contain distribute themselves during cell division. The lack of this information puts the organelle geneticist in much the same position a mendelian geneticist would be in were he not to know about mitosis. What is clearly evident is that it is extremely dangerous to generalize from the average organelle containing an average number of DNA molecules when thinking about genetic mechanisms.

PHYSICAL MAPS OF ORGANELLE DNA

Three techniques are presently in use that permit physical mapping of organelle genomes. The first technique makes use of restriction endonucleases to prepare cleavage maps of DNA. This method has been employed widely only in the last 2 or 3 years. To construct a cleavage map it is necessary to isolate restriction endonuclease fragments, determine their sizes, and order them. The molecular weights of individual fragments are readily established by the agarose-acrylamide gel technique described previously. Ordering of fragments can then

be done by making partial digests with a given enzyme and by successive cleavage with restriction endonucleases that cut the DNA in different nucleotide sequences. An example of this is shown in Fig. 2–13. Suppose that a given organelle DNA is cleaved with an enzyme we will call Endo R▼ , and two fragments of unequal size, such as 1 and 2, are obtained. Now suppose that a second endonuclease, Endo R↑, is used to cleave the DNA, and it yields two fragments of similar size, such as A and B. Thus far we have learned nothing except that the organelle DNA contains two nucleotide sequences that are recognized by each enzyme and that these fall at different points in the genome, as we would have expected. Now, imagine that the DNA is first cleaved with Endo R▼ to yield fragments 1 and 2. Digestion of each of these fragments with Endo R↑ yields two fragments. This must mean that the DNA being studied contains an Endo R↑ restriction site in each of fragments 1 and 2. Similarly, digestion of fragments A and B with Endo R▼ yields two new fragments for each of the original ones; thus there must also be one Endo R▼ restriction site in each of these fragments. These results can be explained only if the DNA in question is a circular molecule with the fragments arranged as shown in Fig. 2–13. Physical maps of organelle DNA us-

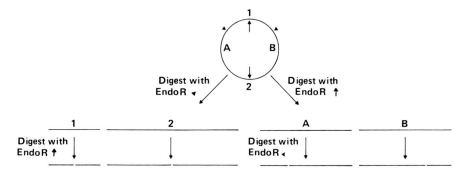

FIG. 2–13. Diagram of results of restriction endonuclease digestion of a circular DNA molecule having two sites (▼) for Endo R ◀ and two sites (arrows) for Endo R ↑. In each case the single digest tells little except that each enzyme recognizes different sites in the molecule as DNA fragments of distinct molecular weight are produced by the two enzymes. Only when the double digests are done does it become evident that the molecule really contains two restriction sites for each enzyme and is, in fact, a circle.

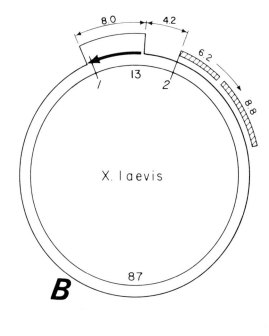

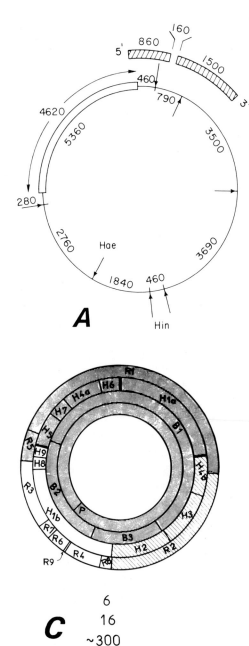

FIG. 2–14. Restriction endonuclease cleavage maps of mtDNA. A: Map of *Drosophila melanogaster* mtDNA. The cleavage sites for the enzymes Hae III and Hind III are indicated by arrows. The early melting region is indicated by an open section of the circle. The rRNA genes are indicated and are separated by a gap of 160 nucleotides. The direction of transcription (i.e., 5'-to-3' polarity of the RNA) is shown. Distances are given in base pairs. The total genome size is 18,400 base pairs. (From Dawid et al., ref. 85, with permission.) B: Map of *Xenopus laevis* mtDNA. The cleavage sites for the enzyme EcoR1 are numbered inside of the circle in italics. The H strand is drawn on the outside and the L strand on the inside. The positions of the rRNA genes and the D-loop are shown, and the polarity of transcription and replication is indicated with an arrow that points from the 5' to the 3' direction on the growing RNA or DNA strand. Distances are given as percentages of total genome, which is 17,100 base pairs. (From Dawid et al., ref. 85, with permission.) C: Map of *S. carlsbergensis* mtDNA. The different rings show the locations of the recognition sites of endonucleases EcoR1 (R fragments), Hind II + Hind III (H fragments), BamH1 (B fragments), and PstI (P fragments). (From Sanders et al., ref. 64, with permission.)

ing restriction enzyme analysis have now been constructed for the mtDNAs of several animals (85–87), yeast (64,88), and *Neurospora* (89–91) and for the clDNAs of spinach (92) and corn (93). Several of these maps are shown in Figs. 2–14 and 2–15. It will be noted that the organellar rRNA genes have been mapped in all of these instances (see Chapter 3).

The second method makes use of DNA-DNA hybridization and so far has been applied only in yeast by Linnane and his colleagues. A description of this method will be deferred until Chapter 7 since an understanding of the method requires an understanding of yeast mitochondrial genetics and in particular how one uses genetically defined petite deletion mutants. The

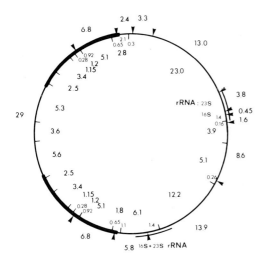

FIG. 2–15. Restriction endonuclease cleavage map of clDNA from *Spinacia oleracea*. Fragment sizes are given × 10⁶ d. Endonuclease Sal fragment sizes are given outside the circle, and endonuclease Sma fragment sizes are given inside. The terminal fragments resulting from the Sal/Sma digestion are given in small numbers. The clDNA contains an inverted repeat indicated by thick lines. The two segments coding for 23S + 16S chloroplast ribosomal RNA are also shown. Key: ▼, Sal cut; |, Sma cut; serial order arbitrary. (From Herrmann et al., ref. 92, with permission.)

establishment of physical maps using restriction enzymes and DNA-DNA hybridization has been of particular significance in the case of yeast because the precise physical location of many mitochondrial genes has now been ascertained (Chapter 7).

The third technique that has been used for physical mapping involves partial denaturation of the DNA. In this method the DNA is heated under conditions such that the A·T-rich regions become denatured but the G·C-rich regions do not. The molecules are then examined by electron microscopy, and the denatured regions are mapped. Denaturation and cleavage mapping can be used in combination to obtain a precise idea of the location of A·T- and G·C-rich sequences (90,94).

CLONING ORGANELLE DNA

Foreign pieces of DNA can be propagated in a bacterial host by introducing them into specific bacterial plasmids (62).

Usually the foreign DNA and the plasmid are cleaved with a restriction endonuclease such as EcoR1 that makes breaks at staggered positions in a given nucleotide sequence. The restricted DNA thus obtained has complementary single-strand ends that can hydrogen-bond with each other or with the complementary sequences in a different DNA cleaved with the same restriction endonuclease. The latter event results in production of a recombinant DNA molecule, and it is this event that makes possible the introduction of a foreign DNA molecule into a plasmid. The enzyme polynucleotide ligase is then used to link the foreign DNA covalently into the plasmid to form a chimera. The plasmid generally contains a marker that will confer antibiotic resistance on a suitable bacterial host such as *E. coli,* and the plasmid DNA can be used to transform *E. coli* to a resistant state. Resistant cells, by definition, contain the plasmid. Similarly, chimeric plasmids can be used to transform *E. coli* to a resistant state. Various methods that we need not detail allow distinction between cells transformed by the plasmid alone and cells transformed by the chimeric plasmid. Resistant cells harboring chimeric plasmids can then be grown in large numbers from a single original isolate, and thus a clone of the DNA fragment is obtained. If a DNA molecule containing a number of restriction sites is cleaved into fragments, it is possible to obtain clones of all the fragments in this fashion.

Clayton and colleagues (95) cloned mouse mtDNA using these methods. They attempted to study transcription and translation of the cloned fragments. Transcription seemed to take place principally off the L-strand fragments, which is not the case *in vivo.* As might be expected from this pattern of transcription, the translation products of these mtDNA fragments did not correspond to known mitochondrial translation products. Cloning is a potentially powerful method for establishing the products of organelle DNA. However, only if the cloned fragments are properly transcribed and their messages translated with fidelity, or if other methods (e.g., hybridization of organelle messengers labeled *in vivo* to cloned fragments) are employed will cloning provide the answers desired.

REPLICATION OF ORGANELLE DNA

Let us consider the replication of organelle DNA in terms of the following three interrelated questions. Can organelle DNA replicate itself autonomously, or is replication dependent on a master template in the nucleus? What is the mode of organelle DNA replication? How is organelle DNA replication timed with respect to replication of nDNA?

A priori one might expect that the way to demonstrate the autonomy of organelle DNA replication would be to develop *in vitro* systems in which isolated organelles would replicate their DNA. Thus far this approach has proved disappointing (15,21, 32,75,96,97). There have been three major problems: although incorporation of DNA precursors can be demonstrated, the efficiency of incorporation is not great; a complete round of replication remains to be demonstrated in any preparation; it is sometimes difficult to disentangle replication and repair. In the case of animal mtDNA, there is evidence that isolated mitochondria do make a structure that has the properties of the D-loop replicative intermediate (15, 75,96). In addition, there is evidence from experiments with rat and chick liver mitochondria that a substantial fraction of the radioactivity is also incorporated into closed circular mtDNA (75). The product made by isolated chloroplasts appears to renature readily following heat denaturation and to hybridize to a much larger extent with clDNA than with nDNA (21). Mitochondria of rat liver, HeLa cells, chick embryo brain, and yeast contain a DNA polymerase distinct from that in the nucleus (75), and a soluble DNA polymerase has been detected in spinach chloroplasts (21).

Although these observations suggest the possibility that isolated chloroplasts and mitochondria may be able to replicate their DNA, they are far from proving it.

An early and elegant suggestion that the nucleus does not generate new copies of clDNA came from the ultraviolet light bleaching experiments of Gibor and Granick (98) with *E. gracilis*. They took advantage of the fact that nonlethal doses of ultraviolet light will permanently bleach this flagellate, resulting in loss of clDNA and plastids as well. Gibor and Granick irradiated the cytoplasms and nuclei of individual *Euglena* cells separately in independent treatments. Irradiation of the cytoplasm resulted in chloroplast loss, but irradiation of the nucleus did not. Their results support the idea that in *Euglena,* clDNA replication is independent of the nuclear genome, and the ability to form clDNA and functional plastids cannot be regenerated from the nucleus.

Perhaps the strongest evidence that organelle DNA replication does not depend on DNA replication comes from studies of the mode of organelle DNA replication *in vivo* and studies of the timing of organelle DNA replication, which brings us to our second and third questions.

The mode of DNA replication in chloroplasts and mitochondria has been examined principally by means of the density labeling method of Meselson and Stahl (99). In the Meselson-Stahl experiment the organisms of choice are usually grown in the heavy isotope of nitrogen (^{15}N) until the DNA is completely labeled with this isotope, after which they are transferred to medium containing the light isotope of nitrogen (^{14}N). The experiment is set up in such a way that ^{14}N effectively replaces ^{15}N as a source of nitrogen in the medium. The classic result of Meselson-Stahl experiments in bacteria, viruses, and the DNA of the eukaryote nucleus is that DNA replication proceeds in a semiconservative fashion, with each parental strand acting as a template for a daughter strand. As a result, all daughter duplexes from the first DNA replication following transfer band in CsCl as hybrid molecules. That is, they band at a position equidistant from completely ^{15}N-labeled and completely ^{14}N-labeled DNA. Furthermore, separation of the strands in the hybrid reveals there is no mixing of isotopes, so that a given strand contains either ^{14}N or ^{15}N, but not both. Nucleotide precursors, such as 5-bromo-deoxyuridine, an analog of thymidine, have also been used in the Meselson-Stahl type of experiment.

Evidence for semiconservative replication of mtDNA was obtained by means of Meselson-Stahl experiments for rat liver mtDNA (100) and the clDNAs of *Chlamydomonas* (101) and *Euglena* (102). Experiments on the replication of mtDNA in *Neurospora* (103) led to a similar conclusion, although the results were equivocal because of the persistence of ^{15}N precursor pools on transfer to a ^{14}N environment.

More detailed investigations into the various intermediates involved in organelle DNA replication have uncovered a veritable paradise for the student of this problem. Before delving into the subject, it will be useful to review briefly the general patterns of DNA replication that are known in nature (104). Perhaps the simplest mode of DNA replication is that characteristic of the DNA bacteriophage T7. The chromosome of this virus is a linear duplex in which DNA replication is bidirectional, beginning at a point of origin 17% from one end of one duplex. The parental duplex strands separate at the origin to form an intermediate called an *eye* form. Replication continues, and the eye keeps enlarging, until finally two new duplexes are formed. As shown by Borst and colleagues (105–107), the linear mtDNA of *Tetrahymena* exhibits the same replication pattern (Fig. 2–16). In *Paramecium,* the only other organism known to have linear mtDNA, replication of mtDNA proceeds unidirectionally from an origin at one end via a lariat type of replicative intermediate to yield a linear

dimer (108) (Fig. 2–16). Linear dimers are not found in *Tetrahymena.*

Two other basic replication patterns have been described: the Cairns pattern and the rolling circle pattern. The replicating intermediate in the Cairns model is rather similar to that seen in T7 or *Tetrahymena* mtDNA, with the important exception that the two ends of the replicating DNA molecule remain covalently sealed during the process. This creates a swivelling problem, because unwinding in the growing points of replication creates a torque that is transmitted forward to the unreplicated portion of the molecule, causing it to become supertwisted. Supertwisting causes conformational strain on the molecule, blocking further replication. The problem is believed to be solved by a swivelling protein that causes transient single-strand nicks in the duplex, allowing parental strands to rotate freely about one another. After rotation, the same protein seals the nick, and replication proceeds.

The circular mtDNA of animals appears to be synthesized via a modified Cairns mode of DNA replication (96) (Fig. 2–16). The first replicative intermediate observed is called displacement loop (D-loop) DNA. This intermediate resembles a circular eye form, except that in the case of mtDNA, synthesis of new DNA takes place unidirectionally along only one parental strand, the L strand. As replication of the new H strand continues, the D-loop DNA expands to form expanded D-loop DNA (Exp-D DNA), until finally a complete new H strand is formed that is complementary to the parental L strand. At the same time, the old H strand becomes separated from the new duplex, and synthesis of a new L strand begins, with the parental H strand serving as the template. Thus semiconservative replication is achieved by these two temporally distinct replication events. Syntheses of the new H and L strands may not necessarily be so completely separated in time as this discussion would indicate. Replication of the new

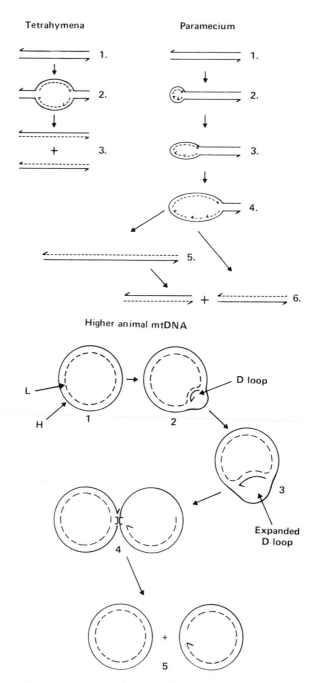

FIG. 2–16. Modes of mitochondrial DNA replication. **Tetrahymena:** Bidirectional synthesis of linear mtDNA molecule. (Adapted from Upholt and Borst, ref. 107.) 1: Unreplicated molecule. 2: Eyeform. 3: Two daughter molecules. **Paramecium:** Unidirectional synthesis of linear mtDNA molecule. (Adapted from Goddard and Cummings, ref. 108.) 1: Unreplicated molecule. 2: Covalent or H bonding occurs between left ends of parental strands, and replication begins. 3 and 4: Replication continues, to yield lariat forms. 5: Dimer that may result from opening out of lariat form at termination of replication. 6: Daughter molecules; they may result either from cleavage of lariat form at point where the two parental strands originally were linked or by cleavage of dimer at the same site. Higher-animal mtDNA: Modified Cairns mode of replication of the circular DNA mtDNA molecules of higher animals. (Adapted from Kasamatsu and Vinograd, ref. 96.) 1: Unreplicated molecule. 2, D-loop DNA in which replication of new H strand has begun. 3: Expanded D-loop DNA in which replication of new H strand continues. 4: Expanded D-loop DNA after duplex synthesis; new H strand is almost completed, and L strand synthesis has begun. 5: H-strand synthesis completed, and L-strand synthesis nearing completion.

L strand may commence, in some instances, before synthesis of the new H strand is completed (109).

In the rolling circle model, one strand of a DNA duplex is nicked at a specific point by a special endonuclease, which creates a linear DNA strand with 3'-OH and 5'-phosphate ends. The 3'-OH end serves as the origin of replication, and the closed strand serves as the template. The 5' end of the linear strand is then displaced into solution. The circular DNA strand can then serve as an endless template for replication. The single-strand tail that emerges from the replicating intermediate functions as a template on which DNA synthesis initiates *de novo*. The movement of the DNA growing point in the rolling circle model does not create a swivelling problem because one of the parental strands is broken, and thus unwinding can take place without transmission of any torque. A rolling circle intermediate has been reported by Kolodner and Tewari (110,111) for corn and pea clDNA, which also appear to have the most complex mode of replication yet described for any organelle DNA. The replication of these DNAs begins with the formation of two D-loops (Fig. 2–17). The two new strands are hydrogen-bonded to opposite parental strands, and as synthesis proceeds, the two loops expand toward each other to form a structure that looks like a Cairns replicative intermediate. Synthesis of the two new strands proceeds until two progeny clDNA molecules are formed. Kolodner and Tewari also observed rolling circle intermediates; they argued, on the basis of denaturation mapping studies, that the progeny molecules produced by the Cairns round of replication then undergo a round of rolling circle replication, with the origin of replication (i.e., the point at which one of the strands is nicked) being at or near the site where the Cairns intermediates terminate replication. In summary, the mode of replication of linear mtDNA molecules resembles that of the linear chromosome of bacteriophage T7, except that replication of mtDNA in *Paramecium* is unidirectional, from one end of the molecule, rather than bidirectional, from an origin at an internal position in the molecule, as in T7 and *Tetrahymena*. The circular mtDNA of animal mitochondria replicates via a modified Cairns mode that is characterized by a specific intermediate called D-loop DNA. The replication of clDNA involves both Cairns-type intermediates and rolling circle intermediates. The mode of replication of clDNA clearly demands detailed investigation in other plants, notably *Chlamydomonas*, in order to deter-

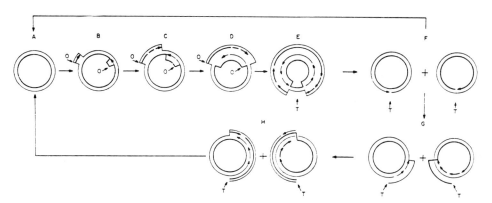

FIG. 2–17. Model of replication of clDNA. A: closed circular parental molecule. B: D-loop-containing molecule. C: Expanded D-loop-containing molecule. D and E: Cairns-type replicative intermediate. F: Nicked progeny molecules. G and H: Rolling circles. Thin and thick lines mark opposite strands of a molecule; lines with arrows are daughter strands; 0 indicates positions of the two origins of D-loop synthesis, which are 5.2% of pea clDNA apart; T indicates the terminus of the Cairns round of replication, which is 180 degrees opposed to the origins of D-loop synthesis. (From Kolodner and Tewari, ref. 110, with permission.)

mine if this rather complex replication pattern is general among green plants.

The timing of clDNA replication was investigated in *Chlamydomonas* by Chiang (101,112,113) using a $^{15}N \rightarrow {}^{14}N$ transfer experiment design on cultures synchronized on a 12-hr-light/12-hr-dark cycle. Chiang found that clDNA replicated primarily in the light period. The first replication (at about 4.5 hr) resulted in the replacement of heavy molecules by hybrid molecules. About 2 hr later, lightweight unlabeled clDNA molecules appeared in the same amounts as hybrid DNA, indicating that a second round of replication had taken place. The nDNA of *C. reinhardtii* showed no density shift during the 12-hr light period or during the first half of the dark period. During the second half of the dark period, two rounds of nDNA replication took place, during which time the cells divided twice. Lee and Jones (114) reported similar results, except that they found a gradual shift toward lighter density for clDNA. However, recent experiments (115) reveal that clDNA replication may not be quite so synchronous as previously supposed. DNA precursors continue to be incorporated into clDNA during the dark period (see Chapter 12). Whether this incorporation represents repair synthesis or replication remains to be seen. Cook (116) reported that cytoplasmic DNA synthesis in *Euglena* cells synchronized by alternating periods of light and dark occurred in two bursts. The first took place early in the light period; the second occurred at about the same time that total cellular DNA increased at the end of the light period. Autoradiography was used to detect synthesis periods, and the DNAs were not characterized by buoyant density or other techniques. A reasonable interpretation of these results would be that clDNA synthesis takes place early in the light period, with a second burst (or perhaps a burst of mtDNA synthesis) occurring late in the light period, when nDNA also replicates. However, such an interpretation will

remain speculative until the replicating DNAs are properly characterized.

Nass (75) distinguished three general patterns of mtDNA synthesis: (a) *Periodicity* of mtDNA and nDNA synthesis has been reported in many mammalian cells, including human HeLa cells (117) and mouse lymphoma cells (118). The maximum rate of mtDNA synthesis in these cases tends to be in G_2, with some synthesis in the S phase. This pattern of synthesis has also been reported in the yeast species *S. cerevisiae* and *Kluyveromyces lactis* (119,120). (b) *Simultaneous* synthesis of mtDNA and nDNA has been reported in *E. gracilis* (121). (c) *Continuous* synthesis throughout the cell cycle has been reported in *Tetrahymena* (122–124), *Physarum* (125, 126), *Pteridium* (127), and mouse fibroblasts (128), as well as *S. cerevisiae* (129). The conflicting results with *Saccharomyces* illustrate the difficulties one can encounter in attempting to fix the timing of mtDNA replication. Suffice it to say that in aggregate the results (with the possible exception of those for *Euglena*) indicate that mtDNA replication takes place either discontinuously at a slightly different time than nDNA replication or continuously throughout the cell cycle. Despite the fact that all the foregoing biochemical data, as well as the genetic evidence (Chapters 5 and 16), indicate that organelle DNA replication does not depend on an intermediary present in the nucleus, it is very difficult to prove that the nucleus does not contain one copy or a few copies of organelle DNA. Practically, the only way that this could be done would be to show by use of molecular hybridization techniques (see Chapter 3) that there is no DNA sequence in the nucleus complementary to organelle DNA. In discussing this problem with respect to vertebrate mtDNA, Borst (15) pointed out that the presence of one copy of mtDNA per haploid nucleus would mean that about 0.001% of the nDNA is complementary to mtDNA. This can barely be detected by hybridization, and therefore it is difficult to determine if any hybridization that is seen results from background noise, from a master copy, or from contamination of the nDNA with traces of mtDNA. The situation in protists is simplified because the ratio of mtDNA to nDNA is higher by two orders of magnitude than in vertebrates, and this is a great aid in the detection of putative master copies. Still, isolation of clean nuclei free of mitochondrial contamination poses a major problem. Borst concluded, in spite of these problems, that early reports of substantial base sequence homology between mtDNA and nDNA are probably incorrect. Tabak and associates (130) set a limit of one to three copies per haploid DNA complement in chick, and Flavell and Trampé (131) found that at most there could be 0.04 copies per haploid genome in *Tetrahymena*. Whether or not these numbers really equal zero remains to be seen.

MINICIRCULAR DNA AND TRYPANOSOME KINETOPLAST

The kinetoplast is part of a highly specialized mitochondrion (Fig. 2–18) found in certain groups of flagellated protozoa that contains an enormous amount of DNA called kinetoplast DNA (kDNA). The structure and function of this organelle, as well as the properties of kDNA, have been dealt with in an excellent review by Simpson (132), and at a recent symposium (133–136) the properties of kDNA were considered in several papers. Here we will review briefly the properties of this remarkable type of organelle DNA.

The kinetoplast represents the definitive taxonomic characteristic by which a large group of flagellated protozoa, the Kinetoplastidae, are grouped together. Both free-living and symbiotic forms (family Bodonidae), as well as parasitic forms (family Trypanosomatidae), are known. The parasitic forms are found in plants, invertebrates, and vertebrates, and they fall into several genera. Protozoa belonging to certain of these genera are parasitic on a single invertebrate host (e.g., *Crithidia*), whereas other genera (e.g., *Leishmania, Trypanosoma*) include medically important human and animal parasites whose life cycles are usually divided between an invertebrate vector (e.g., insect or leech) and a vertebrate host.

The kinetoplast is always located at the base of the flagellum (Fig. 2–18) and appears as a dark purple granule when stained with the ordinary basic dyes used to stain DNA. After reviewing ultrastructural data from a variety of Kinetoplastidae, Simpson

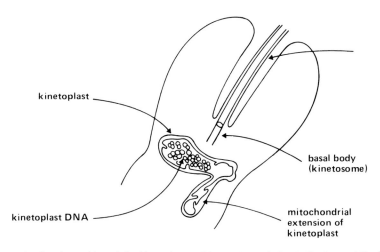

kinetoplast

basal body (kinetosome)

kinetoplast DNA

mitochondrial extension of kinetoplast

FIG. 2–18. Diagram showing the position of the kinetoplast and its structure relative to the base of the flagellum at the anterior end of a leptomonad form of one of the Kinetoplastidae. (Adapted from Trager, ref. 160.)

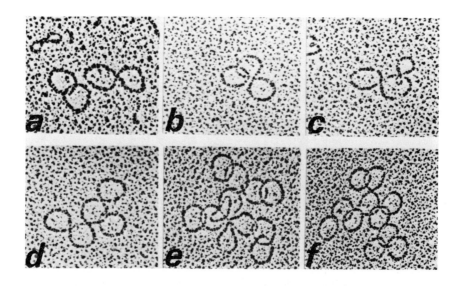

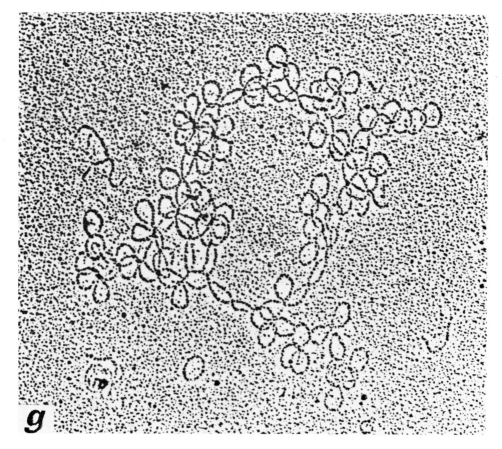

FIG. 2–19. Electron micrographs of several types of molecular configurations seen in purified kDNA from *L. tarentolae*. Minicircles (contour length 0.29 μm) are shown in A–F. A: A twisted circle, a catenated dimer, and a figure 8. B and C: A minicircle catenated with a figure 8. D: An oligomeric figure 8. E and F: Examples of large-scale catenation. G: A small kDNA association. Such circular chains are frequently present in large associations. H: A large association. (From Simpson, ref. 132, with permission.)

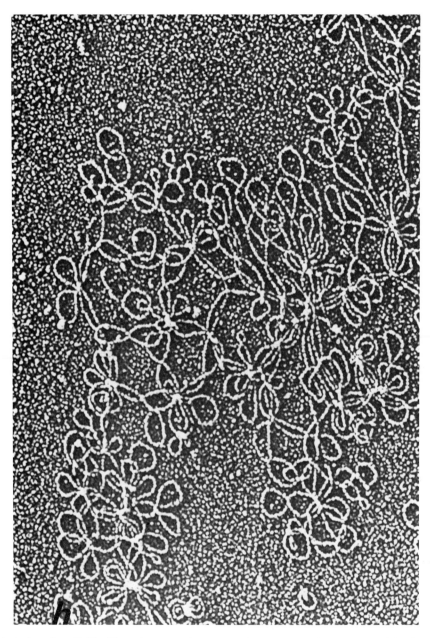

concluded that in all likelihood a member of the Kinetoplastidae contains a single mitochondrion at all stages of its life cycle, the complexity of which depends on the species studied and the stage of the life cycle. The portion of this mitochondrion that surrounds kDNA is defined as the kinetoplast. The kDNA can constitute a substantial fraction (20 to 30% of the total cell DNA [see Simpson's Table 5 (132)]). Purified kDNA has been found to consist of a heterogeneous series of molecular types, including large associations whose structures are difficult to visualize, small associations consisting of small circular DNA molecules (minicircles) catenated with each other, small catenanes of 2 to 10 minicircles, and free minicircles (Fig. 2–19). It has been found that minicircles from any given species form a homogeneous population in terms of size and that between species the range in size corresponds to a molecular weight variation from 0.56×10^6 d (*Leishmania tarentolae*) to 1.49×10^6 d (*Crithidia fasciculata*). Taking 3.3×10^5 d as the molecular weight of DNA required to code for a protein of molecular weight 20,000 d, this would mean that the minicircles could code for somewhere between one and five proteins, depending on source. Clearly this coding capacity is far lower than that of any mtDNA known. Calculations of the number of minicircles per kinetoplast lead to estimates of 10,000 to 24,000, depending on source.

kDNA preparations also include long linear forms, in addition to minicircles. The amount of total kDNA in linear form has been estimated to be as low as 0.8 to 1.0% in *Trypanosoma cruzi* and as high as 33% in *L. tarentolae*. The discrepancies between these estimates may in part be artifactual, and they may in part be representative of species-specific differences. Borst (133) recently reviewed the work of his own group and concluded that these long forms really represent segments of a large circular molecule, which he called a maxicircle. Thus the kDNA network would consist of both minicircles and maxicircles, all of which would be interlocked together. Restriction enzyme analysis leads to the conclusion that the maxicircles in *Crithidia luciliae* would have molecular weights of 20×10^6 d, which is certainly quite respectable for mtDNA molecules. Electron microscopic measurements are in reasonable agreement with this estimate. The maxicircle of *Trypanosoma brucei* appears to be half the size of that found in *C. luciliae*. It will be recalled that a similar mtDNA size variation is encountered among different yeast genera (p. 49). Restriction enzyme analysis of minicircular DNA from *Crithidia* led Borst to conclude that there was some heterogeneity in the sequence of minicircles but that this

was a microheterogeneity; that is, the coding capacity of minicircular DNA was still low, on the order of 1.5×10^6 d. Steinert and associates (134) studied the kinetic complexity of minicircular DNA from three species of *Trypanosomatidae* and found that it is much higher than would be expected from the size of a single minicircle. For *T. brucei* it can be estimated that the kDNA renatures with a complexity equivalent to 300 times the size of a minicircle. Clearly this poses problems with respect to estimates of molecular heterogeneity made on the basis of restriction enzyme analysis.

What can we conclude from these various observations on kDNA? The most likely hypothesis is that the maxicircle is the real mtDNA molecule, with the minicircles fulfilling some unique function. Borst suggested that the minicircle might have a structural function similar to that of the nontranscribed "spacer" genes in nDNA. Clearly the kDNA of the kinetoplast is the most baroque organelle DNA yet found in nature. It seems designed to puzzle, perplex, and delight the student of DNA structure and replication. The real purpose of the minicircle remains to be determined.

CONCLUSIONS

We now know a great deal about the forms of organelle DNAs in different organisms and their potential information content. In the case of mitochondria, it is quite clear that a 5-μm circle that can code 20 to 30 proteins is the minimal mitochondrial genome. Why certain organisms such as the yeast *S. cerevisiae* and higher plants have genomes fivefold to sixfold greater in size and in potential information content is not evident at present. Other species of yeast have mitochondrial genomes ranging in size from 5 to 10 μm. As will be discussed later (see Chapter 6), the mtDNA of *S. cerevisiae* contains long A·T-rich stretches. It is possible that these sequences represent nontranscribed spacers; if so, it may be that the mitochondrial genome of *S. cerevisiae* does not contain a great many more structural genes than that of a mammal. Whether the same is true of higher-plant mtDNA is unknown. The DNA of chloroplasts, with the notable exception of *Acetabularia,* is remarkably uniform in size and information content, and it could code for up to several hundred proteins. As yet we have no idea if all of this DNA is transcribed and if the number of structural genes in clDNA comes anywhere close to the potential number. Much is now known about mtDNA replication in animals, but practically nothing is known in the case of *S. cerevisiae,* where the most work on mitochondrial genetics has been done. In the case of the chloroplast, we know the structure of clDNA replication intermediates in two higher plants, but not in *Chlamydomonas,* which is the organism with the greatest potential for a complete genetic characterization of the chloroplast genome. We know practically nothing of how organelle DNA molecules segregate at the time of organelle division, nor do we know much about the variability in numbers of genomes per organelle. What little we do know with respect to the latter question indicates a rather high degree of variability. As a geneticist the author keenly feels the need for development of a molecular cytogenetics stressing the segregation and distribution of organelle genomes in *S. cerevisiae* and *C. reinhardtii,* which are the organisms of choice for genetic dissection of the mitochondrial and chloroplast genomes, respectively. Despite enormous progress, the organelle geneticist still lacks "cytogenetic" knowledge equivalent to the knowledge that was so central in the development of mendelian genetic theory.

REFERENCES

1. Chiba, Y. (1951): Cytochemical studies on chloroplasts. I. Cytologic demonstration of nucleic acids in chloroplasts. *Cytologia (Tokyo),* 16:259–264.

2. Kirk, J. T. O., and Tilney-Bassett, R. A. E. (1967): *The Plastids.* W. H. Freeman, San Francisco.

3. Ris, H., and Plaut, W. (1962): Ultrastructure of DNA-containing areas in the chloroplast of *Chlamydomonas. J. Cell Biol.,* 13:383–391.

4. Nass, M. M. K., and Nass, S. (1963): Intra-mitochondrial fibers with DNA characteristics. I. Fixation and electron staining reactions. *J. Cell Biol.,* 19:595–611.

5. Nass, M. M. K., and Nass, S. (1963): Intra-mitochondrial fibers with DNA characteristics. II. Enzymatic and other hydrolytic treatments. *J. Cell Biol.,* 19:613–629.

6. Kirk, J. T. O. (1963): Deoxyribonucleic acid of broadbean chloroplasts. *Biochem. J.,* 84:45p.

7. Kirk, J. T. O. (1963): The deoxyribonucleic acid of broadbean chloroplasts. *Biochim. Biophys. Acta,* 76:417–424.

8. Chun, E. H. L., Vaughn, M. H., and Rich, A. (1963): The isolation and characterization of DNA associated with chloroplast preparations. *J. Mol. Biol.,* 7:130–141.

9. Sager, R., and Ishida, M. (1963): Chloroplast DNA in *Chlamydomonas. Proc. Natl. Acad. Sci. U.S.A.,* 50:725–730.

10. Leff, J., Mandel, M., Epstein, H. T., and Schiff, J. A. (1963): DNA satellites from cells of green and aplastidic algae. *Biochem. Biophys. Res. Commun.,* 13:126–130.

11. Luck, D. J. L., and Reich, E. (1964): DNA in mitochondria of *Neurospora crassa. Proc. Natl. Acad. Sci. U.S.A.,* 52:931–938.

12. Rabinowitz, M., Sinclair, J., DeSalle, L., Haselkorn, R., and Swift, H. (1965): Isolation of deoxyribonucleic acid from mitochondria of chick embryo heart and liver. *Proc. Natl. Acad. Sci. U.S.A.,* 53:1126–1133.

13. Ashwell, M., and Work, T. S. (1970): The biogenesis of mitochondria. *Annu. Rev. Biochem.,* 39:251–290.

14. Borst, P., and Kroon, A. M. (1969): Mitochondrial DNA: Physicochemical properties, replication, and genetic function. *Int. Rev. Cytol.,* 26:107–190.

15. Borst, P. (1972): Mitochondrial nucleic acids. *Annu. Rev. Biochem.,* 41:333–376.

16. Borst, P., and Grivell, L. A. (1973): Mitochondrial nucleic acids. *Biochimie,* 55:801–804.

17. Clayton, D. A., and Smith, C. A. (1975): Complex mitochondrial DNA. *Int. Rev. Exp. Pathol.,* 14:1–67.

18. Rabinowitz, M., and Swift, H. (1970): Mitochondrial nucleic acids and their relation to the biogenesis of mitochondria. *Physiol. Rev.,* 50:376–427.

19. Kirk, J. T. O. (1971): Will the real chloroplast DNA please stand up? In: *Autonomy and Biogenesis of Chloroplasts and Mitochondria,* edited by N. K. Boardman, A. W.

Linnane, and R. M. Smillie, pp. 267–276. North Holland, Amsterdam.

20. Green, B. R. (1974): Nucleic acids and their metabolism. In: *Algal Physiology and Biochemistry,* edited by W. D. P. Stewart, Chapter 10. Blackwell, Oxford.

21. Ellis, R. J., and Hartley, M. R. (1974): Nucleic acids of chloroplasts. *MTP International Review of Science,* 6:323–348.

22. Josse, J., Kaiser, A. D., and Kornberg, A. (1961): Enzymatic synthesis of deoxyribonucleic acid. VIII. Frequencies of nearest neighbor sequences in deoxyribonucleic acid. *J. Biol. Chem.,* 236:864–875.

23. Cummins, J. E., Rusch, H. P., and Evans, T. E. (1967): Nearest neighbor frequencies and phylogenetic origin of mitochondrial DNA in *Physarum polycephalum. J. Mol. Biol.,* 23:281–284.

24. Sager, R. (1972): *Cytoplasmic Genes and Organelles.* Academic Press, New York.

25. Kornberg, A. (1974): *DNA Synthesis.* W. H. Freeman, San Francisco.

26. Wang, J. C., Baumgarten, D., and Olivera, B. M. (1967): On the origin of tertiary turns in covalently closed double-stranded cyclic DNA. *Proc. Natl. Acad. Sci. U.S.A.,* 58:1852–1858.

27. Vinograd, J., Lebowitz, J., Radloff, R., Watson, R., and Laipis, P. (1965): The twisted circular form of polyoma viral DNA. *Proc. Natl. Acad. Sci. U.S.A.,* 53:1104–1111.

28. Vinograd, J., and Lebowitz, J. (1966): Physical and topological properties of circular DNA. *J. Gen. Physiol.,* 49:103–125.

29. Bauer, W., and Vinograd, J. (1968): The interaction of closed circular DNA with intercalative dyes. I. The superhelix density of SV40 DNA in the presence and absence of dye. *J. Mol. Biol.,* 33:141–171.

30. Radloff, R., Bauer, W., and Vinograd, J. (1967): A dye-buoyant-density method for the detection and isolation of closed circular duplex DNA: The closed circular duplex DNA of HeLa cells. *Proc. Natl. Acad. Sci. U.S.A.,* 57:1514–1521.

31. Hudson, B., Upholt, W. B., DeVinny, J., and Vinograd, J. (1969): The use of an ethidium bromide analogue in the dye-buoyant density procedure for the isolation of closed circular DNA: The variation of superhelix density of mitochondrial DNA. *Proc. Natl. Acad. Sci. U.S.A.,* 62:813–820.

32. Borst, P. (1970): Mitochondrial DNA: Structure, information content, replication and transcription. *Symp. Soc. Exp. Biol.,* 24:201–226.

33. Hudson, B., Clayton, D. A., and Vinograd, J. (1968): Complex mitochondrial DNA. *Cold Spring Harbor Symp. Quant. Biol.,* 33:435–442.

34. Pikó, L., Blair, D. G., Tyler, A., and Vinograd, J. (1968): Cytoplasmic DNA in unfertilized sea urchin eggs: Physical proper-

ties of circular mitochondrial DNA and the occurrence of catenated forms. *Proc. Natl. Acad. Sci. U.S.A.,* 59:838–845.

35. Goddard, J. M., and Cummings, D. J. (1975): Structure and replication of mitochondrial DNA from *Paramecium aurelia. J. Mol. Biol.,* 97:593–609.

36. Clayton, D. A., and Vinograd, J. (1967): Circular dimer and catenate forms of mitochondrial DNA in human leukaemic leucocytes. *Nature,* 216:652–657.

37. Herrmann, R. G., Bohnert, H.-J., Kowallik, K., and Schmitt, J. G. (1975): Size, conformation and purity of chloroplast DNA of some higher plants. *Biochim. Biophys. Acta,* 378:305–317.

38. Kolodner, R., and Tewari, K. K. (1975): The molecular size and conformation of the chloroplast DNA from higher plants. *Biochim. Biophys. Acta,* 402:372–390.

39. Suyama, Y., and Miura, K. (1968): Size and structural variations of mitochondrial DNA. *Proc. Natl. Acad. Sci. U.S.A.,* 60:235–242.

40. Bohnert, H. J. (1973): Circular mitochondrial DNA from *Acanthamoeba castellani* (Neff-strain). *Biochim. Biophys. Acta,* 324:199–205.

41. Linnane, A. W., and Haslam, J. M. (1971): The biogenesis of yeast mitochondria. *Curr. Top. Cell. Regul.,* 2:101–172.

42. Hollenberg, C. P., Borst, P., and Van Bruggen, E. F. J. (1970): Mitochondrial DNA. V. A 25 µ closed circular duplex DNA molecule in wild-type yeast mitochondria structure and genetic complexity. *Biochim. Biophys. Acta,* 209:1–15.

43. Petes, T. D., Byers, B., and Fangman, W. L. (1973): Size and structure of yeast chromosomal DNA. *Proc. Natl. Acad. Sci. U.S.A.,* 70:3072–3076.

44. Clayton, D. A., and Brambl, R. M. (1972): Detection of circular DNA from mitochondria of *Neurospora crassa. Biochem. Biophys. Res. Commun.,* 46:1477–1482.

45. Agsteribbe, E., Kroon, A. M., and Van Bruggen, E. F. J. (1972): Circular mitochondrial DNA from mitochondria of *Neurospora crassa. Biochim. Biophys. Acta,* 269:299–303.

46. O'Connor, R. M., McArthur, C. R., and Clark-Walker, G. D. (1975): Closed circular DNA from mitochondrial-enriched fractions of four *petite*-negative yeasts. *Eur. J. Biochem.,* 53:137–144.

47. O'Connor, R. M., McArthur, C. R., and Clark-Walker, G. D. (1976): Respiratory-deficient mutants of *Torulopsis glabrata,* a yeast with circular mitochondrial deoxyribonucleic acid of 6 µm. *J. Bacteriol.,* 126:959–968.

48. Lopez Perez, M. J., and Turner, G. (1975): Mitochondrial DNA from *Aspergillus nidulans. FEBS Lett.,* 58:159–163.

49. Wolstenholme, D. R., and Gross, N. J.

(1968): The form and size of mitochondrial DNA of the red bean *Phaseolus vulgaris*. *Proc. Natl. Acad. Sci. U.S.A.,* 61:245–252.

50. Manning, J. E., Wolstenholme, D. R., Ryan, R. S., Hunter, J. A., and Richards, O. C. (1971): Circular chloroplast DNA from *Euglena gracilis. Proc. Natl. Acad. Sci. U.S.A.,* 68:1169–1173.

51. Wong, F. Y., and Wildman, S. G. (1972): Simple procedure for isolation of satellite DNAs from tobacco leaves in high yield and demonstration of minicircles. *Biochim. Biophys. Acta,* 259:5–12.

52. Kolodner, R., and Tewari, K. K. (1972): Physicochemical characterization of mitochondrial DNA from pea leaves. *Proc. Natl. Acad. Sci. U.S.A.,* 69:1830–1834.

53. Kolodner, R., and Tewari, K. K. (1972): Genome sizes of chloroplast and mitochondrial DNAs of higher plants. In: *Proceedings 30th Annual Meeting Electron Microscopy Society of America,* pp. 190–191.

54. Ryan, R. S., Grant, D., Chiang, K.-S., and Swift, H. (1973): Isolation of mitochondria and characterization of mitochondrial DNA of *Chlamydomonas reinhardtii. J. Cell Biol.,* 59:297a.

55. Ryan, R. S., Chiang, K.-S., and Swift, H. (1974): Circular DNA from mitochondria of *Chlamydomonas reinhardtii. J. Cell Biol.,* 63:293a.

56. Kolodner, R., and Tewari, K. K. (1972): Molecular size and conformation of chloroplast deoxyribonucleic acid from pea leaves. *J. Biol. Chem.,* 247:6355–6364.

57. Manning, J. E., Wolstenholme, D. R., and Richards, O. C. (1972): Circular DNA molecules associated with chloroplasts of spinach, *Spinacia oleracea. J. Cell Biol.,* 53:594–601.

58. Behn, W., and Herrmann, R. G. (1977): Circular DNA in the β-satellite DNA of *Chlamydomonas reinhardtii. Mol. Gen. Genet.,* 157:25–30.

59. Britten, R. J., and Kohne, D. E. (1968): Repeated sequences in DNA. *Science,* 161:529–540.

60. Wetmur, J. G., and Davidson, N. (1968): Kinetics of renaturation. *J. Mol. Biol.,* 31:349–370.

61. Kohne, D. E. (1970): Evolution of higher organism DNA. *Q. Rev. Biophys.,* 33:327–375.

62. Nathans, D., and Smith, H. O. (1975): Restriction endonucleases in the analysis and restructuring of DNA molecules. *Annu. Rev. Biochem.,* 44:273–293.

63. Locker, J., Rabinowitz, M., and Getz, G. S. (1974): Electron microscopic and renaturation kinetic analysis of mitochondrial DNA of cytoplasmic petite mutants of *Saccharomyces cerevisiae. J. Mol. Biol.,* 88:489–507.

64. Sanders, J. P. M., Heyting, C., DiFranco, A., Borst, P., and Slonimski, P. P. (1976): The organization of genes in yeast mitochondrial

DNA. In: *The Genetic Function of Mitochondrial DNA,* edited by C. Saccone and A. M. Kroon, pp. 259–272. North Holland, Amsterdam.

65. Bastia, D., Chiang, K.-S., Swift, H., and Siersma, P. (1971): Heterogeneity, complexity, and repetition of the chloroplast DNA from *Chlamydomonas reinhardtii. Proc. Natl. Acad. Sci. U.S.A.,* 68:1157–1161.

66. Wells, R., and Sager, R. (1971): Denaturation and the renaturation kinetics of chloroplast DNA from *Chlamydomonas reinhardtii. J. Mol. Biol.,* 58:611–622.

67. Howell, S. H., and Walker, L. L. (1976): Informational complexity of the nuclear and chloroplast genomes of *Chlamydomonas reinhardtii. Biochim. Biophys. Acta,* 418:249–256.

68. Lambowitz, A. M., Merril, C. R., Wurtz, E. A., Boynton, J. E., and Gillham, N. W. (1976): Restriction enzyme analysis of chloroplast DNA from *Chlamydomonas reinhardtii. J. Cell Biol.,* 70:217.

69. Rochaix, J.-D. (1976): Studies with chloroplast DNA-plasmid hybrids from *Chlamydomonas reinhardtii.* In: *Genetics and Biogenesis of Chloroplasts and Mitochondria,* edited by T. Bücher, W. Neupert, W. Sebald, and S. Werner, pp. 375–378. North Holland, Amsterdam.

70. Green, B. R., Padmanabhan, V., and Muir, B. L. (1975): The kinetic complexity of *Acetabularia* chloroplast DNA. In: *Proceedings 12th International Botanical Congress.* Colloquia Botanicorum, Leningrad.

71. Green, B. R. (1976): Approaches to the genetics of *Acetabularia.* In: *The Genetics of Algae,* edited by R. A. Lewin, Chapter 12. Blackwell, Oxford.

72. Nass, M. M. K. (1966): The circularity of mitochondrial DNA. *Proc. Natl. Acad. Sci. U.S.A.,* 56:1215–1222.

73. Nass, M. M. K. (1967): Circularity and other properties of mitochondrial DNA in animal cells. In: *Organizational Biosynthesis,* edited by H. J. Vogel, J. O. Lampen, and V. Bryson, pp. 503–522. Academic Press, New York.

74. Nass, M. M. K. (1969): Mitochondrial DNA. I. Intramitochondrial distribution and structural relations of single- and double-length circular DNA. *J. Mol. Biol.,* 42:521–528.

75. Nass, M. M. K. (1976): Mitochondrial DNA. In: *Handbook of Genetics, Vol. 5,* edited by R. C. King, Chapter 17. Plenum Press, New York.

76. Gibbs, S. P., Cheng, D., and Slankis, T. (1974): The chloroplast nucleoid in *Ochromonas danica.* I. Three-dimensional morphology in light- and dark-grown cells. *J. Cell Sci.,* 16:557–577.

77. Gibbs, S. P., and Poole, R. J. (1973): Autoradiographic evidence for many segregating

DNA molecules in the chloroplast of *Ochromonas danica. J. Cell Biol.,* 59:318–328.

78. Kowallik, K. V., and Haberkorn, G. (1971): The DNA-structures of the chloroplast of *Prorocentrum micrans* (Dinophyceae). *Arch. Mikrobiol.,* 80:252–261.

79. Woodcock, C. L. F., and Bogorad, L. (1970): Evidence for variation in the quantity of DNA among plastids of *Acetabularia. J. Cell Biol.,* 44:361–375.

80. Herrmann, R. G. (1969): Are chloroplasts polyploid? *Exp. Cell Res.,* 55:414–416.

81. Herrmann, R. G., and Kowallik, K. V. (1970): Selective presentation of DNA-regions and membranes in chloroplasts and mitochondria. *J. Cell Biol.,* 45:198–202.

82. Herrmann, R. G. (1970): Multiple amounts of DNA related to the size of chloroplasts. I. An autoradiographic study. *Planta,* 90:80–96.

83. Herrmann, R. G., and Kowallik, K. V. (1970): Multiple amounts of DNA related to the size of chloroplasts. II. Comparison of electron-microscopic and autoradiographic data. *Protoplasma,* 69:365–372.

84. Kowallik, K. V., and Herrmann, R. G. (1972): Variable amounts of DNA related to the size of chloroplasts. IV. Three-dimensional arrangement of DNA in fully differentiated chloroplasts of *Beta vulgaris* L. *J. Cell Sci.,* 11:357–377.

85. Dawid, I. B., Klukas, C. K., Ohi, S., Ramirez, J. L., and Upholt, W. B. (1976): Structure and evolution of animal mitochondrial DNA. In: *The Genetic Function of Mitochondrial DNA,* edited by C. Saccone and A. M. Kroon, pp. 3–13. North Holland, Amsterdam.

86. Saccone, C., Pepe, G., Cantatore, P., Terpstra, P., and Kroon, A. M. (1976): Mapping of the transcription products of rat-liver mitochondria by hybridization. In: *The Genetic Function of Mitochondrial DNA,* edited by C. Saccone and A. M. Kroon, pp. 27–36. North Holland, Amsterdam.

87. Brown, W. M., and Vinograd, J. (1974): Restriction endonuclease cleavage maps of animal mitochondrial DNAs. *Proc. Natl. Acad. Sci. U.S.A.,* 71:4617–4621.

88. Sanders, J. P. M., Borst, P., and Weijers, P. J. (1975): The organization of genes in yeast mitochondrial DNA. II. The physical map of EcoR1 and Hind II + III fragments. *Mol. Gen. Genet.,* 143:53–64.

89. Küntzel, H., Pühler, A., and Bernard, U. (1976): Physical map of circular mitochondrial DNA from *Neurospora crassa. F.E.B.S. Lett.,* 60:119–121.

90. Bernard, U., and Küntzel, H. (1976): Physical mapping of mitochondrial DNA from *Neurospora crassa.* In: *The Genetic Function of Mitochondrial DNA,* edited by C. Saccone and A. M. Kroon, pp. 105–109. North Holland, Amsterdam.

91. Terpstra, P., Holtrop, M., and Kroon, A. M. (1976): Restriction fragment map of *Neurospora crassa* mitochondrial DNA. In: *The Genetic Function of Mitochondrial DNA,* edited by C. Saccone and A. M. Kroon, pp. 111–118. North Holland, Amsterdam.

92. Herrmann, R. G., Bohnert, H.-J., Driesel, A., and Hobom, G. (1976): The location of rRNA genes on the restriction endonuclease map of the *Spinacia oleracea* chloroplast DNA. In: *Genetics and Biogenesis of Chloroplasts and Mitochondria,* edited by T. Bücher, W. Neupert, W. Sebald, and S. Werner, pp. 351–359. North Holland, Amsterdam.

93. Bedbrook, J. R., and Bogorad, L. (1976): Endonuclease recognition sites mapped on *Zea mays* chloroplast DNA. *Proc. Natl. Acad. Sci. U.S.A.,* 73:4309–4313.

94. Bernard, U., Pühler, A., Mayer, F., and Küntzel, H. (1975): Denaturation map of the circular mitochondrial genome of *Neurospora crassa. Biochim. Biophys. Acta,* 402:270–278.

95. Clayton, D. A. (1976): The expression of mouse mitochondrial DNA in prokaryotic and eukaryotic cells. In: *The Genetic Function of Mitochondrial DNA,* edited by C. Saccone and A. M. Kroon, pp. 47–56. North Holland, Amsterdam.

96. Kasamatsu, H., and Vinograd, J. (1974): Replication of circular DNA in eukaryotic cells. *Annu. Rev. Biochem.,* 43:695–719.

97. Sager, R., and Schlanger, G. (1976): Chloroplast DNA: Physical and genetic studies. In: *Handbook of Genetics, Vol. 5,* edited by R. C. King, Chapter 12. Plenum Press, New York.

98. Gibor, A., and Granick, S. (1962): Ultraviolet sensitive factors in the cytoplasm that affect the differentiation of *Euglena* plastids. *J. Cell Biol.,* 15:599–603.

99. Meselson, M., and Stahl, F. W. (1958): The replication of DNA in *Escherichia coli. Proc. Natl. Acad. Sci. U.S.A.,* 44:671–682.

100. Gross, N. J., and Rabinowitz, M. (1969): Synthesis of new strands of mitochondrial and nuclear deoxyribonucleic acid by semiconservative replication. *J. Biol. Chem.,* 244:1563–1566.

101. Chiang, K.-S., and Sueoka, N. (1967): Replication of chloroplast DNA in *Chlamydomonas reinhardtii:* Its mode and regulation. *Proc. Natl. Acad. Sci. U.S.A.,* 57:1506–1513.

102. Manning, J. E., and Richards, O. C. (1972): Synthesis and turnover of *Euglena gracilis* nuclear and chloroplast deoxyribonucleic acid. *Biochemistry,* 11:2036–2043.

103. Reich, E., and Luck, D. J. L. (1966): Replication and inheritance of mitochondrial DNA. *Proc. Natl. Acad. Sci. U.S.A.,* 55:1600–1608.

104. Dressler, D. (1975): The recent excitement in the DNA growing point problem. *Annu. Rev. Microbiol.,* 29:525–559.

105. Arnberg, A. C., Van Bruggen, E. F. J., Clegg, R. A., Upholt, W. B., and Borst, P. (1974): An analysis by electron microscopy of intermediates in the replication of linear *Tetrahymena* mitochondrial DNA. *Biochim. Biophys. Acta,* 361:266–276.

106. Clegg, R. A., Borst, P., and Weijers, P. J. (1974): Intermediates in the replication of the mitochondrial DNA of *Tetrahymena pyriformis. Biochim. Biophys. Acta,* 361: 277–287.

107. Upholt, W. B., and Borst, P. (1974): Accumulation of replicative intermediates of mitochondrial DNA in *Tetrahymena pyriformis* grown in ethidium bromide. *J. Cell Biol.,* 61:383–397.

108. Goddard, J. M., and Cummings, D. J. (1975): Structure and replication of mitochondrial DNA from *Paramecium aurelia. J. Mol. Biol.,* 97:593–609.

109. Koike, K., and Wolstenholme, D. (1974): Evidence for discontinuous replication of circular mitochondrial DNA molecules from Novikoff rat ascites hepatoma cells. *J. Cell Biol.,* 61:14–25.

110. Kolodner, R. D., and Tewari, K. K. (1975): Chloroplast DNA from higher plants replicates by both the Cairns and rolling circle mechanism. *Nature,* 256:708–711.

111. Tewari, K. K., Kolodner, R. D., and Dobkin, W. (1976): Replication of circular chloroplast DNA. In: *Genetics and Biogenesis of Chloroplasts and Mitochondria,* edited by T. Bücher, W. Neupert, W. Sebald, and S. Werner, pp. 379–386. North Holland, Amsterdam.

112. Chiang, K.-S., and Sueoka, N. (1967): Replication of chromosomal and cytoplasmic DNA during mitosis and meiosis in the eucaryote *Chlamydomonas reinhardi. J. Cell Physiol.* (Suppl 1), 70:89–112.

113. Chiang, K.-S. (1971): Replication, transmission and recombination of cytoplasmic DNAs in *Chlamydomonas reinhardtii.* In: *Autonomy and Biogenesis of Mitochondria and Chloroplasts,* edited by N. K. Boardman, A. W. Linnane, and R. M. Smillie, pp. 235–249. North Holland, Amsterdam.

114. Lee, R. W., and Jones, R. F. (1973): Induction of Mendelian and non-Mendelian streptomycin resistant mutants during the synchronous cell cycle of *Chlamydomonas reinhardtii. Mol. Gen. Genet.,* 121:99–108.

115. Chiang, K.-S. (1976): The nuclear and chloroplast DNA replication mechanisms in *Chlamydomonas reinhardtii:* Their regulation, periodicity and interaction. In: *Les cycles cellulaires et leur blocage chez plusieurs protistes,* edited by M. Lefort-Tran and R. Valencia, pp. 147–157. Editions du Centre National de la Recherche Scientifique, Paris.

116. Cook, J. R. (1966): The synthesis of cytoplasmic DNA in synchronized *Euglena. J. Cell Biol.,* 29:369–372.

117. Pica-Mattoccia, L., and Attardi, G. (1972): Expression of the mitochondrial genome in HeLa cells. IX. Replication of mitochondrial DNA in relationship to the cell cycle in HeLa cells. *J. Mol. Biol.,* 64:465–484.

118. Bosmann, H. B. (1971): Mitochondrial biochemical events in a synchronized mammalian cell population. *J. Biol. Chem.,* 246: 3817–3823.

119. Wells, J. R. (1974): Mitochondrial DNA synthesis during the cell cycle of *Saccharomyces cerevisiae. Exp. Cell Res.,* 85:278–286.

120. Smith, D., Tauro, P., Schweizer, E., and Halvorson, H. O. (1968): The replication of mitochondrial DNA during the cell cycle in *Saccharomyces lactis. Proc. Natl. Acad. Sci. U.S.A.,* 60:936–942.

121. Calvayrac, R., Butow, R., and Lefort-Tran, M. (1972): Cyclic replication of DNA and changes in mitochondrial morphology during the cell cycle of *Euglena gracilis* (Z). *Exp. Cell Res.,* 71:422–432.

122. Parsons, J. A. (1965): Mitochondrial incorporation of tritiated thymidine in *Tetrahymena pyriformis. J. Cell Biol.,* 25:641–646.

123. Parsons, J. A., and Rustad, R. C. (1968): The distribution of DNA among dividing mitochondria of *Tetrahymena pyriformis. J. Cell Biol.,* 37:683–693.

124. Charret, R., and André, J. (1968): La synthèse de l'ADN mitochondrial chez *Tetrahymena pyriformis.* Étude radioautographique quantitative au microscope électronique. *J. Cell Biol.,* 39:369–381.

125. Guttes, E. W., Hanawalt, P. C., and Guttes, S. (1967): Mitochondrial DNA synthesis and the mitotic cycle in *Physarum polycephalum. Biochim. Biophys. Acta,* 142:181–194.

126. Braun, R., and Evans, T. E. (1969): Replication of nuclear satellite and mitochondrial DNA in the mitotic cycle of *Physarum. Biochim. Biophys. Acta,* 182:511–522.

127. Sigee, D. C. (1972): Pattern of cytoplasmic DNA synthesis in somatic cells of *Pteridium aquilinum. Exp. Cell Res.,* 73:481–486.

128. Madreiter, H., Mittermayer, C., and Osieka, R. (1972): ^{3}H-thymidine incorporation into mitochondria of synchronized mouse fibroblasts. *Beitr. Pathol.,* 145:249–255.

129. Williamson, D. H., and Moustacchi, E. (1971): The synthesis of mitochondrial DNA during the cell cycle in the yeast *Saccharomyces cerevisiae. Biochem. Biophys. Res. Commun.,* 42:195–201.

130. Tabak, H. F., Borst, P., and Tabak, A. J. H. (1973): Search for mitochondrial DNA sequences in chick nuclear DNA. *Biochim. Biophys. Acta,* 294:184–191.

131. Flavell, R. A., and Trampé, T. O. (1973): The absence of an integrated copy of mitochondrial DNA in the nuclear genome of *Tetrahymena pyriformis*. *Biochim. Biophys. Acta,* 308:101–105.

132. Simpson, L. (1972): The kinetoplast of hemoflagellates. *Int. Rev. Cytol.,* 32:140–203.

133. Borst, P., Fairlamb, A. H., Fase-Fowler, F., Hoeijmakers, J. H. J., and Weislogel, P. O. (1976): The structure of kinetoplast DNA. In: *The Genetic Function of Mitochondrial DNA,* edited by C. Saccone and A. M. Kroon, pp. 59–69. North Holland, Amsterdam.

134. Steinert, M., Van Assel, S., Borst, P., and Newton, B. A. (1976): Evolution of kinetoplast DNA. In: *The Genetic Function of Mitochondrial DNA,* edited by C. Saccone and A. M. Kroon, pp. 71–81. North Holland, Amsterdam.

135. Price, S. S., DiMaio, D. C., and Englund, P. T. (1976): Kinetoplast DNA from *Leishmania tarentolae*. In: *The Genetic Function of Mitochondrial DNA,* edited by C. Saccone and A. M. Kroon, pp. 83–94. North Holland, Amsterdam.

136. Riou, G. (1976): The kinetoplast DNA from a *Trypanosoma cruzi* strain resistant to ethidium bromide. In: *The Genetic Function of Mitochondrial DNA,* edited by C. Saccone and A. M. Kroon, pp. 95–102. North Holland, Amsterdam.

137. Wu, M., Davidson, N., Attardi, G., and Aloni, Y. (1972): Expression of the mitochondrial genome in HeLa cells. XIV. The relative positions of the 4s RNA genes and the ribosomal RNA genes in mitochondrial DNA. *J. Mol. Biol.,* 71:81–93.

138. Borst, P. (1971): Size, structure and information content of mitochondrial DNA. In: *Autonomy and Biogenesis of Chloroplasts and Mitochondria,* edited by N. K. Boardman, A. W. Linnane, and R. M. Smillie, pp. 260–266. North Holland, Amsterdam.

139. Polan, M. L., Friedman, S., Gall, J. G., and Gehring, W. (1973): Isolation and characterization of mitochondrial DNA from *Drosophila melanogaster*. *J. Cell Biol.,* 56:580–589.

140. Carter, C. E., Wells, J. R., and MacInnis, A. J. (1972): DNA from anaerobic adult *Ascaris lumbricoides* and *Hymenolepis diminuta* mitochondria isolated by zonal centrifugation. *Biochim. Biophys. Acta,* 262:135–144.

141. Dawid, I. B., and Brown, D. D. (1970): The mitochondrial and ribosomal DNA components of oocytes of *Urechis caupo*. *Dev. Biol.,* 22:1–14.

142. Flavell, R. A., and Jones, I. G. (1971): *Paramecium* mitochondrial DNA. Renaturation and hybridization studies. *Biochim. Biophys. Acta,* 232:255–260.

143. Cummings, D. J., Goddard, J. M., and Maki, R. A. (1976): Mitochondrial DNA from *Paramecium aurelia*. In: *The Genetic Function of Mitochondrial DNA,* edited by C. Saccone and A. M. Kroon, pp. 119–130. North Holland, Amsterdam.

144. Christiansen, C., Christiansen, G., and L. Bak, A. (1974): Heterogeneity of mitochondrial DNA from *Saccharomyces carlsbergensis:* Renaturation and sedimentation studies. *J. Mol. Biol.,* 84:65–82.

145. Sanders, J. P. M., Borst, P., and Weijers, P. J. (1975): The organization of genes in yeast mitochondrial DNA. II. The physical map of EcoR1 and Hind II + III fragments. *Mol. Gen. Genet.,* 143:53–64.

146. Wood, D. D., and Luck, D. J. L. (1969): Hybridization of mitochondrial ribosomal RNA. *J. Mol. Biol.,* 41:211–224.

147. Herrmann, R. G. (1973): Number and arrangement of genomes in chloroplasts. *Genetics,* 74:s114.

148. Manning, J. E., and Richards, O. C. (1972): Isolation and molecular weight of circular chloroplast DNA from *Euglena gracilis*. *Biochim. Biophys. Acta,* 259:285–296.

149. Stutz, E. (1970): Kinetic complexity of *Euglena gracilis* chloroplast DNA. *F.E.B.S. Lett.,* 9:25–28.

150. Rawson, J. R. Y. and Boerma, C. (1976): Influence of growth conditions upon the number of chloroplast DNA molecules in *Euglena gracilis*. *Proc. Natl. Acad. Sci. U.S.A.,* 73:2401–2404.

151. Schäfer, K. P., and Küntzel, H. (1972): Mitochondrial genes in *Neurospora:* A single cistron for ribosomal RNA. *Biochem. Biophys. Res. Commun.,* 46:1312–1319.

152. Chua, N.-H., and Luck, D. J. L. (1974): Biosynthesis of organelle ribosomes. In: *Ribosomes,* edited by M. Nomura, A. Tissieres, and P. Lengyel, pp. 519–539. Cold Spring Harbor Laboratory, Cold Spring Harbor, N.Y.

153. Mahler, H. R. (1973): Biogenetic autonomy of mitochondria. *CRC Crit. Rev. Biochem.,* 1:381–460.

154. Loening, U. E. (1968): Molecular weights of ribosomal RNA in relation to evolution. *J. Mol. Biol.,* 38:355–365.

155. Surzycki, S. J., and Rochaix, J. D. (1971): Transcriptional mapping of ribosomal RNA genes of the chloroplast and nucleus of *Chlamydomonas reinhardtii*. *J. Mol. Biol.,* 62:89–109.

156. Gruol, D. J., and Haselkorn, R. (1976): Counting the genes for stable RNA in the nucleus and chloroplasts of *Euglena*. *Biochim. Biophys. Acta,* 447:82–95.

157. Grimes, G. W., Mahler, H. R., and Perlman,

P. S. (1974): Nuclear gene dosage effects on mitochondrial mass and DNA. *J. Cell Biol.,* 61:565–574.

158. Hollenberg, C. P., Borst, P., Thuring, R. W. J., and Van Bruggen, E. F. J. (1969): Size, structure and genetic complexity of yeast mitochondrial DNA. *Biochim. Biophys. Acta,* 186:417–419.

159. Tewari, K. K., and Wildman, S. G. (1970): Information content in the chloroplast DNA. *Symp. Soc. Exp. Biol.,* 24:147–179.

160. Trager, W. (1965): The kinetoplast and differentiation in certain parasitic protozoa. *American Naturalist,* 99:255–266.

161. Chelm, B. K., Hoben, P. J., and Hallick, R. B. (1977): Cellular content of chloroplast DNA and chloroplast ribosomal RNA genes in *Euglena gracilis* during chloroplast development. *Biochemistry,* 16:782–786.

162. Dawid, I. B. (1972): Mitochondrial RNA in *Xenopus laevis* I. Expression of the mitochondrial genome. *J. Mol. Biol.,* 63:201–216.

Chapter 3
Transcription and Translation in Chloroplasts and Mitochondria

We have established that chloroplasts and mitochondria contain DNA that is distinct in several respects from the DNA of the eukaryote cell nucleus. How is the information in organelle DNA expressed? First, it must be transcribed into messenger RNA (mRNA); then the information carried in the mRNA molecules must be translated into the amino acid sequences of specific proteins. This implies that chloroplasts and mitochondria must contain RNA polymerase, the enzyme necessary for transcription of mRNA from a DNA template; it also implies that the messengers synthesized either are exported to the cytoplasm for translation via the cytoplasmic protein-synthesizing system or are translated within the organelle itself. It is now well documented that chloroplasts and mitochondria contain the apparatus necessary to perform both transcription and translation. Thus it is quite possible that chloroplasts and mitochondria promote the complete expression of their own genomes, although this has by no means been proved.

Not only do chloroplasts and mitochondria have the ability to synthesize proteins, but also a portion of the protein-synthesizing machinery of each organelle is coded in the organelle genome. Furthermore, the protein-synthesizing systems of chloroplasts and mitochondria can be distinguished from the system present in the eukaryote cytoplasm. In many respects, particularly in the case of the chloroplast, they resemble the protein-synthesizing systems found in bacteria. In this chapter we will summarize the general characteristics of transcription and translation in chloroplasts and mitochondria, with particular attention to the participating components of these systems; we will consider the specificity of certain inhibitors of organelle transcription and translation; and we will address the question of whether or not mRNAs or proteins or both are imported into and exported out of chloroplasts and mitochondria.

TRANSCRIPTION AND TRANSLATION IN ISOLATED ORGANELLES

The earliest evidence that chloroplasts and mitochondria might be able to promote transcription of their own genomes and translation of the products came from experiments with isolated organelles. In 1964 Kirk (1,2) showed that purified broad bean chloroplasts could form polyribonucleotides, that this synthesis required the presence of all four ribonucleoside triphosphates, and that it could be abolished by either deoxyribonuclease (DNAse) or actinomycin D, the classic inhibitor of RNA polymerase. In the same year, Luck and Reich (3) published essentially similar results for the mitochondria of *Neurospora crassa*. Thus the presence of DNA-dependent RNA-synthesizing systems in chloroplasts and mitochondria was delineated and later confirmed in detail by the experiments of numerous other workers (see 4–8 for reviews).

Incorporation of labeled amino acids by isolated mitochondria was first reported in 1958 by McLean and associates (9); in 1956 it had been reported by Stephenson and associates (10) for chloroplasts. Unfortunately, interpretation of early studies such as these soon became clouded in the case of both chloroplasts and mitochondria because of the presence of contaminating bacteria and, in the case of mitochondria, because incorporation by isolated microsomes also had to be taken into account (see 7,11,12, for reviews). Since then, many other investigators have reported amino acid incorporation into polypeptides by isolated chloroplasts and mitochondria, and there is absolutely no doubt that each of these organelles contains a complete translational apparatus (see 4–8,11–14 for reviews). As Schatz and Mason (12) remarked in the case of amino acid incorporation experiments with isolated mitochondria, the numbers of publications dealing with this phenomenon are awe-inspiring and still increasing. *A priori,* one would have expected these *in vitro* systems to have allowed rapid identification of the products of organelle protein synthesis, but this has not proved to be the case until quite recently (Chapters 19 and 20).

RNA POLYMERASES OF ORGANELLES

The demonstration that isolated chloroplasts and mitochondria could incorporate ribonucleoside triphosphates into RNA in a DNA-dependent reaction immediately indicated that these organelles must contain RNA polymerase. RNA polymerase has since been purified from both organelles (15–20). In Table 3–1 the structures of several organelle RNA polymerases are compared to the much better characterized structures of enzymes of *Escherichia coli* and animal nuclei (21,22). The bacterial and eukaryotic enzymes are complexes of several dissimilar subunits, with the intact enzyme having a very high molecular weight. In contrast, the enzymes obtained from the mitochondria of *Xenopus* (19) and *Neurospora* (16) are very simple in structure, each consisting of a single polypeptide. At low ionic strength, both polymerases aggregate to form structures of much higher molecular weight. Chambon (21) pointed out that the simple structures of these polymerases are reminiscent of the structures of enzymes specific for bacteriophages T3 and T7, each of which consists

of a single subunit of 110,000 d molecular weight. He suggested that this similarity might reflect the fact that each enzyme type transcribes a small genome containing relatively limited structural and regulatory information. The mitochondrial RNA polymerase of the yeast *Saccharomyces cerevisiae* has been studied by several investigators (15,17,23). There is disagreement over the size of the intact enzyme, the number of distinct subunits per enzyme molecule, and the molecular weights of the subunits. Two groups of investigators (15,23) reported that the intact enzyme has a molecular weight of 500,000 d and is composed of two high-molecular-weight subunits. In contrast, another investigator (17) stated that the intact enzyme has a molecular weight of 200,000 d, contains a single subunit of 59,000 to 63,000 d, and forms aggregates similar to those formed by the enzymes of *Xenopus* and *Neurospora* that were discussed previously. In short, it is not clear whether or not the yeast enzyme is really more complex than other mitochondrial RNA polymerases. Bogorad and associates (20,24) studied the chloroplast-specific RNA polymerase of maize. This enzyme contains two high-molecular-weight polypeptides, and it may also contain smaller polypeptides. This enzyme would appear likely to rival bacterial and eukaryotic nuclear RNA polymerases in complexity; in the case of the *Chlamydomonas* chloroplast enzyme, this is certainly true, as shown by Ratcliff and Surzycki (25) (Table 3–1). In fact, Surzycki and Shellenbarger (26) have recently shown that correct recognition of promoter sites for transcription depends on a sigma protein, as it does in *E. coli.*

Studies of inhibitor sensitivities of organelle RNA polymerases have focused principally on rifampicin and α-amanitin. Rifampicin (22) binds tightly to the β subunit of *E. coli* RNA polymerase and blocks RNA chain initiation. Animal nuclear RNA polymerases are resistant to rifampicin (21). However, the B-class enzymes are

TABLE 3–1. Comparison of RNA polymerases from bacteria, from the eukaryote nucleus, and from chloroplasts and mitochondria

Source	Molecular weight (d)		Molar ratio	Inhibitor sensitivity		Reference
	Intact enzyme	Subunits		Rifampicin	α-amanitin	
E. coli	380,000–400,000	150,000–165,000 (β')	1	yes	no	22
		145,000–155,000 (β)	1			
		86,000–95,000 (δ)	1			
		39,000–41,000 (α)	2			
		9,000–12,000 (ω)	2			
Calf thymus (A1)	550,000	197,000 (SA1)	1	no	no	21
		126,000 (SA2)	1			
		51,000 (SA3)	1			
		44,000 (SA4)	1			
		25,000 (SA5)	2			
		16,500 (SA6)	2			
Calf thymus (B1)	600,000	214,000 (SB1)	1	no	yes	21
		140,000 (SB3)	1			
		34,000 (SB4)	1–2			
		25,000 (SB5)	2			
		16,500 (SB6)	3–4			
Calf thymus (B2)	570,000	180,000 (SB2)	1	no	yes	21
		140,000 (SB3)	1			
		34,000 (SB4)	1–2			
		25,000 (SB5)	2			
		16,500 (SB6)	3–4			
Xenopus mitochondria	100,000–150,000 (may be aggregates)	46,000	–	no	no	19
Neurospora mitochondria	64,000 (forms high molecular weight aggregates)	64,000	–	yes	no	16
yeast mitochondria	500,000	190,000	–	no?	no	15,17,23
		150,000	–			
		several smaller subunits	–			
Chlamydomonas chloroplast	–	96,000	1	yes	no	25,26
		93,000	1			
		52,000	2			
		51,000 (2)	–			
		39,000 (1)	–			
Maize chloroplast	500,000	180,000	–	yes?	no	20,24
		140,000	–			
		smaller polypeptides?	–			

sensitive to α-amanitin whereas the A-class enzymes and bacterial RNA polymerases are not (21). Obviously it would be convenient if organelle RNA polymerases, like bacterial enzymes, proved to be sensitive to rifampicin, for then selective inhibition experiments could be performed in which organelle transcription could be blocked without blocking transcription of nDNA. Furthermore, if organelle RNA polymerases should prove to be similar to the bacterial enzymes, they would also be expected to be insensitive to α-amanitin. Thus far, all the mitochondrial enzymes studied have proved to be insensitive to α-amanitin, but rifampicin has yielded highly variable results. For example, Küntzel and Schäfer (16) reported that the mitochondrial RNA polymerase of *Neurospora* that they studied was rifampicin-sensitive, whereas Wintersberger (27) reported that the enzyme he isolated was rifampicin-resistant. Similar uncertainty characterizes the results of experiments with the yeast enzyme. Most investigators have reported that the yeast enzyme is insensitive to rifampicin (15,23,27), but Scragg (17) reported that his enzyme was rifampicin-sensitive. The enzyme from *Xenopus* is clearly rifampicin-insensitive (19). The situation becomes even more confused if one considers the rifampicin sensitivity of RNA synthesis in isolated mitochondria. As Chambon (21) pointed out, "RNA synthesis in isolated mitochondria from various sources was found to be sensitive or insensitive to the antibiotic according to the origin of the mitochondria and to the investigator." Thus caution would dictate that rifampicin not be used in experiments designed to block mitochondrial transcription *in vivo*, since interpretation of any experimental results would most certainly be ambiguous.

Bogorad et al. (20) reported that the chloroplast RNA polymerase of maize is inhibited by the parent compound from which rifampicin is derived, rifamycin SV (28), only in the presence of monovalent cations (NH_4^+, K^+) at high concentrations. Rifamycin SV had no effect on enzyme activity at low concentrations of monovalent cations. Surzycki (29) found that rifampin, which is the same compound as rifampicin (28), blocked RNA synthesis in isolated chloroplasts of *Chlamydomonas*. He also showed that the antibiotic blocked chloroplast RNA synthesis *in vivo*, which is consistent with the fact that chloroplast rRNA is transcribed from clDNA (see pp. 109–113). Since there was no evidence that transcription of nDNA was being affected under

either set of conditions, these results suggested that rifampin could be used to block transcription of clDNA selectively in *Chlamydomonas*. However, it should be noted that in these and later experiments (30), only transcription of rRNA was definitively shown to be blocked, and the inhibitor concentrations that were required were much higher than those in comparable experiments with bacteria.

Bogorad and Woodcock (31) reported similar results with rifampicin on the light-induced synthesis of chloroplast rRNA in maize. Although these results suggest that rifampicin may be a specific transcriptional inhibitor for clDNA, it remains to be established whether or not mRNA transcription is also inhibited *in vivo* by this antibiotic. It could be argued that only the polymerase that transcribes the rRNA cistrons is sensitive to the inhibitor. Also, as Surzycki et al. (32) pointed out, only short-term *in vivo* experiments with rifampicin can be meaningful, since the antibiotic will eventually block chloroplast protein synthesis indirectly by preventing chloroplast ribosome formation.

Other inhibitors have also been tested for their ability to block transcription in organelles. In the case of mitochondria, these include actinomycin D, cordycepin, acriflavine, and ethidium bromide (12,13). Schatz and Mason (12) stated that the results obtained with these inhibitors were not convincing for several reasons. First, actinomycin D usually inhibits amino acid incorporation by isolated mitochondria only partially, probably because mitochondria are quite impermeable to the drug. Second, actinomycin D, as well as cordycepin, may also affect transcription of nDNA, which makes it virtually impossible to interpret the effects of these drugs on protein synthesis *in vivo*. Third, acriflavine and ethidium bromide have been reported to bind to tRNA (33,34), presumably in double-stranded regions, which suggests that these dyes might also behave as translational inhibitors. Grivell and Metz (35), in experiments with isolated mitochondria from *Xenopus,* provided direct evidence that this may be so. These mitochondria, unlike mitochondria from other sources, take up polyuridylic acid (poly-U). The added poly-U causes marked stimulation of phenylalanine incorporation, which is inhibited by ethidium bromide. Since poly-U uptake is not blocked by ethidium bromide, these results provide rather direct demonstration that ethidium bromide can interfere with translation of this artificial message. These results should be kept in mind in interpreting experiments in which ethidium bromide has been employed as a transcriptional inhibitor.

ORGANELLE MESSENGERS:
From whence do they come and whither do they go?

As will be discussed later in this chapter, the mitochondrial genome codes for mitochondrial rRNA and probably codes for a complete set of tRNAs. Chloroplast DNA codes for chloroplast rRNA, and very likely it codes for a set of tRNAs, although information on the latter point is far less complete than in the case of the mitochondrion. There is little reason to believe that any of these stable RNAs are imported from the nucleus. In contrast, there are conflicting views on the origins of the mRNA molecules that are read by organelle protein-synthesizing systems. We can broadly define two general hypotheses concerning the origins of organelle messengers. The first is the import-export hypothesis, which assumes that nuclear transcripts can enter organelles and be translated there and that organelle messengers, in turn, can be translocated into the cytoplasm for translation. The import-export hypothesis enjoyed particular vogue at the beginning of this decade. The second hypothesis, which for want of a better name we will call the barrier hypothesis, assumes that messengers are nei-

ther exported from nor imported into organelles. By definition, this hypothesis implies that all import or export taking place must be at the level of proteins. The barrier hypothesis has enjoyed a much more recent vogue, and most current experimental evidence supports this model.

The issue of RNA transport has been debated most spiritedly and with the greatest experimental rigor in the case of the mitochondrion. In a significant early consideration of the problem, Dawid (36) argued on the side of transport. He contended that mitochondria imported nuclear transcripts that subsequently directed synthesis of their product proteins on mitochondrial ribosomes. This seemed an eminently reasonable suggestion at the time, for it had become evident that the vertebrate mitochondrial genome was very small. Dawid's studies of the *Xenopus* mitochondrial genome revealed that mitochondrial rRNA and 4S RNA were transcripts of mtDNA. Dawid speculated that mitochondrial ribosomal proteins might also be coded by mitochondrial genes. The basis for this speculation was the observation that certain mitochondrial yeast mutants resistant to antibiotics manifested antibiotic-resistant mitochondrial protein synthesis (see Chapter 5). By analogy with bacteria, these mutations might affect specific ribosomal proteins (see Chapter 21). If, in fact, animal mtDNA did code for mitochondrial ribosomal proteins, in addition to rRNA and tRNA, the information in the mtDNA would be exhausted. To explain the finding that certain membrane components were synthesized in mitochondria (see Chapter 19), Dawid suggested that nuclear transcripts were imported into mitochondria and translated there. To explain the evidence that mitochondrial ribosomal proteins were synthesized in the cytoplasm (see Chapter 21), Dawid suggested that the messages for these proteins were exported to the cytoplasm for translation. Initial experimental support for the import-export hypothesis came from Swanson (37), who found that apparently intact mitochondria of *Xenopus* could be stimulated to incorporate phenylalanine by exogenously added poly-U. These results indicated that the poly-U was taken up by the mitochondria and that once it was inside it directed phenylalanine incorporation on mitochondrial ribosomes. Experiments were then performed with radioactive poly-U and several artificial copolymers; the results indicated that the synthetic polynucleotides were incorporated into the mitochondria in a ribonuclease-insensitive form.

Swanson's results gave a significant boost to the idea that polynucleotides might be able to move across the mitochondrial membranes. Dawid (38,39) later suggested that mtDNA sequences not coding for rRNA or tRNA might not code for mitochondrial ribosomal proteins; rather, they might represent spacers. Parts or even all of these sequences might be transcribed *in vivo,* but these spacer RNAs presumably would be degraded and would not be used as mRNA nor as structural RNA. Another piece of suggestive evidence was the finding that mtDNA seemed at that time to code for only 11 to 15 tRNAs. Dawid (38) suggested that this observation could be accounted for if mitochondrial protein synthesis involved only this limited set of tRNAs or if the additional tRNA was made up by molecules imported from the cytoplasm.

Evidence that RNA might be imported into mitochondria also came from a different source. Fukuhara (40) isolated yeast mitochondria as free of contaminating RNA as possible and hybridized the bulk RNA to mtDNA and nDNA. The hybridized RNA was then separated from the DNA to which it hybridized and rehybridized to both mtDNA and nDNA. If both DNAs contain common nucleotide sequences, then the RNA will hybridize with both of them; but if they do not, then the RNA will rehybridize only with that DNA with which it originally hybridized. Fukuhara found that the

latter situation applied in the case of yeast mitochondrial RNA, which showed that there was little, if any, homology between the nuclear and mitochondrial genes from which these RNA transcripts were derived. He interpreted his results as showing that a high proportion of the RNA in his mitochondrial preparation represented nDNA transcripts.

So much for the import-export hypothesis. The barrier hypothesis has found far wider acceptance among students of mitochondrial genetics and biogenesis. Borst (41) addressed the problem specifically in his 1972 review, in which he reviewed the experiments of Swanson critically. His main point was that Swanson had been able to demonstrate uptake of only certain synthetic polynucleotides (e.g., poly-U) with little secondary structure; uptake of RNAs with high degrees of secondary structure (as would be expected for natural mRNAs) was poor in Swanson's experiments. Reijnders and associates also reviewed experiments done in Borst's laboratory involving yeast mitochondrial RNAs (42). These experiments yielded results that differed from those of Fukuhara. Reijnders and associates (42) labeled mitochondrial RNA and then hybridized this RNA with nDNA in the presence of cold, competing cytoplasmic RNA and in its absence. In the absence of this RNA, considerable hybridization occurred, but in the presence of the competing RNA, hybridization to nDNA was reduced almost to background level. An important control was the demonstration that the cytoplasmic RNA fraction was not seriously contaminated with mitochondrial RNA. This was done by establishing that the cytoplasmic RNA did not interfere with hybridization of mitochondrial RNA to mtDNA. These results indicated that the RNA hybridizing with nDNA in the mitochondrial fraction was contaminating cytoplasmic RNA. Since competition experiments of this sort were not reported by Fukuhara, it seems quite possible that his observation that mitochondria contain nuclear transcripts may be accounted for by contamination of his mitochondrial fraction with cytoplasmic RNA. Subsequently, Grivell and Metz (35) repeated Swanson's experiment with poly-U and *Xenopus* mitochondria. However, they were unable to demonstrate poly-U uptake by mitochondria from *Tetrahymena,* yeast, chick liver, or rat liver, which indicates that the poly-U uptake phenomenon may be peculiar to *Xenopus* mitochondria.

One of the most important pieces of experimental evidence in support of the barrier hypothesis was provided by Mahler and Dawidowicz (43). Because this experiment has often been cited in support of the hypothesis that all mitochondrial mRNA, at least in yeast, is transcribed from mtDNA (13,44,45), it deserves special consideration. Mahler and Dawidowicz used a temperature-sensitive mutant (*ts-136*) in which nuclear RNA synthesis is blocked at the restrictive temperature (36°C). When cells, actually spheroplasts, were incubated at this temperature for a short time in the presence of labeled uracil, it was found that labeling of cytoplasmic RNA ceased, but labeling of mitochondrial RNA proceeded at near the control rate. These results showed that although transcription of nDNA was blocked in the mutant at the restrictive temperature, there was little or no effect on mitochondrial RNA synthesis. When ethidium bromide was added to the spheroplast suspension at either permissive or restrictive temperatures, mitochondrial RNA synthesis ceased. Since synthesis of cytoplasmic RNA proceeded normally in the presence of the dye under permissive conditions, it was evident that ethidium bromide was not blocking incorporation of label into transcripts synthesized on nDNA. Thus Mahler and Dawidowicz argued that they were able to shut down either mitochondrial or nuclear RNA synthesis selectively. To assay mRNA specifically, Mahler and Dawidowicz measured the ability of cytoplasmic and mitochondrial RNA to program polyribosomes. They pulse-labeled the polyribosomes with leucine and formate. Labeled leucine was incorporated into nascent protein chains on both cytoplasmic and mitochondrial polysomes, but formate was incorporated specifically into mitochondrial polypeptides in formylmethionyl residues in the N terminal position in nascent chains (13).

The results of these experiments are summarized in Table 3–2. At the restrictive temperature, cytoplasmic RNA synthesis in the mutant is blocked, but mitochondrial RNA synthesis proceeds normally, as described previously. Under these conditions, typical profiles are seen for the mitochondrial polysomes, and formate incorporation proceeds at nearly normal rates; but cytoplasmic polysomes break down, presumably reflecting mRNA depletion, and leucine incorporation into nascent polypeptides declines greatly—all of which is consistent with this interpretation. In the presence of ethidium bromide, functional mitochondrial messages, as assayed by these criteria, also largely vanish. When the mutant is shifted down to the permissive temperature, cytoplasmic RNA transcription and translation recover, and the comparable mitochondrial processes remain at normal levels. However, in the presence of ethidium bromide, mitochondrial polysomes break down, indicating message depletion, but cytoplasmic polysomes are stable. Mahler and Dawidowicz concluded from these results that transcription and translation can proceed at normal levels in mitochondria in the absence of imported transcripts of nDNA. Conversely, their results indicated that blockage of mitochondrial transcription leads to breakdown of mitochondrial polysomes. The only possible difficulty with these experiments has to do with interpretation of the ethidium bromide results. If ethidium bromide, by blocking mitochondrial translation, leads to breakdown of mitochondrial polysomes, then all such polysomes should degrade, whether they carry nuclear or

TABLE 3-2. Synthesis of mRNA in the temperature-sensitive yeast mutant ts-136 at permissive and restrictive temperatures in the presence and absence of ethidium bromide

Temperature (°C)	Ethidium bromide	RNA synthesis		Polysome stability		Protein synthesis		Mitochondrial polypeptide initiation	
		Mitochondrion	Cytoplasm	Mitochondrion	Cytoplasm	Mitochondrion	Cytoplasm	Formate	Formylmethionyl puromycin
36	−	100	1.0	+	−	92	3.5	100	78
	+	0	1.0	−	−	2	not done	29	22
36 → 23	−	30	100	+	+	100	17	100	104
	+	0.2	80	−	+	not done	not done	not done	not done

Adapted from Mahler et al. (45), see (along with text) for details.

mitochondrial transcripts. The problem is one of the experimental level of resolution. Mahler and Dawidowicz showed unequivocally that translation of mitochondrial messages proceeds nearly normally under conditions in which nuclear message production is blocked. However, the converse part of the experiment, which employed ethidium bromide at the permissive temperature, did not yield clear-cut results. That is, if some small fraction of messages translated on mitochondrial polysomes are of nuclear origin, one could be misled by translational inhibition caused by ethidium bromide. After all, the poly-U experiments of Grivell and Metz (35) indicated that ethidium bromide can block translation of artificial messages in mitochondria.

A different line of evidence supporting the barrier hypothesis, which we will develop further in the chapters on mitochondrial genetics and biogenesis in yeast (Chapters 4–7 and 19), has to do with the mitochondrial proteins made in the presence of mitochondrial protein synthesis and in its absence. Virtually all easily measurable mitochondrial proteins (except certain specific inner-membrane proteins) can be made in the absence of mitochondrial protein synthesis; thus importation of the messages for the bulk of mitochondrial proteins is unnecessary for their synthesis. These proteins are also made by respiratory-deficient mutants lacking mtDNA; thus their messages are not exported either. Furthermore, the mitochondrial mutations thus far isolated that affect specific polypeptides affect only polypeptides whose messages are translated on mitochondrial ribosomes (Chapter 19). In short, there is good reason to believe that the barrier hypothesis is largely, if not entirely, correct.

In the case of the chloroplast, we are unable to find rigorous evidence for or against either the barrier hypothesis or the import-export hypothesis. Those working with *Chlamydomonas* have frequently invoked messenger import and export to explain the results of experiments with inhibitors and mutants (46–51) (see Chapters 20 and 21). On the other hand, thus far there has been little need to invoke import and export of messages in *Euglena* (52).

CHARACTERIZATION OF ORGANELLE mRNA

The most complete study of transcription of an organelle genome was carried out by Attardi and his colleagues on the mitochondrial genome of HeLa cells. Aloni and Attardi (53) first established by hybridization experiments that uniformly labeled mitochondrial RNA hybridizes with virtually the entire H strand of mtDNA, indicating complete transcription of this strand *in vivo*. These experiments also indicated that there is very little RNA that hybridizes with the L strand. However, short-term pulse labeling experiments subsequently revealed RNA that does hybridize with the L strand. Aloni and Attardi (54) established that the L strand is transcribed over a majority, if not the entirety, of its length. Thereafter, the L-strand transcripts rapidly disappear from the mitochondrial fraction, either because they are degraded or because they are exported. The results with the H strand are not surprising, since it is known that this strand codes for mitochondrial rRNA and most of the tRNAs, but the L strand codes for only a few tRNAs, as far as is known (pp. 92 to 94). An obvious possibility is that many of the other transcribed sequences serve as spacers or are important for processing, regulation, etc., and have nothing to do with mRNA.

Attardi and his colleagues next set out to determine if the transcripts of the HeLa cell mitochondrial genome included mRNAs. In eukaryotic cells, mRNAs (with the exception of histone mRNAs) have a segment of polyadenylic acid (poly-A) at the 3'-OH end (55,56). The poly-A segment varies in length from 100 to 200 adenylate residues and is added enzymatically following transcription. Since poly-A serves as a convenient handle by which to identify eukaryotic mRNA, the discovery of poly-A covalently linked to mitochondrial RNA would be a good indicator of the presence of mRNA in the mitochondrion. Perlman and associates (57) were the first to report isolation of poly-A-containing RNA from HeLa cell mitochondria. The poly-A sequence was about 50 to 80 bases long, or considerably shorter than the poly-A segment associated with cytoplasmic mRNA. It was possible to release this RNA from mitochondrial ribosomes by use of puromycin, which causes release of nascent polypeptide chains, after which the ribosomes dissociated from the mRNA. Furthermore, it was found that synthesis of this RNA was largely inhibited by ethidium bromide. All of these observations indicated that the mitochondria synthesized an RNA with the properties of eukaryotic mRNA. Substantially similar results were obtained by Ojala and Attardi (58), and an enzyme that catalyzed synthesis of poly-A was reported from rat liver mitochondria (59).

The next step was systematic characterization of poly-A mRNA. Ojala and Attardi (60) found two discrete poly-A-containing RNA components in the polysome region of mitochondrial lysates. One of these, with a sedimentation velocity of 7S and an estimated molecular weight of 9×10^4 d, hybridized with the L strand; the second component, with a sedimentation velocity of 9S, hybridized with the H strand. These results indicated that the L strand coded for a poly-A-containing RNA in addition to several tRNAs. Further study by Ojala and Attardi (61) revealed that although the 7S component hybridizing with the L strand continued to behave as a single species, the poly-A-containing RNAs hybridizing with the H strand could be separated into seven species ranging in molecular weight from 2.6×10^5 d to 5.3×10^5 d. Ojala and Attardi calculated that these RNAs accounted for 2.3×10^6 d of information in the HeLa

cell genome following correction for the poly-A segment, which is not coded by the mitochondrial genome. Since the tRNAs and rRNAs accounted for an additional 1.2×10^6 d, Ojala and Attardi calculated that they had accounted for some 70% of the information in a single strand of HeLa cell mtDNA. The calculation assumes, of course, that in any given segment of the mitochondrial genome only one strand (either H or L) has any information content. Similar results were reported by Hirsch and Penman (62). Most recently, Attardi and associates (63) reported an even greater number of transcripts of HeLa cell mtDNA identified by pulse labeling and a new gel fractionation method. Some 18 poly-A-containing mRNAs have now been identified, and 14 non-poly-A-containing components have also been found, including the two rRNAs. Synthesis of all components is sensitive to ethidium bromide, and the major components have been shown to hybridize with mtDNA. All components tested, except the 7S species described previously, hybridized specifically with the H strand. At the moment it is not clear whether or not the poly-A and non-poly-A components of similar molecular weights are related.

Attardi's work has revealed an interesting problem. The number and size of discrete RNA species reported in his most recent study are such that their sequences cannot be accommodated in the portions of the H strand not already occupied by the rRNA and tRNA genes. These observations imply that at least some tRNA species must be synthesized as part of larger transcripts and that some high-molecular-weight RNA species separated in the gel system must be precursors of others. Messenger-like poly-A-containing RNA has also been associated with hamster, *Drosophila,* and mosquito mitochondria by Hirsch and associates (64).

The poly-A story in yeast mitochondria deserves brief mention because Rabinowitz and associates (65,66) have succeeded in translating this RNA in an *in vitro* system and identifying the products as subunits of cytochrome oxidase. Poly-A-containing RNA from yeast mitochondria was first reported by Cooper and Avers (67). Subsequently, Groot and associates (68) were unable to find this RNA, and they suggested that Cooper and Avers had used preparations contaminated with protoplasts and, hence, cytoplasmic poly-A-containing RNA. They suggested that the absence of poly-A was consistent with the "prokaryotic" origin of mitochondria and that the mitochondria of higher eukaryotes had become further adapted and that their mRNAs had acquired poly-A segments in the course of evolution. Hendler and associates (69) then discovered a yeast mitochondrial RNA fraction containing a short poly-A stretch of 20 to 30 nucleotides. This accounted for 2 to 5% of the total mitochondrial RNA. Translation of this RNA was achieved in an *E. coli* cell-free system (65,66). The products of translation included the three peptides of cytochrome oxidase known to be synthesized in the mitochondrion (see Chapter 19), as shown by precipitation of the labeled peptides by antiserum to cytochrome oxidase. Synthesis of this RNA was almost completely inhibited by ethidium bromide, and at least 55% of this RNA hybridized to mtDNA with essentially no hybridization to nDNA or DNA from *E. coli*. These results strongly suggest that the messages for the three cytochrome oxidase polypeptides are coded by the mitochondrial genome of yeast and that they contain poly-A. These results also support the barrier hypothesis, since it is known that these polypeptides are translated on mitochondrial ribosomes (see Chapter 19).

A great deal less is known about clDNA transcripts with the exception of the mRNA which codes for the large subunit of RuBPCase. Hartley and associates (70) were the first to obtain *in vitro* translation of mRNA for the large subunit of fraction I

protein. They translated total chloroplast RNA using a cell-free extract from *E. coli* and found that two products, one of 52,000 d and the other of 35,000 d were made. The 52,000-d product was slightly smaller (ca. 1,500 d) than the native large subunit of RuBPCase made *in vitro* by isolated pea chloroplasts, and contained seven of the nine chymotryptic peptides of these native large subunits. Wheeler and Hartley (178) then separated the RNA from spinach chloroplasts into poly-A and non-poly-A-containing fractions and repeated the *in vitro* translation experiments using the *E. coli* cell-free system. They found that only the non-poly-A-containing RNA fraction programmed synthesis of the 52,000-d and 35,000-d polypeptides and concluded that the mRNA coding for the large subunit of fraction I protein as well as the mRNA coding for the 35,000-d species were not polyadenylated. Sagher and associates (179) prepared poly-A- and non-poly-A-containing RNA from *Euglena* chloroplasts and assayed these RNAs using a cell-free protein synthesizing system from wheat germ. Like Wheeler and Hartley, they observed that the messenger for the large subunit of fraction I protein was present only in the non-poly-A-containing RNA fraction. Moreover, the product which they obtained was identical with native fraction I large subunit (59,000 d) as determined by a two dimensional gel system employing isoelectric focusing followed by size filtration in SDS-polyacrylamide gels. Their results also revealed that this polypeptide was not synthesized by the non-poly-A-containing RNA fraction from the W_3BUL mutant lacking chloroplast DNA, and that the large subunit message from wildtype sedimented in the 10–20S fraction of a sucrose gradient.

Howell et al. (71) have obtained similar evidence from *Chlamydomonas* that the large subunit mRNA is not polyadenylated. However they found that the *E. coli* translation system only yielded partial large subunit polypeptides. These were identified by a combination of immunoprecipitation and tryptic peptide analysis. The wheat germ system did not translate the large subunit mRNA to form an immunoprecipitable product. Howell et al. (71) argue that the partial polypeptides obtained with the *E. coli* translation system probably did not result from mRNA degradation as the messenger fraction used was large enough to program synthesis of the entire polypeptide. Thus the *E. coli* system appears unable to translate large subunit mRNA in its entirety. Dobberstein and associates (180) have performed essentially complementary experiments on the small subunit of fraction I protein from *Chlamydomonas*. They found that small subunit mRNA is polyadenylated and translated on cytoplasmic ribosomes. These and related experiments will be discussed in more detail in Chapter 20.

The most detailed studies of the transcription pattern of clDNA are those done by Rawson (72,73) and Chelm and Hallick (74) on cells of *Euglena* transferred from dark growth to the light. Dark-grown cells contain proplastid-like structures that differentiate into normal chloroplasts when the cells are transferred to the light (see Chapter 16). Rawson isolated total cell RNA and hybridized it in vast excess to clDNA labeled with ^{125}I. The fraction of clDNA transcribed was calculated as double the amount of single-stranded clDNA in

the form of an RNA-DNA hybrid. Rawson calculated that about 53% of the DNA was transcribed in the dark. After 5 hr of illumination, the fraction of clDNA transcribed increased to 57%. At the completion of chloroplast development, 47% of the clDNA was represented as RNA transcripts. A similar approach was used by Chelm and Hallick (74), except the clDNA was labeled by a different procedure and somewhat different media were used. Results quite similar to those of Rawson were obtained. On transfer of cells to the light, the amount of clDNA-hybridizing RNA declined slightly and then increased. Approximately equal amounts of clDNA were transcribed in the studies of Chelm and Hallick and the studies of Rawson. This approach seems to have much promise. Eventually it should be possible to characterize the transcripts in detail.

ORGANELLE tRNA AND AMINOACYL tRNA SYNTHETASES

The existence of specific tRNAs and corresponding aminoacyl tRNA synthetases in both chloroplasts and mitochondria has been known for some time (75–81). The available evidence indicates that most, if not all, mitochondrial tRNAs are coded by the mitochondrial genome; similar, but far less extensive, evidence exists for the chloroplast (Tables 3–3 and 3–4). The se-

TABLE 3–3. *Number of tRNA cistrons per organelle genome estimated from hybridization of total organelle tRNA to organelle DNA*[a]

Organelle	Source	Number of tRNA cistrons per genome	Method used to determine number	Reference
Mitochondrion	HeLa cells	19	M	63
	Xenopus	22	M	19
	Saccharomyces	20	H	150
Chloroplast	Euglena	26	H	151
	Tobacco	20–30	H	139

[a] In some cases a 4S RNA fraction was used for hybridization; in other cases purified tRNA was used. Numbers of tRNA cistrons are deduced from hybridization experiments (H) and from molecular mapping (M). See text for details.

TABLE 3–4. *Identification of specific tRNAs coded by the mitochondrial genome from hybridization of tRNAs charged with specific amino acids to mtDNA*

Amino acid for which tRNA identified	Source of mtDNA and tRNA		
	HeLa cells[a]	Saccharo- myces[b]	Rat liver[c]
Alanine	+[d]	+	NT
Arginine	+	+ (2)	NT
Asparagine	−	−	NT
Aspartic acid	+	+	NT
Cysteine	+	+ (2)	NT
Formylmethionine	+	+	NT
Glycine	+	+	NT
Glutamine	−	−	NT
Glutamic acid	+	+ (2)	NT
Histidine	−	+	NT
Isoleucine	+	+	NT
Leucine	+	+ (2)	+
Lysine	+	+	NT
Methionine	+	+ (2)	NT
Phenylalanine	+	+	+
Proline	−	+	NT
Serine	+	+	+
Threonine	+	−	NT
Tryptophan	+	−	NT
Tyrosine	+	+ (4)	+
Valine	+	+ (2)	NT

[a] Data from Attardi et al. (63).

[b] Data from Rabinowitz et al. (66).

[c] Data from Nass and Buck (152).

[d] Plus sign indicates that tRNA hybridizes with mtDNA; minus sign indicates that no tRNA is detected that hybridizes with mtDNA; NT indicates not tested. Numbers in parentheses next to a given tRNA indicate the number of isoaccepting tRNAs detected for that amino acid that hybridize to mtDNA.

quence homology and mapping studies of organelle tRNA supporting this conclusion will be discussed later in this chapter and in Chapter 11. In contrast, the corresponding aminoacyl tRNA synthetases are most likely coded by nuclear genes whose messages are translated in the cytoplasm. This rather complex interaction is one of the subjects discussed in Chapter 21, which deals with the biogenesis of organelle protein-synthesizing systems. In this section we will address two questions: What is the mechanism of polypeptide chain initiation in chloroplasts and mitochondria? Are mitochondria-coded tRNAs sufficient to account for mitochondrial protein synthesis without importation of tRNAs from the cytoplasm? Given our present knowledge, it is pointless to address the latter question in the case of chloroplasts.

Polypeptide chains are initiated in bacterial systems by *N*-formylmethionyl tRNA (f-met-tRNA$_F$) (82). Formylmethionyl tRNA is synthesized by formylation of methionyl tRNA (met-tRNA) following charging of this tRNA by a methionine tRNA synthetase. The formylation of the methionine residue is catalyzed by a specific transformylase. In eukaryotic cells two met-tRNAs are known, one of which behaves as the initiator. This met-tRNA is not formylated, but it can be formylated *in vitro* by the transformylase. This initiator tRNA has been termed met-tRNA$_F$ by analogy with *E. coli*. Both the prokaryotic and eukaryotic tRNAs recognize the same initiation codon AUG. If polypeptide chain initiation proceeds in organelles as it does in bacteria, it should be possible to demonstrate the presence of f-met-tRNA$_F$ and a transformylase in both mitochondria and chloroplasts. Formylmethionine tRNA and transformylase activity have been reported in both organelles (see 11, 13, 14 and 83 for reviews); thus it appears that polypeptide synthesis in both organelles is initiated as it is in bacteria. Mahler and Dawidowicz (43) took advantage of this in the experiments described previously in which they used labeled formate as a means of tagging nascent polypeptide chains associated with mitochondrial polysomes.

It is evident from the data presented in Table 3–4 that several mitochondrial tRNAs remain to be detected from both yeast and HeLa cell mitochondria, although they have been sought after. Several years ago, when fewer hybridizable tRNAs were known, the situation was far more confused. Three possible resolutions of this dilemma can be suggested: First, and most simple, the additional tRNAs will eventually be found; or they may already have been detected without their functions having been assigned correctly. For example, Rabinowitz and associates (66) pointed out that their inability to detect glutamine tRNA (tRNAgln) may have been artifactual, since one of the two glutamic acid tRNAs (tRNAgluI and tRNAgluII) may actually be tRNAgln. They pointed out that in some bacterial species tRNAgln may initially be charged with glutamic acid, which is then converted to glutamine while still linked to the tRNA. Second, it may be found that the missing tRNAs are imported from the cytoplasm (also see Chapter 21). Third, it may be found that the proteins synthesized by mitochondria lack the amino acids for which there are no tRNAs. Some support for the last possibility was suggested by the experiments of Costantino and Attardi (84). They assessed the ability of HeLa cell mitochondria to incorporate different amino acids *in vitro* and *in vivo*. They found that certain amino acids were used very little, if at all, for mitochondrial protein synthesis in HeLa cells. These included most of the charged polar amino acids (arginine, aspartic acid, lysine, and possibly glutamic acid) and some of the neutral polar amino acids (cysteine and glutamine). Histidine and some other neutral polar amino acids (tyrosine, serine, and threonine) appeared to be used to a low or moderate extent, but the strongly hydrophobic amino acids such as leucine, isoleucine, valine, phenylalanine, and methionine were actively incorporated. Costantino and Attardi argued that pool effects probably did not account for the variation seen in incorporation of different amino acids. At the time their article was published, hybridization experiments with aminoacylated mitochondrial tRNAs had not been done in HeLa cells and they pointed out that the available information for rat liver mitochondrial tRNAs (Table 3–4) indicated that the tRNAs present generally corresponded to the most actively incorporated amino acids in HeLa cells. Recently, however, it has become evident that HeLa cell mitochondria contain specific tRNAs

for many of the amino acids that were not incorporated into mitochondrial protein in the Costantino and Attardi experiments (Table 3–4). Lynch and Attardi (85) have suggested that pool phenomena may, after all, account for the previously observed marginal incorporation of certain amino acids by HeLa cell mitochondria. As Lynch and Attardi pointed out, their inability to detect tRNAs for all of the amino acids might be accounted for by a number of artifacts (e.g., loss of the synthetases required to charge the amino acids for which tRNAs have not been found).

In summary, it now seems likely that proteins synthesized by mitochondria contain a complete set of amino acids and that the tRNAs corresponding to most if not all of these amino acids are coded by the mitochondrial genome. Whether results reported for *Tetrahymena* (Chapter 21) indicating that some tRNAs are imported from the cytoplasm prove to be generally true, remains to be seen.

STRUCTURE OF ORGANELLE RIBOSOMES

The structure of organelle ribosomes has been reviewed by several authors in recent years (11,14,86–90). Chloroplast ribosomes were the first to be discovered. In 1962 Lyttleton (91) reported the isolation of ribosomes from spinach chloroplasts. He found that these ribosomes had a sedimentation velocity of 66S and could be distinguished by their S value from the cytoplasmic ribosomes of the leaf. True mitochondrial ribosomes were first reported in 1967 by Küntzel and Noll in *Neurospora* (92) and by O'Brien and Kalf in rat liver (93,94). Since the time of these early studies, chloroplast and mitochondrial ribosomes from a variety of sources have been characterized. At the outset, a few general remarks concerning the ribosomes of prokaryote cytoplasm and eukaryote cytoplasm are in order so that they can be compared with organelle ribosomes.

Prokaryotes contain 70S ribosomes; the ribosomes of the eukaryote cytoplasm are 80S (Table 3–5). The structure, assembly, and genetics of these ribosomes have been considered at length in a recent treatise by Nomura and associates (95). Prokaryotic 70S ribosomes dissociate into subunits of 50S and 30S (Table 3–5). The numbers of ribosomal proteins in each subunit are quite accurately known for *E. coli:* 21 for the small subunit and 34 for the large subunit (96) (Table 3–5). The small subunit contains 16S rRNA, with a molecular weight of 0.6×10^6 d. The large subunit contains two species of rRNA. One of these is 23S rRNA, with a molecular weight of 1.1×10^6 d; the other, 5S RNA, is much smaller, with a molecular weight of 0.04×10^6 d (97). Both the 16S and 23S rRNAs appear to arise from a single 30S precursor RNA, which includes 5S RNA and certain tRNAs (98,181). Eukaryotic ribosomes dissociate into subunits with sedimentation velocities of approximately 60S and 40S (Table 3–5). The small subunit contains 18S rRNA, with a molecular weight approximately 1.3×10^6 d and slightly more proteins than its prokaryotic counterpart. The large subunit contains three species of rRNA. The larger of these in lower eukaryotes is 25S rRNA (molecular weight 1.3×10^6 d); in higher eukaryotes this rRNA has an S value of 28 (molecular weight 1.9×10^6 d). The large subunit also contains 5S RNA, and a third species of RNA (5.5S–5.8S) is noncovalently bound to the 25S or 28S rRNA of the large subunit. Formation of the large-subunit and small-subunit rRNAs, with the exception of the 5S rRNA, involves cleaving and trimming of a single large precursor RNA (99).

Chloroplast ribosomes appear to be uniformly similar to prokaryotic ribosomes in terms of sedimentation coefficients of intact

TABLE 3–5. *Some characteristics of eukaryotic and prokaryotic cytoplasmic ribosomes*[a]

Ribosome source		Monomer	Subunits		rRNA		References
			Large	Small	Large	Small	
E. coli	S value	70	50	30	23	16	96,97
	Molecular weight ($\times 10^6$ d)	2.7	1.7	0.9	1.1	0.6	
	Number of proteins	55	34	21	—	—	
Chlamydomonas	S value	83	58	40	25	18	102,153,154
	Molecular weight ($\times 10^6$ d)	4.0	2.4	1.3	1.3	0.7	
	Number of proteins	65	39	26	—	—	
Saccharomyces	S value	82	61	38	25	18	99,155
	Molecular weight ($\times 10^6$ d)	3.6	—	—	1.3	0.7	
Tetrahymena	S value	80	60	40	26	17	107
	Molecular weight ($\times 10^6$ d)	3.4	2.3	1.5	1.4	0.6	
Pea	S value	78	56	36	25	18	156–159
	Molecular weight ($\times 10^6$ d)	3.8	2.4	1.4	1.3	0.7	
	Number of proteins	52	28	24	—	—	
Rat liver or HeLa cells	S value	80	60	40	28	18	160
	Molecular weight ($\times 10^6$ d)	4.5	3.0	1.5	1.7	0.7	
	Number of proteins	70	40	30	—	—	

[a] All of these ribosomes also contain 5S RNA associated with the large subunit, which is not shown. Number-of-proteins row is included only in cases where the number is known with some accuracy.

ribosomes and their subunits (Table 3–6). The two major rRNAs are also similar to their prokaryotic counterparts in terms of sedimentation coefficients and molecular weights. The base compositions of 16S and 23S rRNAs from chloroplasts of two higher plants are compared to those of other rRNAs in Table 3–8. The G + C content is similar to that of *E. coli* with respect to the 16S component, but the 23S component has a higher G + C content than either the 23S rRNA of *E. coli* or the blue-green alga *Anacystis nidulans*. The 23S rRNAs from all chloroplasts thus far studied are notoriously susceptible to cleavage (87,100). The cleavage products are also of high molecular weight, and one of them has a molecular weight not greatly different from that of the rRNA of the small subunit. The large chloroplast ribosomal subunit also contains a low-molecular-weight RNA with a sedimentation coefficient of approximately 5S (87,89). Payne and Dyer (101) established that 5.8S rRNA is present in the cyto-plasmic ribosomes of a wide variety of plants, but no similar component was found in the chloroplast ribosomes of either spinach or broad bean, which reveals another point of similarity between prokaryote and chloroplast ribosomes. The chloroplast ribosomal subunit proteins of *Chlamydomonas* have been separated by two-dimensional gel electrophoresis by Hanson and associates (102) (Fig. 3–1) and Brügger and Boschetti (103). Despite discrepancies between the numbers obtained by these two groups of workers, it is evident that the numbers of proteins in each subunit are in the same range as those found for the comparable subunits of *E. coli* (Tables 3–5 and 3–6). Some data pertaining to chloroplast ribosomal protein composition have also been obtained for *Euglena* and tobacco (Table 3–6).

The similarity between chloroplast and prokaryotic ribosomes is underscored by subunit exchange experiments. Lee and Evans (104) found that the small subunit

TABLE 3–6. *Chloroplast ribosomes of different organisms[a]*

Ribosome source		Monomer	Subunits		rRNA		References
			Large	Small	Large	Small	
Chlamydomonas	S value	68–70	52–54	37–41	23	16	87,89,102,103
	Molecular weight ($\times 10^6$ d)	—	—	—	1.05	0.56	154,161,162
	Number of proteins	48,60	26,35	22,25	—	—	
Euglena	S value	68–70	46–50	29–30	22	17	87,163
	Molecular weight ($\times 10^6$ d)	—	—	—	1.1	0.55	
	Number of proteins	55–58	—	—	—	—	
Spinach	S value	70	50	30	23	16	87
	Molecular weight ($\times 10^6$ d)	—	—	—	1.34	0.7	
Bean	S value	70	—	—	23	16	87
	Molecular weight ($\times 10^6$ d)	—	—	—	—	—	
Pea	S value	70	45	32	23	16	11,87
	Molecular weight ($\times 10^6$ d)	—	—	—	1.06	0.57	
Tobacco	S value	70	50	35	23	16	87,164
	Molecular weight ($\times 10^6$ d)	—	—	—	—	—	
	Number of proteins	—	25	—	—	—	

[a] Number-of-proteins row is included only in cases where numbers are known with some accuracy. For *Chlamydomonas* and *Euglena*, only the best estimates of sedimentation velocity are given.

of the *Euglena* chloroplast ribosome was active in a poly-U-directed phenylalanine-incorporating system in combination with the large subunit from *E. coli* and *E. coli* supernatant factors (i.e., aminoacyl tRNA syntheses and tRNAs). However, the reciprocal combination was inactive. Grivell and Walg (105) repeated this experiment with spinach chloroplast ribosomes and found that both possible hybrid ribosome combinations were active in the poly-U system. In contrast, yeast mitochondrial ribosomes did not form active hybrid combinations with either *E. coli* or chloroplast ribosomes. Despite these findings, it should be emphasized that the poly-U system is highly artificial. Similar experiments should be conducted with natural mRNA.

In contrast to the relative uniformity of chloroplast ribosomes, mitochondrial ribosomes comprise the most diverse class yet identified, as O'Brien and Matthews (90) pointed out in their excellent review. The best-characterized mitochondrial ribosomes among the protists are those from *Euglena gracilis* and *Tetrahymena pyriformis*. *Euglena* mitochondrial ribosomes have a number of similarities to prokaryotic ribosomes in terms of sedimentation velocities of intact monomers, ribosomal subunits, and rRNAs (Table 3–7). However, the mole percent of G + C (27%) found in *Euglena* mitochondrial rRNA is the lowest yet reported for any organism (106). *Tetrahymena* mitochondrial ribosomes have a sedimentation velocity of 80S and do not dissociate into subunits, even in solutions lacking magnesium. Dissociation has been achieved in the presence of EDTA, yielding 55S particles (Table 3–7). These particles include both subunits, which can be differentiated by their densities following buoyant-density centrifugation in CsCl (107). The mitochondrial rRNAs have lower molecular weights than their prokaryotic counterparts (Table 3–7). O'Brien and Matthews (90) pointed out that the G + C content of *Tetrahymena* mitochondrial

rRNA (29%) is nearly as low as that of *Euglena* mitochondrial rRNA; thus there is no obvious evolutionary continuity from bacterial rRNA (52% G + C) through *Tetrahymena* mitochondrial rRNA to the rRNA of higher-animal mitochondrial ribosomes (45% G + C). In view of the high degree of specialization of the ciliates, this is not necessarily a surprising finding; it merely serves to emphasize the paucity of available information about mitochondrial ribosomes from lower eukaryotes.

Accurate characterization of mitochondrial ribosomes from fungi has posed particular problems. As pointed out by O'Brien and Matthews, these problems stem not only from disagreements among different groups of workers about mitochondrial ribosomes from a given organism but also from the variabilities attributed by the same group to a given organism. In *Neurospora,* monomer sedimentation velocities of 73S and 80S have been reported (Table 3–7); the differences reported are probably artifactual. There is better agreement concerning the sedimentation velocities of the separated subunits, and the rRNAs are of slightly higher molecular weight than the rRNAs of *E. coli* (Tables 3–5 and 3–7). The number of mitochondrial ribosomal proteins in *Neurospora crassa* has been accurately determined by Lizardi and Luck (108); it is not greatly different from the number obtained for *E. coli* (Tables 3–5 and 3–7). As in the case of *Neurospora,* there is much confusion concerning the properties of mitochondrial ribosomes from *Saccharomyces.* This is reflected in the variety of disparate values reported for the sedimentation velocities of the intact monomers and their component subunits (Table 3–7). However, the molecular weights of the mitochondrial rRNAs have been accurately determined; as with *Neurospora,* they are slightly greater than those of the comparable rRNAs from *E. coli* (Table 3–7). Mitochondrial ribosomes from several other fungi have also been partially characterized; they have sedimentation velocities ranging from 70S to 80S (90). Mitochondrial ribosomes from higher plants have not been characterized, except in the cases of maize and mung bean (*Phaseolus vulgaris*) (90). These ribosomes have sedimentation velocities in the range of 77S to 80S. The 78S mitochondrial ribosome of maize dissociates into 60S and 44S subunits. The mitochondrial rRNAs of *P. vulgaris* are only slightly different in molecular weight from the rRNAs of the cytoplasmic ribosomes (0.78×10^6 d and 1.15×10^6 d for the mitochondrial rRNAs versus 0.70×10^6 d and 1.30×10^6 d for the cytoplasmic rRNAs), and in the case of maize the two mitochondrial rRNAs appear slightly larger than their cytoplasmic coun-

terparts (0.76×10^6 d and 1.25×10^6 d versus 0.67×10^6 d and 1.19×10^6 d).

Undoubtedly the most interesting mitochondrial ribosomes thus far discovered are those of animals. These ribosomes sediment between 54S and 61S and have the lowest-molecular-weight RNAs of any ribosomes known (90) (Table 3–7). Ribosomes with these properties have been found in mitochondria of all higher animals that have been studied, from mammals to fish; they have also been found in insects (90). These ribosomes were called miniribosomes by Borst and Grivell (86). According to this notion, animal mitochondrial ribosomes were believed to be smaller (lower in molecular weight) than the ribosomes of other

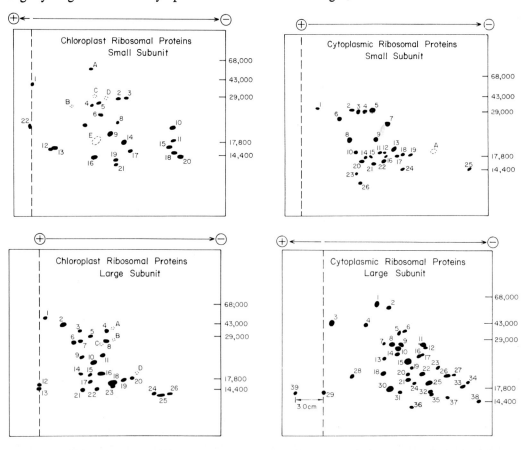

FIG. 3–1. Two-dimensional electrophoretograms of proteins from cytoplasmic and chloroplast ribosomal subunits of *C. reinhardtii*. A: Diagrams showing position of each protein in second (SDS) dimension. True ribosomal proteins are numbered. Proteins that may be contaminants are lettered. Molecular weight scale is given on right-hand side of each diagram. B: Photographs of actual gels of ribosomal proteins. Gel positions correspond to positions of their respective diagrams in part A. (From Hanson et al., ref. 102, with permission.)

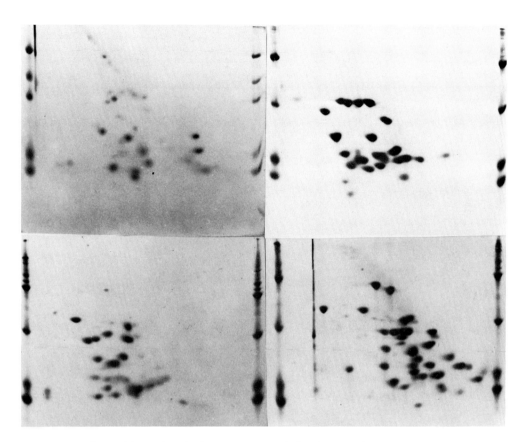

organisms. In fact, this notion is incorrect, as is shown by the data in Tables 3–5 and 3–7. The molecular weights of animal mitochondrial ribosomes are greater than those found in *E. coli*. The question then arises why *E. coli* ribosomes have a sedimentation velocity of 70S and animal ribosomes have sedimentation velocities of 55S to 60S. O'Brien and Matthews argued that the significant difference is the much lower buoyant density of animal mitochondrial ribosomes, which reflects the fact that animal mitochondrial ribosomes are only 30% RNA, with the rest being protein. Bacterial ribosomes are 63% RNA. These differences in protein content are also reflected in numbers of ribosomal proteins, since animal mitochondrial ribosomes have more proteins than any other known ribosomes (Table 3–7). In other words, animal mitochondrial ribosomes have smaller RNAs and more proteins than other ribosomes.

Mitochondrial ribosomes from *Tetrahymena* and higher plants have been reported to contain 5S RNA, but the mitochondrial ribosomes of *Euglena* and *Neurospora* do not appear to contain this component (90). Mitochondrial ribosomes from animal cells also lack 5S RNA, but they contain a 3S component that may be functionally equivalent (90). The presence of 5.8S RNA associated with mitochondrial ribosomes has not been reported.

PROCESSING OF ORGANELLE rRNA

On the basis of pulse labeling experiments at low temperatures, Kuriyama and Luck (109) showed that mitochondrial rRNA in *N. crassa* is transcribed as 32S precursor. They hypothesized that this molecule is cleaved to form precursors of the mature 25S and 19S rRNAs, which are

TABLE 3–7. *Mitochondrial ribosomes of different organisms*[a]

Ribosome source		Monomer	Subunits		rRNA		References
			Large	Small	Large	Small	
Saccharomyces	S value	72,75,80	50–60	37–40	22	15	90
	Molecular weight ($\times 10^6$ d)	—	—	—	1.30	0.70	
Neurospora	S value	73,80	50–52	37–39	23	16	90,108
	Molecular weight ($\times 10^6$ d)	—	—	—	1.28	0.72	
	Number of proteins	53	30	23	—	—	
Euglena	S value	71	50	32	21	16	90,165
	Molecular weight ($\times 10^6$ d)	2.7	—	—	0.93	0.56	
Tetrahymena	S value	80	55[b]	55[b]	21	14	86,90
	Molecular weight ($\times 10^6$ d)	3.2	—	—	0.90	0.47	
Maize	S value	78	60	44	—	—	90
	Molecular weight ($\times 10^6$ d)	—	—	—	1.25	0.76	
Xenopus	S value	60	43	32	21	13	90,166,167
	Molecular weight ($\times 10^6$ d)	3.5	1.9	1.6	0.58	0.32	
	Number of proteins	84	40	44	—	—	
Bovine liver	S value	55	39	28	—	—	90,168
	Molecular weight ($\times 10^6$ d)	2.8	1.65	1.10	0.54	0.36	
	Number of proteins	94	53	41	—	—	
HeLa cells	S value	60	45	35	16	12	90,169
	Molecular weight ($\times 10^6$ d)	—	—	—	0.54	0.35	

[a] Number-of-proteins row is included only in cases where numbers are known with some accuracy.
[b] Subunits can be separated by buoyant-density centrifugation in CsCl (107).

then trimmed to yield the mature rRNA molecules. Experiments to be discussed later (see Chapter 21) have indicated that the 32S rRNA has associated with it some of the ribosomal proteins present in both of the subunits of the mature ribosome, so that a precursor particle is actually formed. Thus far, other mitochondrial precursor rRNAs have not been reported. Chua and Luck (88) noted that the *Neurospora*

TABLE 3–8. *Base composition of rRNA from different sources*

Organism	Source	rRNA	Mole percent					Reference
			A	G	C	U	G + C	
E. coli	Cytoplasm	16S	24.2	32.1	22.3	21.3	54.4	89
		23S	25.5	32.5	21.0	21.0	53.5	
Anacystis nidulans	Cytoplasm	16S	28.7	32.7	20.0	18.6	52.7	89
		23S	29.4	32.0	20.8	17.7	52.8	
Spinach	Chloroplast	16S	23.0	30.6	24.0	22.4	54.6	87
		23S	24.5	34.8	21.7	19.0	56.5	
Beet	Chloroplast	16S	22.8	30.9	23.2	22.9	54.1	87
		23S	24.6	35.2	21.5	18.7	56.7	

pulse labeling experiments were carried out under growth conditions in which the temperature was so low as to increase the doubling time of the culture sixfold. They pointed out that application of this method to slow rRNA maturation might be useful in demonstrating a precursor in other organelles. In this respect it is noteworthy that the mitochondrial rRNAs of a number of animals (e.g., HeLa cells, *Xenopus, Drosophila*) map very close together on the mitochondrial genome, and the same is true of chloroplast rRNA in those cases thus far examined (i.e., maize, spinach, *Chlamydomonas,* (Table 3–11). In these instances one might well expect to find large precursor rRNAs. On the other hand, the mitochondrial rRNAs of *Saccharomyces* are far apart on the mitochondrial genome (Table 3–11); thus there is no reason to suspect large precursors in this instance.

With the aforementioned considerations in mind, we will mention briefly what little is known about chloroplast precursor rRNAs. Stutz and Boschetti (89) recently reviewed the available information. One precursor rRNA of molecular weight 1.2×10^6 d that matures to the 23S rRNA of 1.1×10^6 d and another precursor of molecular weight 0.64×10^6 d have been reported in *Euglena, Chlamydomonas, Chlorella,* and spinach. Until a few years ago, similar small precursor rRNAs had been the only ones reported from *E. coli,* but more recent experiments have indicated that the 16S and 23S rRNAs actually arise from a single RNA transcript, 30S pre-rRNA, with a molecular weight of 2.1×10^6 d (98,181). Clearly, experiments of the kind done by Kuriyama and Luck may be necessary to identify precursor chloroplast rRNAs.

Maturation of rRNA is also accompanied by attachment of ribosomal proteins to form precursor particles in both prokaryotes and eukaryotes (98,99). The only extensive study done on an organelle ribosomal precursor particle is that of Lambowitz and

associates (110) on the mitochondrial ribosomal precursor in *Neurospora.* Their study will be discussed in Chapter 21.

INHIBITOR SENSITIVITY OF ORGANELLE RIBOSOMES

Protein synthesis in mitochondria and chloroplasts tends generally to be sensitive to inhibitors that block protein synthesis on prokaryotic ribosomes, but it tends to be resistant to at least two inhibitors specific for 80S eukaryotic cytoplasmic ribosomes: anisomycin and cycloheximide (Table 3–9). The sensitivity of organelle ribosomes to inhibitors of prokaryotic protein synthesis, coupled with their relative insensitivity to inhibitors acting on eukaryotic cytoplasmic ribosomes, has provided a powerful tool for analysis of the sites of synthesis of specific organelle proteins *in vivo* (see Chapters 19–21). In addition, mutants whose organelle ribosomes are resistant to specific inhibitors have been isolated in both *Chlamydomonas* and yeast (see Chapters 5 and 21). Various methods have been used to assess resistance and sensitivity. These include measurement of the ability of isolated organelles to incorporate amino acids in the presence of an inhibitor (7,11–14,111) and measurement of the ability of isolated organelle ribosomes to bind labeled antibiotics (112–114) or to incorporate labeled amino acids in response to a synthetic mRNA in the presence of an inhibitor (115, 118). Sensitivity of organelle protein synthesis *in vivo* has also been measured in several ways. One method is to incubate cells in the presence of an inhibitor of translation on cytoplasmic ribosomes for a short time, during which incorporation of radioactive amino acids via organelle-dependent protein synthesis is assessed. To determine if a given inhibitor of organelle protein synthesis is effective *in vivo,* different inhibitor concentrations are added during the same incuba-

TABLE 3–9. *Inhibition of protein synthesis in chloroplasts and mitochondria by inhibitors acting on prokaryotic and eukaryotic cytoplasmic ribosomes*

Ribosomal subunit affected	Inhibitor	Organelle	Inhibition	Reference
Prokaryotic large (50S) subunit	Chloramphenicol	Mitochondria	$+^a$	13
	Macrolide group	Chloroplasts	+	11,14,132,133
	Erythromycin	Mitochondria	+	13,170
		Chloroplasts	+	11,50,171
	Spiramycin	Mitochondria	+	13,170
		Chloroplasts	?	
	Carbomycin	Mitochondria	+	13,170
		Chloroplasts	+	115
	Oleandomycin	Mitochondria	+	170
		Chloroplasts	?	
	Lincomycin	Mitochondria	+	13,170
		Chloroplasts	+	11,171
	Cleocin	Mitochondria	?	
		Chloroplasts	+	115
	Thiostrepton group			
	Thiostrepton (Bryamycin)	Mitochondria	+	13
		Chloroplasts	?	
	Streptogramin A group			
	Mikamycin A	Mitochondria	+	13,172
		Chloroplasts	?	
Prokaryotic small (30S) subunit	Aminoglycoside group			
	Kanamycin	Mitochondria	−	13
		Chloroplasts	+	117
	Paromomycin	Mitochondria	+	13,119
		Chloroplasts	?	
	Neamine	Mitochondria	?	
		Chloroplasts	+	50,115,117
	Streptomycin	Mitochondria	−	13
		Chloroplasts	+	50,115,117
	Spectinomycin	Mitochondria	?	
		Chloroplasts	+	50,115,117,171
Eukaryotic large (60S) subunit	Anisomycin	Mitochondria	−	108
		Chloroplasts	−	173
	Glutarimide group			
	Cycloheximide	Mitochondria	−	12,13
		Chloroplasts	−	11,174
50S or 60S large subunits	Puromycin	Mitochondria	+	137
		Chloroplasts	+	132
	Sparsomycin	Mitochondria	+	111
		Chloroplasts	?	
30S or 40S small subunits	Tetracycline group			
	Oxytetracycline	Mitochondria	+	111
		Chloroplasts	?	

a Plus sign indicates inhibition by compound tested; minus sign indicates no inhibition by compound tested; question mark indicates that inhibition experiments were not reported.

tion period, and amino acid incorporation is measured. A second method is to measure the synthesis of a protein known to be made in part or in its entirety on organelle ribosomes in the presence of the inhibitor of organelle protein synthesis. This method has been particularly useful in the case of the chloroplast, since the enzyme RuBPCase, which has a molecular weight of approximately 560,000 d, is composed of two subunits, one of which is made on chloroplast ribosomes and the other in the cytoplasm (see Chapter 20). In *Chlamydomonas,* synthesis of this enzyme in the presence of inhibitors of chloroplast protein synthesis has proved a particularly useful method for establishing the *in vivo* sensitivity or resistance of chloroplast protein synthesis to a specific inhibitor (see Chapter 21). A third method measures the ability of specific inhibitors to prevent runoff of ribosomes from organelle polysomes bound either to the mitochondrial inner membrane (137) or to the thylakoid membranes (50, 116).

Chloramphenicol has undoubtedly been the most popular antibiotic for inhibitor studies with organelle ribosomes (11–14) because it blocks protein synthesis in both chloroplasts and mitochondria (Table 3–9). Erythromycin is another popular antibiotic for this purpose. As will be discussed later, mitochondrial mutations resistant to each of these antibiotics have been isolated in yeast (see Chapter 5), and chloroplast mutations resistant to erythromycin have been obtained in *Chlamydomonas* (see Chapter 12). The target for each of these inhibitors in bacteria is the large ribosomal subunit. The same seems to be true for chloroplast ribosomes, at least in the case of erythromycin, where definitive experiments have been done (Table 3–9). Protein synthesis in chloroplasts and mitochondria is also generally sensitive to macrolide antibiotics related to erythromycin. On the other hand, while streptomycin and related aminoglycoside antibiotics specific for the small subunit of prokaryotic ribosomes clearly inhibit protein synthesis by chloroplast ribosomes (Table 3–9), it is not at all clear that these antibiotics are effective for mitochondrial protein synthesis. Kroon and De Vries (111), for example, report that streptomycin is partially effective in isolated rat liver mitochondria, but Mahler (13) states that streptomycin and kanamycin are generally ineffective as inhibitors of mito-

chondrial protein synthesis. Paromomycin is a notable exception to this statement, at least for yeast, where it is evident that mitochondrial protein synthesis both *in vivo* and *in vitro* is sensitive to this antibiotic (119). The same may also be true for neomycin, but the data are less complete.

In principle, inhibitors should be powerful tools for establishing where messages for various organelle proteins are translated, and this approach has enjoyed great popularity. Since we will have occasion to refer repeatedly to the results of inhibitor experiments in this book, a few cautionary remarks should be made. In practice, there are several reasons why it can be difficult to obtain unequivocal results from inhibitor experiments. First, it is difficult to determine exactly what inhibitor concentration to use *in vivo*. Concentrations that are effective *in vitro* may be too high *in vivo,* and vice versa. On the one hand, there is the risk that the inhibitor is acting as a general metabolic poison; on the other hand, it may be having only a minimal effect on the organelle *in vivo*. Second, an inhibitor that is effective *in vitro* may not get into the organelle *in vivo,* since during isolation for *in vitro* studies the organelle may sustain enough damage to permit entry of the inhibitor. Kroon and De Vries (111), for example, found that several inhibitors of transcription and translation that have no effect on intact rat mitochondria have a marked effect if the mitochondria are swollen (Table 3–10). Third, one of the most popular inhibitors, chloramphenicol, has particular problems of its own, which Ellis and Hartley (120) have called attention to in the case of chloroplasts. Of the four stereoisomers of this antibiotic, only the D-threo isomer inhibits protein synthesis by isolated chloroplasts. However, all four isomers inhibit other processes, such as ion uptake and oxidative phosphorylation. The L-threo isomer will inhibit light-driven protein synthesis in isolated chloroplasts but not ATP-driven protein synthesis. Fourth,

TABLE 3–10. *Effects of various antibiotics on in vitro protein synthesis in rat liver mitochondria*

Antibiotic	Protein synthesis in:	
	Intact mitochondria	Swollen mitochondria
Actinomycin D	$-^a$	+
Rifampicin	+	+
Oxytetracycline	++	++
Streptomycin	+	+
Chloramphenicol	++	++
Lincomycin	−	+
Erythromycin	−	++
Oleandomycin	−	++
Carbomycin	++	++
Sparsomycin	++	++
Puromycin	++	++
Cycloheximide	−	−

^{a} ++ = 50–95% inhibition; + = 20–50% inhibition; − = no inhibition.

Data from Kroon and De Vries (111).

there is the problem of pleitropy. Sythesis of the chloroplast thylakoid membranes and the mitochondrial inner membrane involves cooperation between the protein-synthesizing systems of organelle and cytoplasm. (See Chapters 19 and 20). Inhibition of the synthesis of any one component can easily lead to an array of secondary pleiotropic effects on membrane structure and function as a whole. Hence, special precautions must be taken in distinguishing primary and secondary effects of inhibitors (see Chapters 19 and 20). Fifth, in green plants there is the problem that both chloroplasts and mitochondria may or may not be sensitive to the same set of inhibitors. In *Euglena* (see Chapter 16), antibiotics that cause irreversible bleaching are usually not lethal to the cells. There is good reason to believe that at least some of these antibiotics (e.g., streptomycin) cause bleaching because they block chloroplast protein synthesis. Since streptomycin is not lethal to this organism, except at extremely high concentrations, it may be inferred that mitochondrial protein synthesis is probably insensitive to streptomycin in *Euglena*. In *Chlamydomonas,* on the other hand, which

is sensitive to antibiotics such as streptomycin at very low concentrations, it has been postulated by Gillham, Boynton, and their colleagues that these antibiotics block protein synthesis in both chloroplasts and mitochondria (see Chapter 21).

MEMBRANE-BOUND POLYSOMES AND ORGANELLE BIOGENESIS

If it is assumed that importation of cytoplasmic mRNA for translation on organelle ribosomes is a rare process, if it occurs at all (pp. 85–89), then proteins must be transported into these organelles, since the majority of different proteins found in both chloroplasts and mitochondria are the products of translation of mRNA on cytoplasmic ribosomes. One mechanism by which transport of proteins into chloroplasts and mitochondria might be effected is via cytoplasmic polysomes bound to the outer membranes of these organelles. A second and corollary problem concerns the mechanism by which relatively insoluble membrane polypeptides are inserted during membrane biogenesis. Again cytoplasmic polysomes bound to the outer membranes of chloroplasts and mitochondria could account for the insertion of some of these polypeptides, but other polypeptides of the inner mitochondrial membrane and the thylakoid membrane are evidently synthesized on organelle ribosomes. Therefore, it might also be expected that organelle polysomes would be bound to the inner membranes of both organelles.

Before discussing polysomes bound to organelle membranes, it is necessary to review briefly the most favored current hypothesis of how proteins are translocated across membranes and the evidence for it. Most of the relevant experimental data come from the work of Blobel and Sabatini and their colleagues on ribosomes bound to the rough endoplasmic reticulum of mammalian cells. In a recent version, by Blobel and Dobberstein (121), the hypothesis describing this transfer is called the signal hypothesis. According to the signal hypothesis, a unique sequence of codons is located distal to the AUG initiation codon in those mRNAs whose translation products are to be transferred across membranes (Fig. 3–2). Translation of these signal codons results in a unique sequence of amino acids at the amino terminal end of the growing polypeptide chain emerging from the large ribosomal subunit. This signal sequence triggers attachment of the ribosome to the membrane and fulfills the topologic requirements for transfer of the nascent polypeptide across the membrane. Once the signal sequence is transferred across the membrane, it is removed proteolytically. Vectorial transfer of the polypeptide through the membrane is envisioned as taking place through a "tunnel" that is created by recruitment of specific membrane receptor proteins in response to the signal sequence. The association of the ribosome with the tunnel is stabilized by interaction between exposed sites on the large ribosomal subunit and the membrane receptor sites. After release of the nascent chain into the transmembrane space, the ribosomes become dissociated from the membrane, and the

membrane receptor proteins forming the tunnel are free to diffuse as individual proteins in the plane of the membrane.

There is strong evidence to support many of the features of the signal hypothesis. In 1966 Redman and Sabatini (122) showed that ribosomes bound to the rough endoplasmic reticulum of guinea pig liver discharged the peptides they carried in a vectorial manner into the interior of the endoplasmic reticulum. They labeled microsomes obtained from guinea pig liver with ^{14}C-leucine, after which the nascent polypeptide chains were released from the ribosomes to which they were bound by puromycin, an inhibitor causing chain termination (123). The peptides lost by the ribosomes were not discharged into the medium but were retained by the microsomes in the cisternal space. Subsequently, Sabatini et al. (124) found that the ribosomes were attached to the endoplasmic reticulum by their large subunits. Adelman et al. (125) then reported that these ribosomes were attached to the rough endoplasmic reticulum via two types of interactions: direct interaction, presumably involving direct binding of the large ribosomal subunit to the membrane; indirect

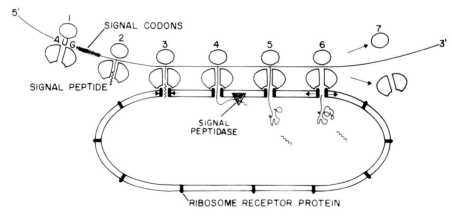

FIG. 3–2. Diagram showing translocation of a polypeptide across a membrane such as that associated with the endoplasmic reticulum according to the signal hypothesis. Ribosomes bind at AUG initiation codon on an mRNA molecule. Signal codons are just distal to initiation codon; they code for an amino acid sequence, the signal peptide, that recognizes a ribosome receptor protein (or group of proteins) in the membrane. The signal sequence proceeds through a tunnel formed at this point, followed by the rest of the polypeptide, which is in the process of being synthesized. A signal peptidase cleaves off the signal peptide as it enters the cisternal space. When the ribosomes reach the chain termination codons, the completed polypeptides are released into the cisternal space, and the ribosomes dissociate into free subunits. (Courtesy of Dr. G. Blobel.)

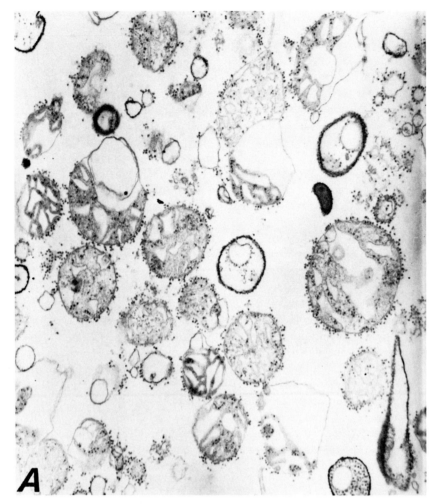

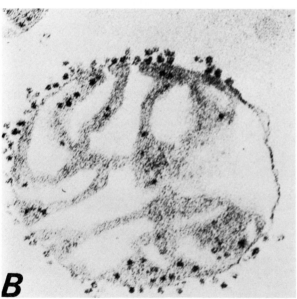

FIG. 3–3. Electron micrographs of mitochondria isolated from growing spheroplasts of *S. cerevisiae* showing cytoplasmic ribosomes attached to mitochondrial outer membrane. A: ×41,000. B: ×91,000. C: ×283,000. (From Butow et al., ref. 131, with permission.)

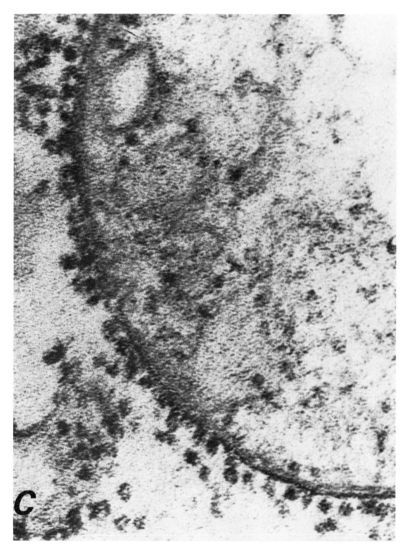

C

interaction, in which the nascent poly-peptide chain anchors the ribosome to the membrane. The direct interaction may be electrostatic, since it can be broken by high concentrations of KC1. The indirect interaction can be terminated by incubating the ribosome membrane preparation with puromycin, which causes premature poly-peptide chain termination. In short, if ribo-somes are released from membranes by a combination of high concentrations of salt and puromycin, this is a rather good indica-tion that they are actively involved in pro-tein synthesis. Support for the idea that there is also a signal sequence in proteins

destined for transmembrane transport comes from the experiments of DeVillers-Thiery and associates (126), who showed that the precursors of several dog pancreas secretory proteins contain similar amino-terminal amino acid sequences that are at least 16 amino acid residues long. These sequences contain an unusually high percentage of hydrophobic amino acids. Evidence that such signal sequences are processed pro-teolytically has also been obtained by Blobel and Dobberstein (127).

Kellems and Butow (128–131) pre-sented virtually the only evidence for bind-ing of cytoplasmic ribosomes to the outer

membranes of either chloroplasts or mitochondria. Their elegant experiments with
yeast mitochondria strongly suggested that
these mitochondria have active cytoplasmic
polysomes bound to their outer membranes.
They showed by electron microscopy that
yeast mitochondria, both *in situ* and in isolated form, have ribosomes closely packed
along their outer membranes (Fig. 3–3).
These ribosomes can be removed by washing the mitochondria with buffer that lacks
magnesium but contains a chelating agent
such as EDTA. Other experiments showed
that cytoplasmic polysomes can be released
from the mitochondria by detergent extraction. In puromycin experiments, Butow and
associates also provided evidence for vectorial release of polypeptides by these ribosomes into the mitochondrial compartment.
Release of about one-third of these bound
cytoplasmic ribosomes was accomplished
by incubation of the mitochondrial membranes in high concentrations of KCl, but
removal of all of the bound ribosomes was
achieved only when puromycin was also
added, which indicates that some two-thirds
of the ribosomes are bound to the outer mitochondrial membrane by nascent chains.
Further electron microscopic study suggested
that cytoplasmic ribosomes are not randomly
distributed over the surface of the outer mitochondrial membrane but are located at
points where the inner and outer membranes
are in close contact. Butow and associates
supposed that it was in these regions of contact that proteins are discharged vectorially

across both membranes and into the matrix
space (Fig. 3–4). These careful experiments
were of great importance; they should be
repeated in other systems for both chloroplasts and mitochondria under suitable experimental conditions, for only then will we
have a satisfactory picture of how generalized the process of vectorial discharge is as
a mechanism for importing proteins into
chloroplasts and mitochondria.

The best evidence that membrane-bound
organelle ribosomes play a role in membrane synthesis comes from experiments
with *Chlamydomonas* (Fig. 3–5). Chua et
al. (132) and Margulies and Michaels
(133) found that in chloramphenicol-treated
cells the isolated thylakoid membranes had
polysomes of chloroplast ribosomes attached
to them. These polysomes could be released
from the membranes by high concentrations
of salt and puromycin, which indicates that
they were actively engaged in polypeptide
synthesis and were bound to the membranes
in much the same way that cytoplasmic
ribosomes are bound to the rough endoplasmic reticulum. Addition of chloramphenicol was necessary in the original
experiments to block runoff of the ribosomes from the membranes. Subsequently,
the same effect was achieved by cooling the
cells rapidly prior to breakage (134). Margulies and associates (135) later presented
evidence that vectorial discharge of the
nascent polypeptides into the thylakoid
membranes was involved; they used much
the same methodology as was originally em-

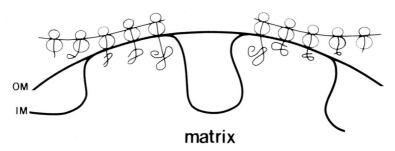

matrix

FIG. 3–4. Model for vectorial translation of nascent polypeptides into mitochondria. Key: OM, outer membrane; IM, inner
membrane. (From Butow et al., ref. 131, with permission.)

ployed by Redman and Sabatini. These authors also showed that solubilization of the membranes with detergents released the polyribosomes (136). Chua et al. (134) studied the variation in proportions of free and bound chloroplast polysomes in light–dark-synchronized cells throughout the cell cycle. During the light part of the cycle, at the time chlorophyll is accumulating and membranes are being synthesized, the fraction of bound chloroplast polysomes increases to about 30% of the total. During the dark period there is no further chlorophyll increase, and the number of membrane-bound chloroplast polysomes declines to a small fraction of the total. The dark period is the time when nDNA synthesis and cell division occur (see Chapters 2 and 12).

All of the foregoing observations support the hypothesis that the membrane-bound chloroplast polysomes are engaged in the synthesis and insertion of thylakoid membrane polypeptides. Similar, but less detailed, evidence for mitochondrial ribosomes has been provided by Kuriyama and Luck (137). These results, which will be discussed in more detail in Chapter 8, indicate that mitochondrial ribosomes that can be released by high concentrations of salt and puromycin are bound to the mitochondrial inner membrane in *Neurospora*.

In summary, there is now good reason to believe that organelle ribosomes engaged in membrane polypeptide biosynthesis probably are bound to these membranes in a manner similar to the manner in which eukaryotic cytoplasmic ribosomes are bound to the endoplasmic reticulum. It seems likely that vectorial discharge of these polypeptides into the membranes occurs. Butow's experiments suggest that cytoplasmic ribosomes bound to the outer mitochondrial membrane probably discharge polypeptides through the outer and inner membranes into the matrix at points of close apposition. Despite these striking findings, it is by no means clear that vectorial discharge is the only mechanism or even the most common

mechanism for transporting polypeptides into and across organelle membranes (see Chapters 19 and 20).

SEQUENCE HOMOLOGY AMONG ORGANELLE rRNA AND tRNA AND ORGANELLE DNA AND nDNA

Molecular DNA-RNA hybridization experiments offer the most direct approach to determining what is coded by organelle DNA. Such experiments have been employed to great advantage to establish that DNA of chloroplasts and mitochondria codes for the rRNA of the ribosomes present in these organelles and for many, if not all, of the tRNAs of chloroplasts and mitochondria. Furthermore, hybridization techniques have made mapping of these RNAs possible, as will be discussed in the next section.

The principle involved in DNA-RNA hybridization is simply that RNA will form hydrogen-bonded molecular hybrids with single-stranded DNA of complementary base composition. RNA that has been transcribed from a given DNA nucleotide sequence will be able to hybridize with that DNA because it contains a complementary sequence of nucleotides. In practice, unlabeled DNA is denatured by heat or alkali, and the single strands are immobilized on a nitrocellulose filter with very small pore size (138). Radioactively labeled RNA is then incubated with the DNA-containing filters under conditions in which hybridization can occur. The filters are then treated with ribonuclease to destroy single-stranded RNA and RNA that has paired imperfectly with the DNA because of partial or false homology. Finally, the filter is washed to remove any unbound radioactive RNA.

The number of identical genes coding for a certain type of RNA in a DNA sample can be ascertained in an experiment where the amount of DNA per filter is held constant while the amount of radioactive RNA is

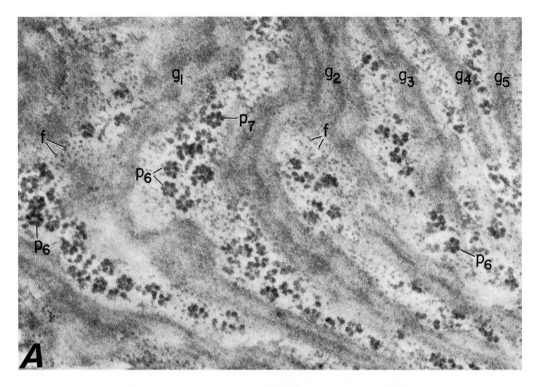

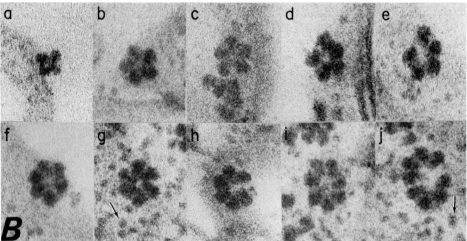

FIG. 3–5. Chloroplast polysomes in *Chlamydomonas*. A: Chloroplast fragment obtained from synchronous cells of *C. reinhardtii* exposed for 10 min to chloramphenicol to prevent runoff of ribosomes from polysomes. Numerous polysomes (p) appear against irregular light-gray bands that represent oblique sections through terminal thylakoids in series of grana (darker composite bands g_1, g_2, g_3, g_4, and g_5 are oblique sections through central part of these grana). Many of the polysomes are circular (either closed or open); they vary in size from pentamers to hexamers (p_6) and heptamers (p_7). Smaller light bodies associated with free surfaces of terminal thylakoids are coupling factor particles (f). ×72,000. B: Gallery of chloroplast polysomes seen in full-face view in grazing sections through terminal thylakoids. A tetramer appears in *a*, closed pentamers in *b* and *c*, open pentamers in *d* and *e*, closed hexamers in *f* and *g*, open hexamer in *h*, a closed heptamer in *i*, and an open octamer in *j*. Coupling factor particles (see Chapter 1) are indicated by arrows. *a*: ×150,000. *b–j*: ×200,000. C: Seven (*a–g*) attached polysomes seen in profile (or side view) on normally sectioned membranes of terminal thylakoids. Attachment of polysomes varies from complete (*a*) to two ends (*b, c,* and *e*) or one end only (*d, f,* and *g*) of the profile. In *h*, a hexamer is seen in full-face view at a distance of 400 Å from nearest (obliquely sectioned) thylakoid membrane (*arrow*). All these appearances can be explained by assuming that these polysomes are assembled in the stroma and attach as groups to the membranes. *a–g*: ×200,000. *h*: ×160,000. (From Chua et al., ref. 134, with permission.)

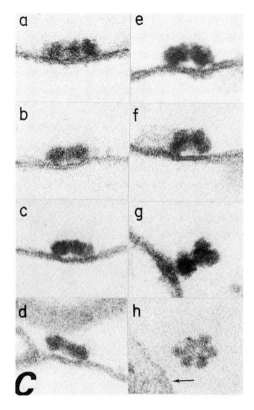

FIG. 3–5. (cont.)

mitochondrial rRNA molecule, the molecular weight of single-strand mtDNA complementary to this rRNA is 4.68×10^5d ($0.04 \times 11.7 \times 10^6$ d). By similar reasoning, the small rRNA molecule hybridizes with mtDNA of approximately 2.9×10^5 d ($0.025 \times 11.7 \times 10^6$ d). This means that there is only one gene for each rRNA molecule per mtDNA molecule, because the molecular weight of the large mitochondrial rRNA molecule in *Xenopus* is 5.3×10^5 d, whereas a small rRNA molecule has a molecular weight of 3.0×10^5 d. Since the two mitochondrial rRNAs hybridize with about 7% of the total mtDNA, this means that about 14% of the sequences in one strand corresponds to mitochondrial rRNA.

Hybridization experiments such as the

varied until no more radioactive RNA hybridizes with the DNA (saturation experiment). An example of a saturation experiment for *Xenopus* mitochondrial rRNA and mtDNA is shown in Fig. 3–6, which is taken from Dawid (38). The percentage hybridization (i.e., micrograms of RNA bound per microgram of DNA) is plotted versus RNA concentration. It can be seen that the mtDNA sites capable of binding the large mitochondrial rRNA molecule saturate at approximately 4% hybridization. Addition of small rRNA molecules to an mtDNA preparation saturated with large rRNA molecules increases the percentage hybridization to 7% (Fig. 3–6); the reciprocal experiment increases it to 6%. The additivity of the two rRNA types in the hybridization experiment shows that they do not compete for the same binding sites (i.e., they are coded by different genes). Since the molecular weight of *Xenopus* mtDNA is 11.7×10^6 d, and 4% hybridizes with the large

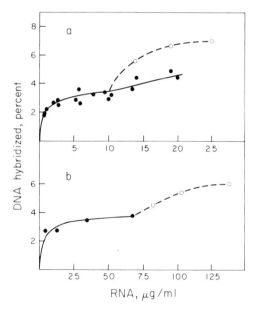

FIG. 3–6. Additive hybridization of large and small mitochondrial rRNA with mtDNA from *Xenopus*. a: Solid circles represent hybridization with ³H-labeled large rRNA; open circles represent hybridization with a constant amount of large rRNA (10 µg/ml) and additional ³H-labeled small rRNA to a total rRNA concentration indicated by the positions of the symbols. Each filter contained 1 µg mtDNA. b: Solid circles represent hybridization with ³H-labeled small rRNA; open circles represent hybridization with 70 µg of small rRNA per milliliter and additional ³H-labeled large rRNA to the concentration indicated. Each filter contained 5 µg DNA. The ratios of RNA to DNA in the two experiments were similar. (From Dawid, ref. 38, with permission from Journal of Molecular Biology. Copyright by Academic Press, Inc. (London) Ltd.)

one described for *Xenopus,* together with molecular mapping experiments (which will be described in the next section), indicate that all mtDNAs characterized to date contain one gene each for the large and small mitochondrial rRNAs (Table 3–11). In most cases these are adjacent, which suggests that they may be transcribed from a single rRNA precursor, but in one case they are far apart on the mitochondrial genome. In contrast, clDNAs contain two sets of rRNA genes per genome. Early reports by Ingle and associates (100) for Swiss chard and Tewari and Wildman (139) for tobacco indicating there is only one cistron for each rRNA molecule per clDNA molecule should be reevaluated. Ingle and associates assumed from kinetic complexity data that the size of the chloroplast genome was approximately 1.5×10^8 d; we now know that 0.95×10^8 d is a better estimate (see Chapter 2). Using the latter estimate, it can be calculated that there are approximately 1.6 cistrons for each rRNA per genome. Tewari and Wildman assumed a kinetic complexity of 1.14×10^8 d for clDNA of tobacco, which is much closer to the true molecular weight for clDNA. They pointed out that their data indicated at least one cistron for each kind of rRNA per clDNA molecule, but they also considered

that to be a minimal estimate. In the case of *Euglena,* Rawson and Haselkorn (140) reported that each clDNA molecule contained a single set of rRNA genes, but more recently Gruol and Haselkorn (141) have revised this number upward to two sets. Surzycki and Rochaix (30), on the basis of transcriptional mapping experiments, suggested that there were two to three pairs of 16S and 23S rRNA genes per chloroplast genome in *Chlamydomonas,* with each pair being arranged in tandem such that the 16S rRNA gene was transcribed first. Bastia and associates (142) calculated about three cistrons per unique sequence of clDNA, based on hybridization experiments reported in abstract only. These estimates need to be checked carefully again, especially in view of the fact that the chloroplast genome of *C. reinhardtii* is smaller than renaturation kinetic analysis first indicated (see Chapter 2).

Three types of experiments have provided evidence that organelle tRNAs are also gene products of organelle DNA. In the first type, labeled 4S RNA (which is primarily tRNA) from the organelle is hybridized with organelle DNA. By performing a saturation experiment, the percentage hybridization can be calculated, and using the knowledge that a tRNA molecule has a molecular

TABLE 3–11. *Numbers and positions of genes coding for large and small organelle rRNAs in organelle DNA*

| | | Genes per DNA molecule | | | |
	Source	Large rRNA	Small rRNA	Position	References
mtDNA	HeLa cells	1	1	Adjacent	63
	Rat liver	1	1	Adjacent	145
	Xenopus	1	1	Adjacent	144
	Drosophila	1	1	Adjacent	144
	Neurospora	1	1	Adjacent	175,176
	Saccharomyces	1	1	Apart	42,146,177
clDNA	Spinach	2	2	Two sets of adjacent 23S plus 16S cistrons separated from each other	147
	Maize	2	2	Same as spinach	148
	Chlamydomonas	2–3	2–3	Probably tandem pairs	30,142
	Euglena	2	2	?	141

weight of approximately 25,000 d, it is possible to calculate the total number of tRNA genes per organelle DNA molecule, just as was done for the rRNA genes. In the second type of experiment, 4S RNA is conjugated with the electron-opaque label ferritin and then hybridized with mtDNA; then the positions of the 4S RNA molecules can be mapped by electron microscopy, as will be described in the next section and in Chapter 11. Data obtained from experiments of these kinds indicate that both the chloroplast genome and the mitochondrial genome have the capacity to code for a complete set of tRNAs (Table 3–3).

Neither of the first two approaches distinguishes between the possibility that organelle DNAs contain repeated sequences coding for only one tRNA or a few tRNAs and the possibility that a complete set of tRNAs is coded in the genomes of chloroplasts and mitochondria. To clarify this point, unlabeled tRNA must be stripped of amino acids and then charged with specific labeled amino acids using synthetase preparations from whole cells or organelles. Hybridization of tRNAs charged with specific amino acids can then be used to determine how many different tRNA species are coded in chloroplast or mitochondrial DNA. Thus far, this kind of analysis has been performed rigorously only with the mtDNA of HeLa cells and *Saccharomyces* (Table 3–4), as discussed previously. It appears that these genomes do contain complete or nearly complete sets of tRNAs. Animal mtDNA has the additional advantage that the two DNA strands can be separated on the basis of buoyant density in alkaline CsCl gradients (see Chapter 2). Attardi et al. (63,85) showed that the H strand of HeLa cell mtDNA hybridizes 12 different tRNAs, whereas the L strand hybridizes 5 tRNAs. All of these tRNAs hybridized to one strand only, except seryl tRNA; thus it appears that there are two mitochondria-coded seryl tRNAs. Multiple tRNA species for a single amino acid coded by the mitochondrial

genome have also been reported for *Saccharomyces* in several instances (66) (Table 3–4). Thus far we do not have a satisfactory picture of the multiplicity of different tRNAs coded by the chloroplast genome.

MOLECULAR MAPPING OF RNA SPECIES

Four methods for mapping specific RNAs are presently in use. Thus far, the most information is available for rRNA, somewhat less for tRNA, and very little for mRNA. In the first (and oldest) method, the rRNA cistrons are mapped by hybridization of rRNA to single-strand DNA, after which the hybrids are examined in the electron microscope. The single-strand DNA collapses on itself, but the duplex hybrid regions remain extended. In a modification of this method, the single-strand DNA remains extended, and the regions of duplex formation are detected by their somewhat thicker appearance in electron micrographs. Neither of these techniques is applicable to the mapping of 4S RNA genes because the regions of duplex formation are too short to be detectable. To map the 4S RNA genes by electron microscopy, the second method is employed. In this method the 4S RNA fraction is conjugated with the electron-opaque label ferritin, which is easily recognized in electron micrographs, and then hybridized to single-strand DNA. The position of each hybridized 4S RNA molecule is indicated by a dark spot adjacent to mtDNA in the electron micrograph. These two methods were pioneered by Attardi and his colleagues; they will be described in more detail in Chapter 11. Attardi's experiments using these mapping methods with HeLa cell mtDNA (63) have resulted in a complete map of the 4S RNA and rRNA genes (Fig. 3–7). Several features of this map should be emphasized. First, there is a 4S RNA binding site (number 2) between the large and small rRNAs on the H strand and one on either side of the rRNA cistrons (num-

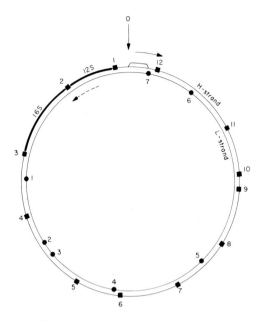

FIG. 3–7. Circular map of positions of complementary sequences for 4S RNAs on H and L strands of HeLa cell mtDNA and 12S and 16S rRNA genes on H strand. (From Attardi et al., ref. 63, with permission.)

The results for the L strand are preliminary, but they indicate five to six binding sites for 4S RNA. Thus in *Xenopus* there are probably about 22 binding sites for 4S RNA, as compared to 19 sites in HeLa cells.

The third method involves mapping of RNA species by hybridization to specific restriction fragments. The positions of different RNA types on the restriction map of the organelle genome (see Chapter 2) are then ascertained. Thus far this method has been applied most successfully to the mapping of the rRNA genes of several mitochondrial and chloroplast genomes (144–148). Southern (149) has adapted a particularly useful technique for this purpose: Restriction fragments are transferred from the agarose gels used for their separation to strips of filter paper, on which hybridization to radioactive RNA is carried out. In this way the specific restriction fragment hybridizing a given RNA can be detected. We may expect that with the development of cloning techniques (see Chapter 2) and techniques for isolation and translation of specific mRNAs, this methodology will soon begin to yield molecular maps of the chloroplast and mitochondrial genomes. In fact, Howell and associates (ref. 71 and Chapter 20) have already used this method to map the mRNA coding for the large subunit of the enzyme RuBPCase in the chloroplast genome of *Chlamydomonas*. This method should make it possible to characterize and map specific organelle genes in systems where the direct genetic approach is impossible.

The fourth method, which thus far has been applicable only to the mitochondrial genome of *Saccharomyces,* involves use of deletions for genetic and physical localization of specific genes. This method, used in combination with the other three, has led to the most complete organelle gene map yet developed. A discussion of these experiments must be preceded by an introduction to yeast mitochondrial genetics and will be deferred until Chapter 7.

bers 1 and 3). Wu et al. (143) speculated that one of these 4S RNAs might actually be the equivalent of ribosomal 5S RNA. Second, the close proximity among the members of this group of genes suggests that they might be formed from a single transcript. Third, 12 binding sites for 4S RNA have been detected on the H strand, as compared to 7 sites on the L strand. Assuming that one of these binding sites corresponds to 5S mitochondrial rRNA, this leaves 18 binding sites for tRNA; this is in remarkably good agreement with the 17 sites detected by hybridization of tRNAs charged with different amino acids (Table 3–4). Dawid and associates (144) reported similar mapping experiments with *Xenopus* mtDNA. They found 15 and possibly 16 sites on the *Xenopus* H strand, which hybridize with 4S RNA. Three of these sites are grouped around the rRNA cistrons in the same way as in HeLa cells (i.e., one between the two rRNA binding sites and one on either side).

CONCLUSIONS

Chloroplasts and mitochondria contain the entire machinery required for transcription of their genomes and translation of the resulting mRNAs. The rRNAs of organelle ribosomes, and probably all of the tRNAs, are transcribed from the organelle genome. In contrast, many of the protein components required for transcription and translation are thought to be coded by nuclear genes, although the information available at the moment is less than satisfactory. This is an area worthy of far more exploration than it has heretofore received, and we will return to it in more detail in Chapter 21.

Current thinking has it that mitochondria do not import RNA, but only proteins. Similarly, it is generally supposed that mRNA is not exported from mitochondria, since evidence is rapidly accumulating that proteins whose messages are transcribed inside the mitochondrion are probably translated there (see Chapter 19). As usual, the question of mRNA transfer is wide open in the case of the chloroplast, which reflects the inverse relationship between the number of workers in the field and the relative sizes of the chloroplast and mitochondrial genomes.

Despite the foregoing comments, it seems likely that the proteins themselves (not their mRNAs) are the entities transported to and fro across organelle membranes. One mechanism for movement of proteins into organelles is vectorial transport as shown by Butow's experiments with yeast, which suggest that proteins are transported into mitochondria in a vectorial manner via polysomes bound to the outside of the outer membrane. Other work has suggested that membrane polypeptide synthesis on organelle ribosomes most likely involves vectorial transport into the membrane. Nevertheless, supporting data for the vectorial transport hypothesis are still limited, and it is now becoming clear that proteins enter organelles by means of other routes as well (Chapters 19 and 20).

As more and more organelle genomes are mapped physically using restriction endonucleases, we may expect emphasis to shift to characterization of the transcripts complementary to each fragment. This information is already available for rRNA and will probably become available (particularly in the case of the mitochondrion) for different tRNAs before too long. But the most exciting results are likely to come from molecular mapping of the mRNAs coded by different fragments. Three general approaches are likely to prove important. In the first, mRNAs will be isolated from organelles for hybridization to restriction fragments and translation *in vitro*. The products of *in vitro* translation will be characterized using antisera against specific organelle proteins and other techniques. In the second approach, cloned restriction fragments will be transcribed and translated *in vitro,* and their products will be identified using appropriate methods. In the third approach, cellular RNA will be pulse-labeled *in vivo*, and organelle polysomes will be precipitated using antisera specific for the nascent chains they carry. The mRNA will then be isolated and hybridized to restriction fragments. In short, molecular mapping will soon become a powerful adjunct to genetic mapping of chloroplast and mitochondrial genomes.

REFERENCES

1. Kirk, J. T. O. (1964): DNA-dependent RNA synthesis in chloroplast preparations. *Biochem. Biophys. Res. Commun.,* 14:393–397.
2. Kirk, J. T. O. (1964): Studies on RNA synthesis in chloroplast preparations. *Biochem. Biophys. Res. Commun.,* 16:233–238.
3. Luck, D. J. L., and Reich, E. (1964): DNA in mitochondria of *Neurospora crassa. Proc. Natl. Acad. Sci. U.S.A.,* 52:931–938.
4. Borst, P., and Kroon, A. M. (1969): Mitochondrial DNA: Physicochemical properties, replication and genetic function. *Int. Rev. Cytol.,* 26:107–189.
5. Ashwell, M., and Work, T. S. (1970): The

biogenesis of mitochondria. *Annu. Rev. Biochem.*, 39:251–290.

6. Rabinowitz, M., and Swift, H. (1970): Mitochondrial nucleic acids and their relation to the biogenesis of mitochondria. *Physiol. Rev.*, 3:376–427.

7. Roodyn, D. B., and Wilkie, D. (1968): *The Biogenesis of Mitochondria*. Methuen, London.

8. Kirk, J. T. O., and Tilney-Bassett, R. A. E. (1967): *The Plastids*. W. H. Freeman, San Francisco.

9. McLean, J. R., Cohn, G. L., Brandt, I. K., and Simpson, M. V. (1958): Incorporation of labelled amino acids into protein of muscle and liver mitochondria. *J. Biol. Chem.*, 233: 657–663.

10. Stephenson, M. L., Thimann, K. V., and Zamecnik, P. C. (1956): Incorporation of ^{14}C-amino acids into proteins of leaf disks and cell-free fractions of tobacco leaves. *Arch. Biochem. Biophys.*, 65:194–209.

11. Boulter, D., Ellis, R. J., and Yarwood, A. (1972): Biochemistry of protein synthesis in plants. *Biol. Rev.*, 47:113–175.

12. Schatz, G., and Mason, T. L. (1974): The biosynthesis of mitochondrial proteins. *Annu. Rev. Biochem.*, 43:51–87.

13. Mahler, H. R. (1973): Biogenetic autonomy of mitochondria. *CRC Crit. Rev. Biochem.*, 1:381–460.

14. Ellis, R. J., Blair, G. E., and Hartley, M. R. (1973): The nature and function of chloroplast protein synthesis. *Biochem. Soc. Symp.*, 38:137–162.

15. Eccleshall, T. R., and Criddle, R. S. (1974): The DNA-dependent RNA polymerases from yeast mitochondria. In: *The Biogenesis of Mitochondria*, edited by A. M. Kroon and C. Saccone, pp. 31–36. Academic Press, New York.

16. Küntzel, H., and Schäfer, K. P. (1971): Mitochondrial RNA polymerase of *Neurospora crassa*. *Nature* [*New Biol.*], 231:265–269.

17. Scragg, A. H. (1974): A mitochondrial DNA-directed RNA polymerase from yeast mitochondria. In: *The Biogenesis of Mitochondria*, edited by A. M. Kroon and C. Saccone, pp. 47–57. Academic Press, New York.

18. Gallerani, R., and Saccone, C. (1974): The DNA-dependent RNA polymerase from rat liver mitochondria. In: *The Biogenesis of Mitochondria*, edited by A. M. Kroon and C. Saccone, pp. 59–69. Academic Press, New York.

19. Dawid, I. B., and Wu, G.-J. (1974): Transcription of mtDNA by mitochondrial RNA polymerase from *Xenopus laevis*. In: *The Biogenesis of Mitochondria*, edited by A. M. Kroon and C. Saccone, pp. 79–84. Academic Press, New York.

20. Bogorad, L., Mets, L. J., Mullinex, K. P.,

Smith, H. J., and Strain, G. C. (1973): Possibilities for intracellular integration: The ribonucleic acid polymerases of chloroplasts and nuclei, and genes specifying chloroplast ribosomal proteins. *Biochem. Soc. Symp.*, 38: 17–41.

21. Chambon, P. (1974): Eucaryotic RNA polymerases. In: *The Enzymes, Vol. 10,* edited by P. D. Boyer, pp. 261–331. Academic Press, New York.

22. Chamberlin, M. J. (1974): Bacterial DNA-dependent RNA polymerase. In: *The Enzymes, Vol. 10,* edited by P. D. Boyer, pp. 333–374. Academic Press, New York.

23. Hallick, R. B., Hager, G. L., and Rutter, W. J. (1973): Transcription in organellular systems. *Fed. Proc.*, 32:abstract #2299.

24. Smith, H. J., and Bogorad, L. (1974): The polypeptide subunit structure of the DNA-dependent RNA polymerase of *Zea mays mays* chloroplasts. *Proc. Natl. Acad. Sci. U.S.A.*, 71:4839–4842.

25. Ratcliff, S., and Surzycki, S. (1974): DNA-dependent RNA polymerases from *Chlamydomonas reinhardtii*. *J. Cell Biol.*, 62:281a.

26. Surzycki, S. J., and Shellenbarger, D. L. (1976): Purification and characterization of a putative sigma factor from *Chlamydomonas reinhardi*. *Proc. Natl. Acad. Sci. U.S.A.*, 73: 3961–3965.

27. Wintersberger, E. (1972): Isolation of a distinct rifampicin-resistant RNA polymerase from mitochondria of yeast, *Neurospora* and liver. *Biochem. Biophys. Res. Commun.*, 48: 1287–1294.

28. Riva, S., and Silvestri, L. G. (1972): Rifamycins: A general view. *Annu. Rev. Microbiol.*, 26:199–224.

29. Surzycki, S. J. (1969): Genetic functions of the chloroplast of *Chlamydomonas reinhardi*: Effect of rifampin on chloroplast DNA-dependent RNA polymerase. *Proc. Natl. Acad. Sci. U.S.A.*, 63:1327–1334.

30. Surzycki, S. J., and Rochaix, J. D. (1971): Transcriptional mapping of ribosomal RNA genes of the chloroplast and nucleus of *Chlamydomonas reinhardi*. *J. Mol. Biol.*, 62:89–109.

31. Bogorad, L., and Woodcock, C. L. F. (1971): Rifamycins: The inhibition of plastid RNA synthesis *in vivo* and *in vitro* and variable effects on chlorophyll formation in maize leaves. In: *Autonomy and Biogenesis of Chloroplasts and Mitochondria*, edited by N. K. Boardman, A. W. Linnane, and R. M. Smillie, pp. 92–97. North Holland, Amsterdam.

32. Surzycki, S. J., Goodenough, V. W., Levine, R. P., and Armstrong, J. J. (1970): Nuclear and chloroplast control of chloroplast structure and function in *Chlamydomonas reinhardi*. *Symp. Soc. Exp. Biol.*, 24:13–37.

33. Urbanke, C., Römer, R., and Maass, G. (1973): The binding of ethidium bromide to

different conformations of tRNA. *Eur J. Biochem.,* 33:511–516.

34. Surovaya, A., and Trubitsyn, S. (1972): Binding isotherms of tRNA-acriflavine complexes. *FEBS Lett.,* 25:349–352.

35. Grivell, L. A., and Metz, V. (1973): Inhibition by ethidium bromide of mitochondrial protein synthesis programmed by imported poly (U). *Biochem. Biophys. Res. Commun.,* 55:125–131.

36. Dawid, I. B. (1970): The nature of mitochondrial RNA in oocytes of *Xenopus laevis* and its relation to mitochondrial DNA. *Symp. Soc. Exp. Biol.,* 24:227–246.

37. Swanson, R. F. (1971): Incorporation of high molecular weight polynucleotides by isolated mitochondria. *Nature,* 231:31–34.

38. Dawid, I. B. (1972): Mitochondrial RNA in *Xenopus laevis.* I. The expression of the mitochondrial genome. *J. Mol. Biol.,* 63:201–216.

39. Dawid, I. B. (1972): Evolution of mitochondrial DNA sequences in *Xenopus. Dev. Biol.,* 29:139–151.

40. Fukuhara, H. (1970): Transcriptional origin of RNA in a mitochondrial fraction of yeast and its bearing on the problem of sequence homology between mitochondrial and nuclear DNA. *Mol. Gen. Genet.,* 107:58–70.

41. Borst, P. (1972): Mitochondrial nucleic acids. *Annu. Rev. Biochem.,* 41:333–376.

42. Reijnders, L., Kleisen, C. M., Grivell, L. A., and Borst, P. (1972): Hybridization studies with yeast mitochondrial RNAs. *Biochim. Biophys. Acta,* 272:396–407.

43. Mahler, H. R., and Dawidowicz, K. (1973): Autonomy of mitochondria in *Saccharomyces cerevisiae* and their production of messenger RNA. *Proc. Natl. Acad. Sci. U.S.A.,* 70:111–114.

44. Mahler, H. R., Feldman, F., Phan, S. H., Hamill, P., and Dawidowicz, K. (1974): Initiation, identification and integration of mitochondrial proteins. In: *The Biogenesis of Mitochondria,* edited by A. M. Kroon and C. Saccone, pp. 423–441. Academic Press, New York.

45. Mahler, H. R., Bastos, R. N., Feldman, F., Flury, U., Lin, C. C., Perlman, P. S., and Phan, S. H. (1975): Biogenetic autonomy of mitochondria and its limits. In: *Membrane Biogenesis,* edited by A. Tzagoloff, Chapter 2. Plenum Press, New York.

46. Hoober, J. K., and Stegeman, W. J. (1973): Control of the synthesis of a major polypeptide of chloroplast membranes in *Chlamydomonas reinhardi. J. Cell Biol.,* 56:1–12.

47. Armstrong, J. J., Surzycki, S. J., Moll, B., and Levine, R. P. (1971): Genetic transcription and translation specifying chloroplast components in *Chlamydomonas reinhardi. Biochemistry,* 10:692–701.

48. Sirevåg, R., and Levine, R. P. (1972): Fatty acid synthetase from *Chlamydomonas reinhardi. J. Biol. Chem.,* 247:2586–2591.

49. Surzycki, S. J., and Gillham, N. W. (1971): Organelle mutations and their expression in *Chlamydomonas reinhardi. Proc. Natl. Acad. Sci. U.S.A.,* 68:1301–1306.

50. Conde, M. F., Boynton, J. E., Gillham, N. W., Harris, E. H., Tingle, C. L., and Wang, W. L. (1975): Chloroplast genes in *Chlamydomonas* affecting organelle ribosomes. *Mol. Gen. Genet.,* 140:183–220.

51. Hoober, J. K., and Stegeman, W. J. (1975): Regulation of chloroplast membrane synthesis. In: *Genetics and Biogenesis of Mitochondria and Chloroplasts,* edited by C. W. Birky, Jr., P. S. Perlman, and T. J. Byers, Chapter 6. Ohio State University Press, Columbus.

52. Schiff, J. A. (1973): The development, inheritance, and origin of the plastid in *Euglena. Adv. Morphogenesis,* 10:265–312.

53. Aloni, Y., and Attardi, G. (1971): Expression of the mitochondrial genome in HeLa cells. II. Evidence for complete transcription of mitochondrial DNA. *J. Mol. Biol.,* 55·251–270.

54. Aloni, Y., and Attardi, G. (1971): Symmetrical *in vivo* transcription of mitochondrial DNA in HeLa cells. *Proc. Natl. Acad Sci. U.S.A.,* 68:1757–1761.

55. Darnell, J. E., Jelinek, W. R., and Molloy, G. R. (1973): Biogenesis of messenger RNA: Genetic regulation in mammalian cells. *Science,* 181:1215–1221.

56. Brawerman, G. (1974): Eukaryotic messenger RNA. *Annu. Rev. Biochem.,* 43:621–642.

57. Perlman, S., Abelson, H. T., and Penman, S. (1973): Mitochondrial protein synthesis: RNA with the properties of eukaryotic messenger RNA. *Proc. Natl. Acad. Sci. U.S.A.,* 70:350–353.

58. Ojala, D., and Attardi, G. (1974): Expression of the mitochondrial genome in HeLa cells. XIX. Occurrence in mitochondria of polyadenylic acid sequences, "free" and covalently linked to mitochondrial DNA-coded RNA. *J. Mol. Biol.,* 82:151–174.

59. Jacob, S. T., Schindler, D. G., and Morris, H. P. (1972): Mitochondrial polyriboadenylate polymerase: Relative lack in hepatomas. *Science,* 178:639–640.

60. Ojala, D., and Attardi, G. (1974): Identification of discrete polyadenylate-containing RNA components transcribed from HeLa cell mitochondrial DNA. *Proc. Natl. Acad. Sci. U.S.A.,* 71:563–567.

61. Ojala, D., and Attardi, G. (1974): Identification and partial characterization of multiple discrete polyadenylic acid-containing RNA components coded for by HeLa cell mitochondrial DNA. *J. Mol. Biol.,* 88:205–219.

62. Hirsch, M., and Penman, S. (1973): Mitochondrial polyadenylic acid-containing RNA: Localization and characterization. *J. Mol. Biol.*, 80:379–381.

63. Attardi, G., Amalric, F., Ching, E., Costantino, P., Gelfand, R., and Lynch, D. (1976): Informational content and gene mapping of mitochondrial DNA from HeLa cells. In: *The Genetic Function of Mitochondrial DNA*, edited by C. Saccone and A. M. Kroon, pp. 37–46. North Holland, Amsterdam.

64. Hirsch, M., Spradling, A., and Penman, S. (1974): The messenger-like poly(A)-containing RNA species from the mitochondria of mammals and insects. *Cell*, 1:31–35.

65. Padmanaban, G., Hendler, F., Patzer, J., Ryan, R., and Rabinowitz, M. (1975): Translation of RNA that contains polyadenylate from yeast mitochondria in an *Escherichia coli* ribosomal system. *Proc. Natl. Acad. Sci. U.S.A.*, 72:4293–4297.

66. Rabinowitz, M., Jakovcic, S., Martin, N., Hendler, F., Halbreich, A., Lewin, A., and Morimoto, R. (1976): Transcription and organization of yeast mitochondrial DNA. In: *The Genetic Function of Mitochondrial DNA*, edited by C. Saccone and A. M. Kroon, pp. 37–46. North Holland, Amsterdam.

67. Cooper, C. S., and Avers, C. J. (1974): Evidence of involvement of mitochondrial polysomes and messenger RNA in synthesis of organelle proteins. In: *The Biogenesis of Mitochondria*, edited by A. M. Kroon and C. Saccone, pp. 289–303. Academic Press, New York.

68. Groot, G. S. P., Flavell, R. A., Van Ommen, G. J. B., and Grivell, L. A. (1974): Yeast mitochondrial RNA does not contain poly(A). *Nature*, 252:167–169.

69. Hendler, F., Padmanaban, G., Patzer, J., Ryan, R., and Rabinowitz, M. (1975): Yeast mitochondrial RNA contains a short polyadenylic acid segment. *Nature*, 258:357–359.

70. Hartley, M. R., Wheeler, A., and Ellis, R. J. (1975): Protein synthesis in chloroplasts. V. Translation of messenger RNA for the large subunit of fraction I protein in a heterologous cell-free system. *J. Mol. Biol.*, 91:67–77.

71. Howell, S. H., Heizmann, P., Gelvin, S., and Walker, L. L. (1977): Identification and properties of the messenger RNA activity in *Chlamydomonas reinhardi* coding for the large subunit of d-ribulose-1,5-bisphosphate carboxylase. *Plant Physiol.*, 59:471–477.

72. Rawson, J. R. Y. (1975): A measurement of the fraction of chloroplast DNA transcribed in *Euglena. Biochem. Biophys. Res. Commun.*, 62:539–545.

73. Rawson, J. R. Y., and Boerma, C. L. (1976): A measurement of the fraction of chloroplast DNA transcribed during chloroplast development in *Euglena gracilis. Biochemistry*, 15:588–592.

74. Chelm, B. K., and Hallick, R. B. (1976): Changes in the expression of the chloroplast genome of *Euglena gracilis* during chloroplast development. *Biochemistry*, 15:593–599.

75. Barnett, W. E., and Brown, D. H. (1967): Mitochondrial transfer ribonucleic acids. *Proc. Natl. Acad. Sci. U.S.A.*, 57:452–458.

76. Barnett, W. E., Brown, D. H., and Epler, J. L. (1967): Mitochondrial-specific aminoacyl-RNA synthetases. *Proc. Natl. Acad. Sci. U.S.A.*, 57:1775–1781.

77. Barnett, W. E., Pennington, C. J., Jr., and Fairfield, S. A. (1969): Induction of *Euglena* transfer RNA's by light. *Proc. Natl. Acad. Sci. U.S.A.*, 63:1261–1268.

78. Epler, J. L. (1969): The mitochondrial and cytoplasmic transfer ribonucleic acids of *Neurospora crassa. Biochemistry*, 8:2285–2290.

79. Burkard, G., Guillamaut, P., and Weil, J. H. (1970): Comparative studies of the tRNAs and the aminoacyl-tRNA synthetases from the cytoplasm and the chloroplasts of *Phaseolus vulgaris. Biochim. Biophys. Acta*, 224:184–198.

80. Leis, J. P., and Keller, E. B. (1970): Protein chain-initiating methionine tRNAs in chloroplasts and cytoplasm of wheat leaves. *Proc. Natl. Acad. Sci. U.S.A.*, 67:1593–1599.

81. Merrick, W. C., and Dure, L. S., III (1971): Specific transformylation of one methionyl-tRNA from cotton seedling chloroplasts by endogenous and *Escherichia coli* transformylases. *Proc. Natl. Acad. Sci. U.S.A.*, 68:641–644.

82. Lewin, B. (1974): *Gene Expression. Vol. I. Bacterial Genomes.* John Wiley & Sons, New York.

83. Burkhard, G., Guillemaut, P., Steinmetz, A., and Weil, J. H. (1973): Transfer ribonucleic acid-recognizing enzymes in bean cytoplasm, chloroplasts, etioplasts and mitochondria. *Biochem. Soc. Symp.*, 38:43–56.

84. Costantino, P., and Attardi, G. (1973): Atypical pattern of utilization of amino acids for mitochondrial protein synthesis in HeLa cells. *Proc. Natl. Acad. Sci. U.S.A.*, 70:1490–1494.

85. Lynch, D. C., and Attardi, G. (1976): Amino acid specificity of the transfer RNA species coded for by HeLa cell mitochondrial DNA. *J. Mol. Biol.*, 102:125–141.

86. Borst, P., and Grivell, L. A. (1971): Mitochondrial ribosomes. *FEBS Lett.*, 13:73–88.

87. Ellis, R. J., and Hartley, M. R. (1974): Nucleic acids of chloroplasts. *MTP Internat. Rev. Science*, 6:323–348.

88. Chua, N.-H., and Luck, D. (1974): Biosynthesis of organelle ribosomes. In: *Ribosomes*, edited by M. Nomura, A. Tissieres, and

P. Lengyel, pp. 519–539. Cold Spring Harbor Laboratory, Cold Spring Harbor, N.Y.

89. Stutz, E., and Boschetti, A. (1976): Chloroplast ribosomes. In: *Handbook of Genetics, Vol. 5,* edited by R. C. King, Chapter 13. Plenum Press, New York.

90. O'Brien, T. W., and Matthews, D. E. (1976): Mitochondrial ribosomes. In: *Handbook of Genetics, Vol. 5,* edited by R. C. King, Chapter 16. Plenum Press, New York.

91. Lyttleton, J. W. (1962): Isolation of ribosomes from spinach chloroplasts. *Exp. Cell Res.,* 26:312–317.

92. Küntzel, H., and Noll, H. (1967): Mitochondrial and cytoplasmic polysomes from *Neurospora crassa. Nature,* 215:1340–1345.

93. O'Brien, T. W., and Kalf, G. F. (1967): Ribosomes from rat liver mitochondria. I. Isolation procedure and contamination studies. *J. Biol. Chem.,* 242:2172–2179.

94. O'Brien, T. W., and Kalf, G. F. (1967): Ribosomes from rat liver mitochondria. II. Partial characterization. *J. Biol. Chem.,* 242: 2180–2185.

95. Nomura, M., Tissieres, A., and Lengyel, P. (editors) (1974): *Ribosomes.* Cold Spring Harbor Laboratory, Cold Spring Harbor, N.Y.

96. Wittmann, H. G. (1974): Purification and identification of *Escherichia coli* ribosomal proteins. In: *Ribosomes,* edited by M. Nomura, A. Tissieres, and P. Lengyel, pp. 93–114. Cold Spring Harbor Laboratory, Cold Spring Harbor, N.Y.

97. Van Holde, K. E., and Hill, W. E. (1974): General physical properties of bacterial ribosomes. In: *Ribosomes,* edited by M. Nomura, A. Tissieres, and P. Lengyel, pp. 53–91. Cold Spring Harbor Laboratory, Cold Spring Harbor, N.Y.

98. Schlessinger, D. (1974): Ribosome formation in *Escherichia coli.* In: *Ribosomes,* edited by M. Nomura, A. Tissieres, and P. Lengyel, pp. 393–416. Cold Spring Harbor Laboratory, Cold Spring Harbor, N.Y.

99. Warner, J. R. (1974): The assembly of ribosomes in eukaryotes. In: *Ribosomes,* edited by M. Nomura, A. Tissieres, and P. Lengyel, pp. 461–488. Cold Spring Harbor Laboratory, Cold Spring Harbor, N.Y.

100. Ingle, J., Possingham, J. V., Wells, R., Leaver, C. J., and Loening, U. E. (1970): The properties of chloroplast ribosomal RNA. *Symp. Soc. Exp. Biol.,* 24:303–325.

101. Payne, P. I., and Dyer, T. A. (1972): Plant 5.8S RNA is a component of 80S but not 70S ribosomes. *Nature [New Biol.],* 235: 145–147.

102. Hanson, M. R., Davidson, J. N., Mets, L. J., and Bogorad, L. (1974): Characterization of chloroplast and cytoplasmic ribosomal proteins of *Chlamydomonas reinhardi* by two-dimensional gel electrophoresis. *Mol. Gen. Genet.,* 132:105–118.

103. Brügger, M., and Boschetti, A. (1975): Two-dimensional gel electrophoresis of ribosomal proteins from streptomycin-sensitive and streptomycin-resistant mutants of *Chlamydomonas reinhardi. Eur. J. Biochem.,* 58:603–610.

104. Lee, S. G., and Evans, W. R. (1971): Hybrid ribosome formation from *Escherichia coli* and chloroplast ribosome subunits. *Science,* 173:241–242.

105. Grivell, L. A., and Walg, H. L. (1972): Subunit homology between *Escherichia coli,* mitochondrial and chloroplast ribosomes. *Biochem. Biophys. Res. Commun.,* 49:1452–1458.

106. Krawiec, S., and Eisenstadt, J. M. (1970): Ribonucleic acids from the mitochondria of bleached *Euglena gracilis* Z. II. Characterization of highly polymeric ribonucleic acids. *Biochim. Biophys. Acta,* 217:132–141.

107. Chi, J. C. H., and Suyama, Y. (1970): Comparative studies on mitochondrial and cytoplasmic ribosomes of *Tetrahymena pyriformis. J. Mol. Biol.,* 53:531–556.

108. Lizardi, P. M., and Luck, D. J. L. (1972): The intracellular site of synthesis of mitochondrial ribosomal proteins in *Neurospora crassa. J. Cell Biol.,* 54:56–74.

109. Kuriyama, Y., and Luck, D. J. L. (1973): Ribosomal RNA synthesis in mitochondria of *Neurospora crassa. J. Mol. Biol.,* 73:425–437.

110. Lambowitz, A., Chua, N.-H., and Luck, D. J. L. (1976): Mitochondrial ribosome assembly in *Neurospora.* Preparation of mitochondrial ribosomal precursor particles, site of synthesis of mitochondrial ribosomal proteins and studies on the poky mutant. *J. Mol. Biol.,* 107:223–253.

111. Kroon, A. M., and De Vries, H. (1970): Antibiotics: A tool in the search for the degree of autonomy of mitochondria in higher organisms. *Symp. Soc. Exp. Biol.,* 24:181–199.

112. Tait, A. (1972): Altered mitochondrial ribosomes in an erythromycin resistant mutant of *Paramecium. FEBS Lett.,* 24:117–120.

113. Mets, L. J., and Bogorad, L. (1971): Mendelian and uniparental alterations in erythromycin binding by plastid ribosomes. *Science,* 174:707–709.

114. Burton, W. G. (1972): Dihydrospectinomycin binding to chloroplast ribosomes from antibiotic-sensitive and -resistant strains of *Chlamydomonas reinhardtii. Biochim. Biophys. Acta,* 272:305–311.

115. Schlanger, G., and Sager, R. (1974): Localization of five antibiotic resistances at the subunit level in chloroplast ribosomes of *Chlamydomonas. Proc. Natl. Acad. Sci. U.S.A.,* 71:1715–1719.

116. Hanson, M. R., and Bogorad, L. (1977): Effects of erythromycin on membrane-bound chloroplast ribosomes from wild-type *chlamydomas reinhardii* and erythromycin-resistant

mutants. *Biochim. Biophys. Acta,* 479:279–289.

117. Harris, E. H., Boynton, J. E., Gillham, N. W., Tingle, C. L., and Fox, S. B. (1977): Mapping of chloroplast genes involved in chloroplast ribosome biogenesis in *Chlamydomonas reinhardtii. Mol. Gen. Genet.,* 155:249–265.

118. Gillham, N. W., Boynton, J. E., Harris, E. H., Fox, S. B., and Bolen, P. (1976): Genetic control of chloroplast ribosome formation in *Chlamydomonas.* In: *Genetics and Biogenesis of Chloroplasts and Mitochondria,* edited by T. Bücher, W. Neupert, W. Sebald, and S. Werner, pp. 69–76. North Holland, Amsterdam.

119. Kutzleb, R., Schweyen, R. J., and Kaudewitz, F. (1973): Extrachromosomal inheritance of paromomycin resistance in *Saccharomyces cerevisiae. Mol. Gen. Genet.,* 125:91–98.

120. Ellis, R. J., and Hartley, M. R. (1971): Sites of synthesis of chloroplast proteins. *Nature [New Biol.],* 233:193–196.

121. Blobel, G., and Dobberstein, B. (1975): Transfer of proteins across membranes. I. Presence proteolytically processed and unprocessed nascent immunoglobulin light chains on membrane-bound ribosomes of murine myeloma. *J. Cell Biol.,* 67:835–851.

122. Redman, C., and Sabatini, D. D. (1966): Vectorial discharge of peptides released by puromycin from attached ribosomes. *Proc. Natl. Acad. Sci. U.S.A.,* 56:608–615.

123. Vazquez, D. (1974): Inhibitors of protein synthesis. *FEBS Lett.* (Suppl.), 40:563–584.

124. Sabatini, D. D., Tashiro, Y., and Palade, G. E. (1966): On the attachment of ribosomes to microsomal membranes. *J. Mol. Biol.,* 19:503–524.

125. Adelman, M. R., Sabatini, D., and Blobel, G. (1973): Ribosome-membrane interaction. Nondestructive assembly of rat liver rough microsomes into ribosomal and membrane components. *J. Cell Biol.,* 56:206–229.

126. DeVillers-Thiery, A., Kindt, T., Schelle, G., and Blobel, G. (1975): Homology in aminoterminal sequence of precursors to pancreatic secretory proteins. *Proc. Natl. Acad. Sci. U.S.A.,* 72:5016–5020.

127. Blobel, G., and Dobberstein, B. (1975): Transfer of proteins across membranes. II. Reconstitution of functional rough microsomes from heterologous components. *J. Cell Biol.,* 67:852–862.

128. Kellems, R. E., and Butow, R. A. (1972): Cytoplasmic-type 80S ribosomes associated with yeast mitochondria. I. Evidence for ribosome binding sites on yeast mitochondria. *J. Biol. Chem.,* 247:8043–8050.

129. Kellems, R. E., Allison, V. F., and Butow, R. A. (1974): Cytoplasmic type 80S ribosomes associated with yeast mitochondria. II. Evidence for the association of cytoplasmic

ribosomes with the outer mitochondrial membrane *in situ. J. Biol. Chem.,* 249:3297–3303.

130. Kellems, R. E., and Butow, R. A. (1974): Cytoplasmic type 80S ribosomes associated with yeast mitochondria. III. Changes in the amount of bound ribosomes in response to changes in metabolic state. *J. Biol. Chem.,* 249:3304–3310.

131. Butow, R. A., Bennett, W. F., Finkelstein, D. B., and Kellems, R. E. (1975): Nuclear-cytoplasmic interactions in the biogenesis of mitochondria in yeast. In: *Membrane Biogenesis,* edited by A. Tzagoloff, Chapter 6. Plenum Press, New York.

132. Chua, N.-H., Blobel, G., Siekevitz, P., and Palade, G. E. (1973): Attachment of chloroplast polysomes to thylakoid membranes in *Chlamydomonas reinhardtii. Proc. Natl. Acad. Sci. U.S.A.,* 70:1554–1558.

133. Margulies, M. M., and Michaels, A. (1974): Ribosomes bound to chloroplast membranes in *Chlamydomonas reinhardtii. J. Cell Biol.,* 60:65–77.

134. Chua, N.-H., Blobel, G., Siekevitz, P., and Palade, G. E. (1976): Periodic variations in the ratio of free to thylakoid-bound chloroplast ribosomes during the cell cycle of *Chlamydomonas reinhardtii. J. Cell Biol.,* 71:497–514.

135. Margulies, M. M., Tiffany, H. L., and Michaels, A. (1975): Vectorial discharge of nascent polypeptides attached to chloroplast thylakoid membranes. *Biochem. Biophys. Res. Commun.,* 64:735–739.

136. Margulies, M. M., and Michaels, A. (1975): Free and membrane-bound chloroplast polyribosomes in *Chlamydomonas reinhardtii. Biochim. Biophys. Acta,* 402:297–308.

137. Kuriyama, Y., and Luck, D. J. L. (1973): Membrane-associated ribosomes in mitochondria of *Neurospora crassa. J. Cell Biol.,* 59:776–784.

138. Gillespie, D., and Spiegelman, S. (1965): A quantitative assay for DNA-RNA hybrids with DNA immobilized on a membrane. *J. Mol. Biol.,* 12:829–842.

139. Tewari, K. K., and Wildman, S. G. (1970): Information content in the chloroplast DNA. *Symp. Soc. Exp. Biol.,* 24:147–179.

140. Rawson, J. R. Y., and Haselkorn, R. (1973): Chloroplast ribosomal RNA genes in the chloroplast DNA of *Euglena gracilis. J. Mol. Biol.,* 77:125–132.

141. Gruol, D. J., and Haselkorn, R. (1976): Counting the genes for stable RNA in the nucleus and chloroplasts of *Euglena. Biochim. Biophys. Acta,* 447:82–95.

142. Bastia, D., Chiang, K.-S., and Swift, H. (1971): Studies on the ribosomal RNA cistrons of chloroplast and nucleus in *Chlamydomonas reinhardtii. Abstracts 11th Annual Meeting American Society of Cell Biology,* p. 25.

143. Wu, M., Davidson, N., Attardi, G., and Aloni, Y. (1972): Expression of the mitochondrial genome in HeLa cells. XIV. The relative positions of the 4S RNA genes and of the ribosomal RNA genes in mitochondrial DNA. *J. Mol. Biol.,* 71:81–93.

144. Dawid, I. B., Klukas, C. K., Ohi, S., Ramirez, J. L., and Upholt, W. B. (1976): Structure and evolution of animal mitochondrial DNA. In: *The Genetic Function of Mitochondrial DNA,* edited by C. Saccone and A. M. Kroon, pp. 3–13. North Holland, Amsterdam.

145. Kroon, A. M., Pepe, G., Bakker, H., Holtrop, M., Bollen, J. E., Van Bruggen, E. F. J., Cantatore, P., Terpstra, P., and Saccone, C. (1977): The restriction fragment map of rat-liver mitochondrial DNA: A reconsideration. *Biochim. Biophys. Acta,* 478:128–145.

146. Sanders, J. P. M., Heyting, C., DiFranco, A., Borst, P., and Slonimski, P. P. (1976): The organization of genes in yeast mitochondrial DNA: In: *The Genetic Function of Mitochondrial DNA,* edited by C. Saccone and A. M. Kroon, pp. 259–272. North Holland, Amsterdam.

147. Herrmann, R. G., Bohnert, H.-G., Driesel, A., and Hobom, G. (1976): The location of rRNA genes on the restriction endonuclease map of the *Spinacia oleracea* chloroplast DNA. In: *Genetics and Biogenesis of Chloroplasts and Mitochondria,* edited by T. Bücher, W. Neupert, W. Sebald, and S. Werner, pp. 351–359. North Holland, Amsterdam.

148. Bedbrook, J. R., and Bogorad, L. (1976): Endonuclease recognition sites mapped on *Zea mays* chloroplast DNA. *Proc. Natl. Acad. Sci. U.S.A.,* 73:4309–4313.

149. Southern, E. M. (1975): Detection of specific sequences among DNA fragments separated by gel electrophoresis. *J. Mol. Biol.,* 98:503–517.

150. Reijnders, L., and Borst, P. (1972): The number of 4-S RNA genes on yeast mitochondrial DNA. *Biochem. Biophys. Res. Commun.,* 47:126–133.

151. Schwartzbach, S., Hecker, L. I., and Barnett, W. E. (1976): Transcriptional origin of *Euglena* chloroplast tRNAs. *Proc. Natl. Acad. Sci. U.S.A.,* 73:1984–1988.

152. Nass, M. M. K., and Buck, C. A. (1970): Studies on mitochondrial tRNA from animal cells. II. Hybridization of aminoacyl-tRNA from rat liver mitochondria with heavy and light complementary strands of mitochondrial DNA. *J. Mol. Biol.,* 54:187–198.

153. Keller, S. J. (1969): The characterization of ribosomes from wild type and streptomycin resistant mutants of *Chlamydomonas reinhardtii.* Ph.D. thesis, SUNY, Stony Brook, N.Y.

154. Bourque, D. P., Boynton, J. E., and Gillham, N. W. (1971): Studies on the structure and cellular location of various ribosome and ribosomal RNA species in the green alga *Chlamydomonas reinhardi. J. Cell Sci.,* 8: 153–183.

155. Hamilton, M. G. (1976): Eukaryotic ribosomes. In: *Handbook of Genetics, Vol. 5,* edited by R. C. King, Chapter 11. Plenum Press, New York.

156. Cammarano, P., Pons, S., Romeo, A., Galdieri, M., and Gualerzi, C. (1972): Characterization of unfolded and compact ribosomal subunits from plants and their relationship to those of lower and higher animals: Evidence for physio-chemical heterogeneity among eucaryotic ribosomes. *Biochim. Biophys. Acta,* 281:571–596.

157. Cammarano, P., Romeo, A., Gentile, M., Felsani, A., and Gualerzi, C. (1972): Size heterogeneity of the large ribosomal subunits and conservation of the small subunits in eucaryote evolution. *Biochim. Biophys. Acta,* 281:597–624.

158. Thomas, H. (1973): Gel electrophoresis of ribosomal components from seeds of *Pisum sativum* L. *Exp. Cell Res.,* 77:298–302.

159. Loening, U. E. (1968): Molecular weights of ribosomal RNA in relation to evolution. *J. Mol. Biol.,* 38:355–365.

160. Wool, I. G., and Stöffler, G. (1974): Structure and function of eukaryotic ribosomes. In: *Ribosomes,* edited by M. Nomura, A. Tissieres, and P. Lengyel, pp. 417–460. Cold Spring Harbor Laboratory, Cold Spring Harbor, N.Y.

161. Hoober, J. K., and Blobel, G. (1969): Characterization of the chloroplastic and cytoplasmic ribosomes of *Chlamydomonas reinhardi. J. Mol. Biol.,* 41:121–138.

162. Siersma, P. W., and Chiang, K.-S. (1971): Conservation and degradation of cytoplasmic and chloroplast ribosomes in *Chlamydomonas reinhardtii. J. Mol. Biol.,* 58:167–185.

163. Freyssinet, G. (1975): Changes in chloroplast ribosomal proteins in a streptomycin-resistant mutant of *Euglena gracilis. Plant Sci. Lett.,* 5:305–311.

164. Bourque, D. P., and Wildman, S. G. (1973): Evidence that nuclear genes code for several chloroplast ribosomal proteins. *Biochem. Biophys. Res. Commun.,* 50:532–537.

165. Avadhani, N. G., and Buetow, D. E. (1972): Isolation of active polyribosomes from the cytoplasm, mitochondria and chloroplasts of *Euglena gracilis. Biochem. J.,* 128:353–365.

166. Dawid, I. B., and Chase, J. W. (1972): Mitochondrial RNA in *Xenopus laevis.* II. Molecular weights and other physical properties of mitochondrial ribosomal and 4S RNA. *J. Mol. Biol.,* 63.217–231.

167. Leister, D. E., and Dawid, I. B. (1974): Physical properties and protein constituents of cytoplasmic and mitochondrial ribosomes of *Xenopus laevis. J. Biol. Chem.,* 249:5108–5118.

168. Hamilton, M. G., and O'Brien, T. W. (1974): Ultracentrifugal characterization of the mitochondrial ribosome and subribosomal particles of bovine liver: Molecular size and composition. *Biochemistry,* 13:5400–5403.

169. Attardi, G., and Ojala, D. (1971): Mitochondrial ribosomes in HeLa cells. *Nature [New Biol.],* 229:133–136.

170. Lamb, A. J., Clark-Walker, G. D., and Linnane, A. W. (1968): The biogenesis of mitochondria. 4. The differentiation of mitochondrial and cytoplasmic protein synthesizing systems *in vitro* by antibiotics. *Biochim. Biophys. Acta,* 161:415–427.

171. Ellis, R. J. (1970): Further similarities between chloroplast and mitochondrial ribosomes. *Planta,* 91:329–335.

172. Bunn, C. L., Mitchell, C. H., Lukins, H. B., and Linnane, A. W. (1970): Biogenesis of mitochondria. XVIII. A new class of cytoplasmically determined antibiotic resistant mutants in *Saccharomyces cerevisiae. Proc. Natl. Acad. Sci. U.S.A.,* 67:1233–1240.

173. Chua, N.-H., and Gillham, N. W. (1977): The sites of synthesis of the principal thylakoid membrane polypeptides in *Chlamydomonas reinhardtii. J. Cell Biol.,* 74:441–452.

174. Ellis, R. J., and MacDonald, I. R. (1970): Specificity of cycloheximide in higher plant systems. *Plant Physiol.,* 46:227–232.

175. Schäfer, K. P., and Küntzel, H. (1972): Mitochondrial genes in *Neurospora:* A single cistron for ribosomal RNA. *Biochem. Biophys. Res. Commun.,* 46:1312–1319.

176. Terpstra, P., Holtrop, M., and Kroon, A. M. (1976): Restriction fragment map of *Neurospora crassa* mitochondrial DNA. In: *The Genetic Function of Mitochondrial DNA,* edited by C. Saccone and A. M. Kroon, pp. 111–118. Elsevier/North-Holland, Amsterdam.

177. Nagley, P., Sriprakash, K. S., Rytka, J., Choo, K. B., Trembath, M. K., Lukins, H. B., and Linnane, A. W. (1976): Physical mapping of genetic markers in the yeast mitochondrial genome. In: *The Genetic Function of Mitochondrial DNA,* edited by C. Saccone and A. M. Kroon, pp. 231–242. Elsevier/North-Holland, Amsterdam.

178. Wheeler, A. M., and Hartley, M. R. (1975): Major mRNA species from spinach chloroplasts do not contain poly(A). *Nature,* 257:66–67.

179. Sagher, D., Grosfeld, H., and Edelman, M. (1976): Large subunit ribulosebisphosphate carboxylase messenger RNA from *Euglena* chloroplasts. *Proc. Natl. Acad. Sci. U.S.A.,* 73:722–726.

180. Dobberstein, B., Blobel, G., and Chua, N.-H. (1977): *In vitro* synthesis and processing of a putative precursor for the subunit of ribulose-1,5-bisphosphate carboxylase of *Chlamydomonas reinhardtii. Proc. Natl. Acad. Sci. U.S.A.,* 74:1082–1085.

181. Nomura, M., and Morgan, E. A. (1977): Genetics of bacterial ribosomes. *Annu. Rev. Genet.,* 11:297–347.

Chapter 4
Mitochondrial Genetics of Bakers' Yeast. I. The Organism, Its Growth, and the Respiration-Deficient Mutants

Over 20 years ago Ephrussi (1) summarized the work that he and his colleagues had done on respiration-deficient mutants in bakers' yeast (*S. cerevisiae*) in a thought-provoking little book entitled *Nucleocytoplasmic Relations in Microorganisms*. They found that respiration-deficient mutants (which they called "petites" because these mutants formed small colonies under the growth conditions used) could be induced quantitatively in both haploid and diploid strains of *S. cerevisiae* with acriflavine and related dyes (2–5). These acriflavine-induced petites had a nonmendelian pattern of inheritance (6).

The survival of mutants impaired in a process as fundamental to life as respiration would be surprising were it not for the fact that *S. cerevisiae* is one of those rare eukaryotic organisms that behave as facultative anaerobes. The ability of this fungus to ferment glucose is the reason that it occupies a central position not only in the baking industry but also in investigations of mitochondrial genetics. In the next several chapters we will see how experiments with *S. cerevisiae* have led to an understanding of the rules that govern the inheritance of mitochondrial genes and have provided a rational basis for deducing the functions of mitochondrial genes from study of mitochondrial gene mutations. In this chapter we will consider the life cycle of *S. cerevisiae*, the effects of growth conditions on mitochondrial phenotypes, and the genetics of petite mutations. The chapter ends with a comparison between *S. cerevisiae* and the so-called petite negative yeasts, in which vegetative petite mutations of the type

studied so extensively in *S. cerevisiae* have thus far not been detected.

LIFE CYCLE AND MENDELIAN GENETICS OF BAKERS' YEAST

The yeast *S. cerevisiae* is a haploid unicellular fungus that reproduces vegetatively

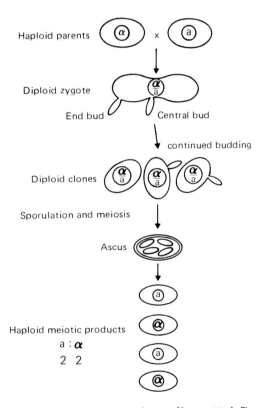

FIG. 4–1. Life cycle of bakers' yeast (*S. cerevisiae*). The mating-type alleles *a* and *α* segregate 2:2 in a normal mendelian fashion at meiosis, as do all other nuclear gene mutations. The diploid zygotes, by budding, form diploid vegetative progeny, which themselves continue to bud off diploid progeny, unless the cells are placed on a special sporulation medium that causes them to undergo meiosis.

by budding off daughter cells (Fig. 4–1). Its sexuality is controlled by a pair of mendelian alleles called *a* and *α*. There is no morphologic distinction between cells of opposite mating types. When haploid *a* and *α* cells are mixed, they fuse pairwise to form diploids. Both haploid and diploid conditions are stable; the diploid cells, like the haploid cells, reproduce vegetatively by budding. Diploid cells will frequently undergo meiosis if their growth medium becomes exhausted or if they are placed in a medium containing either raffinose or acetate as a carbon source (Fig. 4–1). Each diploid zygote undergoing meiosis produces a tetrad of four haploid ascospores. Segregation for any pair of alleles heterozygous in the zygote nucleus (e.g., mating type) is 2:2 (Fig. 4–1).

GROWTH CONDITIONS AND MITOCHONDRIAL DIFFERENTIATION

When *S. cerevisiae* is grown aerobically on nonfermentable carbon sources such as ethanol, glycerol, or lactate, the respiratory phenotype of the mitochondria is fully developed (Table 4–1). Such yeast cells contain mitochondria that are morphologically well developed (Fig. 4–2). All the respiratory cytochromes are present in maximal amounts, as are the attendant respiratory enzymes and the ATP-generating system (Table 4–1). These mitochondria contain mtDNA and carry out mitochondrial protein synthesis.

Exponential growth of *S. cerevisiae* on a readily fermentable carbon source such as

TABLE 4–1. *Mitochondrial phenotype in wild-type cells grown under different conditions and the vegetative petite mutation*

	Wild type				Vegetative petite glucose-repressed
	Aerobic		Anaerobic		
	Nonfermentable substrate	Glucose-repressed	Plus lipids	Minus lipids	
Mitochondrial morphology	Normal	Modified inner membrane	Modified inner membrane	Modified inner membrane	Modified inner membrane
mtDNA	$+^a$	+	+	R	Modified or absent
Mitochondrial protein synthesis	+		+	−	−
Cytochrome a + a$_3$	+	R	−	−	−
Cytochrome b	+	R	−	−	−
Cytochrome c	+	R	−	−	R
Cytochrome c$_1$	+	R	−	−	R
Oligomycin-sensitive ATPase F$_1$	+	R	R	very R	−
Succinate dehydrogenase	+	R	R	R	R
L-malate dehydrogenase	+	R	R	R	R
Succinate cytochrome c reductase	+	R	−		
Polyunsaturated fatty acids	+		R	very R	
Ergosterol	+		R	very R	

a Plus sign indicates the factor is present in normal amounts; R indicates the factor is present in reduced amounts; minus sign indicates the factor is absent or below the level of detection.

Data are derived from various sources (7–9,15–19,22,23,31,32,76–78).

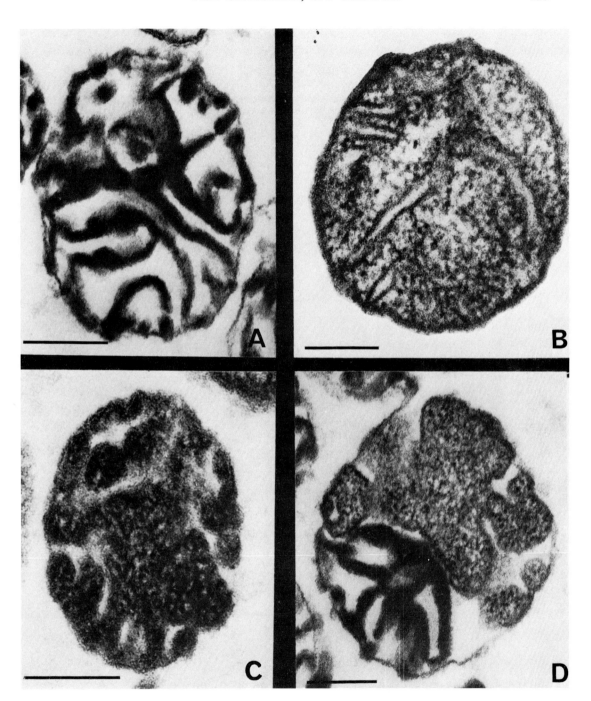

FIG. 4–2. Conformational forms of mitochondria isolated from aerobically grown yeast cells. Bars correspond to 0.25 μm. A: Condensed form. B: Orthodox form. C: Intermediate form. D: Mitochondrion in transitional state. (From Watson et al., ref. 23, with permission.)

glucose under aerobic conditions leads to repression of respiration (reviewed in 7–9,75); instead, the cells ferment glucose aerobically, with production of ethanol and CO_2 (Fig. 4–3). This phenomenon of *glucose repression* was described in detail some years ago by Ephrussi (11) and Slonimski (12,13). Slonimski found that at very low concentrations of glucose (below 6×10^{-3} M), the rate of adaptation to respiration increased with increasing glucose concentration; but at levels greater than 6×10^{-3} M, glucose repression began to set in.

The respiratory capacity of the mitochondria of glucose-repressed cells is greatly reduced as a result of tremendous depletion of respiratory chain components (Tables 4–1 and 4–2). The mitochondria themselves are different in morphology, larger in size, and fewer in number than those of respiring cells (13–15). Glucose-repressed mitochondria contain mtDNA (15,16), but mitochondrial protein synthesis is reported to be inhibited (8,17,18). As the growing culture exhausts the glucose in the medium, the cells enter a stationary phase during which the metabolic pattern changes over from fermentation to respiration (7,8,19). This phase, *glucose derepression,* is accompanied by a rapid increase in the components of the respiratory chain (Table 4–2 and Fig. 4–4). This increase is accompanied by an increase in mitochondrial number and mass, as was shown by the careful electron microscopic observations of Grimes and associates (15) on serially sectioned cells. At the same time, the mitochondria become smaller in size and less highly branched (15). The cells are then able to respire the ethanol they produced fermentatively and enter into the second phase of growth (Fig. 4–3).

Fermentable carbon sources other than glucose also cause repression, but the degree of repression depends on how well the cells ferment the carbon source. For example, galactose is slowly fermented, and cells growing exponentially on this sugar exhibit an intermediate level of repression (Table 4–2). In recent years glucose repression has been referred to as *catabolite repression* (7–8,16,19,75). Whether or not catabolite repression in *S. cerevisiae* is modulated as it is in bacteria (20) through control of the cellular level of cyclic 3′,5′ cyclic adenylic

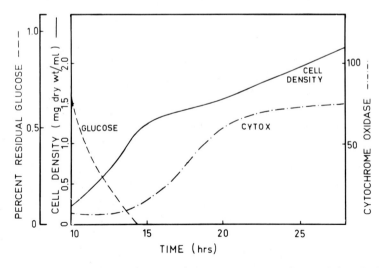

FIG. 4–3. Transition of cells of *S. cerevisiae* from state of glucose repression to aerobic metabolism of ethanol. Cells were grown on medium containing 1% glucose, which they fermented to produce ethanol. On exhaustion of the glucose, there is a lag in growth while the cells undergo respiratory adaptation. Once respiratory adaptation is complete, aerobic growth on ethanol begins. Cytochrome oxidase (CYTOX) is used as a marker for respiratory adaptation. (From Linnane and Haslam, ref. 7, with permission.)

TABLE 4-2. Effects of different levels of glucose repression on mitochondrial respiratory enzyme activities and cytochrome levels in S. cerevisiae

Carbon Source	Stage of growth	Specific activity[a]				Specific absorbance[b]			
		Cytochrome oxidase	NADH oxidase	NADH–cytochrome c reductase	Succinic dehydrogenase	$a + a_3$	b	c_1	c
Lactose (3%)	log	190	145	92	—	38.8	56	20.5	61.6
Galactose (1%)	log	103	143	48	45	—	—	—	—
Glucose (1%)	log	28.8	6.3	1.8	1.3	3.4	18.6	8.2	13.2
Glucose (1%)	stationary	118	114	54.9	46	38.2	83	21.9	77

[a] Nanomoles per minute per milligram.
[b] Absorbance per milligram $\times 10^4$.
Adapted from Perlman and Mahler (19).

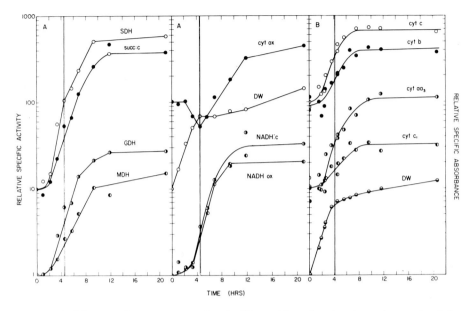

FIG. 4–4. Increases in different respiratory parameters in cells of S. cerevisiae undergoing glucose derepression during growth in 1% glucose. The double vertical lines at 4.5 hr indicate the beginning of stationary phase. Key: SDH, succinic dehydrogenase; succ:c, succinate:cytochrome c oxidoreductase; GDH, L-glutamate dehydrogenase; MDH, L-malate dehydrogenase; cyt ox, cytochrome c oxidase; DW, dry weight; NADH:c, NADH–cytochrome c reductase; NADH ox, NADH oxidase; cyt c, cytochrome c; cyt b, cytochrome b; cyt c_1, cytochrome c_1; cyt aa_3, cytochrome $a + a_3$. A: Enzyme levels. B: Cytochrome levels. (From Perlman and Mahler, ref. 19, with permission.)

acid (cAMP) remains to be determined (75). Although some experimental data suggest that cAMP is involved (21), other experiments have led to ambiguous results (19).

Anaerobic growth of S. cerevisiae on glucose (reviewed in 7–8,75) not only causes great reductions in respiratory cytochromes and other components of the respiratory chain (Table 4–1) but also prevents formation of two important lipid components of mitochondrial membranes, unsaturated fatty acids and ergosterol (22,23,75), the formation of which is oxygen-dependent (24–26). Healthy anaerobic cultures of cells can be obtained only if the growth media are sup-

plemented with ergosterol and a source of unsaturated fatty acids (e.g., oleic acid in the form of Tween 80) (22). Whether or not anaerobically grown cells contained mitochondria at all was a matter of controversy during the 1960s (27–30), but the problem was finally laid to rest by Schatz (22,31,32) and Linnane (23) and their associates. In a series of three definitive papers, Schatz and associates (22,31,32) characterized the mitochondria of anaerobically grown S. cerevisiae biochemically and morphologically. In an earlier paper (33) he had coined the term *promitochondria* for these structures, since they appeared to differentiate into mitochondria on transfer

→

FIG. 4–5. Effects of lipids on promitochondrial structure in anaerobically grown S. cerevisiae. A: Protoplast from cells grown anaerobically in presence of lipids (bar = 0.1 μm). B: Isolated promitochondria from cells grown anaerobically in presence of lipids (bar = 0.1 μm). C: Cells grown anaerobically in absence of lipids. Numerous promitochondria (*arrow*) are seen (bar = 1 μm). D: Protoplast from cell grown anerobically in absence of lipids. After conversion of the cells to protoplasts, the internal structure of the promitochondria can be seen more clearly (*insets*) (bar = 0.1 μm). (From Plattner et al., ref. 36, with permission.)

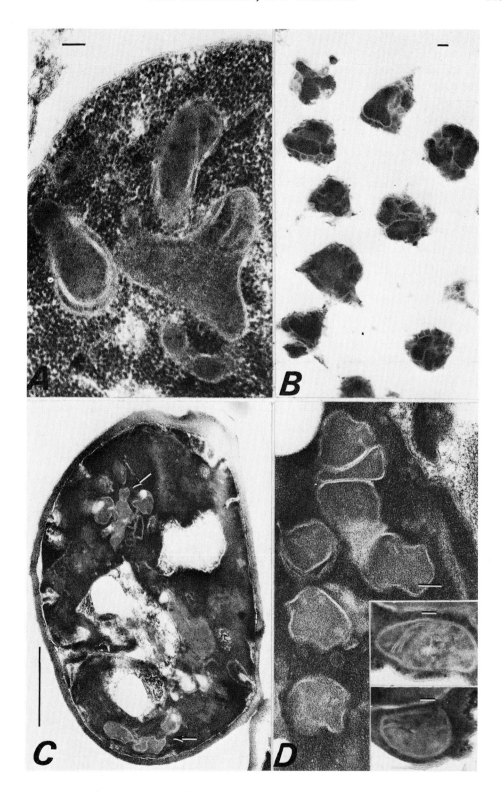

to aerobic conditions in the absence of glucose repression. The term was coined in analogy to the term *proplastid*. The transformation of proplastids into plastids is catalyzed by light (see Chapter 1).

Morphologically the promitochondria of lipid-supplemented cells resemble normal mitochondria; they are the same size, and they have cristae and double-layer envelopes (Fig. 4–5). They also have active protein-synthesizing systems and contain mtDNA (Table 4–1). The promitochondria of lipid-depleted cells are less well developed (23,34); they appear to be inactive in mitochondrial protein synthesis and deficient in mtDNA (23,35). As one might expect, the promitochondria of cells not supplemented with lipids are much poorer in unsaturated fatty acids than those of cells that have received such supplements (Table 4–3). The variation in lipid composition in the promitochondrial membranes appears to have been responsible for much of the earlier confusion concerning the existence of mitochondria in anaerobically grown yeast cells. In anaerobically grown cultures lacking lipid supplements, mitochondria were not

visualized by electron microscopy initially (27), although supplementation of the media with lipids led to visualization of mitochondria in the cells (29). Thus it appeared to be possible that cells grown anaerobically in the absence of lipids might lack mitochondria entirely. We now know that the apparent absence of mitochondria in these cells resulted from technical artifacts of the fixation procedure used in the attempt to visualize these lipid-depleted promitochondrial structures by electron microscopy (22,31,32).

Direct evidence for conversion of promitochondria into mitochondria during respiratory adaptation comes from the elegant experiments of Plattner et al. (36). These workers labeled the promitochondrial protein of anaerobically growing yeast cells with radioactive leucine in the presence of cycloheximide. This antibiotic prevents cytoplasmic protein synthesis in yeast, but it does not affect mitochondrial protein synthesis (see Chapter 3). The cells were then transferred to aerobic growth conditions in the presence of unlabeled leucine, and cycloheximide was omitted from the medium. If

TABLE 4–3. *Fatty acid composition of lipids from promitochondria and mitochondria of S. cerevisiae*

Lipid fraction	Particle preparation	Weight percent of total fatty acids							
		C_{10}	C_{12}	C_{14}	C_{16}	$C_{16:1}$	C_{18}	$C_{18:1}$	$C_{20:1}$
Phospholipids	Mitochondria (from cells grown in presence of lipids)	trace	trace	0.6	17.9	43.7	3.6	34.2	trace
	Promitochondria (from cells grown in presence of lipids)	trace	trace	4.5	20.5	6.5	3.9	61.5	3.1
	Promitochondria (from cells grown in absence of lipids)	14.3	8.9	10.4	33.7	12.0	13.7	7.0	trace
Neutral lipids	Mitochondria (from cells grown in presence of lipids)	trace	trace	1.2	16.1	43.1	3.6	36.0	trace
	Promitochondria (from cells grown in presence of lipids)	trace	trace	4.9	18.0	2.1	10.3	63.0	1.7
	Promitochondria (from cells grown in absence of lipids)	8.5	10.5	14.4	34.0	7.7	17.5	6.4	1.0

Adapted from Paltauf and Schatz (22).

promitochondria differentiate into respiring mitochondria, the labeled amino acid should be found in mitochondrial protein; but if mitochondria are synthesized *de novo* from structures other than promitochondria, the mitochondrial membranes should not contain labeled amino acid (Fig. 4–6). The authors found by biochemical tests that 50% of the label introduced into the promitochondria was present in mitochondrial membranes; they concluded that promitochondria differentiate into mitochondria but that this differentiation process is accom-

panied by some turnover of mitochondrial proteins. Plattner and associates then demonstrated the presence of label in membranes of respiring mitochondria directly by means of electron microscope autoradiography (Fig. 4–7). Thus far, similar evidence for transformation of the large branched mitochondria of glucose-repressed cells into the more numerous and smaller mitochondria of derepressed cells has not been published, but it is assumed that a similar differentiation process occurs (19).

The rapid differentiation of fully respira-

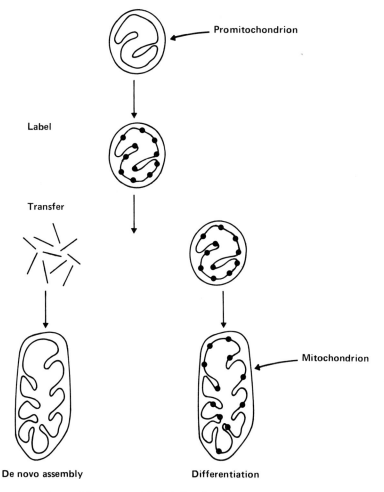

Label

Transfer

Promitochondrion

Mitochondrion

De novo assembly

Differentiation

FIG. 4–6. Diagram of experiment of Plattner et al. (36) designed to determine if promitochondria differentiate into mitochondria or if mitochondria are assembled *de novo*. Promitochondria in anaerobically growing yeast cells were labeled with radioactive leucine in the presence of cycloheximide, following which the cells were transferred to medium containing unlabeled leucine, but lacking cycloheximide, and grown aerobically to induce formation of mitochondria. Labeling is indicated by black circles.

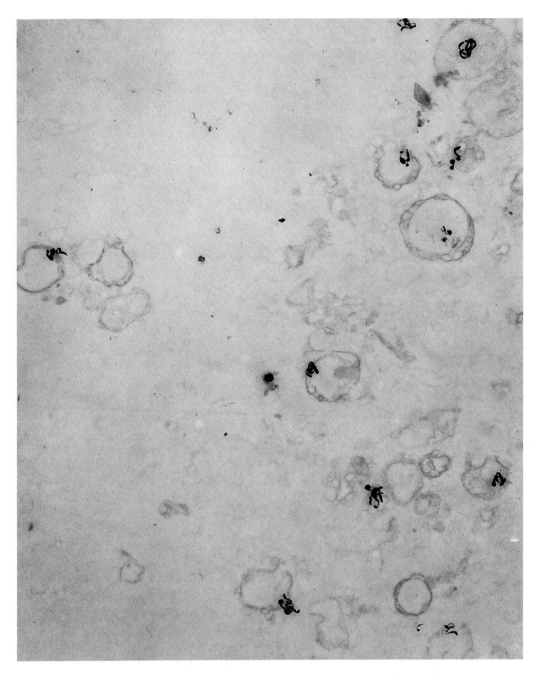

FIGURE 4–7. Autoradiogram of isolated adapted mitochondria that has been stained for cytochrome oxidase. If radiation spread is taken into consideration, the selective association of silver grains with stained vesicles is readily apparent. ×16,000. (From Plattner et al., ref. 36, with permission.)

tion-competent mitochondria that accompanies derepression of glucose-repressed cells and transfer of anaerobically grown cells to aerobic conditions makes these two physiologic transfers ideal for studying the biogenesis of mitochondrial inner-membrane components associated with respiration (see Chapter 19).

PETITE MUTATIONS

The vast majority of spontaneous and induced petite mutations exhibit nonmendelian inheritance and are termed *vegetative petites* (1,2). A minority are inherited in a mendelian fashion and are called *segregational petites* (1,37). Many such mendelian gene mutations have been isolated over the years (see Chapter 19). Vegetative petites are nonreverting pleiotropic mutations that either have grossly altered mtDNA or have no mtDNA at all (see p. 141 and Chapter 6). As a consequence, vegetative petites cannot carry out mitochondrial protein synthesis (Table 4–1) because components of the mitochondrial protein-synthesizing system, notably mitochondrial rRNA and tRNA (see Chapter 3), are coded by mtDNA. This absence of a functional mitochondrial protein-synthesizing system results in pleiotropic defects in the respiratory electron transport chain, since certain components (including cytochromes $a + a_3$, b, and the attendant ATP-generating system) require mitochondrial protein synthesis for their assembly and function (Table 4–1) (see Chapter 19). As might be expected, mitochondrial morphology in vegetative petites is modified, particularly with respect to the inner membrane. One respiratory chain component that is found in vegetative petite mitochondria is cytochrome c. The presence of cytochrome c, the major form of which (iso-*1*-cytochrome c) is known to be coded by a nuclear gene (38) and whose synthesis takes place in the cytoplasm (39–42), illustrates how vegetative petite mutants can be used to determine which

mitochondrial functions are not dependent on mitochondrial genes or protein synthesis. These include the proteins of the outer mitochondrial membrane, the bulk of (possibly all of) the enzymes of the matrix, factor F_1, and many of the polypeptides of the inner membrane itself (see Chapter 19).

Since, by definition, petites must be grown on a fermentable carbon source, the question arises how much of the apparent pleiotropic nature of the mutation results from the mutation itself and how much from glucose repression. Glucose repression–derepression experiments (19) have shown that although those functions repressed in wild type are also repressed in petites, only some of them recover following respiratory adaptation. Among the cytochromes, for example, only cytochrome c synthesis resumes. Thus the phenotypic effects of the vegetative petite mutation on the respiratory electron transport chain (Table 4–1) are those resulting from the mutation itself rather than from glucose repression. During glucose repression or anaerobiosis, wild-type cells become phenocopies of petite mutations in many respects (Table 4–1). It is interesting that this is true not only at the level of fine biochemical similarities and differences but also at the much grosser level of colony size. On 2% glucose medium, petite and wild-type cells form colonies of similar sizes, because both wild-type and petite cells are glucose-repressed and derive their energy from fermentation. On a "differential" medium containing 0.1% glucose and 2% glycerol, wild-type cells form large or *grande* colonies, whereas the colonies formed by petite mutants are small. Under the latter conditions the petite mutant cells are restricted to glucose as the carbon source, since they cannot respire the glycerol.

Segregational petites fall into two general phenotypic classes, as we shall see later in this chapter and in Chapter 19. Some are pleiotropic; either they act as "mutators" to produce vegetative petites, so that the phenotype expressed is that of the double mutant, or they are mutants that themselves cause pleiotropic effects on the respiratory chain. Other segregational petites are not pleiotropic; they affect single components of the respiratory chain.

GENETICS OF VEGETATIVE PETITES

Vegetative petites fall into two subclasses (Table 4–4) called *neutral* and *suppressive* (43). The neutral petite phenotype is not transmitted in crosses to wild type, either among the vegetative diploid progeny of the resulting zygotes or among the haploid meiotic progeny of diploids that have been induced to undergo meiosis (Fig. 4–8). Backcrossing the resulting wild-type progeny to the neutral petite parent does not increase the frequency of transmission of the neutral petite phenotype. In fact, Ephrussi and his colleagues in their initial studies with neutral petites calculated that more than 20 independently segregating nuclear genes would be required to give the rare petite segregations observed (6).

Suppressive petites were first described in 1955 by Ephrussi et al. (43), several years after characterization of the neutral petites. These petites, unlike neutral petites, transmit the petite phenotype to a fraction of the vegetative diploid progeny in crosses (Fig. 4–8). The level of suppressiveness is char-

acteristic of a given petite strain, as has been shown by cloning experiments (44), but it is frequently altered as the result of further mitochondrial mutations (see Chapter 6). Suppressiveness varies from 1% or less petite diploid progeny in some strains to over 99% petite progeny in other strains. Aerobic metabolism seems to be required for meiosis and sporulation in *S. cerevisiae;* thus suppressive petite diploids cannot be induced to undergo meiosis. However, this is not true of the zygotes formed by a cross between a suppressive petite and wild type. Ephrussi et al. (43) found that in a cross between a highly suppressive petite and wild type, the zygotes almost always segregated 0:4 wild type: petite ascospores (Fig. 4–8).

Although the aforementioned characteristics of vegetative petites are consistent with cytoplasmic localization of these mutations, they do not constitute direct proof. Such proof has been provided by Wright and Lederberg (45) in an experiment employing yeast heterokaryons. Heterokaryosis is a common phenomenon in filamentous fungi but rare in unicellular species; it results from fusion of mycelia carrying genetically different nuclei (see Chapters 8 and 9). The resulting mycelium contains both nuclear types and is thus heterokaryotic. Wright and Lederberg used a variety of *S. cerevisiae* called *ellipsoideus* that forms transient heterokaryons. They formed heterokaryons between haploid strains that each carried dif-

TABLE 4–4. Nomenclature and inheritance of the different petite mutants

	Genotype			
	Nuclear	Mitochondrial	mtDNA	Inheritance
Wild type	*PET*	ρ^+	Normal	Not applicable
Vegetative petite				
Neutral	*PET*	ρ^0	Absent	Nonmendelian
Neutral or suppressive	*PET*	ρ^-	Altered	Nonmendelian
Segregational petite	*pet*	ρ^+	Normal	Mendelian
Double mutant	*pet*	ρ^0 or ρ^-	Absent or altered	See text

Nomenclature adopted largely from the *Yeast Genetics* supplement of *Microbial Genetics Bulletin No. 31,* 1969. Oak Ridge National Laboratory, Tenn.

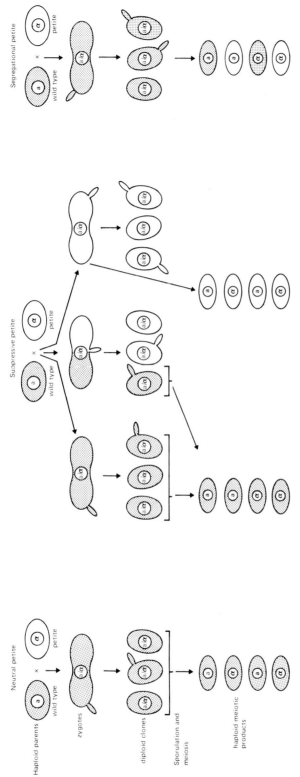

FIG. 4–8. Comparison of inheritance patterns of neutral, suppressive, and segregational petites in crosses to wild type. Neutral petites produce only wild-type diploid progeny in such crosses that segregate 4:0 wild type: petite progeny in crosses to wild type. Zygotes issuing from cross between suppressive petite and wild type may yield wild-type progeny, petite progeny, or a mixture, with the ratio depending on level of suppressivity of the mutant. Petite vegetative diploids will not sporulate, but zygotes destined to yield petite progeny will do so, and they segregate 0:4 wild type: petite progeny. Phenotypically wild-type progeny result from crosses between recessive segregational petites and wild type. On sporulation of these diploids, the petite phenotype is found to segregate 2:2 wild type: petite progeny in a typical mendelian fashion.

ferent nuclear gene mutations (Fig. 4–9). One stock also carried a suppressive petite mutation, whereas the second had a wild-type respiratory phenotype. Buds were separated from the heterokaryons and allowed to form colonies. Haploid buds could be distinguished from each other, as well as from the diploid buds formed by nuclear fusion, on the basis of the nuclear markers they carried (Fig. 4–9). The authors found that some haploid buds of each of the two nuclear types introduced into the cross were petite, whereas others had a wild-type respiratory phenotype (Fig. 4–9). These results could be explained only by cytoplasmic transmission of the petite mutation.

Originally, Ephrussi et al. (3) hypothesized that the specific induction of vegetative petite mutations by acriflavine and the inheritance of the neutral petite mutation could be accounted for in terms of irreversible loss or inactivation of a particulate cytoplasmic autoreproducing factor required for synthesis of respiratory enzymes. The later discovery of suppressive petites (43) necessitated modification of their hypothesis. Suppressiveness was initially explained in terms of a "suppressive factor" that prevented perpetuation of the "normal factor"

by destroying it, interfering with its replication, or preventing its distribution during cell division. Later, Ephrussi et al. (46) suggested that the suppressive factor interfered with replication of the normal factor as a result of "mutual repression." The mechanism underlying suppressiveness has been the subject of much speculation ever since; we will return to the problem in Chapter 6 after developing the necessary background. For the moment, suffice it to say that the property of suppressiveness can be equated with the grossly altered mtDNA found in suppressive petites.

The term rho (ρ) came into general use to designate the cytoplasmic factor required for respiration, in a paper by Sherman and Ephrussi (47), although it had been used earlier in a little-known paper by Marquardt (10). Respiratory-competent grande cells were designated ρ^+ and vegetative petites ρ^-, with the understanding that ρ^- could mean either actual physical absence or gross modification of the factor that made reversion impossible. It is now evident that ρ and mtDNA are equivalent (see p. 141 and Chapter 6) and that some vegetative petites lack mtDNA and others contain grossly altered mtDNA. The former petites, which

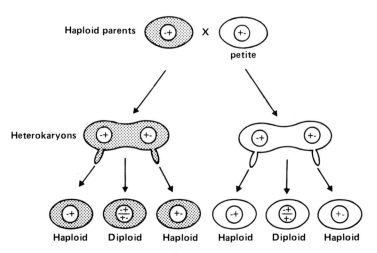

FIG. 4–9. Diagram of heterokaryon test designed by Wright and Lederberg (45) to determine if the suppressive petite mutation is transmitted through the cytoplasm. Each parent carried complementary nuclear gene mutations, designated as +— and —+ to make possible the identification of the nuclei carried by the haploid progeny.

are always neutral, were designated ρ^0 by Linnane and associates (48). All other petites contain mtDNA, and are designated ρ^-. In genetic tests some of these petites approach neutrality, whereas others are highly suppressive (e.g., 49). The convention we will use in this book is to designate all vegetative petites ρ^-, unless an absence of mtDNA has been demonstrated, in which case they will be called ρ^0. This terminology, together with the nomenclature used for segregational petites and double mutants of segregational petites and vegetative petites, is summarized in Table 4–4.

GENETICS OF SEGREGATIONAL PETITES

Segregational petites or *pet* mutants segregate 2:2 in crosses to the wild type (Fig. 4–8), and they "complement" the vegetative petite (PET) mutation; that is, at least some of the diploids produced in a cross of a segregational petite (*pet*ρ^+) and a vegetative petite (*PET*ρ^-) will be of the wild type, with the fraction depending on the suppressivity of the vegetative petite used in the cross (Fig. 4–10). The vegetative petite supplies a wild-type *PET* allele of the nuclear gene mutation, and the segregational petite supplies a wild-type ρ^+ factor. Sporulation of these diploids yields a 2:2 segregation for the nuclear *pet* mutant.

Unfortunately, complementation and allele testing between different *pet* mutants, as well as physiological and biochemical experimentation with individual mutants, are often complicated by a high frequency of *pet*ρ^- double mutants (41). Since segregational petites must be cultured under the

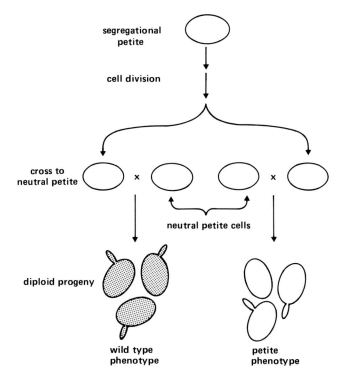

FIG. 4–10. Retention of ρ^+ factor by a segregational petite. Following cell division, the segregational petite is crossed to a neutral petite. A segregational petite cell that has retained the ρ^+ factor will yield phenotypically wild-type diploid progeny in such a cross, because the neutral petite supplies the wild-type nuclear allele of the mutant carried by the segregational petite, and the segregational petite supplies the ρ^+ factor (complementation). If a segregational petite cell happens to be ρ^- or ρ^0, complementation cannot occur, since neither parent supplies the ρ^+ factor.

same conditions that permit growth of vegetative petites, there is no selection against double mutants. Thus it is usually necessary to establish the frequency of double mutants in experiments with *pet* mutants. Sherman and Ephrussi (47) simply took advantage of the complementation test to do this, and their method is a standard procedure. A clone of the *pet* mutant is crossed to a neutral petite, and the frequencies of wild-type and petite diploids are calculated. Those cells in the *pet* stock with the genotype $pet\rho^+$ will complement the neutral petite; those with the genotype $pet\rho^-$ cannot, and they will form petite diploids (Fig. 4–10). We will refer to this test as a *retention test* to distinguish it from the complementation test. The complementation test establishes that the *pet* mutation does not cause a genetic alteration in the ρ factor, but the retention test is used to determine whether a given stock (in this case a *pet* mutant) has retained a functional ρ^+ factor. The retention test can also be used to establish whether a ρ^- stock has retained specific mitochondrial alleles in a functional state, as will be discussed in Chapter 5.

The distinction between the complementation test and the retention test breaks down in an operational sense in the case of those *pet* mutants that cause mutation from ρ^+ to ρ^-. Such mutants will complement rarely, if ever, because they cannot retain a ρ^+ factor. The problem is exemplified by some of the data in Table 4–5, where it is evident that some *pet* mutants retain the ρ^+ factor far more efficiently than others. These data also show that the vegetative petites accumulated by *pet* mutants are usually suppressive and that different double mutants isolated from the same *pet* mutant can have different levels of suppressiveness. Thus it appears that each *pet* mutant clone contains a collection of independently arising vegetative petites.

TABLE 4–5. *Retention of ρ^+ by different pet mutants and suppressiveness of ρ^- in different pet ρ^- clones.*

Mutant	Percentage ρ^+ cells in most stable segregant	Percentage suppressiveness of derived double mutants
pet₁	45.0	34.8
		48.6
		41.4
		47.5
		32.6
		47.0
		23.4
		18.7
		31.4
pet₂	0.1	—
pet₃	0.0	3.4
		5.7
		0.6
		18.4
		0.2
		23.2
		9.8
		83.2
		4.4
		61.6
pet₄	60.0	—
pet₅	99.0	15.1
		1.7
		0.3
		2.0
pet₆	99.0	66.6
		29.4
		22.0
pet₇	95.0	82.2
		70.8
		59.5

Percentage suppressiveness = $[(X - Y)/(1 - Y)] \times 100$ for a $pet\rho^+$ double mutant, where X is the fraction of ρ^- colonies produced by zygotes with a *PET* tester strain and Y is the fraction of ρ^- cells in the normal, haploid tester strain. It is not known whether double mutants were derived from the same segregants.

Data from Sherman and Ephrussi (47) and Sherman (49).

INDUCTION OF VEGETATIVE PETITES WITH SPECIFIC MUTAGENS AND THE NUMBER-OF-TARGETS PARADOX

In their original experiments Ephrussi and his colleagues demonstrated that the acridine dye acriflavine caused quantitative conversion of wild-type cells to vegetative petites (1,6). Later, Marcovich (5) compared the effects of proflavine and euflavine, the acridines present in a 1:2 ratio in acriflavine. He found that whereas proflavine would not induce petites at the highest concentration tested, euflavine was extremely mutagenic and would transform the entire population to petites at nonlethal concentrations. Ephrussi and Hottinguer (50,51) then showed by pedigree analysis that euflavine at concentrations as low as 10^{-6} M produced petite daughter cells with almost 100% efficiency, but the parent cells remained wild type in phenotype and could produce wild-type buds when transferred to a medium lacking the dye.

Many other treatments also induce vegetative petites (for review see 52–54). Exposure of wild-type cells to conventional mutagens used to induce nuclear gene mutations (e.g., ultraviolet light, N-methyl-N'-nitro-N-nitrosoguanidine) is effective, as are specific mutagens (e.g., acridine and phenanthridine dyes and the trypanocide berenil). Currently the most important and most widely used mutagen is the phenanthridine dye ethidium bromide (Fig. 4–11). This dye, whose mechanism of action is detailed in Chapter 6, interacts specifically with mtDNA to block its replication. Blockage of mtDNA replication leads to production of vegetative petites lacking mtDNA (ρ^0). In addition, ethidium bromide treatment also leads to degradation of parental mtDNA molecules, which explains why the dye, unlike euflavine, is mutagenic to mother cells as well as daughter cells. Although the process of parental mtDNA degradation ultimately leads to production of ρ^0 petites, removal of the mutagen after a shorter incubation yields ρ^- petites in which the remaining mtDNA fragments have been replicated. These properties have made ethidium bromide an invaluable tool for probing the mitochondrial genome (see Chapters 5–7).

The kinetics of vegetative petite induction with a variety of mutagens have been used, with the aid of target analysis, to infer the number of ρ^+ factors per cell. Target theory assumes that individual molecules of a

chemical used to produce a specific biologic effect (e.g., the induction of a vegetative petite) collide with a specific target (e.g., a mtDNA molecule) to produce a "hit." A hit on the target is responsible for the biologic effect observed. In the case of radiation, the assumption is that the radiation is delivered in discrete packets (quanta).

The mean number of hits (h) sustained by the target population is directly proportional to the dose of the agent in seconds or minutes (t) multiplied by a constant (k) that depends on the sensitivity of the target molecule to impinging quanta of radiation or molecules of chemical agent. Since hits are assumed to occur at random, they should be distributed among the population of target molecules in a random (Poisson) fashion. The Poisson distribution can be written as follows:

$$P_x = \frac{h^x}{x!} e^{-h} \tag{1}$$

where P_x equals the frequency with which x hits occur, e is the base of natural logarithms (2.718), and $x!$ is factorial x ($3! = 1 \times 2 \times 3 = 6$; $0!$ and $1! = 1$).

The fraction of unhit individuals (S) will be equal to the P_0 term of the distribution; thus

$$S = P_0 = e^{-h} \tag{2}$$

but since $h = kt$, we can rewrite the equation as

$$S = e^{-kt} \tag{3}$$

It is convenient to plot $\log_{10} S$ against time in which case equation 3 becomes

$$\log S = -0.4343kt \tag{4}$$

and a plot of $\log_{10} S$ versus t yields a straight line (one-hit curve) with a slope equal to $-0.4343k$.

When two or more independent units with identical properties must be hit for the biologic effect to be seen, the probability that the effect will be seen depends on the probability that individual units are inactivated simultaneously; thus equation 3 is rewritten as

$$S = 1 - (1 - e^{-kt})^n \qquad (5)$$

where n is the target number. A plot of $\log_{10} S$ versus t in this case yields a straight line with a shoulder (Fig. 4–11). The slope of the straight line is equal to $-0.4343k$ and n is determined by extrapolating the straight-line portion of the curve to its intercept with the ordinate (Fig. 4–11).

In applying target analysis to vegetative petite induction, S is equated with the fraction of wild-type colonies remaining [i.e., $S = \rho^+/(\rho^+ + \rho^-)$], and n is the number of ρ^+ factors. The number-of-targets paradox, as Deutsch et al. (55) called it, arises because the number of ρ^+ factors determined from target analysis ranges between 1 and 20, with 2 to 6 being the most common numbers (55–61), whereas the number of mtDNA molecules per cell is around 50 (15,16). Since there is every reason to believe that ρ and mtDNA are the same (see

p. 141 and see Chapter 6), the paradox is that the target numbers tend to be low.

Several explanations can be offered for the discrepancy. First, for technical reasons the estimates of numbers of ρ^+ factors and mtDNA molecules per cell have inherent inaccuracies that bias them in opposite directions. If these technical problems could be resolved, these estimates would give similar numbers of ρ particles per cell. This explanation seems unlikely, because current estimates of mtDNA molecules per cell seem reasonably accurate (15,16), and most target measurements (no matter what method or mutagen used) give low target numbers.

Second, the shoulders seen on the vegetative petite induction curves reflect the action of repair processes. While repair processes may influence the shapes of these curves, it seems unlikely that they can explain them in full, as is shown by the experiments of Maroudas and Wilkie (57) on the kinetics of induction of vegetative petites with ultraviolet light. Furthermore, it is difficult to see why repair processes for the variety of different mutagens and treatments used would be similar in usually yielding low apparent target numbers.

Third, the number of genetically competent replicating and/or segregating mtDNA molecules is smaller than the actual number of mtDNA molecules (57,62). This hypothesis in its most extreme form is embodied in the notion of a single master template of mtDNA located outside the mitochondrion. This hypothesis was put forward some years ago by Wilkie (58) on the basis of ultraviolet target analysis suggesting that anaerobically grown cells contained one target and aerobic cells contained several. Wilkie suggested that in aerobically grown cells the mitochondria and their mtDNA would replicate normally, but in anaerobic cells mitochondrial formation and replication of mtDNA would be repressed. Under these conditions the master template system would act autocatalytically to ensure genetic continuity during

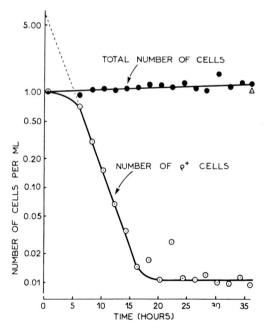

FIG. 4–11. Ethidium bromide transformation of wild-type cells of *S. cerevisiae* (ρ^+) to vegetative petites ($\rho^0 + \rho^-$) under nongrowing conditions. (From Slonimski et al., ref. 60, with permission.)

mitochondrial repression. When the hypothesis was first formulated, it had considerable appeal, since it was thought that anaerobically grown cells lacked mitochondria (see p. 128). Since then, as was discussed previously, both promitochondria and mtDNA have been demonstrated in anaerobically grown cells. Allen and MacQuillan (61), who repeated the ultraviolet experiments and always found the target numbers for anaerobic cells to be three or more, questioned Wilkie's results. The hypothesis was resurrected by Whittaker et al. (62) on different experimental grounds to account for their observation that under certain conditions ethidium bromide treatment leads to a reversible reduction in the amount of mtDNA without causing petite induction. They believed that under these conditions at least a single molecule of mtDNA would be retained. This master copy would be resistant to agents that degrade mtDNA or inhibit its replication under the experimental conditions used and would regenerate mtDNA on removal of the agent. Evidently such a master copy is not present under most experimental conditions, since ethidium bromide usually causes complete conversion of the treated cell population to the vegetative petite phenotype and the mtDNA is either grossly modified or absent entirely (see Chapter 6).

Fourth, it has been hypothesized that a hit on one target leads to inactivation of other targets. Two mechanisms have been proposed. Williamson (16) theorized that an agent such as ultraviolet light would cause a temporary disturbance in the system that regulates synthesis of mtDNA. He postulated that a nuclear operon would be responsible for synthesis of the initiator or initiators of mtDNA synthesis and that the activity of this operon would be controlled by a repressor that would be a gene product of mtDNA. The petite mutation would result in production of a repressor by one of the mitochondrial genomes that would bind irreversibly to the nuclear operator. This "dominant" effect would result in blockage of all mtDNA synthesis in the affected cell, and a genetic copy number lower than the physical copy number would be obtained. This hypothesis seems unlikely, since it does not explain why the target number, although low, is usually greater than one; nor does it explain, without some rather bizarre and strained assumptions, why most vegetative petites contain altered mtDNA rather than no mtDNA. A more plausible mechanism was proposed by Deutsch et al. (55). They suggested that the errors introduced into a fraction of the mtDNA molecules by mutation were spread by recombination to the rest. Hence, the average target number would reflect the average number of mtDNA molecules that would have to be mutated to spread their damages by recombination to the remaining mtDNA molecules.

THE ρ FACTOR IS mtDNA

Despite the discrepancy between numbers of ρ^+ factors and mtDNA molecules, all models equate the ρ factor with mtDNA for excellent reasons. First, neutral petites lacking mtDNA result following prolonged exposure of cells to ethidium bromide. Second, ρ^- petites, whether spontaneous or induced, possess mtDNA distinct in base composition and often distinct in size from wild-type mtDNA, although these petites contain normal amounts of mtDNA. Third, vegetative petites do not revert, which is consistent with the observation that these mutants either lack mtDNA entirely or have mtDNA molecules containing large deletions. This evidence will be discussed in much greater detail in the three chapters that follow.

PETITE NEGATIVE YEASTS

Yeast species in which petite mutations are difficult, if not impossible, to isolate, even following treatment of cells with acridine or phenanthridine dyes, were defined originally by Bulder (63,64) as petite negative yeasts. These yeasts comprise a diverse group whose members were thought to be obligate aerobes not subject to glucose repression (63–67). This is a clear oversimplification, at least in the case of *Schizosaccharomyces pombe,* as experiments to be detailed will show.

Bulder (63,64) found that certain petite negative yeasts produced microcolonies on acriflavine and that these microcolonies lost viability on subculturing. He concluded that these might be lethal petite mutations. Microcolony formation was also noted by Luha and associates (67) in *Kluyveromyces fragilis* during the early phases of growth of this petite negative yeast in ethidium bromide; but this change was temporary, and normal-size colonies later replaced the microcolonies. Whatever the mechanism of adaptation, it did not involve selection of mutants resistant to ethidium bromide. Similar results were reported by Heslot et al. (68) for *S. pombe.*

The petite negative and petite positive yeasts are contrasted in Table 4–6. It is evident that although the two groups generally differ with respect to glucose repression and petite induction, in each group the synthesis of respiratory enzymes is inhibited by euflavine. Measurements of cytochrome levels in the petite negative yeasts (69–71) revealed that, as in *S. cerevisiae,* inhibitors

such as chloramphenicol, ethidium bromide, and acriflavine specifically affect the synthesis of cytochromes $a + a_3$, and b. This is probably the result of blockage of mitochondrial protein synthesis by these drugs in the petite negative yeasts, as it is in *S. cerevisiae.* Long-term growth of petite negative yeasts in acridine or phenanthridine dyes leads to enormous reductions in the levels of the affected cytochromes, and respiration can become almost wholly cyanide-insensitive (69). Nevertheless, the cells continue to grow at reasonable rates as long as they are supplied with a fermentable carbon source (65–68,70). Hence, Bulder's suggestion (64) that petite negative yeasts cannot form viable petite mutants because such species are unable to obtain sufficient energy for fermentative growth seems highly unlikely. The physiologic studies of Goffeau and his colleagues (68,72) with *S. pombe* serve further to blur the distinction between petite negative and petite positive yeasts, since *S. pombe* is subject to glucose repression and derepression, like the petite positive yeasts. These workers also demonstrated that limited growth of *S. pombe* can occur under anaerobic conditions in the presence of a lipid supplement.

Growth of petite negative yeasts in the presence of ethidium bromide or acriflavine may or may not affect mtDNA. Luha et al.

TABLE 4–6. *Petite positive and negative yeast species*

Petite induction	Glucose repression	Inhibition of synthesis of respiratory enzymes by acridines or ethidium	Organisms
Petite positive	yes	yes	*Saccharomyces* spp. *Dabaryomyces globosa* *Torulopsis* spp. *Nematospora coryli* *Brettanomyces lambicus*
Petite negative	no	yes	*Kluyveromyces* spp. *Hansenula* spp. *Schwanniomyces occidentalis* *Candida* spp. *Trichosporon fermentans* *Torulopsis utilis*

Data from various sources (63–70).

(67,70) reported that mtDNA synthesis in *K. fragilis* and *K. lactis* is repressed by both acriflavine and ethidium bromide. In the case of *K. fragilis,* the rate of mtDNA synthesis dropped to less than 1% of the uninhibited rate within 1 hr after ethidium bromide addition. After 11 generations of growth, mtDNA was undetectable; but following 49 generations of growth in ethidium bromide, the cells had recovered their ability to synthesize mtDNA to the extent of 30% of the uninhibited control. Luha et al. (67) suggested that the reestablishment of some mtDNA synthesis might be correlated with the loss of microcolony formation. They also found that removal of ethidium bromide from the medium was followed by resumption of synthesis of mtDNA and respiratory enzymes. In contrast, Bandlow and Kaudewitz (73), working with *S. pombe,* reported that cells labeled with adenine following eight to ten generations of growth in ethidium bromide contained normal amounts of mtDNA whose buoyant density was similar to that found in the wild type.

The return of petite negative yeasts to the wild-type phenotype following growth of cells in acridine dyes or ethidium bromide is the main feature distinguishing them·from petite positive yeasts. Although petite negative yeasts exhibit the same changes in mitochondrial phenotype in the presence of these dyes as are seen with petite induction, these changes are reversible. In short, it seems likely that the acridines and ethidium bromide probably interfere with mitochondrial transcription and possibly translation in both petite negative yeast and petite positive yeast, but only in the latter class are major irreversible changes in mtDNA induced. This is not to say that respiratory-deficient mutants cannot be induced in petite negative yeasts. Goffeau and associates (72) isolated and characterized nuclear respiratory-deficient mutants in *S. pombe.* Wolf et al. (74) also isolated what appear to be mitochondrial respiratory-deficient point mutations in this yeast. Thus, as these authors pointed out, *S. pombe,* at least, is not petite negative either in the sense of mutants equivalent to segregational petites in *S. cerevisiae* or in the sense of mitochondrial point mutations to respiratory deficiency (see Chapters 5, 7, and 19). *S. pombe* is petite negative only in the sense that mutations equivalent to the vegetative petites of *S. cerevisiae* have not been found.

REFERENCES

1. Ephrussi, B. (1953): *Nucleo-cytoplasmic Relations in Micro-organisms.* Clarendon Press, Oxford.
2. Ephrussi, B., Hottinguer, H., and Chimenes, A. M. (1949): Action de l'acriflavine sur les levures I. La mutation "petite colonie." *Ann. Inst. Pasteur (Paris),* 76:351–364.
3. Ephrussi, B., l'Héritier, P., and Hottinguer, H. (1949): Action de l'acriflavine sur les levures. VI. Analyse quantitative de la transformation des populations. *Ann. Inst. Pasteur (Paris),* 77:64–83.
4. Ephrussi, B., and Hottinguer, H. (1950): Direct demonstration of the mutagenic action of euflavine on baker's yeast. *Nature,* 166:56.
5. Marcovich, H. (1951): Action de l'acriflavine sur les levures. VIII. Détermination du composant actif et étude de l'euflavine. *Ann. Inst. Pasteur (Paris),* 81:452–468.
6. Ephrussi, B., Hottinguer, H., and Tavlitzki, J. (1949): Action de l'acriflavine sur les levures. II. Étude génétique du mutant "petite colonie." *Ann. Inst. Pasteur (Paris),* 76:419–450.
7. Linnane, A. W., and Haslam, J. M. (1970): The biogenesis of yeast mitochondria. *Cur. Top. Cell. Regul.,* 2:101–172.
8. Linnane, A. W., Haslam, J. M., Lukins, H. B., and Nagley, P. (1972): The biogenesis of mitochondria in microorganisms. *Annu. Rev. Microbiol.,* 26:163–198.
9. Mahler, H. R., Bastos, R. N., Flury, U., Lin, C. C., and Phan, S. H. (1975): Mitochondrial biogenesis in fungi. In: *Genetics and Biogenesis of Mitochondria and Chloroplasts,* edited by C. W. Birky, Jr., P. S. Perlman, and T. J. Byers, Chapter 2. Ohio State University Press, Columbus.
10. Marquardt, H. (1952): Neue Ergebnisse aus dem Gebiet der Plasmavererbung. *Umschaw. Wiss. Tech.,* 52:545–549.
11. Ephrussi, B., Slonimski, P. P., Yotsuyanagi, Y., and Tavlitzki, J. (1956): Variations physiologiques et cytologiques de la levure au cours du cycle de la croissance aérobie. *Compt. Rend. Trav. Lab. Carlsberg Ser. Physiol.,* 26:87–99.

12. Slonimski, P. P. (1953): *Formation des Enzymes Respiratoires chez la Levure.* Masson, Paris.

13. Slonimski, P. P. (1956): Adaptation respiratoire développement du système hémoprotéique induit par l'oxygène. In: *Proceedings Third International Congress Biochemistry,* p. 242.

14. Yotsuyanagi, Y. (1962): Ètudes sur le chondriome de la levure. I. Variation de l'ultrastructure du chondriome au cours du cycle de la croissance aérobie. *J. Ultrastruct. Res.,* 7: 121–140.

15. Grimes, G. W., Mahler, H. R., and Perlman, P. S. (1974): Nuclear gene dosage effects on mitochondrial mass and DNA. *J. Cell Biol.,* 61:565–574.

16. Williamson, D. H. (1970): The effect of environmental and genetic factors on the replication of mitochondrial DNA in yeast. *Symp. Soc. Exp. Biol.,* 24:247–276.

17. Tzagoloff, A. (1971): Assembly of the mitochondrial membrane system. IV. Role of mitochondrial and cytoplasmic protein synthesis in the biosynthesis of the rutamycin-sensitive adenosine triphosphatase. *J. Biol. Chem.,* 246: 3050–3056.

18. Groot, G. S. P., Rouslin, W., and Schatz, G. (1972): Promitochondria of an anaerobically grown yeast. VI. Effect of oxygen on promitochondrial protein synthesis. *J. Biol. Chem.,* 247:1735–1742.

19. Perlman, P. S., and Mahler, H. R. (1974): Derepression of mitochondria and their enzymes in yeast: Regulatory aspects. *Arch. Biochem. Biophys.,* 162:248–271.

20. Perlman, R. L., and Pastan, I. (1971): The role of cyclic AMP in bacteria. *Curr. Top. Cell. Regul.,* 3:117–134.

21. Fang, M., and Butow, R. A. (1970): Nucleotide reversal of mitochondrial repression in *Saccharomyces cerevisiae. Biochem. Biophys. Res. Commun.,* 41:1579–1583.

22. Paltauf, F., and Schatz, G. (1969): Promitochondria of anaerobically grown yeast. II. Lipid composition. *Biochemistry,* 8:335–338.

23. Watson, K., Haslam, J. M., Veitch, B., and Linnane, A. W. (1971): Mitochondrial precursors in anaerobically grown yeast. In: *Autonomy and Biogenesis of Mitochondria and Chloroplasts,* edited by N. K. Boardman, A. W. Linnane, and R. M. Smillie, pp. 162–174. North Holland, Amsterdam.

24. Andreasen, A. A., and Stier, T. J. B. (1953): Anaerobic nutrition of *Saccharomyces cerevisiae.* I. Ergosterol requirement for growth in a defined medium. *J. Cell. Comp. Physiol.,* 41:23–36.

25. Bloomfield, D. K., and Bloch, K. (1960): The formation of Δ^9-unsaturated fatty acids. *J. Biol. Chem.,* 235:337–345.

26. Frantz, I. D., Jr., and Skroepfer, G. J., Jr. (1967): Sterol biosynthesis. *Annu. Rev. Biochem.,* 36:691–726.

27. Wallace, P. G., and Linnane, A. W. (1964): Oxygen induced synthesis of yeast mitochondria. *Nature,* 201:1191–1194.

28. Polakis, E. S., Bartley, W., and Meek, G. A. (1964): Changes in the structure and enzyme activity of *Saccharomyces cerevisiae* in response to changes in the environment. *Biochem. J.,* 90:369–373.

29. Wallace, P. G., Huang, M., and Linnane, A. W. (1968): The biogenesis of mitochondria. II. The influence of medium composition on the cytology of anaerobically grown *Saccharomyces cerevisiae. J. Cell Biol.,* 37:207–220.

30. Schatz, G. (1970): Biogenesis of mitochondria. In: *Membranes of Mitochondria and Chloroplasts,* edited by E. Racker, pp. 251–314. Van Nostrand Reinhold, New York.

31. Criddle, R. S., and Schatz, G. (1969): Promitochondria of anaerobically grown yeast. I. Isolation and biochemical properties. *Biochemistry,* 8:322–334.

32. Plattner, H., and Schatz, G. (1969): Promitochondria of anaerobically grown yeast. III. Morphology. *Biochemistry,* 8:339–343.

33. Schatz, G. (1965): Subcellular particles carrying mitochondrial enzymes in anaerobically-grown cells of *Saccharomyces cerevisiae. Biochim. Biophys. Acta,* 96:342–345.

34. Watson, K., Haslam, J. M., and Linnane, A. W. (1970): Biogenesis of mitochondria. XIII. The isolation of mitochondrial structures from anaerobically grown *Saccharomyces cerevisiae. J. Cell Biol.,* 46:88–96.

35. Nagley, P., and Linnane, A. W. (1972): Cellular regulation of mitochondrial DNA synthesis in *Saccharomyces cerevisiae. Cell Differ.,* 1:143–148.

36. Plattner, H., Salpeter, M. M., Saltzgaber, J., and Schatz, G. (1970): Promitochondria of anaerobically grown yeast. IV. Conversion into respiring mitochondria. *Proc. Natl. Acad. Sci. U.S.A.,* 66:1252–1259.

37. Chen, S.-Y., Ephrussi, B., and Hottinguer, H. (1950): Nature génétique des mutants à déficience respiratoire de la souche B-II de la levure de boulangerie. *Heredity,* 4:337–351.

38. Sherman, F., Stewart, J. W., Margoliash, E., Parker, J., and Campbell, W. (1966): The structural gene for yeast cytochrome c. *Proc. Natl. Acad. Sci. U.S.A.,* 55:1498–1504.

39. Clark-Walker, G. D., and Linnane, A. W. (1967): The biogenesis of mitochondria in *Saccharomyces cerevisiae. J. Cell Biol.,* 34: 1–14.

40. Mahler, H. R., and Perlman, P. S. (1971): Mitochondriogenesis analyzed by blocks on mitochondrial translation and transcription. *Biochemistry,* 10:2979–2989.

41. Sherman, F., and Slonimski, P. P. (1964): Respiration deficient mutants of yeast. II. Biochemistry. *Biochim. Biophys. Acta,* 90:1–15.

42. Mackler, B., Douglas, H. C., Will, S., Haw-

thorne, D. C., and Mahler, H. R. (1965): Biochemical correlates of respiratory deficiency. IV. Composition and properties of respiratory particles from mutant yeasts. *Biochemistry*, 4:2016–2020.

43. Ephrussi, B., de Margerie-Hottinguer, H., and Roman, H. (1955): Suppressiveness: A new factor in the genetic determinism of the synthesis of respiratory enzymes in yeast. *Proc. Natl. Acad. Sci. U.S.A.*, 41:1065–1071.

44. Ephrussi, B., and Grandchamp, S. (1965): Études sur la suppressivité des mutants à déficience respiratoire de la levure. I. Existence au niveau cellulaire de divers "degrés de suppressivité." *Heredity*, 20:1–7.

45. Wright, R. E., and Lederberg, J. (1957): Extranuclear transmission in yeast heterokaryons. *Proc. Natl. Acad. Sci. U.S.A.*, 43:919–923.

46. Ephrussi, B., Jakob, H., and Grandchamp, S. (1966): Études sur la suppressivité des mutants à déficience respiratoire de la levure. II. Étapes de la mutation grande en petite provoquée par le facteur suppressif. *Genetics*, 54: 1–29.

47. Sherman, F., and Ephrussi, B. (1962): The relationship between respiratory deficiency and suppressiveness in yeast as determined with segregational mutants. *Genetics*, 47:695–700.

48. Nagley, P., and Linnane, A. W. (1972): Biogenesis of mitochondria. XXI. Studies on the nature of the mitochondrial genome in yeast: The degenerative effects of ethidium bromide on mitochondrial genetic information in a respiratory competent strain. *J. Mol. Biol.*, 66:181–193.

49. Sherman, F. (1963): Respiration-deficient mutants of yeast. I. Genetics. *Genetics*, 48:375–385.

50. Ephrussi, B., and Hottinguer, H. (1950): Direct demonstration of the mutagenic action of euflavin on baker's yeast. *Nature*, 166:956.

51. Ephrussi, B., and Hottinguer, H. (1951): Cytoplasmic constituents of heredity. On an unstable state in yeast. *Cold Spring Harbor Symp. Quant. Biol.*, 16:75–85.

52. Nagai, S., Yanagashima, N., and Nagai, H. (1961): Advances in the study of the respiration-deficient (RD) mutation in yeast and other microorganisms. *Bacteriol. Rev.*, 25: 404–426.

53. Sager, R. (1972): *Cytoplasmic Genes and Organelles.* Academic Press, New York.

54. Mahler, H. R. (1973): Biogenetic autonomy of mitochondria. *CRC Crit. Rev. Biochem,* 1:381–460.

55. Deutsch, J., Dujon, B., Netter, P., Petrochilo, E., Slonimski, P. P., Bolotin-Fukuhara, M., and Coen, D. (1974): Mitochondrial genetics. VI. The petite mutation in *Saccharomyces cerevisiae:* Interrelations between the loss of the ρ^+ factor and the loss of the drug resistance mitochondrial genetic markers. *Genetics*, 76: 195–219.

56. Sherman, F. (1959): The effects of elevated

temperatures on yeast. II. Induction of respiratory deficient mutants. *J. Cell. Comp. Physiol.*, 54:37–52.

57. Maroudas, N. G., and Wilkie, D. (1968): Ultraviolet irradiation studies on the cytoplasmic determinant of the yeast mitochondrion. *Biochim. Biophys. Acta*, 166:681–688.

58. Wilkie, D. (1963): The induction by monochromatic U.V. light of respiratory-deficient mutants in aerobic and anaerobic cultures of yeast. *J. Mol. Biol.*, 7:527–533.

59. Sugimura, T., Okabe, K., and Imamura, A. (1966): Number of cytoplasmic factors in yeast cells. *Nature*, 212:304.

60. Slonimski, P. P., Perrodin, G., and Croft, J. H. (1968): Ethidium bromide induced mutation of yeast mitochondria: Complete transformation of cells into respiratory deficient nonchromosomal petites. *Biochem. Biophys. Res. Commun.*, 30:232–239.

61. Allen, N. E., and MacQuillan, A. M. (1969): Target analysis of mitochondrial genetic units in yeast. *J. Bacteriol.*, 97:1142–1148.

62. Whittaker, P. A., Hammond, R. C., and Luha, A. A. (1972): Mechanism of mitochondrial mutation in yeast. *Nature [New Biol.]*, 238: 266–268.

63. Bulder, C. J. E. A. (1964): Induction of petite mutation and inhibition of synthesis of respiratory enzymes in various yeasts. *Antonie van Leeuwenhoek*, 30:1–9.

64. Bulder, C. J. E. A. (1964): Lethality of the petite mutation in petite negative yeasts. *Antonie van Leeuwenhoek*, 30:442–454.

65. DeDeken, R. H. (1966): The crabtree effect and its relation to the petite mutation. *J. Gen. Microbiol.*, 44:157–165.

66. Crandall, M. (1973): Comparison of *Hansenula wingei*, a *petite*-negative, obligately aerobic yeast, to the *petite*-positive yeast *Saccharomyces cerevisiae*. *J. Gen. Microbiol.*, 75: 363–375.

67. Luha, A. A., Whittaker, P. A., and Hammond, R. C. (1974): Biosynthesis of yeast mitochondria: Some effects of ethidium bromide on *Kluyveromyces* (*Saccharomyces*) *fragilis*. *Mol. Gen. Genet.*, 129:311–323.

68. Heslot, H., Louis, C., and Goffeau, A. (1970): Segregational respiratory-deficient mutants of a "petite-negative" yeast *Schizosaccharomyces pombe* 972h⁻. *J. Bacteriol.*, 104:482–491.

69. Kellerman, G. M., Biggs, D. R., and Linnane, A. W. (1969): Biogenesis of mitochondria. XI. A comparison of the effects of growth-limiting oxygen tension, intercalating agents, and antibiotics on the obligate aerobe *Candida parasilopsis*. *J. Cell Biol.*, 42:378–391.

70. Luha, A. A., Sarcoe, L. E., and Whittaker, P. A. (1971): Biosynthesis of yeast mitochondria. Drug effects on the petite negative yeast *Kluyveromyces lactis*. *Biochem. Biophys. Res. Commun.*, 44:396–402.

71. Schwab, R., Sebald, M., and Kaudewitz, F.

(1971): Influence of ethidium bromide on respiration in *Schizosaccharomyces pombe*. *Mol. Gen. Genet.,* 110:361–366.

72. Goffeau, A., Briquet, M., Colson, A. M., Delhez, J., Foury, F., Labaille, F., Landry, Y., Mohar, O., and Mrena, E. (1975): Stable pleiotropic respiratory-deficient mutants of a "petite-negative" yeast. In: *Membrane Biogenesis,* edited by A. Tzagoloff, Chapter 3. Plenum Press, New York.

73. Bandlow, W., and Kaudewitz, F. (1974): Action of ethidium bromide on mitochondrial DNA in the petite-negative yeast *Schizosaccharomyces pombe. Mol. Gen. Genet.,* 131: 333–338.

74. Wolf, K., Lang, B., Burger, G., and Kaudewitz, F. (1976): Extrachromosomal inheritance in *Schizosaccharomyces pombe*. II. Evidence for extrakaryotically inherited respira-tory deficient mutants. *Mol. Gen. Genet.,* 144:75–81.

75. Linnane, A. W., and Crowfoot, P. D. (1975): Biogenesis of yeast mitochondrial membranes. In: *Membrane Biogenesis,* edited by A. Tzagoloff, Chapter 4. Plenum Press, New York.

76. Perlman, P. S. (1975): Cytoplasmic petite mutants in yeast: A model for the study of reiterated genetic sequences. In: *Genetics and Biogenesis of Mitochondria and Chloroplasts,* edited by C. W. Birky, Jr., P. S. Perlman, and T. J. Byers, Chapter 4. Ohio State University Press, Columbus.

77. Tzagoloff, A., Rubin, M. S., and Sierra, M. F. (1973): Biosynthesis of mitochondrial enzymes. *Biochim. Biophys. Acta,* 301:71–104.

78. Schatz, G., and Mason, T. L. (1974): The biosynthesis of mitochondrial proteins. *Annu. Rev. Biochem.,* 43:51–87.

Chapter 5

Mitochondrial Genetics of Bakers' Yeast. II. Point Mutations and the Mitochondrial Genome

For a long time the major impediment to the study of mitochondrial genetics in *S. cerevisiae* was the lack of mutants other than vegetative petites. Although vegetative petites were easily obtainable by the thousands using a variety of mutagens, all of them were phenotypically similar; they lacked the capacity for mitochondrial protein synthesis, and consequently they lacked that constellation of proteins whose production depended on this system. Furthermore, vegetative petites did not revert, recombine, or complement one another. In other words, by themselves they represented a dead end as material with which to study the formal genetics of mitochondria. A way around this dilemma first became apparent because of the work of Linnane and his colleagues. They discovered that antibacterial antibiotics such as chloramphenicol and erythromycin blocked formation of respiratory cytochromes, and they postulated that the site of action was the mitochondrial protein-synthesizing system. They also found that cell division proceeded normally on fermentable substrates but not on nonfermentable substrates. Subsequently Linnane and Wilkie (11) reported isolation of nonmendelian mutations to antibiotic resistance that later proved to be mitochondrial in origin.

The existence of mitochondrial mutations to antibiotic resistance made genetic dissection of the mitochondrial genome feasible for the first time. Intensive formal genetic investigations were launched in several laboratories that yielded a wealth of interesting findings, as we shall see in subsequent chapters. However, it was not possible to go the entire distance with antibiotic-resistant mutations; thus new mutations, particularly point mutations in the mitochondrial genome affecting specific respiratory chain components, were sought. The problem was how to isolate such mutants and distinguish them from vegetative petites. This problem was recently solved by use of several ingenious techniques, and the genetic functions of these so-called *mit⁻* mutants are now being actively investigated.

In this chapter we will begin with a discussion of the nomenclature used to describe mitochondrial mutants in yeast. Although this may seem a boring subject, such a discussion is necessary because the variety of nomenclatures used in the literature can only be described as bewildering. From there we will turn to the isolation and characterization of the antibiotic-resistant mutants and thence to the *mit⁻* mutants. The chapter will end with a short review of the use of molecular weight variants of mitochondrial polypeptides to study mitochondrial genetics.

NOMENCLATURE OF MITOCHONDRIAL MUTATIONS

Three formal proposals have been made regarding nomenclature to describe mitochondrial genes (1–3), and a fourth proposal for mitochondrial gene nomenclature is now under consideration (4). Although this fourth system is likely to come into general use, since it has the tacit approval of most organelle geneticists, it seems appropriate in this book to make use of the major systems currently being employed in order to make the original literature more accessible to the interested reader. There are five general classes of mitochondrial phenotypes other than vegetative petites (Table 5–1). The first consists of a group of three alleles (ω^+, ω^-, ω^N) that map at the ω locus and that influence the transmission of linked mitochondrial genes in crosses. We will have much more to say about the ω locus later; here we will simply note that ω was originally regarded as the mitochondrial "sex factor." Mutations to antibiotic resistance map in a number of loci, and these constitute the second class. We use the symbol *ant* in referring to these genes in a general sense, where *ant^r* and *ant^s* refer to resistance and sensitivity, respectively, and *ant^0* indicates actual loss of the gene from mtDNA. Mitochondrial mutations, other than vegetative petites, that affect either respiratory components or mitochondrial protein synthesis are referred to as *mit$^-$* and *syn$^-$*, respectively. The superscripts plus, minus, and zero indicate, respectively, normal function or altered function or that the gene in question is missing. The *mit$^-$* and *syn$^-$* mutants constitute the third and fourth classes of phenotypes. Finally, by use of techniques that will be described at the end of this chapter, it has proved possible to identify

molecular weight variants of certain polypeptides coded by the mitochondrial genome, and these are called *var-f* and *var-s*. These polypeptides are separable by SDS acrylamide gel electrophoresis on the basis of molecular weight. The *var-f* or fast form moves more rapidly in gels than the *var-s* form, which reflects the fact that *var-f* has the lower molecular weight.

Different mitochondrial genes have been named largely according to nomenclatures developed in the laboratories of P. P. Slonimski and A. W. Linnane (Table 5–2). These workers have independently mapped a large number of mitochondrial gene mutations in *S. cerevisiae*. Although mutants from the two laboratories have not been intercrossed to any great extent, one can assign equivalent genes in the two systems with a reasonable degree of accuracy based on map position and function affected (Table 5–2). The nomenclature devised by Slonimski and his colleagues has enjoyed the widest acceptance, in part because this group published a large volume of very detailed papers on genetic mapping. Their nomenclature began originally as a single-

TABLE 5–1. *General classes of mitochondrial gene mutations excluding vegetative petites*

Function affected	General allelic symbols[a]	Phenotypic manifestation
Polarity of mitochondrial gene transmission and recombination	$\omega^+,\omega^-,\omega^N$	Transmission and recombination of closely linked mitochondrial genes affected in $\omega^+ \times \omega^-$ crosses, but not in $\omega^+ \times \omega^+$, $\omega^- \times \omega^-$, or ω^N by itself or either of the other alleles (see Chapter 7 for details)
Antibiotic resistance	*ant^r* *ant^s* *ant^0*	Antibiotic-resistant Antibiotic-sensitive Gene is lost from mtDNA
Respiratory competence	*mit$^+$* *mit$^-$* *mit^0*	Function normal Function deficient Gene lost from mtDNA
Mitochondrial protein synthesis	*syn$^+$* *syn$^-$* *syn^0*	Function normal Function deficient Gene lost from mtDNA
Mitochondrial translation products	*var*	Variant polypeptides made on mitochondrial ribosomes; may be present in some strains and absent from others, or may show molecular weight variation between strains

[a] With the exception of the alleles at the ω locus that control the polarity of transmission and recombination of linked mitochondrial genes, each gene class is identified by a three-letter symbol. Specific symbols for genes falling in each class are given in Table 5–2.

TABLE 5–2. Mitochondrial loci of S. cerevisiae

Phenotype	Nomenclature	Locus symbol	Typical allele designation	Comments
Polarity of mito-chondrial gene transmission and recombination affected	B, S, L	ω	ω^+, ω^-, ω^N	This nomenclature is in general use by all groups and will be retained
Chloramphenicol resistance	S	*RIB1* or R_I	C^R_{321}	*RIB1* = ribosomal locus 1
	L	cap1	cap1-r	Same locus as *RIB1*
	B	cap1	$cap^r1–1$, etc.	
Erythromycin and/or spiramycin resistance	S	*RIB2* or R_{II}	E^R_{354}, S^R_{352}	*RIB2* = ribosomal locus 2; E^R = erythromycin-resistant or erythromycin- and spiramycin-resistant; S^R = spiramycin-resistant
		RIB3 or R_{III}	E^R_{221}, S^R_{352}	*RIB3* = ribosomal locus 3, with the allele designations having the same meaning as for *RIB2*
	L	ER or ery spi	ery1-r, ery2-r, spi2-r, spi3-r, spi4-r	There appear to be two groups of mutants, possibly corresponding to two genes; they map in the same region as *RIB2* and *RIB3*; ery1-r and ery2-r are resistant to erythromycin and spiramycin and map together with spi2-r, which is resistant only to spiramycin; spi3-r and spi4-r map in a second region and are resistant only to spiramycin
	B	ery1, spi1	$ery^r1–1$, etc.	It is not clear how the locus and cross-resistance problem will be handled in this system; most likely, *RIB2* and *RIB3* would be designated ery1 and ery2
Paromomycin resistance	S	*PAR1* or P_I	P^R_{354}	None
	L	par1	?	Appears to be same locus as *PAR1*
	B	par1	$par^r1–74$, etc.	
Antimycin resistance	S	*ANA1*, A_I	ana^r-101	Mutants map in a cluster in *COB2* (see Chapter 19)
		ANA2, A_{II}	?	The *ANA2* and *FUN1* mutants map in the same
Funiculosin resistance		*FUN1*	FUN^r_5, etc.	cluster in *COB1* (see Chapter 19)
Antimycin resistance	L	ana1 (mik1)	?	Probably the same as one of the *ANA* loci of Slonimski; originally thought to be resistant to mikamycin (see text for discussion)
	B	ana1	$ana^r1–5$, etc.	
Diuron resistance	S	*DIU1*	div1–724	These mutants map in the same cluster as *ANA2*, *FUN1*, and *BOX1* mutants in *COB1* (see Chapter 19)
	B	div1	$div^r1–10$	Presumed nomenclature, but not proposed specifically for these alleles by Birky et al. (4)
	S	*DIU2*	div2–725	These mutants map in the same cluster as *ANA1*, *MUC1*, and *BOX4* mutants in *COB2* (see Chapter 19)
	B	div2	$div^r2–16$	See comment under div1
Mucidin resistance	S	*MUC1*	muc1–771	See comment under *DIU2*
	B	muc1	$muc^r1–17$	See comment under div1
	S	*MUC2*	muc2–772	Maps in a cluster with *BOX6* in *COB1* (see Chapter 19)
	B	muc2	$muc^r2–27$	See comment under div1
Oligomycin resistance	S	*OLI1*, O_I	O^r_1	Mutants resistant to oligomycin and structurally related antibiotics
		OLI2, O_{II}	O^R_{144}	Mutants resistant to oligomycin and structurally related antibiotics

TABLE 5–2. *Mitochondrial loci of* S. cerevisiae *(continued)*

Phenotype	Nomenclature	Locus symbol	Typical allele designation	Comments
		OLI3, O_{III}	V^R_{61}, V^R_{62}	Mutants resistant to oligomycin and the structurally unrelated inhibitor venturicidin; the OLI1 and OLI3 genes are very closely linked
	S	OLI4	O^R_9	This locus is linked to OLI2
	L	OLG2, O_I	oli17	Appears to be same locus as OLI1
		O_{II}	—	Appears to be same locus as OLI2
		OLG1, O_{III}	oli1	Appears to be same locus as OLI3
	B	oli1, oli2, oli3, oli4	oli^r1–2, oli^r2–16, etc.	
Cytochrome oxidase deficiency	S	OXI1	M9–94	Mutants in the OXI and COB loci have yet to be designated by a formal allelic nomenclature.
		OXI2	M15–208	Linked to OXI1
		OXI3	M5–16	
		OXI4	—	Linked to OXI3
	L	cya1	46–3–1	No formal nomenclature as yet for alleles; mapping indicates equivalence with OXI2
		cya2	10–15	Mapping indicates equivalence with OXI3
		cya3	45–3–1	Mapping indicates equivalence with OXI1
	B	oxi1, oxi2, etc.	oxi1–1, etc.	Presumed nomenclature, but not proposed specifically for these genes by Birky et al. (4)
Cytochrome b deficiency	S	COB1	M6–200	No formal nomenclature as yet for alleles
		COB2	M10–152	No formal nomenclature as yet for alleles; linked to COB1
	L	cyb1	41–2–4	Equivalent to either COB1 or COB2; no formal allelic nomenclature as yet
	B	?	?	Nomenclature uncertain
Oxidative phosphorylation deficiency	S	PHO1	?	Maps near OLI2; OLI and PHO mutants could prove to be alleles
		PHO2	?	Maps near OLI1; OLI and PHO mutants could prove to be alleles
	L	?	?	
	B	pho1, pho2, etc.	pho1–1, etc.	
Cytochrome b and cytochrome oxidase deficiency	S	BOX1	BOX1–1	See comment under DIU1
		BOX2	BOX2–1	Maps in COB1 (see Chapter 19)
		BOX3	BOX3–1	Maps in COB2 (see Chapter 19)
		BOX4	BOX4–1	See comment under DIU2
		BOX5	BOX5–1	Maps in COB2 (see Chapter 19)
		BOX6	BOX6–1	See comment under MUC2

Three groups of symbols are given for most loci. Two of these reflect the nomenclatures of Slonimski (S) and Linnane (L) and their colleagues and are in most common use in the literature. The third (B) is the nomenclature proposed by Birky et al. (4); which is likely to come into general use for mitochondrial genes and possibly organelle genes in general. This nomenclature has been compiled from the many sources cited in references for Chapters 5–7 and 19. A much more thorough compilation of all mitochondrial gene mutations in *S. cerevisiae* was published by Dujon et al. (60) as this book was going to press. Because no complementation test exists, it is still not clear what constitutes a gene in mitochondrial genetics, and the term *locus* is used here to designate a group of mutants that are tightly linked and presumably constitute either part of a gene or an entire gene. In the complex III region of the mitochondrial genetic map, for example, clusters of mutants having the ANA, BOX, COB, DIU, FUN, and MUC phenotypes are all found, but it is uncertain how many genes there are (see Chapter 19 for details).

letter system. For example, C^R_{321} referred to a specific mutation to chloramphenicol resistance. Later, as different alleles were found to cluster together, a single-letter system was used to indicate genes as well.

For example, a gene identified by mutations to oligomycin resistance was designated O_I. Currently this system of gene designation is being replaced by a three-letter system; so O_I will become *OLI1*.

Allelic nomenclature also seems to be in a process of transition. The older one-letter system continues in use for designation of alleles at the different *ant* loci, but in the case of the *mit* genes the specific alleles are referred to either by isolation numbers (e.g., M10–152) or by a three-letter system. For example, *BOX1–1* is an allele at the *BOX1* locus. The use of *RIB* as a locus designation deserves specific comment. This notation reflects the supposition that mutations mapping in the *RIB* loci confer antibiotic resistance on mitochondrial ribosomes. Thus mutations resistant to chloramphenicol (C^R) map in the *RIB1* locus, whereas mutations resistant to erythromycin (E^R) and spiramycin (S^R) map in the *RIB2* and *RIB3* loci. These loci were previously designated R_I, R_{II}, and R_{III}.

Since we will refer repeatedly to crosses made by the Slonimski group with mitochondrial mutations to antibiotic resistance, it will be useful to describe here the notation ordinarily employed in such crosses. The symbolism $\omega^- C^R_{321} E^S \times \omega^+ C^S E^R_{514}$ means that one stock carries the ω^- allele of the mitochondrial sex factor and the chloramphenicol resistance mutation 321, but is sensitive to erythromycin, whereas the other stock includes the ω^+ allele, is sensitive to chloramphenicol, and carries the erythromycin resistance mutation 514. This nomenclature can become confusing when phenotypically similar mutations in different genes are considered. For example, the erythromycin-resistant mutations E^R_{553} and E^R_{353} map in the *RIB2* and *RIB3* loci, respectively.

The Linnane system generally has not undergone such complex evolution; thus it is easier to use. Normally, three letters are used to designate mutations (e.g., *ery2-r* and *oli17-r* for resistance to erythromycin and oligomycin, respectively). Usually three letters are also employed to designate loci (e.g., *cap1*, *ery1*, *cya1*, etc.), but here some variations in usage are to be found. For example, the symbols *OLG1*, *OLG2,* and O_{II} are used to designate loci mutations that confer oligomycin resistance (Table 5–2).

The nomenclature system proposed by Birky et al. (4), which very likely will come into general use for organelle genes, is also a three-letter system. Genes for oligomycin resistance are designated oli^r1, oli^s1, oli^r2, oli^s2, etc., to distinguish resistance and sensitivity. Independent mutations at the same locus are designated $oli^r1–1$, $oli^r1–2$, etc. An informal symbolism is also suggested that consists of single upper-case letters; it is usually used throughout an article after the formal designation has been given. For example, the genotype $ery^r3–337$ oli^s1 $par^r1–6$ is informally designated $E^R O^S P^R$. Loss of an allele for antibiotic resistance or sensitivity is designated by the superscript 0 (e.g., C^0, E^0) in the one letter systems of Birky and associates, and Slonimski.

ANTIBIOTIC RESISTANCE AND THE MITOCHONDRIAL GENOME

As mentioned previously, antibiotics such as chloramphenicol and erythromycin block the growth of wild-type *S. cerevisiae* on media containing nonfermentable carbon sources (e.g., glycerol and lactate), but they do not inhibit cell growth on a fermentable carbon source such as glucose (Table 5–3). Linnane and associates (5–8) demonstrated in the mid-1960s that this

TABLE 5–3. Growth of wild-type, vegetative petite, and antibiotic-resistant mutant cells on fermentable and nonfermentable carbon sources in presence and absence of the antibiotic to which the mutant is resistant

| | | Growth | |
| | Carbon | Minus | Plus |
Genotype	source	antibiotic	antibiotic
Wild type	F[a]	+[b]	+
	NF	+	−
Vegetative petite	F	+	+
	NF	−	−
Antibiotic-resistant mutant	F	+	+
	NF	+	+

[a] F = fermentable; NF = nonfermentable.
[b] Plus sign indicates growth; minus sign indicates no growth.

conditionally lethal inhibition of growth of wild-type yeast cells by these antibiotics results because they prevent mitochondrial protein synthesis. Inhibition of mitochondrial protein synthesis, in turn, blocks formation of specific respiratory chain components (Table 5–4) (see Chapter 19). Thus, examination of the cytochrome spectra of antibiotic-inhibited yeast cells reveals that cytochromes a + a_3, and b are present in greatly reduced amounts, but cytochrome c is present in normal amounts (Table 5–4). These biochemical alterations are accompanied by morphologic changes in the mitochondria, including a reduction in the number of cristae.

Not only do these antibiotics affect mitochondrial phenotype, but also they affect those promitochondrial functions dependent on mitochondrial protein synthesis. Unlike normal promitochondria, the promitochondria of cells treated with chloramphenicol or erythromycin lack the F_1 ATPase anaerobic energy-transfer mechanism (9). In short, wild-type cells grown on a fermentable carbon source either anaerobically or aerobically in the presence of chloramphenicol or erythromycin have a mitochondrial phenotype similar in many respects to that of a vegetative petite mutant. However, the antibiotic-treated cells, unlike vegetative petites, contain normal mtDNA and retain the capacity for mitochondrial protein synthesis and resynthesis of the respiratory chain components on removal of antibiotics from the medium (7,8).

It is important to point out, as in the case of vegetative petites (see Chapter 4), that glucose repression is not responsible for the effects of chloramphenicol and erythromycin on wild-type cells. Perlman and Mahler (10) showed that enzymes such as L-malate dehydrogenase and L-glutamate dehydrogenase are synthesized in cells grown on glucose in the presence of chloramphenicol at repressed levels (as they are in vegetative petites) and that derepression occurs as the cells enter the stationary phase (as it does in vegetative petites). Conversely, cytochrome oxidase is not synthesized in the presence of chloramphenicol to any significant degree; thus, as in the case of vegetative petites, derepression of this enzyme cannot occur. In general, it can be said that activities requiring mitochondrial protein synthesis are blocked absolutely in the vegetative petite and are blocked in a concentration-dependent manner in wild-type cells in the presence of inhibitors such as chloramphenicol and erythromycin. Those mitochondrial enzymes synthesized by petites are also synthesized by antibiotic-inhibited cells, and those enzymes that undergo glucose derepression in petites also undergo derepression in wild-type cells in the presence of these antibiotics.

In 1967 Wilkie and associates (11) reported the isolation of mutants capable of growth on glycerol in the presence of chloramphenicol, erythromycin, and tetracycline. In the following year Thomas and Wilkie (12) and Linnane et al. (13)

TABLE 5–4. Effect of chloramphenicol on mitochondrial protein synthesis and phenotype in S. cerevisiae

	Chloramphenicol			Vegetative petite, no inhibitor
	None	Short-term	Long-term	
Mitochondrial protein synthesis	+[a]	−	−	−
Cytochrome a + a_3	+	−	−	−
Cytochrome b	+	R	−	−
Cytochrome c	+	+	+	+
Cristae	+	+	R	R

[a] + = present; − = absent; R = reduced.
Data from various sources.

showed that certain of the erythromycin-resistant mutants were inherited in a nonmendelian manner. They also found that nonmendelian resistance to erythromycin could be eliminated from cells by converting them to the vegetative petite phenotype with acridine dyes. This discovery by Linnane, Wilkie, and their associates ushered in the era of formal mitochondrial genetics, since mutations to antibiotic resistance, unlike vegetative petites, can be phenotypically dissimilar and can behave as point mutants, and recombination studies can be done with them.

The inheritance of a typical mitochon-

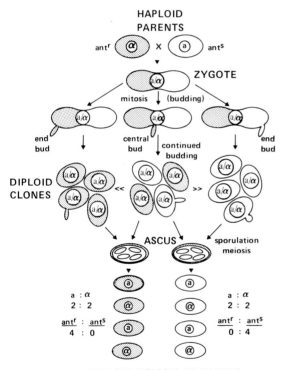

FIG. 5–1. Transmission of mitochondrial antibiotic resistance mutations in *S. cerevisiae*. Regardless of the mating type (α or *a*) of the haploid parent carrying the mutation to antibiotic resistance (*ant*ʳ and shaded cells), the majority of the zygotes produced by the cross give rise to both *ant*ʳ and *ant*ˢ (antibiotic-sensitive and unshaded cells) diploid progeny although the ratio of the two mitochondrial genotypes varies considerably from clone to clone. Sporulation of *ant*ʳ or *ant*ˢ progeny yields only *ant*ʳ or *ant*ˢ meiotic products, respectively, while the nuclear genes α and *a* segregate 2:2 as expected. (Courtesy of Dr. K. Van Winkle-Swift.)

drial *ant*ʳ mutation is reminiscent of the inheritance of a suppressive petite (Fig. 5–1). When haploid cells of an *ant*ʳ mutant are crossed to *ant*ˢ cells, some of the diploids produced are resistant to the antibiotic, and others are sensitive. The proportions of resistant and sensitive diploids are governed by genetic and physiologic factors (14,15) that will be discussed in detail in Chapter 7. Sporulation of a resistant diploid produces 4:0 *ant*ʳ:*ant*ˢ progeny, whereas sporulation of a sensitive diploid yields 0:4 *ant*ʳ:*ant*ˢ progeny. Diploids formed between a mendelian mutation to antibiotic resistance and the wild type are either resistant or sensitive to the antibiotic, depending on the dominance relationships of the alleles; on sporulation of the diploid there is 2:2 segregation of resistance and sensitivity (Fig. 5–2).

There is good reason to believe that the ρ factor is the genetic manifestation of mtDNA (see Chapters 4 and 6). Evidence that nonmendelian mutations to antibiotic resistance are mutations in mtDNA comes from genetic experiments indicating that antibiotic resistance mutations are localized in the ρ factor. Thus the retention of an *ant*ʳ or an *ant*ˢ allele by a given stock is affected when the stock mutates to the vegetative petite phenotype.

Loss of an *ant* gene in a petite mutant cannot be demonstrated directly, since, by definition, the *ant*ʳ and *ant*ˢ phenotypes can be distinguished only when yeast cells are grown on a nonfermentable carbon source, and petite mutants cannot themselves grow on such a carbon source. The method used to determine whether or not an *ant* allele has been lost is indirect; it involves crossing petites isolated from the stock containing the *ant* allele to a respiration-sufficient tester stock and examining the diploid progeny of the cross. For example, suppose an *ant*ʳ stock is treated with ethidium bromide, and the induced petites are isolated and crossed to a respiration-sufficient *ant*ˢ tester (Table 5–5). Theoretically, the cross will yield one of four possible results. If the diploid progeny include a mixture of petite and respira-

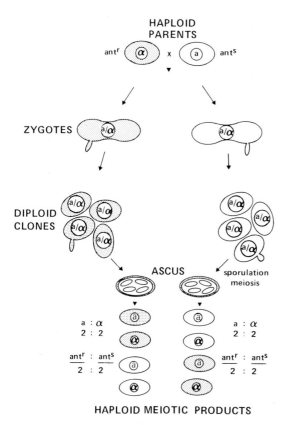

HAPLOID PARENTS

ZYGOTES

DIPLOID CLONES

ASCUS   sporulation meiosis

a : α
2 : 2

a : α
2 : 2

ant^r : ant^s
2 : 2

ant^r : ant^s
2 : 2

HAPLOID MEIOTIC PRODUCTS

FIG. 5–2. Transmission of nuclear antibiotic resistance mutations in *S. cerevisiae*. Regardless of the mating type (a or α) of the haploid parent carrying the mutation to antibiotic resistance (*ant^r* and shaded cells), all of the zygotes produced by the cross give rise to either *ant^r* or *ant^s* (antibiotic-sensitive and unshaded cells) diploid progeny, depending on whether resistance or sensitivity is dominant. Sporulation of either kind of diploid yields *ant^r* and *ant^s* progeny in a 2:2 ratio, as expected, and the same segregation is seen for the nuclear mating type genes.

tion-sufficient progeny and the respiration-sufficient progeny are all *ant^s*, the genotype of the petite parent must be ρ$^-$*ant^0*. That is, the mutation from ρ$^+$ to ρ$^-$ is accompanied by loss of a functional *ant^r* marker. If the diploids include petite progeny and respiration-sufficient progeny, some of which are *ant^r* and some of which are *ant^s*, the petite parent must be ρ$^-$*ant^r*. That is, the *ant^r* allele is retained in the petite parent, although it cannot, by definition, be expressed. If the diploids are all respiration-sufficient, but include *ant^r* and *ant^s* progeny, the petite parent must be ρ^0*ant^r*. Isolation

of a petite of this type means that the *ant* gene does not lie in the ρ factor (and hence mtDNA), since the *ant^r* allele is retained and the ρ factor is lost. Finally, if the diploid progeny are all ρ$^+$*ant^s*, the genotype of the petite parent is either ρ^0*ant^0* or ρ$^+$*ant^0*. The ρ$^+$*ant^0* genotype would be expected to be petite in phenotype, since it is believed that the products made by the *ant* genes either are involved in mitochondrial ribosome biogenesis or are components associated with complex III or the membrane factor of the oligomycin-sensitive F$_1$ ATPase (Table 5–2) (see Chapters 19 and 21). In either case, the *ant^0* mutation will lead to loss of the ability of the mutant cells to grow on a nonfermentable carbon source. Inability to form mitochondrial ribosomes causes the syndrome of respiratory defects described earlier in this chapter for yeast cells grown in the presence of mitochondrial protein synthesis inhibitors, and a defect in complex III or in the membrane factor will block the ability of the cells to respire or to carry out oxidative phosphorylation. The ρ^0*ant^0* and ρ$^+$*ant^0* genotypes can be distinguished genetically by making crosses to ρ$^-$*ant^s* petites (Table 5–5). In such crosses the ρ$^+$*ant^0* genotype will yield respiration-sufficient diploids carrying the *ant^s* allele from the petite parent, whereas the ρ^0*ant^0* genotype will not. The ρ$^+$*ant^0* mutants will qualify as either *mit* or *syn* deletion mutants (Table 5–1), depending on whether loss of the *ant* gene results in loss of functional oligomycin-sensitive F$_1$ ATPase or loss of mitochondrial protein synthesis.

Thomas and Wilkie (12), in their early experiments, found that conversion of non-mendelian erythromycin-resistant (*E^R*) mutants to the ρ$^-$ phenotype with acriflavine resulted in loss of the ability of the petite to transmit the *E^R* allele. They argued, therefore, that the *E^R* determinant was located in the mitochondrion, since the mutation to ρ$^-$ was accompanied by loss of the *E^R* allele. At about the same time, Linnane et al. (13) reported similar results with

TABLE 5–5. *Classification of possible genotypes for ρ and ant^r following mutagenesis of an ant^r stock with ethidium bromide by measuring retention in crosses to wild type ($ρ^+ant^s$) and a petite carrying the ant^s gene ($ρ^-ant^s$)*

Genotype of mutagenized stock	Growth on nonfermentable carbon source	Phenotypes of diploids produced in crosses to wild type ($ρ^+ant^s$)			Phenotypes of diploids produced in crosses to a petite ($ρ^-ant^s$)		
		Respiration-sufficient		Respiration-deficient	Respiration-sufficient		Respiration-deficient
		ant^r	ant^s		ant^r	ant^s	
$ρ^+ant^r$	+	+	+	−	+	+	+
$ρ^+ant^0$	−	−	+	−	−	+	+
$ρ^-ant^r$	−	+	+	+	−	−	+
$ρ^-ant^0$	−	−	+	+	−	−	+
$ρ^0ant^r$	−	+	+	−	−	−	+
$ρ^0ant^0$	−	−	+	−	−	−	+

euflavine, but with an important exception. They found a case in which a petite derived from an erythromycin-sensitive (E^s) stock, when crossed to a $ρ^+E^R$ stock yielded both $ρ^+E^R$ and $ρ^+E^s$ diploid progeny. As a result, Linnane and associates argued that a yeast cell probably carried two separate cytoplasmic factors, one of which was equivalent to ρ and one of which carried the E^R gene. More extensive experiments of this type by Gingold et al. (16) revealed that petites frequently retained the *ant* alleles that they carried. These authors concluded that ρ and the E^R allele were on different cytoplasmic determinants, with ρ generally being required for stability of the E^R determinant, or, alternatively, that ρ and E^R were on the same piece of mtDNA but recombined with high frequency. These experiments indicated the necessity of doing careful kinetic studies on induction of $ρ^-$ mutants and loss of *ant* alleles to establish whether or not the inactivations of ρ and the *ant* markers were really independent.

Later experiments on retention of *ant* markers in ethidium-bromide-induced petites provided strong evidence that these genes were included within $ρ^+$ and hence within mtDNA. The first results were reported by Nagley and Linnane (17), who isolated five $ρ^0$ mutants lacking mtDNA from a $ρ^+$ strain carrying an E^R marker and showed that all of them were E^0. These re-

sults, although limited, suggested that $ρ^0ant^r$ mutants did not exist and that loss of mtDNA was accompanied by loss of the *ant* genes, which was consistent with the localization of these genes in mtDNA.

Subsequently, the careful kinetic experiments of Deutsch and associates (18) provided further evidence that the *ant* genes were localized in $ρ^+$ and hence mtDNA. In these experiments the kinetics of loss of $ρ^+$ and alleles for erythromycin resistance (E^R), chloramphenicol resistance (C^R), and oligomycin resistance (O^R) were measured in nongrowing cells as a function of time of treatment with ethidium bromide. The methods used for measuring retention of $ρ^+$ and the ant^r markers were genetic; they employed tests similar to those summarized in Table 5–5. Since the tests were not designed to measure either suppressiveness or retention of mtDNA, it was not possible to distinguish $ρ^0ant^r$ and $ρ^-ant^r$ genotypes, nor could $ρ^0ant^0$ and $ρ^-ant^0$ genotypes be differentiated.

The data were analyzed by the method of target analysis (see Chapter 4), with S indicating survival of a given marker as measured by the frequency of nonmutated colonies for that marker:

$$S = 1 - (1 - e^{-h})^n$$

where h is the number of effective hits and n is the target number. As discussed in the

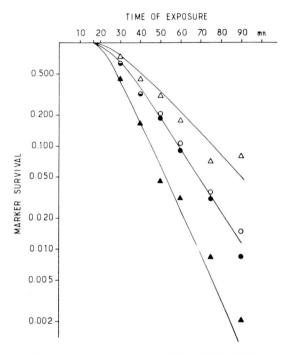

TIME OF EXPOSURE

FIG. 5–3. Kinetics of survival of ρ^+(▲), C^R(●), E^R(○), and O_R(△) genetic markers after ethidium bromide treatment. (From Deutsch et al., ref. 18, with permission.)

preceding chapter, h varies as a linear function of time (t) multiplied by a constant (k) that depends on the sensitivity of the target to the mutagen. Deutsch et al. (18) set $h = k(t - t_0)$, where t_0 is a parameter used to correct for the fact that mutagenesis begins immediately in one strain but not in the other strain until 10 min have elapsed. The interpretation of k in these experiments is particularly important. In the preceding chapter we stated that this parameter de-

pends on the sensitivity of the target to the mutagen. Sensitivity depends not only on the nature of the target in terms of its susceptibility to the mutagen but also on relative target size. In this case, presumably, a single target molecule, mtDNA, is involved, and k can be related to the relative target size of each marker.

The experimental results led to several conclusions. First, the rate at which ethidium bromide converted ρ^+ to ρ^- mutants was greater than the rate at which any of the ant^r markers were converted to ant^0 (Fig. 5–3). This was reflected in the relative target sizes of the markers, with E^R and C^R being 65 to 70% of the size of ρ^+ (Table 5–6). The results also illustrated that although the relative target size for each marker was constant in the two experiments shown, k was influenced by other factors. Thus, k was 10-fold greater for each marker in one strain than in the other (Table 5–6).

Second, if the ρ^+ factor and the ant^r genes represented independent targets, one would expect ρ^+ and ant^r to be converted most frequently by single events either to ρ^+ant^0 or to ρ^-ant^r (ρ^0ant^r), with the double event ρ^-ant^0 (ρ^0ant^0) being the product of the two single events. However, the results showed that ρ^+ant^0 cells were never found, despite an exhaustive search, whereas ρ^-ant^0 (ρ^0ant^0) cells were either as frequent as or more frequent than ρ^-ant^r (ρ^0ant^r) cells (Table 5–7). Hence, ant^r and ρ^+ did not behave as independent targets.

Third, the C^R and E^R markers were usu-

TABLE 5–6. Parameters of survival curves of mitochondrial genetic markers generated as a result of ethidium bromide inactivation

| | | Targets | | | | | | |
| | | Size (K) | | | | Relative size (%) | | |
Genotype	Target number (n)	K_ρ	K_C	K_E	K_0	K_C/K_ρ	K_E/K_ρ	K_0/K_ρ
$\rho^+\omega^+ C^R_{321}E^R_{514}$	2.0	0.65	0.47	0.47	—	72	72	—
$\rho^+\omega^+ C^R_{321}E^R_{221}O^R_1$	1.8	6.3	4.1	4.1	3.3	65	65	52

The parameters of the survival curves were calculated to fit the theoretical equation $S = 1 - (1 - e^{-K(t-t_0)})^n$ where S is survival of a given marker, K is the target size, n is the number of targets, t is exposure time, and t^0 is a delay. The derivation of each parameter is discussed in the text.

Adapted from Deutsch et al. (18).

TABLE 5–7. *Kinetics of survival of mitochondrial genetic markers following ethidium bromide treatment*[a]

Time (min)	Number of colonies scored	ρ^+ $C^RE^RO^R$ (%)	ρ^- $C^RE^RO^R$ (%)	ρ^- $C^RE^RO^0$ (%)	ρ^- $C^RE^0O^R$ (%)	ρ^- $C^RE^0O^0$ (%)	ρ^- $C^0E^RO^R$ (%)	ρ^- $C^0E^RO^0$ (%)	ρ^- $C^0E^0O^R$ (%)	ρ^- $C^0E^0O^0$ (%)
20	394	89.2	5.4	1.8	0.0	0.0	0.4	0.0	2.4	0.7
30	498	44.3	14.6	2.2	0.0	0.2	1.1	0.0	11.5	26.1
40	482	16.2	9.9	5.5	0.4	0.2	0.5	0.0	16.3	51.0
50	572	4.6	10.0	4.4	0.0	0.2	1.6	0.4	14.3	64.5
60	326	3.1	3.2	2.5	0.0	0.3	0.5	0.9	10.8	78.7
75	481	0.8	1.6	0.2	0.0	0.4	0.0	0.8	4.5	91.6
90	491	0.2	0.0	0.2	0.0	0.4	0.0	1.0	7.7	90.4

[a] Kinetics of survival of the ρ^+, C^R, E^R, and O^R genetic markers following treatment of triply resistant cells of *S. cerevisiae* with $2\mu g/ml$ of ethidium bromide. The percentage of each of the possible genotypic classes is shown as a function of time of exposure.
From Deutsch et al. (18).

ally lost or retained as a pair (i.e., C^RE^R + $C^0E^0 >> C^RE^0 + C^0E^R$) among ρ^- mutants, whether the mutation was spontaneous or was induced by ultraviolet light or ethidium bromide (Table 5–7). This pair of markers and the O^R marker represented two different but not independent targets in terms of marker inactivation, since although $C^RE^RO^0$ and $C^0E^0O^R$ petites were frequent, C^0O^0 and E^0O^0 double mutants were in excess of expectation based on the product of the probabilities of inactivation of each marker singly. These findings are consistent with mapping results that will be discussed in Chapter 7 and that show that E^R and C^R are closely linked to each other but not to O^R. These results also ruled out the possibility that ρ^+ represented one part of the mtDNA, with the *ant*^{r} markers being localized in another part. This model predicted that ρ^-ant^0 (ρ^0ant^0) genotypes would be rare for both the C^RE^R and the O^R groups of genes, indicating loose linkage of ρ^+ to both sets of genes, or that ρ^-ant^0 (ρ^0ant^0) genotypes would be frequent for either C^RE^R or O^R, indicating linkage of ρ^+ to one or the other group of markers, but not both. In fact, $\rho^-C^0E^0$ and ρ^-O^0 genotypes were frequent, and the logical problem was created that ρ^+ must be tightly linked to two sets of markers that appear to be very loosely linked to each other on the basis of formal

genetic analysis and marker inactivation studies (see Chapter 7).

In summary, the hypothesis that best explains the results of Deutsch and associates is that the ρ factor is mtDNA and that the *ant*^{r} markers are therefore included within ρ^+. The mutation from ρ^+ to ρ^- is a macrolesion that may or may not include a specific *ant* gene. This hypothesis best accounts for all of the observations, including the fact that ρ^+ has a greater target size than the *ant*^{r} markers. More evidence in support of this hypothesis will be presented in Chapter 6.

A final piece of evidence supporting the theory of localization of the *ant* genes in mtDNA is the finding of Michaelis and associates (19) that genetic recombination between specific C and E alleles is accompanied by formation of recombinant mtDNA molecules. These experiments will be detailed in Chapters 6 and 7.

GENERAL MECHANISMS OF MITOCHONDRIAL ANTIBIOTIC RESISTANCE AND THE PROBLEM OF CROSS-RESISTANCE

Although the mechanisms by which mitochondrial gene mutations confer antibiotic resistance will be explored in more detail later (see Chapter 19), some general remarks on cross-resistance of different mutants to antibiotics are appropriate at this juncture, since cross-resistance essentially puts a limitation on the variety of mutant phenotypes available for genetic analysis. In addition, certain mutants have totally bizarre and unexpected patterns of cross-resistance.

Mutations to antibiotic resistance in *S. cerevisiae* may confer resistance to inhibitors of mitochondrial protein synthesis, inhibitors of respiratory electron transport, or inhibitors of the oligomycin-sensitive F_1 ATPase. The mutation can confer resistance directly on specific mitochondrial components such as mitochondrial ribosomes or the membrane factor of the oligomycin-sensitive F_1 ATPase, or the mutation can cause the cell or mitochondrial membranes to become impermeable to specific antibiotics. Most mitochondrial mutations to antibiotic resistance are cross-resistant only to closely related antibiotics. For example, mutations singly or doubly resistant to erythromycin and spiramycin can be isolated (20). Trembath et al. (21) found that each of three tightly linked and probably allelic mutations resistant to macrolide antibiotics had a different spectrum of resistance. Griffiths (22) reported that oligomycin-resistant mutants mapping in the *OLI1* and *OLI2* genes are generally cross-resistant only to related antibiotics such as rutamycin, ossamycin, and peliomycin. However, Griffiths et al. (23) reported that mutants resistant to venturicidin, a macrolide antibiotic different in structure from oligomycin, are also cross-resistant to oligomycin. These mutants map in a third mitochondrial locus called *OLI3*. In addition, the one venturicidin-resistant mutant not cross-resistant to oligomycin appears to map in the *OLI1* gene. Finally, a class of mutants resistant to venturicidin and to the unrelated inhibitor of the oligomycin-sensitive F_1 ATPase triethyl tin has been isolated; these mutants seem to be localized in a mitochondrial episome (see Chapter 7).

The aforementioned mutants have been called class II mutants by Griffiths (22), and with the exception of the mutants resistant to venturicidin and triethyl tin, they behave as typical mitochondrial mutants and are relatively conservative in terms of their cross-resistance spectra. In contrast, class I mutants have a spectrum of cross-resistance to unrelated antibiotics that is truly impressive. These pleiotropic mutants seem to be of nuclear origin, in part, but they often show more complex genetic behavior (see Chapter 7). For example, class I mutants selected for resistance to oligomycin are also cross-resistant to unrelated inhibitors of the oligomycin-sensitive F_1 ATPase (e.g., aurovertin, Dio 9, venturicidin), as well as to uncoupling agents, electron transport inhibitors (i.e., antimycin), and inhibitors of mitochondrial protein synthesis (e.g., chloramphenicol, erythromycin, mikamycin). Class I mutants almost certainly confer resistance by some generalized indirect mechanism (e.g., mitochondrial membrane permeability). Linnane and his colleagues studied one such mutant in detail; it was originally characterized as being resistant to mikamycin, an inhibitor of bacterial protein synthesis. Since this mutant and its properties have received considerable attention in the literature, it deserves mention here.

The story begins in 1970 with publication of an article by Bunn et al. (24). They hypothesized that mitochondrial protein synthesis could be made antibiotic-resistant either directly through alteration of a mitochondrial ribosomal protein or indirectly by a mutation altering the permeability of the mitochondrial membranes to antibiotic. This hypothesis was formulated to explain the different behaviors of mutants resistant to chloramphenicol, mikamycin, and erythromycin. Linnane et al. (13,25–27) had earlier shown that their erythromycin-resistant mutant behaved genetically like a mitochondrial mutant and that protein synthesis in isolated mitochondria was resistant to the drug. Therefore they postulated that the erythromycin-resistant mutant altered a ribosomal protein, as in certain bacteria. In contrast, Bunn et al. (24) found that the chloramphenicol- and mikamycin-resistant mutants, which they believed to be mitochondrial on the basis of genetic tests, behaved quite differently. When grown on a nonfermentable carbon source, the chloramphenicol-resistant mutant also appeared to be resistant to lincomycin, but the mikamycin-resistant mutant was even more bizarre in that it exhibited cross-resistance to a whole collection of unrelated antibiotics, including chloramphenicol, the macrolides, and oligomycin (24,28). Protein synthesis by isolated mitochondria was sensitive to the antibiotics to which the cells were resistant *in vivo*.

Bunn et al. (24) also examined the effects of the antibiotics on the induction of respiration in anaerobically grown cells of the erythromycin- and mikamycin-resistant mutants. The rationale for these experiments was originally set forth by Thomas and Wilkie (12); they assumed that because anaerobic growth alters the structure of the mitochondrial inner membrane (see Chapter 4), it might also alter the permeability of mitochondria to antibiotics. Thus a mutant that alters the mitochondrial membrane and thus protects the mitochondrial protein-synthesizing system of the mitochondria of aerobically growing cells might not do so in anaerobically growing cells because of the difference in structure of the inner membrane. Bunn et al. (24) reported that the anaerobic–aerobic transition was inhibited by mikamycin in both the erythromycin- and mikamycin-resistant mutants, but they found that this was not true in the erythromycin-resistant mutant with respect to erythromycin.

As a result of their experiments, Bunn et al. (24) suggested that the mikamycin- and chloramphenicol-resistant mutants actually altered the mitochondrial membrane, making it impermeable to antibiotic, and

that during isolation the mitochondria became damaged, so that antibiotics could enter the mitochondria and block protein synthesis. The inhibition of respiration by mikamycin during aerobic adaptation of the mikamycin-resistant mutant was accounted for in terms of differences in membrane structure of aerobic and anaerobic mitochondria that permitted entry of the antibiotic into the latter class of mitochondria, with consequent blockage of mitochondrial protein synthesis. The development of respiration in the erythromycin-resistant mutant in the presence of erythromycin was explained by the fact that the mutant had antibiotic-resistant mitochondrial ribosomes, and thus differences in mitochondrial membrane structure would be without effect on mitochondrial protein synthesis. However, mikamycin concentration curves were not reported; thus it can be argued that a difference in sensitivity between the mikamycin-resistant mutant and the wild type might have been seen at appropriate concentrations. After all, the hypothesis could also imply that the antibiotic sensitivity levels of promitochondria and mitochondria are different because of structural differences between them that are unrelated to genotype.

The membrane barrier hypothesis was elaborated and extended in further articles to include the observation that high concentrations of macrolides, chloramphenicol, and mikamycin also interfere directly with respiration (29,30). Linnane and associates supposed that antibiotics that inhibit mitochondrial protein synthesis interact with an integrated ribosome-membrane system, with consequent secondary effects on other membrane functions (e.g., electron transport in the case of mikamycin).

The membrane-ribosome hypothesis came under suspicion first as it applied to chloramphenicol resistance mutations and later as it was purported to be a satisfactory explanation for the mikamycin-resistant mutants. Grivell et al. (31) found that mitochondrial mutations resistant to chloramphenicol had chloramphenicol-resistant mitochondrial ribosomes. They also showed that chloramphenicol inhibition of protein synthesis in isolated mitochondria of resistant mutants and wild type was clearly concentration-dependent, with wild-type mitochondria being more sensitive than those of either mutant tested, but with the two mutants differing from each other in sensitivity. These findings implied that mutations to chloramphenicol resistance conferred resistance directly on mitochondrial ribosomes; so there was no need to invoke membrane participation. Furthermore, there is no reason to suspect that the mutant studied by Bunn and associates is not allelic with the chloramphenicol-resistant mutants studied in other laboratories, since only one gene has thus far been identified (Table 5–2) (see Chapter 7).

Howell et al. (32) then found that the mikamycin-resistant mutant was really a double mutant containing a dominant nuclear gene mutation and a mitochondrial mutation. The nuclear mutation by itself conferred resistance to oligomycin, chloramphenicol, and intermediate levels of mikamycin (25 μg/ml). That is, the nuclear mutation was a typical class I mutant in the terminology of Griffiths. The mitochondrial mutation conferred resistance only to low levels of mikamycin (5 μg/ml). In the double mutant, however, the two mutations behaved synergistically to produce high levels of mikamycin resistance ($>$ 100 μg/ml). The authors supposed that the nuclear mutation reduced permeability to a group of unrelated antibiotics, but the biochemical basis of mikamycin resistance of the mitochondrial mutation remained obscure. As detailed in a recent article by Groot Obbink et al. (33), the nature of the mitochondrial resistance seems finally to have been resolved. Chromatography of the commercial mikamycin preparation used in these experiments revealed that it contained a minor contaminant in addition to mikamycin A and B. Although both mikamycin A and B appear to block mitochondrial protein synthesis *in vitro,* neither antibiotic blocks cell growth on a nonfermentable carbon source, even at high concentrations; thus it appears that mikamycin is not acting as a mitochondrial protein synthesis inhibitor *in vivo.* However, the minor contaminant does block cell growth on a nonfermentable carbon source, and it has now been determined that mikamycin-resistant mutants are cross-resistant to antimycin A; thus it appears that the contaminant present in the commercial preparations of mikamycin was an antimycin-A-like compound. Small wonder that the history of the mutation to mikamycin resistance is so confusing. Not only is it complex genetically, but also mikamycin blocks mitochondrial protein synthesis *in vitro,* and at one time it appeared that cells were sensitive to mikamycin *in vivo.* In fact, the cells were not sensitive to mikamycin *in vivo,* but to a contaminant in the mikamycin preparation that blocked respiratory electron transport in the manner of antimycin A.

Recently, mutations resistant to inhibitors acting in the complex III region of the respiratory chain (see Chapter 1) have been shown to map in a series of clusters in the cytochrome b (*COB*) region of the mitochondrial genome (see Fig. 19–8). These results are discussed in detail in Chapter 19.

ORIGIN OF MITOCHONDRIAL MUTATIONS TO ANTIBIOTIC RESISTANCE

Because mitochondrial mutations to antibiotic resistance segregate somatically in diploids from crosses between resistant and sensitive cells, it is logical to suppose that new mitochondrial mutants should be readily selected in diploids as well as haploids. Selection in diploids would be particularly useful because the background of recessive nuclear gene mutations to resistance could be eliminated through this method. Rank and Martin (34) made a systematic study of the use of diploids in selection of mitochondrial mutations to erythromycin and oligomycin resistance and found that all of the mutants they obtained showed the characteristic mitochondrial pattern of inheritance. A similar method has been developed in *C. reinhardtii* for the preferential selection of chloroplast mutants (see Chapter 13). However, it should be pointed out that dominant nuclear mutations will also be selected by this method. Thus, the mikamycin-resistant mutant studied so extensively by Linnane and his colleagues was selected using the diploid method by Bunn and associates (24). As was pointed out in the preceding section, it now appears that this mutant is a double mutant containing a dominant nuclear gene mutation plus a mitochondrial mutation.

The experiments with diploids previously discussed support the belief that newly arising mitochondrial mutations undergo somatic segregation prior to becoming homozygous, but they do not distinguish between the possibilities that these mutants arise spontaneously in the absence of a selective agent such as an antibiotic or arise by "adaptation" or "specific mutation" in the presence of antibiotic. This was a crucial question early in the development of bacterial genetics, for it was widely believed that adaptation rather than spontaneous mutation was responsible for the production of bacterial variants. Consequently, various elegant experimental designs were developed to distinguish between the clonal distribution expected for spontaneous mutations arising at random prior to exposure to a selective agent and the random (Poisson) distribution expected for cells that had adapted to the selective agent. The first such test was the fluctuation test designed by Luria and Delbrück (35) (see Chapter 13), who showed that *E. coli* mutants to bacteriophage resistance arose clonally by spontaneous mutation prior to exposure of the cells to virus rather than randomly by adaptation to the presence of virus.

Later, Newcombe (36) designed the respreading experiment, a remarkably simple but definitive technique, and confirmed the fluctuation test results. Birky (37) used the respreading experiment to examine the origin of mitochondrial mutations to erythromycin resistance in *S. cerevisiae*. In this experiment, cells are allowed to grow on solid medium in the absence of the selective agent for some period of time, after which the cells on some of the plates are redistributed. The unspread and respread cells are then exposed to the selective agent. In Birky's experiments, the cells were grown on the surfaces of filters on solid media lacking erythromycin, and the filters were transferred to the selective medium. Then the cells on some of them were respread.

If spontaneous mutation is operative, the respread plates will have vastly more colonies on them than the unspread plates, because mutants arising prior to exposure to the selective agent will have formed clones that will be redistributed into single cells on the respread plates but not the unspread plates. If, on the other hand, cells adapt to the presence of the selective agent or are mutagenized by it, respreading will have no effect, since the mutant cell numbers on the respread plates and unspread plates are the same.

Following transfer of the filters from nonselective medium to selective medium, Birky found that resistant colonies appeared on the respread and unspread plates at precisely the same rate. However, in a control experiment where a small number of resistant cells had purposely been mixed with a great excess of sensitive cells, respreading of the filters following transfer to the selective medium yielded a great excess of resistant colonies. This control indicated that if erythromycin-resistant mutants had arisen prior to respreading, they should have been detected. Crosses showed that most of the

mutants Birky obtained were, in fact, mitochondrial. Birky also obtained the same result with mutations to chloramphenicol resistance. Finally, he did an experiment in
which he exposed cells simultaneously to
chloramphenicol and erythromycin, following which the filters were transferred to
erythromycin, and some of them were respread. The idea was that if erythromycin
were a mutagen, it should induce erythromycin-resistant mutants, but there should be
no selection for these mutants, since the mitochondria should still be chloramphenicol-
sensitive. On transfer, however, an increased
frequency of erythromycin-resistant mutants
should be seen if induction had occurred.
There was no indication of mutant accumulation during the pregrowth period.

Birky considered three explanations for
his results. The first was intercellular selection favoring cells with erythromycin-sensitive mitochondria in the absence of erythromycin and cells with resistant mitochondria
in the presence of the drug. Control experiments with artificial mixtures of resistant
and sensitive cells ruled out this possibility,
although this is not to say that there could
not have been some intercellular selection
even in the control experiments. The second
explanation was mutation induction by the
antibiotic. As Birky pointed out, this explanation is improbable because erythromycin and chloramphenicol are translational inhibitors, not mutagens, as far as is
known. Moreover, there was no evidence
that erythromycin-resistant mutations were
induced in the experiment where the cells
were preincubated with a mixture of chloramphenicol and erythromycin. Finally,
Birky indicated that Slonimski had observed
that erythromycin did not induce mitochondrial mutations to chloramphenicol resistance. It is important to consider these results in light of the claim by Sager that
streptomycin induces chloroplast mutations
in *Chlamydomonas,* including those resistant to streptomycin (see Chapter 13).

The third possible explanation, which

Birky favored, was that there was intracellular selection for erythromycin-resistant
mitochondria in the presence of erythromycin, but against such mitochondria in
the absence of the drug. Intracellular selection at the mitochondrial level involving
antibiotic-sensitive and -resistant mitochondria was demonstrated directly in *Paramecium* (see Chapter 10). Such an explanation was also considered by Gillham and
Levine (38) (see Chapter 13) to explain
the nonclonal distribution of chloroplast
mutations to streptomycin resistance in
Chlamydomonas.

A fourth possibility, which has not been
considered by anyone previously, deserves
mention here; it is that the mode of organelle DNA replication leads to a nonclonal distribution of mutants. A number of
years ago, Luria (39), in a penetrating
analysis, showed that the distribution of
mutants in bacteriophage T2 issuing from
single bacteria (single bursts) was clonal,
and he used his data to support an argument that the mode of replication of the
phage genome is essentially geometric.
However, he considered other possible
replication modes, including one in which
progeny genomes are produced from a
single template by a "stamping-machine"
process. In this mode, if one of the copies
mutates, it will not transmit the mutation
during growth of the virus in the bacterium
because all copies will be made from the
same template. This mechanism of DNA
replication would generate phage mutants
in a random (Poisson) fashion among the
progeny of single bursts.

Denhardt and Silver (40) repeated the
Luria experiment with the single-stranded
bacteriophage ϕX174 and showed that the
distribution of mutants is Poisson and is
consistent with a stamping-machine mode
of replication. This is a particularly interesting result in view of what is known about
ϕX174 replication (41). It is clear that in
this virus DNA replication is by a stamping-
machine process involving double-stranded

replicative intermediates that generate single-stranded progeny. The precise mechanism of φX174 replication is not important here; what is important is the possible implication for organelle DNA. It may be that the nonclonal distribution of organelle mutations can be explained, in part at least, by such a stamping-machine process, with only certain molecules serving as templates for replication. Another hint that such an explanation might not be too far-fetched comes from the disparity between target numbers and physical numbers of mtDNA molecules in *S. cerevisiae* (see Chapter 4) and clDNA molecules in *Euglena* (see Chapter 16). Possibly the low target numbers might be explained by assuming that the physical targets actually measured are the replicative templates and that these are lower in number than the total number of organelle DNA molecules.

ISOLATION AND CHARACTERIZATION OF THE *mit⁻* MUTANTS

A landmark in mitochondrial genetics was the isolation of mitochondrial mutations with defects in the respiratory chain and in oxidative phosphorylation (*mit⁻*). The first mutants qualifying as *mit⁻* mutants were reported in 1974 by Flury et al. (42). These mutants had pleiotropic effects on the respiratory chain and cytochrome oxidase, NADH–cytochrome c reductase, succinate–cytochrome c reductase, and the ATPase were among the components affected. In the following year Tzagoloff and his colleagues (43–45) reported the first really extensive isolations of *mit⁻* mutants, among which were mutants with specific defects in cytochrome oxidase, in cytochrome b, or in the oligomycin-sensitive F_1 ATPase. Since then, other groups, notably those of Linnane and Slonimski, have become actively involved in the search for and characterization of *mit⁻* mutants. Certainly this is currently one of the most active areas of mitochondrial genetics, as is illustrated by the extensive number of papers presented at two recent symposia (46,47).

The isolation of a plethora of desirable mutants after so many years of failure can be attributed to two causes. First, methods were developed for selective isolation of mitochondrial mutations other than vegetative petites. Second, a set of criteria was developed that permitted rapid and effective discrimination between *mit⁻* mutants and vegetative and segregational petites. The discovery by Putrament et al. (48,49) that manganese could be used for specific induction of mitochondrial mutations under conditions where vegetative petite induction was minimized provided a powerful technique for obtaining new mitochondrial mutations in large numbers. The rationale for use of manganese came from experiments with bacteria and bacteriophage in which it was shown that DNA polymerase mutants had a high spontaneous mutation frequency because the mutant polymerases frequently made replication errors. A similar mutagenic mechanism had been proposed by Orgel and Orgel (50) to explain why manganese was mutagenic in phage T4. That is, DNA polymerases have a definite preference for Mg^{2+}, and replacement of Mg^{2+} by Mn^{2+} causes the polymerase to become prone to errors. As Putrament et al. (48) pointed out, the mitochondrial DNA polymerase of yeast is also very sensitive to manganese, which makes this enzyme a good candidate for manganese-induced replication errors. By means of experiments in which the relative amounts of Mn^{2+} and Mg^{2+} in the induction medium were varied, Putrament et al. (48) were able to establish optimal conditions for induction of mitochondrial mutations to antibiotic resistance such that these mutations were induced with high frequency but vegetative petites were not. Kotylak and Slonimski (51) added another useful technical innovation involving use of the nuclear mutant op_1, in which oxidative phosphorylation is impaired (52). Kováčová et al. (53)

found that the double mutant of op_1 and a vegetative petite was lethal. However, Kotylak and Slonimski (51) reported that *mit⁻* mutants were not lethal in combination with op_1. Consequently, *mit⁻* mutants can be obtained by mutagenesis of op_1 without the complication of screening out vegetative petites. Rytka and associates (54) reported yet another method for selecting *mit⁻* mutants. Their selection was based on the observation of Puglisi and Algeri (55) that induction of the enzymes required for galactose metabolism requires a functional mitochondrial protein-synthesizing system that, of course, is lacking in a vegetative petite. Rytka and associates succeeded in selecting *mit⁻* mutants by plating mutagenized cells on galactose medium and picking the small colonies. Under these conditions vegetative petites cannot form colonies, since their lack of a functional mitochondrial protein-synthesizing system prevents them from elaborating the enzymes necessary to metabolize galactose, whereas this constraint does not apply to the *mit⁻* mutants. It is interesting that many of the mutants obtained could not subsequently be grown on galactose, despite the fact that they were initially selected in this way.

Three criteria have commonly been used to distinguish *mit⁻* mutants from vegetative and segregational petites. The first is to determine whether the mutants that are obtained will revert. Since vegetative petites are deletion mutants, they are obviously incapable of reversion. Selection of only those

mutants that revert automatically eliminates those mutants that are *mit⁻* but that contain small deletions in their mtDNA. Such mutants do exist and are distinct from vegetative petites (56). The ·second criterion involves determining if a given mutant is capable of mitochondrial protein synthesis. Tzagoloff and associates (44) miniaturized this technique and made it very efficient so that many mutant isolates could be screened rapidly. In essence, small aliquots of cells are labeled with radioactive leucine in the presence of cycloheximide so that only those proteins made on mitochondrial ribosomes will be labeled. Vegetative petites show no evidence of leucine incorporation, whereas *mit⁻* mutants do. Obviously, use of this criterion eliminates the *syn⁻* mutants, which, like petites, lack mitochondrial protein synthesis. To distinguish *mit⁻* mutants from segregational petites, advantage is taken of the complementation test described in Chapter 4. The putative *mit⁻* mutant is crossed to a ρ^0 mutant, and the resulting diploids are scored for respiration deficiency (Table 5–8). The diploids will be respiration-deficient in the case of a *mit⁻* mutant, since complementation cannot occur because the lesion is in the ρ factor, and this is lacking completely in the ρ^0 mutant. In contrast, complementation will occur for a segregational petite, since it contains a ρ^+ factor and a nuclear *pet* mutation, whereas the ρ^0 mutant carries the wild-type *PET* allele (Fig. 4–10) (see Chapter 4). Other standard tests described earlier in this chap-

TABLE 5–8. *Use of the complementation test with a ρ^0 tester to differentiate between segregational petites and mit⁻ mutants*

Genotype of stock	Growth on nonfermentable carbon source	Phenotype of diploids produced in a cross to a ρ^0 PET mutant	
		Respiration-sufficient	Respiration-deficient
ρ^+mit^+ *PET*	+	+	−
ρ^+mit^+ *pet*	−	+	−
ρ^+mit^- *PET*	−	−	+

ter for the *ant* mutants can also be employed to establish the mitochondrial nature of the *mit⁻* mutants. These include demonstrations that *mit⁻* mutants segregate mitotically in crosses to wild type, that *mit⁻* zygotes segregate 4:0 *mit⁻*:wild-type progeny, and that dyes such as acriflavine and ethidium bromide knock out the ability to transmit the *mit⁻* phenotype in crosses to wild type. With respect to meiotic segregation, it is important to note that because *mit⁻* mutants are, by definition, deficient in respiration or oxidative phosphorylation, vegetative diploids do not sporulate. This means that meiotic segregation of the *mit⁻* phenotype must be studied in newly formed zygotes with wild-type cells, as is the case for suppressive petites (see Chapter 4). In addition, *mit⁻* mutants cannot be selected in diploids if one wishes to analyze them genetically. Mapping of the *mit⁻* mutants has proceeded very rapidly because of the use of ρ⁻ mutants in which the sizes of the deleted mtDNA segments are well established (see Chapter 7). The *mit⁻* mutants thus far described include mutants deficient specifically in cytochrome oxidase, cytochrome b, and oxidative phosphorylation (Table 5–2). In addition, several groups of mutants have been described which lead to simultaneous loss of cytochrome b and cytochrome oxidase activity. We will say more about all the mutants later (see Chapters 7 and 19).

OTHER MITOCHONDRIAL MUTANTS AND THE VARIANT POLYPEPTIDES

The use of mitochondrial protein synthesis to discriminate between *mit⁻* mutants and vegetative petites automatically eliminates point mutants blocked in mitochondrial protein synthesis (*syn⁻*). Since mtDNA codes for a complete set of tRNAs, mitochondrial rRNA (see Chapter 3), and possibly other factors associated with mitochondrial protein synthesis as well, *syn⁻*

mutants are certainly to be expected. The problem is how to distinguish efficiently between *syn⁻* mutants and vegetative petites. Tzagoloff et al. (57) suggested two procedures for distinguishing between *syn⁻* mutants and vegetative petites. The first is simply to select those mutants that revert, which automatically excludes vegetative petites. The second is to cross mutants lacking mitochondrial protein synthesis with a collection of ρ⁻ mutants containing defined deletions (see Chapters 6 and 7). A *syn⁻* mutant will yield wild-type diploids when crossed to those ρ⁻ mutants retaining the *syn⁺* allele, since the *syn⁺* allele can replace the *syn⁻* allele in the otherwise ρ⁺ factor of the mutant through recombination with the ρ⁻ mtDNA. Crosses between ρ⁻ mutants, on the other hand, never yield wild-type diploids. Mahler and associates (58) suggested a different scheme that should lead to isolation of *mit⁻*, *syn⁻*, and perhaps other mitochondrial mutants as well. The idea is to mutagenize a strain carrying distantly linked mitochondrial mutations to antibiotic resistance (i.e., *OLI1, OLI2, PAR1, RIB1*) that essentially span the mitochondrial map (Fig. 7–14). According to Mahler and associates, vegetative petites do not retain even three of these markers in a stable manner. Respiration-deficient mutants retaining these resistance markers are referred to as genomic integrity mutants or *gin⁻* mutants. To distinguish between *gin⁻* mutants and vegetative petites, each mutant clone is mated to a wild-type clone, and the resulting diploids are tested for transmission of three of the four antibiotic resistance markers. If the original mutant clone has retained an intact mitochondrial genome, the three resistance markers will be transmitted to some of the diploid progeny; if it has not, only certain of the markers, or perhaps none of them, will be transmitted. Putative *gin⁻* mutants identified in this way are then distinguished from segregational petites by the complementation test with a ρ⁰ tester. A *gin⁻* mutant, like the *mit⁻* and *syn⁻* mutants, is incapable of complementing a ρ^0 mutant, whereas a segregational petite will complement a ρ^0 mutant, as discussed previously. Mutants identified as *gin⁻* by these tests can then be screened for their ability to carry out mitochondrial protein synthesis and for respiratory defects.

A different approach to the identification of mitochondrial gene products has recently been devised by Douglas and Butow (59). The rationale underlying this scheme is to identify variant (*var*) mitochondrial translation products and then map these. A given variant may be either present or absent in a given strain (e.g., *var2* or *var3*), or molecular weight variants of the same polypeptide (e.g., *var l-s* and *var l-f*) may be present in different strains. To identify mitochondrial translation products specifically, cells are labeled with $^{35}SO_4$ in the presence of cycloheximide, so that cytoplasmic protein synthesis is blocked while mitochondrial protein synthesis continues (see Chapters 3 and 19). The labeled mitochondrial translation products are then separated by SDS-gradient gel electrophoresis and identified on autoradiograms of the gels. Variant polypeptides identified in this manner can be identified as mitochondrial gene products by making diploids between a strain containing a specific variant and a ρ^0 mutant of a strain lacking the variant. If the variant polypeptide is found among the mitochondrial translation products, it must, by definition, be a mitochondrial gene product. The other "standard" tests of mitochondrial inheritance can also be performed, and the variants can also be mapped (see Chapter 7).

CONCLUSIONS

The discovery of the *ant* mutants made systematic analysis of mitochondrial genetics possible. Experiments using these mutants were principally responsible for our present understanding of the processes of segregation and recombination of mitochondrial

genes in *S. cerevisiae* and for establishment of workable methods for genetic mapping of mitochondrial mutants. By use of these mutants it became possible to demonstrate that different ρ^- mutants retained different segments of wild-type mtDNA, and it became apparent that these mutants would be extremely useful for deletion mapping. In short, all of the tools required for genetic dissection of the mitochondrial genome of *S. cerevisiae* existed by the time the *mit⁻* mutants were discovered. The discovery of these mutants, together with the possibility of isolating yet other mitochondrial mutants (*vide supra*), has provided the material necessary for detailed genetic mapping of the mitochondrial genome. At the same time, powerful molecular techniques have been brought to bear on the problem. Efforts toward identification, mapping, and biochemical analysis of new mitochondrial mutations in *S. cerevisiae* have reached a fever pitch, as the many articles in two recent symposium volumes (46,47) attest. We may confidently expect that within a few years the mitochondrial genome of *S. cerevisiae* will be as well characterized as the genomes of many of the coliphages. It will then be possible to analyze the regulatory mechanisms that keep the replication of the mitochondrial genome and its expression attuned to events in the nucleus and the cytoplasm. In the two chapters that follow, we will analyze the nature of the vegetative petite mutation in detail, as well as the segregation, recombination, and mapping of mitochondrial genes in *S. cerevisiae*. Later, in Chapters 19 and 21, we will turn to the functions controlled by these genes.

REFERENCES

1. von Borstel, R. C. (1969): *Microbial Genetics Bulletin No. 31. (Yeast Genetics* supplement). Oak Ridge National Laboratory, Tennessee.
2. Sherman, F., and Lawrence, C. W. (1974): Saccharomyces. In: *Handbook of Genetics, Vol. 1,* edited by R. C. King, pp. 359–393. Plenum Press, New York.
3. Plischke, M. E., von Borstel, R. C., Mortimer, R. K., and Cohn, W. E. (1976): Genetic

markers and associated gene products in *Saccharomyces cerevisiae*. In: *Handbook of Biochemistry and Molecular Biology, Vol. 2, Nucleic Acids* (3rd ed.), edited by G. D. Fasman, pp. 767–832. Chemical Rubber Co. Press, Cleveland.

4. Birky, C. W., Jr., Colson, A.-M., and Perlman, P. S. (1978): A proposal for a uniform nomenclature in mitochondrial genetics. *Mol. Gen. Genet.* (*submitted for publication*).

5. Huang, M., Biggs, D. R., Clark-Walker, G. D., and Linnane, A. W. (1966): Chloroamphenicol inhibition of the formation of particulate mitochondrial enzymes of *Saccharomyces cerevisiae*. *Biochim. Biophys. Acta,* 114:434–436.

6. Linnane, A. W., Biggs, D. R., Huang, M., and Clark-Walker, G. D. (1968): The effect of chloramphenicol on the differentiation of the mitochondrial organelle. In: *Aspects of Yeast Metabolism,* edited by A. K. Mills, pp. 217–242. Blackwell, Oxford.

7. Clark-Walker, G. D., and Linnane, A. W. (1966): *In vivo* differentiation of yeast cytoplasmic and mitochondrial protein synthesis with antibiotics. *Biochem. Biophys. Res. Commun.,* 25:8–13.

8. Clark-Walker, G. D., and Linnane, A. W. (1967): The biogenesis of mitochondria in *Saccharomyces cerevisiae*. A comparison between cytoplasmic respiratory-deficient mutant yeast and chloramphenicol-inhibited wild type cells. *J. Cell Biol.,* 34:1–14.

9. Groot, G. S. P., Kováč, L., and Schatz, G. (1971): Promitochondria of anaerobically grown yeast. V. Energy transfer in the absence of an electron transfer chain. *Proc. Natl. Acad. Sci. U.S.A.,* 68:308–311.

10. Perlman, P. S., and Mahler, H. R. (1974): Depression of mitochondria and their enzymes in yeast: Regulatory aspects. *Arch. Biochem. Biophys.,* 162:248–271.

11. Wilkie, D., Saunders, G., and Linnane, A. W. (1967): Inhibition of respiratory enzyme synthesis in yeast by chloramphenicol: Relationship between chloramphenicol tolerance and resistance to other antibacterial antibiotics. *Genet. Res.* (*Camb.*), 10:199–203.

12. Thomas, D. Y., and Wilkie, D. (1968): Inhibition of mitochondrial synthesis in yeast by erythromycin: Cytoplasmic and nuclear factors controlling resistance. *Genet. Res.* (*Camb.*), 11:33–41.

13. Linnane, A. W., Saunders, G. W., Gingold, E. B., and Lukins, H. B. (1968): The biogenesis of mitochondria. V. Cytoplasmic inheritance of erythromycin resistance in *Saccharomyces cerevisiae*. *Proc. Natl. Acad. Sci. U.S.A.,* 59:903–910.

14. Coen, D., Deutsch, J., Netter, P., Petrochilo, E., and Slonimski, P. P. (1970): Mitochondrial genetics. I. Methodology and phenomenology. *Symp. Soc. Exp. Biol.,* 24:449–496.

15. Birky, C. W., Jr. (1975): Zygote heterogeneity and uniparental inheritance of mitochondrial genes in yeast. *Mol. Gen. Genet.,* 141:41–58.

16. Gingold, E. B., Saunders, G. W., Lukins, H. B., and Linnane, A. W. (1969): Biogenesis of mitochondria. X. Reassortment of the cytoplasmic genetic determinants for respiratory competence and erythromycin resistance in *Saccharomyces cerevisiae*. *Genetics,* 62:735–744.

17. Nagley, P., and Linnane, A. W. (1972): Biogenesis of mitochondria. XXI. Studies on the nature of the mitochondrial genome in yeast: The degenerative effects of ethidium bromide on mitochondrial genetic information in a respiratory competent strain. *J. Mol. Biol.,* 66:181–193.

18. Deutsch, J., Dujon, B., Netter, P., Petrochilo, E., Slonimski, P. P., Bolotin-Fukuhara, M., and Coen, D. (1974): Mitochondrial genetics. VI. The petite mutation in *Saccharomyces cerevisiae*: Interrelations between the loss of the ρ^+ factor and the loss of the drug resistance mitochondrial genetic markers. *Genetics,* 76:195–219.

19. Michaelis, G., Petrochilo, E., and Slonimski, P. P. (1973): Mitochondrial genetics. III. Recombined molecules of mitochondrial DNA obtained from crosses between cytoplasmic *petite* mutants of *Saccharomyces cerevisiae*: Physical and genetic characterization. *Mol. Gen. Genet.,* 123:51–65.

20. Netter, P., Petrochilo, E., Slonimski, P. P., Bolotin-Fukuhara, M., Coen, D., Deutsch, J., and Dujon, B. (1974): Mitochondrial genetics. VII. Allelism and mapping studies of ribosomal mutants resistant to chloramphenicol, erythromycin and spiramycin in *S. cerevisiae*. *Genetics,* 78:1063–1100.

21. Trembath, M. K., Bunn, C. L., Lukins, H. B., and Linnane, A. W. (1973): Biogenesis of mitochondria. 27. Genetic and biochemical characterization of cytoplasmic and nuclear mutations to spiramycin resistance in *Saccharomyces cerevisiae*. *Mol. Gen. Genet.,* 121:35–48.

22. Griffiths, D. E. (1975): Utilization of mutations in the analysis of yeast mitochondrial oxidative phosphorylation. In: *Genetics and Biogenesis of Mitochondria and Chloroplasts,* edited by C. W. Birky, Jr., P. S. Perlman, and T. J. Byers, Chapter 4. Ohio State University Press, Columbus.

23. Griffiths, D. E., Houghton, R. L., Lancashire, W. E., and Meadows, P. A. (1975): Studies on energy linked reactions: Isolation and properties of mitochondrial venturicidin-resistant mutants of *Saccharomyces cerevisiae*. *Eur. J. Biochem.,* 51:393–402.

24. Bunn, C. L., Mitchell, C. H., Lukins, H. B., and Linnane, A. W. (1970): Biogenesis of mitochondria. XVIII. A new class of cytoplasmically determined antibiotic resistant mutants in *Saccharomyces cerevisiae*. *Proc. Natl. Acad. Sci. U.S.A.,* 67:1233–1240.

25. Linnane, A. W. (1968): The nature of mitochondrial RNA and some characteristics of the protein-synthesizing system of mitochondria isolated from antibiotic-sensitive and resistant yeasts. In: *Biochemical Aspects of the Biogenesis of Mitochondria*, edited by E. C. Slater, J. M. Tager, S. Papa, and E. Quagliariello, pp. 333–353. Adriatica Editrice, Bari.

26. Linnane, A. W., Lamb, A. J., Christodoulou, C., and Lukins, H. B. (1968): The biogenesis of mitochondria. VI. Biochemical basis of the resistance of *Saccharomyces cerevisiae* toward antibiotics which specifically inhibit mitochondrial protein synthesis. *Proc. Natl. Acad. Sci. U.S.A.*, 59:1288–1293.

27. Lamb, A. J., Clark-Walker, G. D., and Linnane, A. W. (1968): The biogenesis of mitochondria. 4. The differentiation of mitochondrial and cytoplasmic protein synthesizing systems *in vitro* by antibiotics. *Biochim. Biophys. Acta*, 161:415–427.

28. Mitchell, C. H., Bunn, C. L., Lukins, H. B., and Linnane, A. W. (1973): Biogenesis of mitochondria. 23. The biochemical and genetic characteristics of two different oligomycin resistant mutants of *Saccharomyces cerevisiae* under the influence of cytoplasmic genetic modification. *Bioenergetics*, 4:161–177.

29. Linnane, A. W., and Haslam, J. M. (1970): The biogenesis of yeast mitochondria. *Curr. Top. Cell. Regul.*, 2:101–172.

30. Dixon, H., Kellerman, G. M., Mitchell, C. H., Towers, N. H., and Linnane, A. W. (1971): Mikamycin, an inhibitor of both mitochondrial protein synthesis and respiration. *Biochem. Biophys. Res. Commun.*, 43:780–786.

31. Grivell, L. A., Netter, P., Borst, P., and Slonimski, P. P. (1973): Mitochondrial antibiotic resistance in yeast: Ribosomal mutants resistant to chloramphenicol, erythromycin, and spiramycin. *Biochim. Biophys. Acta*, 312:358–367.

32. Howell, N., Molloy, P. L., Linnane, A. W., and Lukins, H. B. (1974): Biogenesis of mitochondria. 34. The synergistic interaction of nuclear and mitochondrial mutations to produce resistance to high levels of mikamycin in *Saccharomyces cerevisiae*. *Mol. Gen. Genet.*, 128:43–54.

33. Groot Obbink, D. J., Hall, R. M., Linnane, A. W., Lukins, H. B., Monk, B. C., Spithill, T. W., and Trembath, M. K. (1976): Mitochondrial genes involved in the determination of mitochondrial membrane proteins. In: *The Genetic Function of Mitochondrial DNA*, edited by C. Saccone and A. M. Kroon, pp. 163–173. North Holland, Amsterdam.

34. Rank, G. H., and Martin, R. (1972): A selective method for the enrichment of cytoplasmic markers in *Saccharomyces cerevisiae*. *Can. J. Genet. Cytol.*, 14:197–199.

35. Luria, S. E., and Delbrück, M. (1943): Mutations of bacteria from virus sensitivity to virus resistance. *Genetics*, 28:491–511.

36. Newcombe, H. B. (1949): Origin of bacterial variants. *Nature*, 164:150–151.

37. Birky, C. W., Jr. (1973): On the origin of mitochondrial mutants: Evidence for intracellular selection of mitochondria in the origin of antibiotic-resistant cells in yeast. *Genetics*, 74:421–432.

38. Gillham, N. W., and Levine, R. P. (1962): Studies on the origin of streptomycin resistant mutants in *Chlamydomonas reinhardtii*. *Genetics*, 47:1463–1474.

39. Luria, S. E. (1951): The frequency distribution of spontaneous bacteriophage mutants as evidence for the exponential rate of phage reproduction. *Cold Spring Harbor Symp. Quant. Biol.*, 16:463–470.

40. Denhardt, D. T., and Silver, R. B. (1966): An analysis of the clone size distribution of ϕX174 mutants and recombinants. *Virology*, 30:10–19.

41. Sinsheimer, R. I. (1968): Bacteriophage ϕX174 and related viruses. *Prog. Nucleic Acid Res. Mol. Biol.*, 8:115–169.

42. Flury, U., Mahler, H. R., and Feldman, F. (1974): A novel respiration-deficient mutant of *Saccharomyces cerevisiae*. *J. Biol. Chem.*, 249:6130–6137.

43. Tzagoloff, A., Akai, A., and Needleman, R. B. (1975): Properties of cytoplasmic mutants of *Saccharomyces cerevisiae* with specific lesions in cytochrome oxidase. *Proc. Natl. Acad. Sci. U.S.A.*, 72:2054–2057.

44. Tzagoloff, A., Akai, A., and Needleman, R. B. (1975): Assembly of the mitochondrial membrane system: Isolation of nuclear and cytoplasmic mutants of *Saccharomyces cerevisiae* with specific defects in mitochondrial functions. *J. Bacteriol.*, 122:826–831.

45. Tzagoloff, A., Akai, A., Needleman, R. B., and Zulch, G. (1975): Assembly of the mitochondrial membrane system: Cytoplasmic mutants of *Saccharomyces cerevisiae* with lesions in enzymes of the respiratory chain and in the mitochondrial ATPase. *J. Biol. Chem.*, 250:8236–8242.

46. Saccone, C., and Kroon, A. M. (editors) (1976): *The Genetic Function of Mitochondrial DNA*. North Holland, Amsterdam.

47. Bücher, T., Neupert, W., Sebald, W., and Werner, S. (editors) (1976): *Genetics and Biogenesis of Chloroplasts and Mitochondria*. North Holland, Amsterdam.

48. Putrament, A., Baranowska, H., and Prazmo, W. (1973): Induction by manganese of mitochondrial antibiotic resistance mutations in yeast. *Mol. Gen. Genet.*, 126:357–366.

49. Putrament, A., Baranowska, H., Ejchart, A., and Prazmo, W. (1975): Manganese mutagenesis in yeast. IV. The effects of magnesium, protein synthesis inhibitors and hydroxyurea on ant^R induction in mitochondrial DNA. *Mol. Gen. Genet.*, 140:339–347.

50. Orgel, A., and Orgel, L. E. (1965): Induction

of mutations in bacteriophage T4 with divalent manganese. *J. Mol. Biol.,* 14:453–457.

51. Kotylak, Z., and Slonimski, P. P. (1976): Joint control of cytochromes a and b by a unique mitochondrial DNA region comprising four genetic loci. In: *The Genetic Function of Mitochondrial DNA,* edited by C. Saccone and A. M. Kroon, pp. 143–154. North Holland, Amsterdam.

52. Kováč, L., Lachowicz, T. M., and Slonimski, P. P. (1967): Biochemical genetics of oxidative phosphorylation. *Science,* 158:1564–1567.

53. Kováčová, V., Irmlerová, J., and Kováč, L. (1968): Oxidative phosphorylation in yeast. IV. Combination of a nuclear mutation affecting oxidative phosphorylation with a cytoplasmic mutation to respiratory deficiency. *Biochim. Biophys. Acta,* 162:157–163.

54. Rytka, J., English, K. J., Hall, R. M., Linnane, A. W., and Lukins, H. B. (1976): The isolation and simultaneous physical mapping of mitochondrial mutations affecting respiratory complexes. In: *Genetics and Biogenesis of Chloroplasts and Mitochondria,* edited by T. Bücher, W. Neupert, W. Sebald, and S. Werner, pp. 427–434. North Holland, Amsterdam.

55. Puglisi, P. P., and Algeri, A. (1971): Role of the mitochondrion in the regulation of protein synthesis in the eukaryote *Saccharomyces cerevisiae. Mol. Gen. Genet.,* 110:110–117.

56. Slonimski, P. P., and Tzagoloff, A. (1976): Localization in yeast mitochondrial DNA of mutations expressed in a deficiency of cytochrome oxidase and/or coenzyme QH_2-cytochrome c reductase. *Eur. J. Biochem.,* 61:27–41.

57. Tzagoloff, A., Foury, F., and Akai, A. (1976): Resolution of the mitochondrial genome. In: *The Genetic Function of Mitochondrial DNA,* edited by C. Saccone and A. M. Kroon, pp. 155–161. North Holland, Amsterdam.

58. Mahler, H. R., Bilinski, T., Miller, D., Hanson, D., Perlman, P. S., and Demko, C. A. (1976): Respiration deficient mutants with intact mitochondrial genomes: Casting a wider net. In: *Genetics and Biogenesis of Chloroplasts and Mitochondria,* edited by T. Bücher, W. Neupert, W. Sebald, and S. Werner, pp. 857–863. North Holland, Amsterdam.

59. Douglas, M. G., and Butow, R. A. (1976): Variant forms of mitochondrial translation products in yeast: Evidence for location of determinants on mitochondrial DNA. *Proc. Natl. Acad. Sci. U.S.A.,* 73:1083–1086.

60. Dujon, B., Colson, A. M., and Slonimski, P. P. (1977): The mitochondrial genetic map of *Saccharomyces cerevisiae:* Compilation of mutations, genes, genetic and physical maps. In: *Mitochondria 1977,* edited by Schliersee, W. Bandlow, et al. DeGruyter, Berlin (*in press*).

Chapter 6
Mitochondrial Genetics of Bakers' Yeast. III. The Petite Mutation Revisited

In the last few years a great deal has been learned about the mechanism by which vegetative petites arise and the structure of ρ^- mtDNA. This rapid advance in knowledge can be attributed to several factors: the availability of genetic markers for ρ^- mtDNA in the form of mutations to antibiotic resistance; the discovery of a mutagen, ethidium bromide (EtdBr), whose action on mtDNA is remarkably well understood, inducing ρ^- and then ρ^0 mutants in a predictable sequence; and the application of powerful analytical physical-chemical techniques to the study of *S. cerevisiae* mtDNA. As a result, the level of experimental knowledge has reached the point that it should soon be possible to use ρ^- mutants for purification and physical analysis of specific genes and gene sequences. The general interest of such a system of gene purification is obvious.

In this chapter we will consider the induction of ρ^- mutations by specific mutagens, the structure of ρ^- mtDNA, the mechanism of suppressiveness, and models for the origin of ρ^- mutations.

MECHANISM OF PETITE INDUCTION BY EtdBr

Following the discovery by Slonimski et al. (1) that EtdBr causes quantitative conversion of wild-type cells to vegetative petites with no lethality in nongrowing cells or growing cells, the mechanism of action of the dye was studied in detail, particularly by Mahler and his student Perlman, who have written several useful reviews (2–5) in addition to many original research papers. Although EtdBr inhibits mtDNA

transcription in many systems (see Chapter 3) and leads to reversible inhibition of mtDNA replication in certain petite negative yeasts (6,7), in animal cells growing in tissue culture (8–13), and in *Euglena* (14), EtdBr induces vegetative petites because of specific covalent binding of the dye to *S. cerevisiae* mtDNA, which triggers a series of steps during which the mtDNA is progressively degraded into smaller and smaller pieces until it is finally eliminated altogether (Fig. 6–1).

The effects of EtdBr on *S. cerevisiae* mtDNA *in vivo* have been studied by Goldring et al. (15,16) and by Mahler and Perlman (17–20). When native yeast mtDNA is extracted carefully and centrifuged in sucrose gradients, most of the molecules appear to be linear, with molecular weights of 25×10^6 d. Hence they are half the molecular weight of the circular duplex mtDNA of the yeast mitochondrion (see Chapter 2). Exposure of nongrowing, starved cells to EtdBr in buffer results in a shift in the size distribution of mtDNA from these linear half-molecules to quarter-molecules with molecular weights of approximately 12.5×10^6 d. This cleavage of mtDNA in the presence of EtdBr is traceable to the binding, probably covalently, of the dye to mtDNA (4,21–23). Since the dye molecules are known to intercalate between adjacent base pairs, conformational distortion of the helix quite probably occurs at the same time. Following dye binding, double-strand scission takes place, and the quarter-molecules are formed (Fig. 6–1).

No additional mtDNA degradation takes place unless an energy source is provided.

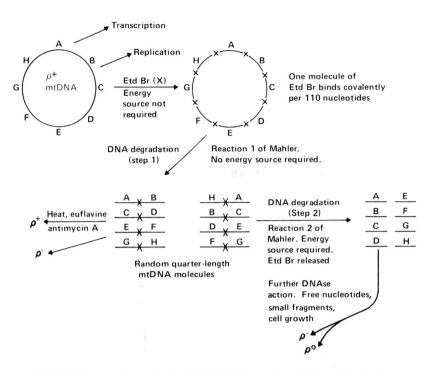

FIG. 6–1. Diagram of steps involved in EtdBr induction of vegetative petite mutations.

When an energy source such as glucose is added to the medium, additional degradation of the mtDNA occurs; this process, if permitted to go to completion, results in the total disappearance of mtDNA and the production of ρ^0 petites (Fig. 6–1). Release of the bound EtdBr accompanies degradation of the mtDNA (22). Removal of dye prior to complete degradation of mtDNA permits the remaining fragments to survive. These mtDNA fragments commence replication, and after a short time the rate of synthesis and the cellular level of these ρ^- mtDNA molecules approach those seen in wild type (24,25). It would appear that at least some of the surviving fragments would have to include the origin of mtDNA replication, assuming that such an origin exists.

Mahler and Bastos (3,4,22,26) succeeded in reconstructing the *in vivo* sequence of EtdBr-induced mtDNA degradation in isolated mitochondria; they recognized four separate reactions. The first reaction yields quarter-molecules and is formulated as follows:

$$mtDNA + mEtdBr \rightarrow mtDNA \cdot EtdBr_m \quad (1)$$

This reaction proceeds with kinetics and stoichiometry virtually identical to those seen for mtDNA degradation *in vivo*. The kinetic course of petite mutagenesis also coincides with that of reaction 1. The value of *m*, which may be expressed as the number of nucleotides for each molecule of EtdBr in the modification product, is 110 $\pm$ 10. Reaction 1 does not require any additional energy source and does not depend on intramitochondrial ATP. This reaction is not inhibited by the ATPase inhibitors oligomycin or Dio 9, but it is inhibited partially by lipophilic uncouplers such as carbonylcyanide-*m*-chlorophenylhydrazone (CCCP) and dinitrophenol (DNP).

Although the modification product is stable *in vitro* ($>$ 2 hr) and *in vivo*, addition of an energy source such as ATP or succinate results in rapid degradation of the modification product *in vitro,* as it does *in vivo.* Coincident with mtDNA degradation *in vitro,* activation of an endogenous mitochondrial ATPase occurs. To account for these findings, Mahler and Bastos formulated the following three reactions:

$$[EtdBr_m \cdot DNA'] + (ATP) \xrightarrow{\text{nuclease(s)}}$$
$$\text{fragments} + \cdots \quad (2)$$

$$mtDNA + mEtdBr + (ATP) \longrightarrow$$
$$\text{fragments} + \cdots \quad (3)$$

$$nATP \xrightarrow{[EtdBr_m \cdot DNA']} nADP + nP_i \quad (4)$$

For reasons that will become apparent later, Mahler and Bastos postulated that the ATPase activity observed was, in fact, that activity associated with the oligomycin-sensitive F_1 ATPase (see Chapter 1) and that there was close coupling between this enzymatic complex and the damage-repair system involved in the preservation of mtDNA integrity.

In order to understand why Mahler and Bastos postulated the aforementioned coupling, we must digress to discuss the effects of physiologic and genetic modulators on the degradation of mtDNA. At the same time, this will permit us to relate the biologic phenomenon of EtdBr induction of vegetative petites to the biochemical steps involved in mtDNA degradation more precisely. The induction of petite mutants by EtdBr can be modulated either genetically and physiologically, as was shown in a series of articles by Mahler and Perlman (5,19,20,27,28). Physiologic modulations of petite induction by EtdBr can be achieved by environmental alterations such as temperature changes or by use of specific reagents whose effects on mitochondria are well established (20,27). Physiologic modulations have been divided operationally into three categories by Mahler and Perlman (18,20), and this classification has been used in attempting to deduce the modes of

action of both EtdBr and the modulating agent in vegetative petite induction. *Protection* occurs when an agent added prior to EtdBr treatment reduces the mutagenic effectiveness of the dye, with *enhancement* being the converse. *Competition* results when a reagent added simultaneously with the dye affects the rate of mutagenesis. *Reversal* or *cure* is effected if the agent reduces the level of vegetative petites when added after the mutagenic exposure has been terminated by removal of all extracellular EtdBr. Modulation of the rate of vegetative petite induction can also be achieved with specific gene mutations.

The results of experiments with a variety of physiologic modulators are summarized in Table 6–1. The inhibitors of ATP transport, the ATPase, and the uncouplers block, at least partially, one or more of the reactions associated with EtdBr-induced degradation of mtDNA. The same is true of the respiratory inhibitor antimycin A and the acridine dye euflavine, which itself induces vegetative petite mutations. Both of these agents have been shown to slow the rate of petite induction under all three conditions of physiologic modulation (2), and both of them inhibit reactions 2 and 4 in the scheme of Mahler and Bastos. Similarly, specific gene mutations modulate the rate of vegetative petite induction, but in different ways (Table 6–2). The mutants $\rho 5$ (which is nuclear) and $\rho 72$ (which may be mitochondrial) modulate the rate of induction of petite mutants in opposite ways. Both mutants were isolated by Moustacchi (29). They are characterized by increased ultraviolet sensitivity to petite formation without any noticeable effects on cell lethality or rate of induction of nuclear gene mutations. They both appear to affect steps involved in repair of mtDNA following ultraviolet irradiation. Mahler and Perlman (20) found that $\rho 5$ protected cells against mutagenesis by EtdBr, whereas $\rho 72$ made cells much more susceptible to mutagenesis by

TABLE 6–1. *Effects of mitochondrial inhibitors and modulators of vegetative petite induction on EtdBr-induced degradation of mtDNA in isolated mitochondria and on the kinetics of vegetative petite induction*

| Agent | | Effects on reaction[a] | | | Modulation of petite induction kinetics | | |
Class	Representative	1	2(+ATP)[b]	4	Pro-tection[c]	Compe-tition	Cure
Inhibitors of AT(D)P transport	EDTA	Complete inhibition[d]			n.d.	n.d.	n.d.
	Atractyloside	None	Inhibits	Inhibits	n.d.	n.d.	n.d.
Inhibitors of ATPase	Dio 9	None	Inhibits	Inhibits	n.d.	n.d.	n.d.
	Oligomycin	None	Inhibits	Inhibits	n.d.	n.d.	n.d.
Uncouplers	Dinitrophenol	Inhibits	Inhibits	Inhibits	n.d.	n.d.	n.d.
	CCCP	Inhibits	Inhibits	Correlation with reaction 1[e]	n.d.	n.d.	n.d.
	Colicin K	Inhibits	Inhibits	Correlation with reaction 1[e]	n.d.	n.d.	n.d.
Respiratory inhibitors	Antimycin A	None	Inhibits	Inhibits	+	+	+
Modulators of mutagenesis	Euflavine	None	Complete inhibition	Complete inhibition	+	+	+
	Cycloheximide	None	n.d.	None	n.d.	+	n.d.
	Galactose	Inhibits	n.d.	None	?	+	?

[a] Dye effects in isolated mitochondria are given in terms of reactions 1, 2, and 4 of Mahler and Bastos.

[b] Reaction 2 was also run with succinate, and it gave the same results, with the following exceptions: atractyloside, none; euflavine, n.d.; galactose, none.

[c] The terms protection, competition, and cure are defined in the text; + = slower rate of petite induction than normal; − = faster rate of petite induction than normal; n.d. = absence of information.

[d] Since reaction 1 is required for reactions 2 and 4, its absence precludes effects on these reactions.

[e] Whenever reaction 4 was checked to see if there was any correlation with inhibition of reaction 1, it was found that reaction 4 was independent of and competitive with the presence of covalent modification product.

Adapted from Mahler et al. (2,4).

the dye. The observations made on the rates of petite mutagenesis in the two mutants are paralleled by those made on the degradation of mtDNA *in vitro* (Table 6–2). Thus, although $\rho 5$ experiences a normal reaction 1, reactions 2 and 4 do not occur in the mutant. In contrast, reactions 1, 2, and 4 are all accelerated in $\rho 72$. Other mutants that decrease the rate of petite mutagenesis also affect one or more of the reactions involved in mtDNA degradation *in vitro*. It is interesting that in petite negative yeast, reactions 1, 2, and 4 are all missing (Table 6–2), as would be expected if mtDNA degradation and petite induction showed obligatory coupling. In summary, all of the foregoing observations support a model that supposes a linkage between functions governed by the oligomycin-sensitive F_1

ATPase of *S. cerevisiae* and enzymes involved in the damage-repair system of mtDNA, as postulated by Mahler and Bastos. EtdBr triggers this system into activity, but damage and eventual loss of mtDNA result, and there is no repair. Thus far, this system seems to operate only in *S. cerevisiae* and presumably other petite positive yeasts.

The phenomenon of reversal deserves special comment before this discussion is closed. The genetic and physiologic correlates accompanying this process have been reviewed by Perlman (5). Antimycin A and euflavine are notably effective in reversing the process of induction of ρ^- mutants in nongrowing cells. Examination of the effects of reversing treatments on mtDNA indicates that reversing treatments primarily

TABLE 6–2. *EtdBr-induced degradation of mtDNA in isolated mitochondria of different mutant strains of S. cerevisiae and in wild-type petite negative yeasts*[a]

Strain tested	Phenotype	Muta-genesis rate	Effects on reactions[b] (relative to wild type)		
			1	2	4
$\rho 5$	ρ-mutation-prone for ultraviolet light	Slow	Normal	Absent	Absent
$\rho 72$	ρ-mutation-prone for ultraviolet light	Fast	Increased in extent	Accelerated	Increased in extent
rec 4	Recombination-deficient	Slow	Decreased in extent but not rate	Normal	Normal
73/1	mit⁻ with oligomycin-resistant ATPase	Slow	Decreased in extent but not rate	Normal in rate but oligomycin-resistant	Reduced in extent and oligomycin-resistant
Hansenula wingei	Petite negative	—	Absent	Absent	Absent
Torulopsis utilis	Petite negative	—	Absent	Absent	Absent
Kluyveromyces lactis	Petite negative	—	Absent	Absent	Absent

[a] Relative rates of vegetative petite induction in *S. cerevisiae* mutants are also indicated.
[b] Dye effects in isolated mitochondria are given in terms of reactions 1, 2, and 4 of Mahler and Bastos. Data from Mahler et al. (4) and Perlman (5).

interferes with the energy-dependent reactions, leading to degradation of the products of reaction 1, but such treatments do not appear to restore the quarter-molecules produced by this reaction to the half-molecules routinely obtained from cells not treated with EtdBr. Genetically, reversal is accompanied by a return in retention of antibiotic resistance markers from less than 50% to 100%. Perlman postulated that reversal involves recombination between the large fragments generated by reaction 1 to generate an intact ρ^+ mtDNA molecule. Presumably this recombination to yield a functional mtDNA molecule would have to involve fragments to which EtdBr was not covalently bound.

INDUCTION OF VEGETATIVE PETITES BY OTHER MUTAGENS

Despite the fact that many other mutagens, some of which have been known for many years, induce vegetative petites with high efficiency (see 30,31 for reviews), little is known of their modes of action. However, Mahler and his colleagues (28, 32,33) have initiated a systematic investigation of the requirements involved in petite induction by acridines and the trypanocidal dye berenil. Whereas acridines like phenanthridines become intercalated into mtDNA, berenil, which binds strongly to kinetoplast (see Chapter 2) and other circular DNAs, does not become intercalated.

The acridine dyes, with the exception of N^{10}-allyl proflavine, induce mutations only in dividing cells. This is consistent with the finding that these dyes cause petite induction only in buds and not in mother cells (see Chapter 4). However, N^{10}-allyl proflavine and berenil will cause petite induction in nongrowing cells in the presence of an energy source, although unlike EtdBr, neither compound will do so in the absence of an energy source. Experiments to determine the fate of mtDNA in cells treated with berenil and N^{10}-allyl proflavine show that these dyes have no effect in the absence of a carbon source but that mtDNA degradation is initiated if a carbon source is present (18,28). Perlman and Mahler (33) also studied the effects of berenil on vegetative petite induction in $\rho 5$ and $\rho 72$ and found that the results were parallel to those observed with EtdBr. That is, $\rho 5$ is relatively resistant to berenil mutagenesis, whereas $\rho 72$ is more sensitive than wild type.

These experiments show that neither N^{10}-allyl proflavine nor berenil can provoke the first step in EtdBr mutagenesis, namely, the scission of mtDNA in the absence of an energy source. However, both compounds initiate mtDNA degradation and petite induction in the absence of growth when an energy source is provided. Whether or not the enzymology of this reaction is the same as in the case of EtdBr mutagenesis remains to be seen.

Modulation of berenil mutagenesis by $\rho 5$ and $\rho 72$ in a manner similar to that seen for EtdBr suggests that at least two steps in the mutagenic pathway are common to berenil and EtdBr. Presumably $\rho 5$ protects cells against mutagenesis, because the mutant is defective in a function involved in recognition and excision of damage caused by both compounds. The $\rho 72$ mutant would affect a different step common to the mutagenic pathways of both compounds that causes mtDNA to become enhanced in its susceptibility to degradation.

Since the acridine dyes, with the exception of N^{10}-allyl proflavine, do not induce petite mutations in nongrowing cells in the presence of a carbon source, it seems likely that mutagenesis by these compounds involves steps different from those present in the EtdBr pathway or that mtDNA replication is required for these dyes to become intercalated and initiate the mutagenic process.

Vegetative petites are also induced with high efficiency by ultraviolet light, which, like EtdBr, acts on nongrowing cells irradiated in buffer (29,34–36). Careful studies of the ultraviolet induction process have yet to be reported, but the fact that ultraviolet light is effective on nongrowing cells in the absence of an energy source suggests that the ultraviolet pathway may share the first step of the EtdBr pathway. Other steps also appear to be common to the two pathways, since $\rho 5$ and $\rho 72$ were

selected as being especially sensitive to vegetative petite induction by ultraviolet light (20,29). The effect of $\rho 5$ is particularly interesting in view of the fact that the mutation *protects* cells *against* petite mutagenesis by berenil and EtdBr. As Mahler and Bastos (22) pointed out, if $\rho 5$ is involved in excision of damage, the rate of expression of the initial mutagenic event must be positively enhanced in the absence of excision in the case of ultraviolet light, but negatively enhanced in its absence in the case of EtdBr and berenil. The $\rho 72$ mutant presumably has the same effect in the case of all three mutagens and leads to excessive degradation of mtDNA.

It seems likely that over the next few years experiments of the type mentioned will allow us to determine if all the diverse mutagens known to cause vegetative petite induction act through some or all of the same steps in a common pathway of mutagenesis or if there are different routes to the same end result.

STABILIZATION OF GENETICALLY DEFINED ρ^- CLONES

Although all ρ^- clones are phenotypically similar, they are not genotypically similar, since they show different levels of suppressiveness, retain different mitochondrial markers, and contain mtDNAs with a variety of distinct base compositions. In order to study the structure of ρ^- mtDNA and relate this in a meaningful way to mitochondrial genotype, it is necessary to have genotypically defined and stable ρ^- mutants. Otherwise, one runs the risk of studying different populations of ρ^- mutants, even though they may have arisen from a single primary ρ^- clone.

The first studies of genotypic stability in ρ^- clones were those of Ephrussi and Grandchamp (37), who showed in cloning experiments that the level of suppressiveness in a clone is heritable. However, these experiments also showed that the level of suppressiveness is frequently altered in specific subclones and that the new level of suppressiveness itself is inherited when these subclones themselves are tested. Hence, although levels of suppressiveness are inherited, alterations occur with relatively high frequency. These observations were subsequently confirmed by Rank and Person (38).

Experiments on the heritability of suppressiveness suffer to some extent from the fact that suppressiveness must be measured essentially as a quantitative trait, with the subclones from a specific primary clone being distributed around some mean level of suppressiveness. The stability of ρ^- clones with respect to retention of a specific mitochondrial allele was first studied systematically by Linnane's group. Saunders et al. (39) showed that spontaneous $\rho^- E^R$ mutants varied widely in suppressiveness. They also examined subclones of a single spontaneous ρ^- mutant for changes in suppressiveness and retention of the E^R allele. They concluded that those ρ^- subclones that had mutated to E^0 were either more or less suppressive than the original clone.

Nagley and Linnane (24) extended these experiments on the stability of spontaneous ρ^- mutants to those induced by EtdBr. They treated $\rho^+ E^R$ cells with a mild dose of EtdBr (10 μg/ml) for 4 hr. Then they washed part of the culture free of EtdBr and allowed growth of the washed cells to continue for eight generations. They compared those cells treated with EtdBr for 4 hr with those that had grown for eight generations in the absence of the dye in terms of amounts of mtDNA, suppressiveness, and retention of the E^R allele.

Superficially, the cultures appeared quite similar in all respects, but clones from the cells grown in the absence of EtdBr proved very heterogeneous. About half the clones were neutral petites lacking mtDNA. Several clones contained mtDNA and were E^0, but in each case the level of suppressiveness was different. In one suppressive petite the

E^R marker had been retained to the level of 20%. When the suppressive petite clones were subcloned, all contained a fraction of neutral petites; they fell into two classes of suppressiveness. In one class, subclones were similarly suppressive. That is, they appeared to have stabilized. The second class, in contrast, had not stabilized, and there was wide variation in suppressiveness of the subclones.

The authors explained the instability of the EtdBr-induced ρ^- mutants in terms of heterogeneity of mtDNA molecules present in the mutant cells. Segregation of mtDNA molecules unable to replicate themselves into buds produced ρ^0 petites. Molecules capable of replicating themselves spawned ρ^- petites, but these molecules formed a heterogeneous collection in terms of retention of the E^R allele and suppressiveness. From among these cells, some were produced that contained relatively homogeneous collections of ρ^- mtDNA molecules, and these cells gave rise to clones of stable petites.

The foregoing experiments show that ρ^- mutants exhibit a certain degree of genotypic instability and must be cloned carefully to obtain stable lines. This problem has been clearly recognized by all groups of workers, and the experiments reported in this chapter have been done largely with genetically characterized ρ^- mutants cloned for stability.

STRUCTURE OF mtDNA IN WILD-TYPE *S. cerevisiae* AND THE ρ^- MUTANTS

Although ρ^+ mtDNA is a 25-μm circle with a molecular weight of 5×10^7 d, as shown by measurement of circle size in lysed mitochondria, as well as by renaturation kinetics and restriction endonuclease mapping (see Chapter 2), intact circles are not obtained on isolation of mtDNA. Instead, the most gentle techniques yield linear half-molecules, as discussed earlier in this chap-

TABLE 6–3. *Some characteristics of mtDNA from wild-type and petite mutants of S. cerevisiae*

Genotype	Origin	Suppressive-ness (%)	mtDNA as percentage of total	Buoyant density (g/cc)	Mole percent G+C determined from:		
					Buoyant density	T_m	Chemical analysis
PET ρ^+	—	—	14	1.683	23.5	13.2	17.4
pet$_1$ ρ^-	Spontaneous	95	10	1.683	23.5	10.2	15.6
pet$_1$ ρ^-	Spontaneous	78	?	1.683	23.5	10.0	15.5
PET ρ^-	Spontaneous	95	?	1.678	19.0	9.8	12.6
PET ρ^-	Ethidium bromide	99	17	1.683	23.5	?	?
PET ρ^-	Ethidium bromide	90	16	1.678	19.0	?	?
PET ρ^-	Ethidium bromide	50	17	1.683	23.5	?	?
PET ρ^-	Ethidium bromide	8	16	1.683	23.5	?	?
PET ρ^0	Ethidium bromide	< 1	< 1	—	—	—	—

Data from Bernardi et al. (43) and Nagley and Linnane (24).

ter. Some years ago, physical and chemical analysis revealed that ρ^+ and ρ^- mtDNAs are peculiar in structure and have exceptionally low G+C contents (40–42). Because of these anomalies, neither buoyant density nor thermal denaturation measurements accurately reflect the base composition of either ρ^+ or ρ^- mtDNA (Table 6–3). The penetrating studies of Bernardi and his colleagues (43–50) revealed much about the structure of ρ^+ mtDNA and resulted in a precise model of its structure (Fig. 6–2). We will begin this account of ρ^+ and ρ^- mtDNA with a brief description of Bernardi's model and its supporting data.

The model of Bernardi supposes that the wild-type mitochondrial genome is divided into A+T-rich spacers and G+C-rich genes (Fig. 6–2). These two sequences represent about half of the mitochondrial genome each; they have an average size greater than 1.6×10^5 d and are interspersed with each other in the mtDNA. The genes have an average G+C content of 26%, whereas the G+C content of the spacer is less than 5%. Associated with the genes are two elements of even richer G+C content. These are called G+C-rich clusters (60% G+C, molecular weight approximately 3×10^4 d) and site clusters (45–62% G+C, molecular weight approximately 2×10^4 d). The site clusters are distinguished from the G+C-rich clusters by their sensitivity to specific restriction endonucleases, as will be discussed later.

The supporting data for Bernardi's model

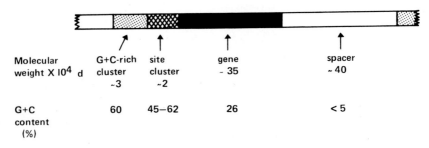

FIG. 6–2. Model of organization of yeast mtDNA (after Bernardi, 50).

have been gathered over the past 8 to 10 years in a series of physical-chemical and biochemical studies. The first evidence for compositional heterogeneity in wild-type yeast mtDNA was published by Bernardi et al. (43). They found that the G+C content of this DNA calculated from buoyant density or from thermal denaturation measurements did not agree with the values obtained by direct chemical analysis (Table 6–3). More important, these authors obtained evidence of striking compositional heterogeneity from analysis of the first derivative of the thermal denaturation curve, which permits a much more precise analysis of the fine structure of this curve (Fig. 6–3). Since yeast mtDNA had an unusually high A+T content (Table 6–3), the differential melting curve for this DNA was compared to the curves obtained for the synthetic polymers poly-(dAT:dAT) and poly-(dA:dT). In the former polymer, A and T alternate regularly in each strand; in the latter polymer, one strand contains only A residues, and its complement contains only T residues. The melting profiles of the

two synthetic polynucleotides were gaussian around their respective T_m values, but yeast mtDNA exhibited a distinctly nongaussian distribution, indicating compositional heterogeneity and the presence of several distinctive melting components. The left-hand part of the curve, amounting to about half of the DNA, consisted of a single component with a very high A+T content; the right-hand part of the curve trailed off over a wide temperature range, indicating the presence of several components with higher, but variable, G+C content.

Bernardi and associates then published three articles (45–47) on the compositional heterogeneity of yeast mtDNA and proposed the working hypothesis that wild-type yeast mtDNA contained A+T-rich and G+C-rich stretches that were intermingled with one another. The latter stretches were presumed to have a "normal" structure, whereas the former stretches were assumed to be responsible for the anomalous physical properties of yeast mtDNA. Since pieces of mtDNA with molecular weights of 2×10^6 d yielded single sym-

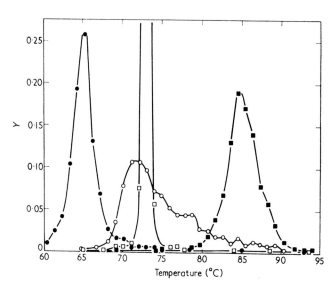

FIG. 6–3. Differential melting curves for synthetic polymers dAT:dAT (*solid circles*) and dA:dT (*open squares*) and for mtDNA (*open circles*) and nDNA (*solid squares*) from wild-type *S. cerevisiae*. Ordinate indicates increment in relative absorbance per degree: $Y = [(A_{t_1} - A_{t_2})/(A_{100} - A_{25})]/(t_1 - t_2)$, where A_{t_1}, A_{t_2}, A_{100}, and A_{25} are absorbances measured at temperatures t_1, t_2, 100°C, and 25°C, respectively. Abscissa values are equal to $(t_1 + t_2)/2$. Y_{max} of poly-(dA:dT) had a value of 0.71. (From Bernardi et al., ref. 43, with permission.)

metrical peaks in buoyant density gradients, while showing clear evidence of intramolecular heterogeneity in thermal denaturation experiments, Bernardi and associates concluded that both types of sequences were present in pieces of this size. Supporting evidence for the existence of two sequence types was obtained through the use of hydroxylapatite chromatography and spleen acid DNAse digestion. Chromatography on hydroxyapatite yielded fragments eluting at low-molarity salt with a G+C content of 22% as opposed to 18% for native mtDNA. These segments comprised at least 1% of the mitochondrial genome and had molecular weights of approximately 1.5×10^6 d. In order to examine this compositional heterogeneity further, it was necessary to fragment the mtDNA more extensively. This was accomplished by controlled degradation of mtDNA with spleen acid DNAse followed by chromatography of the fragments on hydroxyapatite. This treatment yielded fragments having G+C contents of 26% and molecular weights of 1×10^5 d to 2×10^5 d and fragments having G+C contents of 14% and molecular weights of 8×10^5 d. Recurrent degradation of the latter fragments yielded fragments with only 10% G+C content and molecular weights of 2.5×10^5 d. These experiments indicated that the yeast mitochondrial genome consisted of long A+T-rich stretches and G+C-rich stretches with molecular weights of 10^5 to 10^6 d that were intermingled. Finally, oligonucleotide analyses of the 10% G+C fragments revealed that they were probably composed of repeated runs of 10 to 30 nucleotides consisting of nonalternating $dA:dT$ and alternating $dAT:dAT$ sequences, with each run being short (less than four base pairs).

Prunell and Bernardi (48) extended the enzymatic digestion studies to include micrococcal nuclease, an enzyme that preferentially attacks A+T-rich stretches. Following digestion, the macromolecular fragments derived from the G+C-rich regions were separated by gel filtration from the oligonucleotides originating from the A+T-rich spacers. These experiments revealed that the spacers were very homogeneous in base composition, with a G+C content under 5%, but that the G+C-rich sequences were heterogeneous in composition, with G+C content ranging from 25 to 50%, with an average of 32%. Small fragments with molecular weights of 4×10^4 d and G+C contents as high as 65% were also reported.

The micrococcal nuclease digestion studies of Prunell and Bernardi (48), coupled with the earlier experiments, established the existence of A+T-rich "spacers" and G+C rich "genes" in approximately equal amounts and with molecular weights greater than 1×10^5 d. Restriction enzyme analysis (49,50) led to subsequent identification of the restriction site clusters and the G+C-rich clusters (Fig. 6–2). The diagnostic enzymes were HaeIII (Hae) and HpaII (Hpa), which cleave GGCC and CCGG sites, respectively. In a few experiments, HhaI (Hha), which cleaves GCGC sites, was also used. Digestion of yeast mtDNA with Hae and Hpa revealed that the restriction sites for these enzymes were clustered with each other. Digestion with Hae yielded 71 fragments, and digestion with Hpa produced 107 fragments. When the Hae fragments were digested with Hpa, the Hae fragments were shifted down to positions corresponding to only slightly lower molecular weights, and a number of additional bands appeared, mostly in the region of lowest molecular weight. These data indicated clustering of the Hpa restriction sites within the region containing the Hae restriction sites. In the absence of clustering, the Hpa sites should have been located randomly within the Hae fragments. Cleavage of these fragments with Hpa would then have caused most of them to shift to much lower molecular weights. The reciprocal double digest provided more evidence for clustering, since digestion of the Hpa fragments with Hae did not lead to a change in the number of fragments, although some shifted in molecular weight. In other words, the 71 Hae sites were within clusters containing the 107 Hpa sites. The base composition and amount of the oligonucleotides released following digestion with these enzymes were also examined. It was estimated that 2.6 to 3.0% of mtDNA ($1.3–1.5 \times 10^6$ d per genome) was present in the restriction site clusters, and these had G+C contents of approximately 45%.

The association of the restriction site clusters with the G+C-rich clusters was demonstrated by first degrading mtDNA with micrococcal nuclease to eliminate the A+T-rich spacers and then genes that were 26% G+C (Fig. 6–2). Prunell and Bernardi (48) had earlier pointed out that this treatment yielded fragments with molecular weights of 4×10^4 d and G+C contents as high as 65%. Digestion of these fragments with Hae and Hpa released 15% of this material, equivalent to 1.4% of the mtDNA. The remaining material consisted of double-stranded fragments with G+C contents of 60%. This experiment showed that the site clusters and G+C-rich clusters were located in the same fragment and that the latter fragments contained even more G+C than the restriction site clusters. The experiment also showed that the G+C-rich clusters did not contain the quadruplets CCGG, GGCC, or GCGC, which are the targets of Hpa, Hae, and Hha, respectively. The G+C-rich fragments containing the restriction site clusters and the G+C-rich clusters accounted for 9% of the mtDNA. This was slightly greater than that estimated by Prunell and Bernardi (5×10^4 d).

Prunell and Bernardi (48) pointed out that the existence of spacer sequences in yeast mtDNA accounting for 50% of the total mtDNA means that the information content of this DNA is really only 2.5 times that of animal mtDNA, not 5 times, as was originally supposed. The existence of spacer sequences might also explain why there is such heterogeneity in the sizes of fungal mtDNAs (see Table 2–1).

Having discussed the structure of ρ^+ mtDNA, we will turn now to the mtDNA of ρ^- petites. In terms of such usual physical parameters as buoyant density and thermal denaturation characteristics, ρ^- mtDNAs show a wide range of variation (Table 6–3). These differences, in part, reflect differences in base composition, organization of retained sequences, etc. In order to focus this discussion, we will consider the properties of ρ^- mtDNA in terms of five specific questions: What is the evidence for deletion, retention, and amplification of wild-type sequences in ρ^- mtDNA? Does ρ^- mtDNA contain new sequences distinct from those present in ρ^+ mtDNA? What are the form, size, and structure of ρ^- mtDNA? Do ρ^- mutants contain as much mtDNA as the wild type? Is ρ^- mtDNA capable of recombination? Once these questions have been answered as adequately as is possible at present, we will be in a position to consider explanatory models for the origin of ρ^- mutants and the mechanism of suppressiveness.

What is the evidence for deletion, retention, and amplification of wild-type sequences in ρ^- mtDNA? A combination of genetic and molecular methods has been used to establish that ρ^- mtDNAs have large segments of the wild-type mitochondrial genome deleted and that the retained sequences are amplified. Retention or loss of a functional gene in mtDNA can be ascertained genetically (as described in Chapter 5) by crossing a ρ^- mutant derived from a ρ^+ stock carrying a marker of interest (e.g., ant^r) to a ρ^+ stock carrying a different allele (e.g., ant^s) and asking whether the mitochondrial marker present in the parent stock of the ρ^- mutant (i.e., ant^r) is to be found among the resulting diploid progeny. If the marker is transmitted, it is, by definition, retained; if it is not transmitted, one cannot distinguish between the possibility that the whole gene is deleted and the possibility that only a portion of the gene is missing. In any event, application of this test to an array of mutants derived from wild-type stocks carrying different mitochondrial markers has revealed that there is no restriction on which genes may be retained, since libraries of ρ^- mutants, each carrying different mitochondrial markers, have been collected and used for deletion mapping of the mitochondrial genome of *S. cerevisiae* (see Chapter 7).

Molecular evidence for deletion and amplification of specific genes in ρ^- mtDNAs appears to have been reported first by Cohen et al. (51). It will be recalled that mitochondrial tRNAs are transcribed from mtDNA (see Chapter 3). If the sequence for a specific tRNA is deleted in a ρ^- mutant, there will be no significant hybridization of this tRNA to denatured mtDNA from the mutant. On the other hand, if the tRNA genes have been amplified in the ρ^- mutant, a given amount of ρ^- mtDNA will hybridize more tRNA than a comparable amount of ρ^+ mtDNA. Cohen and associates examined the ability of mtDNA from a spontaneous ρ^- mutant to hybridize mitochondrial valyl and leucyl tRNAs. These authors found that the mtDNA of the ρ^- mutant hybridized mitochondrial valyl tRNA from the parent ρ^+ strain very poorly, while hybridizing twice as much mitochondrial leucyl tRNA as the ρ^+ strain. The authors then showed that not only leucyl tRNA but also valyl tRNA could be isolated from the ρ^- mutant and then would hybridize with ρ^+ mtDNA. The results for leucyl tRNA were not unexpected; they showed that the ρ^- mtDNA was capable of transcribing a normal leucyl tRNA molecule. But the presence of mitochondrial valyl tRNA in the ρ^- strain was remarkable in view of the fact that valyl tRNA hybridized so poorly with ρ^- mtDNA. The authors therefore suggested that the ρ^- mutant might be genetically heterogeneous and that, whereas most cells lack the valyl tRNA cistrons, a few contain them.

Subcloning experiments showed that the ρ^- mutant was heterogeneous with respect to retention of the valyl tRNA cistron. Although mtDNA from some subclones hybridized valyl tRNA very poorly, mtDNA from others did hybridize valyl tRNA, but not as well as ρ^+ mtDNA. When leucyl tRNA was hybridized to mtDNA from the subclones, it was found in every case that leucyl tRNA was hybridized to an extent the same as or greater than the corresponding amount of ρ^+ mtDNA. These results suggested that the valyl tRNA cistron was deleted in certain subclones of the ρ^- mutant and retained in others, whereas the leucyl tRNA cistron was retained in all and amplified in some.

Casey et al. (52) then characterized in more detail the mtDNA of four of the ρ^- subclones studied by Cohen and associates. They found that the mtDNA was composed of fast- and slow-renaturing fractions that could be differentially enriched following chromatography on hydroxyapatite. The leucyl tRNA hybridized substantially more to the fast fraction than to the slow fraction, whereas such fractionation had no effect on the hybridization levels of ρ^+ mtDNA. When corrections were made for cross-contamination, it appeared that the leucyl tRNA cistrons were localized exclusively in the fast fractions in the mutants. When the same experiment was done with a clone that also hybridized valyl tRNA, it was found that the latter tRNA was localized in the slow fraction. These results indicate that the cistrons for leucyl tRNA and valyl tRNA are localized in different parts of the mitochondrial genome, with the former being amplified in the fast fraction.

Shortly after Cohen et al. (51) reported deletion and amplification of the valyl and leucyl tRNA genes in mtDNA of their ρ^- mutant and its subclones, other evidence for deletion and amplification began to accumulate rapidly. Nagley et al. (53) reported genetic evidence for amplification of an E^R allele in a study of the effects of EtdBr on ρ^- mtDNA. These authors grew ρ^+ cells and cells of several different petites in the presence of varying concentrations of EtdBr for eight generations and then analyzed the cells for mtDNA and retention of the E^R marker. They found that the amount of mtDNA and retention of the E^R marker decreased in a concentration-dependent manner in all clones. However, in the ρ^- clones, both the decrease in mtDNA and the loss of the E^R marker proceeded at slower rates than in the ρ^+ clones. These results showed that elimination of mtDNA by EtdBr was more resistant in ρ^- clones than in ρ^+ clones, which suggested EtdBr resistance of the ρ^- mtDNA as a possible explanation for the greater retention of the E^R gene by ρ^- clones. The authors then showed that EtdBr resistance was not the explanation by plotting the proportion of E^R genes retained as a fraction of the proportion of mtDNA retained. As amplification of the E^R genes increases, one would expect this ratio to decrease less rapidly with increasing doses of EtdBr. The authors found that this was true of both ρ^- clones tested when compared to ρ^+ clones. Most striking was the observation that although the kinetics of elimination of ρ^- mtDNA were the same for both ρ^- clones, the ratio of E^R genes to mtDNA retained was much greater for one clone than for the other. These results cannot be interpreted in terms of differential susceptibilities of the two ρ^- mtDNAs to EtdBr, but only in terms of greater amplification of the E^R genes in one of the ρ^- mutants than in the other.

Slonimski (54–60) and Rabinowitz (61–67) and their colleagues have also reported on the properties of ρ^- mtDNA in an exhaustive series of articles characterizing the mtDNA of a group of genetically defined stable petites. We will discuss this work in some detail here, as well as in the process of answering the next two of our five questions. In these articles, evidence for deletion of gene sequences, in the general sense, comes from three experimental approaches. First, retention of mitochondrial markers

and kinetic complexity of mtDNA generally decrease in parallel in different ρ^- mutants (Table 6–4). That is, the fewer markers retained in the genetic sense the less complex is the mtDNA in the molecular sense. The decrease in kinetic complexity of mtDNA is also accompanied by a parallel and equivalent decrease in the size of a monomer unit of mtDNA (Table 6–4). As we shall see later, mtDNA in ρ^- mutants increases in size in multiples of this basic monomer unit. Second, DNA-DNA hybridization experiments in which ρ^+ mtDNA in solution is hybridized to varying concentrations of ρ^- mtDNA on filters show that concentrations of ρ^- mtDNA equivalent to those ρ^+ mtDNA amounts sufficient to bind all of the ρ^+ mtDNA in solution only partially exhaust the ρ^+ mtDNA in solution. That is, the ρ^- mtDNAs lack sequences contained in ρ^+ mtDNA, and these sequences cannot be bound on the filters. The retention of sequences in common with ρ^+ mtDNA can be quantitated (Table 6–4). Third, if ρ^- mtDNA contains deletions, digestion of this mtDNA with restriction enzymes should yield fragment patterns different from those seen when ρ^+ mtDNA is digested. For example, if part of a fragment is deleted in a ρ^- mutant between restriction sites, the mutant should have the same number of restriction fragments in its mtDNA as does the wild type, but one of them should move more rapidly in the agarose gels. If a restriction site is deleted, a single fragment should replace two ρ^+ fragments in the ρ^- mtDNA. If the mtDNA of the ρ^- mutant consists of only a small portion of the ρ^+ mitochondrial genome, cleavage of this mtDNA with restriction enzyme should yield a much smaller number of fragments than are found in the wild type. Morimoto et al. (66) obtained results consistent with these predictions following digestion of ρ^- mtDNA with restriction enzymes. For example, cleavage of ρ^+ mtDNA with EcoR1 yielded nine discrete fragments, the sum of whose molecular weights was 51.8×10^6 d,

TABLE 6–4. *Some molecular properties of mtDNA from genetically defined ρ^- mutants*[a]

Clone	Genotype	Retention of sequences (%)	Kinetic complexity relative to ρ^+	Buoyant density (g/cc)	Monomer circle length (μm)	
					Measured	Calculated from kinetic complexity
IL8–8C	$\rho^+ C^R E^R O^S$	100	1	1.684	25	25
D21	$\rho^- C^R E^R O^0$	58	0.36	1.684	—	9.0
F11	$\rho^- C^R E^R O^0$	—	0.24	1.684	4.4	4.2[b]
F21	$\rho^- C^R E^R O^0$	—	0.18	1.684	—	4.5
R53	$\rho^- C^R E^R O^0$	44	0.125	1.685	—	3.1
E41	$\rho^- C^0 E^R O^0$	7	0.015	1.688	0.5	0.45[b]
F12	$\rho^- C^0 E^R O^S$	45	0.25	1.684	5.5	5.5[b]
F22	$\rho^- C^0 E^R O^0$	46	0.125	1.684	—	3.1
R535	$\rho^- C^0 E^R O^0$	28	0.08	1.685	—	2.0
C42	$\rho^- C^R E^0 O^0$	20	0.055	1.681	0.8	0.8[b]
D41	$\rho^- C^R E^0 O^0$	48	0.12	1.682	2.1	2.4[b]
D61	$\rho^- C^R E^0 O^0$	25	0.055	1.681	1.3	1.2[b]
F13	$\rho^- C^0 E^0 O^0$	—	0.0018	1.676	0.13	0.3[b]
F23	$\rho^- C^0 E^0 O^0$	—	0.0165	1.679	—	—

[a] Genotypes are given only with respect to the R_I (C^R), R_{III} (E^R), and O_I (O^S) loci, although in many cases they are also known for the R_{II}, O_{II}, and P_I loci as well (56). All petites were derived from strain IL8–8C.

[b] Kinetic complexities have been estimated independently by Lazowska et al. (56) and Locker et al. (64). Tabulated kinetic complexities are those of Lazowska et al., as are calculated monomer molecular weights, except where actual length measurements have been made. There the calculated values of Locker et al. are used, since they also did the measurements. In general, the kinetic complexities of Locker et al. are somewhat lower than those of Lazowska et al., except in the case of F13, where the value obtained by Locker et al. is much higher (0.012 with respect to ρ^+).

but petite P4 yielded only one major band, with a molecular weight of 2.3×10^6 d, following cleavage of the mtDNA with this enzyme. Restriction enzyme analysis of other ρ^- mtDNAs showed that these mtDNAs contained deletions as well. In some of these a simple contiguous deletion seemed to be involved, but in others the deletion patterns appeared to be more complex.

Coretention and deletion of specific genetic and molecular markers has been examined by studying the ability of genetically marked ρ^- mutants to hybridize mitochondrial rRNA and tRNA. This technique has allowed mapping of these RNAs with respect to specific genetic markers (see Chapter 7). The large (23S) rRNA has been shown by two independent groups of workers to map in the *RIB1–3* region (see Fig. 7–14). Slonimski's group examined the ability of genetically marked ρ^- mutants to synthesize the two rRNAs *in vivo* (54,58) and to hybridize total mitochondrial RNA, most of which is rRNA (55). Faye and associates found that $C^R E^R$ petites retained

the capacity to synthesize 23S rRNA but not 16S rRNA, whereas $C^0 E^0$ petites did not. Fukuhara and associates found that total mitochondrial RNA from a ρ^+ strain hybridized extensively with mtDNA from $C^R E^R$ petites and $C^0 E^R$ petites, while hybridizing less well with $C^R E^0$ petites. Both sets of experiments showed that the *RIB1–3* region of the mitochondrial genome contains genes coding for mitochondrial rRNAs, and the experiments of Faye and associates indicated the only one of these genes present in this region might be the one coding for 23S rRNA. However, since Faye and associates relied on expression of the rRNA genes *in vivo* in ρ^- mutants, it was possible to argue that the 16S rRNA gene was also present in this region but was not expressed. This possibility was ruled out in independent experiments by Linnane and associates (68). These workers separated the two RNAs and labeled them with ^{125}I. Then they hybridized each RNA to mtDNA from ρ^- mutants that had been characterized genetically. Significant hybridization of 23S rRNA occurred when the *cap* and *ery* genes (the equivalent of the *RIB1–3* genes of Slonimski, see Table 5–2) were retained or when *ery* or *cap* alone was retained. Deletion of both *ery* and *cap* resulted in mtDNA incapable of hybridizing 23S rRNA. The 16S rRNA hybridized to the mtDNA of none of the petites, which indicated that the *RIB1–3* region did not contain the 16S rRNA cistron.

The location of the 16S rRNA cistron in the mitochondrial genome was established shortly thereafter by Faye and associates (59) using genetically marked ρ^- mutants and RNA-DNA hybridization techniques. Tritiated 16S and 23S rRNAs were separated and hybridized to mtDNA extracted from different ρ^- mutants. The strains retained markers in one or more of the following loci: *RIB1, RIB3, OLI1, OLI2, PAR1.* These genes are fairly well distributed around the mitochondrial genetic map (see Fig. 7–14). Petites retaining the *RIB1* and *RIB3* genes hybridized 23S rRNA, but not the 16S rRNA as expected. Petites retaining the *PAR1* locus hybridized the 16S rRNA in some cases but not all cases. *In vivo* analysis of the rRNA transcribed in these mutants indicated that two mutants having high levels of 16S rRNA hybridization also were synthesizing this RNA *in vivo,* but in a third petite having a high level of 16S rRNA hybridization, no 16S rRNA was found. These observations show that a negative result using the *in vivo* synthesis criterion can mean either that the gene is not transcribed in the petite or that it is deleted. The *PAR1* gene and the 16S rRNA gene are clearly distinct, although closely linked, since certain ρ^- mutants retaining the *PAR1* locus fail to hybridize 16S rRNA. These experiments provide evidence, confirmed by molecular mapping techniques (see Chapters 2 and 3), that the 16S and 23S rRNA genes are well separated on the mitochondrial genome of *S. cerevisiae.*

Mapping of different mitochondrial tRNAs with respect to mitochondrial genetic markers has also been accomplished by means of coretention experiments. Casey et al. (69) examined hybridization of 11 different mitochondrial tRNAs to a series of ρ^- mutants whose mitochondrial genotypes had been defined with respect to the *RIB1, RIB3,* and *OLI1* genes. These genes are spaced over relatively limited portions of the mitochondrial genetic map of *S. cerevisiae* (see Fig. 7–14). Comparison of three $C^R E^R O^0$ petites revealed that two of them lacked mtDNA sequences complementary to prolyl, valyl, aspartyl, alanyl, tyrosyl, phenylalanyl, and isoleucyl tRNA, but the third contained mtDNA that retained sequences complementary to 9 of the 11 tRNAs tested, including those deleted in the first two strains. Casey and associates concluded from the first two strains that the deleted tRNA genes could not lie between the *RIB1* and *RIB3* genes. The third ρ^- mutation, which did hybridize these tRNAs, had a high kinetic complexity equivalent to one-third of the wild-type mitochondrial genome. Since the *OLI1* gene was also deleted in this petite, these results suggest that the tRNAs in question might map in a region to the right of *RIB1.* The histidyl tRNA gene appeared to be close to *RIB1,* since all petites retaining the C^R marker contained mtDNA that hybridized to a considerable extent with histidyl tRNA. The leucyl and glutamyl tRNAs appeared to be further to the right, since only $C^R E^0 O^0$ and $C^R E^R O^0$ petites in which the mtDNA had a high kinetic complexity showed high levels of hybridization of these tRNAs, whereas $C^0 E^R$ petites did not. These results were confirmed and considerably extended by Martin and Rabinowitz (70) (see Fig. 7–14).

As one might expect, the sequences retained in genetically different ρ^- mutants are distinct from one another. DNA-DNA hybridization experiments have revealed considerable sequence homology between genotypically similar ρ^- mutants but little homology between genotypically different petites (54,56,63). These relationships are also reflected in the RNA synthesized by ρ^- mutants, as has been shown by RNA-DNA hybridization experiments (55). Genotypically different petites also have distinctive mtDNAs in terms of base composition. This can be seen in terms of buoyant densities (Table 6–3), differential melting curves (54,57), and restriction enzyme patterns (66). Thus the mtDNAs of C^0E^R petites have higher buoyant densities than those of C^RE^0 petites; similarly, the latter petites have higher proportions of low-temperature melting regions than the former (Fig. 6–4). Both observations suggest that C^0E^R petites are enriched for G+C sequences, whereas C^RE^0 petites are enriched for A+T sequences.

Evidence for amplification of the retained sequences in genetically marked ρ^- mutants has also been reported (54,68,69). Casey et al. (69) carried out a detailed study of the problem with respect to the tRNA cistrons, and they made the important distinction between amplification as a function of the total mtDNA used in the hybridization experiment and hybridization per mitochondrial genome length retained. They put forth the following argument. If one assumes that the reduction in kinetic complexity of ρ^- mtDNA reflects deletion of a major portion of the wild-type mitochondrial genome, with simple retention of specific sequences, then the hybridization saturation plateau for a given amount of ρ^- mtDNA should be much higher than for a corresponding amount of ρ^+ mtDNA. Thus the elevation in hybridiza-

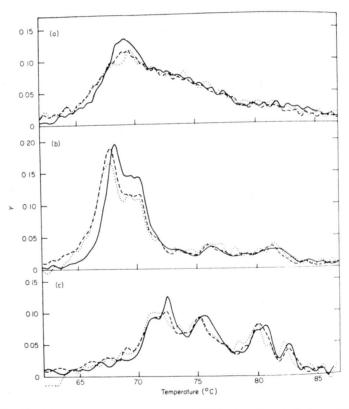

FIG. 6–4. Differential melting profiles of native and reassociated mtDNAs from ρ^+ and ρ^- strains of *S. cerevisiae* (Y is defined in legend of Fig. 6–3, and α is extent of reassociation). a: ρ^+ mtDNA: native (*solid line*), reassociated up to $\alpha = 0.88$ (*dash line*), $\alpha = 0.65$ (*dotted line*). b: ρ^- mtDNA from C^RE^0 mutant (C42): native (*solid line*), reassociated up to $\alpha = 0.92$ (*dash line*), $\alpha = 0.64$ (*dotted line*). c: ρ^- mtDNA from C^0E^R mutant (E41): native (*solid line*), reassociated up to $\alpha = 0.82$ (*dash line*), $\alpha = 0.68$ (*dotted line*). (From Michel et al., ref. 57, with permission. Copyright by Academic Press Inc. (London) Ltd.)

tion level should be directly proportional to the reduction in kinetic complexity.

The results obtained before and after correction for kinetic complexity are shown for some selected data in Table 6–5. Consider mutant D41, for example. The uncorrected levels for hybridization of leucyl or glutamyl tRNA are much higher than those for wild type, but the corrected levels yield between one and two equivalents of the two tRNA cistrons per genome when compared to wild type. The data also show that many tRNAs are retained at much lower levels than one cistron per genome length, even though the ρ^- mutants appear to be genetically homogeneous and stable. These results led Casey and associates to conclude that there was some heterogeneity among the retained mtDNA molecules, despite the apparent phenotypic homogeneity of the ρ^- mutant clones. When corrections were made for kinetic complexity, it became apparent that hybridization of different tRNAs was virtually confined to two levels. These differed by one to two orders of magnitude; few, if any, tRNAs hybridized at intermediate levels. In D21, for example, prolyl and valyl tRNAs hybridized at approximately one-tenth the level of the wild type, whereas all the other tRNAs hybridized at levels suggestive of one or more cistrons per unit genome size (Table 6–5). Casey and associates concluded from these and other data

that the mtDNA responsible for the fractional tRNA hybridization might well contain all of the fractional tRNAs (i.e., those in the hybridization range of approximately 0.2–0.01 cistrons per unit genome) and that these mtDNA molecules constituted no more than a small percentage of the total. Other evidence for low levels of molecular heterogeneity in genetically marked stable ρ^- strains comes from the restriction enzyme studies of Morimoto et al. (66) discussed earlier. In short, genetic stability cannot be taken as *a priori* evidence for genotypic homozygosity of ρ^- mutants. At best, one can be certain only of approximating homozygosity by genetic techniques, and only rigorous molecular characterization can ultimately establish the level of homozygosity. These remarks merely serve to emphasize the pitfalls involved in studying the genetics of multicopy organelle systems.

The results of Casey and associates also show that certain tRNAs (e.g., leucyl tRNA in mutant F11, Table 6–5) appear to be amplified selectively on a per genome length basis. On a per cell basis, or in terms of total mtDNA, retained sequences are amplified extensively, as will be shown in subsequent sections. In other words, although there is not a great deal of gene amplification in ρ^- mutants when corrections are made for the kinetic complexity of ρ^- mtDNA, there is a great deal of amplifica-

TABLE 6–5. *tRNA hybridization levels to petite mtDNA relative to those for wild-type mtDNA*

Strains		Iso	Pro	Val	Phe	Ala	Tyr	Asp	Lys	Glu	Leu	His	C	E	O
						tRNAs								Genetic markers retained	
Wild-type		1	1	1	1	1	1	1	1	1	1	1	+	+	+
mutants															
D21	U[a]	1.1	0.3	0.6	3	3.2	5.2	2.2	1.3	1.5	8	2.6	+	+	0
	C	0.4	0.1	0.2	1.1	1.1	1.9	0.8	0.5	0.5	2.8	0.9			
D41	U	0.2	0.1	0	0.7	0.2	0.1	5.9	3	6.7	14	11.6	+	0	0
	C	0.03	0.01	—	0.1	0.03	0.01	0.8	0.4	0.87	1.8	1.5			
F11	U	0	0	0	0	0	0	0	0.3	4.3	17	7.3	+	+	0
	C	—	—	—	—	—	—	—	0.07	1.0	4.1	1.7			

[a] U = uncorrected; C = corrected for kinetic complexity of the petite mtDNAs.
Data from Casey et al. (69).

tion in terms of total mtDNA and mtDNA per cell, since ρ^- mtDNA probably consists entirely of retained wild-type mtDNA sequences, and since the amount of mtDNA per cell in a ρ^- mutant is equivalent to that present in wild type.

Does ρ^- mtDNA contain new sequences distinct from those present in ρ^+ mtDNA? Conceivably, ρ^- mtDNA could differ from ρ^+ mtDNA by the addition of new sequences not present in ρ^+ mtDNA. Thus far, the only strong case made for the existence of new sequences has been that of Rabinowitz and his colleagues (61,62) for a specific petite called R1–6/1. This petite is a subclone derived from the mutant R1–6, which Cohen et al. (51) originally used in their studies of the heterogeneity of retention of valyl and leucyl tRNA cistrons in this mutant and its subclones (*vide supra*). Gordon and Rabinowitz (61) originally reported, on the basis of DNA-DNA hybridization experiments, that R1–6/1 contained approximately 30% nonhomologous sequences with respect to the wild-type strain, R1, from which R1–6 and its subclones were derived. They compared isohybrid (wild-type mtDNA in solution and wild-type mtDNA on the filter, or petite mtDNA in solution and petite mtDNA on the filter) to heterohybrid (petite mtDNA in solution and wild-type mtDNA on the filter, and the converse) formation and found that the plateau levels for isohybrid formation were always higher than the heterohybrid plateaus, whether or not petite or wild-type mtDNAs were in solution. This result is to be expected when wild-type mtDNA is in solution since we already know that ρ^- mutants contain large deletions; thus, by definition, the ρ^- mtDNA on the filter cannot hybridize as much wild-type mtDNA as can wild-type mtDNA, where all the homologous sequences are present. The surprising result was that heterohybrid formation between ρ^- mtDNA in solution and wild-type mtDNA on the filters was less efficient than isohybrid formation with the ρ^- mtDNAs, even though pla-

teau values were always compared. If the sequences retained in the petite were all homologous to wild type and there were nothing peculiar about the renaturation properties of the mutant mtDNA, one would expect the slopes of the hybridization curves to be different but the final plateau values to be the same. That is, it should take more wild-type mtDNA than petite mtDNA to saturate the petite mtDNA in solution.

The foregoing results suggested that the R1–6/1 petite contained new sequences distinct from those present in wild-type mtDNA. Further support for this notion came from two filter hybridization experiments in which a test filter carrying a fixed amount of mtDNA homologous to the mtDNA in solution was placed in competition against filters containing a graded concentration series of either homologous or heterologous mtDNA. These experiments showed that wild-type mtDNA on the competing filters competed more effectively for wild-type mtDNA in solution at saturation than did petite mtDNA, and the same was true of petite mtDNA on the competing filters for petite mtDNA in solution. This result, again, was consistent with the notion that new sequences were present in the mtDNA of R1–6/1. Gordon et al. (63) subsequently showed that R1–6/1 was unique in its behavior with respect to its kindred subclones R1–6/5, R1–6/6, and R1–6/8, all of which behaved as if they were pure deletion mutants. Experiments with these subclones revealed that when heterohybrids were made between wild-type mtDNA in solution and petite mtDNA on filters, the plateau values obtained were about 60% of the wild-type control values, reflecting extensive deletion in each of the petites. In the converse experiment, 100% plateau values were attained, indicating that the ρ^- mtDNAs contained only wild-type sequences. Again, the sole exception was R1–6/1.

Sanders et al. (71) argued that the results of Gordon et al. (61,62) do not indicate the presence of new sequences. The problem relates to the kinetic complexity of R1–6/1 mtDNA. Although Gordon and Rabinowitz (61) stated that R1–6/1 mtDNA showed a moderate twofold increase in renaturation rate relative to wild-type mtDNA, the experiments of Fauman and Rabinowitz (72), which were quoted as the basis for the statement, indicated that R1–6/1 mtDNA also contained a rapidly renaturing fraction. Sanders and associates argued that this repetitive fraction would hybridize much more effectively in homologous reactions than in heterologous reactions and would, in fact, represent the fraction that was attributed by Gordon and associates to the existence of new sequences. While this explanation may be correct, it relies on the fact that R1–6/1 mtDNA was used by both Gordon and associates and Fauman and Rabinowitz. Gordon and Rabinowitz (61) stated this to be the case, but Fauman and Rabinowitz (72) referred only to R1–6, the parent petite clone, in their article. Lazowska et al. (56) also reported some evidence for new sequences on the basis of filter hybridization experiments; they suggested that the results, although most likely artifactual, could be real. They suggested that new sequences might arise as a consequence of recombination between mutant mtDNAs, leading to reshuffling of the remaining ρ^- sequences, or as a consequence of intramolecular repetition of nondeleted segments to create new local sequences at the junctions (e.g., head to head, tail to tail, etc.).

In summary, it seems likely that ρ^- mtDNA is usually composed entirely of retained ρ^+ mtDNA sequences. The possibility that R1–6/1 is an exception that does contain new sequences is even disputed. So, for purposes of further discussion, we can assume that new sequences do not contribute significantly to the composition of ρ^- mtDNA.

What are the form, size, and structure of ρ^- mtDNA? In answering this question, it seems best to begin with a discussion of the mtDNA of a specific ρ^- mutant, RD1A, and then move on to a consideration of form, size, and structure of ρ^- mtDNA generally. The mtDNA of this mutant was studied in great detail by Borst and his colleagues, and its properties were summarized in a recent article (73). RD1A is an EtdBr-induced petite, and more than 95% of its mtDNA consists of identical repeats of a 68-nucleotide sequence containing two GC and 66 consecutive AT pairs. The repeats are arranged head to tail, and they appear to be faithful copies of a segment of wild-type mtDNA. Two minor tracts that contain unusual bases have also been encountered, but it is not clear if these are present within or between certain of the repeats. The size of this mtDNA is not accurately known, but its sedimentation properties are similar to those of wild-type mtDNA, and at least some molecules are on the order of 7,000 nucleotides long, which is 100 times the repeat length.

Electron microscopic methods have been particularly useful in establishing the size and structure of mtDNA in genetically marked ρ^- mutants as well as the organization of the retained sequences. Locker et al. (64,65) found that circular mtDNA molecules could be obtained readily from ρ^- mutants and that they fell in a multimeric series in which the smallest circle or monomer had a molecular weight estimated from contour length that was in excellent agreement with the kinetic complexity (Table 6–4). They also stated that the linear mtDNA molecules they obtained appeared to be polymers of the same basic monomer unit. These investigators established by use of two methods that the retained sequences in ρ^- mtDNA were organized either as tandem repeats (head to tail) or as inverted tandem repeats (head to head, tail to tail). With the first method, the ρ^- mtDNA was denatured, and the single strands were examined for self-renaturation. If tandem repeats were involved, no self-renaturation occurred, but if the ρ^- mtDNA was organized into inverted

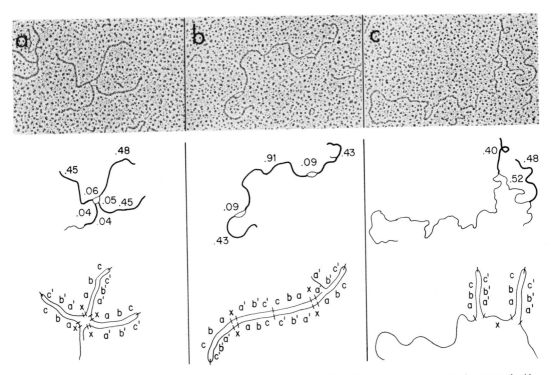

FIG. 6–5. Denatured mtDNA from C^0E^R ρ^- mutant of *S. cerevisiae* E41 illustrating the self-renaturation that occurs in this mtDNA because of the presence of adjacent inverted tandem repeats. Below each electron micrograph are two diagrams; one illustrates the configuration of the molecule (dimensions indicated in micrometers), and the other outlines a hypothetical single-strand molecule containing the sequence abc, its complement a′b′c′, and the nonrenaturing sequence x. a: "Spider" molecule, whose configuration is assumed to be caused by the presence of a regular arrangement of multiple inverted repeats in a single-strand DNA molecule. b: Different configuration formed by the same regular arrangement of multiple neighboring complementary sequences in a single-strand molecule as in a. c: Single-strand molecule showing two self-renaturing complementary sequences, but with an irregular arrangement of sequences, in contrast to molecules of a and b. (From Locker et al., ref. 64, with permission.)

tandems, adjacent repeats self-renatured (Fig. 6–5). The second method employed the partial denaturation technique of Inman (74), in which only the A+T-rich regions become denatured.

Two examples will suffice to show how these techniques have been combined to reveal the structure of ρ^- mtDNA. In the $C^0E^0O^0$ petite F13, the monomer circle size is 0.13 μm (Table 6–4). The denatured mtDNA shows no evidence of self-renaturation, but partial denaturation reveals regularly spaced loops of 0.14 μm separated by undenatured regions 0.03 μm long (Fig. 6–6). It appears from these results that F13 mtDNA is composed of a basic monomer unit 0.14 to 0.17 μm long, with an A+T-rich region comprising the loop and a G+C-

rich segment comprising the undenatured region. These monomers are then tandemly repeated in the multimeric forms. Formation of this type of mtDNA requires excision of a sequence of mtDNA that then becomes amplified. This would account for the loss of mitochondrial genetic information in the ρ^- mutant as well as the regularly repeating tandem sequences seen in molecules larger than the monomer unit.

The E41 petite (Table 6–4) exemplifies the second kind of ρ^- mtDNA structure in which the sequences are arranged in inverted tandems. Denaturation of this mtDNA yielded single strands containing self-complementary duplex regions of 0.4 to 0.5 μm separated by single-strand regions of 0.04 to 0.07 μm (Figs. 6–5 and 6–6).

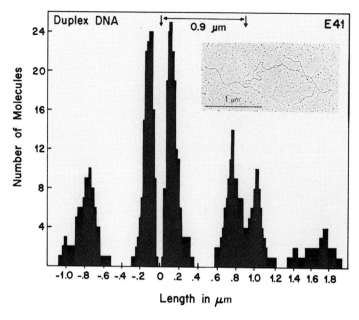

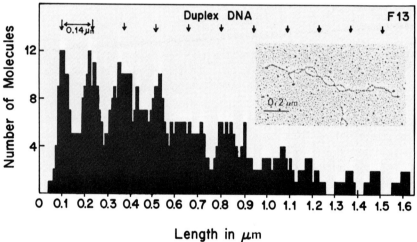

FIG. 6–6. Frequency distributions of duplex regions in mtDNA from two ρ^- mutants following partial denaturation. Insets show typical partially denatured molecules from which data were obtained. A: mtDNA from C^0E^R ρ^- mutant E41. Distribution is plotted on 0.02-μm scale. Spacing of 0.9 μm is apparent for the repeating unit. B: mtDNA from C^0E^0 ρ^- mutant F13. Distribution is plotted on 0.01-μm scale. Regular spacing of about 0.14 μm (arrows) is apparent. (From Locker et al., ref. 64, with permission.)

Partial denaturation of E41 mtDNA revealed a recurring pattern of large and small loops, with the larger loop being 0.9 μm, or twice the size of the duplex region seen in the self-renaturation experiments. These results were interpreted to mean that a single repeat is about 0.5 μm long and that repeats are arranged in inverted tandems. In the self-renaturation experiments these yielded duplex regions 0.5 μm long, and in the partial denaturation experiments they yielded loops 0.9 μm long. It is interesting that not all of the mtDNA molecules were organized in inverted tandems. Some 15% of the molecules contained regular tandem repeats, and another 15% were mixtures of regular and inverted tandems. Assuming that inverted tandems are not present in ρ^+ mtDNA, the genesis of inverted tandems in ρ^- mtDNA would most likely involve at least four steps: excision, duplication leading to head-to-tail tandems, inversion of one of the duplicated segments; and amplification of the inverted tandems.

Locker et al. (65) found by electron microscopic examination that both circular and linear mtDNA molecules were present in their preparations of ρ^- mtDNA. The ratio of the two classes of molecules was different in different mutants. However, the important finding was that in seven of the nine mutants studied the circular forms could be arranged in a single oligomeric length series that was characteristic for each mutant and reflected differences in the sizes of the monomeric forms. The sizes of the monomers were also found to be related to the kinetic complexity of the mtDNA (Table 6-4). That is, the least complex mtDNAs yielded the smallest monomer circles, and the most complex molecules yielded the largest circles. In general, monomers and small oligomers were more frequent than large oligomers. In the other two ρ^- mutants, circles were assignable to more than one oligomeric series. The linear mtDNA molecules obtained in these preparations did not show the same regular size distribution as the circles. One explanation is that these molecules were simply randomly broken forms of larger circles. In any event, denaturation mapping of these molecules indicated that they had repeat spacings similar to those expected from the size of the circular monomers. In short, it is reasonable to conclude that these molecules are made up of monomer units just like the oligomeric circles. Locker and associates did not find small circular molecules in wild-type mtDNA preparations isolated under the same conditions.

Lazowska and Slonimski (60) extended the observations of Locker and associates to another series of genetically characterized ρ^- mutants and drew some general conclusions about the rules governing formation of these oligomeric series of circular mtDNA molecules in ρ^- mutants. The salient observations were as follows: First, four of the nine mutants studied exhibited a single multimeric series of circles ranging from 0.3 to 2.4 μm in monomer length. However, in spreads of two of these mutants, rare molecules of half-monomer size were found. Second, the other five mutants displayed more than one multimeric series, and there were normally three such series of circles. For example, mutant A211 had monomers of 0.68, 2.16, and 2.64 μm. In the first series, molecules up to pentamer size were identified; in the second series there were monomers and dimers; in the third series there were monomers only. Half-monomers were also found in the first series. Third, Lazowska and Slonimski determined the nature of the repeat unit for each mutant (i.e., straight versus inverted).

Lazowska and Slonimski drew several conclusions from these results, three of which are particularly important for this discussion. First, if only one sequence is present and tandem straight repeats are involved, the circles will contain only one, two, or three, etc. fundamental repeat units. In tandem inverted repeats, the circles contain two, four, or six fundamental units. In other words, the monomeric rule of circularization is one unit for straight repeats, but two units for inverted repeats. Occasionally, however, two inverted repeats break apart to form half-monomers. Thus, for all those petites containing inverted tandem repeats in their mtDNA, the fundamental repeat unit length will be half the monomer circle length. Second, the distribution of circular molecules in a multimeric series follows a simple rule in which the frequency of n-meric molecules = $1/n$ × frequency of monomeric molecules, where $n = 1, 2, 3$, etc. This rule has been tested for molecules up to tetramer size, and it appears to fit well. Beyond that size, molecular breakage causes serious problems. If the rule is extrapolated to higher oligomers, the conclusion is that monomers = dimers = trimers = *n*-mers. That is, the fundamental repeat unit, which is equal to the genome size of the ρ^- mutant, has the same probability of constituting one circle or of being part of an *n*-meric circle. Third, for those ρ^- mutants containing more than one multimeric series, two of the sequences are different, and the third is composed of repeats derived from the other two. Evidence that the sequences are different comes from the observation that mtDNAs from ρ^- mutants containing more than one circle series are more complex than is predicted from the repeat length of one or the other series of circles. Evidence against cellular heterogeneity as an explanation is that the ρ^- mutants used have been cloned repeatedly and have remained pure genetically when tested for marker retention. These conclusions are summed up in schematic form in Fig. 6–7.

Borst et al. (73) disputed the frequency rule of Lazowska and Slonimski. They thought it likely, instead, that most ρ^- mtDNA is present in 25-μm circles. In support of this idea they cited the data of Michels and associates (75). These authors selected three genetically characterized ρ^- mutants in terms of their suppressivity and determined the molecular weight of the mtDNA by sedimentation in sucrose gradients. They found that the molecular weight of the mtDNA of these petites appeared to be identical to that of the wild type. Borst and associates also pointed out that for the petite RD1A, which was discussed previously, the repeat unit length is equal to 0.1% of the length of the wild-type mtDNA, and the frequency rule implies that 10% of this DNA should be present as circles with contour lengths of 2.5 μm or smaller. However, this was not found to be true, and small circles were, in fact, rare. They also cited other data that are difficult to interpret

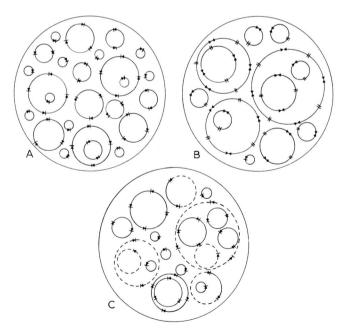

FIG. 6–7. Diagram of various types of arrangements of circular mtDNA molecules in ρ^- mutants. A: Unique fundamental repeat unit is arranged in tandem straight repetitions. B: Unique fundamental repeat unit is arranged in tandem inverted repetitions. C: Two different fundamental repeat units arranged in tandem straight repetition. Three series of circles are present containing one or the other or both types of sequences. In all three schemes the mass of mtDNA is the same, as well as the frequency distribution of monomers, dimers, trimers, etc. (From Lazowska and Slonimski, ref. 60, with permission.)

according to the frequency rule; they came to the conclusion that the frequency rule does not apply to all petites and cannot be extrapolated to higher values of *n*.

We can summarize this discussion of form, size, and structure of ρ^- mtDNA as follows: First, ρ^- mtDNA preparations contain circular and linear mtDNA molecules. The size of the monomer circles seems to be related to the kinetic complexity of the mtDNA. That is, the smaller the monomer circle size, the less complex the DNA. The linear molecules are probably derived from larger circular molecules. Second, ρ^- mtDNA is comprised of both straight tandem and inverted tandem repeats. Third, it is not clear that the frequency rule of Lazowska and Slonimski applies generally; therefore we must reserve judgment on the proportion of ρ^- mtDNA molecules falling into the different categories of the oligomeric series *vis-à-vis* the possibility that the distribution is really discontinuous beyond a certain point. That is, although an oligomeric series does exist, perhaps the smaller circles are generally rare in comparison to 25-μm circles. Fourth, some ρ^- mutants contain up to three classes of circular mtDNA molecules, each having different monomer sizes.

Do ρ^- mutants contain as much mtDNA as wild type? Several early studies (24,76, 77) indicated that ρ^- mutants did contain the same amount of mtDNA as wild-type cells. This question was addressed again in a recent article by Hall et al. (25) in which they reported a careful study of the genetic control of mtDNA levels in wild-type cells of *S. cerevisiae* and in several ρ^- mutants. These investigators found that in wild type, mtDNA constituted from 16 to 25% of the total DNA, depending on the strain. Repeated measurements in two haploid strains yielded levels of 17 and 24%. The genetic control of these differences was shown to be nuclear and to involve at least two genes. When ρ^- mutants were derived from the wild-type strain containing 24% mtDNA, the mtDNA levels were found to range from

20 to 25%, with one exception. The exceptional mutant, LK2.4, contained 16.8% mtDNA. Hall and associates also examined the level of mtDNA in the ρ^- mutant RD1A and its wild-type parent H1. The mutant contained only 6% mtDNA, whereas its parent contained 19%. Borst et al. (73) took note of this observation, pointing out that in their hands the yield of mtDNA from this strain was variable for two reasons. First, under some conditions, RD1A throws off ρ^0 mutants, and if these are not eliminated, mtDNA yields are low. Second, there is a tendency for RD1A mtDNA to form large networks. This leads to variable losses that are difficult to control. In short, with a few possible exceptions that may be explained in other ways, including possible technical artifacts, the levels of mtDNA in ρ^- mutants are as high as in wild-type cells.

Assuming that the amount of mtDNA in ρ^- mutants is normally equal to that in wild type, the level of amplification of the retained sequences in a ρ^- mutant will be related to the size of the monomer unit. In other words, the least complex mtDNAs with the shortest repeat lengths should be amplified the most. This point is illustrated in Table 6–6 for some of the data published by Locker et al. (64).

Is ρ^- mtDNA capable of recombination? Michaelis and associates (78) crossed a C^0E^R petite in which the mtDNA had a buoyant density of 1.688 g/cc with two different C^RE^0 petites, one having mtDNA of buoyant density 1.682 g/cc and the other 1.683 g/cc. The resulting diploid subclones were selected using appropriate nuclear auxotrophic markers and then crossed to a haploid $\rho^+C^SE^S$ tester using yet other selective markers. The resulting ρ^+ triploid progeny were then screened to see if the ρ^- diploid parent had been of the genotype C^0E^R, C^RE^0, C^RE^R, or C^0E^0. The first two genotypes are parental; the third must be recombinant, since it carries a resistance marker from each parent; the fourth could be recombinant or could have arisen be-

TABLE 6-6. Amplification of monomer units of mtDNA in different ρ^- mutants[a]

Strain	Genotype	Monomer circle length (μm)	Monomer units per cell	Amplification of monomer relative to ρ^+
IL8-8C	$\rho^+C^RE^RO^S$	25	50	—
F12	$\rho^-C^0E^RO^S$	5.5	227	4.6
D41	$\rho^-C^RE^0O^0$	2.1	595	11.9
C42	$\rho^-C^RE^0O^0$	0.8	1563	31.3
F13	$\rho^-C^0E^0O^0$	0.13	9615	192.3

[a] The number of 25-μm circles of mtDNA in a cell is taken to be 50 (see Chapter 4), and the sizes of monomer circles in the different ρ^- mutants are taken from the data of Locker et al. (64).

cause the ρ^- mtDNA of one parent or the other had undergone a spontaneous change leading to loss of the resistance marker it originally carried. The buoyant densities of the mtDNA of 23 genotypically defined ρ^- subclones were then compared. The buoyant densities of ρ^- mtDNA extracted from the C^RE^R recombinants were intermediate between those of diploid cells having parental genotypes. The authors concluded that recombination of these *ant*^{r} markers was accompanied by formation of recombinant species of mtDNA of intermediate buoyant density. The C^RE^R recombinants were characterized further in a recent article by Michaelis et al. (79) in terms of kinetic complexity, DNA-DNA hybridization, and differential melting profiles. The one recombinant characterized with respect to kinetic complexity seemed to have a kinetic complexity equivalent to the sum of the complexities of the two parental mtDNAs. In the DNA-DNA hybridization experiments the sequence homologies of the two parental ρ^- mtDNAs were compared to each other and to two different $\rho^-C^RE^R$ recombinants. The two recombinants were also compared to each other. The findings were as follows: The two parental mtDNAs had very little sequence homology with one another. The homology of the two recombinant mtDNAs with each other was complete. Each of the recombinant mtDNAs had considerable sequence homology with each of the parental mtDNAs. Differential

melting profiles also confirmed that the recombinant mtDNAs contained sequences present in both parents.

SUPPRESSIVENESS

In describing suppressive petites for the first time, Ephrussi et al. (80) offered a hypothesis to explain why their behavior is different than that of neutral petites. Ephrussi and associates suggested that in a neutral petite the "normal cytoplasmic factor" is lost or mutated to a functionally inactive form; in a cross of a neutral petite, the normal factor is transmitted to all progeny, which are therefore respiration-sufficient. Suppressiveness was explained by the presence of a "suppressive factor." This factor was supposed to suppress perpetuation of the normal factor in a cross of a suppressive petite to wild type, either by interfering with multiplication of the normal factor or its distribution to diploid buds or by causing actual destruction of the normal factor.

Although this hypothesis accounted in a simple fashion for the occurrence of asci that were either wild type or petite in a cross of a suppressive petite to wild type, it did not account for the existence of petite strains with various degrees of suppressiveness. Ephrussi and associates proposed two alternative hypotheses to account for this phenomenon. First, they suggested that only two types of petites, neutral and completely suppressive, existed. Suppressive strains were mixtures of these two types of petites in various proportions. The smaller the proportion of neutral petites in the strain, the higher its level of suppressiveness. Second, the cells of a suppressive strain all belonged to a single type characterized by a certain efficiency of suppression that had different values in different strains. This hypothesis implied that there were as many cell types as there were degrees of suppressiveness. Obviously, such a hypothesis would be difficult to reconcile with the notion of a discrete suppressive factor.

To evaluate these two hypotheses, Ephrussi and Grandchamp (37) designed an elaborate series of cloning experiments with several suppressive petites. Their procedure was to plate out a haploid suppressive strain, pick individual colonies, allow them to grow for a limited number of divisions, and then divide the cultures in half. Half the clone was subcultured to perpetuate the haploid clone, and the other half was mated to wild type to determine suppressiveness. To measure suppressiveness quantitatively, the zygotes from the mating mixture were plated on appropriate medium and allowed to form colonies, which were then incubated with tetrazolium chloride. Tetrazolium stained wild-type colonies red, but the petite colonies remained white. Percentage suppressiveness was then calculated as $[(X - Y)/(100 - Y)] \times 100$, where X is the percentage of zygotes with petite progeny in the cross and Y is the percentage of spontaneous petites in the wild-type stock.

Ephrussi and Grandchamp found that there was strong correlation between the degree of suppressiveness of a cell and that of its vegetative descendants. However, frequently there arose variants with different levels of suppressiveness that were themselves heritable. The finding that there was strong correlation between the suppressiveness of a cell and that of its vegetative descendants, irrespective of the percentage

suppressiveness of the clone, led Ephrussi and Grandchamp to conclude that suppressiveness was a cellular property of the strain studied and did not result because suppressive strains contained varying mixtures of neutral petites and completely suppressive petites.

Ephrussi and Grandchamp also reported that whereas crosses of neutral petites to the wild type produced mostly red wild-type zygote colonies and a few white petite colonies (presumably resulting from petite mutants in the wild-type stock) following tetrazolium treatment, suppressive petites produced a third type of colony that was a mosaic. Ephrussi et al. (81), as reported in another article, performed more detailed studies of the different types of zygotes arising from crosses of suppressive petites to wild type. They found that the mosaic colonies fell into two principal classes in terms of morphology. Scalloped colonies resulted when the wild-type cells in the colony formed a sector that overgrew the petite portion of the colony. On replating, these colonies yielded high proportions of wild-type cells. At the other extreme were colonies called "abscessed petites." Tetrazolium treatment of these colonies showed them to consist of a white base with a raised red center or papilla. On subculture, abscessed petites produced a mixture of wild-type cells and petite cells, or sometimes petite cells only. Cells forming these mosaic colonies were defined as being in a "premutational state." It was found that cells in the premutational state could be maintained for many cell generations and that the phenotype was transmissible in crosses. Ephrussi and associates also found that the frequencies of the three zygote types could be influenced by temperature and composition of media. Rank and Person (38) extended these results to show that among spontaneously arising petites, not only pure clones but also mosaics could be isolated. On subculture, the mosaics produced wild-type, petite, and mosaic progeny.

Ephrussi and associates suggested that the explanation for their results with crosses of suppressive petites to wild type was that the suppressive factor interfered with replication of the normal factor because of "mutual repression." From this hypothesis of Ephrussi and associates, it was only a short step to one of the most important models for suppressiveness—one that had wide support until a few years ago. According to this model, the suppressive factor and the normal factor were really ρ^- and ρ^+ mtDNA, and the level of suppressiveness reflected the degree to which ρ^- mtDNA had a replicative advantage over ρ^+ mtDNA. This hypothesis was first proposed by Slonimski (82) in 1968; a year later it received support from Carnevali et al. (83) and Rank and Person. The molecular analog that gave rise to the model was the demonstration by Mills et al. (84) that in an *in vitro* system virus $Q\beta$ RNA could be selected for faster and faster replication rate, and the final product of this selection was an RNA that had lost 83% of its original genome and had a very high affinity for the $Q\beta$ replicase.

Rank and associates attempted to test the so-called darwinian selection model of suppressiveness in a series of genetic experiments. Rank and Person (38) dealt with the segregation of ρ^+ and ρ^- cells by colonies of cells in the premutational state. They concluded from their experiments that a cell in the premutational state would arise originally by a spontaneous event in a ρ^+ cell that, in essence, would make the cell heterozygous for ρ^+ and ρ^- factors. Petite cells would be segregated because of the competitive advantage of ρ^- mtDNA over ρ^+ mtDNA. Opposed to this *intracellular* selection would be an *intercellular* selection for those cells that had received ρ^+ mtDNA and were therefore of the wild type. In another article, Rank (85) postulated, again by analogy to the $Q\beta$ selection experiment, that the most highly suppressive mtDNA would contain the least genetic information. Hence, $\rho^- E^R$ clones should give rise to E^0 clones be-

cause the latter mtDNA should have a replicative advantage. His results showed this to be the case, but they should be considered in light of the remarks made at the beginning of this chapter regarding stabilization of genetically defined ρ^- mutants. In his next article, Rank (86) presented the results of crosses between ρ^- mutants of high and low suppressiveness. The diploid progeny of these crosses were selected and, through the use of appropriate nuclear selective markers, mated to ρ^+ diploids. The level of suppressiveness of the parental ρ^- diploids could then be ascertained. Rank found that these diploids were highly suppressive, and he concluded that the highly suppressive ρ^- mtDNA had a competitive advantage over the ρ^- mtDNA of low suppressiveness. The final article in the series (87) showed that nuclear mutations to chloramphenicol resistance were retained in ρ^- cells, whereas mitochondrial mutations to resistance tended to be lost. The authors argued that this result was consistent with darwinian selection and replication of abnormal mtDNA, with the specificity of this effect on mtDNA being emphasized by the retention of phenotypically similar nuclear markers.

More recently, Slonimski and associates (36,88) postulated that recombination between ρ^- and ρ^+ mtDNA molecules plays an important role in suppressiveness. According to this hypothesis, errors (small deletions, insertions, translocations) present in ρ^- mtDNA will be propagated into ρ^+ mtDNA through successive rounds of recombination, converting ρ^+ to ρ^- mtDNA in the process. Michaelis et al. (78), in their article presenting evidence for recombination of ρ^- mtDNA (*vide supra*), also discussed data that they believed were best interpreted by a recombination model of suppressiveness. Michaelis and associates found that in one of their crosses the level of suppressiveness among $C^R E^R$ diploids was low, as it had been in the case of nonrecombinant diploids carrying parental marker configurations (i.e., $C^R E^0$ or $C^0 E^R$); but in the other cross, two classes of recombinants were found with respect to suppressiveness. One class of recombinant diploids had a low level of suppressiveness, as did the diploids carrying the parental marker combinations, but the other class had a high level of suppressiveness. The authors argued that these results indicated that a new level of suppressiveness could arise as a consequence of recombination.

Although the evidence in favor of the recombination model of suppressiveness is still sparse, there are three reasons that it is more attractive than the replication model. First, the results of Michaelis and associates suggest that recombination may be involved in generating new levels of suppressiveness. It is also clear from genetic evidence (see Chapter 5) that ρ^- and ρ^+ mtDNAs recombine; so one can imagine that recombination of these mtDNAs could result in transmission of the suppressive phenotype of the ρ^- parent and perhaps even yield recombinants with new levels of suppressiveness, as seems to be true of certain of the recombinants described previously from crosses of two mutants. Second, by analogy with the $Q\beta$ case, the replication hypothesis would predict that highly suppressive petites would contain small mtDNA molecules that could be replicated rapidly, whereas petites of low suppressiveness would have larger mtDNA molecules that would be replicated more slowly. Such a correlation has not been observed. Michaelis et al. (89) reported no correlation between suppressiveness and the molecular weight of ρ^- mtDNA, but they did not present any data. However, Michels et al. (75) made a careful study of three ρ^- mutants with very different levels of suppressiveness and showed that the molecular weights of mtDNA from all three mutants were similar to that of wild-type mtDNA. Third, Perlman and Birky (90) proposed a rather plausible model for spread of the petite phenotype through recombination between ρ^- and ρ^+ mtDNA molecules. This model suggests that recombination between

ρ^- and ρ^+ mtDNAs can yield one of three general results. First, if there are no deletions within the retained sequences in the ρ^- mutant, recombination will occur as it does between ρ^+ mtDNA molecules, and the mutant will behave as a neutral petite. Second, the ρ^- mutant can retain a single ρ^+ sequence with an internal deletion. During the process of pairing and recombination of this mtDNA with the wild type, a single-stranded piece of wild-type mtDNA becomes looped out in the recombinant. Subsequently, this looped-out segment is clipped by an endonuclease and excised by an exonuclease; the resulting gap is repaired by the sequential action of DNA polymerase and ligase. In short, each recombination event between such a ρ^- mtDNA and ρ^+ mtDNA can yield a ρ^- recombinant. Such a petite will probably be highly suppressive. Third, petites of intermediate suppressiveness could contain sequences lacking internal deletions, as well as a sequence having an internal deletion. Recombination between such ρ^- mtDNA molecules and ρ^+ mtDNA could have one of two consequences. First, if recombination involves a segment of ρ^- mtDNA lacking a deletion, a ρ^+ recombinant will result. Second, if recombination occurs between the ρ^- segment containing the deletion and its homologous ρ^+ segment, a ρ^- recombinant will be produced.

Despite these hopeful signs that recombination may be involved in the phenomenon of suppressiveness, we can summarize with the statement that rigorous evidence for an association of mtDNA recombination and suppressiveness is still lacking. So this old, elusive problem of mitochondrial genetics remains with us.

ORIGIN AND STRUCTURE OF ρ^- mtDNA: A SUMMARY

Perhaps the best way to summarize what we know about the origin and structure of ρ^- mtDNA is by means of the model of Borst and associates (73). Many aspects of the model appear to be plausible. The model assumes that ρ^- mtDNA arises by fragmentation of ρ^+ mtDNA, that these mtDNAs consist of amplified wild-type mtDNA sequences and no new sequences, and that any sequence can be retained. All these assumptions are grounded in strong supporting data. The model also assumes that there is a unique replication sequence (thus far, there are no data to support this assumption) and that ρ^- mtDNA is largely in the form of 25-μm circles, with monomer circles and small oligomers being the exception. That is, the frequency rule of Lazowska and Slonimski (60) is not generally applicable. Thus the steps in formation of ρ^- mtDNA are the following: The ρ^+ mtDNA is first fragmented to yield mtDNA segments, some of which contain the origin of replication and some of which do not. Molecules containing the origin of replication will replicate normally. Molecules not containing the origin of replication can replicate only if they attain an origin through illegitimate recombination. Amplification then occurs for each sequence that contains an origin. Following this, cloning of ρ^- mtDNA molecules occurs to yield genotypically stable ρ^- mutants. There is reasonable genetic and molecular evidence for such a cloning process. Further rearrangements, secondary deletion, and compensatory amplification can occur. Finally, through excision of repeat units, structures are formed that can circularize to form monomers, small circular oligomers, etc.

REFERENCES

1. Slonimski, P. P., Perrodin, G., and Croft, J. H. (1968): Ethidium bromide induced mutation of yeast mitochondria: Complete transformation of cells into respiratory deficient nonchromosomal "petites." *Biochem. Biophys. Res. Commun.*, 30:232–239.
2. Mahler, H. R. (1973): Biogenetic autonomy of mitochondria. *CRC Crit. Rev. Biochem.*, 1:381–460.
3. Mahler, H. R., Bastos, R. N., Flury, U., Lin, C. C., and Phan, S. H. (1975): Mitochondrial biogenesis in fungi. In: *Genetics and Biogenesis of Mitochondria and Chloroplasts,* edited by C. W. Birky, Jr., P. S. Perlman, and T. J. Byers, Chapter 2. Ohio State University Press, Columbus.
4. Mahler, H. R., Bastos, R. N., Feldman, F., Flury, U., Lin, C. C., Perlman, P. S., and Phan, S. H. (1975): Biogenetic autonomy of mitochondria and its limits. In: *Membrane Biogenesis,* edited by A. Tzagoloff, Chapter 2. Plenum Press, New York.
5. Perlman, P. S. (1975): Cytoplasmic petite mutants in yeast: A model for the study of reiterated genetic sequences. In: *Genetics and Biogenesis of Mitochondria and Chloroplasts,* edited by C. W. Birky, Jr., P. S. Perlman, and T. J. Byers, Chapter 4. Ohio State University Press, Columbus.
6. Luha, A. A., Sarcoe, L. E., and Whittaker, P. A. (1971): Biosynthesis of yeast mitochondria. Drug effects on the petite negative yeast *Kluyveromyces lactis. Biochem. Biophys. Res. Commun.*, 44:396–402.
7. Luha, A. A., Whittaker, P. A., and Hammond, R. C. (1974): Biosynthesis of yeast mitochondria. Some effects of ethidium bromide on *Kluyveromyces (Saccharomyces) fragilis. Mol. Gen. Genet.*, 129:311–323.
8. Nass, M. M. K. (1970): Abnormal DNA patterns in mitochondria: Ethidium bromide-induced breakdown of closed circular DNA and conditions leading to oligomer accumulation. *Proc. Natl. Acad. Sci. U.S.A.*, 67:1926–1933.
9. Kleitmann, W., Sato, N., and Nass, M. M. K. (1973): Establishment and characterization of ethidium bromide resistance in simian virus 40-transformed hamster cells. *J. Cell Biol.*, 58: 11–26.
10. Smith, C. A., Jordan, J. M., and Vinograd, J. (1971): *In vivo* effects of intercalating drugs on the superhelix density of mitochondrial DNA isolated from human and mouse cells in culture. *J. Mol. Biol.*, 59:255–272.
11. Ter Schegget, J., and Borst, P. (1971): DNA synthesis by isolated mitochondria. I. Effect of inhibitors and characterization of the product. *Biochim. Biophys. Acta,* 246:239–248.
12. Radsak, K., Kato, K., Sato, N., and Koprowski, H. (1971): Effect of ethidium bromide on mitochondrial DNA and cytochrome synthesis in HeLa cells. *Exp. Cell Res.*, 66:410–416.
13. Leibowitz, J. R. D. (1971): The effect of ethidium bromide on mitochondrial DNA synthesis and mitochondrial DNA structure in HeLa cells. *J. Cell Biol.*, 51:116–122.
14. Nass, M. M. K., and Ben-Shaul, Y. (1973): Effects of ethidium bromide on growth, chlorophyll synthesis, ultrastructure and mitochondrial DNA in green and bleached mutant *Euglena gracilis. J. Cell Sci.*, 13:567–590.
15. Goldring, E. S., Grossman, L. I., Krupnick, D., Cryer, D. R., and Marmur, J. (1970): The petite mutation in yeast. Loss of mitochondrial deoxyribonucleic acid during induction of pe-

tites with ethidium bromide. *J. Mol. Biol.*, 52:323–335.

16. Goldring, E. S., Grossman, L. I., and Marmur, J. (1971): Isolation of mutants containing mitochondrial deoxyribonucleic acid of reduced size. *J. Bacteriol.*, 107:377–381.

17. Perlman, P. S., and Mahler, H. R. (1971): Molecular consequences of ethidium bromide mutagenesis. *Nature [New Biol.]*, 231:12–16.

18. Mahler, H. R. (1973): Genetic autonomy of mitochondrial DNA. In: *Molecular Cytogenetics*, edited by B. A. Hamkalo and J. Papaconstantinou, pp. 181–208. Plenum Press, New York.

19. Mahler, H. R., Mehrotra, B. D., and Perlman, P. S. (1971): Formation of yeast mitochondria. V. Ethidium bromide as a probe for the functions of mitochondrial DNA. *Prog. Mol. Subcell. Biol.*, 2:274–296.

20. Mahler, H. R., and Perlman, P. S. (1972): Mitochondrial membranes and mutagenesis by ethidium bromide. *J. Supramol. Struct.*, 1:105–124.

21. Mahler, H. R., and Bastos, R. N. (1974): A novel reaction of mitochondrial DNA with ethidium bromide. *FEBS Lett.*, 39:27–34.

22. Mahler, H. R., and Bastos, R. N. (1974): Coupling between mitochondrial mutation and energy transduction. *Proc. Natl. Acad. Sci. U.S.A.*, 72:2241–2245.

23. Bastos, R. N., and Mahler, H. R. (1974): A synthesis of labelled ethidium bromide. *Arch. Biochem. Biophys.*, 160:643–646.

24. Nagley, P., and Linnane, A. W. (1972): Biogenesis of mitochondria. XXI. Studies on the nature of the mitochondrial genome in yeast: The degenerative effects of ethidium bromide on mitochondrial genetic information in a respiratory competent strain. *J. Mol. Biol.*, 66:181–193.

25. Hall, R. M., Nagley, P., and Linnane, A. W. (1976): Biogenesis of mitochondria. XLII. Genetic analysis of the control of cellular mitochondrial DNA levels in *Saccharomyces cerevisiae*. *Mol. Gen. Genet.*, 145:169–175.

26. Bastos, R. N., and Mahler, H. R. (1974): Molecular mechanisms of mitochondrial genetic activity: Effects of ethidium bromide on the deoxyribonucleic acid and energetics of isolated mitochondria. *J. Biol. Chem.*, 249:6617–6627.

27. Perlman, P. S., and Mahler, H. R. (1971): A premutational state induced in yeast by ethidium bromide. *Biochem. Biophys. Res. Commun.*, 44:261–267.

28. Mahler, H. R. (1973): Structural requirements for mitochondrial mutagenesis. *J. Supramol. Struct.*, 2:449–460.

29. Moustacchi, E. (1971): Evidence for nucleus independent steps in control of repair of mitochondrial damage. IV. UV-induction of the cytoplasmic petite mutation in UV-sensitive nuclear mutants of *Saccharomyces cerevisiae*. *Mol. Gen. Genet.*, 114:50–58.

30. Nagai, S., Yanagashima, N., and Nagai, H. (1961): Advances in the study of the respiration-deficient (RD) mutation in yeast and other microorganisms. *Bacteriol. Rev.*, 25:404–426.

31. Sager, R. (1972): *Cytoplasmic Genes and Organelles*. Academic Press, New York.

32. Mahler, H. R., and Perlman, P. S. (1973): Induction of respiration deficient mutants in *Saccharomyces cerevisiae* by berenil. I. Berenil, a novel, non-intercalating mutagen. *Mol. Gen. Genet.*, 121:285–294.

33. Perlman, P. S., and Mahler, H. R. (1973): Induction of respiration deficient mutants in *Saccharomyces cerevisiae* by berenil. II. Characteristics of the process. *Mol. Gen. Genet.*, 121:295–306.

34. Wilkie, D. (1963): The induction by monochromatic U.V. light of respiratory-deficient mutants in aerobic and anaerobic cultures of yeast. *J. Mol. Biol.*, 7:527–533.

35. Allen, N. B., and MacQuillan, A. M. (1969): Target analysis of mitochondrial genetic units in yeast. *J. Bacteriol.*, 97:1142–1148.

36. Deutsch, J., Dujon, B., Netter, P., Petrochilo, E., Slonimski, P. P., Bolotin-Fukuhara, M., and Coen, D. (1974): Mitochondrial genetics. VI. The petite mutation in *Saccharomyces cerevisiae:* Interrelations between the loss of the ρ^+ factor and the loss of the drug resistance mitochondrial genetic markers. *Genetics*, 76:195–219.

37. Ephrussi, B., and Grandchamp, S. (1965): Études sur la suppressivité des mutants à déficience respiratoire de la levure. I. Existence au niveau cellulaire de divers degrés de suppressivité. *Heredity*, 20:1–7.

38. Rank, G. H., and Person, C. (1969): Reversion of spontaneously arising respiratory deficiency in *Saccharomyces cerevisiae*. *Can. J. Genet. Cytol.*, 11:716–728.

39. Saunders, G. W., Gingold, E. B., Trembath, M. K., Lukins, H. B., and Linnane, A. W. (1971): Mitochondrial genetics in yeast: Segregation of a cytoplasmic determinant in crosses and its loss or retention in the petite. In: *Autonomy and Biogenesis of Mitochondria and Chloroplasts*, edited by N. K. Boardman, A. W. Linnane, and R. M. Smillie, pp. 185–193. North Holland, Amsterdam.

40. Bernardi, G., Carnevali, F., Nicolaieff, A., Piperno, G., and Tecce, G. (1968): Separation and characterization of a satellite DNA from a yeast cytoplasmic petite mutant. *J. Mol. Biol.*, 37:493–505.

41. Mehrotra, B. D., and Mahler, H. R. (1968): Characterization of some unusual DNAs from the mitochondria from certain "petite" strains of *Saccharomyces cerevisiae*. *Arch. Biochem. Biophys.*, 128:685–703.

42. Van Kreijl, C. F., Borst, P., Flavell, R. A., and Hollenberg, C. P. (1972): Pyrimidine tract analysis of mtDNA from a "low density"

petite mutant of yeast. *Biochim. Biophys. Acta,* 277:61–70.

43. Bernardi, G., Faures, M., Piperno, G., and Slonimski, P. P. (1970): Mitochondrial DNA's from respiratory-sufficient and cytoplasmic respiratory-deficient mutant yeast. *J. Mol. Biol.,* 48:23–42.

44. Bernardi, G., and Timasheff, S. N. (1970): Optical rotatory dispersion and circular dichroism properties of yeast mitochondrial DNA's. *J. Mol. Biol.,* 48:43–52.

45. Bernardi, G., Piperno, G., and Fonty, G. (1972): The mitochondrial genome of wild-type yeast cells. I. Preparation and heterogeneity of mitochondrial DNA. *J. Mol. Biol.,* 65:173–189.

46. Piperno, G., Fonty, G., and Bernardi, G. (1972): The mitochondrial genome of wild-type yeast cells. II. Investigations on the compositional heterogeneity of mitochondrial DNA. *J. Mol. Biol.,* 65:191–205.

47. Ehrlich, S. D., Thiery, J.-P., and Bernardi, G. (1972): The mitochondrial genome of wild-type yeast cells. III. The pyrimidine tracts of mitochondrial DNA. *J. Mol. Biol.,* 65:207–212.

48. Prunell, A., and Bernardi, G. (1974): The mitochondrial genome of wild-type yeast cells. IV. Genes and spacers. *J. Mol. Biol.,* 86:825–841.

49. Bernardi, G., Prunell, A., Fonty, G., Kopecka, H., and Strauss, F. (1976): The mitochondrial genome of yeast: Organization, evolution and the petite mutation. In: *The Genetic Functions of Mitochondrial DNA,* edited by C. Saccone and A. M. Kroon, pp. 185–198. North Holland, Amsterdam.

50. Bernardi, G. (1976): The mitochondrial genome of yeast: Organization and recombination. In: *Genetics and Biogenesis of Chloroplasts and Mitochondria,* edited by T. Bücher, W. Neupert, W. Sebald, and S. Werner, pp. 503–510. North Holland, Amsterdam.

51. Cohen, M., Casey, J., Rabinowitz, M., and Getz, G. S. (1972): Hybridization of mitochondrial transfer RNA and mitochondrial DNA in petite mutants of yeast. *J. Mol. Biol.,* 63:441–451.

52. Casey, J., Gordon, P., and Rabinowitz, M. (1974): Characterization of mitochondrial deoxyribonucleic acid from grande and petite yeasts by renaturation and denaturation analysis and by transfer ribonucleic acid hybridization; evidence for internal repetition or heterogeneity in mitochondrial deoxyribonucleic acid populations. *Biochemistry,* 13:1059–1067.

53. Nagley, P., Gingold, E. B., Lukins, H. B., and Linnane, A. W. (1973): Biogenesis of mitochondria. XXV. Studies of the mitochondrial genomes of petite mutants of yeast using ethidium bromide as a probe. *J. Mol. Biol.,* 78:335–350.

54. Faye, G., Fukuhara, H., Grandchamp, C.,

Lazowska, J., Michel, F., Casey, J., Getz, G. S., Locker, J., Rabinowitz, M., Bolotin-Fukuhara, M., Coen, D., Deutsch, J., Dujon, B., Netter, P., and Slonimski, P. P. (1973): Mitochondrial nucleic acids in the petite colonie mutants: Deletions and repetitions of genes. *Biochimie,* 55:779–792.

55. Fukuhara, H., Faye, G., Michel, F., Lazowska, J., Deutsch, J., Bolotin-Fukuhara, M., and Slonimski, P. P. (1974): Physical and genetic organization of petite and grande yeast mitochondrial DNA. I. Studies by RNA-DNA hybridization. *Mol. Gen. Genet.,* 130:215–238.

56. Lazowska, J., Michel, F., Faye, G., Fukuhara, H., and Slonimski, P. P. (1974): Physical and genetic organization of *petite* and *grande* yeast mitochondrial DNA. II. DNA-DNA hybridization studies and buoyant density determinations. *J. Mol. Biol.,* 85:393–410.

57. Michel, F., Lazowska, J., Faye, G., Fukuhara, H., and Slonimski, P. P. (1974): Physical and genetic organization of *petite* and *grande* yeast mitochondrial DNA. III. High resolution melting and reassociation studies. *J. Mol. Biol.,* 85:411–431.

58. Faye, G., Kujawa, C., and Fukuhara, H. (1974): Physical and genetic organization of *petite* and *grande* yeast mitochondrial DNA. IV. *In vivo* transcription products of mitochondrial DNA and localization of 23 S ribosomal RNA in petite mutants of *Saccharomyces cerevisiae. J. Mol. Biol.,* 88:185–203.

59. Faye, G., Kujawa, C., Dujon, B., Bolotin-Fukuhara, M., Wolf, K., Fukuhara, H., and Slonimski, P. P. (1975): Localisation of the gene coding for the 16 S ribosomal mitochondrial RNA using rho⁻ mutants of *Saccharomyces cerevisiae. J. Mol. Biol.,* 99:203–217.

60. Lazowska, J., and Slonimski, P. P. (1976): Electron microscopy analysis of circular repetitive mitochondrial DNA molecules from genetically characterized rho⁻ mutants of *Saccharomyces cerevisiae. Mol. Gen. Genet.,* 146:61–78.

61. Gordon, P., and Rabinowitz, M. (1973): Evidence for deletion and changed sequence in the mitochondrial deoxyribonucleic acid of a spontaneously generated petite mutant of *Saccharomyces cerevisiae. Biochemistry,* 12:116–123.

62. Fauman, M. A., and Rabinowitz, M. (1974): DNA-DNA hybridization studies of mitochondrial DNA of ethidium-bromide-induced petite mutants of yeast. *Eur. J. Biochem.,* 42:67–71.

63. Gordon, P., Casey, J., and Rabinowitz, M. (1974): Characterization of mitochondrial deoxyribonucleic acid from a series of petite yeast strains by deoxyribonucleic acid-deoxyribonucleic acid hybridization. *Biochemistry,* 13:1067–1075.

64. Locker, J., Rabinowitz, M., and Getz, G. S. (1974): Tandem inverted repeats in mito-

chondrial DNA of petite mutants of *Saccharomyces cerevisiae. Proc. Natl. Acad. Sci. U.S.A.*, 71:1366–1370.

65. Locker, J., Rabinowitz, M., and Getz, G. S. (1974): Electron microscopic and renaturation kinetic analysis of mitochondrial DNA of cytoplasmic petite mutants of *Saccharomyces cerevisiae. J. Mol. Biol.*, 88:489–507.

66. Morimoto, R., Lewin, A., Hsu, H.-J., Rabinowitz, M., and Fukuhara, H. (1975): Restriction endonuclease analysis of mitochondrial DNA from grande and genetically characterized cytoplasmic petite clones of *Saccharomyces cerevisiae. Proc. Natl. Acad. Sci. U.S.A.*, 72:3868–3872.

67. Morimoto, R., Lewin, A., Merten, S., and Rabinowitz, M. (1976): Restriction endonuclease mapping and analysis of grande and mutant yeast mitochondrial DNA. In: *Genetics and Biogenesis of Chloroplasts and Mitochondria*, edited by T. Bücher, W. Neupert, W. Sebald, and S. Werner, pp. 519–524. North Holland, Amsterdam.

68. Nagley, P., Molloy, P. L., Lukins, H. B., and Linnane, A. W. (1974): Studies on mitochondrial gene purification using petite mutants of yeast: Characterization of mutants enriched in ribosomal RNA cistrons. *Biochem. Biophys. Res. Commun.*, 57:232–239.

69. Casey, J. W., Hsu, H.-J., Rabinowitz, M., Getz, G., and Fukuhara, H. (1974): Transfer RNA genes in the mitochondrial DNA of cytoplasmic *petite* mutants of *Saccharomyces cerevisiae. J. Mol. Biol.*, 88:717–733.

70. Martin, N. C., and Rabinowitz, M. (1976): Transfer RNAs of yeast mitochondria. In: *Genetics and Biogenesis of Chloroplasts and Mitochondria*, edited by T. Bücher, W. Neupert, W. Sebald, and S. Werner, pp. 749–754. North Holland, Amsterdam.

71. Sanders, J. P. M., Flavell, R. A., Borst, P., and Mol, J. N. M. (1973): Nature of the base sequence conserved in the mitochondrial DNA of a low-density petite. *Biochim. Biophys. Acta*, 312:441–457.

72. Fauman, M., and Rabinowitz, M. (1972): Analysis of grande and petite mitochondrial DNA by DNA-DNA hybridization. *FEBS Lett.*, 28:317–321.

73. Borst, P., Heyting, C., and Sanders, J. P. M. (1976): The control of mitochondrial DNA synthesis in yeast petite mutants. In: *Genetics and Biogenesis of Chloroplasts and Mitochondria*, edited by T. Bücher, W. Neupert, W. Sebald, and S. Werner, pp. 525–533. North Holland, Amsterdam.

74. Inman, R. B. (1967): Denaturation maps of the left and right sides of the lambda DNA molecule determined by electron microscopy. *J. Mol. Biol.*, 28:103–116.

75. Michels, C. A., Blamire, R., Goldfinger, B., and Marmur, J. (1974): A genetic and biochemical analysis of petite mutations in yeast. *J. Mol. Biol.*, 90:431–449.

76. Hollenberg, C. P., Borst, P., and van Bruggen, E. F. J. (1972): Mitochondrial DNA from cytoplasmic petite mutants of yeast. *Biochim. Biophys. Acta*, 277:35–43.

77. Fukuhara, H. (1969): Relative proportions of mitochondrial and nuclear DNA in yeast under various conditions of growth. *Eur. J. Biochem.*, 11:135–139.

78. Michaelis, B., Petrochilo, E., and Slonimski, P. P. (1973): Mitochondrial genetics. III. Recombined molecules of mitochondrial DNA obtained from crosses between cytoplasmic *petite* mutants of *Saccharomyces cerevisiae:* Physical and genetic characterization. *Mol. Gen. Genet.*, 123:51–65.

79. Michaelis, G., Michel, F., Lazowska, J., and Slonimski, P. P. (1976): Recombined molecules of mitochondrial DNA obtained from crosses between cytoplasmic petite mutants of *Saccharomyces cerevisiae:* The stoichiometry of parental DNA repeats within the recombined molecules. *Mol. Gen. Genet.*, 149:125–130.

80. Ephrussi, B., de Margerie-Hottinguer, H., and Roman, H. (1955): Suppressiveness: A new factor in the genetic determinism of the synthesis of respiratory enzymes in yeast. *Proc. Natl. Acad. Sci. U.S.A.*, 41:1065–1070.

81. Ephrussi, B., Jakob, H., and Grandchamp, S. (1966): Études sur la suppressivité des mutants à déficience respiratoire de la levure. II. Étapes de la mutation grande en petite provoqueé par le facteur suppressif. *Genetics*, 54:1–29.

82. Slonimski, P. P. (1968): Discussion. In: *Biochemical Aspects of the Biogenesis of Mitochondria*, edited by E. C. Slater, J. M. Tager, S. Papa, and E. Quagliariello, pp. 475–477. Adriatica Editrice, Bari.

83. Carnevali, F., Morpurgo, G., and Tecce, G. (1969): Cytoplasmic DNA from petite colonies of *Saccharomyces cerevisiae:* A hypothesis on the nature of the mutation. *Science*, 163:1331–1333.

84. Mills, D. R., Peterson, R. L., and Spiegelman, S. (1967): An extracellular Darwinian experiment with a self-duplicating nucleic acid molecule. *Proc. Natl. Acad. Sci. U.S.A.*, 58:217–224.

85. Rank, G. H. (1970): Genetic evidence for "Darwinian" selection at the molecular level. I. The effect of the suppressive factor on cytoplasmically-inherited erythromycin-resistance in *Saccharomyces cerevisiae. Can. J. Genet. Cytol.*, 12:129–136.

86. Rank, G. H. (1970): Genetic evidence for "Darwinian" selection at the molecular level. II. Genetic analysis of cytoplasmically-inherited high and low suppressivity in *Saccharomyces cerevisiae. Can. J. Genet. Cytol.*, 12:340–346.

87. Rank, G. H., and Bech-Hansen, N. T. (1972): Genetic evidence for "Darwinian" selection at the molecular level. III. The effect of the suppressive factor on nuclearly and cytoplasmi-

cally inherited chloramphenicol resistance in *S. cerevisiae. Can. J. Microbiol.,* 18:1–7.

88. Coen, D., Deutsch, J., Netter, P., Petrochilo, E., and Slonimski, P. P. (1970): Mitochondrial genetics. I. Methodology and phenomenology. *Symp. Soc. Exp. Biol.,* 24:449–496.

89. Michaelis, G., Douglass, S., Tsai, M.-J., and Criddle, R. S. (1971): Mitochondrial DNA and suppressiveness of petite mutants in *Saccharomyces cerevisiae. Biochem. Genet.,* 5:487–495.

90. Perlman, P. S., and Birky, C. W., Jr. (1974): Mitochondrial genetics in baker's yeast: A molecular mechanism for recombinational polarity and suppressiveness. *Proc. Natl. Acad. Sci. U.S.A.,* 71:4617–4621.

Chapter 7

Mitochondrial Genetics of Bakers' Yeast. IV. The Genetic Map of the Mitochondrial Genome and Models for Segregation and Recombination of Mitochondrial Genomes

In his opening remarks to a symposium on cytoplasmic inheritance held in 1958, Boris Ephrussi said that there were two types of inheritance: nuclear and unclear (1). At that time, that was a neat summary of the state of cytoplasmic genetics in general and organelle genetics in particular. As the preceding chapters have shown, that situation no longer exists. Chloroplast and mitochondrial DNA provide the physical entities, the chromosomes, in which organelle genes lie. In yeast, it is certain beyond all reasonable doubt that the vegetative petite mutation and the nonmendelian "point" mutations to antibiotic resistance (ant^r) and respiration deficiency (mit^-, syn^-) are mutations of mtDNA (see Chapters 4–6). Shortly after the discovery of the ant^r mutations, several laboratories in different parts of the world, notably those of Linnane and Slonimski, began attempts to map mitochondrial genes and discover the rules underlying the segregation and recombination of mitochondrial genomes. After an initial period during which a series of conflicting and confusing reports appeared, a general theory of mitochondrial gene recombination began to emerge. At the same time, various methods were developed for the mapping of mitochondrial genes. The result is that the model is now being tested critically, and the mapping of mitochondrial genes in *S. cerevisiae* is proceeding at a rapid pace. In fact, it is not unreasonable to suggest that in a few years the mitochondrial genome of *S. cerevisiae* will be as well characterized in a genetic sense as are those of many bacterial viruses.

Before entering into a specific discussion of the mapping of mitochondrial genes in yeast, some general remarks on the formidable technical problems involved are appropriate, as is some indication about why these obstacles have not proved insurmountable. Perhaps most of the problems that beset the organelle geneticist can be traced to the fact that chloroplasts and mitochondria are parts of cells, not distinct organisms. One cannot score the phenotype of an organelle carrying a specific mutation, only the phenotype of the cell in which the organelle resides. The fact that there are usually several organelles of a given type per cell, and at least that many genomes per organelle, complicates the analysis further. Genotypic purity for a given organelle allele in a cell line can only be inferred, not rigorously proved. The tests of dominance and recessiveness that are routine with mendelian alleles are difficult to apply because, as a rule, both chloroplast (see Chapter 14) and mitochondrial genes segregate rapidly. Consequently, it is difficult to determine what the phenotype of a cell homozygous for an organelle allele ought to be. The processes of organelle fusion and reassortment have the potential to complicate further the analysis of organelle gene segregation and recombination. Suppose, for example, that the organelle genome were broken up into several nonhomologous DNA molecules. Even if these DNA molecules were to assort at random on organelle fusion, formal linkage between markers carried on them would be observed if less than half the organelles present in the zygote and its progeny fused. In two-factor crosses, one could easily confuse the probability of fusion with map distance, but this would be much less likely in a three-factor cross, where true linkage would yield an order irrespective of the probability of fusion. Another possibility is that organelle reassortment could be confused with recombination. The reassortment of similar organelles carrying different alleles might generate a clone of cells with a "recombinant" phenotype that was, in fact, genotypically heterozygous, although this possibility seems remote in view of the apparent rapid segregation of organelle genes. Finally, the input and outflow of organelle genomes in a cross has not been a controllable parameter in the past, although there is now some hope that the input may be subject to systematic manipulation, as will be discussed in this chapter and in Chapter 12. One is tempted to make an analogy to a phage cross, in which the average multiplicity of infection, the average burst size, and the frequency of phenotypically different progeny phages cannot be measured directly.

Fortunately, it has now become evident that the foregoing problems appeared to be more serious than they really were with respect to organelle gene mapping, although they, as well as other obstacles, are likely to impede our understanding of the underlying rules governing organelle gene segregation and recombination. There is one overriding reason why we should be able to map organelle genes. We know that in all genetic systems the mutations that are far apart on a piece of DNA will recombine more frequently than those that are closer together. We also know that mutations falling within the same locus usually recombine very rarely. No matter how many complicating intermediary events occur in the process of getting a pair of organelle DNA molecules together and then segregating them out, we should still be able to order markers based on the recombination frequency (if they are formally linked), although map distances may vary, and we have little idea of what a map unit represents. Similarly, mutants that

are allelic will recombine rarely, irrespective of secondary, tertiary, etc., happenings. These are the facts that must be kept in mind when this complex of problems, real and imagined, threatens to discourage the investigator. Certainly, equally formidable obstacles did not daunt the students of phage and bacterial genetics in the early days, and the study of organelle genetics has only recently emerged from the state those disciplines experienced about 1950.

In the case of *S. cerevisiae*, we can cite a number of specific reasons why it is now possible to map the mitochondrial genome with assurance. First, in the progeny of crosses differing in mitochondrial genotype, marker segregation takes place rapidly in most clones, yielding phenotypically stable and apparently homoplasmic cell lines. Second, the physical structure of yeast mtDNA and the deletion mapping experiments to be discussed later in this chapter suggest that there is only one mitochondrial linkage group. Third, genetic recombination of mitochondrial markers appears to be accompanied by physical exchange between the two parental mtDNAs. Each of these points will be discussed in more detail below, and it should become apparent that the formal mapping of mitochondrial genes in *S. cerevisiae* now rests on very solid ground.

The literature relating to segregation, recombination, and mapping of mitochondrial genes in *S. cerevisiae* has been steadily increasing over the past 7 years, and it is now truly prodigious. The two volumes that resulted from the recent conferences at Bari (2) and Munich (3) attest to this point. Rather than undertaking to review all this literature, this chapter will attempt to define the methodologies used to map mitochondrial genes in *S. cerevisiae* and the present state of thinking about how mitochondrial genomes recombine and segregate. The hope is that this discussion will provide a point of departure from which the interested reader can continue alone. The chapter contains two major themes. The first is the development of formal genetic analysis of mitochondrial gene recombination in *S. cerevisiae*. Basically, this subject is treated historically, culminating with the most widely accepted model for segregation and recombination of mitochondrial genomes, which likens them to the genomes of bacterial viruses. The second major theme concerns the development of deletion mapping techniques employing ρ^- mutants, which has now progressed to the point that mutants having different deletions can be used for precise physical mapping of the mitochondrial genome of *S. cerevisiae*. In the course of the chapter we will also consider a range of other topics relating to mitochondrial gene recombination and segregation. The chapter will end with discussion of a new cytoplasmic DNA in *S. cerevisiae* whose peculiar properties suggest similarities to

bacterial episomes. Mutations in this DNA can cause various cellular membranes to become impermeable to antibiotics that are unrelated in structure and biologic effect.

MITOCHONDRIAL CROSSES: METHODOLOGY AND TERMINOLOGY

Three methods have been used to study segregation and recombination of mitochondrial genes in yeast. In the first, pedigrees of the diploid buds produced by zygotes and their vegetative diploid progeny are analyzed (Fig. 7–1). Pedigree analysis has the advantage of giving detailed information about the pattern of segregation and recombination in the early cell divisions following zygote formation. The disadvantages are that pedigree analysis is time-consuming

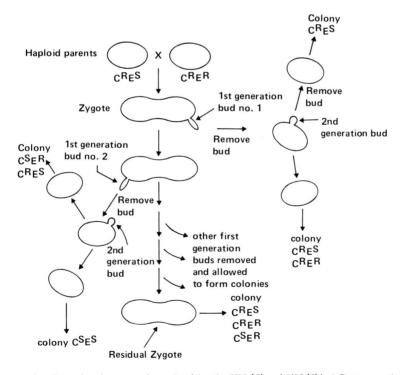

FIG. 7–1. Diagram of pedigree from homosexual cross involving the *RIB1* (C) and *RIB3* (E) loci. First-generation buds formed by the zygote can simply be dissected sequentially, as was done by Forster (14), or they can be allowed to form vegetative diploids, the buds from which (second generation, third generation, etc.) are dissected, as was done by Lukins et al. (9) Cells whose mitochondrial genotypes are to be determined are allowed to form colonies. A sample of the cells in each colony is then tested on appropriate selective media to establish which mitochondrial genotypes are present in the colony and, hence, were present in the original cell.

and only limited numbers of pedigrees are easily done, so that quantitative information concerning recombination frequencies or output ratios of mitochondrial alleles from the two parents is difficult to obtain.

Zygote clone analysis is the second method. Zygote clones can be analyzed qualitatively by replica-plating the entire clone to appropriate selective media to indicate which mitochondrial genotypes are present in the clone (Fig. 7–2), or the clone can be analyzed quantitatively (Fig. 7–3). In the quantitative method, a sample of cells from the clone to be analyzed is plated at appropriate dilution on nonselective medium, and the resulting colonies are then replica-plated to appropriate selective media. Quantitative zygote clone analysis not only indicates which mitochondrial genotypes are present in the clone but also measures their frequency. Zygote clone analysis is much faster than pedigree analysis, and it yields much more extensive quantitative information. It also enables us to determine if individual zygotes usually produce reciprocal recombinants and if these are present in equal frequencies. However, zygote clone analysis can give only indirect information about the timing of recombinational events and segregation patterns.

The third method, referred to as the *standard cross* (4), has yielded the most extensive and most significant data on recombination and linkage of mitochondrial genes. In this method, a mass mating is made and plated on nonselective medium, and zygotes and progeny diploid vegetative cells are allowed to grow for about 20 generations until a confluent lawn of cells is formed (Fig. 7–4). The resulting population is harvested and plated at appropriate dilution on nonselective medium. The colonies formed by the diluted cells are then replica-plated to appropriate selective media. The standard cross yields quantitative information on the genotypic composition of the population as a whole, irrespective of zygote of origin. It is the most efficient method for obtaining large amounts of quantitative information and for making comparisons between different crosses.

Since the most comprehensive model of mitochondrial gene recombination likens the process to phage recombination (pp. 225–239), it is worth noting that zygote clone analysis and the standard cross have their analogs in the single burst experiment and phage cross, respectively. That is, each diploid zygote can be compared to a bacterium in which a number of genomes capable of

QUALITATIVE CLONAL

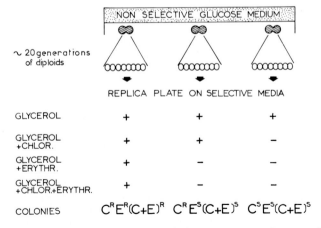

FIG. 7–2. Diagram of qualitative method of analyzing mitochondrial genotypes present in zygote clones. (From Coen et al., ref. 4, with permission.)

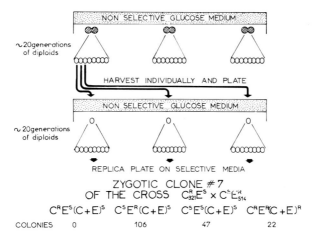

FIG. 7–3. Diagram of quantitative method of analyzing mitochondrial genotypes present in zygote clones. Numerical example gives results obtained with a specific zygote clone from the cross $C^R_{321}E^S \times C^S E^R_{514}$. (From Coen et al., ref. 4, with permission.)

recombination are present. In the case of the bacterium, these would be phage genomes; in the case of the yeast zygote, they are mitochondrial genomes. Thus a zygote clone reveals the genotypic composition of the mitochondrial genomes derived from a single zygote, whereas in the standard cross the population of mitochondrial genomes is analyzed irrespective of the zygote of origin,

in the same way that progeny phage are analyzed in a phage cross irrespective of the bacterium of origin.

With reference principally to the results of standard crosses, Coen et al. (4) introduced the following terminology, to which new words (5,6) have been added as more facts about mitochondrial gene recombination have accumulated.

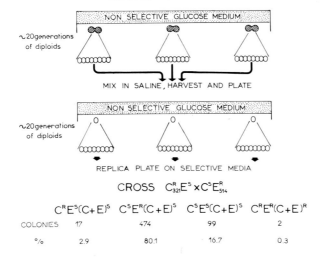

FIG. 7–4. Diagram of procedure used in standard cross for analysis of mitochondrial genotypes among progeny of random diploid cells. Numerical example gives results obtained with a specific zygote clone from the cross $C^R_{321}E^S \times C^S E^R_{514}$. (From Coen et al., ref. 4, with permission.)

Anisochromosomal isomitochondrial cross. Anisochromosomal isomitochondrial crosses (4) are crosses in which the chromosomal genomes are different, but the mitochondrial genomes are the same.

Asymmetry. Asymmetry (5,6) is a situation in which the frequencies of pairs of mitochondrial alleles or reciprocal recombinants for pairs of mitochondrial genes are unequal among the diploid progeny of a cross due to the influence of a nuclear gene.

Bias. The term *bias* (5,6) is applied when the frequencies of pairs of mitochondrial alleles or reciprocal recombinants for pairs of mitochondrial genes are unequal among the diploid progeny of a cross for unknown reasons. The definition of bias should be compared to the definitions of asymmetry and polarity.

Isochromosomal anisomitochondrial cross. Isochromosomal anisomitochondrial crosses (4) are crosses in which the chromosomal genomes of two crosses are the same, but the mitochondrial genomes are different. This term is used for comparison of two or more crosses or strains; it is never applied to a single cross or strain.

Polarity. The term *polarity* (5,6) is applied when the frequencies of pairs of alleles or pairs of reciprocal recombinants are unequal among the diploid progeny of a cross because of the influence of a genetic determinant known to be localized in the mitochondrial genome. The original definition of

the term (4) was sufficiently vague to account in part for the fact that some authors confused asymmetry, bias, and transmission with polarity.

Transmission. In a cross of two pure haploid clones differing by a pair of mitochondrial alleles (e.g., C^R versus C^S), transmission (4) is a measure of the frequency of diploid cells possessing one of the alleles. An allele is highly transmitted if the fraction of cells carrying the allele is high.

The data summarized in Table 7–1 for several crosses of the type $C^R E^S \times C^S E^R$ illustrate the meaning of polarity. It can be seen that these crosses are highly polar in terms of both transmission and recombination. Thus the C^S and E^R alleles carried by one of the parents are much more highly transmitted than the C^R and E^S alleles of the other. Reciprocal recombinants also show polarity, being highly asymmetric in frequency, with the $C^S E^S$ recombinant being in the majority. Interpretation of these results will be discussed later.

The definition of transmission and its relationship to polarity have caused confusion, probably partly because of the original definition of polarity (4) and the rapid accumulation of new and sometimes seemingly contradictory observations by different groups of workers (7,8). As a consequence, the terms *asymmetry* and *bias* were introduced by Avner et al. (5) to distinguish effects on the ratio of reciprocal recombi-

TABLE 7–1. *Analysis of polarity in a series of crosses of the type* $C^R_{351}E^S \times C^S E^R_{514}$ *involving the* R_I *and* R_{III} *genes*

Cross	Percentage of each cell type				Polarity		
					Transmission		Recombination
	$C^R E^S$	$C^S E^R$	$C^S E^S$	$C^R E^R$	C^S/C^R	E^S/E^R	$C^S E^S/C^R E^R$
$aC^R E^S \times \alpha C^S E^R$	2.1	75.0	22.9	0.06	45	0.33	381
$aC^R E^S \times \alpha C^S E^R$	1.5	67.7	30.9	0.04	64	0.48	772
$\alpha C^R E^S \times aC^S E^R$	5.2	56.3	38.5	0.15	18	0.77	256
$\alpha C^R E^S \times aC^S E^R$	1.6	67.4	31.0	0.20	55	0.48	155
$\alpha C^R E^S \times aC^S E^R$	1.4	68.6	29.9	0.16	63	0.45	187
$aC^R E^S \times \alpha C^S E^R$	0.9	82.3	16.7	0.10	99	0.21	167
$aC^R E^S \times \alpha C^S E^R$	1.1	76.6	22.3	0.08	84	0.30	279

Adapted from Bolotin et al. (20).

nants or the transmission of allelic pairs that could not clearly be attributed to polarity. Dujon et al. (6) pointed out that "all crosses which show a strong bias in the frequency of reciprocal recombinants display, thereby, a bias in transmission. The converse is not however true. There exist crosses which show a bias in transmission without showing any bias in the frequency of recombinants." We will consider some examples of asymmetry and bias and their interpretation later in this chapter.

PHYSICAL EVIDENCE FOR MITOCHONDRIAL GENE RECOMBINATION

Evidence that recombination of mitochondrial genes involves physical recombination of mtDNA comes from the experiments of Michaelis and associates, as described in Chapter 6. These workers employed ρ^- mutants having mtDNAs of differing buoyant densities and carrying dissimilar ant^r markers. They found that genetic recombination of these markers was accompanied by changes in the physical properties of the mtDNA in the recombinants such that it appeared to contain sequences derived from the two parents.

TIMING OF MITOCHONDRIAL GENE SEGREGATION AND RECOMBINATION

Largely as a result of pedigree analysis (9–15), two important facts have emerged about mitochondrial gene segregation and recombination in *S. cerevisiae*. First, most mitochondrial gene recombination probably occurs in the zygote. Second, mitochondrial gene segregation proceeds rapidly. Indirect evidence that most recombination takes place in the zygote was first provided by Coen et al. (4) using quantitative zygote clone analysis. These authors examined the clonal distribution of C^sE^s recombinants in a highly polar cross between C^RE^s and C^sE^R

parents in which the C^R allele was very poorly transmitted and C^RE^R recombinants were virtually absent. Coen and associates argued that the proportion of C^sE^s recombinants in each zygote clone could be arranged in the series $(\frac{1}{2})^n$, where $n = 1, 2, 3, 4$, etc., and that a simple model accounting for this distribution would be one in which successive zygote buds had equal probabilities of producing a pure C^sE^s clone. Thus, if the first bud yielded a pure C^sE^s clone, 50% of the cells in the zygote would have the genotype C^sE^s; if the second bud did so, 25% of the cells would be C^sE^s, and so forth. However, the authors were careful to point out that other more complex models could also account for their results.

Pedigree analysis has subsequently shown that although the model of Coen and associates is overly simplistic in its assumptions, the notion of rapid segregation and recombination is essentially correct. To illustrate this, we have summarized some of Callen's pedigree data (11) in Table 7–2. These data, obtained from 15 pedigrees from two-factor mitochondrial crosses, show that 59% of the buds segregated were pure for one parental genotype or the other. The remainder contained one or both recombinant genotypes, but of these only 23% (8

TABLE 7–2. *Mitochondrial genotypes present in buds obtained in the course of analysis of 15 pedigrees from two-factor mitochondrial crosses*

Genotypes present in bud clone	Number of buds	Percentage of total buds
2P + 0R[a]	0	0
2P + 1R	2	2.3
2P + 2R	6	7.0
1P + 0R	51	59.0
1P + 1R	20	23.2
1P + 2R	3	3.5
0P + 1R	4	4.7
0P + 2R	0	0
Total	86	

[a] P = parental genotype; R = recombinant genotype; 0, 1, 2 = number of parental and recombinant genotypes derived from each bud.
Data from Callen (11).

of 35) contained both parental genotypes. The simplest interpretation of these data is that recombination took place in the zygote, following which the recombinant and parental genotypes were parceled out to the buds in small numbers such that only rarely did a bud receive more than two genotypes. This is a notion we will return to later. Forster (14), who summarized her extensive pedigree data in a recent article, found that the rate of mitochondrial marker purification, although rapid, varied between crosses. For example, in one cross 55% of the zygotes were pure from the first bud on, but this was true of only 6% in another cross (Fig. 7–5). However, by bud 17 less than 10% of the zygotes had not purified for one or both of the loci studied in these experiments.

One biologic parameter that may be important in determining both the rate of segregation of homoplasmic mitochondrial genotypes and the recombination frequency between mitochondrial genes is the position of a bud on the zygote. It will be recalled from illustrations in preceding chapters that the yeast zygote is shaped like a dumbbell. Buds may issue from either end of the zygote or from the neck at the center. Callen

(11) found that end buds were often pure for one parental mitochondrial genotype or the other and that (with one exception) daughter buds arising from the same end of the zygote were homoplasmic for the same parental genotype (Table 7–3). His data also suggested that recombinant genotypes were more frequent among center buds than among end buds (Table 7–3). The importance of bud position has also been studied by Perlman and his colleagues (12,13), with similar results. These investigators also presented data showing that the percentage recombination between mitochondrial markers is much higher in center buds than in end buds. This corollary is certainly to be expected in view of the fact that so many end buds are pure for one parental mitochondrial genotype or the other. Forster (14), on the other hand, reported that a bud site on the zygote is randomly associated with the mitochondrial alleles contained within the bud; thus there is no correlation between genetic composition of the bud and its site of origin on the zygote. Until these results can be reconciled with those of Callen and Perlman and associates, the significance of bud position in segregation and recombination of mitochon-

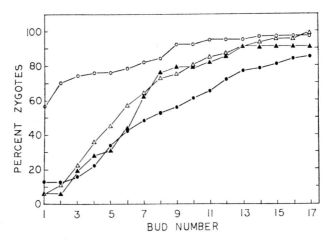

FIG. 7–5. Accumulated percentages of zygotes, from pedigree analysis of specific crosses, that are pure by a given first-generation bud number. Uppermost curve ○—○ denotes crosses between 44 × 51 strains, representing 47 zygotes; ●—● denotes crosses between 41 × 51 strains, representing 152 zygotes; △—△ denotes crosses between 51 × 68, representing 127 zygotes; ▲—▲ denotes 45 × 68 crosses, representing 32 zygotes. (From Forster and Kleese, ref. 14, with permission.)

TABLE 7–3. *Mitochondrial genotypes present in buds obtained by analysis of 15 pedigrees from two-factor mitochondrial crosses as a function of bud position*

Genotypes present in bud clone	Percentage of buds		
	End		
	A	B	Center
Pure parent one	3.3	53.6	7.1
Pure parent two	70.0	0	42.9
Parent(s) plus recombinant(s)	26.7	32.1	50.0
Recombinant(s) alone	0	14.3	0
Total buds	30	28	28

Data from Callen (11).

drial genes must be considered unresolved. If bud position is an important parameter, then crosses between strains that tend to yield end buds will show more rapid mitochondrial gene segregation and lower frequencies of mitochondrial gene recombination than crosses producing a majority of center-budding zygotes.

FORMAL ANALYSIS OF MITOCHONDRIAL GENE RECOMBINATION IN YEAST

Recombination of mitochondrial genes in a three-factor cross in *S. cerevisiae* was first reported by Thomas and Wilkie (16,17). They used mutations resistant to spiramycin (S^R), erythromycin (E^R), and paromomycin (P^R), together with their wild-type alleles. Thomas and Wilkie scored the diploid clones issuing from these crosses to see if they contained parental or recombinant mitochondrial genotypes. The results indicated linkage of the S^R and E^R markers to each other, with the P^R marker showing a high frequency of recombination (approximately 35%) with each of the others. These initial results are in accord with much more extensive data obtained since. The S^R and E^R mutations map in the linked *RIB2* and *RIB3* genes (see Fig. 7–14 and Table 5–2), whereas the *PAR1* locus, in which the P^R mutation would map, is formally unlinked

to the *RIB* gene region, as will be discussed later. In 1970 the Slonimski group (4) published the first of their extensive series of articles on formal mitochondrial genetics in *S. cerevisiae* that led eventually to the formulation of a general theory of mitochondrial gene recombination likening the process to bacteriophage recombination. These articles form the basis for the discussion that will follow. Where appropriate, other articles on the formal analysis of mitochondrial gene recombination, notably those of the Linnane group (8, 18), will also be brought into the discussion.

POLAR GENES AND ω

The polar genes are *RIB1, RIB2,* and *RIB3* (5,19). Although they form but a small part of the mitochondrial genome of *S. cerevisiae* as we now know it (see Fig. 7–14), experiments with mutations in these genes have played a large role in the development of the theory of mitochondrial gene recombination in this organism. Mutations at the *RIB1* locus confer resistance to chloramphenicol (C^R), and those at the *RIB2* and *RIB3* loci can cause resistance to erythromycin and spiramycin together or to either antibiotic alone (Table 5–2). Unfortunately, the erythromycin-resistant mutants at both loci are designated E^R and the spiramycin-resistant mutants S^R. Since these symbols are generally used in the literature, it is frequently necessary to check allele numbers to be certain which locus is being discussed. For example, E^R_{354} is a *RIB2* mutant, but E^R_{221} is a *RIB3* mutant (see Table 5–2). The three loci *RIB1, RIB2,* and *RIB3* are so designated because each is believed to specify a mitochondrial ribosomal function (see Chapter 5).

In the original article on polarity and transmission of mitochondrial genes in *S. cerevisiae,* Coen et al. (4) showed that independently arising C^R and E^R mutations isolated in two α mating type strains were highly transmitted in monofactorial crosses.

whereas similar mutations isolated in an *a* mating type strain showed low or moderate transmission. For example, if the C^R allele was in the α mating type strains, the transmission was always $C^R > C^S$, but if it was in the *a* mating type strain, the transmission was generally $C^R < C^S$ (Table 7–4). When bifactorial crosses were made between an α mating type $C^R E^S$ parent and an *a* mating type $C^S E^R$ parent, the polarity expressed as the ratio of the two recombinant types was $C^S E^S / C^R E^R < 1.0$ (Table 7–4). In reciprocal crosses employing different C^R and E^R mutations, the polarity was $C^S E^S / C^R E^R > 1.0$. The authors believed that the apparent relationship of polarity to cellular mating type was fortuitous; as we shall see, that conclusion has since been well justified.

Coen and associates observed that although the transmission of the C^R allele was always high in their two α mating type strains, it was quite variable in the *a* mating type strain, varying with the mutation under study (Table 7–4). They pointed out that these differences in transmission could be explained by assuming that the C^R mutations were not allelic or by assuming a phenomenon having to do with polarity itself. The authors presented some evidence suggesting that the C^R mutations were allelic, which was amply confirmed in later work (19). They hypothesized that variations in the polarity of transmission could be explained if a polarity point (O) was postulated that varied in its position relative to the C^R gene

in different C^R mutants. Coen and associates used a vectorial representation to describe transmission of different mutants in which each mutant was placed at a distance from O that reflected its relative transmission. Those C^R alleles closest to O showed the highest transmission, and those farthest away showed the lowest. Four groups of mutants were distinguished on the basis of transmission of the C^R alleles, and they were found to correspond to four polarity groups on the basis of the ratios of reciprocal recombinants for the C^R and E^R genes (Table 7–4). Although Coen and associates were careful not to suggest a specific mechanism to explain this correspondence, they did note the great number of analogies that could be drawn with the mechanism of bacterial conjugation. It seems appropriate for illustrative purposes to show how this analogy can be made from the data presented by Coen and associates, especially since the analogy was later made more precisely before being discarded in favor of the phage analogy model. Imagine that the donor mitochondrial chromosome can break at any one of four points, each corresponding to a polarity point (Fig. 7–6). Transmission of the donor chromosome then proceeds in a polar fashion to a recipient mitochondrion, with the segment transmitted being subject to spontaneous breakage (as with a bacterial chromosome) with a certain probability. The transmission of the C^R allele and the ratio of reciprocal recombinants will then depend on the position of the polarity

TABLE 7–4. Summary of data on transmission and polarity on which four polarity groups were defined[a]

Polarity group	Cellular mating type	ω allele	Percentage transmission of C^R allele	Polarity $(C^S E^S / C^R E^R)$
I	α	ω^+	80–100	0.01
II	a	ω^N	30–60	1–10
III	a	ω^-	4–8	100–1,000
IV	a	ω^-	4	>10,000

[a] Defined originally by Coen et al. (4). Assignment of the allele carried at the ω locus is based on the later articles of Bolotin et al. (20) and Dujon et al. (22).

TABLE 7–5. Recombination in homosexual and heterosexual crosses[a]

Mitochondrial cross	Percentage cells of each genotype				Percentage of recombinants		$\dfrac{C\omega^+ E\omega^-}{C\omega^- E\omega^+}$
	$C^R E^S$	$C^S E^R$	$C^S E^S$	$C^R E^R$	$C\omega^+ E\omega^-$	$C\omega^- E\omega^+$	
Heterosexual crosses							
$C^R_{321}E^S\omega^- \times C^S E^R_{514}\omega^+$	8	54	38	0.13	38	0.13	290
$C^R_{36}E^S\omega^+ \times C^S E^R_{221}\omega^-$	79	3.5	0.4	17.5	17.5	0.4	44
$C^R_{321}E^R_{221}\omega^- \times C^S E^S\omega^+$	0.2	23.8	72.6	3.4	23.8	0.2	120
$C^R_{36}E^R_{514}\omega^+ \times C^S E^S\omega^-$	28.8	0.2	14.8	56.3	28.8	0.2	144
$C^R_{321}E^R_{514}\omega^+ \times C^S E^S\omega^-$	25.0	0.2	1.5	73.2	25.0	0.2	125
					$C\omega^+ E\omega^+$	$C\omega^- E\omega^-$	
Homosexual crosses							
$C^R_{321}E^S\omega^- \times C^S E^R_{221}\omega^-$	47.5	41.5	6	5	—	6 5	—
$C^R_{36}E^S\omega^+ \times C^S E^R_{514}\omega^+$	44	44.5	7	4.5	7 4.5	—	—
$C^R_{321}E^R_{221}\omega^- \times C^S E^S\omega^-$	8.2	2.7	46.0	43.1	—	8.2 2.7	—
$C^R_{36}E^R_{514}\omega^+ \times C^S E^S\omega^+$	6.2	4.8	58.9	30.0	6.2 4.8	—	—
$C^R_{321}E^R_{514}\omega^+ \times C^S E^S\omega^+$	4.2	7.6	30.8	57.4	4.2 7.6	—	—

[a] Figures given for each mitochondrial cross for percentage cells of each mitochondrial genotype are usually means derived from several crosses in which the mitochondrial genotypes were the same but the nuclear genotypes varied. Adapted from Bolotin et al. (20).

point. If this is the kind of model the authors had in mind, it is immediately obvious why they were reluctant to invoke cellular sexuality in explaining why α mating type strains tended to transmit their mitochondrial alleles at a higher rate than a mating type strains. A model such as that in Fig. 7–6 automatically assumes that the donor strains in polarity group I are α mating type, but those in polarity groups II–IV are a mating type. Coen and associates suggested that there might, in fact, be a mitochondrial sex factor independent of cellular sex. This concept soon was developed further, as we shall see.

The bacterial sexuality model was extended by Bolotin et al. (20) using specific alleles at the *RIB1, RIB2,* and *RIB3* loci. These authors showed that polarity could not be attributed to cellular sex in a series of crosses in which mitochondrial genotype was reversed with respect to cellular sex (Table 7–1). These crosses revealed that one of the two mitochondrial genotypes used exhibited high polarity regardless of cellular mating type. Bolotin and associates also reported that certain crosses were polar, whereas others were not (Table 7–5). These findings were explained in terms of a mitochondrial sex factor, ω, that had two states designated ω^+ and ω^-. In heterosexual crosses ($\omega^+ \times \omega^-$) polarity was seen, but in homosexual crosses ($\omega^+ \times \omega^+$, $\omega^- \times \omega^-$) there was none.

Using the convention of mitochondrial sex, Bolotin and associates demonstrated that in heterosexual crosses of the type $\omega^- C^R E^S \times \omega^+ C^S E^R$ and $\omega^- C^S E^R \times \omega^+ C^R E^S$ the majority recombinant carried the C allele from the ω^+ parent and the E allele from the ω^- parent, with the converse being true of the minority recombinant (Table 7–5). The convention used to represent this observation was to write the majority recombinant in a heterosexual cross as $C_{\omega^+}E_{\omega^-}$ or C^+E^- and the minority recombinant as $C_{\omega^-}E_{\omega^+}$ or

C^-E^+. The polarity of transmission was then indicated by the ratios C^+/C^-, E^+/E^-, etc., and polarity of recombination by the ratios C^+E^-/C^-E^+, etc.

Bolotin and associates extended their experiments, which had been done with a C^R mutant in the *RIB1* locus and an E^R mutant in the *RIB3* locus, to a *RIB2* mutant resistant to spiramycin (S^R). They proposed that the mutants were linked in the order C^R-S^R-E^R for two reasons. First, in homosexual crosses the sum of recombinant types for the outside genes (i.e., *RIB1* and *RIB3*) was much lower than the sum of non-recombinants. Second, in heterosexual crosses the polarities of recombination were $C^+S^-/C^-S^+ > C^+E^-/C^-E^+$, indicating the order *RIB1, RIB2, RIB3*. Results published by Netter et al. (19), which will be discussed in more detail later in this chapter, appeared to confirm these preliminary findings.

The experiments reported by Bolotin and associates also showed that mitochondrial sex must be transmitted according to the following rules. In $\omega^- \times \omega^-$ homosexual crosses all diploid progeny, whether parental or recombinant, were ω^-. Similarly, in the homosexual cross $\omega^+ \times \omega^+$, all diploid progeny were ω^+. In heterosexual crosses, cells of parental genotype retained their original sex, but recombinants were ω^+. In addition, it was found that ultraviolet irradiation of the ω^+ parent decreased the frequency of diploid cells in a heterosexual cross carrying the ω^+ parental genotype while increasing the frequency of recombinant cells of the type $RIB1^+RIB3^-$ and cells of the ω^- parental genotype. No effect of ultraviolet light was noted on the ω^- parent.

These results were used by the authors in support of the bacterial conjugation model of mitochondrial heredity, which supposed that injection of the ω^+ mitochondrial genome into ω^- mitochondria in heterosexual crosses proceeded in the order ω^+-*RIB1-RIB2-RIB3*. This model seemed to explain why the majority recombinants were $RIB1^+RIB3^-$, $RIB1^+RIB2^-$, etc., why

ultraviolet irradiation only affected transmission of ω^+ genetic markers, and why ultraviolet irradiation had a greater effect on transmission of a *RIB3* allele by the ω^+ parent than on a *RIB1* allele. As the authors pointed out, the latter observation could be explained in terms of the conjugation model by assuming that *RIB3* was farther from ω^+ than *RIB1*, making its probability of transfer more sensitive to ultraviolet irradiation.

Within 3 years of publication of the article by Bolotin and associates, the conjugation analogy model has been discarded in favor of the phage analogy model. The reasons will become apparent as we go along. However, this replacement of one model by another in no way affected the basic observations on polarity that led to definition of ω^+ and ω^- as alternate states or alleles at a locus that governed the frequency of reciprocal recombinants in the *RIB* region of the yeast mitochondrial genome. The original observations made on the three loci comprising this region by Coen and associates and Bolotin and associates were bolstered considerably by the results of a detailed study by Netter et al. (19). These authors defined the three loci and their positions in the *RIB* region more precisely by performing allele tests with selected mutants and by mapping experiments that relied principally on measuring polarity for different pairs of mutants in heterosexual crosses and recombination frequencies in homosexual crosses. Allele tests were done by crossing mutants (normally mutants of similar phenotype) pairwise in homosexual and heterosexual crosses and scoring for wild-type (i.e., antibiotic-sensitive) recombinants. These are, by definition, the only recombinants that can be scored, since the doubly resistant recombinants cannot be distinguished phenotypically from their singly resistant parents. Recombinants should be easily detectable in either $\omega^+ \times \omega^+$ or $\omega^- \times \omega^-$ homosexual crosses, but heterosexual crosses present a special problem.

Only if the *ant*^s allele is proximal to ω⁺ will the doubly sensitive recombinant be detectable. If it is distal to the *ant*^r allele, sensitive recombinants will occur so rarely that they will probably not be detected. In practice, this means that in heterosexual crosses it is necessary to make reciprocal crosses for each pair of mutants with respect to ω⁺ and ω⁻. Allele tests have established the existence of three groups of mutants corresponding to *RIB1*, *RIB2*, and *RIB3* that can be mapped with respect to each other.

Trembath et al. (21) used the frequency of sensitive recombinants to study allelism of their mitochondrial mutations to erythromycin and spiramycin resistance. Their results indicated the existence of two loci, presumably equivalent to *RIB2* and *RIB3* (see Table 5–2). They reported a single recombinant between two of the spiramycin-resistant mutants defining one of these loci. Presumably this recombinant arose between heteroallelic mutants at this locus. The stocks used by Trembath and associates were not defined with respect to ω. Boynton and Gillham and their colleagues used the same methods of allele testing to establish the number of antibiotic resistance loci in the chloroplast genome of *Chlamydomonas* (see Chapter 14). That is, the frequency of sensitive recombinants was measured in pairwise crosses of mutants having the same resistance phenotype. Fortunately, polarity has thus far not proved a complicating factor in the *Chlamydomonas* experiments.

Netter et al. (19) carried out polarity mapping essentially as it had been done by Bolotin et al. (20) using two- and three-factor heterosexual crosses. The two-factor crosses were done with alleles in the trans position (i.e., *ant*^r*ant*^s × *ant*^s*ant*^r), and the polarities using the conventions introduced by Bolotin and associates were as follows: *RIB1*⁺*RIB2*⁻, *RIB1*⁺*RIB3*⁻, *RIB2*⁺*RIB3*⁻, *RIB1*⁻*RIB*⁺, *RIB1*⁻*RIB3*⁺, *RIB2*⁻*RIB3*⁺ ≫ 1. That is, according to the method of polarity mapping, the gene order was ω-

RIB1-RIB2-RIB3. Three-factor heterosexual crosses were also in accord with this order. Two-factor homosexual crosses in which the three loci were mapped by measuring the frequency of reciprocal recombinants in trans-crosses gave the same order, but an unambiguous order could not be deduced in three-factor homosexual crosses because of the rarity of recombinants between *RIB1* and *RIB2*. Despite the fact that polarity mapping consistently yielded the order ω-*RIB1-RIB2-RIB3* we shall see later that the true order based on physical studies is *RIB1-ω-RIB3*.

With the exception of polarity mapping, which is limited to the *RIB* region of the mitochondrial genome of *S. cerevisiae*, the technique of allele testing to define loci by recombination and the technique of recombination analysis in mapping crosses are the two formal genetic tools applicable in common to populations of progeny (as opposed to pedigrees) in the cases of mitochondrial genes in *S. cerevisiae* and chloroplast genes in *Chlamydomonas reinhardtii* (see Chapter 14). Allele testing by recombination has the obvious drawback that two closely linked loci can be mistaken for one in the absence of a satisfactory cis–trans functional test or knowledge of the gene product. Thus far, functional tests do not appear possible in *S. cerevisiae* because stable heteroplasmons are unknown, although they may be possible in *Chlamydomonas*, where persistent heteroplasmons have been reported (see Chapter 14). Furthermore, formal recombination mapping is applicable only to certain clusters of genes in the mitochondrial genome of *S. cerevisiae*. Most mitochondrial genes in this organism behave as if they are unlinked in the formal sense, although they are linked physically, as we shall see shortly. In *Chlamydomonas*, chloroplast gene mapping simply has not proceeded far enough, at least in zygote clones, to indicate whether the same problems will be encountered there.

We now return to the four polarity groups of Coen et al. (4) and a new mutation at the ω locus designated ω^N. Conveniently, the polarity groups had been ignored for some 6 years until a paper by Dujon et al. (22) described the ω^N mutation and provided a plausible explanation for the four polarity groups. The reason that the four polarity groups were ignored for a time was probably as follows: In the article by Coen and associates it was supposed that the polarity point (*O*) had different locations in the three strains used corresponding to the four polarity groups (Table 7–4). That is, all the strains used were acting as donors of mitochondrial markers, but they had different efficiencies of marker donation. However, Bolotin and associates found that the two strains whose mutations constituted polarity group I were ω^+, whereas the other three polarity groups were defined in a strain that was ω^- (i.e., 55R5-3C). In short, the model of Coen and associates assumed that both ω^+ and ω^- strains were acting as donors, but the model of Bolotin and associates assumed that the ω^+ parent was the donor and the ω^- parent the recipient. Even after the conjugation analogy model was abandoned, the four polarity groups posed a problem, for they implied that ω must be able to exist in more than two allelic forms.

Dujon et al. (22) unraveled the mystery with their discovery that mutations could be obtained that affected ω^- and also altered polarity. These were selected as chloramphenicol-sensitive (C^S) revertants of a specific chloramphenicol-resistant mutant (C^R_{336}) at the *RIB1* locus. This mutant was the only member of polarity group IV of Coen and associates (Table 7–4). The polarity exhibited by this mutant was so extreme that no $\rho^+ C^R E^R$ recombinant was ever found. This mutant was used by Dujon and associates to select revertants because it was found that certain strains carrying C^R_{336} grew more slowly on glycerol-containing medium than the C^S revertants.

The revertants obtained were first characterized to see if they behaved as dominant or recessive nuclear suppressor mutations or as unlinked mitochondrial suppressors. The results, which we will not detail here, indicated that the C^S mutations were behaving as true revertants. Dujon and associates then crossed the revertants to ω^+ and ω^- tester stocks and measured the polarity. The revertants isolated in the ω^+ stocks had the same polarity as the parent stocks, but the revertants of the ω^- stocks were of two kinds. The first class of revertants behaved as if they were ω^-, but the second class, designated ω^N, did not. The ω^N revertants exhibited at most a low polarity (called minipolarity by the authors) in crosses to ω^+ testers and no polarity in crosses to ω^- testers or when crossed to each other. Further characterization of the ω^N mutants revealed the following points of significance. First, genetic analysis showed that ω^N was behaving as a typical mitochondrial mutation. Second, $\omega^N C^S$ revertants could accumulate nuclear mutations to chloramphenicol resistance, but C^R mutations at the mitochondrial *RIB1* locus were not found. In contrast, $\omega^- C^S$ revertants did mutate to C^R. This restriction on mutation in the ω^N strains appeared to be limited to the *RIB1* gene, as other mitochondrial genes mutated at normal frequencies. These results suggested that reversion to $\omega^N C^S$ involved a multisite alteration that prevented new C^R mutations from arising, whereas the $\omega^- C^S$ reversion was a point mutation in the *RIB1* gene. Third, crosses of $\rho^- C^R E^0$ isolates obtained from a ω^+ strain to $\omega^N C^S$ revertants yielded virtually no $\rho^+ C^R$ progeny, in contrast to crosses with $\omega^- C^S$ revertants, which gave rise to $\rho^+ C^R$ progeny with a frequency of approximately 1%. This finding was consistent with a multisite change in the mtDNA of the ω *RIB1* region in $\omega^N C^S$ revertants that interfered with recombination involving the homologous region in the ρ^- genome. That this change does not extend to the other loci in the *RIB* region is illustrated by the fact that recombination seems to occur normally for the C^R and E^R mutations in crosses involving ω^N, ω^+, and ω^-. Fourth, not all of the C^R mutations originally obtained by Coen and associates in the ω^- 55R5-3C strain behaved as if they were ω^-. These were classified in polarity group II, having low polarity (Table 7–4). Crosses of several such C^R mutants revealed that each behaved as if it contained a ω^N allele. These results are of interest for two reasons. First, they show that the forward mutation from C^S to C^R can also be accompanied by the ω^- to ω^N mutation. Second, this discovery provides the key to understanding the four polarity groups of Coen and associates, as will now be discussed.

Coen and associates found that the C^R alleles carried by their two α mating type strains showed very high transmission and that the ratio $C^S E^S / C^R E^R$ recombinants was 0.01. The mutants in this group constituted polarity group I (Table 7–4). The two α mating type strains used were ω^+. The a mating type strain was ω^-. In short, this was a normal heterosexual cross in which the C^R mutants were obtained in the ω^+ parent; so the $C^R E^R$ recombinant predominated, and the $C^S E^S / C^R E^R$ recombinant ratio was very low. Polarity group II resulted when ω^- mutated to ω^N together with the C^S to C^R mutation. These mutations have low polarity or no polarity in crosses to ω^+ (Table 7–4). Polarity group III consisted of $\omega^- C^R$ mutants. Crosses of these mutants to the ω^+ α mating type stocks were heterosexual, as were the matings that led to the definition of polarity group I, but this time the C^R mutations were in the ω^- stock (Table 7–4). This means that the $C^S E^S$ recombinant predominated and the ratio $C^S E^S / C^R E^R$ was very high. This leaves polarity group IV, consisting of mutant C^R_{336}, from which the $\omega^N C^S$ and $\omega^- C^S$ revertants were originally derived. This mutant exhibited extremely high polarity of recombination, since $C^R E^R$ recombinants were never obtained at all. Netter et al. (19) demonstrated by some clever crosses that formation of the $C^R E^R$ recombinant in the case of this mutant is accompanied by transformation of the cells from ρ^+ to ρ^-, so that the doubly resistant recombinants cannot be detected except by a marker rescue cross to a $\rho^+ C^S E^S$ strain. Hence they go undetected.

A few final comments on the nature of the ω^N mutation are in order. It appears that ω^N is not a point mutation, but rather a major change in nucleotide sequence organization or a deletion of nucleotides involving both ω and *RIB1*. Almost 50% of the mutations at the *RIB1* locus in ω^- strains are accompanied by the ω^N mutation. The fact that neither ω^- nor ω^N mutations are found in ω^+ strains following induction of C^R mutations or reversion of mutant C^R_{336} suggests that the nucleotide sequences of ω^+ and ω^- must be quite different. Physical evidence that this interpretation is correct has been reported very recently by Jacq et al. (71) who find that the ω^+ DNA sequence consists of an insertion about 1000 base pairs long not present in ω^- mtDNA (see p. 246). Results obtained by these workers and by Heyting and Sanders (48) also show that ω is located between *RIB1* and *RIB3* (Fig. 7–13) which had not been suspected from genetic analysis as we have already described. At the moment, it is not entirely clear whether one or all of these genes correspond to groups of mutations in the large mitochondrial rRNA, nor is it clear how these physical results will eventually be assimilated into the existing genetic framework, assuming they hold up. Dujon and Michel (25) reported some very preliminary physical evidence suggesting that ω^N may be a deletion. Finally, with respect to ω^N, we note that the results of Dujon and associates show that both the forward mutation from C^S to C^R and the reverse mutation from C^R to C^S can be accompanied by a mutation from ω^- to ω^N. If ω^N is the result of a deletion, then we must postulate that this same deletion can be accompanied by mutation to either chloramphenicol resistance or sensitivity with high probability.

Despite claims to the contrary, most other studies of mitochondrial gene recombination involving the polar genes appear to be interpretable within the framework developed by Slonimski and his colleagues. Three articles, in particular, deserve attention, since they challenged the notion that polarity can be explained in terms of alternate states of ω. Rank and Bech-Hansen (7) studied crosses involving mitochondrial markers for chloramphenicol and erythromycin resistance that they designated *chl*^R and *ery*^R, respectively. They found that the markers from one of the parents tended to be much more frequent among the progeny than markers from the other parent. For example, in one cross between a *chl*^s*ery*^R parent and a *chl*^R*ery*^s parent, they found that the ratio of parental genotypes among the diploid progeny was *chl*^s*ery*^R >> *chl*^R*ery*^s, but the reciprocal recombinant types *chl*^R-*ery*^R and *chl*^s*ery*^s were approximately equal in frequency. The result, of course, was that the *chl*^s and *ery*^R markers from one of the parents were high in frequency among the progeny. Rank and Bech-Hansen erroneously concluded from this observation that the *chl*^s*ery*^R parent was ω^+ and that the *chl*^R*ery*^s parent was ω^-. Similar sorts of results were observed in their other crosses. Since there was no polarity of recombination, Rank and Bech-Hansen concluded that the Slonimski model was incorrect. In fact, it appears from the absence of polarity of recombination that Rank and Bech-Hansen were not studying heterosexual crosses, but instead were observing the results of a series of homosexual crosses in which one parental genotype was transmitted in high frequency. The high transmission of one parental genotype in each cross is probably the result of asymmetry in which a nuclear gene or genes common to one of the parents either causes its mitochondrial genotype to be highly transmitted or promotes transmission of the mitochon-

drial genes carried by the other parent. Although the genetic data are insufficient to rule out bias, examination of the strains used by Rank and Bech-Hansen reveal that one strain (62–18a) is common in each cross and that there is an apparent segregation of the tendency to transmit preferentially one parental mitochondrial genotype in tetrads from diploids constructed with this parent.

The article by Howell et al. (8) examined polarity in crosses involving a mutation to oligomycin resistance (*oli1-r*) and a series of other mutations resistant to either erythromycin (*ery1-r*) or spiramycin (*spi2-r, spi3-r, spi4-r*). These authors grouped their results in terms of what they called nonpolar crosses, low-polarity crosses, and high-polarity crosses. They claimed that their results did not support the simple model of Slonimski and associates of two alternate states of ω to explain polarity; rather, they contended that more than two polarity groups were found when several different strains were used. The article by Dujon et al. (22) on ω^N had not been published at that time. However, examination of the data presented by Howell and associates reveals that they do, in fact, follow the simple model of ω. The mutations in the *RIB* region were all isolated in the same strain, which appears from the data to be ω^+. These were coupled by recombination to the *oli1-r* marker, so that all of the derivative strains should, by definition, be ω^+ (p. 216). These doubly resistant ω^+ strains were then crossed to doubly sensitive strains, some of which were ω^+ (nonpolar and low-polarity crosses) and some of which were ω^- (high-polarity crosses). Furthermore, the majority recombinant in the high-polarity crosses carried the *ery* or *spi* allele of the presumptive ω^+ parent and the *oli* allele of the presumptive ω^- parent, as would be expected in a typical heterosexual cross. The difference in frequencies of reciprocal recombinant types in the high-polarity crosses can be used in the same way as it was used by Netter et al. (19) to order the polar markers. The order obtained is ω^+-spi4-spi3-spi2, ery1. The only results that are slightly ambiguous with respect to this interpretation involve the presumptive ω^- strain 652. The polarities of recombination in heterosexual crosses involving this ω^- strain are much weaker than they are in crosses involving the other presumptive ω^- strain D253. Hence, whereas heterosexual crosses with D253 always exhibit high polarity, crosses employing 652 may have either low polarity or high polarity, depending on the markers used. Thus, when 652 was used as a parent in a heterosexual cross involving the least-polar markers, ery1 and spi2, there was low polarity; but when the most-polar markers, spi3 and spi4, were used, there was high polarity. It is also evident from the homosexual crosses ($\omega^+ \times \omega^+$) presented in this article that oli1, as expected (*vide infra*), behaves as if it is not linked to the polar genes. Before leaving the paper by Howell and associates, it is important to note that their results serve to emphasize how the effects of asymmetry and bias can mask and confuse the interpretation of mitochondrial crosses. Thus it is clear from their crosses that different ω^- strains can show various degrees of polarity of recombination in heterosexual crosses and that in homosexual crosses recombination frequencies vary considerably depending on the strains used. Whether some of their isolates may also have carried ω^N mutations is not clear, but such mutations might account for some of the results of the low polarity crosses.

The symposium paper of Linnane et al. (18) summarized a portion of the data presented in the article by Howell and associates and proposed a map based on frequencies of transmission of different mitochondrial markers in crosses. Linnane and associates postulated that in highly polar crosses there is a fixed origin of marker transmission, whereas there are several possible origins in nonpolar crosses. The arguments in support of this model are no more compelling than those used for the conjugation analogy model; this model, like the conjugation analogy model, seems to have little validity in terms of our present understanding of ω and the nature of polarity. Linnane and associates also published a map of their markers based on the frequency of transmission of each marker in crosses showing high polarity. This map is similar to the one already derived from the data of Howell and associates (*vide supra*) for the polar markers taking into account only the polarity of recombination, and it probably reflects the fact that apparent polarity of marker transmission results, at least in part, because reciprocal recombinants are so vastly different in frequency in heterosexual crosses. Linnane and associates also gave transmission map positions for the nonpolar markers oli1-r and mik1-r, which reflected the fact that these markers were transmitted less frequently than the resistance markers in the polar region. This result is probably artifactual with respect to the mapping of these markers for the following reason. The crosses made by Howell and associates and by Linnane and associates always seem to have been done such that the ω^+ parent carried the resistance markers in coupling. This means that the majority recombinant will contain the sensitive allele of the distal marker with respect to ω^+. For example, in the cross ω^+ery1-r oli1-r $\times$ ω^-ery1-s oli1-s the majority recombinant will be ery1-r oli1-s; so transmission of the ery1-r allele will be greater than that of the oli1-r allele. The data from homosexual crosses support the notion that oli1 is actually unlinked to the RIB segment, as has been found by Slonimski and associates (*vide infra*), since the frequency of recombination is 17 to 18%, as compared to the expected 20 to 25% for unlinked mitochondrial markers in *S. cerevisiae*.

NONPOLAR GENES AND THEIR RELATIONSHIPS TO POLAR GENES

Most mitochondrial genes do not exhibit polarity in heterosexual crosses, as was shown first in detailed articles by Avner et al. (5) and Wolf et al. (26) on the mapping of the *OLI1, OLI2,* and *PAR1* loci. They showed that these three genes were nonpolar in both homosexual and heterosexual crosses. Because some of the strains used in these experiments were distinct from those used previously by the Slonimski group, it was first necessary to establish which ω allele they carried. This was done by crossing the new strains, which had the genotype C^SE^S with respect to the *RIB1* and *RIB3* loci, to C^RE^R testers. The mitochondrial ω allele carried by the new strains was then ascertained from the polarity exhibited (Table 7-6). For example, when the new strain D22 was crossed with a ω⁺ tester, the recombinant ratio C^RE^S/C^SE^R was 1, but when D22 was crossed with a ω⁻ tester, this ratio was $\ll 1$. These results showed that D22 must be ω⁺, since polarity was not seen in the cross to the ω⁺ tester but was apparent in the cross to the ω⁻ tester.

The crosses of another new strain, D6, to the appropriate tester not only showed it to be ω⁺ but also illustrated the phenomenon of asymmetry (Table 7-6). In the homosexual cross it was evident that cells of the C^RE^R parental genotype were in great

excess with respect to the C^SE^S parental genotype. Avner and associates showed by tetrad analysis that the tendency to transmit the C^RE^R genotype in excess in homosexual crosses segregated 2:2, suggesting the influence of a nuclear gene or genes. These results, like those of Rank and Bech-Hansen (p. 220), illustrate how in the absence of polarity of recombination there can still be a bias in transmission due to the effects of nuclear genes. These experiments also serve to emphasize the importance of making crosses of new strains to tester strains to establish which ω allele they carry. Until this is done, we cannot disentangle phenomena attributable to bias and asymmetry from polarity; so it is easy to be misled if markers in the polar region are included in crosses, which has usually been the case.

Although only two P^R mutations were used in these studies, and their allelism was established by crossing them and showing that sensitive progeny were not found, as was done by Netter and associates for the polar genes (pp. 216–217), a great many O^R mutations were available. It was established in two ways that the mutants then at hand fell in only two loci. First, allele tests were done, and the resulting diploids were scored for the presence of sensitive progeny. These tests showed that the mutants fell into two mutually exclusive classes. Mutants in the *OLI1* locus did not yield sensitive progeny in crosses to each other, and the same was

TABLE 7-6. *Determination of mitochondrial sex from the polarity of recombination*[a]

Cross		Percentage of total diploid progeny				Total colonies
New strain (C^SE^S)	Tester strain (C^RE^R)	C^RE^S	C^SE^R	C^RE^R	C^SE^S	
D22	IL8–8C ω⁺	5.3	3.5	62.6	28.6	283
D22	IL–126–3A ω⁻	0	31.1	0.5	68.1	405
D6	IL8–8D ω⁺	2.4	6.6	85.3	5.7	211
D6	IL126–1C ω⁻	0	64.1	8.9	26.9	156

[a] Crosses were made between C^RE^R (R_I–R_{III}) testers of known mitochondrial sex and new strains of the genotype C^SE^S, whose mitochondrial sex was unknown. The polarity of recombination was measured among the diploid progeny of each cross.

Adapted from Avner et al. (5).

true when *OLI2* mutants were crossed pairwise; but crosses between *OLI1* and *OLI2* mutants always yielded some sensitive progeny. Second, genetically marked ρ^- mutants were employed as deletion mutants (a methodology we will discuss in great detail in the next section) in what seems to have been the first instance of deletion mapping. The ρ^- mutants were crossed to ρ^+ strains carrying either an O^R_I or an O^R_{II} allele, as deduced from allele testing experiments. If *OLI1* and *OLI2* are not separable loci, crosses made to the ρ^- strains will always yield one of two results, irrespective of whether $\rho^+O^R_I$ or $\rho^+O^R_{II}$ stocks are used (Table 7–7). If the O^s gene is retained in the ρ^- strain, some of the diploid progeny will be O^R and others O^s. If the O^s locus has been lost in the ρ^- strain, all of the diploid progeny will be O^R. On the other hand, if *OLI1* and *OLI2* are distinct loci, one may obtain a ρ^- strain that has, for example, retained the *OLI1* gene while losing the *OLI2* gene. When such a $\rho^-O^s_IO^0_{II}$ stock is crossed to a $\rho^+O^R_IO^s_{II}$ stock, the diploid progeny will be a mixture of oligomycin-resistant and -sensitive cells, but crossing this stock to a $\rho^+O^s_IO^R_{II}$ stock will yield only oligomycin-resistant progeny. That is, the ρ^- stock will give different results with O^R_I and O^R_{II} ρ^+ stocks. Although most of the ρ^- strains tested were $\rho^-O^0_IO^0_{II}$, one was found that was $\rho^-O^s_IO^0_{II}$. The O^R_I and O^R_{II} mutants, as determined by the allele testing method, were then crossed to this ρ^- stock. The allele testing method and the ρ^- deletion method gave fully concordant results, confirming the existence of two genes for oligomycin resistance.

Multifactorial heterosexual crosses established that the *OLI1, OLI2,* and *PAR1* loci exhibited polarity only in combination with a polar gene such as *RIB1* or *RIB3* and were not polar with respect to each other (Table 7–8). In homosexual crosses, where there is no polarity for any pairs of markers, it was possible to obtain accurate estimates of percentage recombination using conventional methods (Tables 7–9 and 7–10). Recombination frequencies in the *RIB1-RIB3* segment in these experiments were between 8 and 12%, but the *OLI1, OLI2,* and *PAR1* genes recombined at frequencies of 20 to 25% with each other and with the genes in the *RIB* segment (Table 7–10). These results were interpreted to mean that whereas the polar genes were formally linked in a genetic sense, the nonpolar genes were not linked to the polar genes or to each other. The reason that 20 to 25% recombination is presumed to indicate absence of linkage will become apparent

TABLE 7–7. *Possible results of $\rho^- \times \rho^+$ crosses if there are one or two mitochondrial genes conferring oligomycin resistance[a]*

Number of loci	Discriminating tester	Genotypes of parents in each cross with respect to OLI loci		Phenotypes of diploid progeny
		ρ^- parent	ρ^+ parent	
One	Not applicable	O^S	O^R	O^S, O^R
		O^0	O^R	O^R
Two	Yes	$O^S_IO^0_{II}$	$O^R_IO^S_{II}$	O^S, O^R
			$O^S_IO^R_{II}$	O^R
	Yes	$O^0_IO^S_{II}$	$O^R_IO^S_{II}$	O^R
			$O^S_IO^R_{II}$	O^S, O^R
	No	$O^0_IO^0_{II}$	$O^R_IO^S_{II}$	O^R
			$O^S_IO^R_{II}$	O^R
	No	$O^S_IO^S_{II}$	$O^R_IO^S_{II}$	O^S, O^R
			$O^S_IO^R_{II}$	O^S, O^R

[a] Two of the possible ρ^- mutants are discriminating testers, and the other two are nondiscriminating testers.

TABLE 7–8. *Three-point heterosexual* ($\omega^+ \times \omega^-$) *mitochondrial crosses involving the polar RIB1 (C) and RIB3 (E) genes and the nonpolar OLI1 (O) and PAR1 (P) genes.*[a]

	Polarity of transmission			
	C^+/C^-	E^+/E^-	O^+/O^-	P^+/P^-
Total colonies scored	6,974/224	3,832/1,783	3,238/3,594	4,082/4,116
Average polarity of transmission	31.1	2.1	0.90	0.94

	Polarity of recombination					
	C^+E^-/C^-E^+	C^+O^-/C^-O^+	C^+P^-/C^-P^+	E^+O^-/E^-O^+	E^+P^-/E^-P^+	O^+P^-/O^-P^+
Total colonies scored	952/7	2,600/18	3,132/26	717/227	1,103/348	855/890
Average polarity of recombination	136	144	120	3.2	3.2	0.96

[a] The data shown are averages taken from many crosses employing different combinations of mitochondrial alleles at these loci. The data are expressed using the conventions described in Chapter 7, according to which polarity of transmission is expressed as the ratio of the ω^+ allele to the ω^- allele (i.e., C^+/C^-, E^+/E^-, etc.) and polarity of recombination is expressed as the ratio of the $\omega^+\omega^-$ recombinant to the $\omega^-\omega^+$ recombinant (i.e., C^+E^-/C^-E^+, C^+P^-/C^-P^+).
Adapted from Wolf et al. (26).

shortly. Netter et al. (19) also examined recombination of the three *RIB* genes with respect to *OLI1*. We need not detail their results, except to say that they add little in terms of mapping this locus with respect to the *RIB* genes and merely serve to point up the limitations of formal recombination analysis in dealing with unlinked mitochondrial genes.

The *OLI1, OLI2,* and *PAR1* genes represent only a small fraction of the nonpolar genes in the yeast mitochondrial genome. Many new genes have recently been described, largely as a result of discovery of the *mit⁻* mutants (2,3). The mapping of these genes, as well as the nonpolar genes already described, would not have been possible without the development of deletion mapping techniques using ρ^- mutants. However, before describing these procedures, it is appropriate to conclude this discussion of formal genetics with a por-

TABLE 7–9. *Transmission of markers in four-point homosexual mitochondrial crosses* ($\omega^+ \times \omega^+$, $\omega^- \times \omega^-$) *involving the polar RIB1 (C) and RIB3 (E) genes and the nonpolar OLI1 (O) and PAR1 (P) genes*[a]

Mitochondrial sex of strains used in cross	Percentage transmission					Total colonies
	C^a	E^a	O^a	P^a	Average	
ω^-	40.7	40.7	42.5	40.4	41.1	339
ω^+	64.4	63.4	60.8	64.3	63.2	664
ω^+	48.1	52.3	45.7	48.6	48.7	545

[a] Data shown are from several of the many crosses made employing different combinations of mitochondrial alleles at these four loci. Since these crosses are homopolar with respect to mitochondrial sex, a different convention has been used for transmission. Transmission is expressed for each pair of alleles in terms of the percentage transmission of the allele carried by the a mating type parent.
Adapted from Wolf et al. (26).

TABLE 7–10. *Recombination of markers in three- and four-point homosexual mitochondrial crosses* $(\omega^+ \times \omega^+, \omega^- \times \omega^-)$ *involving the polar RIB1 (C) and RIB3 (E) genes and the nonpolar OLI1 (O) and PAR1 (P) genes*[a]

	$C^a E^a / C^a E^a$	$C^a O^a / C^a O^a$	$C^a P^a / C^a P^a$	$E^a O^a / E^a O^a$	$E^a P^a / E^a P^a$	$O^a P^a / O^a P^a$
Total colonies scored	130/172	273/181	420/453	277/242	499/490	378/419
Average polarity of recombination	0.8	1.5	0.9	1.1	1.0	0.9
Recombinants for each marker pair as % of total progeny	8.2	17.3	18.3	20.1	21.0	21.7

[a] Data shown are averages taken from many crosses employing different combinations of mitochondrial alleles at these loci. Since these crosses are homopolar for mitochondrial sex, a different convention has been used for recombination. Recombinant ratios like transmission (Table 7–9) are expressed in terms of cellular mating type (i.e., $C^a E^a / C^a E^a$, $C^a O^a / C^a O^a$, $E^a O^a / E^a O^a$, etc.).

Adapted from Wolf et al. (26).

trayal of the phage analogy model that attempts to unify the observations we have thus far described.

PHAGE ANALOGY MODEL OF MITOCHONDRIAL GENE RECOMBINATION

The conjugation analogy model was originally invoked by Bolotin et al. (20) to explain the polarity of recombination of markers in the *RIB* region of the mitochondrial genome by sequential transfer of mtDNA from ω^+ to ω^- mitochondria. Transfer commenced with ω^+, and recombinants were ω^+. The closer a marker was to ω^+, the greater its probability of transfer to ω^- mitochondrion before the mtDNA molecule broke (Fig. 7–6). By this definition, the majority recombinant in a heterosexual cross should carry the ω^+ allele of the marker closest to it (i.e., *RIB1*) and the ω^- allele of the markers farther away (i.e., *RIB2* and *RIB3*). By 1973 the conjugation analogy had become unappealing for two reasons. First, the nonpolar genes had been discovered, and these genes did not exhibit polarity with respect to one another in heterosexual crosses, only with respect to the polar genes. In fact *OLI1, PAR1,* and *RIB3* all showed similar polarities of re-

combination with respect to *RIB1*. Furthermore, Wolf et al. (26) pointed out that when multifactor heterosexual crosses were made and reciprocal recombinants for *OLI1* and *PAR1* were examined to see which alleles of *RIB1* and *RIB3* they carried, polarity was the same as it was in the population as a whole. According to a sequential transfer hypothesis, the *OLI1* and *PAR1* genes would have been distal markers, as they were transmitted less frequently by the ω^+ parent than markers in the *RIB* region. By selecting for recombination of such distal markers, it is ensured that the ω^+ allele of the distal marker is transferred; thus polarity should vanish. The fact that it did not vanish argues against the sequential transfer hypothesis. Second, the conjugation analogy model really did not explain the results of homosexual crosses in which there was no polarity. In short, it began to seem as if the polar genes were merely a special class of mitochondrial genes. At the same time, it was becoming apparent that one might be able to treat recombination of mitochondrial genomes in a formal sense by a population analysis that assumed there were multiple copies of the mitochondrial genome in the zygote, some of which were derived from each parent. The obvious analogy was the phage cross. This analogy was

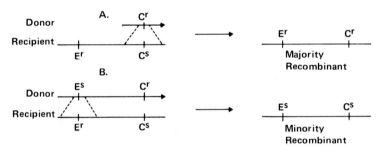

FIG. 7–6. Bacterial conjugation model for recombination of mitochondrial genes in *S. cerevisiae*. Model assumes that donor mtDNA molecule can be broken at any one of four points, corresponding to four polarity groups, following which mtDNA transfer begins from donor to recipient mitochondrion. Markers close to origin are transferred with higher probability than those farther away, because donor mtDNA molecule usually breaks before transfer is completed. This means that majority recombinant will contain donor allele closest to origin, and minority recombinant will contain donor allele farthest from origin. This is shown in the form of an example for polarity group I recombinants.

first suggested by Wolf et al. (26) and shortly thereafter by Linnane and associates (18) and Dujon et al. (6). The latter authors published a detailed quantitative formulation of the model to explain the results of both homosexual and heterosexual crosses, which we will now describe.

The model of Dujon and associates consists of two parts. The first deals with the process of pairing of mtDNA molecules, and the second deals with the mechanism of recombination of mitochondrial genes. The first part of the model makes the fundamental assumption that genetically functional mtDNA molecules are present in the zygote in multiple copies and that genetic

recombination between these molecules can be described by the same sort of model that Visconti and Delbrück (27) proposed for bacteriophage recombination in 1953.

The phage analogy model of mitochondrial gene recombination, as it has been formulated for homosexual crosses, is a general model that is also formally applicable to certain aspects of chloroplast gene recombination in *Chlamydomonas* (see Chapters 14 and 15). We will develop the model principally as it applies to the results of homosexual crosses; later we will turn to the special complications introduced by polarity in heterosexual crosses. In a phage cross the host bacteria are infected with viruses of different genotype at known average multiplicities of infection. The titers of the parent viruses are readily established by plating at appropriate dilution on sensitive indicator bacteria. Thus the average *input* fraction can be measured accurately and varied experimentally with ease. No such technique is available for titrating the relative input fraction of organelle genomes in a cross; therefore the input must be deduced indirectly. One of the important assumptions of the model of Dujon and associates is that in a mitochondrial cross the relative input fraction of mitochondrial genomes from one parent will be proportional to the output, assuming an absence of any selective advantage of mitochondrial genomes at the intracellular level or of specific mitochondrial genotypes at the intercellular level.

If this assumption is correct, the relative input of mitochondrial genomes in a homosexual cross, or nonpolar genes in a heterosexual cross, can be deduced from the ratio of any pair of mitochondrial alleles, with this ratio being similar for all pairs of alleles in terms of parent of origin. For example, if the a mating type parent carries an O^R_I marker and the α mating type parent is O^s_I and the frequency of the O^R allele among the diploid progeny of the cross is 0.3, it can be assumed that the average *input fraction* of mitochondrial genomes from the O^R_I parent is 0.3 and that of the O^s_I parent is $1 - 0.3$, or 0.7. Similarly, any other a mating type mitochondrial allele should have a frequency not significantly different from 0.3, and an α mating type mitochondrial allele should have a frequency of about 0.7. Implicit in this assumption is the notion that all mitochondrial genes are physically linked but not genetically linked, which is, in fact, the case. Evidence that the predicted *coordinate output* of alleles does occur is presented in Fig. 7–7, where it is shown for a number of homosexual crosses that the frequencies of different alleles carried by the same parent are closely correlated.

A second assumption that the model of Dujon and associates has in common with the Visconti-Delbrück model is that the many mtDNA molecules in the yeast zygote are all genetically complete and identical and that they form a panmictic pool in which pairings can take place between any pair of molecules, irrespective of origin or genotype. The authors stated that they were led to the idea of the panmictic pool because the upper limit of recombination for markers that were unlinked genetically appeared to be 20 to 25%, rather than 50%. Such a reduction in percentage recombination of unlinked genes is easily envisaged if homologous as well as heterologous pairings occur. In an equal-input cross where only heterologous pairings are permitted, the observed recombination frequency (R_{ab}) for two unlinked genes will be 50%, but if homologous pairings occur as well, only half of the recombinational events will result in observable recombinants.

A third assumption that is also implicit in the Visconti-Delbrück model is that there are several rounds of mating of mitochondrial genomes that are random in time. Thus there are several pairing events in the line of ancestry (lineage) of an average mtDNA molecule before it is sorted out into a pure line. Therefore the genetic composition of the panmictic pool of mtDNA molecules evolves as a function of the number of mating rounds.

Making these three assumptions, Dujon et al. (6) derived an equation similar to one of the basic Visconti-Delbrück equations. We will derive the equation in a very general way here, following the examples of Stahl (28) and Stent (29). Consider a cross in which the parents are a^+b and ab^+. Let p be the probability of recombination between chromosomes of complementary parental type and w the frequency in the mating pool of genomes carrying allele a. This, in turn, will be equal to the input frequency of parent ab^+. We can now define a lineage for each genome with respect to a given locus

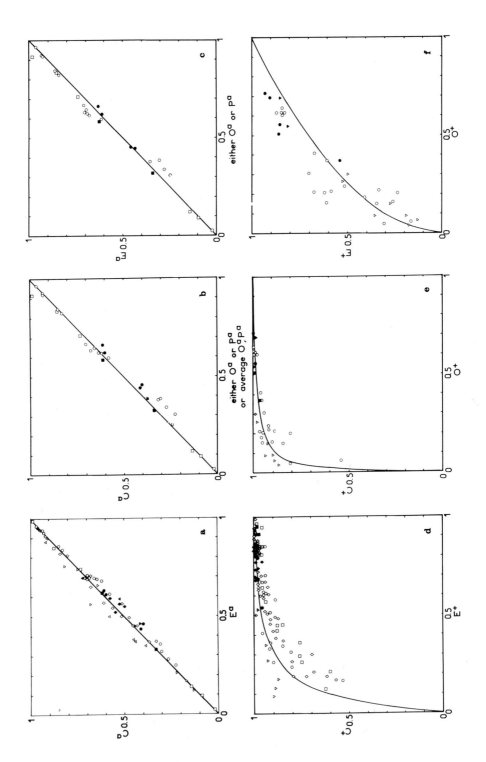

(i.e., *a*) and with a probability *w* that the genome will be of type *a*. Let *m* be the average number of matings per lineage. Then *mp* will be the average number of occurrences per lineage that could lead to recombination for loci *a* and *b*. If matings are randomly distributed among lineages (i.e., Poisson distribution), $1 - e^{-mp}$ is the probability that a lineage has had at least one mating with an exchange pattern that could recombine markers.

The probability that a genome has the recombinant phenotype *ab* is the probability that the last mating in its lineage was with a chromosome carrying marker *b*. This probability is $1 - w$, which is, by definition, the frequency of the *b* allele in the population. Therefore a recombinant will be of the type *ab* with a probability of $w(1 - w)(1 - e^{-mp})$. Since a symmetric argument applies to recombinants of the type a^+b^+, the average frequency of recombinants (R) in the pool is given by the equation

$$R = 2w(1 - w)(1 - e^{-mp})$$

An important prediction of this equation is that the observed frequency of recombination will be a function of the relative inputs of the two parents. That is, at relatively unequal inputs, most recombination events will not be seen, since they result from incestuous matings between similar genomes; at equal inputs, the probability of mating between different genomes is maximized, and so is the observed recombination frequency. The actual function is parabolic, and it is evident that the observed recombination frequencies in different homosexual crosses follow the predicted relationship (Fig. 7–8).

The assumption that homozygous and heterozygous exchanges occur between mtDNA molecules in the panmictic pool makes for an interesting statistical prediction that was first pointed out by Visconti and Delbrück for the phage cross. The prediction is that positive correlations will be observed between pairs of exchanges (i.e., positive coincidence or negative interference) even if these exchanges are independent of each other at the level of individual matings. This point can be illustrated by the following oversimplified example. Suppose *a*, *b*, and *c* are linked in that order and that the probability of an exchange between *a* and *b* (R_{ab}) is 0.1 and the probability of an exchange between *b* and *c* (R_{bc}) is 0.2. Suppose also that double exchanges (R_D) occur at random (i.e., $R_D = R_{ab} \times R_{bc}$). We now make the cross *abc* × +++ such that the input from each parent is 0.5. First we permit only heterozygous exchanges to occur. In this case the coefficient of coincidence is 1; i.e., the coefficient of coincidence = $R_D/R_{ab} \times R_{bc} = 0.02/(0.1)(0.2)$. Now we permit both homozygous

←

FIG. 7–7. Output fraction of one mitochondrial allele as a function of output of other alleles in same cross. In all figures, experimental data are compared with theoretical predictions of model of Dujon et al. (6,30). Experimental: Each point represents results of a single cross. Full symbols refer to control experiments, void symbols to ultraviolet irradiation of one of the parents prior to mating. Shapes of symbols refer to different crosses (see original articles for details). Abscissae and ordinates are output fraction of alleles indicated. a, b, and c: Homosexual crosses. d, e, and f: Heterosexual crosses. Theoretical: Lines relate output fractions (ordinates) of either C or E alleles located in polar region (respectively RIB1 and RIB3 loci) to output fractions (abscissae) of E, O, or P alleles (respectively RIB3, OLI1 or OLI2, and PAR1 loci). They are drawn according to model that assumes the following:

1. The input fractions of all alleles of the same parent carried by the same mtDNA molecule are the same.

2. In homosexual crosses the output fraction of any allele is equal to its input fraction; therefore the output fractions of all alleles must be equal in any given cross.

3. In heterosexual crosses the output fractions of alleles located in the polar region are related to the input fractions by the formula $R_i = w \{(e^m)/[(1 - w) + we^m]\}^{s_i}$, where *m* is the average number of mating rounds, *w* is the input fraction, R_i is the output fraction of a given allele (C^+ or E^+), s_i is the probability of gene conversion (s_C for $C^- \rightarrow C^+$, s_E for $E^- \rightarrow E^+$).

4. In heterosexual crosses the output fractions of alleles located in nonpolar regions is equal to their input fraction. The lines were drawn using the best estimate of the parameters: $m = 4.6$, $s_C = 0.99$, $s_E = 0.50$. (From Dujon et al., ref. 30, with permission.)

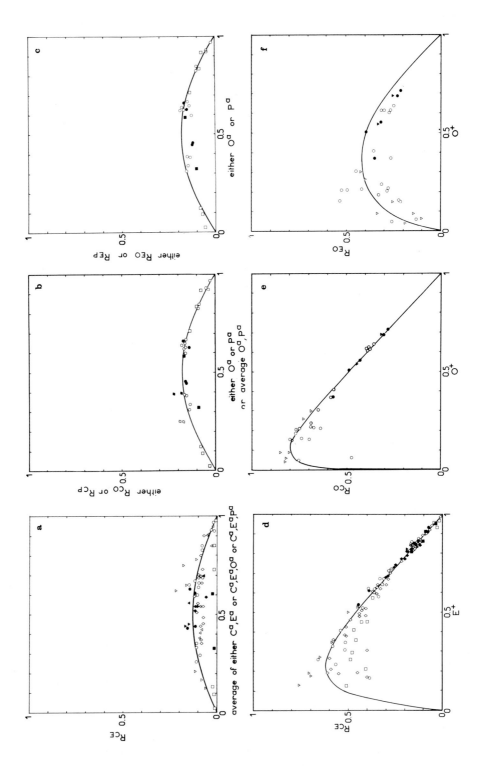

and heterozygous exchanges to take place. At equal input, homozygous exchanges will account for half of the total, and the observed recombination frequencies will be half of the true number (i.e., $R_{ab} = 0.05$, $R_{bc} = 0.1$, $R_D = 0.01$). At the same time, however, the coefficient of coincidence becomes positive and equals 2. As inputs become more unequal, the coefficient of coincidence increases further because homozygous exchanges constitute an ever greater fraction of the total. Dujon et al. (6, 30), using a more sophisticated equation that takes into account not only input but also number of mating rounds, showed that their experimental data fit the predicted values very well.

The second part of the model of Dujon et al. (6) deals with the mechanism of recombination of mtDNA molecules. Their assumption was that mtDNA recombination is fundamentally nonreciprocal, so that one elementary act produces one type or recombinant, another elementary act produces another recombinant, and so forth. Statistically, reciprocal recombinants will be produced at equal frequency, but each elementary act will yield one rather than two recombinants. Dujon and associates believed that in the case of polar genes in a heterosexual cross an obligatory recombinational event takes place every time ω^+ and

ω^- molecules pair. The process is initiated at the ω^+ locus, and the probability of "conversion" of an allele carried by the ω^- parent decreases as one proceeds from ω^- to *RIB3*. Thus all recombinants are ω^+, and the polarity with respect to the *RIB1-RIB3* segment is $RIB1^+ > RIB2^+ > RIB3^+$. This scheme generates one ω^+ parental genome and one recombinant every time ω^+ and ω^- molecules pair. The model postulates that the numbers of mating rounds are similar in homosexual and heterosexual crosses and in both polar and nonpolar regions in heterosexual crosses. Given this assumption, the act of recombination must also be nonreciprocal for nonpolar genes in heterosexual crosses and for both polar and nonpolar genes in homosexual crosses, because reciprocal recombination would generate two rather than one recombinant per elementary act. If reciprocal recombination were to occur for nonpolar genes and for polar genes in homosexual crosses, one would have to postulate additional rounds of mating for polar genes in a heterosexual cross.

Because of polarity, we do not expect coordinate output when the frequency of a polar allele is plotted versus the frequency of a nonpolar allele derived from the same parent in a heterosexual cross (Fig. 7–7). Assuming that polarity does result because of gene conversion, the shape of the trans-

←

FIG. 7–8. Frequency of recombination as a function of output fraction of mitochondrial allele in a cross. Experimental data are compared with theoretical predictions of model of Dujon et al. (6,30). Experimental: Each point represents results of single cross. Full symbols refer to control experiments, void symbols to ultraviolet irradiation of one of the parents prior to mating. Shapes of symbols refer to different crosses (see original articles for details). Ordinates: sums of frequencies of both types of recombinants between loci indicated. Abscissae: output fractions of alleles indicated. Theoretical: Lines relate total frequencies of both types of recombinants (ordinates) between either two alleles located in polar region (CE pair: respectively *RIB1* and *RIB3* loci) or one allele located in polar region and one allele located in nonpolar region (CO, CP, EO, EP pairs: respectively *RIB1* and either *OLI1* or *OLI2*, *RIB1* and *PAR1*, *RIB3* and either *OLI1* or *OLI2*, *RIB3* and *PAR1* loci) to output fractions (abscissae) of E, O, or P alleles (respectively *RIB3*, *OLI1* or *OLI2*, and *PAR1* loci). They are drawn according to model that assumes the following:

1. In homosexual crosses the frequency of recombinants is related to the input fraction by the formula $R_{ij} = 2w(1 - w)(1 - e^{-K'_{ij}m})$, where m is the average number of mating rounds, w is the input fraction, R_{ij} is the sum of frequencies of both types of recombinants between two alleles (CE or CO or CP or EO or EP pairs), K'_{ij} is a factor proportional to the probabilities of genetic exchanges.

2. In heterosexual crosses the frequency of recombinants as a function of the input fraction can be calculated by computer integrations of a complete set of differential equations (Dujon et al., ref. 6).

3. Output fractions are related to input fractions by assumptions given in the legend of Fig. 7–7. Lines were drawn using the best estimates of the parameters $m = 4.6$, $s_C = 0.99$, $s_E = 0.50$. (From Dujon et al., ref. 30, with permission.)

mission plot will be a function of three parameters: number of mating rounds (m), input (w), and the probability of conversion of a ω^- allele in the polar region to its ω^+ counterpart (s_i). Obviously, the latter parameter will be locus-specific, since the model postulates that the probability of gene conversion declines from ω to *RIB3*. Dujon et al. (6,30) developed an equation that takes these parameters into account, and the fit of the experimental data to the theoretical curve is good (Fig. 7–7).

The model of Dujon and associates also predicts that the relationship between frequency of recombination and input in a heterosexual cross will depend on the distance between the markers being considered and ω (Fig. 7–8). When at least one of the markers is located in the polar region, the maximum frequency of recombination occurs for inputs biased in favor of the ω^- parent. The reason for this is evident from the following qualitative considerations. In a heterosexual cross, three sorts of pairings can occur between mtDNA molecules. Two of these ($\omega^+ \times \omega^+$ and $\omega^- \times \omega^-$) are homosexual and do not lead to conversion of ω^- alleles in the *RIB* region to their ω^+ counterparts. The third pairing, $\omega^- \times \omega^+$, leads to conversion of the polar ω^- alleles to their ω^+ alternatives with probabilities estimated at 0.99 for *RIB1* and 0.50 for *RIB3*. There are several mating rounds ($m = 4.6$), and each time $\omega^+ \times \omega^-$ mating occurs the polar ω^- allele is converted to ω^+ with the probabilities indicated, whereas conversion with respect to the nonpolar alleles is random. That is, the frequency of the majority recombinant versus the minority recombinant is largely a function of polarity. When the ω^- genotype is in excess, the frequency of the majority recombinant increases at each mating round as more and more ω^- mtDNA molecules become converted to ω^+ with respect to the polar allele. When the ω^+ genotype is in excess, the frequency of recombination declines because $\omega^+ \times \omega^+$ matings are in excess; so there is an excess of

the ω^+ parental genotype relative to the majority recombinant.

Dujon et al. (30) used ultraviolet light as a probe with which to test the predictions of their model. The basic idea behind these experiments was to vary the input fraction beyond what could be achieved in unirradiated crosses, since the data points gathered from these crosses clustered too closely to provide an adequate test of the theoretical predictions. It was found that ultraviolet irradiation of one of the parents did bias input in favor of the unirradiated parent. When the data points from these experiments were plotted together with the data already obtained from crosses in which neither parent had been irradiated, it became apparent that all the data points fit the theoretical expectations quite well and that a much wider range of inputs had been achieved (Figs. 7–7 and 7–8). Obviously, we would like to imagine that ultraviolet irradiation is effectively withdrawing mtDNA molecules in the irradiated parent from the panmictic pool. Although the formal genetic data appear to be compatible with such a hypothesis, it seems that there are no physical data on the fate of mtDNA molecules following irradiation that bear on the question. It is also apparent that induction of ρ^- mutants is high in the irradiated parent under the experimental conditions used. Since matings of irradiated cells were undertaken in the dark immediately following irradiation, the presence of established petites in the population was not a complicating factor, but the events that led ultimately to petite induction were fully operative in many of these cells.

The model of Dujon and associates also explained the finding of Bolotin et al. (p. 216) that ultraviolet irradiation of the ω^+ parent decreased the frequency of cells of the ω^+ parental genotype while increasing the frequency of recombinant cells of the type *RIB1$^+$RIB3$^-$* and also of the ω^+ genotype, whereas there was no effect of irradiation on the ω^- parent. These observations were interpreted in terms of sequential transfer of ω^+ markers into ω^- mitochondria, a process that ultraviolet irradiation could contravene. According to the model of Dujon and associates, ultraviolet irradiation of the ω^+ parent decreases the input of ω^+ genomes relative to ω^- genomes, resulting in an increase in the majority recombinant, which in this case will be $RIB1^+RIB3^-$ recombinant. Ultraviolet irradiation of the ω^- parent in these experiments did not have a measurable effect on the frequency of the ω^- parental genotype because this genotype was already so rare in the population. However, the effect on this parent was evident in the withdrawal of the majority recombinant from the population relative to the ω^+ parental genotype.

In a molecular sense the model of Dujon and associates assumed that gene conversion resulted because of obligatory excision and degradation of the ω^- allele. The degradation process could then proceed to include closely linked genes, with the probability of degradation being highest for the most closely linked gene $RIB1$ and less for $RIB2$ and $RIB3,$ which are farther away, finally becoming nonexistent for the non-polar genes, which are presumed to be beyond the region of excision. Following degradation, resynthesis proceeded along the ω^+ template, so that genes included in the degraded ω^- segment were resynthesized using the ω^+ template and hence were converted from the genotype they had in the ω^- strain to that of the ω^+ strain. It is not clear from the model of Dujon and associates why degradation should initiate at ω^-, and the molecular mechanism of the degradation and resynthesis process was left vague. Perlman and Birky (31) proposed a precise molecular mechanism to explain these events that seemed to be compatible with the results of Dujon and associates. They suggested that the ω^- mtDNA sequence represented a duplication of ω^+ or that this sequence was deleted in ω^+ strains (Fig. 7–9). In either case, when recombination occurred

between ω^+ and ω^- molecules at or near ω, a heteroduplex was formed in which the extra DNA in the ω^- strand looped out, since there was nothing for it to pair with in the ω^+ strand (Fig. 7–9). Perlman and Birky proposed that this looped-out ω^- DNA was recognized by an endonuclease that made a cut in the ω^- strand near the ω^- sequence, following which the ω^- sequence was excised by an exonuclease. As in the model of Dujon and associates, excision could continue beyond the ω^- sequence to include adjacent genes, and the polarity of degradation was explained in precisely the same way as it was by Dujon and associates. The same was true of resynthesis on the template, so that conversion of the genes in the excised ω^- segment to the ω^+ genotype occurred. The only difference between the models was that the model of Perlman and Birky was precise in molecular terms and supposed that the ω^+ template along which resynthesis occurred was a single-strand region of the ω^+ parental mtDNA and that resynthesis was achieved by a repair polymerase plus ligase (Fig. 7–9). Recent restriction endonuclease studies of ω^+ and ω^- mtDNA by Heyting and Saunders (48) and Jack et al. (71) (p. 246) indicate that the Perlman-Birky model in its original form is wrong in the sense that ω^+ rather than ω^- mtDNA contains the insertion. However, Birky has pointed out to the author that if the endonuclease made its cut in the ω^- strand opposite the inserted sequence in ω^+ and resynthesis included the inserted sequence, unidirectional conversion of ω^- to ω^+ would still be accounted for. To accomodate the fact that ω maps between $RIB1$ and $RIB3$ one would have to assume that conversion via the exonuclease and polymerase can take place along the ω^+ template to the right or left of ω. Birky points out that this observation too is easily accounted for in their model since the direction of conversion in a particular heteroduplex will depend on whether the Watson strand of ω^+ has paired with the Crick strand of ω^- or vice versa. Presumably bidirectional

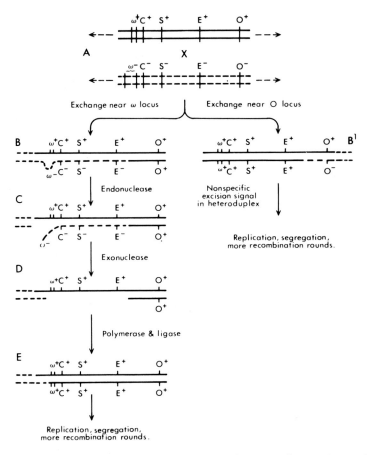

FIG. 7–9. Hypothetical mechanism to explain polarity in yeast mitochondrial crosses. (From Perlman and Birky, ref. 31, with permission.)

gene conversion can also be accomodated in the model of Dujon and associates since the model assumes an independent probability of conversion for each locus. Conversion could then proceed to the left or to the right of ω^- by the mechanism suggested by Birky. Obviously, the model of Dujon and associates could also be modified by the mechanism suggested by Birky such that the ω^- allele rather than being degraded now contains the ω^+ insertion as the result of re-synthesis.

Although the phage analogy model fits the data obtained from population analysis via the standard cross quite well, molecular proof in support of the model is still extremely sparse. In addition, the observations made on populations must ultimately be reconciled with those collected via pedigree and zygote clone analysis. Both of these points deserve further consideration before we move on. The parameter most accessible to experimental manipulation is input. The model makes specific predictions about how variation in input of mtDNA in a cross should affect output of markers and recombination frequencies. Although ultraviolet irradiation seems to be a remarkably versatile tool for varying output, as shown by Dujon and et al. (30), it remains to be shown that input of mtDNA molecules is affected directly. This may prove a difficult problem, since we must distinguish experimentally between the amount of intact undamaged mtDNA entering the zygote and the ability of damaged mtDNA (some of which is undergoing transformation to the ρ^- phenotype) to replicate and recombine. Ideally, we would like to have an agent that varies the amount of mtDNA specifically over a wide range without damaging it or inducing ρ^- mutants. Perlman et al. (13) have attempted to vary the input of mtDNA in crosses by four independent methods that should be nondestructive in their mode of action (Table 7–11). In the first method, nondividing cells of one parent are treated with cycloheximide prior to mating. This treatment is supposed to allow mtDNA to accumulate in nondividing cells. Output of mitochondrial alleles from the treated parent is enhanced (Table 7–11A). Second, the *a* mating type parent is treated with α factor hormone. This treatment arrests the cells in G1 and leads to accumulation of mtDNA. Such treatment also leads to an increase in output of mitochondrial markers by the treated *a* mating type parent (Table 7–11B). Simultaneous incubation of the *a* mating type cells with α factor and hydroxyurea, an inhibitor of DNA synthesis, reduces this increase in output by the *a* mating type parent. Third, crosses between $\rho5$, a mutant with roughly twice the mtDNA of the wild type, yield biased outputs in favor of the mitochondrial markers from the

$\rho5$ parent (Table 7–11C). Fourth, the outputs of diploid-haploid crosses are biased in favor of mitochondrial alleles from the diploid parent (Table 7–11D). Diploids are known to have twice the mtDNA of haploids. All of these results indicate that variation of mtDNA input by nondestructive methods is correlated with variation of mitochondrial marker output such that the parent contributing the most mtDNA has the highest output. Thus far, direct evidence that the treatments are causing variation in the amount of mtDNA has not been published, although the inference is reasonable in view of the four different methodologies used. Unfortunately, none of the treatments used thus far seems to be capable of varying mtDNA content over a wide range. A reagent that does have this potential, 5-fluorodeoxyuridine, has recently been discovered to vary clDNA in *Chlamydomonas* and chloroplast marker output over a considerable range, and the evidence for variation in clDNA input is direct (see Chapter 12).

A second prediction of the phage analogy model is that events must take place in the zygote that lead to panmixia of mtDNA molecules. By definition, this means that mitochondria break down, following which

TABLE 7–11. *Influence of treatments presumed to alter the input of mtDNA on the output of mitochondrial markers*[a]

Part	Remarks on cross	Percentage transmission of markers 1 and 2 by *a* parent		Number of progeny scored
		1	2	
A	Control	41.5 ± 4.5	44.3 ± 4.8	397
	a parent treated with cycloheximide	54.8 ± 5.1	55.6 ± 5.1	365
B	Control	55.5 ± 4.8	63.7 ± 4.7	402
	a parent treated with α factor	83.7 ± 3.9	86.9 ± 3.6	344
	a parent treated with α factor plus hydroxyurea	66.8 ± 4.5	69.0 ± 4.5	413
C	Control	44.2	48.7	3,568
	a parent is $\rho5$ mutant	69.1	69.4	5,016
D	α parent is haploid strain 4810	56.8	56.6	852
	α parent is diploid strain 4810	24.2	25.0	645

[a] In each cross the mitochondrial alleles are designated 1 and 2. In parts A through C of the table the alleles studied were in the *RIB1* and *OLI1* loci, whereas in part D they were in one of the *RIB* loci and the *PAR1* locus. All crosses were homosexual.

Adapted from Perlman et al. (13).

their mtDNA molecules commingle and are repackaged into genetically different assemblages than were present in the parent cells, or that mitochondria fuse in the zygote, following which a sorting out of mitochondrial genomes occurs during mitochondrial division in the progeny. Smith et al. (32) searched for physical evidence in support of mitochondrial genome recombination in the zygote, and based on electron microscopic observations, they argued that mitochondria dedifferentiated in the young zygote and redifferentiated in older zygotes. They suggested that the disorganized condition of the mitochondria in young zygotes might correspond both to the period of mitochondrial genome recombination and to the time when a heterogeneous complement of mitochondrial genomes was transmitted to the buds. However, the published electron micrographs did little to inspire confidence in their interpretation. A definitive study on the fate of mitochondria in zygotes of *S. cerevisiae* has yet to be published. It is interesting that in *Chlamydomonas* chloroplast fusion does occur in the zygote (see Chapter 12). Certainly it would appear from the pedigree studies of Callen (11) and Perlman and associates (12,13) discussed earlier in this chapter (p. 212) that panmixia is not always complete and may be strongly dependent on the strain and the cross. They reported that end buds were usually homoplasmic for one parental mitochondrial genotype or the other and that the center buds usually contained the recombinant genotypes. Forster (14), on the other hand, found that bud position and mitochondrial genotype were randomly associated. Clearly, this issue needs to be settled, particularly with respect to the strains used by Dujon and Slonimski in constructing the phage analogy model.

Birky's studies on zygote clones (13,33, 34) have yielded other observations of importance that must eventually be incorporated into any general model of mitochondrial gene recombination and segregation.

He observed that in certain crosses high frequencies of uniparental zygotes were produced in which mitochondrial genes from only one parent seemed to be transmitted. These zygotes were found in homosexual crosses and were uniparental with respect to nonpolar as well as polar genes; thus polarity did not appear to be a complicating factor. Birky argued that uniparental inheritance could not be accounted for by biased output, since this would require input ratios of 9:1 or even more extreme; thus if one parent had an average of 45 mtDNA molecules, a number that may be high (see p. 239), the other parent would have to have 5 mtDNA molecules or less. Of particular interest were the homosexual crosses involving $\rho 5$. Birky found that when glucose was used as the carbon source, so that the cells were glucose-repressed, there were no distinct classes of zygotes with respect to mitochondrial gene transmission; but when glycerol was used as the carbon source, so that the cells were derepressed, there was a striking shift toward marker transmission by the $\rho 5$ parent such that almost half of the zygotes were uniparental and transmitted only mitochondrial genes from the $\rho 5$ parent. These are remarkable results in view of the fact that $\rho 5$ is thought to have only twice the mtDNA of wild-type cells. Birky argued that the phenomenon he was observing was strikingly similar to that seen with presumptive chloroplast genes in *Chlamydomonas* (see Chapter 12) and *Pelargonium* (see Chapter 17). In *Chlamydomonas*, chloroplast gene transmission is normally maternal, but it can be perturbed by treatment of the maternal parent with ultraviolet irradiation prior to mating or by growth of this parent in fluorodeoxyuridine prior to mating, which reduces the amount of chloroplast DNA present. Both treatments lead to marked increases in chloroplast gene transmission by the paternal parent, and the latter treatment, as far as is known, affects only input. In the case of *Pelargonium,* the same sorts of effects are

apparent in crosses between different cultivars. Birky argued that perhaps uniparental
inheritance in yeast can be explained by
preferential replication of mtDNA from one
parent, possibly coupled with destruction of
mtDNA from the other parent. Whether or
not this mechanism will also explain the fate
of clDNA in *Chlamydomonas* and *Pelargonium,* as Birky suggested, remains to be
seen. In any event, it would seem important
for the phage analogy model to determine if
any of the crosses studied or the ultraviolet
irradiation treatments used by Dujon and
associates led to an increase in the frequency of uniparental zygotes. For example,
in crosses where output is biased strongly in
favor of one parent or the other, it is conceivable that uniparental zygotes are frequent. If so, the frequency of recombination
will diminish as a function of the exclusive
transmission of one parental mitochondrial
genotype, rather than as a function of the
frequency of homozygous and heterozygous
pairings between mtDNA molecules.

Van Winkle-Swift and Birky (35,69)
have provided direct evidence that mitochondrial gene recombination is formally
nonreciprocal in biparental zygote clones of
S. cerevisiae using a method of analysis
which bears some resemblance to that employed many years earlier by Hershey and
Rotman (70) to demonstrate the nonreci-
procity of recombination in single burst experiments with bacteriophage. Van Winkle-
Swift and Birky provided similar evidence
for chloroplast gene recombination in *Chlamydomonas* in both vegetative diploids and
the progeny of sexual zygotes.

Pedigree analysis has proved to be a
much more powerful tool for analyzing
chloroplast gene recombination and segregation in *Chlamydomonas* than has been the
case thus far in *Saccharomyces;* this has
been because of the elegant system devised
by Sager (see Chapter 14). This system
permits reconstruction of the timing of
segregation of each chloroplast gene from
the meiotic divisions up to the second post-

meiotic mitotic division, by which time most
segregation events have occurred. Sager interpreted her results as indicating that
chloroplast genes can exhibit both nonreciprocal (gene conversion) and reciprocal
recombination. The evidence for nonreciprocal recombination is that linked chloroplast genes frequently do not segregate at
the same division. That is, if we consider
two linked markers *a* and *b* in the cross
$ab \times a^+b^+$, one daughter cell in a pedigree
may behave as if it is heterozygous for both
genes, whereas the other may behave as if it
is homozygous for *a,* but heterozygous for
the two alleles at the *b* locus. This event is
assumed to result because the *a*⁺ allele has
become converted to *a* in the second daughter cell. Reciprocal recombination occurs
for a gene when both daughters become
homozygous for a pair of alleles at the same
division (e.g., one cell is homozygous *a* and
the other is homozygous *a*⁺). Sager also reported that this pattern of segregation was
polar such that "reciprocal" events might
occur in the order $a > b > c$. This ordered
segregation was one of the methods she
used to map chloroplast genes. Since, according to Sager, nonreciprocal events are
constant per locus, the segregation rate of
each gene in liquid culture reflects the rate
of reciprocal recombination of each gene.
These segregations also seem to show a polarity, which can be used for mapping.
These remarks are by way of introduction
to two pertinent pedigree and segregation
studies in *S. cerevisiae* that will now be considered.

Before discussing specific pedigree data,
it is important to note that the modes of cell
division in *C. reinhardtii* and *S. cerevisiae*
differ, and this has important implications
for the sort of data that can be obtained. In
Chlamydomonas the entire contents of each
mother cell are distributed relatively equi-
tably among two to four daughter cells, depending on whether one or two mitotic
divisions take place. Therefore, one cannot
directly sample the content of choloroplast

genotypes present in the mother cell but must infer it from the daughter cells. In *S. cerevisiae*, repeated mitoses take place in the zygote to yield buds that can then be dissected off. These buds can then be allowed to form mother cells from which further buds can be isolated. The original zygote and the mother cells, as well as the buds, can then be grown up into clones, and the mitochondrial genotypes present in each clone can be ascertained. In short, the sample of mitochondrial genotypes present in the buds can be compared directly to the sample that was present in the zygote or the mother cell. Lukins et al. (9) published pedigrees from three zygotes in which zygotic buds were dissected, following which the buds were allowed to form mother cells, and the buds formed by these mother cells were dissected. Buds, mother cells, and zygotes were then allowed to form colonies, which were then characterized as to the mitochondrial genotypes present. The results of Lukins and associates showed that buds often did purify for one marker before another, but the genotypic composition of the mother cells was such that the results could be accounted for in terms of sampling. In other words, the mother cell might contain the mitochondrial genotypes $E^R O^S$, $E^R O^R$, and $E^S O^S$, but two buds obtained from this mother cell might be pure for the E^R gene but heterozygous at O. That is, the genotypes sampled in the buds would be $E^R O^S$ and $E^R O^R$. Forster (14) dissected buds sequentially from zygotes in experiments discussed earlier. She found that in a two-factor cross involving C^R and E^R mutations, the two loci did not purify among the sequential buds at the same rate. In most zygotes the C alleles purified first. These findings do not appear to be connected with polarity, as her crosses appear to have been homosexual.

Callen (36) attempted to map mitochondrial genes in *S. cerevisiae* in terms of their segregation rates in liquid cultures. The method was formally analogous to one of the methods used by Sager for chloroplast genes in *Chlamydomonas,* as mentioned previously. The segregation of four different antibiotic resistance markers, two resistant to oligomycin (*OLG1* and *OLG2*) and two resistant to spiramycin (*SPR1* and *SPR2*), was followed in different experiments. Homoplasmic cell frequency was calculated from the mean percentage of sensitive clones for each locus normalized to the final percentage of sensitive clones at different times following mating. Resistant clones could not be used because both homoplasmic resistant and heteroplasmic resistant clones would have scored as being resistant. Callen ordered his markers on the basis of their relative segregation rates in liquid, as Sager had done for *Chlamydomonas*. He obtained the order *OLG1, SPR2, SPR1, OLG2.* Unfortunately, we do not know the relationships of these markers to mitochondrial markers of similar phenotype that have been mapped by the other methods described in this chapter. Sager postulated that reciprocal recombination was the underlying mechanism responsible for polarity of chloroplast gene segregation in *Chlamydomonas* (see Chapter 14), but thus far there is no evidence for a similar process of recombination for mitochondrial genes in *S. cerevisiae.* Despite this, Callen's results are of interest, and clearly these experiments need repetition. If they are reproducible with markers mapped by other methods, it could be established that the polarity-of-segregation map does in fact reflect the true map.

In closing this discussion, it should be pointed out that Dujon and Slonimski (6, 37) clearly recognized that their multicopy phage analogy model must also account for the apparent rapidity of segregation of mitochondrial genes that we have considered here and earlier. They made calculations from their own data that are consistent with a model that supposes that the number of independent genetic units entering each bud is small compared to the total and that these are sampled from the zygote or mother cell in a random and nonexhaustive manner. Such a model would certainly seem to be consistent with pedigree data similar to those of Lukins et al. (9) However, this model is inconsistent with the physical data without modification as Dujon and Slonimski realized. Thus Sena et al. (38) measured mtDNA amounts in synchronously formed zygotes, as well as in the unbudded parent cells and the first diploid buds formed by the zygotes. They calculated that each parental cell contained about 28 mtDNA molecules and that the young zygote contained 112 mtDNA molecules. They estimated that after the first zygotic bud maturation, the zygote had within it 235 to 307 mtDNA molecules. The first diploid buds contained 44 to 67 mtDNA molecules. These numbers are much too high to agree with the numbers of units suggested by the genetic data. Dujon and Slonimski (37) suggested that the solution is that a single segregating unit corresponds to several mtDNA molecules. Other solutions to be mentioned include master copies and the notion that not all copies express their genetic information, at least with respect to the phenotypes studied. This problem of small numbers of segregating genetic units and large numbers of organelle DNA molecules is one of the great paradoxes of organelle genetics. We will return to it with reference to *Chlamydomonas* in Chapters 14 and 15. Although hypotheses abound, the experimental data that will discriminate among them have yet to appear.

DELETION MAPPING OF MITOCHONDRIAL GENES WITH ρ^- MUTANTS

Two genetic strategies have emerged for mapping mitochondrial genes using ρ^- mutants. The first is to measure retention or deletion of pairs of markers in such mutants. The basic idea is that if most petites contain simple deletions, then physically adjacent markers should be retained or de-

leted together more often than those farther apart. This method has been used to map mitochondrial markers unlinked by recombination analysis, as well as linked markers. The method can also be used to map specific genetic markers relative to the mitochondrial tRNA and rRNA genes, as was discussed in Chapter 6. The second strategy is more akin to classic deletion mapping as it is normally used in genetic analysis. The idea is to use deletions having differing endpoints to localize a point mutation to a given segment of the map by recombination analysis (Fig. 7–10). For example, if a particular point mutant recombines with one deletion, but not with another that overlaps the first deletion and is slightly longer, the point mutant must fall in the region between the two deletion endpoints (Fig. 7–10). This method has been used in mapping both closely linked markers and distantly linked markers. Development of these genetic strategies has been accompanied by adoption and refinement of sophisticated physical techniques for determining which segments of wild-type ρ^+ mtDNA are deleted and retained in specific ρ^- mutants. Perhaps the most pleasing aspect of most of the deletion mapping data published thus far is that, with minor discrepancies, there has been reasonably good agreement among laboratories on the positions of different markers. We will now describe some of the relevant findings in more detail, beginning with the purely genetic aspects of petite deletion mapping and ending with a discussion of physical mapping methods and the physical map itself. Any student interested in this general subject is referred to the recent Bari symposium (2), which contains a wealth of papers on the various aspects of petite deletion mapping.

Marker retention as a means of mapping unlinked mitochondrial genes seems to have been reported first by Avner et al. (5). The complete genotype of the ρ^- mutant described previously (p. 223), which was used to distinguish between the *OLI1* and *OLI2*

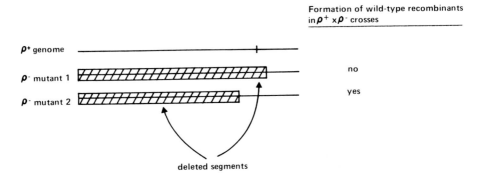

FIG. 7–10. Use of ρ^- deletion mutants for mapping mitochondrial genes in S. *cerevisiae.*

loci by deletion mapping, is $\rho^- R^0{}_I R^R{}_{III} O^S{}_I$- $O^0{}_{II}$. If it is assumed that the retained sequences in a ρ^- mutant are adjacent, then *RIB3* and *OLI1* must be adjacent, and *OLI2* cannot fall between them. The order must be either ω-*RIB1-RIB2-RIB3-OLI1-OLI2* or *OLI2*-ω-*RIB1-RIB2-RIB3-OLI1*. It cannot be ω-*RIB1-OLI2-RIB3-OLI1*, since *RIB1* and *RIB3* behave as if they are linked in homosexual crosses, whereas *OLI2* does not. Avner and associates also argued that the order could not be ω-*OLI2-RIB1-RIB2-RIB3-OLI1* because this order would yield $O^+{}_{II} R^-{}_I$ as the majority recombinant in heterosexual crosses, but it is the minority recombinant.

Linnane and his colleagues (24,39), having run into the same sorts of problems in their attempts to map mitochondrial genes by formal recombination analysis which we described earlier in this chapter (pp. 222–225) turned to marker retention mapping experiments with ρ^- mutants. They examined a large number of spontaneous ρ^- mutants for retention of the *oli1*, O_{II}, *mik1*, *ery1*, and *cap1* loci (their nomenclature). The results of these experiments indicated that *oli1* and *mik1* were retained together more often than either locus was retained with *ery1* or *cap1*. However, *oli1* could be coretained with *ery1*. From these results these investigators concluded that the order of the markers was *ery1-oli1-mik1*. Similarly, whenever *par1* and *mik1* were retained or deleted, the O_{II} locus also was always re-

tained or deleted. These results suggest that O_{II} must normally be retained within the segment of the wild-type genome present in ρ^- *mik1 par1* mutants. However, because the mitochondrial genome of S. *cerevisiae* is circular, ρ^- *mik1 par1* mutants could be obtained, in theory, lacking the O_{II} locus; this could also be the result of a single deletion. The fact that such deletions were not obtained indicates that the segment *mik1-O_{II}-par1* is shorter and less subject to deletion than the other arc of the circle, which, if retained, would yield ρ^- *mik1 par1* mutants. That is indeed the case (Fig. 7–14). In summary, the results obtained by Molloy et al. (39) were consistent with the map order *ery1-oli1-mik1-O_{II}-par1*, which is precisely the same order obtained for these markers by other workers.

Schweyen et al. (40) also employed coretention and deletion to order genetically unlinked markers in the circular mitochondrial genome of S. *cerevisiae*. Their method was to make use of a stock carrying four markers at the *RIB1, OLI1, PAR1,* and *TSM1* loci and examine the frequencies of each of the three possible pairs of double-deletion/double-retention ρ^- mutants (Fig. 7–11). The mutant at the latter locus is temperature-sensitive (T^s) for petite induction, but wild-type cells (T^R) are not. Two of these complementary pairs can be explained in terms of single deletions, but the third would have to result from two independent deletions. Obviously, the first two

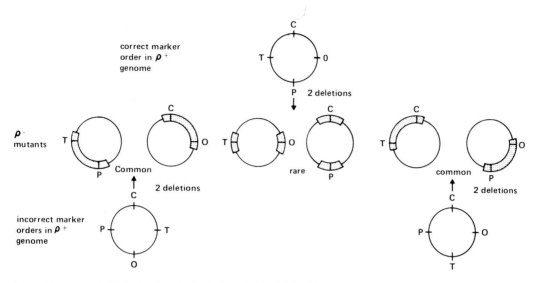

FIG. 7–11. Diagram to illustrate principles involved in double-deletion/double-retention mapping of mitochondrial markers using ρ^- mutants, as employed by Schweyen et al. (40,41). Correct order for genes *RIB1* (C), *OLI1* (O), *PAR1* (P), and *TSM1* (T) in ρ^+ mtDNA is shown at top of figure. Below, three possible pairs of double-deletion/double-retention ρ^- genotypes are shown. Each member of two of these pairs can be generated by a single deletion, but in the case of the final pair, two deletions are required to generate the two complementary double-retention ρ^- genotypes. Retained markers in each ρ^- mutant are included in the thick arc of each circle, with the deleted segment being indicated by the thin arc. At bottom of diagram, two other possible orders of these markers in mitochondrial genome are shown. In each case, these orders would generate one of the common classes of ρ^- mutants as the result of two deletions.

classes ought to be frequent relative to the third. The correct order will be the one that yields the rarest complementary pair as a double deletion. The order that does this is *RIB1-OLI1-PAR1-TSM1* (Fig. 7–11).

Double retention/double deletion can be used not only for ordering markers but also for obtaining relative map distances. Consider the P^R and O^R markers. Each of these markers will be deleted or retained with a probability that can be calculated from the fraction of ρ^- mutants of each kind (i.e., P^R, P^0, O^R, O^0) in the sample studied. If x is the fraction of ρ^-P^R mutants and y is the fraction of ρ^-O^0 mutants, the probability of obtaining a $\rho^-P^RO^0$ mutant will be xy if these markers are always deleted independently. To obtain map distance between two mutants, we take the ratio of the observed number of ρ^- mutants carrying one marker of a pair (i.e., $P^RO^0 + P^0O^R$) to the number calculated, assuming that each marker is lost independently of the other. A ratio of 1 means that the markers are behaving independently of one another; a ratio of < 1 means that

they are not. That is, they are being deleted or retained together more frequently than would be expected by chance. The lower the value, the tighter the linkage. Some representative data are given in Table 7–12.

Schweyen et al. (41) recently extended these mapping methods to several of the genes that are marked by *mit* mutants. One of the nice features of the method is that it yields reasonable additivity of distances. In short, coretention-codeletion mapping seems to have much promise as a purely formal genetic mapping method that gets around the problem of mapping mitochondrial genes that are formally unlinked by recombination analysis.

Classic deletion mapping with ρ^- mutants is now in full swing. both as a means of establishing the physical locations of different genes on the mtDNA molecule and as a tool for dissecting the fine structure of mitochondrial genes. Two different methods of physical mapping are currently in use. The first, developed by Linnane and his colleagues (24,42,43), makes use of DNA-

TABLE 7-12. *Frequency of separation of mitochondrial markers in double-deletion double-retention experiments*[a]

Genotype pair	Total colonies		Observed/expected
	Observed	Expected	
$T^R C^0 + T^0 C^R$	369	4,934	0.07
$T^R O^0 + T^0 O^R$	1,330	4,672	0.28
$C^R O^0 + C^0 O^R$	1,165	4,753	0.25
$T^R p^0 + T^0 p^R$	3,058	3,628	0.84
$C^R p^0 + C^0 p^R$	3,313	3,797	0.87
$O^R p^0 + O^0 p^R$	3,492	3,314	1.05

[a] The frequency of separation is the frequency of genotypes showing retention of one and loss of the other of a pair of markers (i.e., $X^R Y^0 + X^0 Y^R$). Expected values are equal to the products of the frequencies of the single markers considered. Abbreviations: T = TSM1, C = RIB1, O = OLI1, P = PAR1.
Adapted from Schweyen et al. (40).

DNA hybridization to establish the sequences of ρ^+ mtDNA in a genetically defined stable collection of ρ^- mutants. The mtDNA of each ρ^- mutant is characterized to establish the fraction of the ρ^+ genome retained and the fraction held in common with a reference ρ^- mutant, U4, which has retained about 36% of the ρ^+ genome together with markers spaced over about a quarter of the map (i.e., *cap1, ery1, oli1,* and *ana1*). The physical positions of different genes in the mtDNA molecule are then deduced from inclusion and exclusion principles. The inclusion principle is that loci retained in common between two or more petites lie within the mtDNA segment retained in common (i.e., within the overlap region). The exclusion principle is that a locus not found to be present in a petite is assumed to lie outside the entire ρ^+ mtDNA segment retained in the petite.

The basis of the physical mapping method is as follows: It is assumed that a complete ρ^+ sequence is equivalent to 100 units (i.e., 100% hybridization). Each ρ^- mutant has some number of ρ^+ units that depends on its level of hybridization with ρ^+ mtDNA. To ascertain the amount of ρ^+ mtDNA retained by a given ρ^- mutant, filter-bound denatured ρ^- mtDNA in excess is incubated with denatured ρ^+ mtDNA labeled to high specific activity and sheared to an

average molecular weight of 130,000 d in solution. Assuming that a given ρ^- mutant contains a continuous segment of ρ^+ mtDNA, the fraction of ρ^- mtDNA hybridized will be proportional to the amount of ρ^+ mtDNA in that segment. That is, if ρ^+ mtDNA hybridizes to the extent of 10% with a certain ρ^- mutant, it will be assumed that the mutant contains 10 units of ρ^+ mtDNA. In order to obtain a reference library of overlapping petite deletion mutants, it was first necessary to select a reference ρ^- mutant. As mentioned previously, the mutant selected, U4, retains 36 units of ρ^+ mtDNA and four markers spanning about one-quarter of the map. The right-hand end of the segment retained by this mutant is arbitrarily assumed to correspond to 0 units in the ρ^+ mitochondrial genome, so that the left-hand end corresponds to 36 units (Fig. 7-12). The segment is also oriented so that *cap1* is the most distal marker to the right and *ana1* is the most distal marker to the left. Accordingly, the maximum distance between these markers is 36 units, although the minimum distance, based on these data alone, could be much less. To position these and other markers accurately on the physical map, it is necessary to make use of other ρ^- mutants retaining different markers and different sequences of ρ^+ mtDNA. Each ρ^- mutant to be used is subjected to a

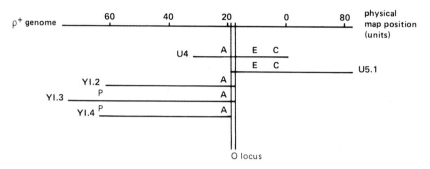

FIG. 7–12. Physical mapping of *oli1* (O) locus using ρ^- mutants retaining different mitochondrial genes and different amounts of ρ^+ mitochondrial genome as determined by DNA-DNA hybridization. Petite U4, which retains *cap1* (C), *ery1* (E), *oli1* (O), and *ana1* (A) loci, is reference petite. Its mtDNA hybridizes to the extent of 36% with ρ^+ mtDNA. This number is set equal to 36 units on the physical map, with 0 and the C gene being set arbitrarily at the right end of U4 mtDNA. The sequences retained by other ρ^- mutants in common with ρ^+ mtDNA and that of U4 are determined by establishing the degree of mtDNA sequence homology between each ρ^- mutant and these two genotypes, as well as by determining which markers are coretained by each mutant and U4 (see text for details). Sequences retained by each ρ^- mutant are indicated by horizontal lines beneath the total physical map of the ρ^+ mitochondrial genome. (Adapted from Nagley et al., ref. 42.)

two-step characterization. First, the number of units of ρ^+ mtDNA retained is determined, as was described previously. Second, the number of units of ρ^+ mtDNA retained in common with U4 must be determined. Linnane et al. (42) pointed out two reasons that this could not be determined directly using labeled U4 ρ^- mtDNA in solution and filter-bound mtDNA of the ρ^- mutant being tested. First, the mtDNA of U4 might contain new sequences not present in either ρ^+ mtDNA or the mtDNA of the ρ^- mutants being tested. Their presence would lead to serious errors in estimates of sequence homology between the DNA on the filter and that in solution. That is, some of the counts in solution would be in DNA incapable of hybridization because of the presence of new sequences, and the level of homology between the ρ^+ segments retained in U4 and the ρ^- mutant being tested would be underestimated. Second, differential amplification of certain ρ^+ sequences in the U4 mutant versus the mutant being tested could lead to erroneous estimates of homology. If the labeled U4 DNA were to contain differentially amplified sequences within the region of homology to another ρ^- mtDNA, the fraction of label bound would overestimate the proportion of sequences in common. Conversely, if the dif-

ferentially amplified labeled sequences were outside the region of homology, the sequence homology would be underestimated by the hybridization test.

In an attempt to obviate these problems, Linnane and associates made use of "petite equivalent" ρ^+ mtDNA. Labeled ρ^+ mtDNA in solution was first hybridized to U4 mtDNA in excess on filters. The filters were washed free of unbound mtDNA, and the DNA hybrids were denatured to yield the labeled petite equivalent ρ^+ mtDNA for hybridization to other ρ^- mutants. Such DNA obviously did not contain new sequences. The authors also argued that this labeled DNA contained all sequences the petite had in common with ρ^+, but in the proportion present in the ρ^+ mitochondrial genome.

Once the fraction of ρ^+ mtDNA retained by each ρ^- mutant and the degree of homology between U4 and each mutant have been determined, construction of the physical map can begin. For example, petite U5.1 lacks the *ana1* locus but retains *oli1, ery1,* and *cap1* (Fig. 7–12). This mutant retains 46 units of ρ^+ mtDNA, of which 19 units are in common with U4. Since U4 has arbitrarily been assigned units 0 to 36 with the *ana1* marker, the most distal marker to the left, U5.1 must extend 27 units to the right of 0. That is, U5.1 embraces units 1 to 19 and 73 to 100 (Fig. 7–12). Because U5.1 retains the *oli1* locus, it can be estimated that *oli1* is no more than 19 units from 0, but to fix the position of this locus precisely requires the use of other ρ^- mutants. Mutants Y1.2 and Y1.3 retain the *ana1* locus and the *oli1* locus (Fig. 7–12). Each of these mutants retains a large sequence of ρ^+ mtDNA, of which 16 to 17 units are in common with U4. Since these mutants do not retain *cap1* or *ery1*, the remaining sequences retained must be to the left of the sequences present in U4. From these results, *oli1* can be positioned at 18 to 19 units on the map as the following logic shows. Petites U5.1, Y1.2, and Y1.3 all retain the *oli1* locus. Petite U5.1 contains 19 units of ρ^+ mtDNA in common with U4 and extends beyond the right end of U4, but petites Y1.2 and Y1.3 retain about 18 units of ρ^+ mtDNA in common with U4, and their ρ^+ sequences extend beyond the left end of U4. In order that *oli1* be retained in both sets of ρ^- mutants, the locus must be at the left end of the retained sequences in U5.1 and at the right end of these sequences in Y1.2 and Y1.3 (Fig. 7–12). This places the locus 18 to 19 units from 0. The *oli1* locus is also excluded from mutant Y1.4, which extends to the left of U4, but this mutant retains *ana1*. Since Y1.4 retains 17 units in common with the left end of U4, but lacks *oli1*, this also places *oli1* at about 18 to 19 units (i.e., $36 - 17 = 19$). Using similar arguments, Linnane and associates (24,42) also mapped a number of other mitochondrial markers using this physical technique. Rytka and associates (43) put their standard collection of physically defined ρ^- mutants to work in order to map newly isolated *mit⁻* mutants to specific segments of the map rapidly. Each mutant was simply tested for its ability to yield wild-type recombinants with the standard set of ρ^- testers. Localization to a specific segment proceeded as described at the beginning of this section (Fig. 7–10).

The only major potential difficulty that might be encountered in the sort of mapping being done by Linnane and associates relates to the presence of multiple deletions in certain petites, as shown by Rabinowitz and his colleagues (44,45). In cases such as these, the retained ρ^+ sequences in the ρ^- mutant might not be adjacent in the ρ^+ mitochondrial genome. One sort of problem that could be encountered because of this is illustrated by a mutant such as U5.1, which retains a sequence in common with U4 and a sequence that extends to the right. The sequence shared with U4 and the one that extends to the right need not be contiguous physically in ρ^+ mtDNA. A different sort of problem could arise with respect to mapping. Suppose Y1.3 retains a segment of 7 units homologous with U4 that contains *oli1* and *ana1*, a deletion of 11 units corresponding to the left end of U4, and a second segment of ρ^+ mtDNA not homologous with the mtDNA of U4. This petite would give a maximum distance from the left end of U4 of 7 units for *oli1*. That would place this locus at unit 29, whereas results obtained with U5.1 would place *oli1* at a maximum of 19 units from 0. Fortunately, it seems unlikely that the ρ^- mutants used by Linnane and associates contained multiple deletions, since they were able to map different genes rather precisely and the order they obtained seems in good agreement with that observed by others (Fig. 7–14) (see Table 5–2). In short, it appears that Linnane and associates selected ρ^- mutants that retained continuous segments of ρ^+ mtDNA or contained small internal deletions at most. To distinguish these possibilities, it would be important to examine the tester ρ^- mutants with restriction endonucleases.

The second physical mapping method establishes the segments of ρ^+ mtDNA retained in a set of stable genetically marked ρ^- mutants to ρ^+ mtDNA by comparison of restriction endonuclease fragments. This method is now in widespread use, as is indicated by numerous papers in the Bari and Munich symposia (2,3). Two general approaches are in use (46). In the first, endonuclease restriction fragment patterns are compared between a set of ρ^- mutants, all of which retain the same sets of mitochondrial genes, and the wild type. The idea is to establish which restriction endonuclease fragments are common to all of the mutants. Ideally, only one such fragment will be found, and the genes being studied must therefore be in that fragment. In practice, most petites have sequences on either side of the fragment in question, in addition to the fragment itself; so while it is possible to place a given gene very close to the fragment, and quite possibly in it, there can be some ambiguity. It is also usually important to use more than one set of enzymes to identify fragments unambiguously. Second, the mtDNA of the ρ^- mutants being studied can be compared to that of ρ^+ mtDNA through the use of RNA probes. For example, if the petites in question carry genes mapping close to one of the rRNA genes, the fragments containing the rRNA gene in both ρ^- and ρ^+ mtDNAs can be determined by RNA-DNA hybridization using the technique of Southern (47). Similarly, the mtDNA of a ρ^- mutant can be used to transcribe complementary RNA (cRNA) *in vitro,* and this RNA can be hybridized to ρ^+ mtDNA fragments to determine homology.

Heyting and Sanders (48) used the foregoing methods to map the *RIB* region. Their experiments not only serve as a good example of how these techniques are applied but also illustrate why ω is now being placed between *RIB1* and *RIB3*. Heyting and Sanders made use of a group of stable ρ^- mutants retaining different segments of the *RIB* region. The primary restriction enzymes used to compare ρ^+ mtDNA with each of the petites were Hind II + III.

These enzymes yield a limited number of fragments (< 15) when they are used to digest ρ^+ mtDNA, and these separate well on gels. The ρ^- mtDNAs were first restricted with these enzymes to allow preliminary identification of the retained ρ^+ sequences. Confirmation of the preliminary identification was then achieved principally by use of two methods. First, the Hind II + III fragments were split with a second enzyme, in this case Hap II, to establish that they yielded the same fragment pattern as the putative equivalent fragments from ρ^+ mtDNA. Second, the 21S mitochondrial rRNA was used to determine if fragment H6 produced by Hind II + III had been retained, since this is the fragment in ρ^+ mtDNA that hybridizes with this RNA.

The results obtained in these experiments are summarized in Fig. 7–13. The region of the ρ^+ genome of interest contains four large Hind II + III fragments (H7, H4a, H6, H1a), a group of small fragments localized between H6 and H1a, and one small fragment (not shown) between H4a and H6. The F41 mutant, which retains all three *RIB* loci, ω^+, and the gene coding for 21S mitochondrial rRNA, contains fragments H4a and H6 and a novel fragment called J. Fragment J contains segments derived from H1a and H7 that are not adjacent in ρ^+ mtDNA but are at either end of the region of interest (Fig. 7–13). Presumably, fragment J arose because the deletion in F41 had endpoints in fragments H1a and H7; following deletion, these fragments rejoined to form fragment J and the basic circular monomer unit of this ρ^- mtDNA.

To pinpoint the location of each *RIB* gene more precisely, other ρ^- mutants retaining only one or two of the genes in this region were examined, and the Hap II digest patterns obtained with these mutants assumed much significance. Among the cleavage products of Hap II digestion are two fragments designated 1 and 5 (Fig. 7–13). Fragment 1 appears to embrace most of the

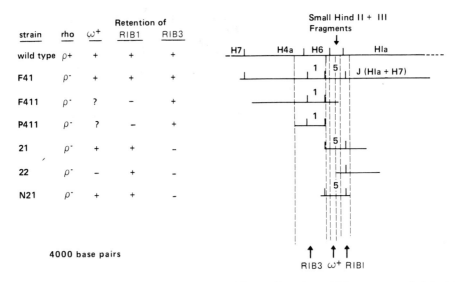

FIG. 7–13. Mapping of *RIB* region of mitochondrial genome of *S. cerevisiae* using restriction enzyme analysis to determine which ρ^+ fragments are present in different ρ^- mutants that have retained various combinations of markers in this region. On the left the genotype of each of the ρ^- mutants is given with respect to the *RIB1* and *RIB3* genes and ω. On the right the restriction fragments present in each strain are indicated. The H fragments are those obtained following Hind II + III digestion, and fragments 1 and 5 are the result of Hap II digestion. For the sake of simplicity, we have omitted other Hap II fragments falling to the right and left of these two adjacent fragments and also the one small Hind II + III fragment between H6 and H4a and the several small fragments between H6 and H1a. The 21S mitochondrial rRNA hybridizes to the mtDNA region embracing H6 and fragments 1 and 5. (Adapted from Heyting and Sanders, ref. 48.)

H6 fragment, whereas fragment 5 includes the region between H6 and H1a that yields the small Hind II + III fragments. Petite P411 retains the *RIB3* locus and only fragment 1 plus a short piece to the left. Therefore, *RIB3* is localized either within this fragment or very close to it. The *RIB1* locus appears to be localized in the right half of fragment 5. Petite 411 retains fragment 1 and the left half of fragment 5, but it retains only the *RIB3* locus. Petite 21 retains all of fragment 5, but fragment 1 is totally missing. This petite retains the *RIB1* locus, but not *RIB3*. Petite N21 retains the same gene, but only fragment 5, plus very short pieces on either side. Therefore, *RIB1* must be localized in fragment 5 or in the very short piece to the right. All petite strains retaining *RIB1* also retain ω^+, except petite 22, which retains the right half of fragment 5 but not the left half. Another petite designated 23 has slightly more of this fragment, but it is

ω^+. From these data it appeared that ω lay between *RIB1* and *RIB3*. Very recently Jacq et al. (71) have published a paper which confirms and extends the results reported by Heyting and Sanders. Using appropriate petites and restriction enzyme analysis, they confirmed the order *RIB1*-ω-*RIB1*-*RIB2*. More important, however, they showed that ω^+ contained a sequence referred to as △ about 1000 base pairs long which was absent from ω^- and in the right position for the ω locus.

Rabinowitz and his colleagues have also been intensively engaged in mapping the mitochondrial genome of *S. cerevisiae* using petite deletion mutants and restriction fragment analysis. From their most recent results (Fig. 7–14) it is apparent that the molecular mapping of this genome is proceeding rapidly and with great precision. The reader should note that although most maps in the literature are drawn with the same orientation shown in Fig. 7–14, this is not true of all. Nevertheless, virtually all published map orders are consistent with the one shown here.

FINE-STRUCTURE MAPPING

Fine-structure mapping is now proceeding with a number of yeast mitochondrial genes, notably those defined by the *mit⁻* phenotype. In the initial article on the mapping of the *mit⁻* mutants, Slonimski and Tzagoloff (49) described three mapping methods. The first method makes use of ρ⁺ stocks carrying mutations in the *RIB1*, *PAR1*, *OLI1*, and *OLI2* genes to localize representative *mit⁻* mutants to specific intervals in the genome. This ingenious method will not be described here because it is being supplanted by the use of genetically and physically defined tester ρ⁻ mutants for localization of new genes through deletion mapping.

The second method makes use of genetically marked ρ⁻ mutants to group the *mit⁻* mutants into clusters. Pairwise crosses are made between ρ⁻ mutants and *mit⁻* mutants. If a particular ρ⁻ mutant can rescue a *mit⁻* mutant by recombination, the recombinant will be able to grow on a nonfermentable carbon source. For this to happen, the *mit⁻* mutant must be within the ρ⁺ segment retained in the ρ⁻ mutant. We can then take different *mit⁻* mutants and classify them with respect to rescue by a set of ρ⁻ mutants that have retained different segments of the mitochondrial genome. If two *mit⁻* mutants are closely linked, they should be rescued by the same set of ρ⁻ mutants.

Through the use of this method, Slonimski and Tzagoloff demonstrated that the *mit⁻* mutants they had were grouped in four clusters (Table 7–13). The first cluster consisted of mutants lacking cytochrome b (*COB*). Of the 65 ρ⁻ mutants that rescued these *mit⁻* mutants, 62 were *nondiscriminating* testers that restored all members of the cluster, and three were *discriminating* testers that restored some, but not all, members of the cluster. These results indicated tight linkage of the mutants belonging in the *COB* cluster, but they also indicated that the fine structure of this region might be subject to investigation with the proper set of discriminating testers. The other three clusters that Slonimski and Tzagoloff identified, *OXI1*, *OXI2*, and *OXI3*, all included mutants deficient in cytochrome oxidase (see Chapters 5 and 19). The *OXI1* and *OXI2* clusters were of intermediate homogeneity in that 36% of the ρ⁻ mutants used to define the *OXI1* cluster discriminated between mutants in that cluster, whereas this was true of 18% in the case of the *OXI2* cluster (Table 7–13). The *OXI3* mutants constituted a very disperse cluster in that very few ρ⁻ mutants restored all members of this group.

Certain ρ⁻ mutants can rescue *mit⁻* mutants belonging to more than one cluster. This is true, for example, in the case of

TABLE 7–13. Separation of *mit⁻* mutants into clusters based on their ability to recombine and yield wild-type progeny in crosses with independently isolated ρ⁻ mutants

Mutant cluster	Number of *mit⁻* mutants tested	Number of ρ⁻ mutants that recombine with:		
		All cluster members	No cluster members	Some cluster members
COB	6	65	499	3
OXI1	6	34	142	19
OXI2	4	61	121	13
OXI3	5	3	181	11

Adapted from Slonimski and Tzagoloff (49).

mutants belonging to the *OXI1* and *OXI2* clusters, which are fairly closely linked (Fig. 7–14). Slonimski and Tzagoloff attempted to quantify the degree of linkage by means of a parameter they called the *disjunction coefficient,* which is defined as follows: If *x* is the number of ρ^- mutants that restore one mutant and *y* is the number restoring a second mutant and *z* is the number restoring both, then the disjunction coefficient is xy/zn, where *n* is the total number of ρ^- strains screened. If the disjunction coefficient is less than 1, the two sites are prefer-

entially associated in the retained wild-type sequences of the ρ^- tester mutants, which would tend to indicate close linkage. If the disjunction coefficient is 1, the probability of association is random, and nothing can be said concerning linkage. If the disjunction coefficient is much greater than 1, the two *mit⁻* mutants are in sequences of wild-type mtDNA, which tend not to be retained together in ρ^- mutants. Examination of the disjunction coefficients indicated that *COB* was closely linked to *OLI1,* that *OXI1* and *OXI2* were close together, and that both

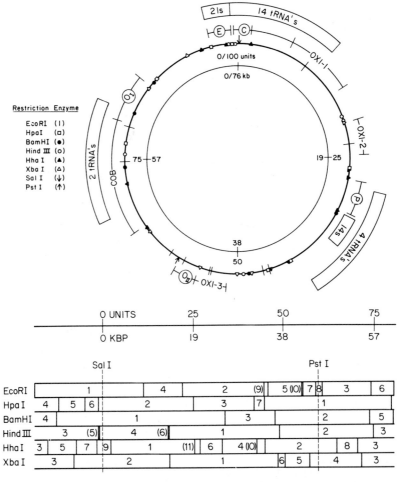

FIG. 7–14. Physical and genetic map of mtDNA of *S. cerevisiae* as determined by restriction fragment analysis of mtDNA from marker-retaining ρ^- mutants and by hybridization of complementary RNA probes made on ρ^- mtDNA to ρ^+ mtDNA. The fragment map units are thousands of base pairs (KBP), and the RNA is given in units. Gene designations are as follows: 21S and 14S are large and small mitochondrial rRNAs, respectively; C and E are *RIB1* and *RIB3* genes, respectively; O_I and O_{II} are *OLI1* and *OLI2*; P is *PAR1*; *OXI1*, *OXI2*, and *OXI3* are three cytochrome oxidase genes; *COB* embraces two cytochrome b genes; mitochondrial tRNAs map in three separate regions. (Courtesy of Dr. M. Rabinowitz.)

clusters were close to *RIB1* and *PAR1,* all of which have proved to be the case (Fig. 7–14).

The third method involved the use pairwise crosses of *mit⁻* mutants to see which pairs yielded wild-type recombinants. In addition to positive and negative responses, papillae growing on an essentially negative background were sometimes obtained. The *OXI1* mutants, when crossed by each other, produced negative responses or papillae. When these mutants were crossed to *OXI2* and *OXI3* mutants, only positive responses were obtained. Thus, *OXI1* mutants form a distinct cluster using this method of allele testing. The data were not so clear-cut in the case of the *OXI2* mutants. None of the mutants in this group yielded completely negative responses with each other. Either papillae were obtained or the responses were completely positive, suggesting that no two mutants fell at the same site in this cluster. However, the sample size was only five. All three possible responses were obtained in the analysis of the *OXI3* mutants. One mutant, M5–16, did not restore any of the other mutants, although it, like any other *OXI3* mutant, gave a positive response in crosses with all *OXI1* and *OXI2* mutants. Thus the M5–16 mutant defined the cluster, since it appeared to be a deletion (but, interestingly enough, not a petite deletion). Other mutants in this cluster behaved as if they were shorter deletions, so that a deletion "map" could be constructed for the locus. Using restriction enzyme analysis, Rabinowitz et al. (45) showed that M5–16 is, in fact, a deletion of about 3.5×10^6 d. Three other putative deletions were also tested, and one of these showed a change in restriction enzyme pattern compatible with the idea that it was a deletion. To summarize, the results of both the allele tests and the ρ^- petite restoration tests were concordant in indicating that the *mit⁻* mutants studied by Slonimski and Tzagoloff fell into four discrete groups. However, the *OXI2* results, in particular,

might have been difficult to interpret had it not been for the petite restoration test. Using petite restoration tests and allele tests, the fine structure of the *mit* genes is now under active study, as papers dealing with the subject in the case of the complex III region of the mitochondrial genetic map (50–52 and Chapter 19) attest.

PLEIOTROPIC RESISTANCE MUTATIONS, THE 2-μm CIRCULAR DNA, AND THE QUESTION OF YEAST EPISOMES

In Chapter 5 we mentioned that Griffiths (53–56) found that mutations resistant to inhibitors of oxidative phosphorylation fell into two classes phenotypically. Class I mutations resistant to oligomycin were resistant to a variety of ATPase inhibitors, as well as unrelated inhibitors of oxidative phosphorylation (e.g., uncouplers), inhibitors of mitochondrial protein synthesis (chloramphenicol, erythromycin), and cycloheximide. Similar pleiotropy was seen among class I mutants selected for resistance to triethyl tin or venturicidin (53, 57–60). The genetic basis of this complex resistance pattern characteristic of class I mutants appeared to involve both a nuclear gene or genes and a cytoplasmic element.

Class II mutants to oligomycin resistance were resistant specifically to related antibiotics (e.g., rutamycin, ossamycin, peliomycin) and were not cross-resistant to other ATPase inhibitors or other agents that block oxidative phosphorylation. These class II mutants proved to be mitochondrial and fell in the *OLI1* and *OLI2* loci (see Table 5–2) (Fig. 7–14). Class II mutations to triethyl tin resistance, on the other hand, while exhibiting cytoplasmic inheritance, did not appear to be mitochondrial. These mutants, like the class II mutations to oligomycin resistance, were specific in phenotype and conferred cross-resistance only to venturicidin, an ATPase inhibitor, and in some cases to 1799, an uncoupler. Two lines of evidence indicated that the class II mutations to triethyl tin and venturicidin resistance ($VEN^R TET^R$), although cytoplasmic in inheritance, were not mitochondrial. First, these mutants assorted randomly with other mitochondrial genes. That is, recombination frequencies were around 45%, as compared to 20 to 25% for unlinked mitochondrial genes. Second, although ethidium bromide induced loss of the $VEN^R TET^R$ determinant, this determinant was clearly not a part of the ρ factor, since ρ^0 $VEN^R TET^R$ isolates could be obtained that lacked mtDNA (60). In summary, Griffiths' findings indicated that resistance in class I mutants involved both a nuclear gene or genes and a cytoplasmic element and that resistance in the class II $VEN^R TET^R$ mutants resided in a cytoplasmic element distinct from mtDNA.

The results obtained by Griffiths can be explained in part by a complex interaction between cytoplasmic and nuclear genes, as described by Guerineau et al. (61,62), the genetics of which are still not entirely clear. These authors postulated the existence of two cytoplasmic factors and three nuclear genes to explain the behavior of oligomycin-resistant mutants and pleiotropic mutants resistant to triethyl tin, venturicidin, chloramphenicol, and cycloheximide, which have some of the properties of the class I mutants and class II $VEN^R TET^R$ mutants of Griffiths. The two cytoplasmic elements are designated τ and π, and the three nuclear genes can have either R or S alleles (for resistance and sensitivity, respectively). The τ element seems to be most readily equated with the $VEN^R TET^R$ determinant of Griffiths. Like the element described by Griffiths, the factor can be eliminated by ethidium bromide, but this process requires higher concentrations of the dye and proceeds with slower kinetics than ρ factor elimination. The τ element has yet to be identified physically, although its behavior is consistent with a cytoplasmic pattern of inheritance distinct from the ρ factor. The π

element is identical with a covalently closed circular 2-μm DNA molecule that has been known for several years in yeast (63–66). According to Guerineau and associates (62), the various possible interactions between the three nuclear genes and the two cytoplasmic elements can be depicted as shown in Table 7–14. Resistance to oligomycin and venturicidin alone is conferred when gene 1 and π are resistant. Genes 2 and 3 and τ can be sensitive. However, it is not clear from the published data if resistance arises because both π and gene 1 carry resistance alleles, if the nuclear allele and the cytoplasmic factor act synergistically to cause resistance, or if one facilitates the expression of the other. Also, the phenotype seen is not identical to that found by Griffiths with respect to his class I oligomycin-resistant mutants. These mutants are cross-resistant not only to oligomycin and venturicidin but also to low levels of chloramphenicol and cycloheximide, whereas resistance to these antibiotics was not reported by Guerineau et al. (62) for this genetic combination.

The genotype $1^S 2^R 3^S$ is sensitive to oligomycin, as is the genotype $1^S 2^R 3^R$. These two genotypes are also characterized by absence of the π factor. In short, the combination of 1^S with 2^R seems to lead simultaneous loss of π and oligomycin resistance. However, the $1^S 2^R 3^R$ genotype is still resistant to triethyl tin, chloramphenicol, and cycloheximide; so the τ element is still present even though π DNA has been lost. Expression of τ-mediated resistance seems to require that gene 3 plus either gene 1 or 2 carry R alleles. However, as in the case of the π factor, it is unclear if both the nuclear genes and cytoplasmic factor carry resistance genes or if one acts synergistically with the other to confer resistance.

Venturicidin resistance seems to require, as a minimum, either of two distinct genotypic combinations. First, the 1^R allele in combination with π^R confers resistance to oligomycin and venturicidin simultaneously,

TABLE 7–14. Genotypes for three mendelian genes (1, 2, 3) and two cytoplasmic factors (π, τ) postulated by Guerineau and associates (62) to explain various patterns of multiple drug resistance in S. cerevisiae[a]

Genotype		Resistance phenotype					
Nuclear	Cytoplasmic	OLI	VEN	CAP	TET	CYC	2-μm DNA
$1^R 2^S 3^S$	$\pi^R \tau^S$	+	+	−	−	−	+
$1^S 2^R 3^S$	$\pi^- \tau^S$	−	+	−	−	−	−
$1^R 2^S 3^R$	$\pi^R \tau^R$	+	+	+	+	+	+
$1^S 2^R 3^R$	$\pi^- \tau^R$	−	+	+	+	+	−
$1^S 2^S 3^R$	$\pi^S \tau^R$	−	−	−	−	−	+
$1^S 2^S 3^S$	$\pi^S \tau^S$	−	−	−	−	−	+

[a] Superscripts indicate that the nuclear genes and cytoplasmic factors confer resistance (R) or sensitivity (S). The designation π^- means that the π factor is absent. Resistant and sensitive phenotypes with respect to a given antibiotic are indicated by + and −. Presence or absence of the 2-μm circular DNA is indicated similarly. Abbreviations: OLI, oligomycin; VEN, venturicidin; CAP, chloramphenicol; TET, triethyl tin; CYC, cycloheximide.

as mentioned previously. Second, the $1^S 2^R 3^S$ genotype, which lacks π and carries a sensitive τ factor, is resistant to venturicidin. Whether or not the τ factor can mediate venturicidin resistance is not clear. because venturicidin resistance is expressed only in stocks containing a resistant τ factor if either the first or the second genotypic combination just described is also present. This consideration, together with the observation that τ-factor-mediated resistance includes resistance to chloramphenicol and cycloheximide, suggests that the τ factor and the $VEN^R TET^R$ determinants of Griffiths are not identical. His determinants do not confer resistance to chloramphenicol and cycloheximide simultaneously, but they do just that with respect to triethyl tin and venturicidin resistance.

Guerineau et al. (61,62) suggested that these complex findings might be accounted for by an episome hypothesis that postulates that π, in particular, can exist either in a cytoplasmic state or integrated into one of the chromosomes. They suggested that mutations in this DNA might cause changes in membrane permeability leading to multiple drug resistance. However, the genetics and possible cross-resistance patterns are still rather ill-defined; so there is still much latitude for new interpretations and hypotheses as new discoveries are made. One fact

that is established beyond doubt is that π is a circular DNA molecule that contains two inverted repeats of 600 bases situated opposite each other on the perimeter of the circle (67,68). The circular monomer itself contains 6,000 bases. However, four EcoR1 endonuclease fragments are generated in 1:1:1:1 stoichiometry that sum to a total length of 12,000 bases. Guerineau et al. (67) proposed that this difference could be accounted for by a model that assumes there are two classes of monomers that can be generated by intramolecular recombination within the inverted repeats. These two monomers differ by the relative orientation of the nonrepeated sequences. Perhaps recombination between different monomers can lead to different combinations of drug resistance patterns. Guerineau et al. (67) also pointed out that π DNA may prove to be useful as a eukaryotic cloning vehicle, in view of its unique molecular structure. This does seem quite likely.

CONCLUSIONS

The length and complexity of this chapter reflect the very rapid rate at which knowledge concerning segregation, recombination, and mapping of mitochondrial genes in S. cerevisiae has accumulated in the past 7 years. There are many papers in the Bari and Munich symposia (2,3) that have been neglected here, even though they bear on the topics discussed in this chapter. They are not left unmentioned because they are unimportant, but rather because this chapter must have an end and must attempt to focus on the principal methods, problems, and achievements in this area of research. Study of this literature has prompted two conclusions: First, the mapping of mitochondrial genes in *S. cerevisiae* has now reached a high level of sophistication, and soon we may expect the map of the mitochondrial genome in this organism to rival those of bacteriophages such as λ and T4 in completeness. The development of fine-structure mapping techniques, including allele testing by recombination and the use of small-deletion mutants that are not ρ^- in phenotype, coupled with peptide analysis, will soon lead to identification of specific structural genes in mtDNA. At the same time, mitochondrial gene regulation is beginning to become amenable to the same powerful genetic probes that have been used for years with bacteria and bacterial viruses. This is a subject we will discuss briefly in Chapter 19. Similarly, mitochondrial gene mutants will come into use for genetic dissection of mitochondrial inner membrane assembly. In short, the basic genetic tools are now at hand; relying on the availability of these tools, the next generation of questions can now be approached.

Second, our understanding of the rules governing mitochondrial gene recombination and segregation is still primitive. The phage analogy model, despite its conceptual elegance and predictive capabilities, fails to deal in an intellectually satisfying manner (at least as far as this author is concerned) with data obtained from pedigrees indicating that mitochondrial genes segregate rapidly, as well as with the results of zygote clone analysis indicating inhomogeneity of zygote clones with respect to mitochondrial gene transmission, which at its most extreme yields high frequencies of uniparental

zygotes. Also, there is practically no consideration of the molecular biology of the events governing transmission and replication of mtDNA in the zygote and its buds, and one has no idea of how the panmictic pool of mtDNA molecules is actually realized in the zygote. These deficiencies in no way affect the mapping of mitochondrial genes, only our ability to conceptualize the events surrounding mitochondrial gene segregation and recombination.

In the case of the chloroplast genome of *Chlamydomonas,* mapping and model building have become intertwined to an unfortunate degree (see Chapters 14 and 15). In the future it will be important in this organism simply to continue development of workable mapping methods irrespective of underlying events and models. The success of this approach in yeast in the last few years, coupled with a much larger group of interested workers, has led to a much more complete map of the mitochondrial genome of *S. cerevisiae* than we presently have for the chloroplast genome of *Chlamydomonas.*

REFERENCES

1. Lewis, D. (1970): Conclusions: Organelles as membrane complexes. *Symp. Soc. Exp. Biol.,* 24:497–501.
2. Saccone, C., and Kroon, A. M. (editors) (1976): *The Genetic Function of Mitochondrial DNA.* North Holland, Amsterdam.
3. Bücher, T., Neupert, W., Sebald, W., and Werner, S. (editors) (1976): *Genetics and Biogenesis of Chloroplasts and Mitochondria.* North Holland, Amsterdam.
4. Coen, D., Deutsch, J., Netter, P., Petrochilo, E., and Slonimski, P. P. (1970): Mitochondrial genetics. I. Methodology and phenomenology. *Symp. Soc. Exp. Biol.,* 24:449–496.
5. Avner, P. R., Coen, D., Dujon, B., and Slonimski, P. P. (1973): Mitochondrial genetics. IV. Allelism and mapping studies of oligomycin mutants in *S. cerevisiae. Mol. Gen. Genet.,* 125:9–52.
6. Dujon, B., Slonimski, P. P., and Weill, L. (1974): Mitochondrial genetics. IX. A model for recombination and segregation of mitochondrial genomes in *Saccharomyces cerevisiae. Genetics,* 78:415–437.
7. Rank, G. H., and Bech-Hansen, N. T. (1972): Somatic segregation, recombination, asym-

metrical distribution and complementation tests of cytoplasmically-inherited antibiotic-resistance mitochondrial markers in *S. cerevisiae. Genetics,* 72:1–15.

8. Howell, N., Trembath, M. K., Linnane, A. W., and Lukins, H. B. (1973): The biogenesis of mitochondria. 30. An analysis of polarity of mitochondrial gene recombination and transmission. *Mol. Gen. Genet.,* 122:37–51.

9. Lukins, H. B., Tate, J. R., Saunders, G. W., and Linnane, A. W. (1973): The biogenesis of mitochondria. 26. Mitochondrial recombination: The segregation of parental and recombinant mitochondrial genotypes during vegetative division of yeast. *Mol. Gen. Genet.,* 120:17–25.

10. Wilkie, D., and Thomas, D. Y. (1973): Mitochondrial genetic analysis by zygote cell lineages in *Saccharomyces cerevisiae. Genetics,* 73:367–377.

11. Callen, D. F. (1974): Segregation of mitochondrially inherited antibiotic resistance genes in zygote cell lineages of *Saccharomyces cerevisiae. Mol. Gen. Genet.,* 134:65–76.

12. Strausberg, R. L., and Perlman, P. S. (1974): Cellular analysis of mitochondrial inheritance in *Saccharomyces cerevisiae. Genetics,* 77: s62–63.

13. Perlman, P. S., Birky, C. W., Jr., Demko, C. A., and Strausberg, R. L. (1976): Confirmations and exceptions to the phage analogy model: Input bias, bud position effects, zygote heterogeneity, and uniparental inheritance. In: *Genetics and Biogenesis of Chloroplasts and Mitochondria,* edited by T. Bücher, W. Neupert, W. Sebald, and S. Werner, pp. 405–414. North Holland, Amsterdam.

14. Forster, J., and Kleese, R. A. (1975): The segregation of mitochondrial genes in yeast. II. Analysis of zygote pedigrees of drug-resistant × drug-sensitive crosses. *Mol. Gen. Genet.,* 139:341–355.

15. Waxman, M., and Eaton, N. (1973): Effect of antibiotics on the transmission of mitochondrial factors in *Saccharomyces cerevisiae. Mol. Gen. Genet.,* 127:277–284.

16. Thomas, D. Y., and Wilkie, D. (1968): Recombination of mitochondrial drug resistance factors in *Saccharomyces cerevisiae. Biochem. Biophys. Res. Commun.,* 30:368–372.

17. Wilkie, D. (1970): Analysis of mitochondrial drug resistance in *Saccharomyces cerevisiae. Symp. Soc. Exp. Biol.,* 24:71–83.

18. Linnane, A. W., Howell, N., and Lukins, H. B. (1974): Mitochondrial genetics. In: *The Biogenesis of Mitochondria,* edited by A. M. Kroon and C. Saccone, pp. 193–213. Academic Press, New York.

19. Netter, P., Petrochilo, E., Slonimski, P. P., Bolotin-Fukuhara, M., Coen, D., Deutsch, J., and Dujon, B. (1974): Mitochondrial genetics. VII. Allelism and mapping studies of ribosomal mutants resistant to chloramphenicol, erythromycin and spiramycin in *S. cerevisiae. Genetics,* 78:1063–1100.

20. Bolotin, M., Coen, D., Deutsch, J., Dujon, B., Netter, P., Petrochilo, E., and Slonimski, P. P. (1971): La recombinaison des mitochondries chez *Saccharomyces cerevisiae. Bull. Inst. Pasteur,* 69:215–239.

21. Trembath, M. K., Bunn, C. L., Lukins, H. B., and Linnane, A. W. (1973): The biogenesis of mitochondria. 27. Genetic and biochemical characterization of cytoplasmic and nuclear mutations to spiramycin resistance in *Saccharomyces cerevisiae. Mol. Gen. Genet.,* 121: 35–48.

22. Dujon, B., Bolotin-Fukuhara, M., Coen, D., Deutsch, P., Slonimski, P. P., and Weill, L. (1976): Mitochondrial genetics. XI. Mutations at the mitochondrial locus ω affecting the recombination of mitochondrial genes in *Saccharomyces cerevisiae. Mol. Gen. Genet.,* 143:131–165.

23. Sanders, J. P. M., Heyting, C., DiFranco, A., Borst, P., and Slonimski, P. P. (1976): The organization of genes in yeast mitochondrial DNA. In: *The Genetic Function of Mitochondrial DNA,* edited by C. Saccone and A. M. Kroon, pp. 259–272. North Holland, Amsterdam.

24. Linnane, A. W., Lukins, H. B., Molloy, P. L., Nagley, P., Rytka, J., Sriprakash, K. S., and Trembath, M. K. (1976): Biogenesis of mitochondria: Molecular mapping of the mitochondrial genome of yeast. *Proc. Natl. Acad. Sci. U.S.A.,* 73:2082–2085.

25. Dujon, B., and Michel, F. (1976): Genetic and physical characterization of a segment of the mitochondrial DNA involved in the control of genetic recombination. In: *The Genetic Function of Mitochondrial DNA,* edited by C. Saccone and A. M. Kroon, pp. 175–184. North Holland, Amsterdam.

26. Wolf, K., Dujon, B., and Slonimski, P. P. (1973): Mitochondrial Genetics. V. Multifactorial mitochondrial crosses involving a mutation conferring paromomycin-resistance in *Saccharomyces cerevisiae. Mol. Gen. Genet.,* 125:53–90.

27. Visconti, N., and Delbrück, M. (1953): The mechanism of genetic recombination in phage. *Genetics,* 38:5–33.

28. Stahl, F. W. (1965): *The Mechanics of Inheritance.* Prentice-Hall, Englewood Cliffs, N.J.

29. Stent, G. S. (1963): *Molecular Biology of Bacterial Viruses.* W. H. Freeman, San Francisco.

30. Dujon, B., Kruszewska, A., Slonimski, P. P., Bolotin-Fukuhara, M., Coen, D., Deutsch, J., Netter, P., and Weill, L. (1975): Mitochondrial genetics. X. Effects of UV irradiation on transmission and recombination of mitochondrial genes in *Saccharomyces cerevisiae. Mol. Gen. Genet.,* 137:29–72.

31. Perlman, P. S., and Birky, C. W., Jr. (1974):

Mitochondrial genetics in baker's yeast: A molecular mechanism for recombinational polarity and suppressiveness. *Proc. Natl. Acad. Sci. U.S.A.,* 71:4612–4616.

32. Smith, D. G., Wilkie, D., and Srivastava, K. C. (1972): Ultrastructural changes in mitochondria of zygotes in *Saccharomyces cerevisiae. Microbios,* 6:231–238.

33. Birky, C. W., Jr. (1975): Effects of glucose repression on the transmission and recombination of mitochondrial genes in yeast (*Saccharomyces cerevisiae*). *Genetics,* 80:695–709.

34. Birky, C. W., Jr. (1975): Zygote heterogeneity and uniparental inheritance of mitochondrial genes in yeast. *Mol. Gen. Genet.,* 141: 41–58.

35. Van Winkle-Swift, K. P., and Birky, C. W., Jr. (1977): The non-reciprocality of organelle gene recombination in *Chlamydomonas reinhardtii* and *Saccharomyces cerevisiae. Genetics,* 86:s67.

36. Callen, D. F. (1974): Recombination and segregation of mitochondrial genes in *Saccharomyces cerevisiae. Mol. Gen. Genet.,* 134: 49–63.

37. Dujon, B., and Slonimski, P. P. (1976): Mechanisms and rules for transmission, recombination and segregation of mitochondrial genes in *Saccharomyces cerevisiae.* In: *Genetics and Biogenesis of Chloroplasts and Mitochondria,* edited by T. Bücher, W. Neupert, W. Sebald, and S. Werner, pp. 393–403. North Holland, Amsterdam.

38. Sena, E. P., Welch, J., and Fogel, S. (1976): Nuclear and mitochondrial DNA replication during synchronous mating in *Saccharomyces cerevisiae. Science,* 194:433–435.

39. Molloy, P. L., Linnane, A. W., and Lukins, H. B. (1975): Biogenesis of mitochondria: Analysis of deletion of mitochondrial antibiotic resistance markers in petite mutants of *Saccharomyces cerevisiae. J. Bacteriol.,* 122: 7–18.

40. Schweyen, R. J., Steyrer, U., Kaudewitz, F., Dujon, B., and Slonimski, P. P. (1976): Mapping of mitochondrial genes in *Saccharomyces cerevisiae:* Population and pedigree analysis of retention or loss of four genetic markers in rho⁻ cells. *Mol. Gen. Genet.,* 146:117–132.

41. Schweyen, R. J., Weiss-Brummer, B., and Backhaus, B. (1976): Localization of seven gene loci on a circular map of the mitochondrial genome of *Saccharomyces cerevisiae.* In: *The Genetic Function of Mitochondrial DNA,* edited by C. Saccone and A. M. Kroon, pp. 251–258. North Holland, Amsterdam.

42. Nagley, P., Sriprakash, K. S., Rytka, J., Choo, K. B., Trembath, M. K., Lukins, H. B., and Linnane, A. W. (1976): Physical mapping of genetic markers in the yeast mitochondrial genome. In: *The Genetic Function of Mitochondrial DNA,* edited by C. Saccone and A. M. Kroon, pp. 231–242. North Holland, Amsterdam.

43. Rytka, J., English, K. J., Hall, R. M., Linnane, A. W., and Lukins, H. B. (1976): The isolation and simultaneous physical mapping of mitochondrial mutations affecting respiratory complexes. In: *Genetics and Biogenesis of Chloroplasts and Mitochondria,* edited by T. Bücher, W. Neupert, W. Sebald, and S. Werner, pp. 427–434. North Holland, Amsterdam.

44. Morimoto, R., Lewin, A., Hsu, H.-J., Rabinowitz, M., and Fukuhara, H. (1975): Restriction endonuclease analysis of mitochondrial DNA from grande and genetically characterized cytoplasmic petite clones of *Saccharomyces cerevisiae. Proc. Natl. Acad. Sci. U.S.A.,* 72:3868–3872.

45. Rabinowitz, M., Jakovic, S., Martin, N., Hendler, F., Halbreich, A., Lewin, A., and Morimoto, R. (1976): Transcription and organization of yeast mitochondrial DNA. In: *The Genetic Function of Mitochondrial DNA,* edited by C. Saccone and A. M. Kroon, pp. 219–230. North Holland, Amsterdam.

46. Borst, P., Sanders, J. P. M., and Heyting, C. (1977): Biochemical methods to locate genes on the physical map of yeast mitochondrial DNA. *Methods Enzymol. (in press).*

47. Southern, E. M. (1975): Detection of specific sequences among DNA fragments separated by gel electrophoresis. *J. Mol. Biol.,* 98:503–517.

48. Heyting, C., and Sanders, J. P. M. (1976): The physical mapping of some genetic markers in the 21S ribosomal region of the mitochondrial DNA of yeast. In: *The Genetic Function of Mitochondrial DNA,* edited by C. Saccone and A. M. Kroon, pp. 273–280. North Holland, Amsterdam.

49. Slonimski, P. P., and Tzagoloff, A. (1976): Localization in yeast mitochondrial DNA of mutations expressed in a deficiency of cytochrome oxidase and/or coenzyme QH₂-cytochrome c reductase. *Eur. J. Biochem.,* 61:27–41.

50. Tzagoloff, A., Foury, F., and Akai, A. (1976): Genetic determination of mitochondrial cytochrome b. In: *Genetics and Biogenesis of Chloroplasts and Mitochondria,* edited by T. Bücher, W. Neupert, W. Sebald, and S. Werner, pp. 419–426. North Holland, Amsterdam.

51. Kotylak, Z., and Slonimski, P. P. (1976): Joint control of cytochromes a and b by a unique mitochondrial DNA region comprising four genetic loci. In: ,*The Genetic Function of Mitochondrial DNA,* edited by C. Saccone and A. M. Kroon, pp. 143–154. North Holland, Amsterdam.

52. Pajot, P., Wambier-Kluppel, M. L., Kotylak, Z., and Slonimski, P. P. (1976): Regulation of cytochrome oxidase formation by mutations in a mitochondrial gene for cytochrome b. In: *Genetics and Biogenesis of Chloroplasts and Mitochondria,* edited by T. Bücher, W. Neu-

pert, W. Sebald, and S. Werner, pp. 443–451. North Holland, Amsterdam.

53. Griffiths, D. E. (1975): Utilization of mutations in the analysis of yeast mitochondrial oxidative phosphorylation. In: *Genetics and Biogenesis of Mitochondria and Chloroplasts,* edited by C. W. Birky, Jr., P. S. Perlman, and T. J. Byers, Chapter 3. Ohio State University Press, Columbus.

54. Avner, P. R., and Griffiths, D. E. (1970): Oligomycin resistant mutants in yeast. *FEBS Lett.,* 10:202–207.

55. Avner, P. R., and Griffiths, D. E. (1973): Studies on energy-linked reactions. Isolation and characterization of oligomycin-resistant mutants of *Saccharomyces cerevisiae. Eur. J. Biochem.,* 32:301–311.

56. Avner, P. R., and Griffiths, D. E. (1973): Studies on energy-linked reactions. Genetic analysis of oligomycin-resistant mutants of *Saccharomyces cerevisiae. Eur. J. Biochem.,* 32:312–321.

57. Lancashire, W. E., and Griffiths, D. E. (1975): Studies on energy-linked reactions: Isolation, characterization and genetic analysis of trialkyl-tin resistant mutants of *Saccharomyces cerevisiae. Eur. J. Biochem.,* 51:377–392.

58. Griffiths, D. E., Houghton, R. L., Lancashire, W. E., and Meadows, P. Λ. (1975): Studies on energy-linked reactions: Isolation and properties of mitochondrial venturicidin-resistant mutants of *Saccharomyces cerevisiae. Eur. J. Biochem.,* 51:393–402.

59. Lancashire, W. E., and Griffiths, D. E. (1975): Studies on energy-linked reactions: Genetic analysis of venturicidin-resistant mutants. *Eur. J. Biochem.,* 51:403–413.

60. Griffiths, D. E., Lancashire, W. E., and Zanders, E. D. (1975): Evidence for an extra-chromosomal element involved in mitochondrial function: A mitochondrial episome? *FEBS Lett.,* 53:126–130.

61. Guerineau, M., Slonimski, P. P., and Avner, P. R. (1974): Yeast episome: Oligomycin resistance associated with a small covalently closed non-mitochondrial circular DNA. *Biochem. Biophys. Res. Commun.,* 61:412–419.

62. Guerineau, M., Grandchamp, C., and Slonimski, P. (1976): Structure and genetics of the

2 μM circular DNA in yeast. In: *Genetics and Biogenesis of Chloroplasts and Mitochondria,* edited by T. Bücher, W. Neupert, W. Sebald, and S. Werner, pp. 557–564. North Holland, Amsterdam.

63. Guerineau, M., Grandchamp, C., Paoletti, C., and Slonimski, P. P. (1971): Characterization of a new class of circular DNA molecules in yeast. *Biochem. Biophys. Res. Commun.,* 42: 550–557.

64. Clark-Walker, G. D. (1972): Isolation of a circular DNA from a mitochondrial fraction from yeast. *Proc. Natl. Acad. Sci. U.S.A.,* 69:388–392.

65. Clark-Walker, G. D. (1973): Size distribution of circular DNA from petite-mutant yeast lacking ρ DNA. *Eur. J. Biochem.,* 32:263–267.

66. Zeman, L., and Lusena, C. V. (1974): Closed circular DNA associated with yeast mitochondria. *FEBS Lett.,* 38:171–174.

67. Guerineau, M., Grandchamp, C., and Slonimski, P. P. (1976): Circular DNA of a yeast episome with two inverted repeats: Structural analysis by a restriction enzyme and electron microscopy. *Proc. Natl. Acad. Sci. U.S.A.,* 73:3030–3034.

68. Hollenberg, C. P., and Royer, H.-D. (1976): Electron microscopical analysis of native and cloned 2-μm DNA from *Saccharomyces cerevisiae.* In: *Genetics and Biogenesis of Chloroplasts and Mitochondria,* edited by T. Bücher, W. Neupert, W. Sebald, and S. Werner, pp. 565–568. North Holland, Amsterdam.

69. Van Winkle-Swift, K. P., and Birky, C. W., Jr. (1978): The non-reciprocality of organelle gene recombination in *Chlamydomonas reinhardtii* and *Saccharomyces cerevisiae.* Manuscript to be submitted.

70. Hershey, A. D., and Rotman, R. (1949): Genetic recombination between host-range and plaque-type mutants of bacteriophage in single bacterial cells. *Genetics,* 34:44–71.

71. Jacq, W., Kujawa, C., Grandchamp, C. and Netter, P. (1977): Physical characterization of the difference between yeast mtDNA alleles. In: *Mitochondria 1977, Proceedings of the Colloquium on Genetics and Biogenesis of Mitochondria,* edited by W. Bandlow et al. DeGruyter, Berlin.

Chapter 8
Mitochondrial Genetics in *Neurospora*

In 1952, three years after the discovery of petite mutations by Ephrussi and his colleagues, Mitchell and Mitchell (1) described a maternally inherited mutant in *Neurospora* that they called *poky* because it grew slowly. Shortly thereafter, the Mitchells showed that, like the vegetative petite mutants, *poky* was deficient in cytochromes a + a$_3$ and b, but not c (2,3). However, unlike the vegetative petites, *poky* did not lack these cytochromes completely; in older cultures of the *poky* mutant, concentrations of cytochromes a + a$_3$ and b were increased. In other words, *poky* differed from the vegetative petites in being leaky, and for reasons that will become apparent later, it is probably best compared to one of the *syn⁻* mutants of *S. cerevisiae* (see Chapter 5). Since then, other maternally inherited cytochrome-deficient mutants have been isolated in *Neurospora crassa,* and much of the discussion in this chapter will concern the genetics and physiology of these mutants, as well as the indirect evidence that they may be mitochondrial in origin.

LIFE CYCLE AND MENDELIAN GENETICS OF *Neurospora crassa*

The vegetative mycelium of *N. crassa* is normally haploid and can be propagated indefinitely by means of transfer of mycelial fragments or asexual spores to fresh medium. The mycelium forms two kinds of asexual spores, the macroconidia, which contain four or five nuclei on the average, and the uninucleate microconidia. The two mating types, designated *A* and *a,* are controlled by a pair of mendelian alleles. Mycelia of either mating type will elaborate fruiting bodies (protoperithecia) on an appropriate medium. After protoperithecia are produced, the sexual cycle is initiated by fertilization of the protoperithecia with conidia or mycelial fragments of opposite mating type (Fig. 8–1).

Each protoperithecium puts forth structures called trichogynes; these thin filaments act as the female receptors through which the haploid male nucleus donated by mycelial fragment or conidium moves into the protoperithecium. After migration of the male nucleus, a large number of equational divisions of nuclei of opposite mating type take place side by side with subsequent fusion of pairs. Each fused nucleus then undergoes meiosis and produces four haploid nuclei. These divide mitotically and the eight ascospores are then formed. The nuclei of sister spore pairs are genetically identical. Mendelian genes segregate as they do in yeast; so in a cross employing two different mendelian alleles (for example, the mating type alleles), four of the ascospores carry one allele, and the other four ascospores carry the second allele.

Each ascospore produces a mycelium containing haploid nuclei. This mycelium remains homokaryotic (i.e., containing a single genotypically distinct nuclear type) until it comes in contact with a mycelium carrying a different nuclear type, whereupon the two mycelia may fuse to form a *heterokaryon* in which two genotypically distinct nuclear types are present (Fig. 8–2). Obviously, the cytoplasms of the two *homoplasmons* fuse too; so a *heteroplasmon* is created at the same time. In the heterokaryon, nuclear identity is conserved, but cytoplasmic mixing occurs.

Heterokaryons are easily obtained in *N. crassa,* and the presence of cytoplasmic mixing in the absence of nuclear fusion makes heterokaryons extremely useful in determin-

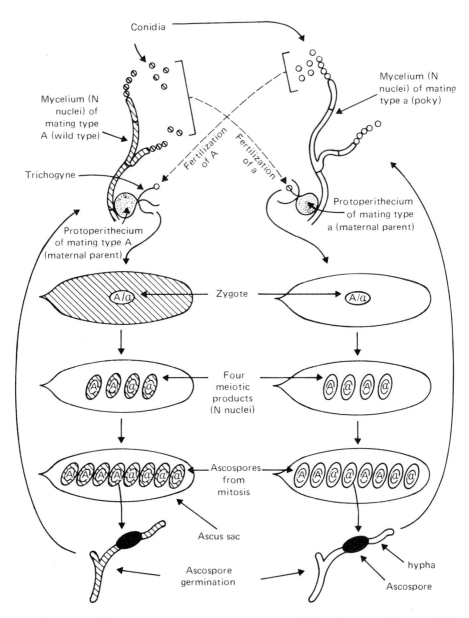

FIG. 8–1. Life cycle of *N. crassa*, also showing inheritance of *poky* mutant. Haploid mycelium reproduces asexually by conidia (macro and micro) that germinate to give more haploid mycelium, as well as simple proliferation of existent mycelium. Sexual cycle is illustrated. Two mating types (*A* and *a*) are required for sexual reproduction. Fertilization takes place by passage of nuclei of conidia or mycelia into protoperithecia of opposite mating type through trichogynes. Fusion of nuclei of opposite mating types takes place within protoperithecia, as does meiosis and ascospore formation. In order to illustrate maternal inheritance of *poky*, asci are drawn out of scale in relation to perithecia containing them. If protoperithecial parent is wild type (*shaded*), all eight ascospores have wild-type phenotype; but if it is *poky* (*clear*), all eight are *poky*. Nuclear mating type alleles segregate in normal mendelian fashion.

ing if a certain trait has a cytoplasmic basis (Fig. 8–2). A homokaryon carrying a suspected cytoplasmic mutant (which we will call *mi-x*) and a readily identified nuclear gene mutation (*a*) is allowed to form a heterokaryon with a second homokaryon that is wild type with respect to the suspected cytoplasmic mutant (i.e., *mi-x*⁺) and

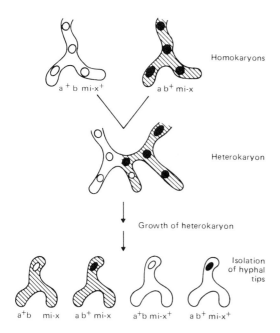

FIG. 8–2. Diagrammatic representation of heterokaryon test in *N. crassa* designed to determine if suspected cytoplasmic mutant, *mi-x*, is indeed transmitted via cytoplasm.

carries a mutation in a different nuclear gene (*b*). After a suitable amount of growth, hyphal tips or conidia are isolated and characterized as to the mutations they carry. If only the parental combinations are found (i.e., ab^+ *mi-x* and a^+b *mi-x*$^+$), then *mi-x* is probably nuclear in origin; but if the recombinant combinations ab^+ *mi-x*$^+$ and a^+b *mi-x* are also found, the *mi-x* mutation and its wild-type allele have recombined under conditions where there is no recombination of nuclear genes. That is, *mi-x* and its wild-type allele are transmitted through the cytoplasm.

GENETICS OF CYTOCHROME DEFICIENCY IN *Neurospora*

In the course of describing the *poky* mutation, Mitchell and Mitchell (1) demonstrated that in reciprocal crosses between *poky* and wild type, the progeny, with rare exceptions, have the phenotype of the protoperithecial parent (Fig. 8–1). In 1953, to emphasize this maternal pattern of inheritance, Mitchell and associates (2) redesignated *poky* as *mi-1* (for maternal inheritance) in an article that described the genetics and physiology of three new respiration-deficient mutants. Two of these, C115 and C117, were mendelian, whereas the third, *mi-3*, was maternally inherited. Their article compared the cytochrome spectra of all four mutants and showed that *mi-3* differed from *poky* in its ability to synthesize cytochrome b in addition to cytochrome c. The two mendelian mutants also differed from each other, with C115 lacking only cytochrome a+a$_3$, and C117 lacking these cytochromes as well as cytochrome c. However, each mutant was slightly leaky, as might be expected because of the obligate aerobic nature of *N. crassa*.

Since the discovery of *poky* and *mi-3*, various other nonmendelian slow-growing mutants have been described in *N. crassa*, most of which, if not all of which, have cytochrome defects. Bertrand and Pittenger (4), using a rather ingenious classification system, have placed these in three groups (Table 8–1). Group I mutants include *poky* and other mutants of similar phenotype, such as *SG1* and *SG3* (for slow growth) (5,6) and *exn-1* through *exn-4* (for extranuclear) (4). These mutants are all fertile as female parents, and their maternal inheritance has been demonstrated in reciprocal crosses. Of particular note, in this respect, are the experiments of Srb (5) with the *SG1* mutant, which he showed to be strictly maternally inherited over 20 generations of backcrossing, thus ruling out any possible effects of nuclear genes in determining maternal inheritance.

Group 1 mutants are all deficient in cytochromes a+a$_3$ and b, but they contain an excess of cytochrome c (Table 8–1). They are also uniformly suppressed with respect to growth rate by a nuclear suppressor mutation called *f*. This suppressor was described by Mitchell and Mitchell (7), who showed that although *f* causes a marked

TABLE 8–1. *Characteristics of extranuclear mutants affecting mitochondria in N. crassa*

Group	Mutant	Growth phenotype	Female fertility	Cytochromes			Suppression	
				$a + a_3$	b	c	by f	by su-1
I	poky	poky	+	deficient	deficient	excess	+	−
	SG-1	poky	+	deficient	deficient	excess	+	−
	SG-3	poky	+	deficient	deficient	excess	+	−
	stp-B1	poky	+	deficient	deficient	excess	+	−
	exn-1	poky	+	deficient	deficient	excess	+	−
	exn-2	poky	+	deficient	deficient	excess	+	−
	exn-3	poky	+	deficient	deficient	excess	+	−
	exn-4	poky	+	deficient	deficient	excess	+	−
II	mi-3	intermediate	+	deficient	normal	excess	−	+
III	stp	stopper	−	deficient	deficient	excess	−	−
	stp-A	stopper	−	deficient	deficient	excess	−	−
	stp-A18	stopper	−	deficient	deficient	excess	−	−
	stp-B2	stopper	−	deficient	deficient	excess	−	−
	stp-C	stopper	−	deficient	deficient	excess	−	−
	abn-1	stopper	−	deficient	deficient	excess	?	?
	abn-2	stopper	−	deficient	deficient	excess	?	?

Adapted from Bertrand and Pittenger (4).

increase in growth rate of *poky,* it does not restore the mutant cytochrome spectrum to normal.

The *mi-3* mutant is the sole representative of group II. It has a growth rate intermediate between those of *poky* and wild type (2); it is deficient in cytochrome $a + a_3$, but it has normal amounts of cytochrome b and an excess of cytochrome c. This mutation is not suppressed by *f,* but it is suppressed by a different nuclear gene mutation, *su-1,* which is specific for *mi-3* and suppresses no other cytoplasmic mutation in *Neurospora* (4,6). The *su-1* suppressor, in combination with *mi-3,* restores not only normal growth but also a wild-type cytochrome spectrum (8).

The group III mutants have a stop–start growth phenotype in which mycelial growth begins, then stops, then starts again (9,10). This phenotype is referred to as *stopper* by Pittenger and his colleagues (9,10). The group III mutants will yield viable progeny in a cross only when used as paternal parents, and in such crosses to wild type the progeny are wild type. Thus, female sterility prevents direct demonstration of maternal transmission of the group III mutants. Like the group I mutants, these mutants are deficient in cytochromes $a + a_3$ and b, but they contain excess cytochrome c. They are not suppressed by either *f* or *su-1.* They include the stopper (*stp*) mutants described by Pittenger (8–10), the abnormal (*abn*) mutants (about which more will be said later) (11), and probably *mi-4,* a mutant described by Pittenger (12) that has since been lost.

There are also several mutants that either have not been classified or do not fit the classification. Among this group is one mutant, *AC-7,* described by Srb (5), that is of particular interest. This mutant is not transmitted by either parent in reciprocal crosses to wild type. Furthermore, in crosses to *SG-1,* which shows very strict maternal inheritance, it is found that *SG-1* is transmitted by the paternal as well as the maternal parent. If it is assumed that *AC-7* and *SG-1* are mitochondrial mutants and that their phenotypes can be distinguished accurately among the progeny of reciprocal crosses, these results indicate that whereas mitochondria (or mtDNA) are normally maternally inherited in *Neurospora* they can be transmitted by the paternal parent under appropriate conditions. Because of the behavior of *AC-7,* it may be speculated that elimination of paternal mitochondrial genomes may be controlled, at least in part, by some property of the maternal mitochondrial genome. Thus the presence of the *AC-7* mutation in the maternal parent permits paternal transmission of *SG-1.* It is unfortunate that the crosses between *AC-7* and *SG-1* have not been described in detail and that a more complete description of *AC-7* and its interesting properties has not been published.

In conclusion, we should note that conidia germinate asexually to yield viable mycelia; so they obviously contain mitochondria. That is, maternal inheritance of presumptive mitochondrial mutants must be determined at some point following fertilization.

EVIDENCE THAT CYTOPLASMICALLY INHERITED MUTANTS IN *Neurospora* ARE LOCALIZED IN MITOCHONDRIA

Support for the idea that the cytoplasmically inherited cytochrome-deficient mutants described in the preceding section are localized in the mitochondria comes not only from observations that they affect specific mitochondrial components but also from two other pieces of experimental evidence: the observation of Reich and Luck (13) that mtDNA tends to be maternally inherited in *Neurospora* and the finding that the phenotypic characteristics of the *abn-1* mutation can be transferred to wild-type mycelia by injection of mitochondria from the mutant strains (14,15).

Reich and Luck (13) reported that mtDNA from wild-type and *poky* strains of *N. crassa* could be separated into components with buoyant densities of 1.698 and 1.702 g/cc, whereas in a second species, *N. sitophila,* a third component of mtDNA with a buoyant density of 1.692 g/cc was present in addition to the other two. Needless to say, the existence of mtDNA com-

ponents of varying buoyant densities is surprising in view of the apparent homogeneity of mtDNA in other organisms, and more recent reports mention only a single form, with the buoyant density being given as 1.698 g/cc (16) or 1.701 g/cc (17). Furthermore, *N. crassa* mtDNA is a single circular molecule with molecular weight of 40×10^6 d (18). These considerations suggest that Reich and Luck must somehow have succeeded in shearing *N. crassa* and *N. sitophila* mtDNA into two or three classes of fragments with distinct buoyant densities. Whatever the explanation, the authors were able to use the 1.692-g/cc fragment as a marker to investigate the inheritance of mtDNA in *Neurospora* in the same way that restriction enzymes would be used for this purpose many years later in other systems (see Chapters 10 and 11).

Reich and Luck used an *N. crassa poky* stock as the maternal (protoperithecial) parent in a cross with wild-type *N. sitophila* (1.692 g/cc) (cross 1). They analyzed the progeny of 19 asci for the transmission of the *poky* mutation and showed that all progeny were *poky,* as expected. None of the four asci analyzed had the 1.692-g/cc mtDNA derived from the *N. sitophila* parent. Although these results are consistent with the expected maternal transmission of *poky,* they show only that the 1.692-g/cc mtDNA is not transmitted by the paternal parent; they do not establish that it is transmitted by the maternal parent.

Since the reciprocal cross was infertile, Reich and Luck made a second cross using the same *N. sitophila* strain used in cross 1, but a new strain designated 6–486–2a (also *SG-2A*) was used in place of the *N. crassa* parent. This strain had been synthesized by Puhalla and Srb (19) by backcrossing the *SG-1* strain of *N. crassa* as the protoperithecial parent to *N. sitophila* over 10 generations. The resulting 6–486–2a strain, as a consequence, carried the *SG-1* mutation from *N. crassa* but had *N. sitophila* nucleoplasm. The 6–486–2a strain also lacked the

1.692-g/cc mtDNA characteristic of *N. si-tophilia,* suggesting that selection for the *SG-1* mutation from *N. crassa* was accompanied by selection for *N. crassa* mtDNA even in an *N. sitophila* nuclear background.

In the second cross (cross 2), *N. sito-phila* (1.692 g/cc) was the maternal parent and 6–486–2a was the paternal parent. The progeny of four asci were examined for transmission of the 1.692-g/cc mtDNA by the maternal parent; in one ascus all four products of meiosis produced clones containing the 1.692-g/cc mtDNA, but this was true in none of the other three asci.

Two F_1 isolates were selected from cross 2, and reciprocal crosses were made between them (crosses 3 and 4). One of the isolates contained 1.692-g/cc mtDNA, and the other did not. In cross 3 the isolate containing the 1.692-g/cc mtDNA was used as the maternal parent. In this cross only two of the five asci analyzed transmitted the 1.692-g/cc mtDNA. In the reciprocal cross (cross 4) the 1.692-g/cc mtDNA was not transmitted by the paternal parent to any of the four asci examined.

In aggregate, the results of the four crosses show that the 1.692-g/cc mtDNA characteristic of *N. sitophila* is not transmitted by the paternal parent but is transmitted by the maternal parent, but only to less than one-third of the progeny. The pattern of inheritance seen for this mtDNA is intermediate between those observed for the *poky* mutation (maternal transmission) and the *AC-7* mutation (no transmission by either parent). Furthermore, the transmission of this form of mtDNA in crosses need not be related to the transmission pattern of any of the cytoplasmically inherited mutants isolated in *N. crassa* and described in the previous section, for *N. crassa* lacks this form of mtDNA. In short, these elegant experiments on the transmission of mtDNA in crosses in *Neurospora* leave many questions unanswered and can only be taken as suggesting that maternal inheritance in *Neu-rospora* results from maternal inheritance of mtDNA.

Perhaps the strongest evidence that the cytoplasmically inherited mutants affecting mitochondrial phenotype in *N. crassa* are, indeed, mitochondrial comes from the microinjection experiments reported in two articles by Tatum and his colleagues (11, 14). The mutants used, *abn-1* and *abn-2,* were described in the first article by Garnjobst et al. (11). These mutants were deficient in cytochromes $a + a_3$ and b, but they contained excess cytochrome c. They had a typical "stopper" growth phenotype and were female sterile. They were placed in group III by Bertrand and Pittenger (Table 8–1). Garnjobst et al. (11) also reported that both mutants characteristically produced lethal segregants at a high rate. For example, of the 113 *abn-1* colonies and germinated conidia isolated, only a few became more or less normal, and only seven of the remainder survived an extensive number of transfers. Two of those seven then died, but five cultures were maintained successfully. The basis for the lethality of these mutants is not known, but three possible explanations suggest themselves. First, the mutations themselves are semilethal, and only under the most propitious of circumstances can they be maintained in culture. Second, those cultures that survive are heteroplasmic and contain a small fraction of normal mitochondria. Third, there is no selection against lethal mitochondrial mutations in these stocks, and such mutations tend to accumulate at a high rate, much as vegetative petite mutations accumulate in segregational petite stocks.

In spite of their lethality, it was possible to maintain *abn-1* and *abn-2* in culture, and Tatum and his colleagues showed that they were transmitted cytoplasmically in both natural heterokaryons and heterokaryons formed artificially by microinjection. In the first set of experiments, Garnjobst et al. (11) attempted transfer of the *abn-2* trait using microinjection methods devised by Wilson (20–22). The method involves removing cytoplasm and nuclei from donor cells with a pipette and then injecting them into a recipient hypha. The injection is accomplished by piercing a cell close to the septum, separating it from the adjacent cell, and then injecting the donor material through the septal pore into the adjacent cell. Within a few seconds the septal pore usually becomes plugged; thus the recipient cell retains the donated material. Following injection, a hyphal segment of two to three cells, including the injected cell, is cut out and cultured on appropriate medium.

Garnjobst and associates injected a normal strain carrying a nuclear mutation to tryptophan dependence (*tryp*) with cytoplasm and nuclei from the *abn-2* mutant. The *abn-2* character exhibited suppressiveness in these artificial heterokaryons and could be found associated with recipient *tryp* nuclei as well as donor nuclei in the heterokaryons. More important, however, was the finding that some cultures that became *abn-2* contained only recipient nuclei. These experiments suggested that only cytoplasmic transfer was required to transmit the *abn-2* mutation. The reciprocal transfer experiments were also attempted, but for technical reasons and because of the propensity of the *abn-2* mutant to die, they were much less successful.

Diacumakos et al. (14) then attempted to determine if the mitochondria were the elements responsible for transfer of the *abn* phenotype. In these experiments mitochondria from the *abn-1* mutant were purified and injected into recipient strains. After several transfers the injected cultures frequently exhibited the *abn* phenotype, both in terms of growth and also, to a large extent, in terms of cytochrome spectra, However, particularly in the early experiments, there were problems with the uninjected controls, which also often showed abnormal growth after several transfers. These problems were overcome in the later experiments; it appears that they were the result of a nuclear gene mutation. However, they do illustrate one of the principal problems in work with *Neurospora* mutants: the phenotypes of mutants such as *abn* or *poky* are complex and often difficult to score. Thus growth rate and cytochrome spectrum as phenotypes are much harder to work with than ability or inability to grow under specific conditions. This is why the petite mutants or mitochondrial mutations to antibiotic resistance are much easier to work with. In the case of the mitochondrial injection experiments, this point becomes particularly apparent when similar experiments with *Paramecium* using mitochondrial antibiotic resistance mutations are considered (see Chapter 10).

Experiments by Tatum and Luck (15) attempting to show directly that mtDNA is responsible for transformation of the wild type to the *abn-1* phenotype by injection were unsuccessful for technical reasons. Thus it is clear that injection of *abn-1* mitochondria into normal mycelia effects transformation of the recipient to the *abn* phenotype, but direct proof that mtDNA is the causative agent is lacking. In spite of this, it is difficult to imagine where else the mutation could be localized. It seems a safe assumption that *abn-1* and the other cytoplasmically inherited mutants affecting mitochondrial phenotype in *N. crassa* are probably mutations of the mitochondrial genome.

SUPPRESSION AND COMPLEMENTATION OF MITOCHONDRIAL MUTANTS IN HETEROKARYONS

When a heteroplasmon is created between a mitochondrial mutant and wild type through the use of appropriately marked homokaryons to force a heterokaryon, the phenotype of the heteroplasmon often becomes mutant. This phenomenon of *suppression* is found not only in *Neurospora* mitochondrial mutants but also in many other cytoplasmically inherited, possibly mitochondrial, mutants in other filamentous fungi (see Chapter 9).

In *Neurospora* the kinetics of suppression have been studied in a controlled manner by Luck and Wilson for the *abn-1* mutant (15). In these experiments normal hyphae were injected with *abn-1* mitochondria, and the resulting cultures were examined over a number of serial transfers for their ability to conidiate and form subcultures with normal growth morphology, as well as for their cytochrome spectra (Table 8–2). Luck and Wilson found that up until the 10th transfer the injected cultures remained reasonably normal; thereafter, the following were observed: conidial plating efficiency declined, a steadily increasing fraction of conidia produced abnormal cultures, and the cytochrome spectrum at each transfer shifted steadily from that of the wild type to that of *abn-1*. By the 16th transfer the cultures had become almost indistinguishable from *abn-1*. Tatum and Luck (15) pointed out that these results were consistent with "heterogeneity of determinants," presumably mitochondria or mtDNA, and gradual predominance of the mutant determinants. Thus, suppression in *Neurospora* and other

filamentous fungi presumably results from competition at the intracellular level between mutant and normal cytoplasmic determinants, which in the case of *Neurospora* are almost certainly mitochondria. In this respect, *Neurospora* more closely resembles *Paramecium* (see Chapter 10) than yeast. Suppressiveness of vegetative petites in yeast quite possibly is the result of recombination (see Chapter 6).

It is only to be expected that mutants such as the *Neurospora* mitochondrial mutants, which have no obvious selective value, will be suppressive, as Jinks (23) pointed out some years ago. When a mitochondrial mutation occurs, it will affect only one member of a population of mitochondria or mtDNA molecules. Therefore the mutation will normally pass unnoticed, unless the mutant determinant is able to become dominant to the wild type, either because of the nature of the mutation itself or because the mutant determinant can outcompete the wild type.

Complementation of mitochondrial mutants in heteroplasmons can be assessed by using nuclear markers to force the appropriate heterokaryons. Mitchell et al. (2), in their paper describing *mi-3*, briefly noted that *mi-3* did not appear to complement *poky* in terms of growth in heteroplasmons. Gowdridge (24) made a much more detailed study of the same pair of mutants

TABLE 8–2. *Transformation of normal cultures injected with mitochondria from abn-1 to the abnormal phenotype in serial transfer experiments*

Transfer number[a]	Conidia			Cytochromes (mμmoles/mg protein)		
	Plating efficiency (%)	Wild type	Intermediate, slow, or failed to grow	$a + a_3$	b	$c + c_1$
10	100	30	20	0.46	0.81	1.04
12	94	28	22	0.41	0.82	1.06
13	91	17	33	0.33	0.86	1.33
14	62	5	30	0.29	0.82	1.41
15	42	0	18	0.01	0.46	1.54
16	—	—	—	0.01	0.29	2.01

[a] Cytochrome contents of controls were as follows: wild type: cytochromes $a + a_3$, 0.45; cytochrome b, 0.81; cytochromes $c + c_1$, 0.96; *abn-1*: cytochromes $a + a_3$, 0.01; cytochrome b, 0.29; cytochrome c, 2.01.
Adapted from Tatum and Luck (15).

and reached the same conclusion. However, in addition, she examined heteroplasmons formed between wild type and *poky* and wild type and *mi-3*. She reported that the *poky*/wild-type combination was wild type in phenotype, but that the *mi-3*/wild-type heteroplasmon sometimes had the wild-type phenotype and sometimes had the phenotype of *mi-3*. Her results point out the complicating nature of suppression in any attempt to assess the presence or absence of complementation in heteroplasmons. If the heteroplasmon has a wild-type phenotype, the result is reasonably unequivocal; but where no complementation is seen, the problem of suppression by one component or the other of the heteroplasmon always exists.

In 1956, the same year that Gowridge reported the absence of complementation between *mi-3* and *poky*, Pittenger (12) showed that *poky* did complement a new mutant in heteroplasmons that he had isolated and designated *mi-4*. An interesting feature of this complementation (to which we will return in a moment) was that it affected only growth rate, not the cytochrome spectrum, which remained mutant. Pittenger also showed that the heteroplasmons between *poky* and wild type and between wild type and *mi-4* were wild type at first and then became mutant and that the same was true of the *mi-4*/*poky* heteroplasmon. He clearly recognized that suppression was probably responsible for the eventual expression of the mutant phenotype in every heteroplasmon.

Bertrand and Pittenger (25) reexamined the question of complementation between mitochondrial mutants in *N. crassa* in terms of their classification of these mutants into three groups (Table 8–1). Their results essentially confirmed and expanded the earlier findings. In terms of growth, all group I mutants, of which *poky* is one, complemented all group III mutants, of which *mi-4* is one (Table 8–3). The single group II mutant, *mi-3*, did not complement any

group I mutant, including *poky*, which confirmed the earlier findings of Mitchell et al. (2) and Gowdridge (24); but *mi-3* did complement the group III mutants (Table 8–3). Within groups I and III there was no complementation between mutants. Thus the classification of Bertrand and Pittenger seems to hold up reasonably well in terms of complementation.

Like the heteroplasmon between *poky* and *mi-4* studied by Pittenger, the complementing heteroplasmons examined by Bertrand and Pittenger exhibited complementation only in terms of growth rate, not in terms of normalization of the cytochrome spectrum. Bertrand and Pittenger pointed out that the nuclear suppressor of *poky*, *f*, also increased growth rate without suppressing the abnormal cytochrome spectrum of *poky*. An explanation for the latter situation, which may also apply to the complementing heteroplasmons, is suggested by the recent experiments of Slayman et al. (26), which will be discussed later in this chapter. These authors found a small but significant increase in respiration via the cytochrome chain in *poky f*, as compared to *poky* alone, that was not evident from the spectral measurements but was sufficient to account for the increased growth rate.

Bertrand and Pittenger introduced some much needed order into the classification of respiration-deficient mitochondrial mutants in *Neurospora* that came at a propitious time in view of the introduction of more powerful techniques for selection and identification of respiration-deficient mutations in this fungus (6,27,28). For example, tetrazolium dyes can be used to detect respiration-deficient mutants in the same way they are used in *S. cerevisiae* (see Chapter 6). On the other hand, the Pittenger and Bertrand classification also points up the phenotypic complexity of the *Neurospora* mitochondrial mutants. For purposes of genetic experiments, there are defined largely in terms of growth parameters, aberrant morphology, lethality, etc.

TABLE 8–3. *Positive heteroplasmon complementation in terms of growth rate between extranuclear mutants of N. crassa*

	Group I							Group II
	poky	stp-B1	SG-1	SG-3	exn-1	exn-3	exn-4	mi-3
Group III								
stp	+[a]	+	+	+	+	+	+	+
stp-A	+	+	+	+	+	+	+	+
stp-A18	+	+	+	+	+	+	+	+
stp-B2	+	+	+	+	+	+	+	+
stp-C	+	+	+	+	+	+	+	+

[a] Indicates positive heteroplasmon complementation.
Adapted from Bertrand and Pittenger (25).

Thus the mutants do not easily lend themselves to use as genetic markers. As a result, there are no accurate estimates of the frequency of transmission of paternal mitochondrial genomes in crosses, and it is not possible at present to map mitochondrial genes in *Neurospora* or even to know if they recombine in heteroplasmons. Thus far, mitochondrial mutations to antibiotic resistance have not been reported, despite the availability of a chloramphenicol-sensitive strain of *Neurospora* (29) that could be used for isolation of mutants resistant to chloramphenicol, the relative sensitivity of the fungus to erythromycin (30), and the description of presumptive mitochondrial mutations resistant to chloramphenicol and oligomycin in *Aspergillus,* a related fungus (see Chapter 9). Until such clearly defined mutants become available, it will be difficult to gain further insight into the mechanism of suppression and the significance of heteroplasmon complementation. The latter phenomenon, for example, might not result from complementation at all, but rather from recombination between genotypically different mtDNA molecules, followed by selection for mitochondria containing the recombinant molecules. This possibility could be investigated through the use of appropriate mitochondrial markers.

Neurospora COMPARED TO PETITE NEGATIVE YEASTS

It appears that *N. crassa* resembles petite negative yeasts (see Chapter 4) far more closely than petite positive yeasts, as exemplified by *S. cerevisiae,* in terms of its mitochondrial genetics and physiology. *N. crassa,* like the petite negative yeasts, is an obligate aerobe. Howell et al. (31) succeeded in isolating a facultatively anaerobic strain (An^+) of *N. crassa,* but they found that the mutant, like wild type, was not glucose-repressed, although anaerobic growth led to production of mitochondria with less cristae and lower cytochrome oxi-

→

FIG. 8–3. Electron micrograph of ethidium-bromide-treated culture of *N. crassa* cells. Samples were taken 48 hr after inoculation of conidia and 44 hr after addition of ethidium bromide (25 μM). Typical paracrystals (P) are visible in nucleus and much more frequently in cytoplasm. Here, cytoplasmic paracrystal is surrounded by glycogen deposits (G) and ribosomes. Intranuclear structure is smaller and appears to consist of only a few fibers. Mitochondrial profiles (M) have strikingly different appearance than that of wild-type organelles. Numbers of cristae per profile are small, and in some cases no cristae are seen. When present, cristae often show a fixation artifact not encountered in wild type, namely, dilatation of intracristal space. In structural form these mitochondria closely resemble those seen in *abn-1.* A number of osmiophilic inclusions (O) are visible; in many cases (e.g., lower left corner of micrograph) it is clear that they are within mitochondria. These large osmiophilic inclusions are seen in ethidium-bromide- and euflavine-treated cultures but not in *abn-1.* ✕45,000. (From Wood and Luck, ref. 32, with permission.)

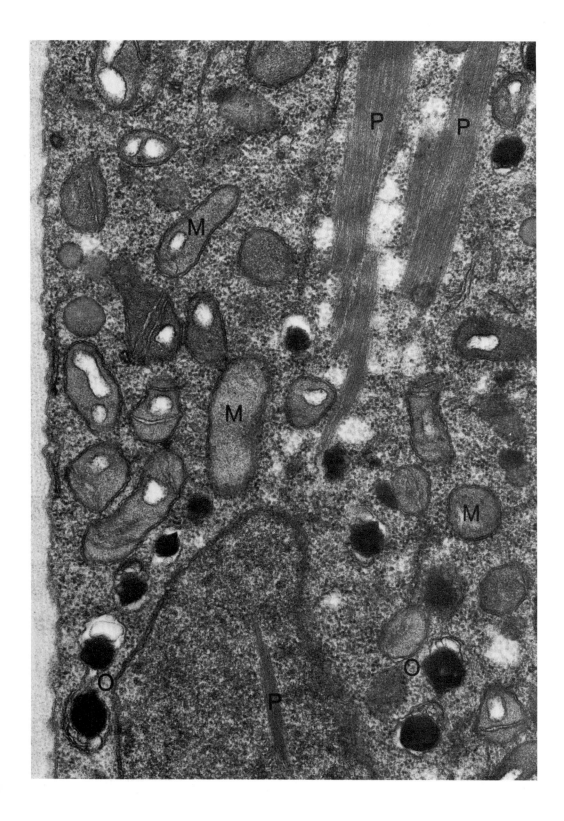

dase content than normal. Thus the mutant differed from *S. cerevisiae* and other petite positive yeasts which are subject to glucose repression, and in this respect it resembled certain of the petite negative yeasts. The extent to which the mitochondria of the *An⁺* mutant grown anaerobically resemble yeast promitochondria remains to be established.

It also appears that *N. crassa* resembles the petite negative yeasts in its responses to euflavine and ethidium bromide, as shown by the experiments of Wood and Luck (32). These dyes do not induce respiration-deficient mutants at appreciable frequencies, but treated mycelia do show large decreases in cytochromes a+a$_3$ and b, with an increase in cytochrome c and increased accumulation of paracrystals (*vide infra*). Mitochondrial ultrastructure also becomes markedly altered, with cristae being few in number and abnormal when present at all (Fig. 8–3). When the mycelia are shifted to dye-free medium, they become normal once again. Presumably, euflavine and ethidium bromide interfere with mitochondrial transcription and/or translation, as they appear to do in petite negative yeasts, without causing irreversible loss or mutation of mtDNA. It would be interesting to know if the dyes also suppress mtDNA synthesis, as they do in petite negative yeasts.

POKY DISEASE: STUDY IN BIOCHEMICAL PLEIOTROPY

Poky is the most thoroughly characterized of the *Neurospora* mitochondrial mutants; yet the primary defect in the mutant is still not established. Past investigations of this mutant can be grouped into several phases. With each phase, new facts were added about the amazing diversity of effects that the *poky* mutant has on *Neurospora* mitochondria.

Phase one began with the description of *poky* and its genetics by the Mitchells in 1952 and continued in the largely biochemical experiments carried out by the Mitchells

and their colleagues throughout the decade. These experiments, commenced with the report by Haskins et al. (33) who found that whereas young *poky* cultures were very deficient in cytochromes a + a$_3$ and b and contained cytochrome c in great excess, this was not true of older cultures. It was shown that cytochrome levels approached those seen in the wild type as cultures aged. Evidence was also reported for a "cytochromase" that destroyed cytochrome c in older *poky* cultures. This cytochromase appeared to be absent from the wild type, but wild-type cytochrome absorption bands disappeared in the presence of a *poky* extract. Later, Mitchell and Herzenberg (34) reported a procedure for purifying the cytochromase from *poky* and described how it could be used to degrade cytochrome c from *Neurospora* and beef heart. They reported indirect evidence that the enzyme also acts on cytochromes a and b.

Tissieres et al. (3) characterized respiration in wild type and *poky* and found that whereas respiration in wild type was largely inhibited by cyanide and azide, there was little inhibition of respiration by these reagents in *poky*. These results demonstrated that *poky* respiration was largely independent of the terminal oxidase system involving cytochrome a + a$_3$ and that this was not true of wild type. Tissieres and associates also observed that there were twice the amounts of flavin and niacin in *poky* as in wild type; they postulated that the alternate oxidase system in *poky* might be mediated by a flavoprotein such as flavin adenine dinucleotide.

A different aspect of the biochemical pleiotropy caused by *poky* was discussed by Hardesty and Mitchell (35), who reported that *poky* cultures, unlike wild type, accumulated large amounts of free fatty acids. In short, the experiments of the Mitchells and their colleagues not only established the general features of respiration deficiency in *poky* but also established other effects of the mutation not obviously related to respiration deficiency. The Mitchells' description (7) of the nuclear suppressor of *poky, f* (for fast *poky*), served to emphasize further the complexity of *poky,* since the suppressor seemed able to correct one aspect of the phenotype (growth rate) without affecting another aspect (cytochrome spectrum).

The second phase of investigations into the nature of the *poky* defect began with the dramatic announcement by Woodward and Munkres (36) that *poky,* and also *mi-3,* differed from wild type by a single amino acid in what they termed "mitochondrial structural protein (MSP)." This protein fraction had been described earlier by Criddle et al. (37) from beef heart mitochondria and was assumed to be homogenous. Woodward and Munkres claimed that amino acid analysis of MSP isolated from *poky* and *mi-3* revealed that the two mutants differed from wild type in manifesting an absence of a single tryptophan residue that was replaced in *poky,* but not in *mi-3,* by a cysteine residue. Woodward and Munkres also presented evidence that the mutant MSPs bound malate dehydrogenase, NADH, and ATP less tightly than did wild type. Therefore they conjectured that the lower affinity of mutant MSPs for these respiratory chain components could account for the pleiotropic effects of *poky* and *mi-3* on the respiratory chain. This appealing hypothesis was extended in a second article by Woodward and Munkres (38) in which they claimed that structural protein isolated from nuclei, microsomes, and the soluble fraction had the same amino acid composition as MSP and showed the same cysteine-for-tryptophan substitution in every case in *mi-1.* Woodward and Munkres pointed out that their results implied that one role of mtDNA might be to code for a protein important not only in organizing mitochondrial membranes but also in organizing other cellular membranes.

The experiments of Woodward and Munkres·were soon questioned, as was all work on MSP, for it became apparent that MSP was not a single protein but rather a group of proteins (39,40). Obviously, measurement of the amino acid composition of such a mixture would be meaningless in terms of the primary defect present in the mutant, since if proteins were present in different proportions in the MSP fractions obtained from *poky* and from wild type, there would be a difference in total amino acid composition. In fact, this appears to be the case, as Zollinger and Woodward (41) showed in an article that retracted the claim that *poky* MSP differed from wild type by a single amino acid substitution. The MSP fraction in *poky* is richer in cysteine-containing material than the comparable fraction in wild type. In other words, another pleiotropic effect of the *poky* mutation is to change the composition of a protein fraction that at first was thought to be homogeneous structural protein but now is known to be a heterogeneous mixture of denatured proteins (40).

Recent phases of the investigation into the nature of the *poky* disease have led to a much better understanding of the defects in electron transport in *poky* and to the finding that most of these defects can be accounted for by a faulty mitochondrial protein-synthesizing system in the mutant. Lambowitz and his colleagues published a series of articles that provide us with a detailed comparative picture of electron transport in *poky* and in wild type (42–46). Their work revealed that electron transport in *N. crassa* proceeds via a branched system similar to that found in higher plant mitochondria (47,48) in which one branch is represented by the classic cytochrome system and the other by a cyanide-insensitive alternate oxidase (Fig. 1–8). The cyanide-insensitive pathway is inhibited by hydroxamic acids such as salicylhydroxamic acid (SHAM) in both *N. crassa* and higher plant mitochondria, whereas these inhibitors do not block cytochrome-mediated respiration. In wild-type *N. crassa* the alternate oxidase is uninduced under normal conditions, and over 90% of respiration proceeds via the cytochrome-mediated branch; in *poky,* induction of the alternate oxidase occurs in response to the cytochrome deficiency, and most electron flux occurs through the cyanide-insensitive system.

Careful spectrophotometric and biochemical studies of the electron transport components present in *poky* and wild-type mitochondria by Lambowitz et al. (44) confirmed the results of earlier work showing the expected deficiencies in cytochromes $a+a_3$ and b and excess cytochrome c in *poky* (Table 8–4). These experiments also showed the following: the cytochromes $a+a_3$ α absorption peak was slightly shifted in *poky,* possibly because of altered interaction between heme and protein moieties (49,50); there were two forms of succinate-reducible cytochrome b in wild type and *poky* that were present at equal concentrations, although there were much smaller amounts in *poky* than in wild type; there were two forms of cytochrome c, with one component being in excess in *poky* and the other (probably corresponding to cytochrome c_1 of animal mitochondria) being deficient in *poky.* The authors also noted that flavoprotein was present in excess in *poky;* they suggested, as had Tissieres et al. (3) years earlier, that a flavoprotein might be the alternate oxidase for the cyanide-resistant respiration pathway, pointing out

that there was evidence for flavoprotein participation in cyanide-resistant respiration in higher plants (51).

In another article Lambowitz et al. (45) examined oxidative phosphorylation in wild-type and *poky* mitochondria. It will be recalled that there are three phosphorylation sites in the respiratory electron transport chain (see Fig. 1–7). Oxidative phosphorylation occurs at sites two and three regardless of whether succinate or NADH is used as the electron donor, since they involve parts of the electron transport chain after coenzyme Q, to which both substrates donate electrons. Site 1 phosphorylation, on the other hand, occurs prior to coenzyme Q in the NADH-mediated branch of respiratory electron transport.

Lambowitz and associates found that oxidative phosphorylation was less efficient at all three sites in *poky* mitochondria. The first phosphorylation site seemed to be nonfunctional in *poky* and quite inefficient even in wild type, since phosphorylation efficiencies in the presence of NADH and succinate were similar (Table 8–5). Furthermore, the phosphorylation efficiencies of *poky* mitochondria were reduced to zero by cyanide or antimycin A and increased in the presence of SHAM (Table 8–5), suggesting that all oxidative phosphorylation in the mutant was taking place through what remained of the cytochrome pathway.

Von Jagow et al. (52) published results

TABLE 8–4. *Concentrations of electron transport components in wild-type and poky mitochondria*

	Wild-type mitochondria		Poky mitochondria	
Components	Concentration (nmoles/mg protein)	Concentration relative to cytochromes $a + a_3$	Concentration (nmoles/mg protein)	Concentration relative to cytochromes $a + a_3$
Cytochromes $a + a_3$	0.38	1.0	0.05	1.0
Cytochrome b	0.38	1.0	0.16	3.2
Cytochrome c	1.6	4.2	2.16	43
Flavoprotein	1.1	2.9	1.6	32
$NAD^+ + NADH$	12	32	5.3	106
$NADP^+ + NADPH$	1.0	2.6	0.36	7.2

Adapted from Lambowitz et al. (43).

substantially in agreement with those of Lambowitz and his colleagues; however, there were two points of difference between the results of Lambowitz and those of von Jagow that deserve mention. First, von Jagow and associates observed that under certain conditions there was a small amount of oxidative phosphorylation associated with site 1 in the cyanide-resistant pathway in *poky*. Perhaps this observation can be attributed to differences in growth conditions used by Lambowitz and associates and von Jagow and associates. Second, von Jagow and associates, using dithionite reduction, reported that the b-type cytochromes in *poky* varied widely from a 1:1 stoichiometric ratio, with the ratio b_{561}: b_{566} being equal to six. They suggested that their results disproved the concept of fixed molar ratios of these cytochromes. In fact, it seems that von Jagow and associates simply observed another misleading aspect of the pleiotropic expression of this mutant. Lambowitz and Bonner (46) showed that there is a third cytochrome-b-like component in *Neurospora* mitochondria that is reduced only by dithionite and has an α absorption maximum of 558 nm, which is very close to the absorption maximum of b_{561}. This b-like component does not appear to be an essential component of the respiratory chain; it is found in excess in *poky* mitochondria. Lambowitz and Bonner suggested that the b-like component might represent a nascent or damaged form of cytochrome b.

Slayman et al. (26) reexamined the question of whether or not there is any ATP synthesis associated with the alternate oxidase in *Neurospora* by comparing wild type (in which cyanide-insensitive respiration is largely missing) with a series of cytochrome-deficient mutants (in which the cyanide-insensitive pathway accounts for a large proportion of total respiration). These experiments revealed that there was a small but significant amount of ATP synthesis (partly substrate level and partly oxidative phosphorylation at site 1) that occurred during respiration via the alternate oxidase. In *poky f* the alternate oxidase was about 20% as efficient as the residual cytochrome chain in this double mutant in producing ATP. Most interesting was the observation that *poky f* would grow slowly in the presence of antimycin A, where respiration through the cytochrome chain is completely blocked, which led to the important prediction that stringent as well as leaky respiration-deficient mutants should be obtainable in *Neurospora*. The results of Slayman and associates also indicated that the differences in growth rate seen among the various cytochrome-deficient mutants and between *poky*

TABLE 8–5. *Phosphorylation efficiencies in wild-type, poky f⁺, and chloramphenicol-induced wild-type mitochondria measured as either ADP:O or P:O ratios*[a]

	Phosphorylation efficiencies				
	Wild-type mitochondria	Poky mitochondria		Chloramphenicol-induced mitochondria	
Substrate		−SHAM	+SHAM	−SHAM	+SHAM
TMPD[b] + ascorbate	0.77	0.44	—	—	—
NADH	1.44	0.24	0.71	0.70	1.04
Succinate	1.31	0.64	0.75	1.00	1.10
Pyruvate + malate	1.59	0.41	0.75	0.78	1.03

[a] In the case of poky f⁺ and chloramphenicol-induced wild-type mitochondria, phosphorylation efficiencies have also been measured in the presence of SHAM, which specifically inhibits cyanide-resistant respiration.

[b] TMPD = N,N,N′,N′-tetramethyl-p-phenylenediamine dihydrochloride.

Adapted from Lambowitz et al. (45).

and *poky f* could be accounted for quantitatively in terms of the amount of respiration occurring through the residual cytochrome chain versus that taking place through the cyanide-insensitive pathway.

It is apparent from a variety of experimental data that the phenotypic effects of *poky* on the respiratory chain can be mimicked by treatment of the wild type with the mitochondrial protein synthesis inhibitor chloramphenicol. Growth of *N. crassa* in chloramphenicol results in a high level of cyanide-resistant respiration (42,53), and the mitochondria produced have excess cytochrome c and little detectable cytocromes a+a$_3$ (53,54). Oxidative phosphorylation of mitochondria from cells grown in chloramphenicol is similar to that of *poky* mitochondria (Table 8–5). Therefore, it was very satisfying when Rifkin and Luck (55), in an elegant study, showed that young cultures of *poky* contained very few small mitochondrial ribosomal subunits, although they accumulated normal amounts of large subunits (Fig. 8–4). Thus the number of mitochondrial ribosomes was greatly reduced in young *poky* cultures as compared to wild type. However, as the cultures aged, more small subunits and consequently more ribosomes were formed, and the mutant began to approach the wild type. This partial normalization of mitochondrial ribosomal phenotype was accompanied by increases in amounts of deficient respiratory chain components (Table 8–6). The mitochondrial ribosomal defects seen by Rifkin and Luck in *poky* were not found in *abn-1, mi-3,* or the nuclear mutant C117 (Table 8–6); so Rifkin and Luck made the significant deduction that the slow growth rate and abnormal cytochrome spectrum did not cause the abnormal mitochondrial ribosomal pattern in *poky*. Instead, the cytochrome defects observed in *poky* were most probably direct consequences of the ribosome deficiency and the resulting constraints on mitochondrial protein synthesis. Rifkin and Luck also found that the nuclear suppressor *f*

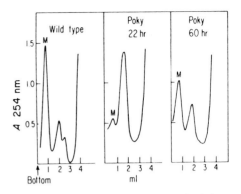

FIG. 8–4. Sedimentation profiles of mitochondrial ribosomes from mitochondrial lysates of *N. crassa*. Following centrifugation on sucrose gradients, ultraviolet absorbance (A 254 nm) of each gradient was analyzed with density-gradient fractionator. M indicates mitochondrial ribosomal monomer. Peak next to M is large mitochondrial ribosomal subunit, and peak next to it is small subunit. (From Rifkin and Luck, ref. 55 with permission.)

did not affect the small ribosomal subunit deficiency in *poky*. Neupert et al. (56) independently confirmed many of the results of Rifkin and Luck.

Although the experiments of Rifkin and Luck and Neupert et al. strongly indicated that the mitochondrial ribosomal deficiency in *poky* led to reduction or loss of mitochondrial protein synthesis, they did not prove this point directly. Evidence that mitochondrial protein synthesis was greatly reduced in *poky* mitochondria was first reported by Sebald et al. (57), who observed that amino acid incorporation by isolated mitochondria was only 10% of that in wild type. However, it was not clear from this observation if there was a real defect in mitochondrial protein synthesis in *poky* or if the mitochondria themselves responded differently than wild-type mitochondria *in vitro*. Kuriyama and Luck (58) demonstrated directly that *poky* mitochondrial ribosomes were actively engaged in protein synthesis *in vivo* and that protein synthesis was a function of the concentration of the small ribosomal subunit in the mutant. They found that in wild-type *N. crassa* mitochondria the large and small mitochondrial ribosomal subunits were present both in the

TABLE 8–6. *Mitochondrial cytochrome content and mitochondrial ratio of ribosomal monomers to large subunits in various mutants of N. crassa*

Strain	Cytochromes (nmoles/mg mitochondrial protein)			Ribosomes (monomer/ large subunit)
	a	b	$c + c_1$	
Wild type	0.40	0.84	1.07	$(4.1–7.4)^a$
poky (22 hr)	0.01	0.30	1.60	0.27
poky (60 hr)	0.03	0.92	1.38	1.25
mi-3 (20 hr)	0.03	0.78	1.89	4.98
abn-1	0.01	0.63	2.44	2.02
C117 (47 hr)	0.16	1.11	0.34	3.32
lys-5 (17 hr)	0.05	0.34	1.84	2.24

[a] Although in wild type this ratio is generally between 4 and 7, it may be as low as 2.7. From Rifkin and Luck (55).

mitochondrial supernatant and bound to the mitochondrial inner membrane. The bound ribosomes were mostly releasable by puromycin and high concentrations of salt (i.e., they were actively engaged in protein synthesis, see Chapter 3), and the ribosomal subunits were present in approximately equal amounts in the supernatant and membrane fractions. In *poky,* on the other hand, the membrane fraction contained equal numbers of large and small subunits, but large subunits were in fivefold excess in the supernatant fraction. That is, in the membrane fraction from *poky* the small subunits were actively engaged in protein synthesis with equivalent numbers of large subunits, with the excess large subunits being found in the mitochondrial supernatant.

The experiments of Luck and his colleagues showed convincingly that many of the pleiotropic effects of *poky* on the synthesis of respiratory chain components were attributable to a deficiency in mitochondrial protein synthesis, but they did not explain why *poky* could not make normal amounts of small mitochondrial ribosomal subunits. The two obvious possibilities were that the *poky* defect resulted because a mutant ribosomal protein was produced that interfered with normal small subunit assembly or that there was something wrong with the small rRNA or its processing that led to a paucity

of small subunits. Kuriyama and Luck (61) believed that the latter possibility was the more likely, since it seemed at the time, from the work of Lizardi and Luck (62) (see Chapter 21), that all proteins of the small mitochondrial ribosomal subunit were synthesized in the cytoplasm. Kuriyama and Luck assumed, excepting possible export of mRNA from the mitochondrion (see Chapter 3), that this observation probably meant that the ribosomal proteins of the small subunit were coded by nuclear genes and hence would be normal in *poky.*

Mitochondrial rRNA synthesis in *N. crassa* begins with the formation of a 32S rRNA precursor molecule that is cleaved to 25S and 19S precursors, which are themselves processed to form mature 25S and 19S rRNAs (63) (see Chapter 3). Kuriyama and Luck decided to compare rRNA methylation in wild type and in *poky,* pointing out that processing of the 45S precursor of eukaryotic cytoplasmic rRNA is accompanied by methylation and that in HeLa cells deprived of methionine ribosome production is abolished (64). Two other aspects of the *poky* disease also suggested that *poky* might be traced to some generalized defects in mitochondrial methylation. Brambl and Woodward (65) reported differences in amino acylated mitochondrial tRNAs between *poky* and wild type with respect to four amino acids. They involved either marked reductions in certain tRNA species in *poky* or additions of new components. Scott and Mitchell (66) showed that *N. crassa* had two kinds of cytochrome c. The C_I form had a trimethyl lysine residue in position 72, whereas the amino acid at that position in the form C_{II} was lysine. Pulse chase experiments showed that C_{II} was probably converted to C_I by specific methylation. C_{II} proved to be a minor component in wild type, but in young *poky* cultures C_{II} accumulated in the cytosol, as it was not converted through methylation to C_I, which appears to be the species bound to the mitochondrial inner membrane. Kuriyama and Luck argued that the results of Brambl and Woodward could be explained in terms of different degrees of tRNA methylation in *poky* as compared to wild type, and it was obvious that methylation was involved in conversion of the C_{II} form of cytochrome c to C_I.

Kuriyama and Luck found in pulse labeling experiments with uracil that, compared to wild type, *poky* mitochondrial RNA was very heterogeneous. Thus, whereas only 32S, 25S, and 19S rRNAs were seen in wild type, there was a highly heterogeneous distribution of high-molecular-weight RNAs in *poky,* with only the 32S and 25S rRNAs being clearly resolved in the mutant. These authors then did long-term labeling experiments with methionine to measure methylation of *poky* and wild-type rRNAs. They found that both the 25S and 19S rRNAs of *poky* were undermethylated and that this was true of both base methylation and methylation of the 2′-OH group of ribose. Kuriyama and Luck proposed that undermethylation of rRNA was responsible for the low rates of assembly of small subunits, accounting for the apparent degradation of most of the putative 19S rRNA.

In the course of the same experiments, Kuriyama and Luck also reported levels of methylation for *N. crassa* cytoplasmic rRNAs. The apparent proportions of base methylations in these RNAs (about 50% of

the total) were unusually high, which suggested that the ³H-methyl label in their methionine might be randomized directly into the purine carbon skeleton rather than being added on through methylation. To test this possibility, Lambowitz and Luck (67) added unlabeled formate to the growth medium, as had been done in other similar studies on rRNA methylation, and repeated the methionine labeling experiments. Under these conditions the methylation of both cytoplasmic and mitochondrial rRNAs declined markedly. The methylation levels found for mitochondrial rRNA were found to be extremely low, and no significant differences between *poky* and wild-type mitochondrial rRNA methylation could be detected.

These findings still did not rule out the possibility that the primary defect in *poky* involved mitochondrial rRNA, since it was conceivable that the base sequence in the small subunit rRNA might be altered in *poky* through deletion, insertion, misprocessing, etc., of certain nucleotides. To examine this possibility, Lambowitz and Luck (68) fingerprinted *poky* and wild-type mitochondrial rRNAs using three different complementary fingerprint systems. The RNAs were first labeled with ³²P and digested with appropriate enzymes (e.g., T_1 RNAse plus phosphatase), and then the oligonucleotides were separated in two dimensions using a combination of electrophoretic and chromatographic techniques. The oligonucleotide positions were then identified by autoradiography. No differences were found between *poky* and wild-type rRNAs in the positions of specific oligonucleotides or in their stoichiometries; in other experiments, levels and patterns of methylation were also compared without differences being noted. Although these results indicated that nucleotide additions and deletions in the 19S mitochondrial rRNA of *poky* were probably not the primary defects in the mutant, they did not rule out the possibility that single base changes or mutations in the unconserved regions of the 32S precursor molecule were the primary defects. However, by that time it was becoming increasingly clear from other work of Lambowitz and Luck (69) that a specific protein in the small mitochondrial ribosomal subunit, S4A, might be the culprit in *poky*. The arguments leading to this conclusion are not yet rigorous, only suggestive; they will be discussed late in this book (see Chapter 21), since an understanding of them requires some description of experiments relating to the sites of synthesis of mitochondrial ribosomal proteins and the assembly of the mitochondrial ribosomal precursor particle in *Neurospora*. To sum up, the primary defect in this tantalizingly complex mutant continues to elude detection, although it seems clearly to involve some aspect of small mitochondrial ribosomal subunit assembly.

IS *mi-3* A MITOCHONDRIAL POINT MUTANT?

Surprisingly little work seems to have been done with *mi-3*, the only class II mutant in the Bertrand and Pittenger classification (Table 8–1). Mitchell et al. (2), in their original description of *mi-3*, showed that the mutant lacked detectable amounts of cytochrome $a+a_3$, but contained cytochrome b and excess cytochrome c. Tissieres and Mitchell (70) characterized *mi-3* further; they stated that whereas cytochrome $a+a_3$ appeared to be missing, the α band of another cytochrome identified as a_1, which replaced cytochrome a in *Acetobacter,* was present. Despite the absence of the typical α bands of cytochromes $a+a_3$, Tissieres and Mitchell found that cytochrome oxidase and succinic oxidase could be assayed enzymatically in the mutant. The authors suggested that cytochrome a_1 was the probable oxidase in the system.

Von Jagow et al. (52), in an article discussed earlier in connection with *poky,* made a detailed reexamination of the cytochromes and other respiratory components in isolated mitochondria of *mi-3*. It is clear from their spectra that the component

identified as a_1 by Tissieres and Mitchell, with an α absorption peak at 590 nm, is absent from mitochondria and that cytochrome $a+a_3$ are present, but in reduced amounts as compared to wild type. It seems likely from these experiments that the cytochrome a_1 observed by Tissieres and Mitchell was nonmitochondrial, or else the α band of cytochromes $a+a_3$ was grossly shifted *in vivo,* for Tissieres and Mitchell made their spectroscopic measurements on intact mycelia. Thus it seems probable that the cytochrome oxidase activity reflected the presence of cytochromes $a+a_3$ rather than a novel new cytochrome. Von Jagow et al. (52) also found two types of b-type cytochromes in *mi-3,* probably corresponding to the electron transport components b_{561} and b_{566}.

Unraveling the primary defect in *mi-3* should prove an easier job than it has in *poky*. The *mi-3* phenotype is apparently not complicated by defects in mitochondrial protein synthesis as *poky* is. Thus the accumulation of b-type cytochromes is not inhibited as it is in wild-type cells treated with chloramphenicol or in *poky,* which has deficient amounts of small mitochondrial ribosomal subunits. It also seems likely that *mi-3* is a point mutation, since the nuclear gene suppressor of *mi-3,* designated *su-1,* isolated by Gillie (6), not only restores the growth rate to normal but also restores the cytochrome spectrum, as shown by Bertrand (8). In fact, the major defect in *mi-3* appears to be its inability to synthesize normal amounts of cytochrome oxidase, and it is not too unreasonable to suggest that *mi-3* is a point mutation in a gene that codes either for a cytochrome oxidase polypeptide or for a polypeptide involved in the processing of cytochrome oxidase or its binding to the mitochondrial inner membrane.

PARACRYSTALS AND VIRUS-LIKE PARTICLES IN *abn-1*

In describing the phenotype of the class III mutant (Table 8–1) *abn-1,* Garnjobst et al. (11) pointed out the accumulation of very long crystals in the hyphae that moved out of the hyphae along with the cytoplasm following hyphal puncture with a micro-needle. Wood and Luck (32) attempted to deduce the nature of these inclusions, which they called paracrystals. They observed that treatment of wild-type *Neurospora* mycelia over long periods with ethidium bromide or euflavine, which presumably inhibited mito-chondrial transcription and possibly transla-tion (see Chapter 2), led to accumulation of paracrystals similar to those found in *abn-1* (Fig. 8–3) and to production of a wild-type phenocopy of *abn-1* in terms of cytochrome spectra. Wood and Luck (32) chose to characterize the paracrystals accumulated by wild type treated with ethidium bromide rather than *abn-1* because wild type pro-vided a richer and more reproducible source of the paracrystals. They found that the paracrystals were composed of a single polypeptide (molecular weight 68,000 d) that when aggregated, formed oligomers arranged end to end to form the fibers of which the paracrystals were composed. The paracrystals were found only in the cyto-plasm and could not be isolated from the mitochondrial fraction and antibodies pre-pared against the wild-type paracrystal poly-peptide also reacted with *abn-1* polypeptide. To explain the fact that paracrystal protein accumulated in cells of *abn-1* and wild type treated with ethidium bromide, the authors suggested that the paracrystal protein was a nuclear gene product that crystallized in the cytoplasm in response to the lack of a mito-chondrial gene product. This mitochondrial gene product would not be made by *abn-1* or its wild-type phenocopy.

A possible identity for the paracrystal protein with mitochondrial RNA polymer-ase was reported by Barath and Küntzel (71). This protein was a single polypeptide of molecular weight 64,000 d that had strong affinity for mtDNA and, like the bac-terial enzyme, was resistant to α-amanitin but sensitive to rifampicin (72). Barath and

Küntzel found that growth of wild-type *N. crassa* in the presence of ethidium bromide resulted in a large increase in the rifampicin-sensitive RNA polymerase activity of *N. crassa* and that this increase in mitochon-drial RNA polymerase was accompanied by an increase in a polypeptide of the same molecular weight in the cytoplasm. Frac-tionation studies indicated that this poly-peptide and the mitochondrial RNA poly-merase were identical. Barath and Küntzel also isolated paracrystals from the ethidium-bromide-treated cultures by the method of Wood and Luck and showed that the para-crystal polypeptide comigrated as a single sharp band with the mitochondrial RNA polymerase during SDS gel electrophoresis. However, Barath and Küntzel did not estab-lish immunologic identity between the para-crystal protein and RNA polymerase.

On the basis of their results, Barath and Küntzel (71) proposed a control model that assumes that mtDNA codes for a repressor or repressors that bind to nuclear DNA cistrons coding for certain mitochondrial proteins such as the RNA polymerase. In the presence of inhibitors of mitochondrial transcription and translation, synthesis of these structural proteins would become de-repressed because repressor synthesis would be blocked. Hence the affected polypeptides, like mitochondrial RNA polymerase, would be overproduced. The model was expanded in a second article by Barath and Küntzel (73) that suggested similar derepression of a mitochondrial peptide chain elongation factor and methionyl tRNA transformylase. This interesting model has been extended and made more specific for the mitochon-drial leucyl tRNA synthetase of *N. crassa* by Gross and his colleagues (Chapter 21).

A very different aspect of the *abn-1* phenotype was first brought to light by Tuveson and Peterson (74), who reported the presence of structures they termed virus-like particles (VLP) in *abn-1* and *poky* that were absent from wild-type *N. crassa*. Simi-lar particles had been reported earlier in

Aspergillus and *Penicillium* (75,76). The VLP particles accumulated by *abn-1* were isolated and characterized by Küntzel et al. (77). They found that the particles contained single-stranded RNA with a sedimentation velocity of 33S; on denaturation, it separated into two RNA species with sedimentation velocities of 9S and 7S. The RNA was associated with two proteins (one a lipoprotein and the other a glycoprotein) to form a particle that was 10% RNA, 85% protein, and about 5% phospholipid. About 40% of the VLPs were separable from the mitochondria by differential centrifugation, but the rest were associated with the mitochondrial fraction such that, of the remaining 60%, some 40% could be released only by mitochondrial lysis.

More recently, Küntzel et al. (78) found that the 33S RNA of the VLPs hybridized with mtDNA and that this DNA segment was different from the cistrons coding for mitochondrial ribosomal RNA and was present in wild-type mtDNA even though VLPs were not detectable in wild type. It also appeared that this RNA had message-like activity *in vitro* for a polypeptide the authors believed to be part of the lipoprotein of the VLP. The production of this protein appeared to take place in the mitochondrion, since its synthesis was chloramphenicol-sensitive. On the basis of their results, Küntzel and associates concluded that VLPs are composed largely, if not entirely, of mitochondrial gene products that in *abn-1* are made in large amounts and are exported to the cytoplasm. The relationships of VLPs to suppressiveness and senescence in filamentous fungi are of great interest and merit further study. Perhaps they can infect and alter normal mitochondria to the abnormal phenotype in heteroplasmons.

CONCLUSIONS

Initially it seemed that *N. crassa* would prove the equal of *S. cerevisiae* as a model in which to study mitochondrial genetics.

This has proved not to be the case, principally because mitochondrial mutants with distinctive and easily scored phenotypes have not been isolated and because deletion mutants equivalent to vegetative petites are probably lethal in this obligate aerobic fungus. Consequently, we have no idea whether or not mitochondrial genes recombine in heteroplasmons of *N. crassa,* and thus no genetic mapping of the mitochondrial genome has been possible. In short, the genetic tools necessary for systematic probing of the mitochondrial genome of this fungus are lacking. On the other hand, *N. crassa* does lend itself to study of a number of rather interesting mitochondrial genetic problems, including the genetic basis of suppressiveness in filamentous fungi, fungal senescence (see Chapter 9), and the mechanism of heteroplasmon complementation. In addition, this fungus is particularly well suited to the study of biogenesis of mitochondrial ribosomes, primarily because of the work of Luck, Lambowitz, and their colleagues (see Chapter 21), and it is not unreasonable to suppose that all of the group I mutants of Bertrand and Pittenger (Table 8–1) are in some way blocked in mitochondrial protein synthesis and perhaps even in mitochondrial ribosome assembly, as is *poky*. Ultimately, however, our ability to make use of these mutants in a systematic fashion to dissect the genetics of mitochondrial ribosome biogenesis will depend on development of a satisfactory mitochondrial genetic system. This stricture, of course, does not apply to nuclear mutants affecting biogenesis of mitochondrial ribosomes in *N. crassa* (see Chapter 21).

REFERENCES

1. Mitchell, M. B., and Mitchell, H. K. (1952): A case of "maternal" inheritance in *Neurospora crassa. Proc. Natl. Acad. Sci. U.S.A.,* 38:442–449.
2. Mitchell, M. B., Mitchell, H. K., and Tissieres, A. (1953): Mendelian and non-Mendelian factors affecting the cytochrome system in

Neurospora crassa. Proc. Natl. Acad. Sci. U.S.A., 39:606–613.

3. Tissieres, A., Mitchell, H. K., and Haskins, F. A. (1953): Studies on the respiratory system of the *poky* strain of *Neurospora. J. Biol. Chem.,* 205:423–433.

4. Bertrand, H., and Pittenger, T. H. (1972): Isolation and classification of extranuclear mutants of *Neurospora crassa. Genetics,* 71:521–533.

5. Srb, A. M. (1963): Extrachromosomal factors in the genetic differentiation of *Neurospora. Symp. Soc. Exp. Biol.,* 17:175–187.

6. Gillie, O. J. (1970): Methods for the study of nuclear and cytoplasmic variation in respiratory activity of *Neurospora crassa,* and the discovery of three new genes. *J. Gen. Microbiol.,* 61:379–395.

7. Mitchell, M. B., and Mitchell, H. K. (1956): A nuclear gene suppressor of a cytoplasmically inherited character in *Neurospora crassa. J. Gen. Microbiol.,* 14:84–89.

8. Bertrand, H. (1971): A nuclear suppressor of the cytochrome defects of the [mi-3] extranuclear mutant of *Neurospora crassa. Can. J. Genet. Cytol.,* 13:626–627.

9. McDougall, K. J., and Pittenger, T. H. (1966): A cytoplasmic variant of *Neurospora crassa. Genetics,* 54:551–565.

10. Bertrand, H., and Pittenger, T. H. (1969): Cytoplasmic mutants selected from continuously growing cultures of *Neurospora crassa. Genetics,* 61:643–659.

11. Garnjobst, L., Wilson, J. F., and Tatum, E. L. (1965): Studies on a cytoplasmic character in *Neurospora crassa. J. Cell Biol.,* 26:413–425.

12. Pittenger, T. H. (1956): Synergism of two cytoplasmically inherited mutants of *Neurospora crassa. Proc. Natl. Acad. Sci. U.S.A.,* 42:747–752.

13. Reich, E., and Luck, D. J. L. (1966): Replication and inheritance of mitochondrial DNA. *Proc. Natl. Acad. Sci. U.S.A.,* 55:1600–1608.

14. Diacumakos, E. G., Garnjobst, L., and Tatum, E. L. (1965): A cytoplasmic character in *Neurospora crassa:* The role of nuclei and mitochondria. *J. Cell Biol.,* 26:427–443.

15. Tatum, E. L., and Luck, D. J. L. (1967): Nuclear and cytoplasmic control of morphology in *Neurospora. Symp. Soc. Dev. Biol.,* 26:32–42.

16. Schäfer, K. P., Bugge, G., Grandi, M., and Küntzel, H. (1971): Transcription of mitochondrial DNA *in vitro* from *Neurospora crassa. Eur. J. Biochem.,* 21:478–488.

17. Wood, D. D., and Luck, D. J. L. (1969): Hybridization of mitochondrial ribosomal RNA. *J. Mol. Biol.,* 41:211–224.

18. Bernard, U., and Küntzel, H. (1976): Physical mapping of mitochondrial DNA from *Neurospora crassa.* In: *The Genetic Function of Mitochondrial DNA,* edited by C. Saccone and A. M. Kroon, pp. 105–109. North Holland, Amsterdam.

19. Puhalla, J., and Srb, A. (1967): Heterokaryon studies of the cytoplasmic mutant SG in *Neurospora. Genet. Res.,* 10:185–194.

20. Wilson, J. F. (1961): Micrurgical techniques for *Neurospora. Am. J. Bot.,* 48:46–51.

21. Wilson, J. F. (1963): Transplantation of nuclei in *Neurospora crassa. Am. J. Bot.,* 50:780–786.

22. Wilson, J. F., Garnjobst, L., and Tatum, E. L. (1961): Heterokaryon incompatibility in *Neurospora crassa:* Microinjection studies. *Am. J. Bot.,* 48:299–305.

23. Jinks, J. L. (1964): *Extrachromosomal Inheritance.* Prentice-Hall, Englewood Cliffs, N.J.

24. Gowdridge, B. M. (1956): Heterokaryons between strains of *Neurospora crassa* with different cytoplasms. *Genetics,* 41:780–789.

25. Bertrand, H., and Pittenger, T. H. (1972): Complementation among cytoplasmic mutants of *Neurospora crassa. Mol. Gen. Genet.,* 117:82–90.

26. Slayman, C. W., Rees, D. C., Orchard, P. P., and Slayman, C. L. (1975): Generation of adenosine triphosphate in cytochrome-deficient mutants of *Neurospora. J. Biol. Chem.,* 250:396–408.

27. Edwards, D. L., Kwiecinski, F., and Horstmann, J. (1973): Selection of respiratory mutants of *Neurospora crassa. J. Bacteriol.,* 114:164–168.

28. Edwards, D. L., and Kwiecinski, F. (1973): Altered mitochondrial respiration in a chromosomal mutant of *Neurospora crassa. J. Bacteriol.,* 116:610–618.

29. Chalmers, J. (1977): Unpublished data.

30. Luck, D. J. L. (1977): Personal communication.

31. Howell, N., Zuiches, C., and Munkres, K. D. (1971): Mitochondrial biogenesis in *Neurospora crassa.* I. An ultrastructural and biochemical investigation of the effects of anaerobiosis and chloramphenicol inhibition. *J. Cell Biol.,* 50:721–736.

32. Wood, D. D., and Luck, D. J. L. (1971): A paracrystalline inclusion in *Neurospora crassa:* Induction by ethidium and acridine, isolation, and characterization. *J. Cell Biol.,* 51:249–264.

33. Haskins, F. A., Tissieres, A., Mitchell, H. K., and Mitchell, M. B. (1953): Cytochromes and the succinic acid oxidase system of *poky* strains of *Neurospora. J. Biol. Chem.,* 200:819–826.

34. Mitchell, H. K., and Herzenberg, L. A. (1955): Enzymatic degradation of cytochrome c. In: *Methods in Enzymology, Vol. 2,* edited by S. P. Colowick and N. O. Kaplan, pp. 167–169. Academic Press, New York.

35. Hardesty, B. A., and Mitchell, H. K. (1963): The accumulation of free fatty acids in *poky,* a maternally inherited mutant of *Neurospora crassa. Arch. Biochem. Biophys.,* 100:330–334.

36. Woodward, D. O., and Munkres, K. D. (1966): Alterations of a maternally inherited mitochondrial structural protein in respiratory-deficient strains of Neurospora. Proc. Natl. Acad. Sci. U.S.A., 55:872–880.

37. Criddle, R. S., Bock, R. M., Green, D. E., and Tisdale, H. D. (1962): Physical characteristics of proteins of the electron transfer system and interpretation of the structure of the mitochondrion. Biochemistry, 1:872–842.

38. Woodward, D. O., and Munkres, K. D. (1967): Genetic control, function, and assembly of a structural protein. In: Organizational Biosynthesis, edited by H. J. Vogel, J. O. Lampen, and V. Bryson, pp. 489–502. Academic Press, New York.

39. Ashwell, M., and Work, T. S. (1970): The Biogenesis of Mitochondria. Annu. Rev. Biochem., 39:251–290.

40. Schatz, G., and Mason, T. L. (1974): The biosynthesis of mitochondrial proteins. Annu. Rev. Biochem., 43:51–87.

41. Zollinger, W. D., and Woodward, D. D. (1972): Comparison of cysteine and tryptophan content of insoluble proteins derived from wild type and mi-1 strains of Neurospora crassa. J. Bacteriol., 109:1001–1013.

42. Lambowitz, A. M., and Slayman, C. W. (1971): Cyanide-resistant respiration in Neurospora crassa. J. Bacteriol., 108:1087–1096.

43. Lambowitz, A. M., Slayman, C. W., Slayman, C. L., and Bonner, W. D. (1972): The electron transport components of wild type and poky strains of Neurospora crassa. J. Biol. Chem., 247:1536–1545.

44. Lambowitz, A. M., Smith, E. W., and Slayman, C. W. (1972): Electron transport in Neurospora mitochondria: Studies on wild type and poky. J. Biol. Chem., 247:4850–4858.

45. Lambowitz, A. M., Smith, E. W., and Slayman, C. W. (1972): Oxidative phosphorylation in Neurospora mitochondria: Studies on wild type, poky, and chloramphenicol-induced wild type. J. Biol. Chem., 247:4859–4865.

46. Lambowitz, A. M., and Bonner, W. D., Jr. (1974): The mitochondrial b-cytochromes of wild type and poky strains of Neurospora crassa: Evidence for a component reduced only by dithionite. J. Biol. Chem., 249:2886–2890.

47. Ikuma, H. (1972): Electron transport in plant respiration. Annu. Rev. Plant Physiol. 23:419–436.

48. Henry, M.-F., and Nyns, E. J. (1975): Cyanide-insensitive respiration. An alternative mitochondrial pathway. Subcell. Biochem., 4:1–65.

49. Edwards, D. L., and Woodward, D. O. (1969): An altered cytochrome oxidase in a respiratory deficient mutant of Neurospora. FEBS Lett., 4:193–196.

50. Woodward, D. O., Edwards, D. L., and

Flavell, R. B. (1970): Nucleocytoplasmic interactions in the control of mitochondrial structure and function in Neurospora. Symp. Soc. Exp. Biol., 24:55–59.

51. Erecinska, M., and Storey, B. T. (1970): The respiratory chain of plant mitochondria. VII. Kinetics of flavoprotein oxidation in skunk cabbage mitochondria. Plant Physiol., 46:618–624.

52. von Jagow, G., Weiss, H., and Klingenberg, M. (1973): Comparison of the respiratory chain of Neurospora crassa wild type and the mi-mutants mi-1 and mi-3. Eur. J. Biochem., 33:140–157.

53. Edwards, D. L., Rosenberg, E., and Maroney, P. A. (1974): Induction of cyanide-insensitive respiration in Neurospora crassa. J. Biol. Chem., 249:3551–3556.

54. von Jagow, G., and Klingenberg, M. (1972): Close correlation between antimycin titer and cytochrome b_t content in mitochondria of chloramphenicol-treated Neurospora crassa. FEBS Lett., 24:278–282.

55. Rifkin, M. R., and Luck, D. J. L. (1971): Defective production of mitochondrial ribosomes in the poky mutant of Neurospora crassa. Proc. Natl. Acad. Sci. U.S.A., 68:287–290.

56. Neupert, W., Massinger, P., and Pfaller, A. (1971): Amino acid incorporation into mitochondrial ribosomes of Neurospora crassa wild type and MI-1. In: Anatomy and Biogenesis of Mitochondria and Chloroplasts, edited by N. K. Boardman, A. W. Linnane, and R. M. Smillie, pp. 328–338. North Holland, Amsterdam.

57. Sebald, W., Bücher, T., Olbrich, B., and Kaudewitz, F. (1968): Electrophoretic pattern of and amino acid incorporation in vitro into the insoluble mitochondrial protein of Neurospora crassa wild type and mi-1 mutant. F.E.B.S. Lett., 1:235–240.

58. Kuriyama, Y., and Luck, D. J. L. (1973): Membrane-associated ribosomes in mitochondria of Neurospora crassa. J. Cell Biol., 59:776–784.

59. Chua, N., Blobel, G., Siekevitz, P., and Palade, G. E. (1973): Attachment of chloroplast polysomes to thylakoid membranes in Chlamydomonas reinhardi. Proc. Natl. Acad. Sci. U.S.A., 70:1554–1558.

60. Adelman, M. R., Sabatini, D. D., and Blobel, G. (1973): Ribosome-membrane interaction: Nondestructive disassembly of rat liver rough microsomes into ribosomal and membranous components. J. Cell Biol., 56:206–229.

61. Kuriyama, Y., and Luck, D. J. L. (1974): Methylation and processing of mitochondrial ribosomal RNAs in poky and wild-type Neurospora crassa. J. Mol. Biol., 83:253–266.

62. Lizardi, P. M., and Luck, D. J. L. (1972): The intracellular site of synthesis of mitochondrial ribosomal proteins in Neurospora crassa. J. Cell Biol., 54:56–74.

63. Kuriyama, Y., and Luck, D. J. L. (1973): Ribosomal RNA synthesis in mitochondria of *Neurospora crassa. J. Mol. Biol.,* 73:425–437.

64. Vaughan, M. H., Jr., Soeiro, J. R., and Darnell, J. E., Jr. (1967): The effects of methionine deprivation on ribosome synthesis in HeLa cells. *Proc. Natl. Acad. Sci. U.S.A.,* 58:1527–1534.

65. Brambl, R. M., and Woodward, D. O. (1972): Altered species of mitochondrial transfer RNA associated with the *mi-1* cytoplasmic mutation in *Neurospora crassa. Nature [New Biol.],* 238:198–200.

66. Scott, W. A., and Mitchell, H. K. (1969): Secondary modification of cytochrome c by *Neurospora crassa. Biochemistry,* 8:4282–4289.

67. Lambowitz, A., and Luck, D. J. L. (1975): Methylation of mitochondrial RNA species in the wild-type and *poky* strains of *Neurospora crassa. J. Mol. Biol.,* 96:207–214.

68. Lambowitz, A. M., and Luck, D. J. L. (1976): Studies on the *poky* mutant of *Neurospora crassa.* Fingerprint analysis of mitochondrial ribosomal RNA. *J. Biol. Chem.,* 251:3081–3095.

69. Lambowitz, A. M., Chua, N.-H., and Luck, D. J. L. (1976): Mitochondrial ribosome assembly in *Neurospora:* Preparation of *mit* ribosomal precursor particles, site of synthesis of *mit* ribosomal proteins and studies on the *poky* mutant. *J. Mol. Biol.,* 107:223–253.

70. Tissieres, A., and Mitchell, H. K. (1954): Cytochromes and respiratory activities in some slow growing strains of *Neurospora. J. Biol. Chem.,* 208:241–249.

71. Barath, Z., and Küntzel, H. (1972): Induction of mitochondrial RNA polymerase in *Neurospora crassa. Nature [New Biol.],* 240:195–197.

72. Küntzel, H., and Schäfer, K. P. (1971): Mitochondrial RNA polymerase from *Neurospora crassa. Nature [New Biol.],* 231:265–269.

73. Barath, Z., and Küntzel, H. (1972): Cooperation of mitochondrial and nuclear genes specifying the mitochondrial genetic apparatus in *Neurospora crassa. Proc. Natl. Acad. Sci. U.S.A.,* 69:1371–1374.

74. Tuveson, R. W., and Peterson, J. F. (1971): Virus-like particles in certain slow-growing strains of *Neurospora crassa. Virology,* 47:527–531.

75. Banks, G. T., Buck, K. W., Chain, E. B., Himmelweit, F., Marks, J. E., Tyler, J. M., Hollings, M., and Last, F. F. (1970): Antiviral activity of double stranded RNA from a virus isolated from *Aspergillus foetidus. Nature* 227:505–507.

76. Wood, H. A., Bozarth, R. F., and Mislivec, P. B. (1971): Virus like particles associated with an isolate of *Penicillium brevi-compactum. Virology,* 44:592–598.

77. Küntzel, H., Barath, Z., Ali, I., Kind, J., and Althaus, H.-H. (1973): Virus-like particles in an extranuclear mutant of *Neurospora crassa. Proc. Natl. Acad. Sci. U.S.A.,* 70:1574–1578.

78. Küntzel, H., Ali, I., and Blossey, C. (1974): Expression of the mitochondrial genome in wild type and in an extranuclear mutant of *Neurospora crassa.* In: *The Biogenesis of Mitochondria,* edited by A. M. Kroon and C. Saccone, pp. 70–78. Academic Press, New York.

Chapter 9
Cytoplasmic Inheritance and Mitochondria in Other Fungi

Ephrussi's discovery of cytoplasmically inherited petite mutations in yeast in 1949 marked the beginning of a period during which a whole host of genetic phenomena attributable to cytoplasmic inheritance were uncovered in fungi. During the next 15 years, the cytoplasmic respiratory mutants of yeast and *Neurospora* were characterized genetically and to some extent biochemically, and Jinks in England and Rizet in France and their students, working with *Aspergillus* and *Podospora,* respectively, discovered a variety of new cytoplasmically inherited traits. Unfortunately, the elegant genetic analyses carried out by these workers were largely unaccompanied by biochemical observations; thus, for the most part, we remain ignorant to this day of the nature of the cytoplasmic genetic elements involved. However, it seems likely that some of the traits studied (e.g., those involved in senescence) may be mitochondrial, and these genetic experiments with *Aspergillus* and *Podospora* will be described here partly in order to contrast their findings with those in yeast and *Neurospora* and partly because interest in the phenomena involved has arisen recently again and they are being investigated both in *Aspergillus* and *Podospora* with the aid of probable mitochondrial gene mutations.

We shall begin this chapter with a description of the nucleocytoplasmic incompatibility interaction known as the barrage phenomenon in *Podospora* following which we shall turn to discussions of the cytoplasmic variants of *Aspergillus* and the phenomenon of senescence. We also consider how the experiments of Rowlands and Turner with *Aspergillus* and Belcour with *Podospora* employing well defined mutants whose origins are probably mitochondrial

are leading to a new definition of these old problems in fungal cytoplasmic inheritance.

BARRAGE PHENOMENON IN *Podospora anserina*

If two fungal strains of the same species are placed on agar medium, the growing mycelia usually intermingle in the zone of contact to form innumerable hyphal fusions, so that the border between the two mycelia eventually becomes unrecognizable. However, a number of instances are known in both Ascomycetes and Basidiomycetes in which the two mycelia exhibit antagonism in the contact zone (1). The two types of hyphae form abnormal fusions that are often lethal, and hyphal tips branch profusely. A clear line of separation in the zone of contact becomes evident as the age of the culture increases; in the case of *P. anserina,* this is characterized by absence of the melanic pigment found in the other parts of the mycelium. This was named the barrage phenomenon by Vandendries (2).

In most fungi, formation of a barrage is determined by mendelian genes, and the same is true of *P. anserina* (1), with one notable exception. In 1952 Rizet (3) showed that barrage formation in strains *S* and *s* of *P. anserina* was governed not only by a pair of nuclear alleles but also by cytoplasmic factors. However, before discussing the barrage phenomenon in these strains, it will be useful to digress briefly on the life cycle of *P. anserina* and the genetics of mating and incompatibility.

P. anserina is an Ascomycete closely related to *Neurospora.* Wild-type mycelia of *P. anserina* contain both female elements (ascogonia) and male elements (spermatia) for sexual reproduction. The spermatia, which are also referred to as microconidia, are microscopic uninucleate cells containing very little cytoplasm. Unlike the microconidia of *Neurospora,* the spermatia of *P. anserina* are able to germinate only to a very limited extent for asexual reproduction (4). Fertilization is controlled by a pair of mating type alleles (called + and −); fertilization occurs only if a spermatium of one mating type becomes attached to an ascogonial trichogyne of opposite mating type.

Meiosis consists of the usual two nuclear divisions and is followed by a postmeiotic mitotic division, as in *N. crassa.* Although some of the ascospores are uninucleate, most of them are binucleate, with each containing two nonsister nuclei. The existence of binucleate ascospores, of course, complicates genetic analysis, since the result is that many ascospores are heterokaryotic for the mendelian gene markers they carry. Nevertheless, once the problem is recognized, it can be dealt with effectively in analyzing the results of crosses. Since the complications introduced by binucleate ascospores are only marginally germane to a general understanding of the barrage phenomenon, we will ignore them here and treat the results as if all ascospores were, in fact, uninucleate.

P. anserina has been isolated from a number of different geographic locations, mostly in France and Germany (1), and the descendants of spores or mycelia isolated from a particular location have been used to define geographic races of the fungus. Although + and − mycelia within any one race will intermingle and form perithecia, a barrage is frequently formed between strains belonging to different races, whether they are of the same or opposite mating type. Esser (1,5) reviewed the genetics of barrage formation *P. anserina* in some detail. In most races barrage formation is strictly under the control of a series of nuclear genes, and it results in vegetative incompatibility that is accompanied by sexual incompatibility in one or both members of a pair of reciprocal crosses. In contrast, bar-

rage formation between races S and s leads only to vegetative incompatibility; it is under the control of cytoplasmic factors as well as a pair of nuclear alleles, which will now be described.

Rizet (3) found that both S and s strains were perfectly stable in vegetative culture; but regardless of the direction in which the cross was made, the s character underwent a profound change. Basically, the results showed that in reciprocal crosses of S by s there was $2:2$ segregation for the ability of the progeny to form a barrage (Fig. 9–1). Two of the spores behaved as if they were $S,$ but the other two produced mycelia that did not form barrages with either S or s testers. Thus it appeared that the progeny, which should have been $s,$ had been modified to a new form, which was called s^s.

A simple interpretation of these results would be that under the influence of S there is a directional mutation of the s allele to the new form s^s. However, other experiments by Rizet (3) showed that this was not the case. Instead, the change from s to s^s appeared to involve an alteration in a cytoplasmic factor. When Rizet made reciprocal crosses between s and s^s stocks, he found that the progeny always resembled the maternal parent (Fig. 9–1). That is, they were all either s or s^s. Since the nuclear genes contributed in both crosses were the same, the two crosses should have given identical results if the s and s^s phenotypes were solely determined by nuclear genes. However, they did not. The results are what would be expected if the maternal parent contributed a cytoplasmic factor, either s or s^s, to the progeny. Possibly this maternal inheritance is a result of the spermatia not contributing the cytoplasmic factor because they contain very little cytoplasm.

Rizet showed that the change from s to s^s does not involve a loss of $s,$ since s^s reverts to $s.$ Whenever this happens, the reversion spreads rapidly through the mycelium. The spread can actually be measured by taking fragments of the originally s^s mycelium and

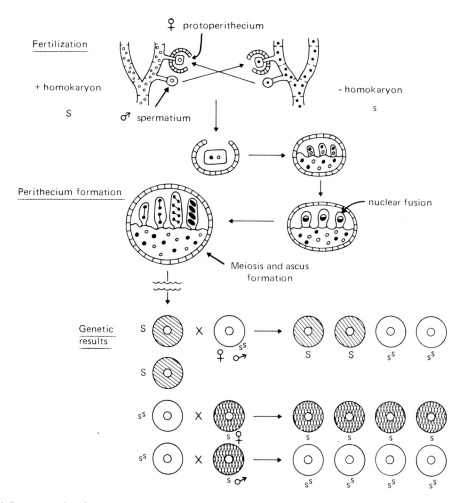

FIG. 9–1. Top two-thirds of figure depict sexual cycle of *P. anserina,* starting with two homokaryotic strains of opposite mating type (+ and −) and culminating with meiosis and ascus formation. Most ascospores are binucleate. Lower one-third of figure diagrams results of crosses of S by s and crosses of the resulting progeny to s^s and s testers. In this part of the diagram, the complications introduced into the actual results because of binucleate ascospore formation are ignored. (Adapted from Ephrussi, ref. 57, and Esser and Kuenen, ref. 1.)

testing them for barrage formation with *S.* This type of experiment shows that the reversion progresses at a speed of 1 cm/day. Rizet also demonstrated that the rate of reversion is constant (10^7/cell) and independent of time; so the mutation from s^s to *s* behaves as a true reversion of a point mutation. Obviously, the reverse change would be very difficult to demonstrate in mycelia because the *s* factor has a tremendous selective advantage over s^s. However, Beisson-Schecroun (4) succeeded in demonstrating the reverse change by plating sper-

matia, which presumably contain few cytoplasmic factors. Only a few of these germinated, but among those that did, she succeeded in showing reversion from *s* to s^s. Presumably these were spermatia containing only a reverted element. In short, the mutation is reversible in both directions, but in mycelia it is normally seen only from s^s to *s* because of the tremendous selective advantage of *s* over s^s. An interesting problem raised by these experiments is whether or not differential cytoplasmic contribution is really the mechanism that results in the maternal transmission of *s* and *s*ˢ. Certainly those spermatia that germinate contain one or the other element. We would like to know if the reason for nongermination of the large majority of spermatia is related to the presence or absence of the cytoplasmic factor or if it results for some entirely different reason.

Experiments by Rizet (3) and Beisson-Schecroun (4) also showed that fusion between *s* and *s*ˢ hyphae leads to spread of the *s* factor into the *s*ˢ hyphae. Under these conditions nuclear migration does not occur, as is shown by the use of marker genes, but cytoplasmic exchange can be clearly seen in the microscope. If the cytoplasmic bridge is broken after a few hours, a stable *s* mycelium grows out of the *s*ˢ hyphae. Filaments can be cut out from the *s*ˢ mycelium following such a fusion experiment and tested for reversion to *s* and appropriate nuclear markers. When this is done, it is seen that the hyphal fragments rapidly become converted to *s* (Table 9–1). The conversion is temperature-dependent and occurs twice as rapidly at 24°C as it does at 18°C.

Finally, there is the question of whether or not strain *S* itself makes a cytoplasmic factor that interacts with *s*. Beisson-Schecroun investigated this question in a series of elegant hyphal micromanipulation experiments. She first sought to determine if a cytoplasmic factor from *S* could be transmitted through an *s*ˢ hypha to cause an incompatibility reaction with an *s* hypha attached to the *s*ˢ hypha at a different point

(Fig. 9–2). At various times following fusion of the *S* and *s*ˢ hyphae, *s* hyphae were allowed to fuse with the *s*ˢ hypha, and the intersection point was scored for the incompatibility reaction. It was found that a cytoplasmic factor was, indeed, transmitted from the *S* strain via the *s*ˢ hypha to cause an incompatibility reaction with *s*. However, these experiments also showed that production of the *S* factor depended on the presence of an *S* nucleus, even though migration of *S* nuclei into the *s*ˢ hypha had not occurred. In order to examine the interaction between the *S* nuclear gene and *s* cytoplasmic factor, Beisson-Schecroun allowed *S* and *s*ˢ hyphae to form a connection and then cut out the resulting **H**-shaped structure. The tips of the **H** then began to grow, and an *s* hypha was allowed to fuse to the *s*ˢ side. As mentioned previously, *s* rapidly replaces *s*ˢ in heteroplasmons; but if the nuclear gene *S* prevents propagation of the *s* factor, these heteroplasmons should retain the *s*ˢ phenotype, and after separating off the connection to the *S* mycelium, *s* sectors should be regenerated. That is, in the *s*ˢ mycelium both *s* and *s*ˢ factors should be present, but the ability of *s* to predominate should be inhibited as long as the inhibitory gene product of the *S* nucleus is present. This was found to be the case. In other experiments using the *s*ˢ hypha as a conduit, it was clear that there was marked antagonism between the *S* and *s* cytoplasmic factors leading to hyphal death. These experiments suggest

TABLE 9–1. *Spread of s cytoplasmic factor through sˢ mycelia as function of time of contact of s and sˢ mycelia*

Time before contact broken (hr)	Mycelial phenotype	Test of hyphal fragments		
		s	Total	Percentage
6	*s*ˢ	0	89	0
11	*s*ˢ	0	190	0
24	*s*	8	177	4
36	*s*	12	113	10
48	*s*	80	147	54
57	*s*	124	149	80

Adapted from Beisson-Schecroun (4).

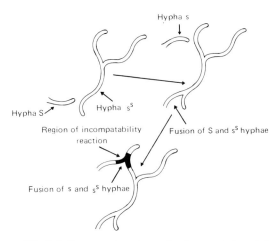

Hypha s

Hypha S

Hypha s^S

Region of incompatability reaction

Fusion of S and s^S hyphae

Fusion of s and s^S hyphae

FIG. 9–2. Demonstration of cytoplasmic transmission of S gene product through s^s hypha, which results in incompatibility reaction in fusion zone when s hypha is brought into contact with s^s hypha. (Adapted from Beisson-Schecroun, ref. 4.)

that the barrage results because of interaction between a stable cytoplasmic factor *s* and an unstable factor produced by the *S* nucleus.

It appears that practically no work has been done on the barrage phenomenon since 1961; thus all such data antedate the discovery of mtDNA and mitochondrial genes. Consequently, as would be expected, mitochondria have not been implicated in models attempting to explain the phenomenon (4,6). Nevertheless, there is no reason why mitochondria might not be the cytoplasmic elements involved, since phenomena such as maternal inheritance and suppressiveness are reminiscent of those we have already encountered in the case of mitochondrial inheritance in *Neurospora* (see Chapter 8). In any event, resolution of this matter should be forthcoming now that Belcour (7,8) has succeeded in isolating mutants in *P. anserina* that are very likely of mitochondrial origin, one of which is resistant to chloramphenicol. Such antibiotic-resistant mutants could be used in repetitions of the elegant hyphal fusion experiments of Beisson-Schecroun. For example, an *s* hypha carrying the chloramphenicol resistance marker (*cap*) could be fused with

an *s*ˢ mycelium carrying a mutation to oligomycin resistance. If *s* and *cap* are both mitochondrial markers, the mycelium should rapidly become chloramphenicol-resistant and oligomycin-sensitive and should acquire the *s* phenotype. Fusion of an *S* hypha to the *s*ˢ mycelium should block this transformation. It is already known that the *cap* marker is maternally inherited, as are *s* and *s*ˢ.

RED AND OTHER CYTOPLASMIC VARIANTS IN *Aspergillus*

Beginning in 1954, Jinks (9–21) and his students published an extensive series of articles on cytoplasmic inheritance in *Aspergillus* that culminated a decade later in several methodologic articles (22,23) and a book (24) on methods of analysis of cytoplasmic inheritance, particularly as they apply to filamentous fungi. Unfortunately, as in the case of *Podospora*, there was no accompanying biochemical work, and until the recent experiments of Rowlands and Turner (pp. 297–301) on probable mitochondrial mutants in *A. nidulans*, the system was largely abandoned for studies of cytoplasmic inheritance. We shall discuss this work here partly because the genetic analysis itself is most interesting and partly because we believe that at least some of the traits studied are likely to be mitochondrial. This discussion will be divided into two parts. In this section we shall consider Jinks's experiments with certain morphologic and color variants exhibiting cytoplasmic inheritance (Table 9–2). In the next section we shall consider phenomena such as vegetative death in *A. glaucus* (16) in the context of the general topic of the cytoplasmic basis of senescence in fungi.

Aspergillus, like *Neurospora* and *Podospora*, is a filamentous Ascomycete, but it belongs to the subgroup Plectomycetes rather than Pyrenomycetes. The mendelian genetics of *A. nidulans* had become well understood by the early 1950s as a result

TABLE 9–2. *Cytoplasmic variants of A. nidulans exclusive of those causing senescence and vegetative death*[a]

Name of variant	Origin of variant	Asexual segregation of variant		Suppressiveness[b]	Nonmendelian segregation in crosses	References
		In conidia	In heterokaryons			
Red	Ultraviolet	+	+	$m > n$	Red not transmitted	13,17,20
Purple	Spontaneous	+	+	$m < n$	Purple transmitted; segregation nonmendelian	20
Minute	Acriflavine	+	+	$m < n$	Minute not transmitted	19
Compact	High temperature		?	$m < n$	?	23
Alba	Spontaneous	+	+	$m < n$	Alba transmitted; segregation nonmendelian	27
Mycelial	Acriflavine	–	+	$m > n$	Minute transmitted; segregation nonmendelian	26
Nonsexual	Spontaneous and high temperature	–	+	$m < n$	Nonsexual not transmitted	9,18,23
Low sexual	Spontaneous	+	?	?	Low sexual transmitted; segregation nonmendelian	9,10,23
Oligomycin resistance (*oliA1*)	Spontaneous	+	+	$m < n$ or $m > n$	Perithecia yield resistant or sensitive progeny	50–52
Chloramphenicol resistance (*camA1*)	N-methyl-N′-nitro-N-nitrosoguanidine	+	+	?	Perithecia yield resistant or sensitive progeny	52,55
Cold sensitivity (*cs67*)	N-methyl-N′-nitro-N-nitrosoguanidine	–	+	?	Perithecia yield resistant or sensitive progeny	51–53

[a] Table includes most of the known variants, except a few that have been reported in the literature but not characterized in any detail.
[b] $m < n$ indicates normal is suppressive to mutant; $m > n$ indicates mutant is suppressive to normal.

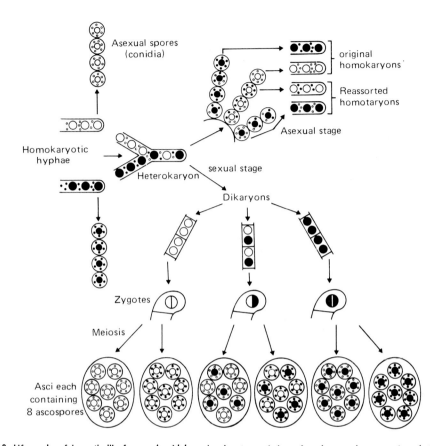

FIG. 9–3. Life cycle of homothallic fungus A. *nidulans* showing transmission of nuclear and extranuclear (mitochondrial?) markers. Both homokaryons and heterokaryons can reproduce asexually by means of uninucleate conidia, which are born in chains. Each conidium contains a single haploid nucleus. Among the conidia formed by a heterokaryon, extranuclear markers such as those described in this chapter can reassort with respect to nuclei to yield new genetic combinations. The sexual cycle is initiated by dikaryon formation, with each cell containing two nuclei. In a heterokaryon three kinds of dikaryons can be formed, two of which contain nuclei contributed by one or the other parent, with the third containing nuclei derived from both parents. In most cases one dikaryotic cell gives rise to all of the dikaryotic cells present in a single fruiting body (cleistothecium or perithecium). Diploid zygotic nuclei are formed, meiosis follows, and asci containing ascospores are formed. Some extranuclear variants such as red are not transmitted sexually, probably because of sterility, but other variants are (e.g., the putative mitochondrial markers to oligomycin and chloramphenicol resistance). In these cases a given cleistothecium derived from a heterokaryon will produce asci that all contain the extranuclear factor derived from one or the other parent (e.g., oligomycin resistance or sensitivity). In this diagram the homokaryotic nuclei derived from each parent are indicated by the large circles and the extranuclear factors by the small circles.

of the work of Pontecorvo and his students (25). The following brief account of the life cycle of *A. nidulans* is based on the descriptions of Pontecorvo (25) and Jinks (24).

A. nidulans is a homothallic species (i.e., it is self-fertile). When two homokaryons come together, the vegetative hyphae anastomose to form a heterokaryon, and the haploid nuclei continue to divide without

any particular balance with respect to one another (Fig. 9–3). The first step in the sexual cycle is a change from the heterokaryotic condition to a dikaryon in which the cells contain two nuclei. In a heterokaryon three kinds of dikaryons can arise: two in which both nuclei are contributed by the same homokaryotic component and one in which both homokaryotic components contribute a single nucleus (Fig. 9–3). In the majority of cases, one initial dikaryotic cell gives rise to all the dikaryotic cells enclosed within a single fruiting body (cleistothecium or perithecium). Hence a heterokaryon will produce three kinds of cleistothecia. The diploid zygotic nuclei are formed by the fusion of the two haploid nuclei in a dikaryotic cell. An immediate meiotic division followed by mitosis yields eight haploid ascospores. Only in those cleistothecia in which the dikaryon contains one nucleus from each of the original homokaryons is there any segregation for different mendelian markers, and in these cleistothecia there is the usual 1:1 segregation in every ascus. In the other two types of cleistothecia, the chromosome complements of the ascospores are uniformly like those of one or the other homokaryotic parent.

Heterokaryotic cleistothecia can be identified by the use of appropriate nuclear markers. For example, *A. nidulans* forms asexual spores called conidia, as well as ascospores that result from meiosis (Fig. 9–3). Wild-type conidia are green, but a mutant that forms yellow condidia (y) is known. A heterokaryotic cleistothecium derived from nuclei containing y and its wild-type allele (y^+) will produce both yellow- and green-spored progeny, but a homokaryotic cleistothecium will produce either yellow or green cleistothecia.

The conidia themselves are uninucleate and are born in chains on a conidiophore (Fig. 9–3). A heterokaryotic conidiophore containing y and y^+ nuclei will produce single chains that are either all yellow or all green, but within the conidiophore, chains of both colors are found. Thus it appears that each conidial chain is derived from a single nucleus but that both nuclear components of the conidiophore are able to produce chains. An important point that should be stressed before we proceed to discuss the mutants themselves is the lack of sexual differentiation (either genetic or morphologic) in *A. nidulans,* as compared

to yeast, which has mating types but no morphologically differentiated sex organs, and *Neurospora* and *Podospora,* which have both.

The most thoroughly studied of the cytoplasmic variants is the red variant isolated by Arlett (13). The red variant was obtained in an ultraviolet light mutagenesis experiment; so it seems likely that it arose because of mutation of a preexisting genetic element in the cell. As we shall see, loss of a preexisting element, as opposed to mutation, can be eliminated because the red variant is suppressive to normal in heteroplasmons. We will spend some time discussing this variant because it provides an interesting example of the sorts of models that can be made for the segregation of cytoplasmic mutants based on genetic data; we will have occasion to return to these later in considering the segregation patterns of presumptive plastid mutations in *Chlamydomonas* and higher plants (see Chapters 15 and 17).

The reasons for thinking that the red variant had a cytoplasmic basis were enumerated in a second article by Arlett et al. (17). The most striking reason was that although the initial variant colony arose from a single irradiated conidium, the colony itself was a mosaic consisting of relatively normal sectors and sectors characterized by reduced density of cleistothecia (which contained no viable spores), increased rate of growth, and marked red coloration, some of which was secreted into the medium. Conidia obtained from subsequent hyphal transfers of the variant invariably segregated into two major phenotypic classes, one apparently normal (*r*-normal) and the other forming abnormal red sectors like the parent colony (*r*-red). Since the conidia of *A. nidulans* are haploid and uninucleate, persistent segregation of this type is presumptive evidence for an extranuclear genetic factor. Further evidence for an extranuclear genetic basis of the red variant came from the demonstration by Arlett and associates that the variant was transmissible in heterokaryon tests. They confronted an *r*-red colony that contained a wild-type nucleus with an *r*-normal colony carrying a nuclear gene mutation that caused the conidia to be white (*w*) rather than green (*W*). Hyphal anastamoses were achieved at the interface, with production of heterokaryotic conidiophores. The *r*-red phenotype was found in association with *W* nuclei as well as *w* nuclei among the conidial progeny, thus indicating its cytoplasmic transmission.

On the basis of these results, Arlett and associates concluded that the red variant had a cytoplasmic basis, and they set about trying to explain the basis of its persistent segregation. They found that whereas the red variant occasionally gave rise to *r*-normal segregants that were stable, this was never true of the *r*-red segregants. Thus it became a matter of deciding if red continuously backmutated to normal or if homoplasmons for the red determinant were lethal. If the latter were true, it would have to be supposed that red was suppressive to normal, or else it would quickly be replaced by the normal determinant in heteroplasmons.

Arlett and associates argued that red homoplasmon lethality was probably the basis for persistent segregation by the red variant. They pointed to a strong parent–offspring correlation in proportions of *r*-red and *r*-normal progeny produced by a given conidium indicating a balance between determinants rather than a random change resulting from backmutation. They also noted that the red variant conidiated poorly and that there was a negative correlation between viability and percentage *r*-red segregants. Furthermore, on aging, conidia that were eventually to give rise to *r*-red colonies lost viability at five times the rate of conidia that were destined to produce *r*-normal colonies. Finally, Arlett and associates pre-

sented evidence suggesting that *r*-red was suppressive to *r*-normal. All these results support the hypothesis that persistent segregation in the red variant results because the responsible factor is lethal in the homoplasmic state but is also suppressive with respect to the normal factor.

On the basis of this hypothesis, Arlett and associates constructed a model to describe the segregation of the normal and variant determinants; this was elaborated in greater detail by Jinks (24). According to their model, production of conidia in *Aspergillus* occurs by the budding off of daughter cells from a mother cell. Therefore each conidium contains a small cytoplasmic sample cut off from a much larger cytoplasmic mass. The model supposes that both normal (+) and red (*r*) cytoplasmic determinants are present in the conidial mother cell and that they are distributed randomly to the conidia. If it is assumed that the number of determinants in the mother cell is infinite ($m = \infty$) and that each conidium receives an equal number of determinants, the model can be described mathematically in terms of a binomial distribution in which *p* is the frequency of + determinants, *q* is the frequency of *r* determinants, and *n* is the number of determinants per spore. Thus, if there were only two homologs per conidium and *p = q,* the constitution of the resulting conidia would be ¼ ++, ½ +*r*, ¼ *rr*.

In practice, *m* will be finite, and so the more complex hypergeometic distribution is more appropriate than the binomial. However, if *m* is large in relation to *n,* the binomial is a good approximation. For example, with $n = 2$ and $m = 50$, the deviation is less than 1%.

The relative frequencies of red segregants fall into at least seven significantly different classes whose means are 0, 13, 40, 57, 66, 79, and 92% red segregants. If the model is correct, the eighth class, which yields only red segregants, is lethal. In the binomial distribution, the number of classes (*k*) is

TABLE 9–3. *Frequency with which different conidial constitutions will be found with respect to normal and mutant homologs in red variant in random sample of 25 conidia[a]*

Proposed ratio of mutant (r) to normal (+) homologs in conidium	Frequency of conidia	
	Observed	Expected
7+:0r	0	0.20
6+:1r	1	1.37
5+:2r	3	4.10
4+:3r	10	6.83
3+:4r	7	6.83
2+:5r	2	4.10
1+:6r	2	1.37
0+:7r	0	0.20 (will be lethal)

[a] The conidium containing only mutant homologs is presumed to be lethal. Calculation of observed and expected classes is described in the text.

Adapted from Jinks (22).

equal to $n+1$; this means that $k = 8$ and $n = 7$. The constitutions of the seven recovered classes of spores will therefore be 7:0, 6:1, 5:2, 4:3, 3:4, 2:5, and 1:6 normal-to-mutant determinants.

Let us now apply the model to some sample data from a collection of 25 conidia (Table 9–3). If $n = 7$, the total number of determinants in a sample of 25 spores will be 25×7 or 175. If we then make the simple assumption that $p = q$, we will have 88 normal and 87 mutant determinants. If we expand the binomial accordingly, we can calculate the expected frequency of conidia in the eight classes. The agreement between the model and this expectation is good (Table 9–3). Therefore the model is an adequate representation (but not necessarily the only adequate representation) of persistent segregation in the red variant.

A second way of estimating n for the red determinant alone is considered by Arlett and associates. It does not depend on making the distinction between the different frequency classes of r-red segregants; instead, it makes use of the fact that some of the conidia produced by the red variant yield only r-normal progeny. In this method the fraction of conidia that are r-normal is set equal to the zero term of a Poisson distribution (i.e., these are the conidia that received no r determinants). This will then yield an estimate of the mean number of mutant determinants per conidium, which we will call n_r. That is, $P_{r(0)} =$ fraction of r-normal conidia $= e^{-n_r}$. Arlett and associates applied this method to the asexual progeny of an r-red colony giving 60% r-red segregants and found that 0.3% of the total progeny were pure r-normal (i.e., wild type). Therefore $P_{r(0)} = 0.003 = e^{-n_r}$ and $n_r = 5.8$. According to the binomial method, a colony producing 60% pure red segregants would have somewhere between three and four mutant determinants to begin with; thus the agreement between the two estimates is not bad. This method cannot be used to determine the mean number of normal particles per conidium because the class that would be used to make the P_0 estimate for this class would, by definition, contain only mutant determinants and would never be recovered.

The analysis of Arlett and associates bears some formal resemblance to the analysis of the average multiplicity of infection in a cross involving two genotypically different bacteriophages. In a virus cross it is assumed that the two genotypes will be distributed in a Poisson fashion with respect to the bacteria they infect. Thus a certain fraction of the bacteria will be infected only by viruses of one genotype. These, together with any uninfected bacteria, can be used to estimate the average multiplicity of infection of the remaining bacteria with viruses of the other genotype. In this case, to estimate the average multiplicity of infection by the other genotype, which we will call b, the number of uninfected bacteria plus those infected only with viruses of genotype a will be taken as a fraction of the total and set equal to the zero term of a Poisson distribution whose mean will be equal to the mean multiplicity of infection by virus b. Of course, in the case of a virus, this estimate can be arrived at more simply by titrating the two viruses prior to making the cross. Unfortunately, such direct methods cannot be applied in estimating multiplicities of infection either in the case of the cytoplasmic determinants involved in persistent segregation of the red variant or in the case of mitochondrial genes in *Saccharomyces* (see Chapter 7) or chloroplast genes in *Chlamydomonas* (see Chapter 15).

Grindle (20) studied the effects of different nuclear gene mutations on the red variant and found that they had profoundly different effects on the frequencies of *r*-red and *r*-normal progeny produced. Some mutations had no effect, whereas others increased the fraction of *r*-normal segregants; but the most interesting were those that increased the frequency of *r*-red segregants, for among these were found apparently pure homoplasmic red segregants. These findings show that pure mutant conidia are not lethal, given the appropriate genetic background. In addition, the pattern of persistent segregation broke down rapidly in the presence of nuclear gene mutations favoring *r*-red segregants, which suggests that given appropriate circumstances *r*-red is indeed suppressive to *r*-normal. Grindle also observed that in those few instances where viable ascospores were obtained from red variant mycelia, they always produced *r*-normal progeny. We do not know if this result was obtained because heteroplasmic and homoplasmic *r*-red ascospores were inviable, because there was postzygotic exclusion of *r*-red determinants, or because there was prezygotic selection of *r*-normal determinants in the cleistothecia. Some suggestive evidence that the latter mechanism may be involved comes from Grindle's finding that conidiophores from cleistothecial regions yield conidia that produce lower proportions of *r*-red progeny than those from noncleistothecial regions.

The characteristics of the other *Aspergillus* cytoplasmic variants (except those causing senescence and vegetative death) are summarized in Table 9–2. None of them seems to have received as much detailed study as the red variant, and they will be discussed here only by way of contrast. Recently, several putative mitochondrial markers to oligomycin resistance, chloramphenicol resistance, and cold sensitivity in *Aspergillus* have been studied in some detail, and their behaviors are contrasted with those of the older variants in Table 9–2. We will discuss these mutants in detail in the last part of this chapter.

The minute and purple variants are probably the most interesting by way of comparison to the red variant. Both these mutants show persistent segregation among conidial progeny, and neither has been recovered in pure breeding form (19,20). Cleistothecia produced by the minute variant yield only wild-type ascospores; the reason appears to be that the cleistothecia are not produced at random but tend to be found in mycelial regions that behave largely as wild types in terms of conidiation (19). Thus nontransmission, as in the case of the red variant, may be the result of a prezygotic exclusion mechanism in which cleistothecia tend to be formed in nonmutant regions of mycelium. In the case of the purple variant, in contrast to the red and minute variants, the mutant determinants are transmitted to the ascospores, but in nonmendelian ratios. In terms of suppressiveness, it appears that in the presence of wild-type nuclei, red is suppressive to normal, but the normal determinants are suppressive to the mutant determinants in the other two cases, with the minute determinants being at the greatest disadvantage.

The mycelial variants isolated by Roper (26) after acriflavine treatment were unlike the variants described thus far in that they did not appear to show persistent segregation. Thus conidia from the mycelial mutant produced only mycelial progeny. However, cytoplasmic transmission of the mycelial trait was shown in heterokaryons. Transmission of the trait was always unidirectional, from mutant into normal; thus it appeared that the mycelial determinants were suppressive to wild type. The mycelial trait was also found to be transmitted in the sexual cycle, but the expression of the mycelial trait was under the control of nuclear genes. Thus, given the appropriate nuclear genotype, the mycelial phenotype may not be expressed, despite the presence of the mutant cytoplasmic factor. Other cytoplasmic variants that affect perithecium formation, such as alba, nonsexual, and low sexual variants (Table 9–2), tend to behave much as the mutants already described. They will not be discussed in detail.

Now that presumptive mitochondrial mutants have been isolated in *Aspergillus* (Table 9–2) and the potential for mapping them is at hand (pp. 297–301) it should be of interest to reinvestigate the older cytoplasmic variants. Among the questions that come to mind are the following: Is the red variant a lethal suppressive mitochondrial mutant that can under normal circumstances be maintained only as a heteroplasmon? If so, then how is it possible for certain nuclear genes to ameliorate the effect of the mutation in such a way that homoplasmic red segregants become viable? Mutants such as the minute and mycelial variants, which are induced by acriflavine, may be mitochondrial. Are they equivalent to petites? If so, must they be maintained in the heteroplasmic state? The experiments discussed suggest that the answer is affirmative in the case of the minute variant but negative in the case of the mycelial variant. Mutants such as the alba, nonsexual, and low sexual variants all elaborate cleistothecia poorly, and the nontransmission of the minute and red variants sexually can largely be explained in terms of the hypothesis that cleistothecia are elaborated primarily in mycelial regions rich in normal determinants. Does this mean, as appears to be the case for *Saccharomyces* (see Chapter 4) and possibly *Neurospora* (see Chapter 8), that mitochondria with intact respiratory chains are required for completion of the sexual cycle? If so, can we account for the fact that senescence and vegetative death in *Aspergillus* and *Podospora* can be successfully circumvented only by sexual reproduction? Only time and more experiments will provide answers to these questions; they are perplexing and fascinating problems that merit attention.

SENESCENCE

Indefinite asexual propagation is not possible in a number of fungi because the mycelium eventually sickens and dies. This phenomenon of senescence has been found to be under the control of cytoplasmic genes in *A. glaucus* (12,16), *P. anserina* (28–38), and *Pestalozzia annulata* (39–46). It can be avoided only by introducing cycles of sexual propagation. In this section we will discuss senescence in *Aspergillus* and *Podospora* and then make comparisons of these phenomena to the suppressive mitochondrial mutants of *Neurospora,* such as *abn-1,* that are eventually lethal.

Jinks (10,12,16) studied senescence, which he called vegetative death, in *A. glaucus*. He showed by means of heterokaryon tests that the responsible factor was transmitted cytoplasmically and was suppressive to the wild type, spreading rapidly into the normal homokaryon. Mather and Jinks (12) also studied the successive stages of degeneration in an aging clone of *A. glaucus* and showed that degeneration usually follows a sequence of events in which the sexual stage is lost first, followed by the asexual stage; finally there is an abrupt and irreversible change in which vegetative death spreads rapidly through the colony. This sequence can be reversed at any time prior to loss of the sexual stage by ascospore propagation, and it is never seen in strains propagated exclusively by sexual means. Mather and Jinks (12) called this rejuvenation process that occurs in the sexual cycle cytoplasmic restandardization.

Senescence in *P. anserina* is analogous to vegetative death. It has been studied in detail by Rizet (28–31) and his student Marcou (32–35) and more recently by Smith and Rubenstein (36–38). When a mycelium of any geographic race of *P. anserina* is kept in continuous culture, it will grow at a rate of about 7 mm/day for a variable distance (5–100 cm), depending on the geographic race. The growth rate then suddenly decreases, and 1 or 2 days later the mycelium ceases to grow, with the terminal hyphae becoming small and vacuolized. Mycelial fragments transferred from the region of abnormal morphology are incapable of further growth. This ensemble of growth and morphologic defects is referred to as the senescence syndrome.

Mycelia derived from single ascospores obtained by crossing isolates of the same race will grow for various distances before becoming senescent. Since different mycelia from an apparently homogeneous population are capable of different amounts of growth, the length of growth of a single mycelium is not a very useful parameter. The median length of growth (MLG) of a population of mycelia is much more useful. The MLG of a population of mycelia is the length of growth that half the members of the population fail to reach and that half the members reach or exceed. The use of this parameter is illustrated in Fig. 9–4, where it is seen that the MLG for geographic race A is much less (15 cm) than that for race S (170 cm). For the moment we will not discuss the interpretation of the shape of this curve further; rather, we will present the evidence that senescence is caused by a cytoplasmic factor. Once this is done, we can return to the models by which Marcou (34) and Smith and Rubenstein (36,37) endeavored to explain the onset of senescence.

Two lines of evidence support the conclusion that senescence is caused by a cytoplasmic factor. First, as Marcou (34) showed by the use of mycelial grafts, senescence is infectious, and it converts normal mycelia to senescent mycelia in the absence of nuclear transfer. These elegant experiments were similar to those of Beisson-Schecroun (4) regarding the barrage phenomenon that were described earlier in this chapter. Second, sexual reproduction is unaffected by the senescence syndrome; thus, in contradistinction to the situation with vegetative death, the inheritance of senescence can be studied in reciprocal crosses in *Podospora*.

When a senescent strain is used as the paternal parent, all of the perithecia yield normal asci and ascospores; but in the reciprocal cross, normal as well as senescent progeny are recovered. In some strains, entire perithecia from senescent females produce either all normal or all senescent progeny. In other strains, the perithecia are heterogeneous, containing normal and senescent asci, each producing all normal or all senescent progeny. Since senescence is not transmitted by the male parent, and since in heterokaryons it is invasive, it seems likely that normal as well as senescent asci result when the female parent is senescent because the senescent mycelium contains both normal and senescent factors (i.e., lethality results when all factors are senescent). That is, inheritance in both crosses is probably strictly maternal.

These results have been extended to the normal factor itself by Smith and Rubenstein (38) in interracial crosses between races A and S, where barrage formation due to nuclear genes does not seem to be a problem. The patterns of onset of senescence in these two races are quite different, as was mentioned previously; thus it is possible to measure the onset of senescence in reciprocal crosses and determine whether the pattern resembles that for race A or that for race S or neither. The pattern of senescence onset in interracial crosses resembles quite precisely the pattern seen for the maternal parent. Backcrosses have also been used to further rule out any important nuclear gene effects. These results indicate that the normal factors themselves are maternally inherited. It is interesting that heterokaryons synthesized between races A and S have an MLG similar to that of A homokaryons, which suggests that the heterokaryon normally has the senescence phenotype of the parent that, by itself, becomes diseased the earliest.

Let us now return to interpretation of the curves shown in Fig. 9–4 in terms of a cytoplasmic senescence factor. It will be seen that the plots bear a strong resemblance to the survival curves used in target analysis. On the ordinate is plotted the log survival (i.e., proportion of nonsenescent colonies), and on the abscissa is plotted length of growth (in centimeters) on a linear scale. Length is not strictly equivalent to time or dose, but since it is proportional to time, it can be used in a similar fashion to calculate a transformation rate (TR). The length of the shoulder is referred to as the incubation distance (ID), and it can be seen that the MLG becomes equivalent to a dose at which 50% of the population is inactivated (LD_{50}) in target analysis.

By analogy with target analysis, we might look at these curves as multiple-hit curves in which the fraction of nonsenescent mycelia is equated to the surviving fraction (S/S_0), and

$$S/S_0 = 1 - (1 - e^{-kt})^n$$

where k is the transformation rate or constant (determined from the linear portion of the survival curve), t is distance (which is proportional to time), and n is the target number (in this case the number of factors that must become senescent before senescent morphology is observed).

Marcou (34), in the experiment that will be described next, showed that this was the wrong interpretation of these curves. If multiple hits are involved in transformation from the nonsenescent state to the senescent state, the MLG will be expected to increase in a continuous fashion as the mycelium is sampled farther and farther away from the point of growth stoppage (i.e., the concentration of hits or senescence factors declines as one proceeds along the shoulder of the curve from the point where the decline in survival begins). However, Marcou found that close to the front the MLG was approximately equal to the distance between the region of sampling and the position of growth stoppage; also, the TR was greater

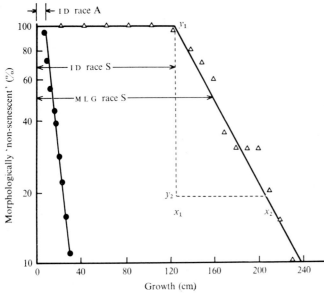

FIG. 9–4. Percentages of mycelial populations of races A and S of *P. anserina* morphologically nonsenescent plotted as a function of distance grown. Each mycelial population originated from a binucleate ascospore. MLG = median length of growth; ID = incubation distance; TR = transformation rate (log y_1 − log y_2)/(x_2 − x_1); ● = race A; △ = race S. (From Smith and Rubenstein, ref. 37, with permission.)

than for a nonsenescent strain, and the ID was shorter. However, when samples were taken farther away from the point of growth stoppage, the MLG was the same for all groups of samples, and the TR and ID were the same as those found in a nonsenescent culture. That is, the distribution of senescence factors was discontinuous.

Marcou (34) therefore interpreted the curves in Fig. 9–4 in terms of the following model. A mycelium can exist in either of two states: nonsenescent or senescent. A mycelium in the nonsenescent state can be transformed to the senescent state by a single random event (the first appearance of a senescence factor) (Fig. 9–5). The probability of an event occurring in a given unit growth has a constant value for each unit of growth. After the occurrence of an event (i.e., the first appearance of the senescence factor), an amount of growth equivalent to the ID is necessary before senescent morphology is manifest. This interpretation explains the pseudo-multiple-hit nature of the curves shown in Fig. 9–4; it also constitutes a cautionary fable on why one must exercise great care in the use of target kinetic analysis. An interesting point that is also illustrated by these data is that the ID and TR differ for races A and S, just as does the MLG (Fig. 9–4), with the TR

being steeper for strain A and the ID being very short for strain A. These results show that transformation to the senescent state is more rapid in race A than in race S, and the senescence factor expresses itself much more rapidly in the former than in the latter. This interpretation is supported by the results of heterokaryon tests, from which it will be recalled that senescence follows kinetics similar to those for an A homokaryon (i.e., it occurs rapidly) rather than those for an S homokaryon.

It might be supposed, as Marcou (34) postulated, that the concentration of the senescence factor would increase during growth of the mycelium in the senescent state. Smith and Rubenstein (37) devised a method for testing this theory. During growth, they took transverse slices of mycelium containing 1 to 20 cells and then measured growth of these fragments. They assumed that any fragment that grew a distance at least as great as the ID contained no senescence factors at the time of isolation. They also assumed that senescence factors were distributed in a random (Poisson) fashion throughout the mycelium. Therefore, the fraction of hyphal segments containing no senescence factors could be set equal to the zero term of a Poisson distribution: fraction of nonsenescent hyphae =

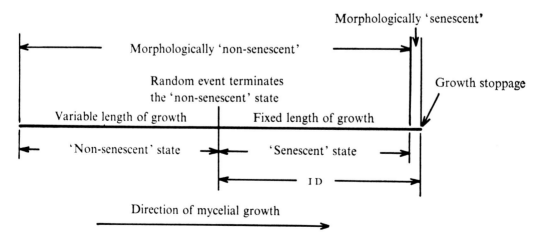

FIG. 9–5. Schematic representation of the two physiologic and morphologic states of the mycelium of *P. anserina*. ID = incubation distance. (From Smith and Rubenstein, ref. 37, with permission.)

e^{-m}, where m is the mean number of senescence factors per hypha. Smith and Rubenstein found that the concentration of senescence factors estimated by this method could indeed be related to the distance from the point of growth stoppage; it increased logarithmically to a concentration of about 1 to 5 senescence factors per cell at the point of growth stoppage (Fig. 9–6).

Marcou (34) discussed several models that might account for senescence in *P. anserina:* (a) A cytoplasmic gene mutates. This mutation will exert a deleterious effect on the mycelium, and the relative number of mutant genes will increase as growth proceeds. (b) An episome or provirus is released from nuclear DNA. The episome or virus in its free form will exert a deleterious effect on the mycelium and will increase in numbers as growth proceeds. (c) A self-maintaining change occurs in some cytoplasmic structure. For example, if B is the normal factor and C the senescence factor, the transformation B → C can occur only at a low frequency when C is absent; but if C is present, the rate of transformation B → C will increase in a manner dependent on the concentration of C. When the concentration of C becomes high enough, senescent morphology will result. (d) The mycelium can exist in two or more physiologic states, each of which is stable. The mycelium can be converted from one state to another by transient changes in the concentrations of intracellular constituents. Neighboring hyphae will be homogeneous with respect to the concentrations of these constituents. This will account for the fact that large areas of the mycelium appear to be senescent at the same time.

In addition to these models, there have been other suggestions. Holliday (47) proposed that senescence might result from errors in protein synthesis of the type discussed by Orgel (48). For example, if an error occurs that alters the specificity of an RNA polymerase or an aminoacyl tRNA synthetase molecule, errors in protein synthesis can increase exponentially and result in death of the organism.

Each of these models was reviewed by Smith and Rubenstein (37) in light of their own data. They rejected all but the first and third models of Marcou. They pointed out that from Marcou's second model and the model proposed by Holliday, nuclear genes would be expected to control the time of expression of senescence; however, the interracial crossing data presented by Smith and Rubenstein show that the time of ex-

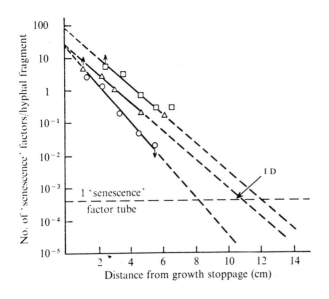

FIG. 9–6. Average number of senescence factors per hyphal fragment in a mycelium of race A of *P. anserina* plotted as a function of distance from the manifestation of senescent morphology. Different symbols represent independent experiments performed on different mycelia of race A (*circles* and *triangles*) or interracial hybrid A (AS$_{10}$) (*squares*). (From Smith and Rubenstein, ref. 37, with permission.)

pression is under the control of cytoplasmic genes. Nevertheless, it could be argued that the provirus or episome was associated with the cytoplasmic genome or that the altered enzyme was a product of the cytoplasmic genome. After all, Küntzel claimed to have found virus-like particles associated with the mitochondria of the *abn-1* mutant of *N. crassa* (see Chapter 8).

According to the fourth model proposed by Marcou, the transformation rate should decrease or remain constant as the number of hyphae per culture increases. However, Smith and Rubenstein found that the transformation rate increased linearly with the number of hyphae per culture. This narrows the possibilities to the first and third models of Marcou, both of which are consistent with the results of Smith and Rubenstein. It is clear that the latter authors preferred the first model (i.e., mutation of a cytoplasmic gene). With this in mind, they made calculations of the transformation rate per cell division to determine if it was consistent with known spontaneous mutation rates. The transformation rate for race A was 1×10^{-7} per cell division; it was 1×10^{-8} per cell division for race S. These values are clearly compatible with the spontaneous mutation hypothesis. Thus a difference in spontaneous mutation rates could account, in part, for the rapid onset of senescence in race A as compared to race S. This is not an entirely adequate explanation, because the incubation distance is much shorter for race A than for race S, which suggests that once the mutation occurs, it spreads much more rapidly in race A than in race S.

Smith and Rubenstein considered the possibility that the mutant cytoplasmic factor causing senescence could be a mitochondrion, but thus far neither they nor others have published more than vaguely suggestive evidence. In view of the elegant methods devised by Marcou and by Smith and Rubenstein to "titrate" the senescence factor, it would seem profitable to establish the molecular nature of this factor. If the senescence factor is a mutant mitochondrion, these experiments should provide information on such things as mitochondrial mutation rates and the kinetics by which a suppressive mitochondrial mutant spreads throughout a mycelium.

Bertrand et al. (49) looked for the senescence phenomenon in *N. crassa* by means of serial transfer experiments of mycelia in growth tubes. These experiments showed that although senescence is not an obligatory feature of the asexual cycle in *Neurospora* (as it is in *A. glaucus* and *Podospora*), suppressive extranuclear mutants do arise under conditions of prolonged growth that in *N. crassa* produce the stop–start or stopper phenotype (see Chapter 8).

MITOCHONDRIAL GENETICS OF *Aspergillus*

Rowlands and Turner (50–52) recently made an important breakthrough in mitochondrial genetics that involved the use of markers with clearly defined phenotypes in *Aspergillus*. Because of their work, we should soon know a great deal about the rules governing the transmission and inheritance of mitochondrial genes in this fungus. Their experiments are also interesting by way of contrast with the earlier experiments of Jinks and his students on cytoplasmic variants of *A. nidulans*. It seems likely that the methods used by Rowlands and Turner could provide definitive evidence about whether variants such as the red and minute variants are mitochondrial mutations or mutations in some other cytoplasmic element.

In their first article, Rowlands and Turner (50) reported the isolation and characterization of a nonmendelian mutation to oligomycin resistance that they first referred to as O^R6 and later called *oliA1* (51). This mutant is almost certainly mitochondrial, as we shall see later. The *oliA1* mutant was 1 among some 77 isolated, most of which proved to be mendelian. In these experiments, but not in later ones (52), hybrid perithecia derived from heterokaryons formed between the *oliA1* mutant and the wild type produced only oligomycin-sensitive ascospore progeny, despite the fact that nuclear markers segregated normally. The lack of transmission of the *oliA1* phenotype was not the result of ascospore sterility, as Rowlands and Turner showed by making selfings of *oliA1*. These selfings produced resistant ascospore progeny, which showed that these progeny were viable. Heterokaryon analysis demonstrated that the nonmendelian behavior of *oliA1* could be explained in terms of a cytoplasmic particle (Table 9–4). In these experiments heterokaryons were synthesized in reciprocal directions with respect to oligomycin resistance and sensitivity and the nuclear gene markers. Conidia were plated from these heterokaryons, and in each case it was possible to show that the resistant marker had become associated with nuclei carrying the markers of the sensitive strain as well as the resistant strain (Table 9–4). Similarly, it was possible to associate sensitivity with nuclei of the resistant strain. In contrast, the mendelian oligomycin-resistant mutants always remained in association with nuclear markers of the resistant strain in heterokaryon tests (Table 9–4).

Oligomycin resistance in the *oliA1* mutant was also found to be associated with impaired growth ability. This proved to be a useful characteristic of the mutant, because resistant conidia formed small colonies, whereas sensitive conidia formed large colonies. As a result, by plating conidia on minimal medium and looking for large colonies, it was possible to determine if the *oliA1* mutant reverted. Reconstruction experiments in which small numbers of wild-type conidia were mixed with large numbers of resistant conidia proved the feasibility of the method. The reversion experiments showed that resistance was a very stable phenotype, and no reversions were found in over 1 million conidia that were plated.

TABLE 9–4. *Demonstration of transmission of oliA1 cytoplasmically in heterokaryon test in A. nidulans and nontransmission of a mendelian mutant* O_2^{Ra}

| | Number of conidial progeny | | | |
| | Oligomycin-resistant | | Oligomycin-sensitive | |
Heterokaryon	paba y	w pyro	paba y	w pyro
oliA1 paba y/ w pyro	93	15	22	57
paba y/oliA1[b] w pyro	2	17	139	3
O_2^R paba y/w pyro	310	0	0	213

[a] One of the homokaryons carries the mendelian markers *paba* and *y*, whereas the other carries the markers *w* and *pyro*.

[b] There appears to be an error in the original article; the *oliA1* stock is incorrectly designated O_2^R.

Adapted from Rowlands and Turner (50).

Having established the cytoplasmic nature of the *oliA1* marker and its phenotypic stability, Rowlands and Turner proceeded to examine the behavior of the mutant in heterokaryons with sensitive strains. Two methods were used. In the first method the heterokaryotic mycelium was repeatedly subcultured on minimal medium, and conidia were sampled to determine the frequencies of resistant and sensitive phenotypes in the population. These experiments showed that the heterokaryons sooner or later became pure for either resistance or sensitivity, and they provided no evidence for suppressiveness of one or the other phenotype. As might be expected, transfers made on oligomycin-containing plates always led to loss of the sensitive character.

In the second method used by Rowlands and Turner, heterokaryons formed between resistant and sensitive strains were allowed to grow out radially from a small inoculum in the center of a plate of minimal medium. These circular mycelial masses were scored for resistant and sensitive sectors. Both kinds of sectors were found, and these were independent of sectors formed with respect to nuclear markers where part of the mycelium had become homokaryotic.

At about the same time that these experiments were reported, Waldron and Roberts (53) described a new nonmendelian mutant of *A. nidulans*. This was one of many cold-sensitive (*cs*) mutants isolated by these authors in an attempt to obtain mutants defective in ribosome assembly. Such cold-sensitive ribosome assembly mutants had previously been reported in *Escherichia coli* by Guthrie and associates (54). The new nonmendelian mutant was designated *cs67*. At 37°C *cs67* had a wild-type growth rate, but at 20°C the colonies formed by *cs67* were much smaller than the wild type.

The nonmendelian inheritance of *cs67* was confirmed by Waldron and Roberts through an examination of hybrid perithecia formed by heterokaryons with a cold-tolerant strain. These perithecia produced exclusively either cold-tolerant ascospore progeny or cold-sensitive ascospore progeny. Nuclear markers segregated normally.

Waldron and Roberts were unable to demonstrate heterokaryotic transmission of *cs67*, but this was accomplished by Rowlands and Turner (51), who reported the first case of recombination between putative mitochondrial genes in filamentous fungi. Heterokaryons were synthesized between *oliA1* and *cs67*, and the conidia were tested to see if recombinant colonies could be obtained. Young heterokaryons were used to minimize the problems resulting from sectoring out mitochondrial genotypes within the mycelium and to reduce the possible rounds of mitochondrial gene recombination.

The conidia obtained from the heterokaryons were plated according to either of two protocols. In the first, or selective, method the frequencies of ++ and +*cs* genotypes were estimated by plating a known number of conidia on nonselective medium at 20°C. Under these conditions the former genotype formed large colonies, and the latter genotype formed small colonies. Obviously *oli+* and *olics* genotypes will also be present on these plates. Fortu-

nately, the four genotypes can be distinguished in terms of mean colony size 5 days after plating, with the relative sizes of the four genotypes being as follows: ++ > *oliA+* > +*cs* > *oliAcs*. The frequencies of *oli+* and *olics* progeny were then estimated by plating conidia at high temperature on plates of rutamycin (an inhibitor related to oligomycin, see Chapter 4) and then distinguishing the two genotypes on the basis of colony size at 20°C. In the second, or nonselective, method, conidia were plated at 37°C on medium lacking inhibitor, and the single colonies were retested on oligomycin-containing medium at 37°C and on inhibitor-free medium at 20°C. This allowed all possible genotypes to be scored.

The two approaches yielded similar results, and recombinants were clearly less frequent than parentals. However, it was also clear that reproducible recombination frequencies could not be obtained by expressing recombination frequencies as fractions of the total. The reason appeared to be that the amounts of cytoplasmic mixing in the heterokaryons were variable. Obviously, if mixing were poor, the frequency of recombination would be underestimated. In order to obtain a more nearly true frequency of recombination, only those conidia in which the mitochondrial markers of one parent had become combined with the nuclear genotype of the other parent were used in the calculations. In these "reassorted" types, cytoplasmic mixing presumably had occurred. By definition, all of the recombinants were included among the reassorted types; thus this estimate should maximize recombination frequencies and could lead to an overestimate if the basic assumptions about reassortment and cytoplasmic mixing were inadequate descriptions of the true situation. In any event, this method gave far more consistent recombination frequencies. The recombinants themselves were stable on retesting, and thus they appeared to be true recombinants.

Mason and Turner (52) extended the

analysis of transmission and recombination of these presumptive mitochondrial markers in sexual crosses. In their experiments a third mutation resistant to chloramphenicol *camA1* was included, in addition to the two described previously. This mutation was isolated by Gunatilleke et al. (55). These crosses confirmed the earlier observation that the progeny of a given perithecium carry only one extranuclear genotype, but in this case for as many as three markers at a time. The crosses also revealed that different extranuclear genotypes were transmitted with different efficiencies or "transmission strengths," with the transmission strengths for each genotype being as follows: (*camA1*) > (*cs67, camA1*) > (+) = (*cs67*) > (*oliA1, cs67*) > (*oliA1*) > (*oliA1, camA1*). In contrast to the earlier report of Rowlands and Turner (50), these experiments showed that *oliA1* was, in fact, transmitted in monofactorial crosses to the wild type, but with low efficiency (approximately 10%). Both transmission and recombination were analyzed using the methods developed by Rowlands and Turner (51) for asexual conidia, which correct for incomplete cytoplasmic mixing by dealing only with genotypes, in this case ascospore progeny, in which extranuclear and nuclear genotypes have reassorted. Once again recombinants were found for all possible genotypes, except in the case of *oliA1, camA1* which seems to be sterile or nearly so. Most recently Rowlands and Turner (58) have been able to establish using appropriate genetic techniques that the inheritance of these putative mitochondrial gene mutations in *A. nidulans* is truly maternal and that a fertilization mechanism of some kind exists in this fungus. Therefore, the all or none inheritance of these mutations in sexual crosses is the consequence of maternal inheritance and the results imply that the recombinants isolated by Mason and Turner from the progeny of sexual crosses must have arisen prior to the sexual stage of the life cycle.

In the course of their genetic studies,

Rowlands and Turner (51) also encountered another phenomenon that deserves mention. It will be recalled that the *oliA1* mutant has impaired growth ability, but among the conidial isolates obtained in the recombination experiments, some were observed to be oligomycin-resistant and to have a wild-type growth rate. It was found that after a maximum of two subcultures in the absence of oligomycin, resistance to the drug was lost, but after two subcultures on oligomycin-containing medium, wild-type growth ability was lost. By alternating the cultures between the two kinds of media, both characters could be retained indefinitely. When conidia from these strains possessing oligomycin resistance and wild-type growth ability were allowed to grow radially, it was found that about half the plates produced mycelia with oligomycin-resistant and oligomycin-sensitive sectors and that the others were either all sensitive or all resistant. It was further noted that on oligomycin these strains gave smaller colonies that the parent *oliA1* strains, indicating an intermediate level of resistance.

In view of these results, it was concluded that these strains were heteroplasmic, possessing both wild-type and oligomycin-resistant mitochondrial genomes. These results show that at least two mitochondrial genomes can be transmitted by a conidium, and they provide a possible molecular basis for the somatic segregation of cytoplasmic variants during conidial propagation observed by Jinks. By alternating these strains between media lacking and containing oligomycin, the authors were able to mimic the persistent segregation seen by Jinks and others for the red, purple, and minute variants. The heteroplasmons may also be useful for doing complementation studies of mitochondrial markers. Obviously their existence could make the estimation of recombination frequencies difficult, but they can be differentiated from true recombinants, which are stable, by the subculture methods just described.

Biochemical evidence that *oliA1* and

cs67 are mitochondrial mutants is limited and somewhat indirect, but nevertheless strongly suggestive. Rowlands and Turner (56) found that *in vitro oliA1* mitochondria have an ATPase that is twice as resistant to oligomycin as that of wild-type mitochondria. Although this difference is small, it is probably real, since it is based not on a single point but on concentration curves. The *oliA1* mutant also has about twice the amount of cytochrome c present in wild type, but the other cytochromes are normal in amount. However, since the spectra were done on whole mycelia, it is not certain that the excess cytochrome c is associated with the mitochondria. Rowlands and Turner (56) suggested that the mechanism of resistance might involve an inner-membrane change that leads to exclusion of oligomycin from the mitochondrion and at the same time renders the ATPase slightly more resistant to oligomycin. An inner-membrane alteration might also be related to the increased level of cytochrome c seen in the *oliA1* mutant. Rowlands and Turner (56) also pointed out that the mitochondrial mutations to oligomycin resistance in yeast cause only small increases in mitochondrial resistance and that these can be attributed specifically to the membrane factor associated with the ATPase (see Chapters 5 and 19). All that has been reported for *cs67* so far is that Rowlands and Turner found an altered cytochrome spectrum at the restrictive temperature (53).

CONCLUSION

The experiments described in this chapter on cytoplasmic variants in filamentous fungi provide much food for thought. Let us suppose that the variants described are, for the most part, mitochondrial. If so, some interesting problems arise regarding transmission of mitochondria (or mtDNA) in relation to sexual differentiation. *A. nidulans* is homothallic, and neither male nor female sex organs have been identified. In this system mitochondrial genotypes are now

known to be transmitted maternally, but as yet we do not know the morphologic basis of maternal inheritance in this homothallic fungus. In *N. crassa* sex is controlled by a pair of nuclear mating type alleles, and male elements (conidia) fertilize female elements (protoperithecia). Here mitochondrial inheritance appears to be maternal. Yet the conidia themselves are able to form colonies; this indicates that they must contain mitochondria, since *N. crassa* is an obligate aerobe, and crosses of the extranuclear mutant *AC-7* to the maternally inherited mutant *SG-1* show that in the presence of *AC-7* the *SG-1* mutation is transmitted by the paternal parent as well. These results suggest that elimination of paternal mitochondria (or mtDNA) is a zygotic or postzygotic phenomenon that is not related to the differential cytoplasmic contributions of the male and female parents. In *P. anserina* the pattern of inheritance of the cytoplasmic factors involved in barrage and senescence is strictly maternal, and once again there is a pair of mating type alleles and morphologic sexual differentiation. However, in *P. anserina* the male fertilizing elements, the microconidia, contain very little cytoplasm and germinate very poorly. It is possible that in this case mitochondrial elimination in the male parent is normally prezygotic. Hence, in these three fungi, mitochondrial transmission or elimination could involve as many as three mechanisms, and obviously in any one case more than one mechanism could be involved.

Different sets of problems arise in the case of suppressiveness: Is the suppressiveness of the *s* factor to the *sˢ* factor in the barrage phenomenon in *P. anserina* related to suppression of the normal factor by the senescence factor? Can senescence in *Podospora* and *Aspergillus* be explained by a high rate of occurrence of suppressive mitochondrial mutants of the *abn-1* type in *Neurospora*? Are there different mechanisms of suppression or only one? Among the possibilities that we can imagine are mitochondrial selection, mtDNA selection,

spreading of mitochondrial mutants by recombination as was postulated in yeast
(see Chapter 6), and cytoplasmic transmission of fungal viruses that cause the mycelium, perhaps the mitochondria themselves, to become sick.

Answers to all these questions await the
isolation and characterization of phenotypically well-defined mitochondrial mutants, as
well as careful biochemical and molecular
biological studies. In this respect, it is quite
remarkable how much progress Rowlands
and Turner have already made with these
questions in a few years by the use of three
phenotypically well-characterized *Aspergillus* mutants that are probably mitochondrial. Similarly, the recent discovery by
Belcour (7,8) of apparent mitochondrial
mutants in *Podospora,* one of which is resistant to chloramphenicol, promises to
open new avenues by which the old problems of barrage and senescence can be
approached.

REFERENCES

1. Esser, K., and Kuenen, R. (1967): *Genetics of Fungi.* Springer-Verlag, New York.
2. Vandendries, R. (1932): La tétrapolarité sexuelle de *Pleurotus colombinus. Cellule,* 41: 267–278.
3. Rizet, G. (1952): Les phénomènes de barrage chez *Polospora anserina.* I. Analyse génétique des barrages entre souches *S* et *S̄. Rev. Cytol. Biol. Végét.,* 13:51–92.
4. Beisson-Schecroun, J. (1962): Incompatibilité cellulaire et interactions nucléocytoplasmiques dans les phénomènes de "barrage" chez le *"Podospora anserina." Ann. Genet.* (Paris), 4:1–50.
5. Esser, K. (1966): Incompatibility. In: *The Fungi, Vol. 2,* edited by G. C. Ainsworth and A. S. Sussman, pp. 661–676. Academic Press, New York.
6. Rizet, G., and Schecroun, J. (1959): Sur les facteurs cytoplasmiques associés ou couple des gènes *S-S* chez *Podospora anserina. C. R. Acad. Sci. [D] (Paris),* 249:2392–2394.
7. Belcour, L. (1975): Cytoplasmic mutations isolated from protoplasts of *Podospora anserina. Genet. Res.,* 25:155–161.
8. Belcour, L., and Begel, O. (1977): Mito-chondrial genes in *Podospora anserina:* Re-combination and linkage. *Mol. Gen. Genet.,* 153:11–21.
9. Jinks, J. L. (1954): Somatic selection in fungi. *Nature,* 174:409–410.
10. Jinks, J. L. (1956): Naturally, occurring cytoplasmic changes in fungi. *Comp. Rend. Lab. Carlsberg, Ser. Physiol.,* 26:183–203.
11. Jinks, J. L. (1957): Selection for cytoplasmic differences. *Proc. R. Soc. Lond. [Biol.],* 146: 527–540.
12. Mather, K., and Jinks, J. L. (1958): Cyto-plasm in sexual reproduction. *Nature,* 182: 1188–1190.
13. Arlett, C. F. (1957): Induction of cytoplasmic mutations in *Aspergillus nidulans. Nature,* 179:1250–1251.
14. Subak Sharpe, H. (1958): A closed system of cytoplasmic variation in *Aspergillus glau-cus. Proc. R. Soc. Lond. [Biol.],* 148:355–359.
15. Jinks, J. L. (1958): Cytoplasmic differentia-tion in fungi. *Proc. R. Soc. Lond. [Biol.],* 148: 314–321.
16. Jinks, J. L. (1959): Lethal suppressive cyto-plasms in aged clones of *Aspergillus glaucus. J. Gen. Microbiol.,* 21:397–409.
17. Arlett, C. F., Grindle, M., and Jinks, J. L. (1962): The "red" cytoplasmic variant of *As-pergillus nidulans. Heredity,* 17:197–209.
18. Arlett, C. F. (1960): A system of cytoplasmic variation in *Aspergillus nidulans. Heredity,* 15: 377–388.
19. Faulkner, B. M., and Arlett, C. F. (1964): The minute cytoplasmic variant of *Aspergillus nidulans. Heredity,* 19:63–73.
20. Grindle, M. (1964): Nucleo cytoplasmic in-teractions in the "red" cytoplasmic variant of *Aspergillus nidulans. Heredity,* 19:75–95.
21. Croft, J. H. (1966): A reciprocal phenotypic instability affecting development in *Aspergillus nidulans. Heredity,* 21:565–579.
22. Jinks, J. L. (1963): Cytoplasmic inheritance in fungi. In: *Methodology in Basic Genetics,* edited by W. J. Burdette, pp. 325–354. Holden-Day, San Francisco.
23. Jinks, J. L. (1966): Mechanisms of inheri-tance. 4. Extranuclear inheritance. In: *The Fungi, Vol. 2,* edited by G. C. Ainsworth and A. S. Sussman, pp. 619–660. Academic Press, New York.
24. Jinks, J. L. (1964): *Extrachromosomal In-heritance.* Prentice-Hall, Englewood Cliffs, N.J.
25. Pontecorvo, G. (1953): The genetics of *As-pergillus nidulans. Adv. Genet.,* 5:142–238.
26. Roper, J. A. (1958): Nucleo cytoplasmic in-teractions in *Aspergillus nidulans. Cold Spring Harbor Symp. Quant. Biol.,* 23:141–154.
27. Mahoney, M., and Wilkie, D. (1962): Nucleo-cytoplasmic control of perithecial formation in *Aspergillus nidulans. Proc. R. Soc. Lond. [Biol.],* 156:524–532.
28. Rizet, G. (1953): Sur l'impossibilité d'obtenir la multiplication végétative interrompue et illimitée de l'ascomycète *Podospora anserina. C.R. Acad. Sci. [D] (Paris).* 237:838–840.
29. Rizet, G. (1953): Sur la longévité des souches

de *Podospora anserina. C.R. Acad. Sci.* [D] (*Paris*), 244:1106–1109.

30. Rizet, G. (1957): Les modifications qui conduisent à la sénescence chez *Podospora* sontelles de nature cytoplasmique? *C.R. Acad. Sci.* [D] (*Paris*), 244:663–665.

31. Rizet, G., and Marcou, D. (1954): Longévité et sénescence chez l'ascomycète *Podospora anserina. Compt. Rend. VIII Cong. Intern. Bot. Sect.,* 10:121–128.

32. Marcou, D. (1954): Sur la longévité des souches de *Podospora anserina* cultivées à divers températures. *C.R. Acad. Sci.* [D] (*Paris*), 239:895–897.

33. Marcou, D. (1954): Rajeunissement et arret de croissance chez *Podospora anserina. C.R. Acad. Sci.* [D] (*Paris*), 244:661–663.

34. Marcou, D. (1961): Notion de longévité et nature cytoplasmique de déterminent de la sénescence. *Ann. Sci. Nat. Bot.,* 2:653–764.

35. Marcou, D., and Schecroun, J. (1959): La sénescence chez *Podospora anserina* pourrait etre due à des particules cytoplasmiques infectantes. *C.R. Acad. Sci.* [D] (*Paris*), 248: 280–283.

36. Smith, J. R. (1970): A genetic study of the development of senescence in *Podospora anserina.* Ph.D. thesis, Yale University, New Haven, Conn.

37. Smith, J. R., and Rubenstein, I. (1973): The development of senescence in *Podospora anserina. J. Gen. Microbiol.,* 76:283–296.

38. Smith, J. R., and Rubenstein, I. (1973): Cytoplasmic inheritance of the timing of "senescence" in *Podospora anserina. J. Gen. Microbiol.,* 76:297–304.

39. Chevaugeon, J., and Lefort, C. (1960): Sur l'apparition régulière d'un "mutant": Infectant chez un champignon du genre *Pestalozzia. C.R. Acad. Sci.* [D] (*Paris*), 250:2247–2249.

40. Chevaugeon, J., and Digbeu, S. (1960): Un second facteur cytoplasmique infectant chez *Pestalozzia annulata. C.R. Acad. Sci.* [D] (*Paris*), 251:3043–3045.

41. Chevaugeon, J. (1962): Modification extra-chromosomique et age du thalle chez le *Pestalozzia annulata. C. R. Acad. Sci.* [D] (*Paris*), 255:1980–1982.

42. Chevaugeon, J. (1962): Conditions de la différenciation de mycélium modifié chez le *Pestalozzia annulata. C.R. Acad. Sci.* [D] (*Paris*), 255:3450–3452.

43. Chevaugeon, J. (1963): Une période d'incubation s'interpose entre deux événements impliques dans la modification extra-chromosomique du *Pestalozzia annulata. C.R. Acad. Sci.* [D] (*Paris*), 257:217–220.

44. Chevaugeon, J. (1966): Mise en évidence de mécanismes de répression de la variation extra-chromosomique chez *Pestalozzia annulata. C.R. Acad. Sci.* [D] (*Paris*), 263:120–123.

45. Chevaugeon, J., and Clouet, L. (1963): Temps et lieu des deux événements aléatoires impliqués dans la modification extra-chromosomique du *Pestalozzia annulata. C.R. Acad. Sci.* [D] (*Paris*), 256:4068–4071.

46. Chevaugeon, J., Clouet, L., and Michel, G. (1965): Nutrition, croissance et modification extra-chromosomique du *Pestalozzia annulata. C.R. Acad. Sci.* [D] (*Paris*), 261:517–520.

47. Holliday, R. (1969): Errors in protein synthesis and clonal senescence in fungi. *Nature,* 221:1224–1228.

48. Orgel, L. E. (1963): The maintenance of the accuracy of protein synthesis and its relevance to aging. *Proc. Natl. Acad. Sci. U.S.A.,* 49: 517–521.

49. Bertrand, H., McDougall, K. H., and Pittenger, T. H. (1968): Somatic cell variation during uninterrupted growth of *Neurospora crassa* in continuous growth tubes. *J. Gen. Microbiol.,* 50:337–350.

50. Rowlands, R. T., and Turner, G. (1973): Nuclear and extranuclear inheritance of oligomycin resistance in *Aspergillus nidulans. Mol. Gen. Genet.,* 126:201–216.

51. Rowlands, R. T., and Turner, G. (1974): Recombination between the extranuclear genes conferring oligomycin resistance and cold sensitivity in *Aspergillus nidulans. Mol. Gen. Genet.,* 133:151–161.

52. Mason, J. R., and Turner, G. (1975): Transmission and recombination of extranuclear genes during sexual crosses in *Aspergillus nidulans. Mol. Gen. Genet.,* 143:93–99.

53. Waldron, C., and Roberts, C. F. (1973): Cytoplasmic inheritance of a cold sensitive mutant in *Aspergillus nidulans. J. Gen. Microbiol.,* 78:379–381.

54. Guthrie, C., Nashimoto, H., and Nomura, M. (1969): Structure and function of *E. coli* ribosomes. VIII. Cold sensitive mutants defective in ribosome assembly. *Proc. Natl. Acad. Sci. U.S.A.,* 57:384–391.

55. Gunatilleke, I. A. U. N., Scazzocchio, C., and Arst, H. N., Jr. (1975): Cytoplasmic and nuclear mutations to chloramphenicol resistance in *Aspergillus nidulans. Mol. Gen. Genet.,* 137:269–276.

56. Rowlands, R. T., and Turner, G. (1974): Physiological and biochemical studies of nuclear and extranuclear oligomycin-resistant mutants of *Aspergillus nidulans. Mol. Gen. Genet.,* 132:73–88.

57. Ephrussi, B. (1953): *Nucleo-cytoplasmic Relations in Microorganisms.* Clarendon Press, Oxford.

58. Rowlands, R. T., and Turner, G. (1976): Maternal inheritance of mitochondrial markers in *Aspergillus nidulans. Genet. Res., Camb.* 28:281–290.

Chapter 10
Mitochondrial Genetics of *Paramecium*

In 1938 Sonneborn described a peculiar phenomenon in the ciliate protozoan *Paramecium aurelia* that occurred when animals from different stocks were allowed to swim together in the same culture: animals of certain stocks (called killers) had the ability to kill those of other stocks (called sensitives). In 1943, in a classic study of extrachromosomal inheritance that is known to every student of elementary genetics, Sonneborn showed that the killer phenomenon was the result of nucleocytoplasmic interaction. The killer factor, later called *kappa* and now identified as a bacterial endosymbiont, was localized in the cytoplasm, but its maintenance was dependent on the presence of the dominant nuclear gene *K*. These experiments and many others established *Paramecium* as a premier organism for the study of cytoplasmic inheritance and nucleocytoplasmic interactions; thus it was natural that this ciliate would become a model for the study of mitochondrial genetics. Experiments with *Paramecium* have provided insight into three important problems of mitochondrial heredity: intracellular mitochondrial selection, compatibility of genetically different mitochondria and nuclei, and identification of specific mitochondrial components as gene products of the nuclear or mitochondrial genome through the use of interspecific hybrids. The reader who would delve further into other aspects of cytoplasmic inheritance in *Paramecium* is referred to review articles by Preer (1–3) and Sonneborn (4–8) and the excellent book on the genetics of *Paramecium* by Beale (9), from which much of the following account of *Paramecium* biology is drawn.

BIOLOGY AND LIFE CYCLE OF *Paramecium aurelia*

A stock of *P. aurelia* consists of the descendants of a single animal isolated in nature. Sonneborn showed that *P. aurelia* stocks can be grouped into 14 varieties, each of which contains two mating types (6). For example, variety 1 contains mating types I and II, variety 2 contains mating types III and IV, and so forth. Matings between stocks of the same variety are usually fertile, but matings between stocks belonging to different varieties, when they occur at all, are usually inviable (6). Thus the *P. aurelia* varieties are reproductively isolated from one another in the sense of true biological species. Sonneborn (6) coined the term *syngen* as an alternative to the term *variety* in a classic article in which he used *Paramecium* to show that morphological species are by no means equivalent to biological species, since the varieties of *P. aurelia* are all morphologically similar. More recently, however, Sonneborn (10) applied the binomial nomenclature to these sibling species; so *P. aurelia* variety 1 or syngen 1 now becomes *P. primaurelia, P. tetraurelia* replaces the designation variety 4 or syngen 4, and so forth. In this chapter we will use the name *P. aurelia* when discussing material relevant to the species complex as a whole, but the specific binomial will be used in considering individual experiments.

P. aurelia is a diploid organism that reproduces asexually by binary fission. Each animal contains a large macronucleus and two micronuclei (Fig. 10–1). Prior to each fission, the two micronuclei and the macronucleus in each cell divide. When animals of different mating type but the same variety are mixed together under appropriate conditions, the cells form clumps, after which pairs of animals swim off together, with each pair consisting of one animal of each mating type. The animals are first attached at their anterior ends; later they are also attached in the region of their mouths (Fig.

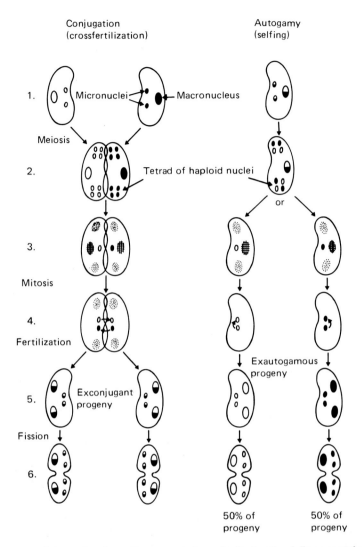

FIG. 10–1. Nuclear changes during conjugation and autogamy in *P. aurelia*. Conjugation: 1. Two parental animals of the same species, but belonging to different mating types, each with one macronucleus and two micronuclei. Nuclear differences with respect to any pair of nuclear alleles are indicated by open (○) and closed (●) nuclear areas. 2. The animals pair, and each micronucleus undergoes meiosis to give four haploid nuclei. 3. Seven of the haploid nuclei in each conjugant disintegrate, and the macronucleus breaks up. 4. The remaining haploid nucleus in each conjugant divides mitotically; one nucleus, the migratory nucleus, moves into its mate, where it fuses with the other, the stationary nucleus. 5. Each fusion nucleus divides twice mitotically. Two products give rise to macronuclei, and the other two become new micronuclei. The nuclei of the exconjugants are then heterozygous for any pair of nuclear alleles that differ between the two parents (indicated by half-filled nuclear areas). 6. Mitotic division of the micronuclei, accompanied by fission, initiates an exconjugant clone in which each animal once more contains one macronucleus and two micronuclei. Autogamy: All stages are identical with conjugation, except that only one animal is involved; hence the two mitotic nuclei at stage 4 fuse to give the fusion nucleus at stage 5. Because only a single haploid nucleus is involved in fusion, all exautogamous progeny become isogenic for any nuclear allelic differences present. For any pair of nuclear alleles, half the exautogamous progeny will carry one member of the pair, and the other half will carry the second. (Adapted from Jinks, ref. 38.)

10–1). Conjugation normally lasts 5 to 6 hr at 27 to 29°C.

In order to understand why *P. aurelia* is such a good subject for studies of cytoplas- mic inheritance, it is necessary to consider the cytologic and genetic changes that ac- company conjugation (Fig. 10–1). During conjugation the macronucleus disintegrates into fragments; the fragments become distributed (without multiplying) to the products of the successive divisions of the exconjugants and may disappear completely. Each pair of micronuclei undergoes a pair of meiotic divisions; then each conjugant contains eight haploid meiotic products, of which seven subsequently disintegrate. This leaves a single haploid nucleus in each conjugant, which then divides mitotically. Each parent donates one of these haploid nuclei to its mate. The donor and recipient nuclei fuse, and each fusion nucleus divides twice mitotically. Two of the products of the fusion nucleus eventually differentiate into new macronuclei, and the other two form the micronuclei. At the time of the first fission, the two micronuclei divide again, but the macronucleus does not; so each cell once again contains one macronucleus and two micronuclei. In some varieties a bridge is formed between certain pairs of conjugants, and through this bridge cytoplasm is exchanged. If the bridge persists long enough, the animals may eventually fuse to form double animals, which reproduce as such thereafter (8). Bridge formation can also be induced by treatment of the conjugating pairs with a dilute antiserum (11). As will be seen later, cytoplasmic exchange is important in proving that presumptive mitochondrial genes are cytoplasmically inherited and in studying competition between genotypically different mitochondria.

P. aurelia can also undergo a process of self-fertilization (called *autogamy*) under appropriate conditions (Fig. 10–1). In autogamy the two micronuclei in a single animal divide meiotically to form eight meiotic products, and the macronucleus disintegrates. Seven of the resulting meiotic products break down, and the remaining product divides once mitotically. These two genetically identical haploid nuclei then fuse to form a diploid fusion nucleus. The fusion nucleus, like that of a conjugating animal, divides twice; two of the resulting nuclei form macronuclei, and the other two form

micronuclei. The two new micronuclei then divide once mitotically, but there is no macronuclear division. Thus after the first fission the *exautogamous* animals, like the exconjugant animals, contain a single macronucleus and two micronuclei.

INHERITANCE OF ERYTHROMYCIN RESISTANCE IN *P. aurelia*

In order to understand why these rather complex sexual processes make *P. aurelia* such a good subject for the study of cytoplasmic inheritance, we will discuss the genetics of presumptive mitochondrial mutations to erythromycin resistance (E^R). Adoutte and Beisson (12) isolated a series of E^R mutations in *P. tetraurelia*. By definition, these mutants cannot be recessive nuclear mutations, since *P. aurelia* is a diploid animal; thus they must be either dominant nuclear mutations or somatically segregating cytoplasmic mutations. Diploids are useful for the selection of cytoplasmically inherited organelle mutations in other systems as well (see Chapters 5 and 13).

Adoutte and Beisson crossed an erythromycin-sensitive (E^S) stock homozygous for a recessive nuclear mutation to temperature sensitivity (*ts*) to each of three E^R mutations homozygous for the wild-type allele of temperature sensitivity (*ts$^+$*) isolated in a stock of opposite mating type (Fig. 10–2). Following conjugation in the absence of cytoplasmic exchange, it was found that all exconjugants expressed a wild-type phenotype with respect to the *ts* allele and that the exconjugant clone derived from the E^R parent remained E^R, whereas that derived from the E^S parent remained E^S. When exconjugants of both kinds were put through autogamy, it was found that half the clones were *ts* and the other half *ts$^+$*.

These results can be explained as follows. During conjugation the E^R conjugant donates a *ts$^+$* nucleus to its mate, and the E^S conjugant donates a *ts* nucleus in the opposite direction. Therefore the fusion nuclei of

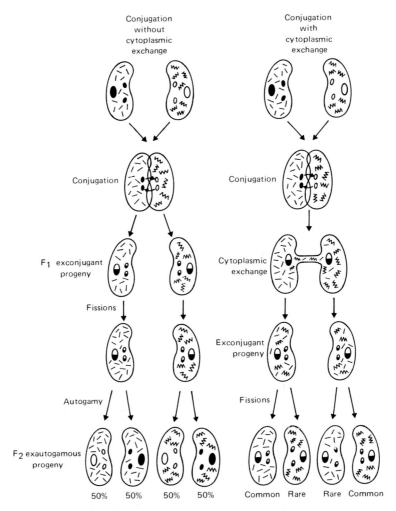

FIG. 10–2. Diagram illustrating how cytoplasmic inheritance of a mitochondrial trait such as erythromycin resistance can be demonstrated in *P. aurelia*. **Conjugation without cytoplasmic exchange:** Animals differing in mitochondrial phenotype (indicated by cytoplasmic symbols / and ⋀⋀) and by a pair of nuclear alleles (indicated by ● and ○ nuclear areas) are crossed. Each F₁ exconjugant resembles its parent with respect to the mitochondrial marker but is heterozygous for the nuclear alleles. These phenotypes are maintained stably through vegetative fissions. Following induction of autogamy, half the progeny of each exconjugant clone carry one nuclear allele of the pair, and the other half carry the other allele; but in any one clone all F₂ exautogamous progeny carry the mitochondrial allele present in the original parent. **Conjugation with cytoplasmic exchange:** Behavior of nuclear markers is the same as in conjugation without cytoplasmic exchange, except that mitochondria from each parent are transferred across the cytoplasmic bridge, so that the exconjugants contain a mixed mitochondrial population. The mitochondria segregate out during subsequent fissions, so that each exconjugant clone usually contains animals pure for both mitochondrial phenotypes. The most frequent mitochondrial phenotype in each exconjugant clone is generally that of the parent, but it is also controlled by intercellular and intracellular selection of specific mitochondrial phenotypes during vegetative growth.

both F₁ exconjugants will be *ts⁺/ts*, and all the exconjugant progeny of both parents will exhibit the wild-type phenotype with respect to temperature sensitivity. However, to prove that this is the correct genotype, it is necessary to cause the animals to undergo

autogamy. Following meiosis, the seven meiotic products that degenerate would be expected to do so at random with respect to the *ts⁺* and *ts* alleles. Thus half the animals should contain a *ts* nucleus and the other half a *ts⁺* nucleus. On completion of autogamy, half of the F_2 exautogamous clones should be temperature-sensitive, and the other half should not. Since that is the result observed, nuclear exchange must have occurred during conjugation, and meiosis must have proceeded normally in both conjugation and autogamy.

The E^R trait, on the other hand, behaved in an entirely different manner (Fig. 10–2). The exconjugant progeny of the E^R parent remained erythromycin-resistance, whereas those of the E^s parent were sensitive to erythromycin. The same was true of the exautogamous clones. Thus the E^R and E^s phenotypes are inherited clonally, and there

is no meiotic segregation, which shows that they are not inherited as nuclear alleles are.

To prove the cytoplasmic basis of inheritance of the E^s and E^R traits, Adoutte and Beisson also examined the progeny of conjugating pairs of animals in which cytoplasmic exchange had occurred (Fig. 10–2). By growing the exconjugant clone derived from the E^s parent in erythromycin-containing medium, they readily demonstrated that the sensitive cells gradually became resistant, following which they began to divide in the erythromycin-containing medium. The phenotypic transformation from sensitivity (dark color, slow swimming) to

TABLE 10–1. Antibiotic-resistant mitochondrial mutants of P. aurelia[a]

Species	Phenotype	Genotype	Cross-resistance		
			Erythro-mycin	Mika-mycin	Chloram-phenicol
P. primaurelia	Low resistance to erythromycin (125 μg/ml)	$513E^R$–018	+	+	?
	High resistance to erythromycin (250 μg/ml)	$513E^R$–48	+	−	?
	Mikamycin-resistant	$513M^R$–2	Low levels (125 μg/ml)	+	?
		$513M^R$–3	Low levels (125 μg/ml)	+	?
		$513M^R$–4	Low levels (125 μg/ml)	+	?
	Chloramphenicol-resistant	C^R	?	?	+
P. tetraurelia	Very weak resistance to erythromycin (slow growth at 100 μg/ml)	E_{50}^R	+	+	+
	Weak resistance to erythromycin (faster growth at 100 μg/ml)	E_1^R	+	−	−
	Moderate resistance	E_{111}^R	+	?	+/−
	to erythromycin	E_{37}^R	+	?	+/−
	(slow growth at	E_{104}^R	+	?	−
	400 μg/ml)	E_{110}^R	+	+	−
	Spiramycin resistance	S_7^R	+/− at 100 μg/ml	?	?
	Chloramphenicol resistance	C_4^R	+/− at 100 μg/ml	−	+

[a]This classification is adapted mainly from the articles of Beale (18) and Adoutte (21). Further details about individual mutants are to be found in these and other references.

resistance (clear color, rapid swimming) could actually be seen taking place within the living cell.

These experiments showed that the E^R determinant was transferred via the cytoplasm to the E^s conjugant, but the reciprocal transfer was at first more difficult to detect, since the E^s phenotype could not be selected for. Nevertheless, Adoutte and Beisson reported one instance in which an E^R exconjugant clone grown in nonselective medium and tested periodically for resistance first became heterogeneously E^R and E^s after 40 fissions, but by the 45th fission it became pure for E^s. These results not only demonstrated reciprocal cytoplasmic transfer of the E^s determinant into the E^R exconjugant but also formed the basis for the mitochondrial competition experiments that will be described shortly.

The experiments of Adoutte and Beisson demonstrated that cytoplasmically inherited mutations to erythromycin resistance could be selected in *P. aurelia,* and they showed that these segregated somatically following cytoplasmic exchange. Using slightly different methods, Beale (13) was able to demonstrate cytoplasmic inheritance of erythromycin resistance in *P. primaurelia* at about the same time. Since then, a number of other cytoplasmically inherited mutations to antibiotic resistance have been isolated in *P. primaurelia* and *P. tetraurelia* (Table 10–1). We shall now discuss the evidence that these are mitochondrial mutations.

NATURE OF CYTOPLASMICALLY INHERITED MUTATIONS TO ANTIBIOTIC RESISTANCE IN *P. aurelia*

Evidence that the mitochondria of sensitive (but not resistant) cells were drastically altered in ultrastructure by chloramphenicol and erythromycin was published by Adoutte et al. (14). These authors found that the mitochondria of sensitive cells became elongated and that the number of cristae decreased considerably. At the same time, certain abnormal cristal structures appeared. These changes were not observed in resistant mutants. Adoutte and associates pointed out that such changes were known to accompany blockage of mitochondrial protein synthesis in other organisms, and they suggested that mitochondrial protein synthesis was probably blocked by erythromycin and chloramphenicol in sensitive (but not resistant) cells of *P. aurelia*. At about the same time, Tait (15) published supporting evidence for this hypothesis. He found that mitochondrial ribosomes from sensitive (but not resistant) cells bound ^{14}C-labeled erythromycin.

One problem that arises in mitochondrial isolation experiments with *P. aurelia* stems from the conditions used to grow the animals. In nature, *P. aurelia* feed on other microorganisms such as bacteria and algae. In the laboratory, the animals are usually fed on a grass infusion or dried lettuce extract in which a pure culture of *Klebsiella aerogenes* has been growing. Consequently, in purifying mitochondria there is always the risk of bacterial contamination of the mitochondrial pellet. It would be expected that bacterial ribosomes would behave in much the same manner as mitochondrial ribosomes with respect to antibiotic sensitivity and resistance, as was discussed previously (see Chapter 3).

Tait considered this problem; bacterial contamination as an explanation for his results was ruled out on the following grounds: (a) The bacteria used for culture were the same for both resistant and sensitive animals, and the bacteria were themselves sensitive to erythromycin. (b) Bacteria were not lysed under the conditions used to lyse the mitochondria, and bacteria should not have been present in the postmitochondrial supernatant used for ribosome analysis. (c) The mitochondrial ribosomes sedimented at 80S, whereas the bacterial ribosomes sediment at 70S. There was no evidence for 70S peaks in Tait's gradients. In another article, Beale et al. (16) reported that the change from sensitivity to resistance was accompanied by a change in one or more mitochondrial ribosomal proteins. However, the data were unconvincing, since the mitochondrial ribosomal subunits were not separated, and the resolution of the one-dimensional gels used to distinguish the ribosomal proteins was poor. Only 16 bands were resolved, which is far too few (see Chapter 3).

The foregoing observations were consistent with the notion that the E^R mutations were probably mitochondrial, but direct evidence was still lacking. Beale et al. (16) attempted to provide this proof by means of mitochondrial injection experiments. They prepared cell fractions from a cytoplasmically inherited E^R mutant of P. primaurelia and injected them into sensitive animals. The method used was originally devised by Koizumi and Preer (17) for the injection of kappa and related P. aurelia endosymbionts from one stock to another. The authors found that a cell fraction from the E^R mutant greatly enriched in mitochondria could bring about transformation of the E^S recipient to the E^R phenotype. They concluded that the E^R determinant was localized in the mitochondria. These results were extended to double mutants by Beale (18). By successive mutations he obtained a double mutant resistant to erythromycin and mikamycin ($E^R M^R$) in P. primaurelia, and in mitochondrial injection experiments using recipients sensitive to both antibiotics, he showed that the two markers were transferred together.

The ultrastructural events accompanying transformation of E^S cells to the E^R phenotype following mitochondrial injection were studied by Knowles (19). He injected sensitive cells with a mitochondrial fraction from E^R cells and then placed the injected cells in erythromycin-containing medium. He observed that such injected cells developed resistance between 4 and 15 days following injection. However, it appeared that before

becoming completely resistant, the transforming cells passed through an intermediate stage in which they changed from slow swimming to fast swimming. These were the cells Knowles compared to sensitive and resistant animals in terms of ultrastructure.

Like Adoutte et al. (14), Knowles observed that the mitochondria of sensitive cells lost cristae in the presence of erythromycin, and the mitochondria became irregular in shape. Also, bundles of round "tubules" were seen in these mitochondria. None of these changes were seen in E^R cells in the presence of erythromycin, and the mitochondria of these cells were indistinguishable from those of untreated cells. Knowles found that cells in the process of transformation from sensitivity to resistance contained three kinds of mitochondria: those that were unaffected in structure by erythromycin and presumably were resistant to the antibiotic, those that were similar in appearance to the mitochondria of sensitive cells exposed to erythromycin, and those with intermediate mitochondrial profiles that had some of the characteristics of both sensitive and resistant mitochondria. Knowles's main conclusion was that within a single transforming cell, two and often three classes of mitochondria can exist in a common cytoplasmic environment.

Although these mitochondrial injection experiments provided strong suggestive evidence that the cytoplasmically inherited antibiotic resistance markers were localized in the mitochondrion, they were subject to the criticism made earlier concerning bacterial contamination. Beale et al. (16) reported that 80% of the particulate components in the pellet used for injection appeared to be mitochondria and 10% looked like bacteria. Two questions immediately arose: Where did the bacteria come from? Did the bacteria rather than the mitochondria cause the transformation from sensitivity to resistance? The answer to the first question was that the bacteria could have come from either of two places. Since P.

aurelia cells were cultured on infusions containing *K. aerogenes,* it was possible that organisms belonging to this species were contaminants. The second possibility was that the bacteria were endosymbionts existing within the cells. Preer et al. (3) pointed out in a comprehensive review that *kappa* and related endosymbionts are almost certainly Gram-negative bacteria. They also mentioned that endosymbionts called *mu* are found in *P. primaurelia,* although they were not reported by Preer and associates specifically for the stocks used by Beale et al. (16) in their microinjection experiments. The second question could be answered only on logical grounds at the time. It could be argued that contaminating bacteria could not account for the results of the injection experiments, but this would be difficult to prove. Cummings et al. (20) subsequently provided rigorous proof that transformation of recipient animals from erythromycin sensitivity to resistance by donor mitochondria was accompanied by transformation of the mtDNA of these animals from that originally present in the recipient to that found in the donor. This demonstration was possible because interspecific mitochondrial transfer can be achieved in both directions between *P. primaurelia* and *P. pentaurelia,* and both species can also donate their mitochondria to *P. septaurelia* (see pp. 319–321 for further discussion). If the donor cells are E^R and the recipient cells are E^S, erythromycin-resistant hybrid animals can be selected by injecting the recipient cells with the mitochondrial fraction obtained from the donor cells and then growing the injected animals on erythromycin-containing medium. Cummings and associates examined the EcoR1 restriction enzyme patterns of mtDNA from all three species and found that they were all different. The mtDNA of hybrid cells in which the E^R marker of the donor was selected following injection of donor mitochondria had the same restriction enzyme pattern as mtDNA obtained directly from

the donor, rather than that of the recipient; thus selection of the donor E^R marker was accompanied by selection for donor mtDNA.

INTRACELLULAR MITOCHONDRIAL SELECTION AND EXPRESSION OF MITOCHONDRIAL MUTATIONS

Cells of *P. aurelia* contain several thousand mitochondria (21). A mixed mitochondrial population can be created either by isolating exconjugant animals from pairs in which cytoplasmic exchange has occurred or by the use of mitochondrial injection. As a result, *P. aurelia* is the most favorable organism currently available in which to study the expression of mitochondrial gene mutations and mitochondrial selection (suppressiveness?) in mixed populations.

Adoutte and Beisson (22) made use of the cytoplasmic exchange technique to study both problems in *P. tetraurelia.* They used both E^R and chloramphenicol-resistant (C^R) mitochondrial mutations. Certain of the E^R mutations die after prolonged exposure to high temperature, but these are clearly distinguishable from the previously described nuclear *ts* mutation that in the homozygous condition causes rapid death at high temperature.

Adoutte and Beisson made crosses between animals carrying the E^R or C^R markers and sensitive cells or between C^R and E^R animals. Conjugating pairs in which cytoplasmic bridges had formed were isolated. In most such pairs the bridge is of short duration; in a few pairs it persists, and eventually the two mates fuse to form a double animal, which then reproduces clonally as such. The exconjugant animals from such pairs can be studied under selective conditions or nonselective conditions. Under selective conditions the expression of a mitochondrial mutation to resistance can be studied; under nonselective conditions intracellular selection of different mitochondrial genotypes can be studied. If a cross is made between E^R and E^s animals, and the exconjugant cell derived from the E^s parent is taken, it will contain a majority of E^s mitochondria and a minority of E^R mitochondria. If such cells are put into selective erythromycin-containing medium, they will divide two or three times and then stop for a period of 2 to 4 days. During this period the cells express the E^s phenotype and are dark in color and slow swimming. They then progressively become transformed to the E^R phenotype, and the animals are characterized by a clear color and fast swimming.

The transformation from sensitivity to resistance can be assayed genetically by removing the cells undergoing transformation from their selective medium, allowing them to divide under nonselective conditions, and again testing the clonal progeny for resistance and sensitivity under selective conditions. These experiments show that the transformation takes place very rapidly, so that by the time the cells begin to divide and exhibit the E^R phenotype, they are also pure genotypically for the E^R mutation. The lag between the time the sensitive exconjugants are placed in erythromycin-containing medium and the time the cells begin to express resistance and divide again is proportional to the character and size of the cytoplasmic bridge. These results suggest that the rate at which resistance is expressed depends on the "dosage" of mitochondria or mitochondrial genes contributed by the resistant parent. As might be expected, the resistant exconjugant divides rapidly in erythromycin-containing medium, since it contains few sensitive mitochondrial genomes.

Perasso and Adoutte (23) examined the mitochondrial ultrastructural changes accompanying the transformation from erythromycin sensitivity to resistance and at the same time assayed genotypic purification. They found that the transformation process could be divided roughly into three stages. In the first stage, cells divide two or three times and then stop. These cells have sensi-

tive-looking mitochondria and produce mostly sensitive progeny on subculture under nonselective conditions. In the second stage the cells assume a healthier appearance and begin to divide. Three types of mitochondria can be distinguished in these cells: clearly sensitive-looking mitochondria with dense matrix and few or no cristae, normal resistant-appearing mitochondria, and intermediate mitochondria resembling resistant mitochondria in terms of the clear matrix but resembling sensitive mitochondria in terms of the wavy arrangement of cristae. On subculture under nonselective conditions, these cells are found to be highly enriched in E^R mitochondria. It should be noted that the ultrastructural changes described by Perasso and Adoutte for transforming cells obtained as a result of cytoplasmic exchange are reminiscent of those seen by Knowles (19) in similar cells obtained by mitochondrial injection.

Perasso and Adoutte found that in the final stage of transformation the cells became fully resistant in terms of ultrastructural phenotype and in terms of genotype, as assayed by the genotypic purification technique. As these authors pointed out, their experiments suggest an active multiplication of resistant genomes under selective conditions where the cells are dividing slowly or not at all. At the same time, sensitive genomes must disappear. Although the mechanism by which transformation takes place is unknown, these remarkable experiments may provide a key to one of the great mysteries of organelle genetics, for they demonstrate directly that an organelle genotype that is greatly in the minority within a cell can rapidly replace the majority type under selective conditions. In a sense, these experiments serve as "reconstructions" with respect to how a new organelle mutation becomes expressed. In short, it may not be necessary to invoke notions such as master copies or small numbers of genetically functional organelle genomes to explain why mutations can be obtained even though there are so many copies of chloroplast or mitochondrial DNA per organelle or per cell. Rather, a different process may be operative in which intracellular replacement of one organelle genome by another occurs under selective conditions in the absence of significant cell division.

As was mentioned at the beginning of this section, intracellular competition or selection for different mitochondrial genotypes can be studied by allowing cytoplasmic exchange to take place and then culturing the exconjugant clones under nonselective conditions for varying numbers of fissions, following which some of the cells are challenged with antibiotic. Adoutte and Beisson (22) used this method to examine the competition between various combinations of resistant and sensitive mitochondria and between E^R and C^R mitochondria. Two types of results were obtained (Table 10–2). Either the two mitochondrial genotypes were relatively stable, and cells remained mixed for a great number of fissions (more than 80 for C^R/C^S combinations), or one of the two mitochondrial genotypes (always the same for a given combination) was progressively eliminated, and cells pure for the favored mitochondrial genotype were recovered between the 20th and 40th fissions. In the latter case, the wild-type mitochondrial genotype was always favored. Moreover, among the various mutants studied, a classification could be made according to their relative degrees of stability in what Adoutte and Beisson (22) termed "heteromitochondriotes." Thus in terms of stability, $C^R_2 > E^R_{102} > E^R_1 > E^R_{38} > C^R_2 E^R$. This classification paralleled the classification of the mutants in terms of increasing thermosensitivity, with C^R mutants being thermoresistant and the $C^R E^R$ genotype being the most thermosensitive tested. At the intercellular level, Adoutte et al. (24) reported the same relationship for mixed populations of genotypically pure cells; this can be related to cell generation time at 26°C.

TABLE 10–2. *Evolution of various mixed mitochondrial populations in* P. tetraurelia *during growth on nonselective medium*[a]

Mechanism of cytoplasmic exchange	Mitochondrial genotypes	Derivation of exconjugant	Percentage of resistant cells at fission number:[b]							
			10	20	30	40	50	60	70	80
Bridge formation during conjugation	$E_1^R \times E^S$	Sensitive parent	50	—	13	—	—	0	—	—
		Resistant parent	100	—	100	100	100	0	—	—
	$E_{102}^R \times E^S$	Sensitive parent	R[c]	R	24	—	—	—	—	—
		Resistant parent	R	R	100	—	—	—	—	—
	$E_{38}^R \times E^S$	Sensitive parent	100	—	7	—	0	—	—	—
		Resistant parent	100	—	100	—	0	—	—	—
	$C_2^R \times C^S$	Sensitive parent	R	R	93	60	60	100	11	0
		Resistant parent	R	R	100	100	100	100	100	100
	$C_2^R E^R \times C^S E^S$	Sensitive parent	0	—	0	—	—	—	—	—
		Resistant parent	100	—	0	—	—	—	—	—
Double-animal formation[d]	$E_1^R \times E^S$	Not applicable	100	11	—	—	—	—	—	—
	$E_{102}^R \times E^S$	Not applicable	100	90	30	0	—	—	—	—
	$C_2^R \times C^S$	Not applicable	100	100	100	—	—	32	—	R

[a] It will be noted that many of the allele numbers for erythromycin resistance (E^R) and chloramphenicol resistance (C^R) do not correspond to those given in Table 10–1. A complete classification of the alleles being used in the *Paramecium* experiments has not yet been published.

[b] In the original experiments, clones were not tested for the fraction of resistant animals at precisely 10, 20, etc., fissions. For ease of comparison, we have modified the original data in this respect.

[c] R indicates that only one to three cells were tested and were found to be resistant.

[d] Except for the first cross, the percentages represent pooled data from several double animals.

Adapted from Adoutte and Beisson (22).

Thus wild type has a generation time of 6 hr, E^R_1 6.5 hr, and $C^R E^R$ 10 hr. Thus selection operates in the same direction at both intracellular and intercellular levels with respect to mitochondrial genotype. It is perhaps worth noting that since mitochondrial gene recombination is rare in *P. aurelia* (*vide infra*), these experiments show that the phenomenon of suppressiveness, at least in this ciliate, can be explained in terms of competition between different mitochondrial genomes and mitochondrial phenotypes. Thus wild-type mitochondria are suppressive to each of the mutants tested (except possibly C^R), C^R is suppressive to E^R, and so forth.

Limited mitochondrial competition experiments have also been reported in *P. primaurelia* by Beale (18). In these experiments Beale induced cytoplasmic exchange with dilute antiserum in conjugating pairs of animals; in each pair, one member carried a mutation causing resistance to high levels of erythromycin (E^R-48), and the other was resistant to mikamycin (M^R-4). Both markers can be transferred by mitochondrial injection. Beale obtained five pairs in which small cytoplasmic bridges were formed. One exconjugant clone produced cells resistant only to erythromycin; the other yielded progeny that if selected on mikamycin became mikamycin-resistant, but if selected on erythromycin became erythromycin-resistant. These results imply a competitive advantage of E^R to M^R mitochondria, so that mikamycin resistance is never expressed among the exconjugant progeny of the E^R parent, but erythromycin resistance is expressed among the exconjugant progeny of the M^R parent.

MITOCHONDRIAL COMPATIBILITY: ROLE OF NUCLEAR MUTATION *cl-1* IN *P. tetraurelia*

Sainsard and associates (25–27) reported on a recessive nuclear mutation, *cl-1,* in which the mitochondrial genome was also altered. These articles are particularly interesting in that they illustrate how an organelle incompatibility mechanism can arise with respect to the nuclear genome as a result of a minimal number of genetic alterations. One of the articles (27) also elegantly demonstrated how genetic analysis can be used to reveal that a specific suppressor mutation is localized in the mitochondrion rather than in the nucleus.

Homozygous *cl-1/cl-1* cells have a slower growth rate than wild-type cells, and they are very deficient in cytochrome a + a_3 (25,28). Although the mutant cells have normal respiration rates, 80% of the respiration is cyanide-insensitive; in the wild type this level is 10 to 20%. The *cl-1* mutation is also weakly thermosensitive at high temperatures; this is important because the thermosensitivity characteristic of *cl-1* has been used in some instances to score for the mutation (27). In other instances a different and apparently more stringent nuclear mutation to temperature sensitivity (*ts-401*) has been scored in crosses involving *cl-1* (25). That is, cells homozygous for *cl-1, ts-401,* and the double mutant are apparently distinguishable.

The genetics of the system were first defined by Sainsard and associates (25). These authors crossed cells homozygous for *cl-1* with cells homozygous for the unlinked *ts-401* (*ts*) marker (Fig. 10–3). The F_1 exconjugants were therefore heterozygous for both genes, but they maintained their original cytoplasmic parentage, since there was no cytoplasmic exchange (Fig. 10–3). Phenotypically, the F_1 was thermoresistant and had normal growth rate, indicating the dominance of the two wild-type alleles (i.e., *cl-1⁺* and *ts⁺*).

In the exautogamous F_2, the wild-type and mutant alleles of each gene segregated 1:1, proving the nuclear heterozygosity of the F_1. However, the striking observation was that the exautogamous progeny homozygous for *cl-1* differed with respect to which of the two original conjugants had produced them, even though 20 fissions or more had elapsed between conjugation and

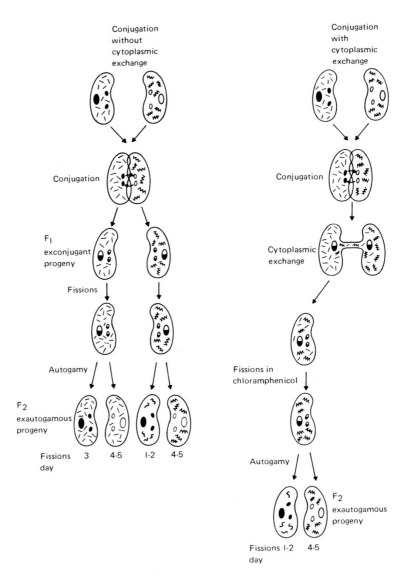

FIG. 10–3. Diagram illustrating inheritance of *cl-1* mutation and M^+ and M^{cl} mitochondria in *P. tetraurelia*. **Conjugation without cytoplasmic exchange:** Animals homozygous for the *cl-1* mutation (indicated by ● nuclear areas) and containing M^{cl} mitochondria (indicated by / cytoplasmic symbols) are allowed to conjugate with animals homozygous for the *cl-1*$^+$ allele (indicated by ○ nuclear areas) and containing M^+ mitochondria (indicated by ⋀⋀ cytoplasmic symbols). Each F_1 exconjugant and its descendants resemble its parent with respect to mitochondrial phenotype, but both exconjugant clones are then heterozygous for *cl-1* and its wild-type allele (indicated by half-filled nuclear areas). Following autogamy, homozygous *cl-1*$^+$ progeny divide at the rate of 4 to 5 fissions per day whether they contain M^+ or M^{cl} mitochondria; homozygous *cl-1* progeny containing M^{cl} mitochondria undergo 3 fissions per day, but homozygous *cl-1* progeny containing M^+ divide only once or twice per day, and the mitochondria assume a very abnormal morphology (indicated by the cytoplasmic symbols ∼). **Conjugation with cytoplasmic exchange:** In these experiments the homozygous *cl-1*$^+$ parent also carried a mitochondrial mutation conferring chloramphenicol resistance, and the *cl-1* parent was chloramphenicol-sensitive. The exconjugant progeny produced by the *cl-1* parent were selected in chloramphenicol-containing medium in order to replace the M^{cl} mitochondria with M^+ mitochondria. Following autogamy, the homozygous *cl-1*$^+$ progeny divided at the rate of 4 to 5 fissions per day, but the *cl-1* progeny divided only once or twice a day, once again reflecting the incompatibility of M^+ mitochondria with this genotype. The symbols used in this part of the figure have the same meaning as those used in the first part.

autogamy (Fig. 10–3). Homozygotes for the *cl-1* gene derived from the *cl-1* exconjugant clones following autogamy had the same growth rate as the *cl-1* parent, but *cl-1* homozygotes derived from the exconjugant clones produced by the *ts* parent had a drastically reduced growth rate. This reduction in growth rate was reflected in mitochondrial phenotype. Although the exconjugant clones derived from the *ts* parent had normal mitochondria (M^+), the mitochondria of the exautogamous *cl-1* F_2 progeny of these clones were disorganized and had few cristae. Thus the cytoplasmic difference associated with *cl-1* mitochondria (M^{cl}) and M^+ mitochondria was stable for at least 20 fissions between conjugation and autogamy, even though both exconjugant clones were heterozygous for the *cl-1* allele. On autogamy, there was a drastic interaction between M^+ mitochondria and the *cl-1* gene, but M^{cl} mitochondria were unaffected by this gene. No detectable interaction was noted between the *cl-1*+ homozygote and M^{cl} mitochondria.

In the presence of the homozygous *cl-1* allele, M^+ mitochondria gradually became transformed over a period of 15 to 25 generations to the M^{cl} phenotype. It is not known whether physiologic or genetic transformation is involved in this change from incompatible to compatible mitochondrial phenotype.

To demonstrate more rigorously that the incompatible reaction really involved mitochondria, and not some other cytoplasmic factor, Sainsard et al. (25) crossed a strain with normal mitochondria carrying a mitochondrial mutation to chloramphenicol resistance (M^+C^R) to a strain homozygous for *cl-1* and chloramphenicol-sensitive ($M^{cl}C^S$) (Fig. 10–3). Conjugating pairs in which cytoplasmic exchange had occurred were selected and the F_1 progeny derived from the *cl-1* parent were exposed to chloramphenicol. In this way M^+C^R mitochondria transferred by cytoplasmic exchange were selected, so that the *cl-1* exconjugants then

had M^+C^R mitochondria instead of $M^{cl}C^S$ mitochondria (Fig. 10–3). When these cells were caused to undergo autogamy, the exautogamous *cl-1* homozygotes carrying M^+C^R mitochondria behaved as *cl-1* homozygotes carrying M^+ mitochondria in the original cross where no cytoplasmic exchange had taken place (Fig. 10–3). That is, the mitochondria became disorganized, and the cells divided poorly. In other words, the incompatibility reaction will occur either when a *cl-1* nucleus is substituted into a strain carrying M^+ mitochondria or when M^+ mitochondria are introduced into a strain that originally contained M^{cl} mitochondria. The reaction in either case is specific for M^+ mitochondria and the *cl-1* gene in homozygous condition. The cytoplasmic distinction between M^+ and M^{cl} mitochondria is maintained for at least 20 fissions in clones heterozygous for the *cl-1* gene.

The mitochondrial transfer experiments also revealed that transformation of the *cl-1* exconjugant cells by M^+C^R mitochondria was a slow and inefficient process in contrast to crosses in which *cl-1* was absent. This observation suggested that the mutant phenotype of the cytoplasm persisted for a time in these cells, despite the fact that they were then heterozygous for the *cl-1* gene. This observation was confirmed in more detail by Sainsard-Chanet (26) using E^R and E^S alleles as mitochondrial markers. In essence, these experiments showed that M^{cl} mitochondria become expressed efficiently in the exconjugant clone derived from the *cl-1*+/*cl-1*+ parent as well as in that derived from the *cl-1*/*cl-1* parent. However, M^+ mitochondria are rarely selected and expressed in exconjugant animals derived from the *cl-1*/*cl-1* parent. In this same series of experiments, Sainsard-Chanet also succeeded in obtaining some exconjugant progeny of mixed mitochondrial genotype that were then induced to undergo autogamy. In most of the exautogamous progeny, M^+ mitochondria were rejected in favor of M^{cl} mitochondria if the nuclear genotype of the animals was *cl-1/cl-1;* the converse was true of *cl-1⁺/cl-1⁺* exautogamous progeny. That is, in homozygotes there is a selection for "self" with regard to mitochondria.

Finally, Sainsard (27) reported on the genetics of an apparent mitochondrial suppressor (*su*) of the *cl-1* gene. This mutant was isolated as a spontaneous fast-growing revertant of *cl-1*. However, *su* did not appear to have significantly more cytochrome a + a_3 than *cl-1*, nor had it lost the property of thermosensitivity associated with *cl-1*. When *cl-1 su* animals were allowed to conjugate with *cl-1⁺ su⁺* animals under conditions where there was no cytoplasmic exchange, both *cl-1 su* and *cl-1⁺ su⁺* were thermoresistant with a short generation time. That is, the *cl-1⁺* allele behaved as if it were dominant; as expected, there was no nuclear mitochondrial interaction in either exconjugant clone, which led to a slower rate of growth. The F_2 exautogamous progeny derived from the *cl-1 su* exconjugants all had short generation times whether they were genotypically homozygous for *cl-1* or for its wild-type allele, but the exautogamous progeny derived from the *cl-1⁺ su⁺* exconjugants grew rapidly only if they were homozygous for the *cl-1⁺* allele. Exautogamous progeny homozygous for *cl-1* grew very slowly and manifested the typical reaction expected if they had inherited *M⁺* mitochondria.

Sainsard considered two hypotheses to explain these results. According to the first, *su* was a nuclear suppressor closely linked to *cl-1* that would suppress the slow growth of *cl-1* homozygotes only when they contained M^{cl} mitochondria. The second hypothesis was that *su* was a cytoplasmic, presumably mitochondrial, suppressor. To determine whether or not *su* was a mitochondrial suppressor mutation, Sainsard made use of homozygous *cl-1⁺* exautogamous progeny derived from the *cl-1 su* exconjugant clones from the previous cross.

These animals were also *E*ˢ and would contain M^{cl} *su* mitochondria assuming that *su* is a mitochondrial suppressor.

These homozygous *cl-1⁺* $E^s M^{cl}$ *su* animals were allowed to conjugate with homozygous *cl-1* $E^R M^{cl}$ *su⁺* animals with and without cytoplasmic exchange. If *su* is a mitochondrial mutation, then in the absence of cytoplasmic exchange, all exautogamous F_2 progeny of the *cl-1⁺* exconjugant should have the mitochondrial genotype $E^s M^{cl}$ *su*, and these animals should be fast growing, whether they are *cl-1* or *cl-1⁺* homozygotes. This result was observed. Conversely, the exautogamous progeny of the *cl-1* exconjugant should have the mitochondrial genotype $E^R M^{cl}$ *su⁺*, and whereas *cl-1⁺* homozygotes containing these mitochondria ought to divide rapidly, the *cl-1* homozygotes should divide with the longer generation time that characterizes M^{cl} *cl-1* homozygotes. This result was also observed.

If *su* is a mitochondrial mutation, it should be possible to transfer it cytoplasmically in conjugating pairs in which cytoplasmic exchange has occurred. Sainsard showed that this was possible in the following way. Erythromycin resistance was selected for among exconjugant progeny from the *cl-1⁺* parent following cytoplasmic exchange. If *su* is a mitochondrial mutation, these animals should then be of the mitochondrial genotype $E^R M^{cl}$ *su⁺* instead of $E^s M^{cl}$ *su*. Therefore, exautogamous F_2 progeny homozygous for the *cl-1* allele should no longer have the same generation time as exautogamous homozygous *cl-1⁺* progeny because *su* is absent. That is, the *cl-1* homozygotes should divide slowly, and the *cl-1⁺* homozygotes should divide rapidly. This prediction was realized. Similarly, an erythromycin-sensitive *cl-1* exconjugant clone was obtained following cytoplasmic exchange. These cells should then have the mitochondrial genotype $E^s M^{cl}$ *su* instead of the genotype $E^R M^{cl}$ *su⁺*. Therefore, homozygous *cl-1* F_2 exautogamous progeny should then have the same rapid generation time as those homozygous for the *cl-1⁺* allele. This result was obtained.

Sainsard concluded from these intricate genetic analyses that *su* was a mitochondrial mutation linked to *Eˢ*. She also concluded that *su* could suppress the interaction between *M⁺* mitochondria and the homozygous *cl-1* nucleus, but there is no reason to assume in the critical crosses just discussed that *M⁺* mitochondria were ever involved (*vide supra*). It appears that Sainsard's results warrant only the conclusion that *su* suppresses the interaction between *Mᶜˡ* mitochondria and the homozygous *cl-1* nuclear genotype, which results in a reduction in the growth rate.

Sainsard et al. (25) considered two kinds of hypotheses that might explain the incompatibility reaction and the differences between *M⁺* and *Mᶜˡ* mitochondria. The first assumes that compatibility is determined by a difference in the mitochondrial genomes of *M⁺* and *Mᶜˡ* mitochondria. According to this hypothesis, transformation of *M⁺* mitochondria to the *Mᶜˡ* phenotype in *cl-1* homozygotes is the result of mutation from *M⁺* to a form similar to *Mᶜˡ*, following which the mutant mitochondria replicate selectively and the cells begin to divide normally. This hypothesis leads to the interesting prediction that the *M⁺ cl-1/cl-1* genotype can be used for selection of new mitochondrial mutants. The second hypothesis assumes that compatibility is determined by physiologic differences. Sainsard and associates considered two possibilities. The first was that the *cl-1* mutation affects the "pattern" of organization of the mitochondrial membrane and that the latter is endowed with a certain "structural inheritance" similar to the cortical inheritance described in *Paramecium* by Beisson and Sonneborn (8, 29). These elegant experiments showed that the inheritance of various structural properties of the cell cortex depended on neither nuclear nor cytoplasmic genes, but rather on

the structural properties of the cortex itself. In this hypothesis the *cl-1* gene would be responsible for the *Mᶜˡ* structural pattern, and the mutant mitochondrial pattern would persist in cells heterozygous for the *cl-1* mutation. If cells containing *M⁺* mitochondria were rendered homozygous for *cl-1*, the conversion from *M⁺* to *Mᶜˡ* would be a long and difficult process because the mutant gene product would not be easily inserted into the wild-type mitochondrial organization pattern.

The second possibility was that the difference between *M⁺* and *Mᶜˡ* mitochondria can be correlated in some way with mitochondrial metabolism. If respiration proceeds mainly through the cytochrome chain in *M⁺* mitochondria, but via the cyanide-insensitive pathway in *Mᶜˡ* mitochondria (which are also deficient in cytochrome $a+a_3$) the transformation will involve conversion of mitochondria with normal levels of cytochrome oxidase and normal respiration to mitochondria that are deficient in cytochrome oxidase and have high levels of cyanide-insensitive respiration. This hypothesis does not adequately explain the slowness of the transformation of *M⁺* mitochondria in *cl-1* homozygotes, nor does it explain why *Mᶜˡ* mitochondria are maintained for so many fissions in *cl-1* heterozygotes.

For the moment, the hypotheses of mutation selection and structural inheritance of pattern must be regarded as having equal merit, since they have not been subjected to rigorous experimental test.

MITOCHONDRIAL COMPATIBILITY: INTERSPECIFIC MITOCHONDRIAL INJECTION EXPERIMENTS

Beisson and associates (30), and more recently Beale and Knowles (31), reported the results of experiments on mitochondrial compatibility in which mitochondria were transferred between species by injection of

E^R donor mitochondria into an E^S host, following which the E^R mitochondria were selected for by attempting to grow the recipient paramecia on erythromycin-containing medium. The three species used in these experiments were *P. primaurelia, P. pentaurelia,* and *P. septaurelia,* which will be designated species 1, 5, and 7, respectively, for the sake of brevity, as was done by Beale and Knowles. These are probably the most closely related species of the *P. aurelia* group, with 1 and 5 having the closest kinship.

There were three major conclusions from these mitochondrial transfer experiments. First, mitochondria from species 1 and 5 can readily be transferred to any of the three species 1, 5, and 7, but species 7 mitochondria have never been transferred successfully to species 1 or 5. Second, Beale and Knowles reported that there are indications that interspecific mitochondrial transfers, where they can be made, are sometimes less successful than intraspecific mitochondrial transfers. Third, mitochondria from nuclear/mitochondrial "hybrids" sometimes behave in a manner differing markedly from that of mitochondria of the "pure" species. Thus species 5 mitochondria, after having been injected and established in cells of species 7, were never successfully injected back into species 5 and only rarely into species 1. Similarly, species 1 mitochondria, once established in species 7 animals, could only rarely be transferred back to species 1 and never to species 5. Therefore, when species 1 or 5 mitochondria are injected into species 7, they acquire the compatibility characteristics of species 7 mitochondria. Similar but less striking results are observed in transfers from species 1 to species 5 and the reverse. Thus species 1 mitochondria that have been established in species 5 animals can be transferred to species 5 cells with greater ease than mitochondria taken directly from species 1.

All these results favor nuclear control of mitochondrial compatibility. In this respect it is important to note that Beisson et al. (30) earlier reported some provisional evidence favoring the hypothesis that mitochondrial genetic determinants are responsible for certain compatibility differences between species 1 and 5 mitochondria. These authors reported that species 1 mitochondria could be transferred to species 7 with only low efficiency, whereas species 5 mitochondria could be transferred to species 7 with high efficiency. When species 1 mitochondria were first injected into species 5, and mitochondria from these hybrids were then injected into species 7, the same low efficiency of transfer characteristic of species 1 mitochondria, as opposed to species 5 mitochondria, was seen again. In short, the results suggested that species 1 mitochondria had certain heritable properties not modifiable by growth in species 5 that inhibited their efficiency of transfer into species 7. The more recent experiments of Beale and Knowles (31) revealed a great deal of uncontrolled variation among mitochondrial preparations from the same species in terms of transferability; thus the original results reported by Beisson and associates can no longer be regarded as proving the existence of mitochondrial factors controlling compatibility even if such factors do exist.

Several mechanisms have been considered in the attempt to explain the results of these interspecific mitochondrial transfer experiments. Among them are restriction and modification of mtDNA by nuclear genes and a structural compatibility model similar to that already set forth for the *cl-1* mutation (p. 319). One hypothesis that can be eliminated is that the mtDNA of the donor mitochondrion recombines with the recipient to yield E^R recombinant mitochondria, which are then selected; this hypothesis was ruled out because restriction enzyme analysis of the mtDNA of the hybrid animals revealed that they contained donor mtDNA (p. 312). Whatever the mechanism of mitochondrial compatibility, the discovery of the phenomenon opens up an interesting area for exploration in mitochondrial heredity. Perhaps *P. aurelia* hybrids can serve as models to help us understand the sorts of mechanisms governing the inheritance of mitochondria (or mtDNA) in sexual crosses in other organisms.

MITOCHONDRIAL GENE RECOMBINATION IN *Paramecium*

Adoutte (21) made several efforts to isolate mitochondrial recombinants in *P. tetraurelia* and found them to be very rare. He began by attempting to isolate $C^R E^R$ recombinants from various $C^R E^S$ by $C^S E^R$ crosses, but no conclusions could be drawn from these experiments because of the high spontaneous frequency of E^R mutations in C^R strains. Therefore Adoutte turned his attention to isolation of $C^S E^S$ recombinants from such crosses, pointing out that C^R and E^R mutants did not revert to sensitivity at an appreciable frequency.

Adoutte studied a cross involving the alleles E^R_6 and C^R_4 using conjugating pairs in which cytoplasmic exchange had taken place. Clones were obtained from each exconjugant, and these were subcloned every 10 fissions using a single cell. C^R mitochondria showed a selective advantage over E^R mitochondria; so the E^R exconjugant clones became progressively enriched with C^R mitochondria during vegetative growth. At the 60th fission the subclones derived from the original E^R exconjugant appeared to be almost pure C^R. However, with further subcloning these cells began yielding double-sensitive $C^S E^S$ animals. These did not result from sudden transformation of $C^R E^S$ animals to the $C^S E^S$ genotype but from progressive dilution of the former mitochondrial genotype by the latter, as is shown by the steadily increasing lag before the cells began to grow in chloramphenicol-containing medium as the number of fissions increased. The $C^S E^S$ isolates were stable and were never found in control crosses of the two parents; thus it appears that they arose as a result of mitochondrial gene recombination.

It is not clear why mitochondrial recombinants are so rare in *P. aurelia*. One possibility is that the numbers of mitochondria are great in *P. aurelia* and that extensive mitochondrial sorting-out is required for recombinant expression. Although this might help to explain the rarity of double-sensitive recombinants and the long time required for their expression, it does not explain why doubly resistant recombinants have not been found. It should be easy to detect these recombinants if they occur, since the cytoplasmic exchange experiments described earlier showed that resistant mitochondrial genotypes become expressed rapidly in a sensitive recipient when the latter is placed in antibiotic-containing medium. Perhaps the opportunities for recombination are themselves very limited in *P. aurelia*.

USE OF CYTOPLASMIC EXCHANGE AND MITOCHONDRIAL INJECTION TO STUDY GENETIC CONTROL OF SPECIFIC MITOCHONDRIAL PROTEINS

Because nuclear exchange normally occurs without cytoplasmic exchange in *P. aurelia,* it is possible to determine whether a specific mitochondrial protein is coded by nuclear or cytoplasmic (mitochondrial) genes, provided that identifiable differences with respect to such a protein can be found in different stocks of the same syngen. Tait, who devised the method, showed that the outer mitochondrial membrane enzyme β-hydroxybutyric acid dehydrogenase (HBD) in *P. novaurelia* and the mitochondrial form of NADP isocitrate dehydrogenase in *P. biaurelia* were under the control of nuclear genes (32,33). Since the analysis is based on the same principle in both cases, a single example will suffice.

Two electrophoretically different forms

of HBD were identified among clones isolated from a natural population of *P. novaurelia*. Cells were allowed to conjugate without cytoplasmic exchange, and mitochondria were isolated from the exconjugant clones. If the two forms of HBD (HBD-1 and HBD-2) are controlled by mitochondrial alleles, each exconjugant clone will resemble its parent; but if the forms of HBD are controlled by nuclear alleles, both HBD-1 and HBD-2 should be present in each exconjugant clone, excluding dominance of one allele to the other, since the exconjugants will be heterozygous for the two alleles. The heterozygotes contained both forms of HBD and, in addition, three new forms. The new forms were explained by Tait (32) on the assumption that HBD contains four polypeptides that assort at random with respect to each other, thus yielding three new forms of the enzyme in addition to the two original forms.

These results suggested that HBD-1 and HBD-2 were under the control of a pair of nuclear alleles. This was confirmed by causing the heterozygotes to undergo autogamy, whereupon a 1:1 segregation was found among the exautogamous clones for the two forms of HBD originally observed.

The method of Tait is applicable only within species, since crosses between species are generally incompatible. To overcome this problem, Knowles and Tait (34) constructed mitochondrial hybrids between species using the mitochondrial injection methods described previously (p. 311). The hybrids were constructed using *P. primaurelia* mitochondria and *P. septaurelia* recipient cells. These two species have electrophoretically different forms of fumarase. If fumarase formation is controlled by a nuclear gene, the fumarase formed should resemble that present in *P. septaurelia;* but if the enzyme is coded by the mitochondrial genome, it should resemble the enzyme synthesized by *P. primaurelia*. The authors found that the enzyme formed by the injected animals resembled that of *P. septau-*

relia, which indicated that its formation was controlled by nuclear genes rather than mitochondrial genes. Tait et al. (35,36) recently extended this approach to include mitochondrial ribosomal proteins using the same two species and the interspecific mitochondrial hybrid containing *P. primaurelia* mitochondria and a *P. septaurelia* nuclear genome. Antisera were raised to pure mitochondrial ribosomal proteins from both species, and these were tested against mitochondrial ribosomal proteins isolated from the two parent species and the hybrid. In the controls it was evident that homologous antisera reacted more strongly than heterologous antisera. That is, there existed immunologic differences between the mitochondrial ribosomal proteins of the two parent species. The reaction seen for the mitochondrial ribosomal proteins from the hybrid suggested some homology with both of the parent species. That is, some proteins, probably the majority, seemed to be determined by the *P. septaurelia* nuclear genome, but others were determined by the *P. primaurelia* mitochondrial genome.

These experiments are important in that they provide the first evidence that mitochondrial ribosomal proteins might be coded in part by the mitochondrial genome as well as the nuclear genome. Experiments on the sites of synthesis of mitochondrial ribosomal proteins (see Chapter 21) indicate that virtually all of them are synthesized on cytoplasmic ribosomes in *Neurospora* and *Saccharomyces* but that some are synthesized on mitochondrial ribosomes in *Tetrahymena*. In view of the relatively close relationship between *Paramecium* and *Tetrahymena*, it seems likely that ribosomal proteins coded by the mitochondrial genome of *Paramecium* are synthesized in the mitochondrion. Had this result been obtained with *Neurospora* or *Saccharomyces,* it would have raised serious questions concerning the concordance of the location of the structural genes coding for these polypeptides and their sites of synthesis.

CONCLUSIONS

Experiments with *P. aurelia* have made it possible to investigate some unusual problems in mitochondrial heredity that are not readily approachable in other systems. These include the rate and mechanism of expression of mitochondrial genes that are a minority within the cell, selection at the mitochondrial level under conditions that render mitochondrial gene recombination rare, and mitochondrial compatibility with different nuclear genotypes. Already it is clear that when selective conditions are applied, a mitochondrial mutation can be expressed rapidly within a cell even though it is in great minority. It is also evident in *P. aurelia* that mitochondrial selection (suppressiveness, if you will) is operative at the level of the mitochondrion itself. Studies of mitochondrial compatibility have revealed new aspects of mitochondrial heredity and generated a host of interesting questions. Does mitochondrial heredity involve the inheritance of preformed structure as well as genes localized in mtDNA and nuclear DNA? Are restriction modification systems involved in the recognition and maintenance of mtDNA? The unique attributes of the life cycle of *P. aurelia,* coupled with the availability of mitochondrial injection techniques, also provide ingenious new methods for ascertaining where different mitochondrial proteins are coded.

REFERENCES

1. Preer, J. R., Jr. (1969): Genetics of protozoa. In: *Research in Protozoology, Vol. 3,* edited by T.-T. Chen, pp. 129–278. Pergamon Press, Oxford.
2. Preer, J. R., Jr. (1971): Extrachromosomal inheritance: Hereditary symbionts, mitochondria, chloroplasts. *Annu. Rev. Genet.,* 5:361–406.
3. Preer, J. R., Jr., Preer, L. B., and Jurand, A. (1974): Kappa and other endosymbionts in *Paramecium aurelia. Bacteriol. Rev.,* 38:113–163.
4. Sonneborn, T. M. (1947): Recent advances in the genetics of *Paramecium* and *Euplotes. Adv. Genet.,* 1:263–358.
5. Sonneborn, T. M. (1959): Kappa and related particles in *Paramecium. Adv. Virus Res.,* 6:229–356.
6. Sonneborn, T. M. (1957): Breeding systems, reproductive methods, and species problems in protozoa. In: *The Species Problem,* edited by E. Mayr, pp. 155–324. American Association for the Advancement of Science, Washington, D.C.
7. Sonneborn, T. M. (1970): Methods in Paramecium research. *Meth. Cell Physiol.,* 4:241–33g.
8. Sonneborn, T. M. (1963): Does preformed cell structure play an essential role in cell heredity? In: *The Nature of Biological Diversity,* edited by J. M. Allen, pp. 165–221. McGraw-Hill, New York.
9. Beale, G. H. (1954): *The Genetics of Paramecium aurelia.* Cambridge University Press, Cambridge.
10. Sonneborn, T. M. (1975): The *Paramecium aurelia* complex of 14 sibling species. *Trans. Am. Microsc. Soc.,* 94:155–178.
11. Sonneborn, T. M. (1950): Methods in the general biology and genetics of *Paramecium aurelia. J. Exp. Zool.,* 113:87–147.
12. Adoutte, A., and Beisson, J. (1970): Cytoplasmic inheritance of erythromycin resistant mutations in *Paramecium aurelia. Mol. Gen. Genet.,* 108:70–77.
13. Beale, G. H. (1969): A note on the inheritance of erythromycin-resistance in *Paramecium aurelia. Genet. Res. (Camb.),* 14:341–342.
14. Adoutte, A., Balmefrézol, M., Beisson, J., and André, J. (1972): The effects of erythromycin and chloramphenicol on the ultrastructure of mitochondria in sensitive and resistant strains of *Paramecium. J. Cell Biol.,* 54:8–19.
15. Tait, A. (1972): Altered mitochondrial ribosomes in an erythromycin resistant mutant of *Paramecium. FEBS Lett.,* 24:117–120.
16. Beale, G. H., Knowles, J. K. C., and Tait, A. (1972): Mitochondrial genetics in *Paramecium. Nature,* 235:396–397.
17. Koizumi, S., and Preer, J. R., Jr. (1966): Transfer of cytoplasm by microinjection in *Paramecium. J. Protozool. (Suppl.),* 13:27.
18. Beale, G. H. (1973): Genetic studies on mitochondrially inherited mikamycin-resistance in *Paramecium aurelia. Mol. Gen. Genet.,* 127:241–248.
19. Knowles, J. K. C. (1972): Observations on two mitochondrial phenotypes in single *Paramecium* cells. *Exp. Cell Res.,* 70:223–226.
20. Cummings, D. J., Goddard, J. M., and Maki, R. A. (1976): Mitochondrial DNA from *Paramecium aurelia.* In: *The Genetic Function of Mitochondrial DNA,* edited by C. Saccone and A. M. Kroon, pp. 119–130. North Holland, Amsterdam.
21. Adoutte, A. (1974): Mitochondrial mutations in *Paramecium:* Phenotypical characterization and recombination. In: *The Biogenesis of Mi-*

tochondria, edited by A. M. Kroon and C. Saccone, pp. 263–271. Academic Press, New York.

22. Adoutte, A., and Beisson, J. (1972): Evolution of mixed populations of genetically different mitochondria in *Paramecium aurelia. Nature,* 235:393–395.

23. Perasso, R., and Adoutte, A. (1974): The process of selection of erythromycin-resistant mitochondria by erythromycin in *Paramecium. J. Cell Sci.,* 14:475–497.

24. Adoutte, A., Sainsard, A., Rossignol, M., and Beisson, J. (1973): Aspects génétiques de la biogenèse des mitochondries chez la *Paramécie. Biochimie,* 55:793–799.

25. Sainsard, A., Claisse, M., and Balméfrézol, M. (1974): A nuclear mutation affecting structure and function of mitochondria in *Paramecium. Mol. Gen. Genet.,* 130:113–125.

26. Sainsard-Chanet, A. (1976): Gene controlled selection of mitochondria in *Paramecium. Mol. Gen. Genet.,* 145:23–30.

27. Sainsard, A. (1975): Mitochondrial suppressor of a nuclear gene in *Paramecium. Nature,* 257:312–314.

28. Doussiere, J., Adoutte, A., Sainsard, A., Ruiz, F., Beisson, J., and Vignais, P. (1976): Physiological and genetical analysis of the respiratory chain of *Paramecium.* In: *Genetics and Biogenesis of Chloroplasts and Mitochondria,* edited by T. Bücher, W. Neupert, W. Sebald, and S. Werner, pp. 873–880. North Holland, Amsterdam.

29. Beisson, J., and Sonneborn, T. M. (1965): Cytoplasmic inheritance of the organization of the cell cortex in *Paramecium aurelia. Proc. Natl. Acad. Sci. U.S.A.,* 53:275–282.

30. Beisson, J., Sainsard, A., Adoutte, A., Beale,

G. H. Knowles, J., and Tait, A. (1974): Genetic control of mitochondria in *Paramecium. Genetics,* 78:403–413.

31. Beale, G. H., and Knowles, J. K. C. (1976): Interspecies transfer of mitochondria in *Paramecium aurelia. Mol. Gen. Genet.,* 143:197–201.

32. Tait, A. (1968): Genetic control of β-hydroxybutyric dehydrogenase in *Paramecium aurelia. Nature,* 219:941.

33. Tait, A. (1970): Genetics of NADP isocitrate dehydrogenase in *Paramecium aurelia. Nature,* 225:181–182.

34. Knowles, J. K. C., and Tait, A. (1972): A new method for studying the genetic control of specific mitochondrial proteins in *Paramecium aurelia. Mol. Gen. Genet.,* 117:53–59.

35. Tait, A., Knowles, J. K. C., and Hardy, J. C. (1976): The genetic control of mitochondrial ribosomal proteins in *Paramecium.* In: *The Genetic Function of Mitochondrial DNA,* C. Saccone and A. M. Kroon, pp. 131–136. North Holland, Amsterdam.

36. Tait, A., Knowles, J. K. C., Hardy, J. C., and Lipps, H. (1976): The study of the genetic function of *Paramecium* mitochondrial DNA using species hybrids. In: *Genetics and Biogenesis of Chloroplasts and Mitochondria,* edited by T. Bücher, W. Neupert, W. Sebald, and S. Werner, pp. 569–572. North Holland, Amsterdam.

37. Arber, W., and Linn, S. (1969): DNA modification and restriction. *Annu. Rev. Biochem.,* 38:467–500.

38. Jinks, J. L. (1964): *Extrachromosomal Inheritance.* Prentice-Hall, Englewood Cliffs, N.J.

Chapter 11

Experiments in Mitochondrial Heredity in Toads, Horses, and Tissue Culture Cells

No "system" has been developed that will permit study of the formal aspects of mitochondrial genetics among the higher animals, equivalent to those in use in the fungi and *Paramecium*. Until recently, investigators concentrated on molecular approaches to mitochondrial genetics that involved mammalian cells in tissue culture. Those experiments showed that cellular hybrids derived from mammals as distantly related as mouse and man often contain mtDNA derived from each parent cell line and that these DNAs can even recombine. As we shall see, the techniques used to make these demonstrations involve hybridization of RNAs synthesized *in vitro* on each of the parental mtDNAs. The base sequences of these complementary RNAs (cRNAs) are distinctly different in even closely related species; they will hybridize only with the mtDNAs on which they were synthesized. Hence they can be used for accurate measurement of the composition of mtDNA in an interspecific hybrid. We shall also discuss the classic experiments of Attardi and his colleagues that have established the precise positions of the mitochondrial ribosomal RNA genes and the genes for 4S RNA on the HeLa cell mitochondrial genome. These experiments involved a combination of molecular hybridization and electron microscopy. In short, investigation of the molecular cytogenetics of the mitochondrial genomes of tissue culture cells is at an advanced stage, but formal genetic analysis of these genomes is just beginning. It is to be hoped that a merger of the two approaches will soon be achieved, for tissue culture cells have many unusual attributes that make them suitable for the study of

mitochondrial genetics, and they are our best hope for achieving genetic dissection of the mitochondrial genome in higher organisms. As was said previously, this mitochondrial genome is considerably smaller than the mitochondrial genomes of fungi and ciliate protozoa, both in physical size and in information content. From the evolutionary point of view, it will be interesting to discover the meaning of this reduction in genome size.

One problem that experiments with tissue culture cells cannot illuminate concerns inheritance of mtDNA in crosses. Is it maternal, as the disparity in size between egg and sperm might suggest? How does the mechanism of mitochondrial genome transmission in vertebrates relate to those in other organisms? This problem was studied in a series of elegant molecular hybridization experiments using F_1 hybrids derived from crosses between the closely related clawed toads *Xenopus laevis* and *X. mulleri* by Dawid and by Hutchison and Edgell who used restriction enzymes to analyze the mtDNAs of the horse, the donkey, and the two reciprocal F_1 hybrids between them. We shall begin this chapter by discussing their experiments.

INHERITANCE OF mtDNA IN TOADS AND HORSES

The feasibility of studying the inheritance of mtDNA in *Xenopus* depended on two observations. First, Blackler and Gecking (1) found that *X. laevis* and *X. mulleri* [now designated *X. borealis* (36)] could be crossed to yield viable hybrid progeny. Second, Dawid (2) found that there were sequence differences between the mtDNAs of the two species. These sequence differences became apparent during reannealing experiments (see Chapter 2); it was found that whereas homoduplexes of mtDNA from either *X. laevis* or *X. mulleri* reannealed with a sharp thermal transition similar to that seen when the DNAs were originally melted, the same was not true of interspecific heteroduplexes. The heteroduplexes had a very broad thermal transition, and they began to melt at much lower temperatures than the homoduplexes. From quantitative analysis of differential plots made during the melting experiments (see Chapter 6), Dawid was able to conclude that only about 20% of the mtDNA of the two species showed a high degree of cross homology. The rest of the mtDNA from the two species showed much less homology.

The discovery of mtDNA sequence differences between *X. laevis* and *X. mulleri* did not provide the key to probing the inheritance of mtDNA in interspecific hybrids; it only showed that given the right technique such a study would be feasible. The key was provided by another technique that will be described in some detail, since it is the same technique that was used by Dawid and his colleagues to study mtDNA in tissue culture cell hybrids between different species.

Native mtDNA of *Xenopus* can be transcribed *in vitro* using *E. coli* RNA polymerase to yield an RNA product that is complementary to a portion of the mitochondrial genome and is called cRNA. It will be recalled that vertebrate mtDNA can be separated into light (L) and heavy (H) strands on the basis of buoyant density differences in cesium chloride and that *in vivo* both rRNA and most 4S RNA molecules are transcribed from the H strand. *In vitro* the *E. coli* enzyme with or without sigma factor preferentially transcribes the L strand to the extent of 70 to 90%. In addition, about one-half of the cRNA synthesized *in vitro* forms duplexes with mitochondrial RNA made *in vivo*. Since the *in vivo* RNA con-

sists mostly of rRNA and 4S RNA that are transcribed from about 20% of the base pairs in mtDNA, this means that the *E. coli* polymerase preferentially transcribes the "antisensesequences" of the ribosomal and 4S RNA sequences. The other half of the cRNA contains transcripts of sequences of mtDNA other than the antisensesequences for the stable RNAs, but it is not known if all other mtDNA sequences are represented among these transcripts.

Dawid (2) examined interspecific hybridization of cRNA and RNA synthesized *in vivo,* mostly rRNA and 4S RNA, to mtDNA (Table 11–1). In these experiments *X. laevis* RNA labeled with ^{32}P was mixed with ^{3}H-labeled *X. mulleri* RNA and then hybridized with mtDNA obtained from each species. The way the method works is best illustrated by an ideal example. Suppose that equal amounts of RNA having similar radioactivities are mixed. If there is no difference between the coding sequences for a given type of RNA between the mtDNAs of *X. mulleri* and *X. laevis,* the ^{3}H/^{32}P ratio will be 1.0. If there is a difference, the ^{3}H/^{32}P ratio will be greater than 1.0 when hybridization is done using *X. mulleri* mtDNA and less than 1.0 when *X. laevis* mtDNA is used.

It can be seen from Table 11–1 that the isotope ratios for *in vivo* synthesized rRNA and 4S RNA do not vary greatly with the type of mtDNA being used for hybridization. Therefore, these RNAs have not diverged greatly between *X. laevis* and *X. mulleri.* The cRNA, on the other hand, showed more pronounced variations in isotope ratios, depending on the mtDNA to which it was hybridized. Thus, when *X. laevis* mtDNA was used for hybridization, the isotope ratio was 0.22, indicating that more *X. laevis* cRNA was being hybridized by *X. laevis* mtDNA than *X. mulleri* cRNA. The isotope ratio was 22 when *X. mulleri* mtDNA was used for hybridization, indicating preferential hybridization of *X. mulleri* cRNA. Since about half the cRNA se-

TABLE 11–1. *Hybridization between X. laevis and X. mulleri*
mitochondrial RNA and DNA[a]

Type of mito-chondrial RNA	Source of mtDNA	RNA hybridized $^3H/^{32}P$
rRNA	X. laevis	0.5
	X. mulleri	2.4
4S RNA	X. laevis	0.4
	X. mulleri	2.8
cRNA	X. laevis	0.22
	X. mulleri	22

[a] Each hybridization mixture contained ^{32}P-labeled *X. laevis* RNA and ^{3}H-labeled *X. mulleri* RNA. Filters containing mtDNA of *X. laevis* and filters containing mtDNA of *X. mulleri* were hybridized together in these RNA solutions.
From Dawid (2), with permission.

quences are known to be complementary to antisensesequences of rRNA and 4S RNA genes that do not appear to differ greatly between the mtDNAs of the two species (Table 11–1), these striking interspecific differences in sequence homology for cRNA must be attributed to mtDNA sequences other than those coding for the stable RNAs. This prediction was borne out in competition experiments in which the antisensesequence cRNA was allowed to form duplexes with stable RNA formed *in vivo*, leaving only that cRNA in single-stranded

form that did not correspond to stable RNA antisequences. This cRNA had a very low level of interspecific homology, as revealed by RNA-DNA hybridization experiments.

The virtual absence of homology between the mtDNA sequences coding for cRNA in *X. laevis* and *X. mulleri,* other than the antisensesequences for the stable RNAs, provided the analytical tool for studying the inheritance of mtDNA in interspecific crosses. Dawid and Blackler (3) made reciprocal crosses between *X. mulleri* and *X. laevis* and compared the mtDNAs of the F_1 hybrids and the parent stocks using the cRNA hybridization technique. Using the same technique, cRNA synthesized on nuclear DNA of each of the parents was hybridized to nuclear DNA of the parents and the F_1 hybrids. These experiments showed that mtDNA extracted from the F_1 hybrids was virtually identical in its hybridization properties to that of the maternal parent (Table 11–2). On the other hand, cRNA synthesized on nuclear DNA showed a pattern of hybridization intermediate between that of the two parents, indicating that the F_1 progeny contained nuclear DNA from both parents.

TABLE 11–2. *Inheritance of mtDNA in Xenopus[a]*

Source of mtDNA		cRNA bound to DNA	
		Nuclear cRNA nuclear DNA $(^3H/^{32}P)$	Mitochondrial cRNA mitochondrial DNA $(^3H/^{32}P)$
X. laevis ovary		1.0	9.8
X. mulleri ovary		0.23	0.083
X. laevis/X. mulleri artificial mixtures	100:1	—	8.4
	20:1	—	4.7
X. laevis ♀ × X. mulleri ♂ F₁ whole tadpole		0.51	11
		—	9.8
X. mulleri ♀ × X. laevis ♂ F₁ whole tadpole		0.44	0.12
		0.47	0.12
		0.50	0.10
F₁ liver mtDNA		0.52	0.09
X. laevis ♀ × X. mulleri ♂ oocytes (F₂) of F₁ hybrid		0.51	10.5
X. mulleri ♀ × X. laevis ♂ F₂ whole tadpole DNA from F₁ hybrid backcrossed to X. mulleri		—	0.085

[a] Adapted from Table 1 of Dawid and Blackler (3), which should be consulted with respect to experimental details, sample sizes, etc. In the cRNA mixture, *X. laevis* cRNA was labeled with ^{3}H and *X. mulleri* cRNA with ^{32}P. Tissue sources of mtDNA, but not nuclear DNA are specified in the original table.

Although these experiments demonstrated maternal transmission of mtDNA over one generation, they did not show that mtDNA would exhibit cytoplasmic inheritance independent of the nuclear gene complement over succeeding generations. It could be imagined, for example, that in F_1 hybrids the nuclear genes present in one or the other parent would control the species of mtDNA transmitted to the ovaries. Although the F_1 hybrids obtained from these crosses usually formed no gonads or contained sterile gonads, Dawid and Blackler were able to obtain two second-generation examples. In one case they obtained an ovary with many large oocytes from an F_1 hybrid female and showed that it contained *X. laevis* mtDNA like that of the maternal parent. This observation suggests, but does not prove, that *X. laevis* mtDNA would have been transmitted to the F_2. In the second instance, Dawid and Blackler were able to obtain backcross progeny from a hybrid whose female and male parents were *X. mulleri* and *X. laevis,* respectively, and that contained *X. mulleri* mtDNA. The backcross was made between the hybrid, which was a female, and an *X. mulleri* male. The progeny contained *X. mulleri* mtDNA, which is not surprising, since this was true of both parents. In short, only one of the two second-generation examples was relevant to the question of whether or not mtDNA is maternally transmitted to the F_2, and in this case mtDNA did seem to be maternally transmitted to the F_2.

The fate of sperm mitochondria following fertilization is not entirely clear. The mitochondria are contained in the sperm midpiece. In tunicates, the midpiece does not enter the egg (4). In mammals, it is reported that the midpiece usually enters the egg (5), following which the mitochondria may degenerate (6) or may become scattered through the egg cytoplasm (5). A conservative interpretation of both the *Xenopus* and *Equus* interspecific crosses (*vide infra*) would be that the apparent ma-

ternal inheritance of mtDNA can be accounted for simply in terms of dilution of paternal mitochondrial genomes by maternal mitochondrial genomes in the fertilized egg. Dawid and Blackler (3) calculated that each *Xenopus* egg contains 10^8 mtDNA molecules, whereas each sperm contains only 100. If paternal and maternal mitochondrial genomes were to replicate at equal rates following fertilization, only one in a million would be of paternal origin, which is surely well below the resolution of the experiments of Dawid and Blackler. In addition, it may be that paternal mitochondria are actually excluded from the *Xenopus* egg. In any event, this second mechanism would merely serve to accentuate the pattern of maternal inheritance that must necessarily result because of the first mechanism.

Dawid and Blackler (3) also used cRNA synthesized on mtDNA to determine if the nucleus contains a master copy of mtDNA. To be successful, such experiments must have a level of resolution high enough to detect as little as one mtDNA genome per nuclear genome. Obviously, cross-contamination of the nuclear DNA fraction with even a very small amount of mtDNA could confuse the picture. Dawid and Blackler were able to attain the necessary sensitivity in the assay because the binding of cRNA to homologous mtDNA sequences is a linear function over a wide range of DNA concentrations down to an amount equivalent to less than one copy of mtDNA per haploid genome.

When different nuclear DNA preparations were hybridized with mitochondrial cRNA, significant levels of hybridization were observed in all cases. Dawid and Blackler were able to show that this hybridization was caused by contamination of the nuclear DNA fraction with mtDNA (rather than being caused by master copies in the nuclear genome) by making use of sequence differences in *X. laevis* and *X. mulleri* mtDNA with respect to cRNA and the maternal inheritance of mtDNA in interspecific crosses. They first showed that the isotope ratio following hybridization of a mixture of ³H-labeled *X. laevis* and ³²P-labeled *X. mulleri* cRNA to nuclear DNA preparations from the two species closely resembled the isotope ratios from hybridization experiments with pure mtDNA from each species, as would be expected if the nucleus contained master copies (Table 11–3). However, they then showed that in the F₁ hybrids the isotope ratios following hybridization to nuclear DNA resembled those of the maternal parent (Table 11–3). Since these frogs are nuclear hybrids between *X. laevis* and *X. mulleri,* intermediate isotope ratios would be expected because both paternal and maternal master copies should be present in the nuclear DNA of the hybrids. Only if the nuclear master copy itself were maternally inherited could these results be explained by a master copy model.

These experiments serve to emphasize the care in experimental design required to obviate cross-contamination between DNA fractions in molecular hybridization experiments designed to detect rare copies in one fraction or the other. The radiochemical hybridization assay is so sensitive that a small amount of cross-contamination yields a rather high estimate of master copies of mtDNA per nuclear genome in *Xenopus* (Table 11–3). These experiments of Dawid and Blackler should be brought to mind each time molecular hybridization experiments designed to detect rare copies are undertaken. They make use of a unique observation (that mtDNA is maternally inherited in interspecific hybrids of *Xenopus*) as a control for cross-contamination of the nuclear DNA fraction with mtDNA. Without this or other appropriate and relatively independent controls, all hybridization experiments relying on detection of very small amounts of organelle DNA in nuclear DNA fractions or vice versa must be regarded as suspect because of the possibility of low levels of cross-contamination. The importance of this observation cannot be overemphasized in view of the role that nuclear master copies have played in models of mitochondrial heredity.

TABLE 11–3. *Hybridization of mitochondrial cRNA with nuclear DNA preparations in Xenopus. Evidence for contamination by maternal mtDNA.*[a]

DNA	Mitochondrial cRNA bound ³H/³²P	mtDNA present in nuclear preparation	
		Species	Copies per haploid genome
Experiment A			
X. laevis mtDNA	17.9	—	—
X. mulleri mtDNA	0.13	—	—
X. laevis nuclear DNA	9.6	X. laevis	9
X. mulleri nuclear DNA	0.23	X. mulleri	15
F₁ (female X. laevis × male X. mulleri) nuclear DNA	9.9	X. laevis	14
Experiment B			
X. laevis mtDNA	26	—	—
X. mulleri mtDNA	0.127	—	—
X. laevis nuclear DNA	16	X. laevis	5.3
X. mulleri nuclear DNA	0.18	X. mulleri	20
Hybrid (female X. mulleri × male X. laevis) nuclear DNA	0.18	X. mulleri	16
Hybrid, plus one equivalent of X. laevis mtDNA	0.29	—	—

[a] Experiments A and B tested different DNA preparations. Each experiment was carried out with one mixture of ³H cRNA transcribed from *X. laevis* mtDNA and ³²P cRNA transcribed from *X. mulleri* mtDNA.
From Dawid and Blackler (3), with permission.

Hutchison et al. (7) examined the pattern of inheritance of mtDNA in reciprocal hybrids between the horse and the donkey. The progeny of crosses between female horses and male donkeys are mules; the progeny of the reciprocal cross are referred to as hinnies. Mules and hinnies may be of either sex, but they are generally sterile. Purified mtDNAs from the parent animals and the F_1 hybrids were digested with restriction endonuclease Hae III, and the resulting digests were analyzed by gel electrophoresis. Mules have the same mtDNA cleavage pattern as their maternal parents (horses), and hinnies have the mtDNA cleavage pattern of their maternal parents (donkeys). Thus mtDNA is maternally inherited in horses and donkeys. Hutchison and associates considered the mechanism of maternal inheritance of mtDNA in these crosses and concluded that simple dilution of the type already discussed for *Xenopus* was sufficient to account for their results.

USES OF INTERSPECIFIC SOMATIC CELL HYBRIDS IN DISSECTING MAMMALIAN MITOCHONDRIAL GENOMES

Experiments with interspecific somatic cell hybrids have proved useful in investigating several aspects of mitochondrial heredity in mammals. The first studies focused on the nature of the mtDNA retained in such hybrids. Clayton et al. (8) and Attardi and Attardi (9) examined the species of mtDNA present in established mouse–human hybrid cell lines, using cesium chloride buoyant-density centrifugation to distinguish between the two kinds of mtDNA. Such cell hybrids typically lose human chromosomes, especially during the early divisions, following which the hybrid chromosome complement becomes fairly stable (10–12). Both groups of investigators reported that only mouse mtDNA was present in the hybrid cell lines, even though some human chromosomes (up to 23 in one case) had been retained. Several explanations for

these results were considered, including the possibility that certain human chromosomes required for the maintenance of human mtDNA are lost in the hybrids. Why the same argument would not also apply to mouse mtDNA is not clear.

More recent experiments by Eliceiri (13) and particularly by Coon et al. (14) revealed that irrevocable loss of one mitochondrial genome is not characteristic of all interspecific cell hybrids. Eliceiri showed that hybrid cell lines derived from mouse and hamster contained mitochondrial rRNAs characteristic of both species, with the preponderent rRNA being derived from the species that had contributed the most chromosomes to the nuclear genome. The rRNAs from each species were distinguishable on polyacrylamide gels.

The experiments of Coon et al. (14) using rat–human and mouse–human hybrid lines were the most careful and exhaustive studies reported thus far; they will now be discussed in some detail. These authors synthesized the hybrid lines using the established human cell lines VA2 and D98/AH2 and freshly dissociated cells of either rat or mouse embryos. The human cell lines were both deficient in hypoxanthine guanine phosphoribosyl transferase activity (HGPRT), which, as we shall see shortly, is important for hybrid cell selection.

Cell fusion was induced by use of Sendai virus inactivated by β-propiolactone, with human cells being present in 10-fold excess. Cell fusing activity is associated with a specific glycoprotein present in the cell envelope (15). The fusing mixture was then placed in normal growth medium for a day, following which it was transferred to HAT medium; HAT medium, which was designed by Littlefield (16), contains hypoxanthine, thymidine, and aminopterin. Aminopterin blocks *de novo* synthesis of hypoxanthine and thymidine in the cell, but as long as HGPRT activity and thymidine kinase (TK) activity are present, a given cell line will be able to use exogenously supplied hypoxanthine and thymidine as sources of purine and pyrimidine nucleotides, respectively. In this instance, this means that the HGPRT⁻ human cells will be unable to grow, although the rodent cells and hybrid cells can grow because they are not deficient in HGPRT. By using an excess of human cells in the fusion mixture, it was possible to reduce greatly the fraction of parental rodent cells that had not undergone fusion.

Once the hybrid cell clones had been established, Coon et al. (14) made use of the cRNA hybridization technique described previously for *Xenopus* to measure the relative amounts of nuclear DNA and mtDNA of each species per hybrid clone. Labeled cRNA was prepared *in vitro* using mtDNA or nuclear DNA of each parent as a primer for *E. coli* RNA polymerase. As in the case of the *Xenopus* experiments, the cRNA from one species was labeled with ³H, and that from the other species was labeled with ³²P. Separate mixtures were then made between the cRNAs transcribed from mtDNA and nuclear DNA.

In order to use these cRNA mixtures to estimate relative amounts of mtDNA or nuclear DNA from each species present in a given hybrid clone, it was first necessary to establish calibration curves using artificial mixtures of DNA from cells of the two species for hybridization of cRNA. Such a calibration curve for different mixtures of rat and human mtDNA establishes how the ³H/³²P ratio varies as the input of mtDNA is varied. Following cRNA hybridization, one can then calculate the relative amounts of the two kinds of mtDNA present in a hybrid clone from the ³H/³²P ratio. Similar curves can be constructed for nuclear DNA.

Once the calibration curves were available, the amount of mtDNA or nuclear DNA derived from each species could be estimated for any hybrid cell line using the appropriate cRNA mixture. Hybrid cell lines were analyzed for DNA after 40 to 60 cell doublings. Of 33 hybrid lines analyzed, 19 contained nuclear DNA and

mtDNA derived from both parents, and 14 were pure human for both kinds of DNA (Fig. 11–1). Subclones from the lines containing mixtures of human and rat mtDNA and nuclear DNA were found to be tightly clustered, which indicated that these hybrids were relatively stable once a particular proportion of each type of DNA had been established (Fig. 11–2). Provided that those cells that are virtually pure human segregants are excluded from consideration, the rat–human hybrids fall into two groups: one containing approximately 40% rat nuclear DNA and 20% rat mtDNA, the other containing approximately 60% rat nuclear DNA and 90% rat mtDNA. An axis connecting these points is inclined more steeply than a 45-degree angle. If the 45-degree diagonals in Fig. 11–1 are considered, it is seen that all but one of the hybrid cell strains fall into either the top quarter or bottom quarter of the field. This means that in general the proportions of nuclear DNA are more balanced than the proportions of mtDNA. These facts suggested the following interpretations to the authors: (a) There is a tendency, especially in the mtDNA, toward segregation away from equal proportions. This development is relatively slow, so that many cell strains remain hybrid in nuclear DNA and mtDNA after 40 to 150 generations of growth. (b) The segregation of the nuclear and mitochondrial genomes is correlated, but not strictly so. The possibility that propagation of mtDNA of one parental species in the hybrid cell depends on a particular chromosome or set of chromosomes cannot be evaluated from these experiments; more detailed karyotype analysis will be required. (c) Apparent segregation of mtDNA proceeds further and more rapidly than segregation of nuclear DNA. Obviously it will be important to establish whether the more rapid segregation of mtDNA is dependent on or independent of specific chromosomes.

Coon et al. (14) also examined the stability of mtDNA from both parents in

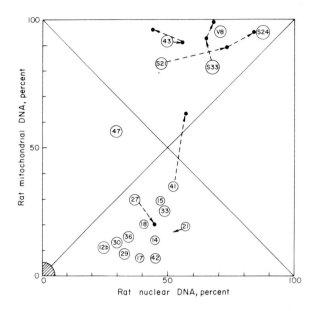

FIG. 11-1. DNA composition of rat–human hybrid cell strains. For each strain, a point is plotted at the position corresponding to its content of rat nuclear DNA and rat mtDNA (the contents of human DNA in each case are 100% minus the rat DNA content). The open circles signify hybrid cell strains tested for the first time 40 to 60 population doublings after fusion. The numbers in the circles identify the strains for correlation with Table 2 of Coon et al. Solid circles are average values of percentages of rat DNAs in sets of subclones derived from some of the hybrid cell strains (see Fig. 11–2); arrows connect these points with the populations from which they were derived. The subclones of hybrid strain (VRE) 21 contained virtually no rat nuclear or mitochondrial DNA, and this is indicated by an arrow pointing toward the origin. Fourteen hybrid cell strains contained no measurable rat nuclear or mitochondrial DNA when first tested; these strains are indicated by the shaded area at the origin. (From Coon et al., ref. 14, with permission.)

mouse–human hybrids. It is important to emphasize again that these were newly formed hybrids employing fresh rodent embryo cells, as opposed to the established hybrid cell lines used by Clayton et al. (8) and Attardi and Attardi (9). Coon and associates found mtDNA of both species in 7 of 23 lines tested. Of these 7 lines, 4 lines contained no measurable mouse nuclear DNA. Several of these lost much of their mouse mtDNA after additional doublings. Among the remaining 16 lines, 7 lines were hybrid with respect to their nuclear DNA (and chromosomes) but contained virtually pure mouse mtDNA, and the remaining 9 lines were essentially pure human with respect to both nuclear DNA and mtDNA.

The authors pointed out that mouse–human cell hybrids presented a rather different picture than did rat–human cell hybrids with respect to mtDNA. Loss of nuclear DNA proceeded in either direction.

However, as in the case of rat–human hybrids, loss of rodent nuclear DNA was frequent, but no instance occurred in which all human nuclear DNA had disappeared. Coon and associates also found that there was a strong tendency for the hybrids to lose human mtDNA. Hybrids with a predominance of mouse nuclear DNA contained only mouse mtDNA, but hybrids with very little mouse nuclear DNA still contained mouse mtDNA. The fact that the lines that lost mouse nuclear DNA also lost mouse mtDNA suggests that at least some mouse chromosomes may be required for long-term maintenance of mouse mtDNA.

The experiments of Coon and associates confirmed and extended those of Clayton et al. (8) and Attardi and Attardi (9) because they showed that human mtDNA is not necessarily absent from mouse–human hybrid cells but rather that there is strong selection for mouse mtDNA. The mecha-

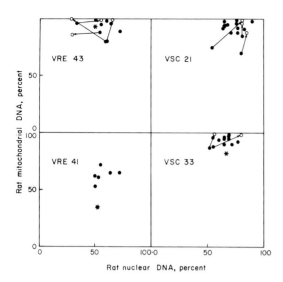

FIG. 11-2. Presence of both parental nuclear and mitochondrial DNAs in four sets of subclones of rat–human hybrid cell strains. Original population is shown as an asterisk and subclones as solid circles. Average values of these subclone sets are also presented in Fig. 11-1. Open circles represent subclonal populations tested again after additional growth for approximately 25 doublings; these populations are connected by arrows with the original subclones from which they were derived. (From Coon et al., ref. 14, with permission.)

nism of this selection is not understood. As we shall see later in this section, the existence of mouse–human cell lines containing nuclear DNA from both species, but only mouse mtDNA, permits use of a method similar to that used in *Paramecium* for determining whether specific mitochondrial enzymes, which are electrophoretically distinguishable in the two species, are coded by the nuclear or mitochondrial genomes.

Since somatic cell hybrids can contain mtDNA from both parents, it is desirable to know if this mtDNA is being maintained in its parental state or if interspecific mtDNA recombination has taken place or both. Horak et al. (17) set out to examine this question in subclones of the rat–human and mouse–human cell lines used by Coon et al. (14). The specific problem they addressed was whether or not interspecific recombinant mtDNA molecules could be detected in the hybrid cells. The test Horak and associates used was based on two find-

ings. First, the buoyant-density difference in cesium chloride between human mtDNA (1.707 g/cc) and mouse (1.700 g/cc) or rat (1.701 g/cc) mtDNA is great enough so that the mtDNAs of human and rodent cells are easily separated in cesium chloride gradients. Second, under stringent conditions, cross-hybridization between human and rodent mitochondrial cRNA is only 1 or 2% of the homologous hybridization level. Therefore, in a mixture, human and rodent mtDNAs can be completely resolved following cesium chloride centrifugation by use of the appropriate mitochondrial cRNA mixture (Fig. 11-3A). If recombination occurs between mtDNA molecules, a peak of intermediate buoyant density will be expected to appear that will hybridize with mitochondrial cRNA derived from both human and rodent species.

By use of this test, Horak and associates found that the mtDNA from many of the hybrid lines was indeed recombinant. Such rodent–human recombinant mtDNAs are illustrated in Fig. 11-3B–E. Fig. 11-3B depicts mtDNA from a mouse–human hybrid clone. It can be seen in this hybrid that one major peak of mtDNA sequences is at a density close to or slightly lower than that of pure human mtDNA. However, mtDNA in this region also hybridizes with mouse mitochondrial cRNA. A second peak is evident near the density of mouse mtDNA, which hybridizes mouse mitochondrial cRNA much more extensively than human mitochondrial cRNA. Therefore, this hybrid cell line appears to contain at least two kinds of hybrid mtDNA molecules: some with a high human-to-mouse ratio of mtDNA sequences and some containing a high mouse-to-human ratio. In addition, pure mtDNA molecules of each parent may be present in the hybrid. Figure 11-3C depicts recombinant mtDNA from another mouse–human hybrid cell line, and Fig. 11-3D shows recombinant mtDNA from a rat–human hybrid cell line.

One possible interpretation of the fore-

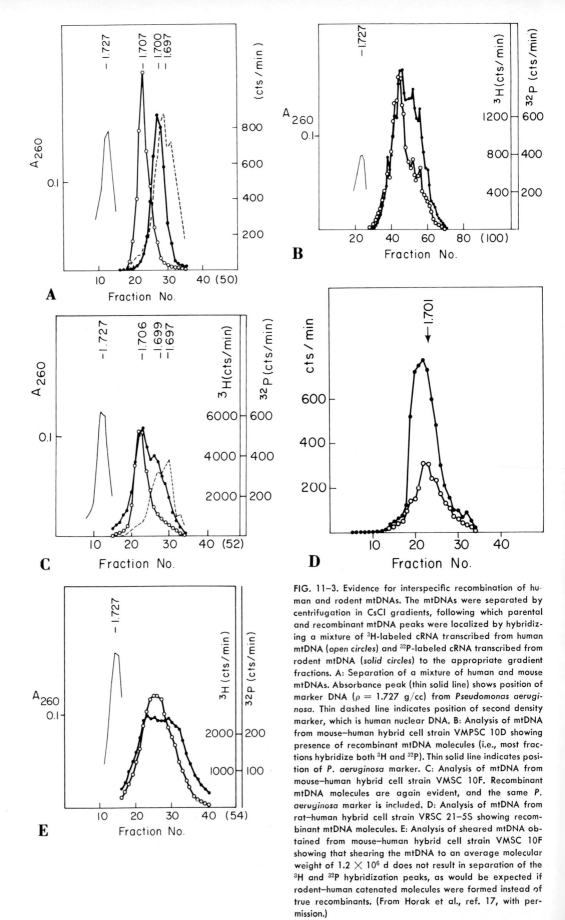

FIG. 11–3. Evidence for interspecific recombination of human and rodent mtDNAs. The mtDNAs were separated by centrifugation in CsCl gradients, following which parental and recombinant mtDNA peaks were localized by hybridizing a mixture of ^{3}H-labeled cRNA transcribed from human mtDNA (*open circles*) and ^{32}P-labeled cRNA transcribed from rodent mtDNA (*solid circles*) to the appropriate gradient fractions. A: Separation of a mixture of human and mouse mtDNAs. Absorbance peak (thin solid line) shows position of marker DNA (ρ = 1.727 g/cc) from *Pseudomonas aeruginosa*. Thin dashed line indicates position of second density marker, which is human nuclear DNA. B: Analysis of mtDNA from mouse–human hybrid cell strain VMPSC 10D showing presence of recombinant mtDNA molecules (i.e., most fractions hybridize both ^{3}H and ^{32}P). Thin solid line indicates position of *P. aeruginosa* marker. C: Analysis of mtDNA from mouse–human hybrid cell strain VMSC 10F. Recombinant mtDNA molecules are again evident, and the same *P. aeruginosa* marker is included. D: Analysis of mtDNA from rat–human hybrid cell strain VRSC 21–5S showing recombinant mtDNA molecules. E: Analysis of sheared mtDNA obtained from mouse–human hybrid cell strain VMSC 10F showing that shearing the mtDNA to an average molecular weight of 1.2 × 10[6] d does not result in separation of the ^{3}H and ^{32}P hybridization peaks, as would be expected if rodent–human catenated molecules were formed instead of true recombinants. (From Horak et al., ref. 17, with permission.)

going gradient profiles is that human and rodent mtDNA sequences are linked by catenation (see Chapter 2) rather than covalently. To test this possibility, Horak and associates sheared the mtDNA to fragments one-eighth the size of intact mtDNA. As Fig. 11–3E indicates, shearing does not separate mouse sequences from human sequences; thus apparently they are covalently linked as a result of recombination rather than being catenated.

Horak and associates examined a number of hybrid cell lines containing rodent and human mtDNA; their results are summarized in Table 11–4 and Fig. 11–4. From these results, Horak and associates concluded the following: (a) Hybrid cell lines containing mtDNA from two animal species frequently contain high proportions of re-

combinant mtDNA molecules. (b) Recombined mtDNA molecules occur in hybrid cells with predominantly rodent or human mtDNA sequences. (c) Primary clones or subclones may or may not contain a high proportion of recombinant mtDNA molecules. (d) Recombinant mtDNA molecules can be found in hybrid cell populations that have undergone between 40 and 150 doublings following fusion.

To these conclusions might be added the observation that it would be most interesting to know if the hybrid mtDNA molecules are functional in a biologic sense. The fact that the hybrid cells containing these recombinant mtDNA molecules are themselves viable suggests either that the recombinant mtDNA molecules are functional or that there are enough parental mtDNA

TABLE 11–4. Recombinant and nonrecombinant mtDNA molecules in human–rodent hybrid cell strains[a]

| Cell strain | Subcloned | Percentage of mouse | | Recombinant mtDNA |
		mtDNA	Nuclear DNA	
VMSC 10F	Yes	7	N.D.	Yes
VMSG 16	No	20	1	Yes
VME 15	No	27	32	Yes
VMSC 10 G	Yes	28	N.D.	Yes
VMPOC 15F	Yes	39	49	Yes
VMPSC 10D	Yes	39	35	Yes
VMPE 18F	Yes	40	39	Yes
VMPOC 18E	Yes	46	38	Yes
VMPth 18	No	48	9	Yes
VMPOC 15L	Yes	69	32	No
VMPOC 15B	Yes	73	20	No
VMSC 10E	Yes	71	13	No
VMSC 10B	Yes	73	19	No

| Cell strain | Subcloned | Percentage of rat | | Recombinant mtDNA |
		mtDNA	Nuclear DNA	
VRE 36A	Yes	15	58	Yes
VRSC 21–5S	Yes (2×)	84	89	Yes
VRSC 21–5N	Yes (2×)	89	90	Yes
VRSC 21–8	Yes	95	81	Yes
VRSC 33	No	24	8	No

[a] Strain designations allow comparison with Fig. 11–3. All rodent cells were derived from 15-day to 18-day embryos, and human cells were the line VA2 (WI-18 VA2 SV 40 transformed, HGPRT⁻); V = VA2, M = mouse, R = rat, SC = spinal cord, SG = spinal ganglion, E = neural retina, OC = optic cortex, th = thalamus; P designates a separate series. Cell strains not subjected to subcloning had undergone 40 to 60 doublings after fusion. Subcloned strains had undergone 60 to 150 doublings. N.D. = not determined.
From Horak et al. (17), with permission.

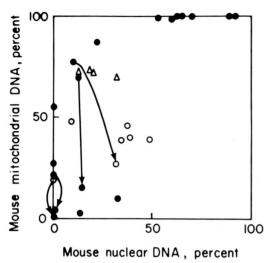

FIG. 11–4. Summary of DNA compositions of mouse–human hybrid cell strains. Solid points refer to hybrids that have doubled 40 to 60 times, except that arrows lead from several points to DNA values measured in same strains after additional 25 doublings. Open circles indicate hybrid strains containing recombinant mtDNA molecules; open triangles indicate strains that did not contain such molecules. (From Dawid et al., ref. 18, with permission.)

molecules retained in the hybrid cells to preserve biologic function and hence viability. It would also be of interest to know more about the molecular nature of the recombinants. Perhaps recombination al-

ways occurs in regions of molecular homology that have been conserved during the course of mammalian evolution.

It should be noted that much of the foregoing work by Dawid and his colleagues was also reviewed in two symposium papers (18,19).

Bodmer and his colleagues (20–22) took advantage of the fact that mouse–human hybrid cell lines frequently contain only mouse mtDNA to establish that several enzymes of the mitochondrial matrix are coded by nuclear DNA. The method depends on being able to make a distinction between mouse and human enzymes either electrophoretically or immunologically. If a hybrid cell containing only mouse mtDNA also contains human enzyme, the enzyme must be coded by nuclear DNA. On the other hand, absence of human enzyme activity from the hybrid may simply mean that the chromosome carrying the structural gene for that enzyme or a component polypeptide has segregated out of the hybrid.

The results for mouse–human cell hybrid lines for the enzymes citrate synthetase (CS), mitochondrial NAD malate dehydrogenase (mMOR), and mitochondrial

TABLE 11–5. Distribution of human enzymes in different man–mouse hybrid cell lines[a]

Hybrid line	LDHB/Pep B	CS	mMOR	mAAT	mtDNA	sMOR	sAAT
1W1	−	−	−	−	NT	−	−
2W1	−	−	+	−	NT	−	−
3W4	+	+	+	−	−	−	−
4W10	−	−	−	−	−	−	−
HORL	−	−	−	−	NT	−	−
HORP	+	+	+	−	NT	+	+
4.12Z	+	±[b]	+	+	−	+	−
4.31Z	+	+	+	−	NT	−	−
4.42.7Z	−	−	−	−	NT	+	−

[a] Mouse activity is always seen to be present, except in the case of the immunologic technique, which scores only for human activity. Key: mMOR = mitochondrial NAD malate dehydrogenase, sMOR = cytoplasmic NAD malate dehydrogenase, mAAT = mitochondrial aspartate aminotransferase, sAAT = cytoplasmic aspartate aminotransferase, CS = citrate synthetase, LDHB = lactate dehydrogenase B, Pep B = peptidase B. Plus sign indicates human activity is present; minus sign indicates no human activity; NT indicates not tested.

[b] Weak activity, possibly because of extreme gene dosage imbalance; 4.12Z has a doubled chromosome complement from the mouse parent, 3T3.

From Van Heyningen et al. (22), with permission.

aspartate aminotransferase (mAAT) are shown in Table 11–5. It is clear that human activity or immunologically cross-reacting material for each enzyme was detectable in one or more cell lines. These included lines known to contain only mouse mtDNA. Therefore these human mitochondrial matrix enzymes are coded by nuclear DNA. It was also found that the presence or absence of human CS activity was strongly correlated with the presence or absence of human forms of the unrelated enzymes lactate dehydrogenase and peptidase B. The latter loci have been shown to be linked and are probably on chromosome 12. Hence the gene for human CS activity must also be in the same linkage group.

MAPPING OF MITOCHONDRIAL rRNA AND 4S RNA CISTRONS IN HeLa CELLS BY MOLECULAR HYBRIDIZATION

Animal mtDNA can be separated into its two complementary single strands on the basis of buoyant density in alkaline cesium chloride gradients (see Chapter 2). The strand with the higher buoyant density in such an equilibrium gradient is known as the heavy (H) strand, and that with the lower buoyant density is the light (L) strand. Once the mtDNA strands are separated, they can be used in molecular hybridization experiments; such experiments reveal that mitochondrial rRNA and most mitochondrial tRNAs hybridize with the H strand (see Chapter 3). However, few tRNAs do hybridize specifically with the L strand; so it is clear that this strand is not composed entirely of antisense sequences.

Attardi and his colleagues established the molecular positions of the rRNA cistrons and the 4S RNA (presumably tRNA) cistrons in HeLa cell mtDNA in a sophisticated combination of molecular hybridization and DNA electron microscopy. The first cistrons to be mapped were those for the 12S and 16S rRNAs. Robberson et al.

(23) established by length measurements that the 12S and 16S rRNAs were 0.27 and 0.42 μm long, corresponding respectively to molecular weights of 0.35×10^6 d and 0.54×10^6 d. In a second article, Robberson et al. (24) showed that the two mitochondrial rRNA cistrons were very close together (< 500 nucleotides apart) on the H strand. They prepared hybrids of the mtDNA H strand and a mixture of the two mitochondrial rRNAs by an aqueous technique, following which the hybrids were spread on a basic protein film and examined by electron microscopy (25). Under these conditions single-stranded DNA becomes collapsed into a recognizable bush, and duplex regions remained extended as gently curved filaments. Experiments with a deletion mutant of bacteriophage ϕX174 in which a single strand from the deletion mutant was allowed to anneal to a complementary strand of the wild type revealed a single-stranded bush of wild-type DNA approximately 500 nucleotides long on electron microscopy.

When hybrids of the H strand of HeLa cell mtDNA and a mixture of the two mitochondrial rRNAs were examined, bushes were usually seen at one or both ends of the duplex hybrid region, which was linear in conformation; but bushes were seen only very rarely in the middle of the linear region. Length measurement showed the linear region to be equal to the sum of the lengths of the linear regions when 12S and 16S rRNA were hybridized separately. These results led Robberson and associates to conclude that the 12S and 16S rRNA cistrons are very close together on the H strand and that any spacer between them could not be greater than 500 nucleotides long.

Wu et al. (26) went on to examine the relative positions of the rRNA and 4S RNA cistrons on the mtDNA using even more powerful analytical techniques. In these experiments the mtDNA strands were prepared by formamide modification of the basic protein film method for electron microscopy (25,27). The important difference between this method and the aqueous method is that in this method both single-strand and duplex DNA appear as extended filaments following formamide treatment. That is, there is no collapse of the single-stranded regions into bushes. Regions of the H strand that hybridize with the 12S and 16S rRNA molecules can be distinguished from single-stranded regions because they appear somewhat thicker under the electron microscope. These experiments confirmed the findings of Robberson et al. (24) that the mtDNA sequences coding for 12S and 16S mitochondrial rRNA were close together on the H strand, with length measurements of the spacer region showing it to be very short (approximately 160 nucleotides long).

Since single 4S RNA molecules, when hybridized with complementary mtDNA sequences, yielded duplex regions too short to be detected by electron microscopy, Wu and associates used a different technique to locate these hybrids. The electron-opaque label ferritin, which is easily recognized in the electron microscope using the basic protein film technique, was covalently coupled with 4S RNA. The 4S RNA fraction was then hybridized separately to preparations of the H and L strands using the formamide method. Wu and associates identified 4S RNA–ferritin conjugates at nine different positions on the H strand and three different positions on the L strand (see Fig. 3–7). The H strand contained one 4S site (H2) within the spacer region between the 12S and 16S rRNA coding sequences; another (H1) was immediately adjacent to the 12S sequence; a third (H3) was adjacent to the 16S sequence; two other sites (H7, H8) were close together; and the remaining sites were evenly spaced throughout the heavy strand.

Although the experiments of Wu and associates succeeded in revealing the binding sites for 12 4S RNA molecules, the number

of sites was clearly insufficient to account for a complete set of tRNAs. Costantino and Attardi presented data indicating that proteins synthesized in HeLa cell mitochondria might actually lack certain amino acids; later it became apparent that these results were probably artifactual, since Attardi and his colleagues succeeded in demonstrating by hybridization experiments with aminoacylated mitochondrial tRNAs (see Chapter 3 for discussion and Table 3–4) that HeLa cell mtDNA coded for at least 17 different tRNAs. Clearly, then, the experiments of Wu and associates did not succeed in revealing all the tRNA binding sites in HeLa cell mtDNA. Angerer et al. (28) set about identifying the remaining sites using a modification of the ferritin labeling technique in which the vitamin biotin was first coupled covalently to the oxidized 3′-OH termini of the 4S RNA molecules, which were then hybridized to mtDNA. Following hybridization, ferritin conjugated with the protein avidin (which forms a rapid noncovalent association with biotin) was used to label the bound 4S RNAs. Angerer and associates confirmed the nine original map positions on the H strand and also located three new 4S RNA binding sites (see Fig. 3–7). The number of L-strand binding sites was also increased from three to seven, bringing the total number of 4S RNA sites mapped up from 12 to 19, which is sufficient to account for virtually all the tRNAs needed to transfer all the amino acids (see Fig. 3–7). Angerer and associates also reviewed the disadvantages and advantages of the new procedure as compared to that used previously by Wu and associates. They will not be discussed here, except to point out that much more reproducible results have been obtained with the new procedure.

Some examples of ferritin-labeled L strands and H strands taken from the article by Angerer and associates are shown in Figs. 11–5 and 11–6. In these preparations the ferritin-labeled tRNAs are seen as black dots, and the regions where rRNA or a

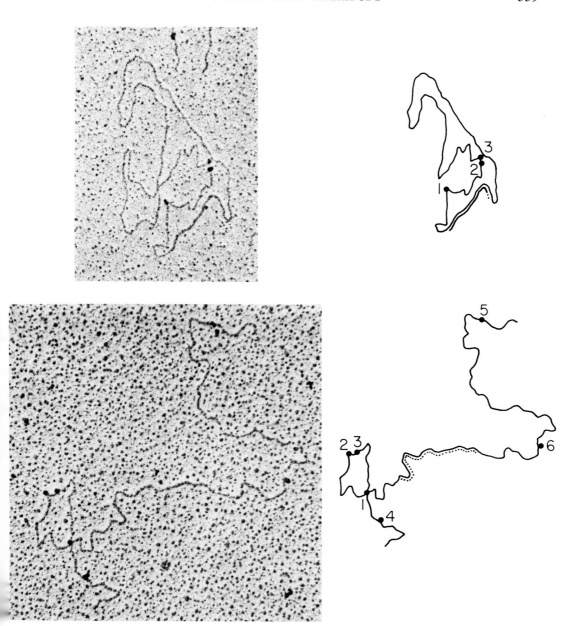

FIG. 11–5. Electron micrographs of HeLa cell L-strand mtDNA hybrids with HpA II fragment 3 and 4S RNA-biotin labeled with ferritin-avidin. HpA II fragment identifies mtDNA region coding for mitochondrial rRNA (see text for details). Explanatory tracings indicate positions of 4S RNA genes and reference rDNA duplex segment identified by hybridization of the HpA II mtDNA fragment with the L strand. An ambiguous duplex region is indicated by dotted lines. (From Angerer et al., ref. 28, with permission.)

restriction endonuclease fragment from the rDNA region have bound to the single-stranded DNA are seen to be thickened. It is evident that not all the 4S RNA binding sites are revealed through examination of any one mtDNA molecule; so one must examine a number of L strands and H strands, always using the rDNA region of the molecule as an orientation marker. Once this is done, the number of 4S RNA binding

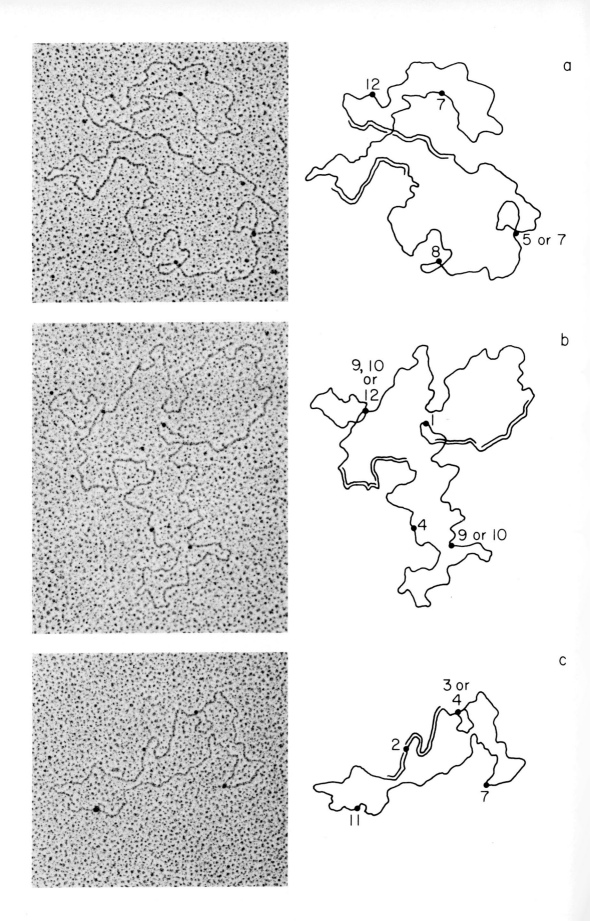

sites and their positions relative to the rDNA region can be established for both strands.

The H-strand preparation shown in Fig. 11–6 contained about 30% dimers. These can be seen to be arranged in head-to-tail fashion, with some of the ferritin labeling sites being diametrically opposed.

The experiments just described provided a molecular map of the HeLa cell mitochondrial genome in terms of the stable RNA species (see Fig. 3–7). Obviously, coupling these mapping techniques with techniques for mapping mRNAs (see Chapter 3) and determining their function, together with the use of genetically defined mitochondrial mutants, should eventually allow mapping of the entire mitochondrial genome of this mammalian cell line. It will be most interesting to compare this map with that being constructed for yeast (see Fig. 7–14) to see if the content of the mitochondrial genome has really diminished during the course of evolution or if the mtDNA of *Saccharomyces,* as opposed to other yeast species (see Chapter 2), is merely very rich in spacers (see Chapter 6).

FORMAL MITOCHONDRIAL GENETICS OF TISSUE CULTURE CELLS

Study of the formal mitochondrial genetics of tissue culture cells is getting under way, but at only a slow pace thus far. In 1972 Spolsky and Eisenstadt (29) reported the isolation of chloramphenicol-resistant mutants in HeLa cells; they showed, by use of isolated mitochondria, that mitochondrial protein synthesis in these mutants had become resistant to chloramphenicol. In a second article, Kislev et al. (30) reported

that chloramphenicol also caused gross alterations in mitochondrial ultrastructure in the sensitive strain but not in a mutant manifesting chloramphenicol-resistant mitochondrial protein synthesis. These observations are noteworthy because Storrie and Attardi (31) reported very little change in mitochondrial ultrastructure in chloramphenicol-sensitive HeLa cells grown in the presence of drug concentrations even higher than those used by Kislev et al. (30). The discrepancy between these reports has not been resolved.

In any event, the experiments of Spolsky and Eisenstadt (29) and Kislev et al. (30) indicated that mutants whose mitochondria had become chloramphenicol-resistant could be isolated in tissue culture cells, and in an article that could become a model for this sort of analysis in tissue culture cells, Bunn et al. (32) set out to determine whether or not resistance was cytoplasmically inherited. Bunn and associates used two subclones of mouse L cells designated A9 and LMTK⁻. The A9 strain is deficient in hypoxanthine guanine phosphoribosyl transferase (HGPRT) and is unable to use exogenously supplied hypoxanthine as a purine source, and as a result of the mutation the A9 strain is also resistant to the purine analog 8-azaguanine. The LMTK⁻ strain is deficient in thymidine kinase activity (TK⁻); so it is unable to use exogenously supplied thymidine as a pyrimidine source. The deficiency in TK activity also causes LMTK⁻ to become resistant to the thymidine analog 5-bromodeoxyuridine (BrdUrd).

Both A9 and LMTK⁻ are chloramphenicol-sensitive. Using methods similar to those used previously by Spolsky and Eisenstadt (29) for HeLa cells, Bunn and associates isolated a chloramphenicol-resistant deriva-

←

FIG. 11–6. Electron micrographs of HeLa cell H-strand mtDNA hybrids with 12S and 16S mitochondrial rRNA and 4S RNA-biotin labeled with ferritin-avidin. Explanatory tracings are given. Molecules shown in a and b constitute dimers. Note that identical sequences are diametrically opposed, confirming a head-to-tail arrangement of the monomer genomes. c: Monomer H strand is shown. (From Angerer et al., ref. 28, with permission.)

tive of A9 that they called 501–1. To determine whether the mutation to chloramphenicol resistance was located in a nuclear or cytoplasmic gene, Bunn and associates treated 501–1 cells with cytochalasin B, which induces enucleation in mammalian cells in tissue culture (33). They fused the resulting enucleated cytoplasts with LMTK⁻ cells using inactivated Sendai virus (Fig. 11–7). The cell fusion mixture was then plated on a medium containing BrdUrd and chloramphenicol. This medium selects not only against the two parental strains but also against nuclear hybrids of the two strains (Fig. 11–7). The 501–1 strain cannot grow on BrdUrd because it contains a normal thymidine kinase; the same is true

of nuclear hybrids between 501–1 and LMTK⁻, since the mutation resulting in thymidine kinase deficiency and BrdU resistance in LMTK⁻ is recessive. Similarly, LMTK⁻ cells cannot grow on the selective medium because it contains chloramphenicol. Therefore, excluding the possibility of a mutation to chloramphenicol resistance in LMTK⁻, three conditions must be satisfied for growth on the selective medium: chloramphenicol resistance must occur because of mutation in a cytoplasmic gene in 501–1; a mutant cell must have been converted to a cytoplast at the time of fusion; the mutant cytoplast must fuse with an LMTK⁻ cell to form a *cybrid*.

When Bunn and associates performed the

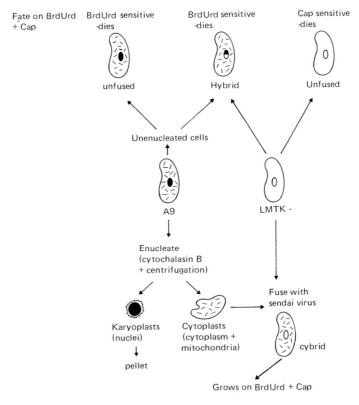

FIG. 11–7. Schematic diagram of methods used by Bunn et al. (32) to select mouse cell cybrids. Strain A9 has normal thymidine kinase; it can incorporate exogenous pyrimidines and is sensitive to BrdUrd for that reason (● = wild-type nucleus), but it carries a cytoplasmically inherited mutation to chloramphenicol (cap) resistance (/ = mutant cytoplasm). Strain LMTK⁻ is deficient in thymidine kinase because of a nuclear gene mutation, and it cannot metabolize exogenous pyrimidines, which causes this strain to be resistant to BrdUrd (○ = mutant nucleus). This strain is also sensitive to cap (indicated by clear cytoplasm). Only cybrids are able to grow on medium containing cap plus BrdUrd. Cell hybrids like the A9 parent are BrdUrd-sensitive because the mutation in thymidine kinase carried by the LMTK⁻ parent is recessive to its wild-type allele.

fusion experiments following enucleation, they found colonies growing on the BrdUrd-chloramphenicol medium (Table 11–6). Control experiments established that a mutation to chloramphenicol resistance in LMTK⁻ was not involved. Therefore, the three aforementioned conditions were fulfilled, and chloramphenicol resistance must have resulted from mutation in a cytoplasmic gene. An elaborate series of control experiments, together with cytologic and biochemical characterization of the cybrids, bolstered this conclusion even further.

Similar results have since been obtained by Wallace et al. (34) working with human cells. Using methods generally similar to those of Bunn and associates, these authors demonstrated cytoplasmic transfer of chloramphenicol resistance in HeLa cells. They also showed that mitochondrial protein synthesis in the resulting chloramphenicol-resistant cybrids was resistant to the antibiotic, which indicated that the site of phenotypic expression of the mutation was the mitochondrion. Most recently, Bunn et al. (35) followed segregation of chloramphenicol resistance and sensitivity in nuclear hybrids between cell lines and in cybrids. In each case chloramphenicol resistance was known to be cytoplasmically transmitted, and both mouse and human fusions were studied. The principal findings were as follows: First, in cybrids or hybrids between homologous mouse or human cell lines, the resulting cells retained a high level of chloramphenicol resistance and did not segregate very many chloramphenicol-sensitive cells. Second, in hybrids resulting from fusion of mouse cells of heterologous nuclear origin, chloramphenicol resistance of the cell population as a whole dropped very rapidly in the doublings following fusion, which indicated segregation of a high proportion of chloramphenicol-sensitive cells. When clones not selected on chloramphenicol were isolated, grown up, and retested to determine the proportions of resistant and sensitive cells, it was found that the proportion of resistant cells, although quite variable from clone to clone, was generally low. Third, cybrids between heterologous human cell lines (which must, by definition, be selected on medium containing chloramphenicol) exhibited a slow initial growth

TABLE 11–6. *Transfer of chloramphenicol resistance by L-cell cytoplasts*[a]

Cells	Number of cells plated	Average number of colonies per flask[b]	Colonies per 10⁶ cells
en501–1 × LMTK⁻	2 × 10⁶	Confluent (2)	
	1 × 10⁶	Confluent (2)	
	5 × 10⁵	Confluent (3)	
	2.5 × 10⁵	70 (3)	280
	1 × 10⁵	18 (3)	180
	8 × 10⁴	14 (1)	175
	5 × 10⁴	5 (2)	100
en501–1	2 × 10⁶	0	0
LMTK⁻	2 × 10⁶	0	0

[a] Cells were fused and plated in selective medium containing BrdUrd and chloramphenicol. The prefix en denotes a culture previously treated cytochalasin B to induce enucleation. This treatment produced 93% enucleation of 501–1 cells. Cells of LMTK⁻ (5 × 10⁶) and cells of the enucleated preparation of 501–1 (5 × 10⁶) were fused. Colonies growing on flasks were counted when clearly visible. Otherwise the flasks were simply scored as confluent, since the number of colonies was so high that they fused to produce confluent growth.

[b] Numbers in parentheses represent numbers of flasks counted at that particular cell number.

Adapted from Bunn et al. (32).

phase followed by a faster, normal rate of growth. During the slow phase the cells exhibited oxygen sensitivity and excess acid production, which suggested a reduction in ATP formation by oxidative phosphorylation coupled with a parallel increase in glycolysis. These changes in cell phenotype were not evident in the fast phase. Cells cloned from the slow phase in the absence of chloramphenicol segregated both resistant and sensitive cells, with the latter predominating, but cells cloned from the fast phase segregated a high proportion of chloramphenicol-resistant progeny. Heterologous mouse cybrids did not show the initial slow growth phase, but they did exhibit wide variation in chloramphenicol resistance following growth in the absence of chloramphenicol. These results are reminiscent of similar observations made on *Paramecium* (see Chapter 10) in that they suggest nuclear–mitochondrial incompatibility in heterologous combinations.

CONCLUSIONS

Molecular genetic study of the mtDNA of tissue culture cells is at an advanced stage. The stable RNAs have been mapped in HeLa cells, and it is clear that, given the right conditions, mtDNA from two species can be maintained in interspecific cell hybrids and that recombinant mtDNA molecules can even be formed between mtDNAs of different species. The tissue culture system is now ready for the study of formal mitochondrial genetics, and Eisenstadt and his colleagues have made an auspicious beginning. It is to be hoped that such work will now be concentrated on a limited number of mammalian cell lines and that formal and molecular genetic analysis will be coordinated. It would, for example, seem feasible to map the position of the chloramphenicol resistance gene in HeLa cell mitochondria with respect to the stable RNA genes. One possible approach is suggested by Dawid's interspecific mtDNA recombinants. It might be possible to use interspecific mtDNA recombinants to map in the following way: Suppose the mutation to chloramphenicol resistance in HeLa cell mtDNA is close to the rRNA cistrons. Hybrid cells resistant to chloramphenicol could be selected using HeLa and mouse cells on HAT medium. The clones could then be tested by Dawid's complementary cRNA technique to see if they contain recombinant mtDNA. Once recombinants have been detected, hybridization experiments can be carried out using mitochondrial rRNA from mouse and HeLa cells. If the marker for chloramphenicol resistance is close to the rRNA cistrons in HeLa mtDNA, most recombinant mtDNA will hybridize with HeLa mitochondrial rRNA, but not mouse mitochondrial rRNA.

REFERENCES

1. Blackler, A. W., and Gecking, C. A. (1972): Transmission of sex cells of one species through the body of a second species in the genus *Xenopus*. II. Interspecific matings. *Dev. Biol.*, 27:385–394.
2. Dawid, I. B. (1972): Evolution of mitochondrial DNA sequences in *Xenopus. Dev. Biol.*, 29:139–151.
3. Dawid, I. B., and Blackler, A. W. (1972): Maternal and cytoplasmic inheritance of mitochondrial DNA in *Xenopus. Dev. Biol.*, 29:152–161.
4. Ursprung, H., and Schabtach, E. (1965): Fertilization in tunicates: Loss of the paternal mitochondrion prior to sperm entry. *J. Exp. Zool.*, 159:379–384.
5. Austin, C. R. (1965): *Fertilization*. Prentice-Hall, Englewood Cliffs, N.J.
6. Szollosi, D. G. (1965): The fate of the sperm middle-piece mitochondria in the rat egg. *J. Exp. Zool.*, 159:367–378.
7. Hutchison, C. A., III, Newbold, J. E., Potter, S. S., and Edgell, M. H. (1974): Maternal inheritance of mammalian mitochondrial DNA. *Nature*, 251:536–538.
8. Clayton, D. A., Teplitz, R. L., Nabholz, M., Dovey, H., and Bodmer, W. (1971): Mitochondrial DNA of human-mouse cell hybrids. *Nature*, 234:560–562.
9. Attardi, B., and Attardi, G. (1972): Fate of mitochondrial DNA in human-mouse somatic cell hybrids. *Proc. Natl. Acad. Sci. U.S.A.*, 69:129–133.
10. Davidson, R. L. (1973): *Somatic Cell Hybridization: Studies on Genetics and Development*. Addison-Wesley, Reading, Mass.

11. Davidson, R., and de la Cruz, F. (editors) (1974): *Somatic Cell Hybridization.* Raven Press, New York.

12. Davidson, R. (1974): Gene expression in somatic cell hybrids. *Annu. Rev. Genet.,* 8: 195–218.

13. Eliceiri, G. L. (1973): Synthesis of mitochondrial RNA in hamster-mouse hybrid cells. *Nature [New Biol.],* 241:233–234.

14. Coon, H. G., Horak, I., and Dawid, I. B. (1973): Propagation of both parental mitochondrial DNAs in rat-human and mouse-human hybrid cells. *J. Mol. Biol.,* 81:285–298.

15. Scheid, A., and Choppin, P. W. (1974): Identification of biological activities of paramyxovirus glyproteins. Activation of cell fusion, hemolysis and infectivity by proteolytic cleavage of an inactive precursor protein of Sendai virus. *Virology,* 57:475–490.

16. Littlefield, V. W. (1964): Selection of hybrids from matings of fibroblasts *in vitro* and their presumed recombinants. *Science,* 145:709–710.

17. Horak, I., Coon, H. G., and Dawid, I. B. (1974): Interspecific recombination of mitochondrial DNA in hybrid somatic cells. *Proc. Natl. Acad. Sci. U.S.A.,* 71:1828–1832.

18. Dawid, I. B., Horak, I., and Coon, H. G. (1974): The use of hybrid somatic cells as an approach to mitochondrial genetics in animals. *Genetics,* 78:459–471.

19. Dawid, I. B., Horak, I., and Coon, H. G. (1974): Propagation and recombination of parental mtDNAs in hybrid cells. In: *The Biogenesis of Mitochondria,* edited by A. M. Kroon and C. Saccone, pp. 255–262. Academic Press, New York.

20. Craig, I. (1973): A procedure for the analysis of citrate synthase (EC 4.1.3.7) in somatic cell hybrids. *Biochem. Genet.,* 9:351–358.

21. Van Heyningen, V., Craig, I., and Bodmer, W. (1973): Genetic control of mitochondrial enzymes in human-mouse somatic cell hybrids. *Nature,* 242:509–512.

22. Van Heyningen, V., Craig, V. W., and Bodmer, W. (1974): Mitochondrial enzymes in man-mouse hybrid cells. In: *The Biogenesis of Mitochondria,* edited by A. M. Kroon and C. Saccone, pp. 231–244. Academic Press, New York.

23. Robberson, D., Aloni, Y., Attardi, G., and Davidson, N. (1972): Expression of the mitochondrial genome in HeLa cells. VI. Size determination of mitochondrial ribosomal RNA by electron microscopy. *J. Mol. Biol.,* 60:473–484.

24. Robberson, D., Aloni, Y., Attardi, G., and Davidson, N. (1972): Expression of the mitochondrial genome in HeLa cells. VII. The relative position of ribosomal RNA genes in mitochondrial DNA. *J. Mol. Biol.,* 64:313–317.

25. Davis, R. W., Simon, N., and Davidson, N. (1971): Electron microscope heteroduplex methods for mapping regions of base sequence homology in nucleic acids. *Methods Enzymol.,* 210:413–428.

26. Wu, M., Davidson, N., Attardi, G., and Aloni, Y. (1972): Expression of the mitochondrial genome in HeLa cells. XIV. The relative positions of the 4S RNA genes and of the ribosomal RNA genes in mitochondrial DNA. *J. Mol. Biol.,* 71:81–93.

27. Westmoreland, B. C., Szybalski, W., and Ris, H. (1969): Mapping of deletions and substitution of heteroduplex DNA molecules of bacteriophage λ by electron microscopy. *Science,* 163:1343–1348.

28. Angerer, L., Davidson, N., Murphy, W., Lynch, D., and Attardi, G. (1976): An electron microscope study of the relative positions of the 4S and ribosomal RNA genes in HeLa cell mitochondrial DNA. *Cell,* 9:81–90.

29. Spolsky, C. M., and Eisenstadt, J. M. (1972): Chloramphenicol-resistant mutants of human HeLa cells. *FEBS Lett.,* 25:319–324.

30. Kislev, N., Spolsky, C. M., and Eisenstadt, J. M. (1973): Effect of chloramphenicol on the ultrastructure of mitochondria in sensitive and resistant strains of HeLa. *J. Cell Biol.,* 57:571–579.

31. Storrie, B., and Attardi, G. (1973): Expression of the mitochondrial genome in HeLa cells. XV. Effect of inhibition of mitochondrial protein synthesis on mitochondrial formation. *J. Cell Biol.,* 56:819–831.

32. Bunn, C. L., Wallace, D. C., and Eisenstadt, J. M. (1974): Cytoplasmic inheritance of chloramphenicol resistance in mouse tissue culture cells. *Proc. Natl. Acad. Sci. U.S.A.,* 71: 1681–1685.

33. Croce, C. M., Tomassini, N., and Koprowski, H. (1974): Enucleation of somatic cells with cytochalasin B. *Methods Cell Biol.,* 8:145–150.

34. Wallace, D. C., Bunn, C. L., and Eisenstadt, J. M. (1975): Cytoplasmic transfer of chloramphenicol resistance in human tissue culture cells. *J. Cell Biol.,* 67:174–188.

35. Bunn, C. L., Wallace, D. C., and Eisenstadt, J. M. (1976): The behavior of cytoplasmic genes in mammalian cells. In: *The Genetic Function of Mitochondrial DNA,* edited by C. Saccone and A. M. Kroon, pp. 15–25. North Holland, Amsterdam.

36. Brown, D. D., Dawid, I. B., and Reeder, R. H. (1977): *Xenopus borealis* misidentified as *Xenopus mulleri. Dev. Biol.,* 59:266–267.

Chapter 12
Chloroplast Genetics of *Chlamydomonas*. I. Nonmendelian Inheritance and the Chloroplast

The green alga *Chlamydomonas reinhardtii* occupies a central position in the study of chloroplast genetics, much as has *S. cerevisiae* in the study of mitochondrial genetics. The story of plastid inheritance in *Chlamydomonas* began in 1954 with a report by Sager (1) that two mutations to streptomycin resistance in the alga exhibited different patterns of inheritance (Fig. 12–1). One of the mutations, designated *sr-1*, was mendelian in inheritance and was resistant to low levels of streptomycin, but the second, called *sr-2,* was maternally inherited and was resistant to high levels of streptomycin. Since that time many other mutations with the same maternal pattern of inheritance as *sr-2* have been isolated (Table 12–1), and it appears likely that these mutations are localized in the chloroplast genome. We shall begin this account of plastid inheritance in *C. reinhardtii* with a description of the organism, its genetics, and its growth, together with a summary of the

evidence that supports the hypothesis that maternally inherited genes in *Chlamydomonas* are localized in the chloroplast. In Chapters 13 and 14 we shall consider mutation, segregation, recombination, and mapping of chloroplast genes in this alga. In Chapter 15 we shall consider the hypotheses that have been suggested to account for the inheritance of chloroplast genes in *Chlamydomonas.*

CELL STRUCTURE

C. reinhardtii is a heterothallic, unicellular green alga with all the convenient handling characteristics of other unicellular microorganisms such as bacteria and yeast. Vegetative cells are haploid, and the chromosome number has variously been reported as 8 to 16 (2–5). Eight is probably the correct number, according to a very recent and detailed study by Maguire (4). However, the number of linkage groups is

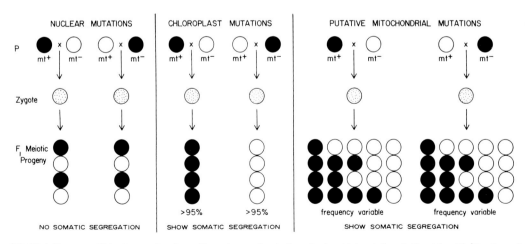

FIG. 12–1. Patterns of inheritance of nuclear, chloroplast, and putative mitochondrial mutations in *C. reinhardtii*. (Courtesy of Dr. J. E. Boynton.)

TABLE 12–1. *Summary of known chloroplast gene mutations in C. reinhardtii*[a]

Phenotype	Gene symbol	Other designations	Origin induced by	Comments
Acetate-requiring	ac-u-1	ac_1, ac1	STREP	Small, flat, light-green colonies on minimal medium (54); mapped by several procedures (84–86)
	ac-u-2	ac_2, ac2	STREP	Tiny, high, dark-green colonies on minimal medium (54); mapped by several procedures (84–86)
	ac-u-3	ac3	STREP	Same phenotype as ac-u-1 (54); mapping claimed (68, 83), but no supporting data published
	ac-u-4	ac4	STREP	Same phenotype as ac-u-2 (54); mapping claimed (68, 83), but no supporting data published
	ac-u-Cl	Cl	?	Lacks chloroplast-protein complex 1 and polypeptide 2 (87, and see Chapter 20)
Carbomycin-resistant	car-u-1	car1, car, car-r	STREP	Mapping claimed, and some supporting data published (85)
Cleocin-resistant	clr-u-1	cle, cle1	STREP	Mapping claimed, and some supporting data published (85)
Erythromycin-resistant	er-u-1a	ery-U1a	EMS	Not mapped (89).
	er-u-11	er-ll, eryll, ery1, ery	STREP	Mapped by several methods (84–86,88,89); cross-resistant to CARB, OLE, and SPI (83); same mutant as ery1 (68,76)
	er-u-3	ery3	STREP	Not mapped (68,83); same cross-resistance pattern as er-u-ll
	er-u-37	ery-u-37	Spontaneous	Mapping claimed, and supporting data published (88,89)
	er-u-AW-17, 48, 82, and 84	None	MNNG	Mapping claimed, and supporting data published (88,89)
	er-u-2y	ery-2y, Ery-2-y	MNNG	Uniparental inheritance claimed (90), but inheritance pattern is more complex (91)
	er-u-3-6	ery-3-6	MNNG	No mapping data available; has been described (92)
Kanamycin-resistant	knr-u-1	kan1	STREP	Not mapped (68,83)
	knr-u-2	kan-1	Spontaneous	Not mapped (90)
Neamine-dependent	nd-u-1	nd-a, nd-1	MNNG	Not mapped; cross-dependent on streptomycin (78)
	nd-u-2	nd-b, nd-2	MNNG	See nd-u-1
	nd-u-3	nd-c, nd-3	MNNG	See nd-u-3
Neamine-resistant	nr-u-2-1	nr-2, nr-2-1, Nea^r, nr, nea, nea-r, nea-2-1	MNNG	Mapping claimed, and supporting data published (68,84,88, 89); cross-resistant to kanamycin (89)
Oleandomycin-resistant	olr-u-1, 2, and 3	oli1, 2, and 3	STREP	Mapping claimed (68,83), and limited supporting data published (85)
Streptomycin-dependent	sd-u-sm4	sd, sm_4, sm4	STREP	Mapping claimed (68,83), and some supporting data published (84)
	sd-u-3-18	sd, sd-3, sd-3-18	MNNG	Not mapped; cross-dependent on neamine (78)
Conditional streptomycin-dependent	sd_{cd}-u-tm2	tm2	STREP	Not mapped; requires streptomycin at 35°C (68,83)
	sdR_{cd}-u-sm4-D-769	csd	Growth of sm4 without STREP	Mapping claimed (68,83), and some supporting data published (84)

TABLE 12–1. *(Continued)*

Phenotype	Gene symbol	Other designations	Origin induced by	Comments
Reversion from strep-tomycin dependence	*sdR*r*-u-sm4-D-310*	*D-310*	Growth of *sm4* without STREP	Streptomycin-resistant (68,83)
	*sdR*r*-u-sm4-D* (4 mutants)	*D*	Growth of *sm4* without STREP	Resistant to 500 μg/ml strepto-mycin; segregates like persist-ent hets (*sd/sr*) (68,83, and see Chapter 14)
	*sdR*r*-u-sm4-D* (11 mutants)	*D*	Growth of *sm4* without STREP	Resistant to various low levels of streptomycin; segregates like persistent hets (*sd/low sr*) (68,83, and see Chapter 14)
	*sdR*cd*-u-sm4-D* (3 mutants)	*D*	Growth of *sm4* without STREP	Conditional streptomycin-depen-dent; segregates like persistent hets (*sd/cond. sd*) (68,83, and see Chapter 14)
	*sdR*s*-u-3-18R1*	*R1*	Growth of *sd-u-3-18* without STREP	Streptomycin-sensitive (93)
	*sdR*s*-u-3-18R20*	*R20, sd-3-18R20*	Growth of *sd-u-3-18* without STREP	Streptomycin-sensitive; segre-gates *sd* progeny over long pe-riod of time (94,95), and was incorrectly designated strepto-mycin-resistant (92)
Spiramycin-resistant	*spir-u-1* through *spir-u-5*	*spi1* through *spi5*	STREP	Mapping claimed (68,83), and limited supporting data pub-lished (85)
Spectinomycin-resistant	*spr-u-1-27-3*	*spr-1-27-3, spr-1-27, sp-2-73, Spe*r*, spr-1**	MNNG	Mapping claimed, and support-ing data published (88,89); re-sistant to high spectinomycin lev-els only on acetate (80)
	spr-u-1-6-2	*spr-1*	MNNG	Mapping claimed, and support-ing data published (88,89)
	spr-u-sp-23	*sp-23, spc1, spc*	STREP	Mapping claimed, and support-ing data published (85,86,88,89)
	spr-u-1-H-4	None	Spontaneous	Mapping claimed, and support-ing data published (88,89)
Streptomycin-resistant	*sr-u-2-1*	*sr-2-1*	Spontaneous	Not mapped; resistant to 500 μg/ml streptomycin (51,92)
	sr-u-2-60	*sr-2-60, sr-2, Str*r	Spontaneous	Mapping claimed, and support-ing data published (88,89); re-sistant to 500 μg/ml streptomycin
	sr-u-2-235	*sr-2-235*	Spontaneous	Not mapped; resistant to 500 μg/ml streptomycin (96)
	sr-u-2-241	*sr-2-241*	Spontaneous	See *sr-u-2-235*
	sr-u-2-280	*sr-2-280*	Spontaneous	See *sr-u-2-60*
	sr-u-2-281	*sr-2-281*	Spontaneous	See *sr-u-2-60*
	sr-u-2-23	None	MNNG	Resistant to high streptomycin levels only in the presence of acetate (80); mapping claimed, and supporting data published (88,89)
	sr-u-2-NG-1	None	MNNG	See *sr-u-2-60*
	sr-u-sm2	*sr-2, sr, sr-500, sm-2*	STREP	Resistant to 500 μg/ml strepto-mycin; mapping claimed, and supporting data published (84–86,88,89)
	sr-u-sm3	*sm3*	STREP	Resistant to 50 μg/ml streptomy-cin; mapping claimed, and sup-porting data published (84,85, 88,89)

TABLE 12–1. *(Continued)*

Phenotype	Gene symbol	Other designations	Origin induced by	Comments
	sr-u-sm3a sr-u-sm3c	sm-3a sm-3c	MNNG	Resistant to 100 μg/ml strepto-mycin; mapping claimed, and supporting data published (88, 89,97)
	sr-u-sm3b sr-u-sm3d	sm-3b sm-3d	MNNG	Not mapped; resistant to 100 μg/ml streptomycin (97)
	sr-u-sm5	sm5	STREP	Resistant to 500 μg/ml strepto-mycin; mapping claimed, and supporting data published (85, 88,89)
	sr-u-sr$_{35}$	sr$_{35}$	STREP	Resistant to high levels of strep-tomycin (98)
Tiny colonies	ti-u-1 through ti-u-5	ti1 through ti5	MNNG	Not mapped (68)
Temperature-sensitive	tm-u-1	tm1, tr1	STREP	Cannot grow at 35°C; mapping claimed, and supporting data published (85,86)
	tm-u (7 mutants)	tm	STREP	Not mapped; cannot grow at 35°C (68,83)
Thylakoid membrane polypeptide	thm-u-1	None	Spontaneous	Altered mobility of one of thyla-koid membrane polypeptides in SDS gels (99)

[a] The nomenclature used is that introduced by Conde et al. (80) and applied broadly to mutants from all laboratories for the first time by Adams et al. (81). Summaries of the known chloroplast gene mutations in *C. reinhardtii* are also to be found in the book by Sager (68), as well as in her more recent review articles (26,82,83). In the nomenclature adopted by Conde et al. each mutant is given a symbol descriptive of its phenotype followed by *u* (for uniparentally inherited) and by its previously published designation or isolation number. In the case of mutants with altered sensitivity to antibiotics, this symbol indicates the name of the antibiotic and whether the mutant is resistant (*r*), dependent (*d*), or sensitive (*s*). Conditional mutants are designated by a subscript c. Reversions are indicated by the original mutant symbol followed by R and a subscript desig-nating the new phenotype. A great many reversions of streptomycin-dependent mutants have been isolated by Sager (see above references) and by Arnold et al. (see Chapter 13 references). Only a few of these are indicated here to show how the nomenclature works. Several mutations allelic with sr-u-2-23 and one allelic with sr-u-2-60 have been omitted (89). Key: CARB = carbomycin, CLE = cleocin, EMS = ethyl methane sulfonate, ERY = erythromycin, KAN = kanamycin, NEA = neamine, MNNG = N-methyl-N′-nitro-N-nitrosoguanidine, OLE = oleandomycin, SPI = spiramycin, SPEC = spectinomycin STREP = streptomycin.

reported to be 16 (6). The reasons for this discrepancy are not presently evident, and they probably will not be revealed until further systematic and extensive mapping of mendelian genes is done in *C. reinhardtii*. The cell contains a single, large cup-shaped chloroplast occupying about 40% of the cell volume (Table 12–2; see Fig. 1–13). Mito-chondria are also evident; they vary in shape and probably in number depending on growth conditions. Arnold and Schötz (7–9) reported, on the basis of serial sec-tion electron microscopy of gametes, that there are 9 to 14 mitochondria per cell. The mitochondria occupy a much smaller por-tion of the total cell volume than the chloro-plast (Table 12–2). The nucleus contains a

prominent nucleolus, and the anterior end of the cell contains a pair of flagella (see Fig. 1–13).

The structure of the chloroplast itself de-serves some additional comment. The chlo-roplast contains a prominent pyrenoid sur-rounded by starch plates (see Fig. 1–13). DNA-containing areas lie adjacent to the pyrenoid, as was shown by Ris and Plaut (10). Thylakoid membranes occupy about 43% of the volume of the chloroplast, and the lamellae are usually paired two by two (Table 12–2; see Fig. 1–13). The granular appearance of the chloroplast stroma is caused in part by the presence of chloroplast ribosomes, representing 30 to 40% of the ribosomes of the cell (Table 12–2), and

TABLE 12–2. *Quantification of cell organization in* C. reinhardtii[a]

		Percentage of total[b]	
		Area	Volume
Whole cell			
Area of median section (μm^2)	43.4 ± 1.8	100	—
Volume including cell wall (μm^3)			
Cell 1	56.1	—	—
Cell 2	80.4	—	—
Volume within plasma membrane (μm^3)[c]			
Cell 1	44.1	—	100
Cell 2	59.1	—	100
Nucleus			
Volume (μm^3)			
Cell 1	4.4	—	10
Cell 2	4.3	—	7
Chloroplast			
Area of median section (μm^2)	14.7 ± 1.1	34	—
Net stroma	7.0 ± 0.7	16	—
Lamellar system	6.3 ± 0.3	15	—
Volume (μm^3)			
Cell 1	17.1	—	40
Cell 2	25.2	—	43
Mitochondria			
Area of median section (μm^2)	1.46 ± 0.26	4	—
Volume			
Cell 1	1.4	—	3
Cell 2	2.0	—	3
Number			
Cell 1	9	—	—
Cell 2	14	—	—
Ribosomes[d]			
Cytoplasm			
Number per μm^2 of median section	161 ± 7	—	—
Number estimated per cell × 10^5	0.98		
Chloroplast			
Number per μm^2 of median section	115 ± 4	—	—
Number estimated per cell × 10^5	0.53	—	—

[a] Area measurements and estimates of ribosome number are from Bourque et al. (100), who used cells growing exponentially under mixotrophic conditions. The numbers represent means and standard errors derived by quantification of median sections of 10 cells. The volume measurements and estimates of mitochondrial number were made by Schötz et al. (9) and Arnold et al. (8) on the basis of serial sections of two gametes derived from cells grown phototrophically. Both sets of measurements were made on wild-type cells of strain 137c.

[b] Percentages are rounded off to nearest whole number.

[c] Schötz et al. (9) normalized all their volume percentages to the volume of the cell within the plasma membrane, and excluding the cell wall.

[d] Calculation of number of ribosomes per cell is described by Bourque et al. (100).

ribulose bisphosphate carboxylase (RuBP-Case), or fraction I protein, which has a molecular weight of approximately 560,000 d and constitutes at least 10% of the soluble protein of the cell (11). This enzyme is a convenient marker for measuring chloro-plast protein synthesis *in vivo,* since one of the two component polypeptides of which the enzyme aggregate is formed is synthesized on chloroplast ribosomes (see Chapter 20). Also found within the chloroplast is the eyespot (see Fig. 1–13). This organelle, which is rich in β-carotene, is important in the phototactic behavior of the cell.

DNA SPECIES OF *C. reinhardtii*

Four species of DNA have been identified in vegetative cells of *C. reinhardtii* on the basis of their buoyant densities in cesium chloride (Fig. 12–2). The major species, α-DNA, which comprises about 85% of the total cell DNA, has a buoyant density of 1.723 g/cc and is assumed to be nuclear (Table 12–3). This DNA has a kinetic complexity of 5×10^{10} d and is composed largely of unique sequences (12). The satellite at 1.715 g/cc (γ), which amounts to 1% of the total (Table 12–3), was originally thought to be mtDNA (13) but has now been reported to hybridize with cytoplasmic rRNA (14). Hence $\dot{\gamma}$-DNA is probably a nuclear satellite composed of the cytoplasmic ribosomal DNA cistrons.

Studies of isolated mitochondrial preparations from *C. reinhardtii* by Chiang and his colleagues (15,16) showed that they are enriched for a satellite DNA with a buoyant density of 1.706 g/cc. This mtDNA amounts to 0.5 to 1.0% of the total DNA of the cell; it is in the form of 4.6-μm circles and has a kinetic complexity of 1.6×10^7 d (Table 12–3). The kinetic complexity and electron microscopic measurements are in reasonably good agreement, and they indicate that the mtDNA of *C. reinhardtii* is small, like that

of many animals (see Chapter 2). There are about 20 to 40 copies of mtDNA per cell.

In 1963 Sager and Ishida (17) showed that chloroplast preparations from *C. reinhardtii* were greatly enriched for the satellite with a buoyant density of 1.695 g/cc (β). This clDNA comprises 14% of the total in vegetative cells and 7% in gametes. Molecular weight estimates based on kinetic complexity, restriction fragment analysis, and measurement of intact circles do not agree, as was discussed in Chapter 2. Circular molecules 62 μm in length have been obtained from clDNA preparations of *C. reinhardtii* (18). Using T7 (2.07×10^6 d/μm) as a standard, these would have a molecular weight of 1.29×10^8 d. This value is in good agreement with the molecular weight estimate obtained by summing the molecular weights of EcoR1 restriction fragments (19,20). All kinetic complexity estimates reported thus far (12,21,22) have indicated a complexity equivalent to 2×10^8 d. In two of these studies *E. coli* was used as the standard (12,21), and in the third study T4 (22) was the standard. When T4 was used as the standard, the molecular weight of its DNA was assumed to be 2×10^8 d. In one of the two studies in which *E. coli* was used as the standard, the kinetic complexity of T4 DNA was calculated under the same conditions (12), and an overestimate of the molecular weight

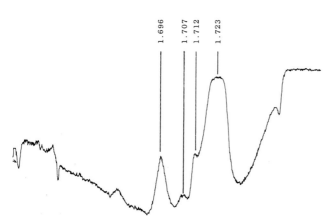

FIG. 12–2. Densitometer tracing of CsCl equilibrium gradient containing DNA extracted from whole cells of C. *reinhardtii.* Gradient was purposely overloaded to display the minor DNA species, thereby sacrificing resolution of the major component; clDNA (β) = 1.696 g/cc; mtDNA (δ) = 1.707 g/cc; cytoplasmic ribosomal DNA (γ) = 1.712 g/cc; nDNA (α) = 1.723 g/cc. (From Adams et al., ref. 81, with permission.)

TABLE 12–3. *Properties of DNA components of C. reinhardtii[a]*

Growth stage	DNA/cell ($\times 10^7 \mu$g)	Component	Buoyant density (g/cc)	Percentage of total DNA	Unique sequences	
					Molecular weight (d)	Number per cell
Vegetative cells	1.24 ± 0.081^b	Nuclear (α)	1.723	85	5×10^{10}	1
		Chloroplast (β)	1.695	14	1.3×10^8	80
		Mitochondrial (δ)	1.706	1	1.6×10^7	40^c
		γ	1.715	1	—	—
Gametes	1.23 ± 0.064^b	Nuclear (α)	1.723	89	—	1
		Chloroplast (β)	1.695	7	—	40
		γ	1.715	4	—	—

[a] See various references (12,15–17,19,20,22,24). Molecular weights are based on kinetic complexity estimates in the case of mtDNA and nDNA, but on restriction fragment analysis in the case of clDNA.

[b] Standard deviation.

[c] R. S. Ryan, D. Grant, K.-S. Chiang, and H. Swift, personal communication.

of the DNA was obtained. It is reasonable to suspect that such an overestimate would also have been obtained in the third study (21) where *E. coli* DNA was taken as the standard for kinetic values. Kolodner and Tewari (23) pointed out that if a lower and more accurate estimate of the molecular weight of T4 DNA is used, the kinetic complexity of *C. reinhardtii* clDNA will reduce to the equivalent of 9.9×10^7 d. In view of these uncertainties about the kinetic complexity measurements, it seems that the best estimate for the molecular size of clDNA in *C. reinhardtii* is about 1.3×10^8 d, which is the estimate obtained by molecular measurement and restriction fragment analysis. It should also be noted that Wells and Sager (12) reported a second fraction in clDNA, since they found an initial period of nonlinearity in their renaturation kinetic experiments. This fast-renaturating fraction comprised 10% of the clDNA, and it had a kinetic complexity of 10^6 to 10^7 d. It was estimated that the fast-renaturing fraction was 2 to 20 times more redundant than the major fraction of clDNA. This fast-renaturing fraction was not reported in either of the other studies of kinetic complexity. Since the actual amount of clDNA in gametes has been estimated at 5×10^9 d (24), clDNA molecules of 1.3×10^8 d

molecular weight would have to be present on the average in 40 copies per gamete chloroplast and 80 copies per vegetative cell.

Sager (25,26) suggested, on the basis of genetic evidence (see Chapter 14), that the chloroplast of *C. reinhardtii* may also contain a very slowly renaturing DNA. She suggested that this DNA would be present in one or two copies per plastid and would have a genomic content of about 8×10^8 d. She argued that this DNA would not have been detected in the renaturation experiments thus far published and that this is the DNA fraction that contains the chloroplast genes studied in *C. reinhardtii*. Although there is no physical evidence of any kind to support this model, neither is there evidence that rigorously rules it out.

In concluding this discussion of *Chlamydomonas* DNA, it is important to note that cell fractionation in this alga is still at a very primitive stage. Thus α-DNA is assumed to be nuclear, since it represents most of the cellular DNA, but this has not been proved by a demonstration that α-DNA is enriched in isolated nuclear preparations. Worse yet is the situation with respect to the chloroplast. Despite reports to the contrary, it is the author's opinion that no convincing report of intact chloroplast isolation has yet appeared for *Chlamydomonas*. This does not mean that thylakoid membrane preparations lacking the outer chloroplast membrane are not easily obtained. Such preparations are readily made, and they are enriched for clDNA. The point is that such preparations also contain a considerable amount of DNA having the buoyant density of α-DNA. In fact, in the chloroplast preparations of Sager and Ishida (17), clDNA accounted for only 40% of the total. The remaining DNA had the buoyant density of the α component. It seems likely that this represented nDNA trapped in the chloroplast preparations, since the chloroplasts were probably not intact. However, the possibility was not ruled out then (nor has it been ruled out since) that part of the clDNA has the same buoyant density as the α species.

GROWTH CONDITIONS

Strain 137c of *C. reinhardtii*, which was isolated by G. M. Smith and is the organism used in most genetic investigations, can be grown either on complex media containing supplements such as yeast extract or on simple inorganic media containing a few salts and trace elements (27–34). *C. reinhardtii* is one of the so-called acetate flagellates

(35); it will use acetate as well as carbon dioxide as a carbon source.

C. reinhardtii can be grown photosynthetically in the light with carbon dioxide as the sole source of carbon, or it can be grown in light or dark when supplemented with acetate; this provides the basis for defining three fundamentally different sets of growth conditions (Table 12–4). *Phototrophically* grown cells are cultured in the light with carbon dioxide as the sole source of carbon. Under these growth conditions the organism must be able to generate the necessary ATP and "reducing power" to permit reduction of carbon dioxide to carbohydrate via photosynthesis (see Chapter 1). Thus phototrophic growth conditions are restrictive for those chloroplast functions having to do with both the light and dark reactions of photosynthesis. Any chemical agent or mutation that blocks photosynthesis is lethal under phototrophic growth conditions.

Fortunately, photosynthetic functions are dispensable when cells are grown either in the light (*mixotrophic growth*) or in the dark (*heterotrophic* growth) in the presence of acetate. Photosynthetic mutants can be maintained under these permissive conditions. These mutants form part of a very large class broadly defined as acetate-requiring (*ac*) mutants. Many of these mutants

TABLE 12–4. *Diagnostic growth conditions for C. reinhardtii*[a]

Growth condition			Specific inhibitors	References	Specific classes of mutants	References
Phototrophic	Light	+	Rifampicin	38,90,101	Restrictive for acetate (*ac*) mutants, many of which have photosynthetic defects	6,36,37
	CO₂	+				
	Acetate	−				
Heterotrophic	Light	−	Oligomycin, acriflavine, ethidium bromide	59,60	Permissive for *dk* mutants	44,62
	CO₂	+			Permissive for *ac* mutants	6,36,37
	Acetate	+			Restrictive for obligate photoautotrophic mutants (*dk*), some of which have mitochondrial defects	44,62
Mixotrophic	Light	+	None		Permissive for *ac* and *dk* mutants	
	CO₂	+				
	Acetate	+				

[a] The table also indicates inhibitors known to be specific for a given mode of growth and general phenotypic classes of mutants lethal under a specific growth condition.

have been mapped, and they affect a large array of different functions associated with photosynthesis (for reviews see 6,36,37). One inhibitor, rifampicin, appears to block phototrophic growth preferentially, since it specifically inhibits transcription of the cistrons for 23S and 16S chloroplast rRNA and thus blocks the synthesis of chloroplast ribosomes (38).

Although *C. reinhardtii* is an obligate aerobe, as are most other algae (39–41), it can dispense with the mitochondrial respiratory electron transport chain and oxidative phosphorylation when it is growing in the light. *C. reinhardtii,* like fungi and higher plants (see Chapters 1 and 8), has a cyanide-insensitive alternate oxidase as well as the classic electron transport chain (see Fig. 1–8). This oxidase can serve to transport electrons and protons from the tricarboxylic acid intermediates to molecular oxygen.

As was discussed by Walker (42) in a recent review, not all the ATP produced photosynthetically may be required to drive the dark reactions of photosynthesis. Some of this ATP might be used to drive other reactions in the cell outside of the chloroplast. The major problem with this idea is that evidence for direct transfer of ATP or ADP across the chloroplast envelope is contradictory. Despite this, Walker pointed out that an indirect mechanism may exist for ATP transfer. The proposed shuttle mechanism involves export of dihydroxyacetone phosphate (DHAP) to the cytoplasm where it is oxidized via glyceraldehyde 3-phosphate (G3P) and 1,3-diphosphoglyceric acid (DPGA) to phosphoglyceric acid (PGA) via the normal glycolytic sequence. In the process, NADH and ATP are also produced. The PGA then reenters the chloroplast and is reduced back to triose phosphate at the expense of photosynthetically generated NADPH and ATP. The net effect is export of ATP and reducing equivalents. Walker pointed out that this shuttle mechanism could account for the observed increase in cytoplasmic ATP in the light and

its decrease in the dark and that it could also explain several observations that implicate the use of photosynthetically derived ATP in cytoplasmic events.

Direct evidence that mitochondrial electron transport and ATP production can be bypassed comes from studies of obligate photoautotrophic mutants or dark dier (*dk*) mutants, and inhibitors specific for heterotrophic growth in *C. reinhardtii* (Table 12–4). Some years ago Lewin (43) isolated an obligate photoautotrophic mutant of *C. dysosmos* that was unable to grow heterotrophically with acetate as a carbon source. He obtained experimental evidence suggesting that the mutant could not obtain energy by degrading acetate. Prompted by this observation, Wiseman et al. (44) set out to isolate *dk* mutants of *C. reinhardtii* in the hope of obtaining mutants that had defects in mitochondrial electron transport and/or ATP production. Their experiments proved successful; they characterized nine mendelian *dk* mutants of *C. reinhardtii*. Although none of the mutants was able to grow on acetate in the dark, all of them took up and metabolized acetate normally. The nine mutants mapped in eight genes, and all of them had pleiotropic effects on mitochondrial structure and function. Class I mutants had gross alterations in ultrastructure in their mitochondrial inner membranes, together with deficiencies in cytochrome oxidase activity and antimycin-A/rotenone-sensitive NADH–cytochrome c reductase activity. Class II mutants had a variety of less severe alterations in mitochondrial ultrastructure and deficiencies in cytochrome oxidase activity. Class III mutants had normal or nearly normal mitochondrial ultrastructure and reduced cytochrome oxidase activity. In short, all of the mutants died in the dark because respiratory electron transport and, presumably, oxidative phosphorylation were blocked. Support for the notion that oxidative phosphorylation is dispensable in the light comes from the observation (N. W. Gillham and J. E.

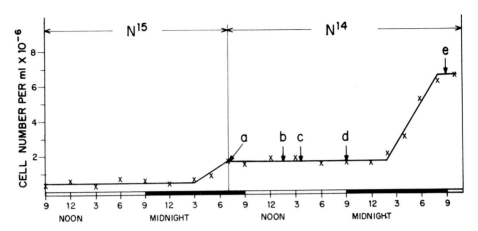

FIGURE 12–3. Demonstration of time and nature of nDNA (α) and clDNA (β) replication during vegetative cell division in synchronous cultures of C. *reinhardtii* after transfer from ^{15}N to ^{14}N medium. Suspension of mt^+ cells was subjected to light–dark cycle to maintain synchronous cell division. Samples were taken at various times (indicated by a, b, c, d, and e), and DNA was extracted and centrifuged (see Fig. 12–12 for densitometer tracings). (From Chiang and Sueoka, ref. 24, with permission.)

Boynton, unpublished data) that oligomycin apparently blocks cell growth preferentially under heterotrophic conditions (Table 12–4). This inhibitor uncouples respiratory ATP formation by the mitochondrial F_1 ATPase, but it does not appear to inhibit ATP formation by the CF_1 ATPase of the chloroplast (see Chapter 1). Other inhibitors that preferentially block heterotrophic growth are acriflavine and ethidium bromide (Table 12–4). These dyes also induce a class of mutations similar in some aspects to vegetative petites in yeast, as we shall see later in this chapter.

In summary, phototrophic growth is restrictive for photosynthetic mutants, but heterotrophic and mixotrophic conditions are permissive for such mutants. Heterotrophic growth is restrictive for mutants blocked in mitochondrial electron transport and probably in mitochondrial ATP formation, whereas phototrophic and mixotrophic conditions are permissive for such mutants. Thus by use of the appropriate combination of growth conditions, one can select for mutants affecting either the chloroplast or the mitochondrion in C. *reinhardtii*.

C. *reinhardtii* can be grown synchronously or asynchronously (34,45–47). Vegetative cell synchrony is achieved by growing cells with carbon dioxide as the sole carbon source on a cycle of alternating 12-hr periods of light and darkness (Fig. 12–3). Under such conditions, each cell divides into four daughter cells near the end of the dark period. The timing of a whole series of different cellular events has been investigated in synchronous culture. We have already had occasion to refer to synchronous cultures in connection with the timing of clDNA replication (see Chapter 2). Other experiments employing synchronous cultures will be discussed in this and later chapters.

LIFE CYCLE

Cells of opposite mating types in C. *reinhardtii* are morphologically identical, and mating is controlled by a pair of alleles, more likely different but tightly linked genes (see Chapter 15), in linkage group VI. Sager and Granick (28) showed that vegetative cells of either mating type can be converted into gametes by nitrogen starvation (Fig. 12–4). On mixing, gametes of opposite mating type pair at the flagellar tips. The gametes then shed their cell walls, and a cytoplasmic bridge is formed, following which the cells fuse completely to form a diploid zygote (Fig. 12–4). Not only the nuclei but also the chloroplasts fuse during

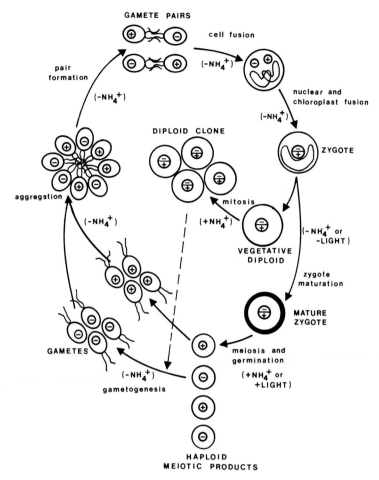

FIG. 12–4. Life cycle of *C. reinhardtii.* Vegetative cells are normally haploid, and mating type is controlled by a pair of nuclear gene alleles (designated by plus and minus). Following nitrogen deprivation (designated $-NH_4^+$), vegetative cells differentiate into gametes. Gametes of opposite mating types first aggregate and then pair off, following which the pairs fuse to form diploid zygotes. The majority of zygotes are meiotic zygotes, but a few mitotic zygotes are also formed in each cross, and they then divide as vegetative diploids. Germination of meiotic zygotes is triggered by nitrogen and light. Each zygote yields four haploid meiotic products, two of which belong to each mating type. (Courtesy of Dr. K. Van Winkle-Swift.)

the first 3 hr after mating (Fig. 12–5). Rarely, a triple fusion occurs in which two gametes of one mating type fuse with one of the opposite mating type (Fig. 12–6). Fortunately, such fusions are not frequent enough to interfere with routine genetic analysis.

Zygotes can be matured on an alternating cycle of light (approximately 20 hr), dark (approximately 5 days), and light (Fig. 12–4). During the second light period meiosis occurs (approximately 10 hr), and four haploid meiotic products are formed. In each tetrad two meiotic products are mating

type plus (*mt⁺*), and the other two are mating type minus (*mt⁻*), reflecting the mendelian segregation of these genes (Fig. 12–4). The clones produced by each meiotic product can be maintained either on solid medium or in liquid and can be induced to undergo gametic differentiation at any time by nitrogen withdrawal.

The entire zygote maturation cycle can also be effected in the light (Fig. 12–4), provided the zygotes are deprived of nitrogen, as was shown by Van Winkle-Swift (48,49). Germination is triggered by transferring the zygotes to nitrogen-containing

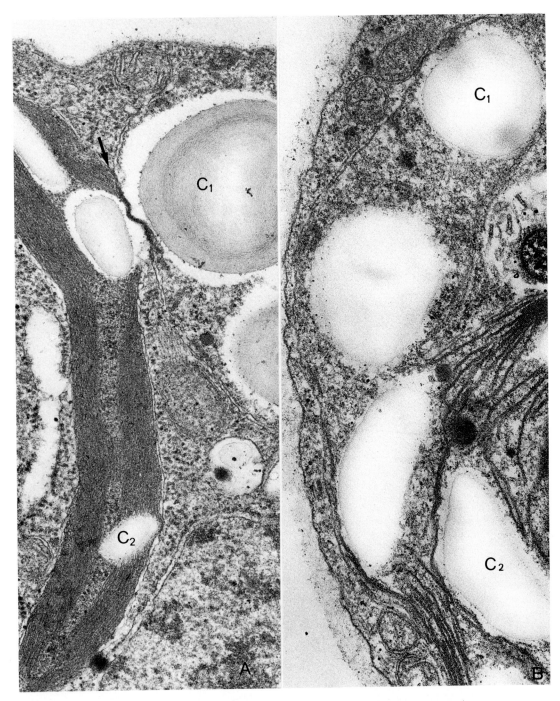

FIG. 12–5. Chloroplast fusion in a young zygote of C. *reinhardtii*. ×54,000. A: Ten minutes after mixing mt^+ and mt^- gametes together. B: Three hours after mating. Key: C_1 indicates chloroplast of br_{s-1} mt^+ gamete, which lacks chloroplast lamellae; C_2 indicates chloroplast of wild-type mt^- gamete with normal chloroplast lamellae; arrow indicates deposition of osmiophilic material in region where the two chloroplast envelopes come in contact. (Courtesy of Dr. J. E. Boynton.)

medium. Light maturation has proved invaluable in the study of certain mutant types (e.g., the *dk* mutants).

A small fraction of the zygotes formed in a cross divide mitotically as vegetative diploids (Fig. 12–4). These diploids are readily selected using complementary and closely linked auxotrophic markers such as the arginine-requiring mutants *arg-1, arg-2,* and *arg-7* located close together in linkage group I. Matings are made between stocks carrying complementary *arg* markers, and the mating mixture is plated on arginine-free medium. The colonies that first grow up usually prove to be diploids. This can be checked by measuring cell size, by measuring the amount of DNA per cell, or by testing for mating type. Diploids, although heterozygous for the mating type genes, are phenotypically *mt⁻* (50). Thus far, no haploidization process has been described for diploid *C. reinhardtii;* so markers can be extracted from diploids only by mating them to haploid *mt⁺* gametes. The triploid zygotes formed undergo meiosis, which is accompanied by much product lethality, as might be expected. However, a certain fraction of the resulting clones do grow well and carry markers that were present in the diploid. On crossing, these clones show normal marker segregation patterns. Diploids have proved useful in routine complementation analysis of mendelian genes, in the isolation of chloroplast gene mutants (see Chapter 13), and in the study of segregation and recombination of chloroplast genes (see Chapter 14).

GENETIC SYSTEMS

Three distinct genetic systems have now been identified in *C. reinhardtii.* One of these is the classic mendelian system in which genes segregate 2:2 (Figs. 12–1 and 12–4). Mutations belonging to the second genetic system exhibit maternal inheritance (Fig. 12–1). In the literature these mutations have been referred to by a variety of names, including cytoplasmic, nonmendelian, nonchromosomal, uniparental, and chloroplast. There is considerable indirect evidence (which will be summarized later in this chapter) that these mutations are localized in the chloroplast; here they will be referred to as chloroplast mutations with the proviso that a rigorous, unambiguous experiment proving their location remains to be done.

High-level mutations to streptomycin resistance (*sr-u-2*) exemplify the pattern of transmission of chloroplast genes in *C. reinhardtii.* When an *sr-u-2 mt⁺* stock is crossed to a wild-type *mt⁻* stock, all four of the meiotic products (more than 95%) are streptomycin-resistant. In the reciprocal cross, most of the zygotes produce only streptomycin-sensitive meiotic products. In the backcrosses of the F₁ progeny the same pattern of mating-type-dependent maternal inheritance is retained. Thus if the *mt⁺* F₁ progeny are streptomycin-resistant, they will transmit resistance to all four meiotic products in backcrosses to the sensitive parent. Similarly, if the *mt⁺* F₁ progeny are streptomycin-sensitive, they will transmit sensitivity in backcrosses to the resistant parent. The *mt⁻* F₁ progeny, on the other hand, do not transmit resistance or sensitivity, even though the marker in each case is, by definition, derived from the *mt⁺* parent. Thus the pattern of inheritance of *sr-u-2* and other chloroplast gene mutations is maternal (Table 12–1).

If 100% of the zygotes showed maternal inheritance of chloroplast genes, it would be impossible to study segregation and recombination of these genes. Fortunately, in addition to *maternal zygotes,* small percentages of *exceptional zygotes* are obtained in every cross. These are of two types. *Biparental zygotes* transmit chloroplast genes from both parents to their meiotic products. In the progeny of such zygotes, chloroplast genes, unlike mendelian genes, continue to segregate during the postmeiotic mitotic divisions, as well as during the initial meiotic

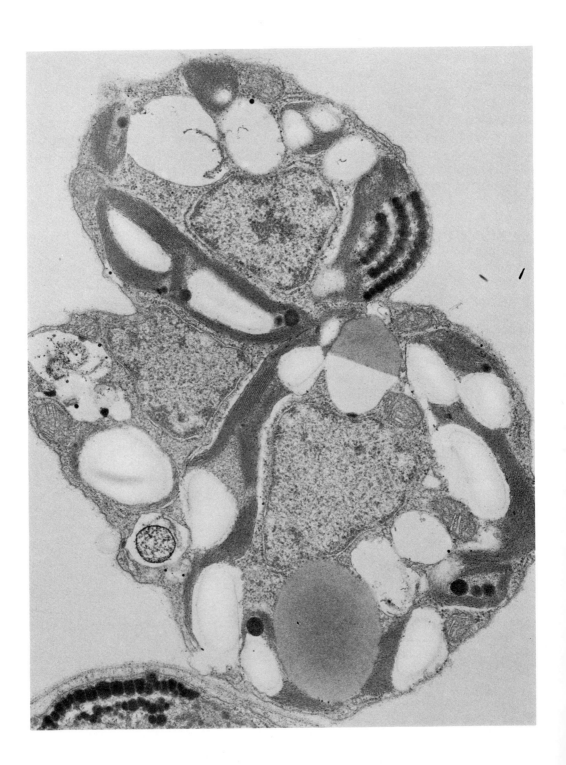

divisions (51,52). These zygotes are of critical importance in the study of chloroplast genetics in *C. reinhardtii,* for it is among their progeny that chloroplast gene recombination and segregation are studied (see Chapter 14). *Paternal zygotes* are the second group of exceptional zygotes. As their name implies, paternal zygotes transmit only chloroplast genes derived from the mt^- parent.

The frequency of spontaneous exceptional zygotes is variable from cross to cross, but it is almost always less than 10% (51). The frequencies of exceptional zygotes reported by Sager and Ramanis (53, 54) have thus far proved to be much lower (approximately 0.1% or less) than those observed in the laboratory of Gillham and Boynton (51,56). In any event, the frequency of exceptional zygotes can be increased greatly by irradiation of the mt^+ parent with ultraviolet light, as was shown by Sager and Ramanis (53). That discovery made possible the systematic study of chloroplast gene recombination and segregation in *C. reinhardtii* (see Chapter 14).

Gillham (51,57) found that the behavior of chloroplast genes could also be analyzed in the vegetative diploids arising in a cross. Like the sexual zygotes, diploids fall into maternal, biparental, and paternal classes with respect to chloroplast gene transmission (Fig. 12–7). When the appropriate mendelian markers (e.g., *arg-1, arg-2,* or *arg-7*) were used to select these diploids, it was found that biparental and paternal diploids were much more frequent than their zygotic counterparts. Chloroplast genes, unlike mendelian genes, segregate mitotically among the vegetative progeny of biparental diploids (Fig. 12–7). This is a useful property with respect to the isolation of chloroplast mutants in *Chlamydomonas* (see Chapter 13), as it is for mitochondrial mutations in other systems (see Chapters 5 and 10). Segregation and recombination of chloroplast genes in diploids have now been studied in detail by Van Winkle-Swift (48, 58); we shall return to these experiments in Chapter 14.

Mutations belonging to the third genetic system are believed to be mitochondrial in origin and to belong to two distinct phenotypic classes. We shall characterize these mutations in some detail here because they will not be discussed again in this book. The first class of mutations, called *minutes,* is induced with virtually 100% efficiency by either acriflavine or ethidium bromide (EtdBr) (59,60). The kinetics of induction are reminiscent of those seen for petite induction in yeast, and the same low target number (approximately 4) is observed (Fig. 12–8). Like petite mutants, the minutes form small colonies (Fig. 12–9), but for a different reason. Minutes are lethal after seven to nine generations of growth in the light. Nevertheless, minute mutants are obligate photoautotrophs. If cells are plated under heterotrophic conditions following induction, there is no cell division at all. Since minute mutants are capable of limited growth in the light following induction, and since the induction process operates with almost 100% efficiency, it is possible to study the phenotypic effect of the minute mutation as well as its inheritance. To study the phenotypic effects of the minute mutation, cells induced to form a very high frequency of minutes are transferred to medium lacking the inducing agent (i.e., acriflavine or EtdBr) and allowed to divide until growth ceases after seven to nine cell generations. Under these conditions it can be shown that the cells fix carbon dioxide at normal rates and contain normal amounts of chloroplast ribosomes. These rather crude measurements suggest that photosynthesis

FIG. 12–6. Fusion of one br_{s-1} mt^+ gamete with two mt^- wild-type gametes to give a triploid zygote. Cells were fixed 10 min. after mating. The br_{s-1} gamete can be identified by its undifferentiated chloroplast that lacks lamellae. ×27,000. (Courtesy of Dr. J. E. Boynton.)

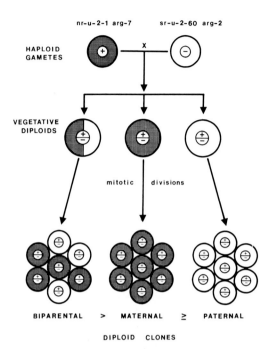

FIG. 12–7. Transmission of chloroplast genes in vegetative diploids of C. reinhardtii. Each parent carries a chloroplast mutation to antibiotic resistance (nr-u-2-1 or sr-u-2-60) and a nuclear mutation conferring arginine dependence (arg-2 or arg-7). The two arginine mutations complement one another in vegetative diploids and are used to select the diploids against a background of unmated gametes and meiotic zygotes (the mutants are closely linked and recombine infrequently) by plating the whole mating mixture on medium lacking arginine. Each diploid clone is then scored with respect to the chloroplast mutations to antibiotic resistance to determine if it is maternal, paternal, or biparental. (Courtesy of Dr. K. Van Winkle-Swift.)

and *Neurospora* (see Chapter 19), then the minutes may be analogs of vegetative petites. If so, treatment of *C. reinhardtii* with acriflavine or EtdBr would cause gross lesions in mtDNA or its loss, with resulting loss of mitochondrial protein synthesis and loss of the ability of the cells to synthesize essential components of the respiratory chain, including specific cytochrome oxidase polypeptides. In any event, the inability of the cells to synthesize cytochrome oxidase probably explains in part why they cannot grow in the dark, as in the case of the mendelian *dk* mutants discussed earlier in this chapter. However, it is not clear why these mutants die after seven to nine generations in the light, nor is it known if there are permissive conditions under which cell growth might be continued.

The pattern of inheritance of the minute mutants is nonmendelian (Fig. 12–10). In a cross of a minute mutant to wild type, the

and clDNA transcription are proceeding normally, since any defect in photosynthetic electron transport or the dark reactions of photosynthesis would lead to a reduction in the rate of carbon dioxide fixation and since chloroplast rRNA is coded by clDNA (see Chapter 3). On the other hand, cytochrome oxidase activity declines at each generation, and minute mutants show abnormalities in mitochondrial ultrastructure, whereas chloroplast ultrastructure appears normal. If it is assumed that mtDNA in *C. reinhardtii* codes for mitochondrial rRNA and that certain of the cytochrome oxidase polypeptides are translated on mitochondrial ribosomes, as is the case in both yeast

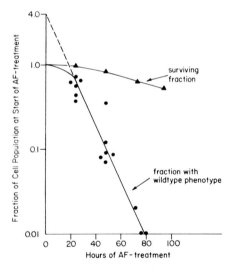

FIG. 12–8. Kinetics of transformation of wild-type mt^- cells to minute colony-forming cells during treatment with acriflavine (AF) (5 μg/ml). Cells were grown heterotrophically, and known numbers of cells were plated on acriflavine-free media at times indicated. Plates were scored for wild-type and minute colonies; viability was measured as number of wild type colonies plus number of minute colonies divided by total cells plated. The target number of 4 was estimated by extrapolating the linear part of the decay curve for wild-type colonies to the ordinate. (From Alexander et al., ref. 59, with permission.)

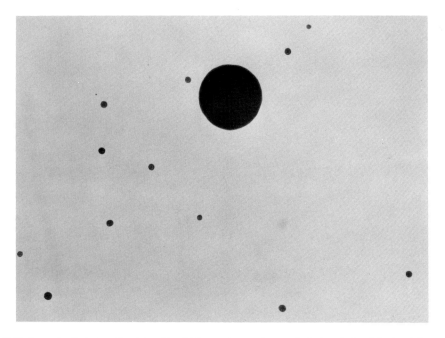

FIG. 12–9. Photograph of mixotrophic plate with wild-type colony and several minute colonies. Prior to plating, cells were grown heterotrophically in acriflavine (5 μg/ml) for 48 hr. (From Alexander et al., ref. 59, with permission.)

resulting tetrads segregate either 0:4 wild type:minutes or the reverse. Mixed tetrads account for less than 2% of the total. Rather striking differences are seen in the patterns of inheritance of EtdBr-induced and arcriflavine-induced minutes. When minutes induced in an *mt⁻* stock by EtdBr are crossed to wild-type *mt⁺* cells, almost all of the tetrads segregate 0:4 wild type:minute for the first several days, following which tetrads segregating 4:0 wild type:minute are observed. These increase to about 20% of the total. However, at the same time germination is declining. Thus it seems likely that the observed increase in wild-type tetrads is an artifact that results because zygotes produced by matings of unmutated *mt⁻* gametes germinate normally, whereas those resulting from matings of minute *mt⁻* gametes lose the capacity to germinate with time. In the reciprocal cross, where the *mt⁺* stock is converted to the minute phenotype and is mated to untreated *mt⁻* cells, germination does not decline with time, and virtually all tetrads segregate 4:0 wild type:

minute. In short, if one had only the EtdBr results to go by, one would say that the minute phenotype is *paternally* inherited, which is just the converse of the inheritance pattern seen for chloroplast genes.

When minute mutants induced by acriflavine in an *mt⁻* stock are mated to wild-type *mt⁺* cells, zygote germination declines with time and tetrads segregate 0:4 wild type:minute for the first few days, whereupon tetrads segregating 4:0 wild type:minute increase to about 20% of the total (Fig. 12–10). In other words, the results are just like those obtained for EtdBr. However, when *mt⁺* cells are converted to the minute phenotype and mated to untreated *mt⁻* cells, the results obtained are very different from those described above for EtdBr (Fig. 12–10). Zygotes matured for only a short time mostly segregate 0:4 wild type:minute. That is, the progeny have the phenotype of the maternal *mt⁺* parent. Zygotes matured for longer times segregate 4:0 wild type:minute. That is, the progeny then have the phenotype of the paternal *mt⁻* parent. As in

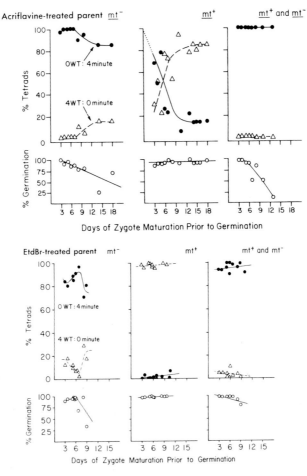

FIGURE 12–10. Effect of zygote age on tetrad types observed after treating one or both parents with acriflavine or EtdBr prior to gametogenesis. The 4:0 wild-type (WT):minute tetrad (*triangles*) and 0:4 WT:minute tetrad (*filled circles*) were the most prevalent types. Mixed tetrads were very rare. Zygote germination rates (*open circles*) are given for each time point. In all experiments, induction of minutes in treated parent(s) was measured (85%). Top: Inheritance of acriflavine-induced minute mutants. (From Alexander et al., ref. 59, with permission.) Bottom: Inheritance of EtdBr-induced minute mutants. (From Alexander, ref. 60, with permission.)

the case of EtdBr, there is no drop in germination when the minute mutation is induced by acriflavine in the mt^+ parent. In summary, the pattern of inheritance of the minute mutation is paternal in all crosses save the one in which minutes are induced by acriflavine in the mt^+ stock. In this cross the pattern of inheritance is at first predominantly maternal, but later it becomes paternal. This pattern of inheritance is very different from that seen for the maternally inherited chloroplast genes of *C. reinhardtii*.

Thus far, only one study has been published dealing with the effects of either acriflavine or EtdBr on DNA in *C. reinhardtii*. Flechtner and Sager (61) reported that EtdBr had a reversible effect on clDNA replication. Incorporation of labeled adenine into clDNA was blocked in cells that had been grown phototrophically in the presence of EtdBr for one to two generations. If the cells were washed free of EtdBr incorporation of labeled adenine into clDNA resumed. The effect of EtdBr on mtDNA was not studied by Flechtner and Sager, nor was minute induction reported. In any event, a reversible effect of the dye on clDNA in no way compromises the hypothesis that the primary mutagenic effect of EtdBr is on mtDNA.

Flechtner and Sager also found that incorporation of adenine into nDNA as well as clDNA was blocked by EtdBr under heterotrophic conditions. Alexander et al. (59) argued that the differential effects of EtdBr on nDNA synthesis under phototrophic and heterotrophic conditions suggested that the dye was blocking mitochondrial transcription and that mitochondrial function was required to generate the energy necessary for DNA synthesis under heterotrophic conditions. Indirect evidence for this hypothesis has been obtained for both acriflavine and EtdBr by Alexander (59,60). If the hypothesis is correct, it means that the dyes should have much the same effect on *Chlamydomonas* that they have on yeast (see Chapters 3 and 5). Because of the affinity of the two dyes for mtDNA, transcription of mtDNA should be blocked, and at the same time the mtDNA should be mutagenized. It may be that translation on mitochondrial ribosomes is also blocked by one or both of the dyes. The next step in

the characterization of the minute mutants would appear to be systematic examination of mtDNA and clDNA following mutant induction by acriflavine or EtdBr.

Like the *dk* mutants, the minute mutants are obligate photoautotrophs that apparently acquire this phenotype because respiratory electron transport is blocked. However, the two classes of mutants differ in two significant respects: First, the *dk* mutants are mendelian in inheritance, but the minute mutants have a nonmendelian pattern of inheritance. Second, the *dk* mutants are conditional lethals that die only under heterotrophic conditions, but the minute mutants eventually die in the light as well as in the dark. Ideally, one would like to be able to isolate conditional *dk* mutants with a nonmendelian pattern of inheritance. Two such mutants were isolated by Wiseman et al. (62). At first these mutations exhibited a nonmendelian, but biparental, pattern of inheritance (Fig. 12–11). The F_1 progeny included clones having *dk,* wild-type, and intermediate phenotypes. Examination of phenotypically wild-type clones revealed that they usually contained some *dk* cells as well. When F_1 progeny with *dk,* intermediate, or wild-type phenotype were backcrossed to wild type, the *dk* phenotype con-

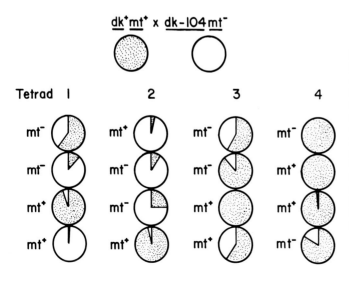

FIG. 12–11. Distributions of wild-type (dk^+), intermediate ($dk^{+/-}$), and obligate photoautotrophic or dark dier (dk^-) phenotypes in zoospore clones from a cross between the wild type and the *dk-104* mutant. Distributions of these phenotypes are presented in pie diagrams, with each pie representing the distribution for one zoospore. Frequencies of cells with the dk^- phenotype are indicated by light sectors; combined frequencies of dk^+ and $dk^{+/-}$ phenotypes are indicated by shaded sectors. These results are indicative of the somatic segregation of the dk^+, $dk^{+/-}$, and dk^- phenotypes. (From Wiseman et al., ref. 62, with permission.)

tinued to be inherited in a nonmendelian and apparently random fashion. On selection, neither mutant formed stable clones, producing only *dk* progeny at first. Eventually, stable clones were selected in both cases, but these then segregated 2:2 in crosses, as do ordinary mendelian *dk* mutants. Wiseman et al. (62) suggested that the nonmendelian *dk* mutation might be lethal in the homozygous condition and that the stable *dk* clones obtained through repeated selection had acquired new mendelian *dk* mutations while losing the original nonmendelian *dk* factor. Because of the propensity of the nonmendelian *dk* mutants to segregate wild-type cells, no attempt was made to do biochemical or ultrastructural studies of these mutants; so it is not known if the mutation in either case had specific effects on mitochondrial structure and function.

In summary, it is evident that obligate photoautotrophic mutants usually affect components of the mitochondrial respiratory electron transport chain such as cytochrome oxidase. This is true of both the mendelian *dk* mutants and the *minute* mutants, but it is not known if the nonmendelian *dk* mutants have mitochondrial defects. If, in fact, both the *minute* and nonmendelian *dk* mutants are mitochondrial in origin then it is clear that the pattern of inheritance of both groups of mutants is very different from that of the maternally inherited chloroplast mutants. Inheritance of the minute mutant phenotype is paternal, except when these mutants are induced in the *mt*+ parent by acriflavine, in which case the pattern of inheritance is at first maternal and later paternal. The nonmendelian *dk* mutants exhibit a biparental pattern of inheritance. These groups of mutants differ from each other in one very important respect. The progeny of a cross between minute mutants and wild type are pure. That is, with very few exceptions, all four progeny are either minute or wild type. Conversely, the progeny of a cross between

a nonmendelian *dk* mutant and wild type are almost always mixed, and progeny clones segregate cells having *dk,* intermediate, and wild-type phenotypes. It remains to be seen if these disparate patterns of inheritance can be reconciled into a unifying model for the inheritance of the mitochondrial genome in *Chlamydomonas.*

EVIDENCE FOR CHLOROPLAST LOCALIZATION OF MATERNALLY INHERITED GENES IN *Chlamydomonas*

The evidence that maternally inherited genes in *Chlamydomonas* are in the chloroplast derives from certain experiments suggesting that clDNA shows the same maternal pattern of inheritance as putative chloroplast mutants, that these mutants can be induced selectively at the time of clDNA replication, that experimentally induced alterations in the amount of clDNA lead to alterations in the pattern of transmission of presumptive chloroplast genes, that presumptive mitochondrial mutations show a pattern of inheritance distinct from that of chloroplast mutations, and that chloroplast mutations to antibiotic resistance cause chloroplast ribosomes to become resistant to antibiotics. The rest of this chapter will be devoted to detailed discussion of the first three lines of evidence; the fourth has just been considered, and the fifth is treated in Chapter 21.

INHERITANCE OF clDNA

Since presumptive chloroplast genes in *C. reinhardtii* are maternally inherited, and since fusion of gametes of opposite mating types is accompanied by chloroplast fusion as well as nuclear fusion without detectable elimination of any cytoplasm (63,64) (Fig. 12–5), the mechanism of maternal inheritance must be sought at the molecular level. An obvious possibility is that clDNA in *C.*

reinhardtii is maternally inherited, so that the clDNA from the mt^+ parent but not the mt^- parent is normally transmitted in a cross.

The physical fate of clDNA from the mt^+ and mt^- parents in the zygote was examined by Chiang (13,65,66) and by Sager and co-workers (26,67,68) who employed somewhat similar methods but drew dramatically different conclusions. Chiang initially concluded that clDNA from both parents was conserved in the zygote (13,65), but later he reported extensive degradation of clDNA from both parents in the zygote (66,71). Sager found that the clDNA of the mt^+ parent was conserved in the zygote (26,67,68). Although the latter possibility is most attractive in view of the pattern of inheritance observed for chloroplast genes, it is irreconcilable with Chiang's most recent findings. Thus once again we are in the dark as to the true pattern of inheritance of clDNA in *Chlamydomonas,* which of course makes the formulation of models to explain the inheritance of chloroplast genes in *Chlamydomonas* (see Chapter 15) imprecise at best.

In an attempt to understand why different results were obtained by Chiang and Sager, it should be useful first to review what is known about clDNA replication in three situations: in vegetative cells, during gametogenesis, and in the zygote. The experiments of Chiang and Sager will then be interpreted in light of these observations. Chiang and Sueoka (24) reported that in exponentially growing vegetative cells 14% of the total DNA is clDNA. In the same article they discussed the replication pattern of clDNA as shown by $^{15}N \rightarrow {}^{14}N$ transfer experiments using synchronously growing cells. These experiments revealed that the clDNA replicated twice semiconservatively between 4 and 7 hr during the light period (Fig. 12–3 and 12–12). Replication of nDNA was temporally separated from clDNA replication and did not occur until toward the end of the dark period. The cell

number then doubled twice. These experiments suggested there were no problems with precursor pools of ^{15}N, since the first clDNA replication yielded hybrid clDNA, and the second hybrid and ^{14}N-labeled clDNA in the proportions expected. Lee and Jones (70), using essentially the same methods, reported the same timing for clDNA replication under their conditions, but they observed only hybrid clDNA. That is, a second semiconservative round of clDNA replication yielding ^{14}N-labeled clDNA was not observed. Two semiconservative rounds of nDNA replication were seen during the dark period, and then the cells divided to four.

Chiang's most recent experiments (69, 71) suggested that control of clDNA replication in synchronous cultures may not be as stringent as it appeared originally. Swinton and Hanawalt (72) found that labeled thymidine was incorporated preferentially into clDNA of *Chlamydomonas*. Chiang et al. (69) confirmed that observation and showed further that the thymidine analog 5-bromodeoxyuridine (BrdUrd) was also incorporated specifically into clDNA and could be used as a density label for clDNA. Subsequently Chiang (71) followed the pattern of thymidine incorporation into clDNA and found that incorporation occurred throughout the entire life cycle, although the major incorporation peak was still in the light period, as predicted by the earlier density labeling experiments. At present, the significance of thymidine incorporation into clDNA during the dark phase of the synchronous cycle remains to be elucidated. This incorporation could result from a second period of clDNA synthesis or from extensive repair synthesis of clDNA. In those same experiments, nDNA continued to show a very sharp peak of synthesis during the dark phase of the cycle.

In order to follow the fate of clDNA in the zygote, both Sager and Chiang induced gametogenesis by nitrogen withdrawal during the light period using synchronous cul-

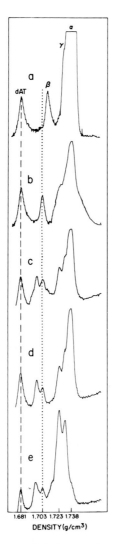

FIG. 12–12. Demonstration of time and nature of nDNA (α) and clDNA (β) replication during vegetative cell division in synchronous cultures of C. reinhardtii after transfer from ^{15}N to ^{14}N medium (see Fig. 12–3 for sampling times). Analyses by CsCl density-gradient centrifugation showing shifts of buoyant densities of α-DNA and β-DNA indicating their respective times of replication. (From Chiang and Sueoka, ref. 24, with permission.)

small fraction of the newly synthesized DNA can be accounted for by γ, most of the increase results from synthesis of clDNA. If the contribution of γ is ignored, two rounds of clDNA replication will be required to produce 0.18 $\mu g \times 10^{-7}$ of clDNA per cell, thus prior to clDNA replication, each cell must contain 0.045 $\mu g \times 10^{-7}$ clDNA per cell, or about 4% of the total. For gametogenesis, Chiang's group (73) picked cells about midway through the light period; such cells should already have replicated their clDNA but not their nDNA. Therefore the resulting gametes should replicate their clDNA once and their nDNA twice during gametogenesis, since the amount of nDNA per gamete is the same as in a vegetative cell and the amount of clDNA is half that in a vegetative cell (Table 12–3). Because these DNA replications take place in the absence of an external nitrogen source, it is obvious that the precursors for gametogenic DNA synthesis must be derived from within the cell itself. Furthermore, if equal amounts of clDNA are to be contributed to the zygote, the requisite round of replication must take place in gametes of both mating types. If it does not, it might appear that clDNA of one mating type or the other was partially eliminated in the zygote, whereas differential gametic DNA replication could also account for the observation. From Chiang's most recent experiments (66,69,71) it would appear that clDNA synthesis occurs in gametes of both mating types, since labeled thymidine is incorporated into the clDNA of both mating types during gametogenesis. However, it is not clear from these experiments how much of this incorporation represents replication of new clDNA, nor is it clear if the amounts of new clDNA synthesized are the same in gametes of both mating types.

So much for clDNA replication prior to zygote formation. In the zygote itself Chiang and Sueoka (73,74) followed replication of the various DNA species using the $^{15}N \rightarrow$ ^{14}N density transfer experiment. Vegetative cells were grown synchronously using ^{15}N

tures. Under these conditions the gametic cell number doubles twice, and the nDNA content remains constant, but the gametic clDNA value falls to 7% of the total (Table 12–3). Chiang and Sueoka (24) calculated that there is a 15% increase in the amount of DNA during the light period (from 1.19 $\times$ 10^{-7} to 1.37 $\times 10^{-7} \mu g$/cell). Although a as the nitrogen source and then transferred to nitrogen-free medium; the zygotes were plated out on ^{14}N-containing medium. At various times during zygote maturation, DNA was isolated and examined by buoyant-density centrifugation in the analytical ultracentrifuge.

Chiang and Sueoka observed that the clDNA replicated once semiconservatively during the first 24-hr light period of zygote maturation. During that period a new DNA species called M-band DNA, with a density between that of α and γ, was also synthesized to the extent that it accounted for almost 50% of the total cell DNA. Replication of nDNA was shown to be temporally separated from replication of clDNA and M-band DNA, and it did not occur until after the light was turned on, approximately 7 to 8 hr after onset of germination. Only one DNA replication was observed (that is, all the nDNA was at the buoyant density of hybrid DNA), but eight zoospores were produced. This observation suggested that the zoospores have one-half the nDNA of the gametes; needless to say, that suggestion sparked controversy. It was also evident that by the time of germination the clDNA had replicated several more times since it had shifted from the hybrid position assumed during the first 24 hr of maturation; it had become entirely light.

Since these experiments did not reveal the pattern of inheritance of clDNA in zygotes, Chiang (13) examined meiotic transmission of DNA in *C. reinhardtii* by labeling synchronously growing cells of opposite mating types with ^{14}C- and ^{3}H-adenine. Cells were harvested and resuspended in medium lacking inorganic nitrogen but containing cold adenine. Following gametogenesis, equal numbers of gametes of opposite mating types were mixed, and for maturation the zygotes were transferred to minimal medium supplemented with adenine. It was found that radioactive adenine was not incorporated during gametogenesis, which implied that a chase with cold adenine during game-

togenesis would be ineffectual, but it appeared that a chase with cold adenine was effective in the zygote, since M-band DNA synthesized in these zygotes was unlabeled. Samples were taken for DNA analysis from gamete pairs, mature zygotes, and zoospores. Radioactivity appeared to be conserved in all species, including clDNA, although this was difficult to estimate in the mature zygotes because the DNA was badly sheared. Later Chiang (65) combined density methods and radioactive labeling of DNA in the same experiment. In the first such experiment Chiang labeled the mt^+ parent with ^{3}H-adenine and ^{14}N and the mt^- parent with ^{14}C-adenine and ^{15}N and matured the zygotes on ^{14}N-containing medium (i.e., $^{3}H^{14}N$ $mt^+ \times {}^{14}C^{15}N$ $mt^- \rightarrow {}^{14}N$). In each experiment the mt^- parent also contained three chloroplast mutations to antibiotic resistance, and the mt^+ parent was sensitive to all three antibiotics. The markers were transmitted to only a small fraction of the progeny, as was expected, since they were carried by the mt^- parent.

The banding profiles of the labeled DNA in preparative cesium chloride gradients were examined for gamete pairs and germinated zoospores at a time when a fourfold cell increase had taken place. Chiang found that the distribution of nDNA was consistent with the hypothesis that replication and recombination had occurred. Thus the mt^- nDNA had moved from its heavy position to the hybrid position, indicating that it had replicated once, as was expected incorporating ^{14}N from the maturation medium; the mt^+ nDNA remained at the light position, since it had been labeled with ^{14}N to begin with. Chiang also found that a considerable fraction of the ^{14}C counts originally associated with the mt^- nDNA were found to be associated with ^{14}N-labeled nDNA as well as hybrid nDNA. Similarly, some of the ^{3}H counts from the ^{14}N-labeled mt^+ nDNA had become associated with hybrid nDNA derived from the mt^- parent. These results were interpreted as physical evidence for recombination between the two nuclear genomes.

Chiang also noted that the zoospores showed a single clDNA peak that contained both radioactive labels and was at the light (^{14}N) density position. These results were interpreted as indicating that clDNAs from both parents not only were conserved but also recombined extensively. To substantiate this conclusion, Chiang labeled the same two stocks reciprocally with respect to density (but not radioactive labels) and crossed them (i.e., ^{14}N^{14}C mt^+ × ^{15}N^{3}H mt^- and ^{15}N^{14}C mt^+ × ^{14}N^{3}H mt^-). The zygotes were plated on ^{14}N medium. DNAs from gamete pairs, young zygotes (30 hr), and (in one experiment) mature zygotes were banded in cesium chloride gradients. Chiang found that by 30 hr there was a single clDNA density peak containing both radioactive labels. This seemed to indicate that replication and recombination of clDNA had occurred in both experiments. By comparison, two nDNA peaks were seen; thus nDNA replication had not then occurred, as was expected. Chiang interpreted his results as supporting the view that clDNAs from both parents were conserved in the zygote and that extensive recombination took place between these DNAs. He pointed out that these results were not reconcilable with the apparent maternal transmission of chloroplast genes in *C. reinhardtii*.

Interpretation of Chiang's results depends on the fate of the radioactivity incorporated in the form of adenine in vegetative cells in young zygotes. Since radioactive adenine is not taken up during gametogenesis, it appears likely that a cold adenine chase during that period should be ineffective. Therefore it is important that the cold adenine chase be effective in the young zygotes. The chase was assumed to be effective because the M-band DNA, which was synthesized in large amounts in the young zygotes, was unlabeled.

Siersma and Chiang (75), using labeled adenine, then investigated the fate of chloro-plast and cytoplasmic ribosomes during gametogenesis in order to determine if rRNA provided the precursors for gametogenic replication of nDNA. They found that approximately 90% of the chloroplast and cytoplasmic ribosomes were degraded during gametogenesis. Furthermore, the DNA synthesized during gametogenesis became radioactive, which indicated that the degradation products of rRNA were probably being used to synthesize DNA. Siersma and Chiang (75) also observed that although much of the radioactivity (34%) was found in the medium, there was a considerable acid-soluble pool of radioactivity (15%) in the cells themselves at the time gametogenic differentiation was completed. This pool presumably would be available for early zygotic DNA replication.

Since the experiments of Chiang and Sueoka indicated that clDNA, as well as M-band DNA, replicates during the first 24 hr following zygote formation, it seems plausible that the radioactive precursor pools present in the gametes could become mixed following gamete fusion and zygote formation. Clearly the chloroplast pools should mix, because chloroplast fusion takes place in the young zygote. Therefore the apparent conservation of clDNA from both parents in the zygote likely reflects this mixing of radioactive precursor pools, coupled with clDNA replication, which is known to occur during the first 22 hr. Thus, even if only the clDNA from one parent is conserved and replicated, it will be labeled from the pools contributed by both parents. Thus it would appear that both parental clDNAs have been conserved.

The fact that M-band DNA was not labeled under these conditions could be explained in at least two ways. The first possibility is that the cold adenine used for the chase was extensively incorporated into M-band DNA, a new DNA component, but the gametogenic pattern of DNA replication continued in the chloroplast. That is, the precursors derived from rRNA degradation might be incorporated preferentially into clDNA. After all, we know only that radioactive adenine is an effective label for the RNA and DNA of vegetative cells; we do not know the nature of the precursors used for gametogenic DNA replication following ribosome degradation. Perhaps these are used in preference to exogenously added adenine. A second possibility is that there is some temporal separation between clDNA and M-band DNA replication such that clDNA replicates first, using the radioactive precursor pool, which becomes diminished and diluted by cold adenine by the time of M-band DNA replication. In a recent article that interprets his earlier experiments, Chiang (71) considered the problems posed by radioactive precursor pools.

Following the initial attempts by Chiang to elucidate the fate of clDNA in the zygote of *Chlamydomonas,* Sager and Lane (67, 68) attempted to determine the fate of clDNA in the zygote using density labeling techniques. As in the experiments of Chiang, cells were grown synchronously until the middle of the light cycle; then they were centrifuged and resuspended in nitrogen-free medium to permit gametogenesis. During gametogenesis the cells divided to form four gametes. The zygotes thus formed, unlike those used by Chiang, were kept in liquid medium, rather than being plated onto solid medium. Extensive M-band DNA synthesis was not seen in any of these experiments.

The presence or absence of M-band DNA synthesis probably depends on the method used for zygote maturation. The zygotes used by Sager and Lane were matured in liquid, and for the first 24 hr no nitrogen source was present in the medium. In the experiments of Chiang and Sueoka (73, 74), zygotes were plated on agar medium containing a nitrogen source, and M-band DNA synthesis took place during the first 24 hr of maturation when the zygotes were exposed to light. The zygotes used by both groups of investigators were then placed in the dark for further maturation. If M-band

DNA synthesis is triggered by light exposure during the first phase of zygote maturation, then the extent of synthesis probably was very limited in the experiments of Sager and Lane, since they omitted nitrogen from the medium during that period.

In any event, the absence of extensive M-band DNA synthesis made it possible for Sager and Lane to label vegetative cells with the density isotopes ^{14}N and ^{15}N prior to gametogenesis and mating, as well as to distinguish clDNA carrying each label. Thus they avoided the complications inherent in the sensitive radiochemical assay used by Chiang. The two mating types were labeled reciprocally in different experiments (i.e., ^{15}N $mt^+ \times$ ^{14}N mt^- and ^{14}N $mt^+ \times$ ^{15}N mt^-).

In their first set of experiments, Sager and Lane examined the fate of clDNA in mature zygotes (68). From these experiments it appeared that most of the clDNA in the zygote had a buoyant density similar to that of the clDNA of the mt^+ parent. This was found to be the case even if the mt^+ parent was labeled with ^{14}N and then matured on ^{15}N or vice versa. That is, in their experiments, contrary to the findings of Chiang (73,74), there did not appear to be significant zygotic clDNA replication. Sager and Lane also noted that during the first 24 hr the clDNA of the zygote underwent a small density shift (e.g., 1.694 to 1.689 g/cc). The published densitometer tracings suggest that this density shift was maintained in mature zygotes in some cases but not in others.

Sager and Lane (67) then looked specifically at the fate of clDNA in young zygotes (6 hr old) using the density labeling techniques (Fig. 12–13). Again, a single density species of clDNA predominated, and it had the buoyant density expected for the density-shifted clDNA of the mt^+ parent.

The experiments of Sager and Lane supported the hypothesis that clDNA is transmitted largely if not entirely by the mt^+ parent, as are chloroplast genes. One potential complication with such experiments is that

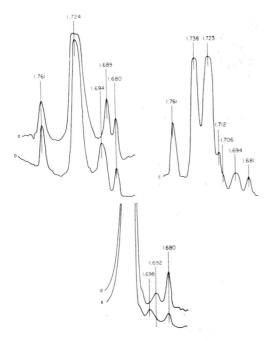

FIG. 12–13. Inheritance of clDNA in *C. reinhardtii* crosses as indicated by centrifugation of whole-cell DNA to equilibrium in CsCl gradients. a: DNA from 24-hr zygotes from cross of $^{14}N \times ^{14}N$ (unlabeled) gametes showing shift of clDNA to lighter density (1.689 g/cc) than is normally found in vegetative cells and gametes. b: DNA from one of ^{14}N gamete types used as parents in the cross shown in a. The chloroplast DNA has a density of 1.694 g/cc. c: DNA from a 1:1 mixture of ^{14}N and ^{15}N gametes showing resolution of both light and heavy DNAs (1.723 and 1.738 g/cc). d: DNA from 6-hr zygotes from cross of $^{14}N \times ^{15}N$ gametes showing single clDNA peak at 1.692 g/cc. e: DNA from 6-hr zygotes from cross of $^{15}N \times ^{14}N$ gametes showing single clDNA peak at 1.698 g/cc. DNAs used as density markers: SP-15 DNA at 1.761 g/cc and poly (dAT) at 1.680 or 1.681 g/cc. (From Sager and Lane, ref. 67, with permission.)

clDNA might replicate differentially in the gametes under Sager's experimental conditions. If clDNA from the *mt⁺* parent replicates during gametogenesis and that from the *mt⁻* parent does not, there will be an excess of clDNA with the buoyant density characteristic of the *mt⁺* parent in the zygote. Sager (76) indicated that this was not a problem and that gametes of opposite mating types had similar amounts of clDNA.

Sager (68) also reported that when ^{3}H- and ^{14}C-adenine were used as labels, the results were less clear-cut. For example, after 24 hr, when the clDNA had shifted to its new position, both maternal and paternal labels were found in the clDNA peak. However, the ratio of paternal label to maternal label was low compared to that in the nuclear region, which indicated that most of the counts were from the maternal parent. Later this ratio gradually increased and approached that of the nuclear region. Sager interpreted this finding as indicating considerable turnover or repair synthesis of clDNA in the zygote. She discussed Chiang's experiments in light of her own observations and suggested that Chiang had missed the differential fates of the clDNAs from the two parents because of his exclusive use of adenine labeling and because he had looked only at late times in zygote development, although in fact he had looked at zygote DNA as early as 30 hr after gamete fusion.

Schlanger and Sager (26,77) then reported some potentially important results, but the findings were published in abbreviated form. Sager and Ramanis (p. 361) found that ultraviolet irradiation of the *mt⁺* parent greatly increased the frequencies of biparental and paternal zygotes. Therefore it would be expected that following ultraviolet irradiation of the *mt⁺* parent, transmission of *mt⁻* as well as *mt⁺* clDNA would be observed in the zygote. In experiments in which ^{3}H- and ^{14}C-adenine were used to distinguish the clDNAs from the two parents, Schlanger and Sager reported disappearance of radioactive label from the *mt⁻* parent from clDNA in the zygotes, as was previously reported for the density labeling experiments. However, when the *mt⁺* parent was pretreated with ultraviolet light, clDNA from the *mt⁺* parent disappeared partially or entirely. They reported similar results when *mt⁻* cells containing the *mat-1* mutation, which greatly enhances the frequency of exceptional zygotes in a cross (see Chapter 15), were mated to wild-type *mt⁺* cells. It appears that these experiments provide important confirmation of the original experiments, and their details are anxiously awaited. However, one wonders why the authors chose to use radioisotopes in view of the difficulties that they themselves discussed.

The complicated story of the inheritance of clDNA in zygotes of *C. reinhardtii* has come full circle with the recent report of Chiang (66,71) that clDNAs from both parents are extensively degraded in the zygote. In these experiments total DNA of one of the two parents was labeled during gametogenesis with ^{14}C-adenine, but the clDNA was labeled specifically with ^{3}H-thymidine. Cold thymidine was then used as a chase for the labeled gametes, which subsequently were mated to unlabeled gametes of opposite mating type. This labeling protocol was carried out in separate experiments with mt^- and mt^+ gametes. The banding patterns of the radioactivity were then examined in gamete pairs, in mature zygotes just prior to germination, and in zoospores following zygote germination (Fig. 12–14). The results obtained are consistent with substantial degradation of clDNAs from both parents in the zygote. Chiang also presented evidence that prior to this degradation in the young zygote, both parental clDNAs replicated. At least two rounds of semiconservative replication of clDNA appeared to have been completed by 10 hr after zygote formation. Chiang (66) also reported evidence for extensive recombination between the two parental clDNAs, but his results can also be explained by repair replication. The observations are as follows: Maternal gametes labeled with ^{3}H-thymidine were mated with unlabeled paternal gametes, after which ^{14}C-thymidine and BrdUrd were added to the mating mixture. Newly synthesized clDNA was labeled with ^{14}C-thymidine and was density-shifted toward the nDNA peak because of incorporation of the density label BrdUrd. Nevertheless, substantial ^{14}C label was found, together with a parental clDNA fraction that lacked the density label but was heavily labeled with ^{3}H-thymidine. Chiang assumed that the association of ^{14}C-thymidine with this light clDNA peak signified

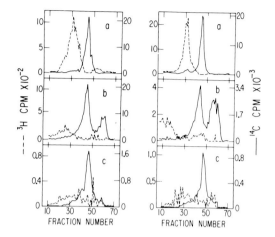

FIG. 12–14. Inheritance of clDNA in *Chlamydomonas* as indicated by centrifugation of whole-cell DNA to equilibrium in CsCl gradients. Left: mt^- cells were induced to form gametes in presence of ^{14}C-adenine to label whole-cell DNA and ^{3}H-thymidine, which is incorporated specifically into clDNA. A cold chase was then used to eliminate unincorporated label. These labeled gametes were then mixed with unlabeled mt^+ gametes, and the homogeneous population of young zygotes thus obtained was allowed to mature and germinate. a: Equal numbers of male and female gametes. b: Mature zygotes. c: Zoospores. Right: Same as left panel, except that female gametes were labeled with ^{3}H-thymidine and ^{14}C-adenine, whereas male gametes were unlabeled. (From Chiang, ref. 66, with permission.)

recombination, but the results could also be explained by repair replication.

We can summarize this discussion of a complex set of experiments with the simple statement that at the moment we do not know how clDNA is inherited in the *Chlamydomonas* zygote, nor will we have a solution to this problem until the results of Sager and Chiang are reconciled.

INDUCTION OF MENDELIAN AND CHLOROPLAST MUTATIONS AT DIFFERENT TIMES IN SYNCHRONOUS CELL CYCLES

Lee and Jones (70) used selective mutagenesis of replicating DNA as a probe for localizing chloroplast and mendelian genes to their respective DNAs. They mutagenized synchronously growing cells of *C. reinhardtii* with *N*-methyl-*N'*-nitro-*N*-nitrosoguanidine (MNNG), which is an effective mutagen for both mendelian and chloroplast genes in this alga (78). In *E. coli* MNNG is believed to induce mutations selectively at the replication fork of DNA. As an assay, Lee and Jones measured the frequency of mendelian (*sr-1*) and chloroplast (*sr-u-2*) mutations to streptomycin resistance. These mutations are easily distinguished on the basis of resistance level (p. 347).

Lee and Jones observed that *sr-1* mutations increased 15- to 30-fold when cells were mutagenized at the time of nDNA replication in the dark period of the synchronous cycle. There was no concomitant increase in the frequency of *sr-u-2* mutants at that time. When MNNG was added at the time of the major replication of clDNA, both *sr-1* and *sr-u-2* mutations showed small (twofold) but significant increases in frequency.

The explanation for the increase in frequency of *sr-1* mutants following MNNG treatment at the time of clDNA replication appears to be that between 2 and 3% of the cells are out of synchrony and are replicating nDNA while the rest are replicating clDNA. Since *sr-1* mutants increase 15- to 30-fold following MNNG treatment during the normal period of nDNA replication, the twofold increase in this class of mutants at the time of clDNA replication is not surprising.

It is not known why the increase in frequency of *sr-u-2* mutants is so small at the time of clDNA replication. It may be that only small fractions of potential mutants are ever expressed. In any event, the results are consistent with the hypothesis that the *sr-u-2* mutations originate in the chloroplast.

EFFECTS OF 5-FLUORODEOXYURIDINE ON AMOUNTS OF clDNA AND ON CHLOROPLAST GENE TRANSMISSION

Since clDNA is present in many copies per chloroplast in *C. reinhardtii*, it is not

unreasonable to suppose that perturbation of the amount of clDNA per cell might lead to alteration in the transmission pattern of chloroplast genes. As was discussed previously, radioactive thymidine labels clDNA in *C. reinhardtii* but does not label nDNA in vegetative cells or cells in the process of gametogenesis. The existence of a chloroplast-specific thymidine kinase in *C. reinhardtii* is believed to account for this incorporation. Since BrdUrd is also incorporated specifically into *Chlamydomonas* clDNA (69,72), Wurtz et al. (79) reasoned that the cell and chloroplast envelopes must also be permeable to thymidine base analogs. In *E. coli* these base analogs are phosphorylated by thymidine kinase. 5-Fluorodeoxyuridylate acts as an analog of deoxyuridylate and becomes covalently bound to thymidylate synthetase, thus inhibiting *de novo* thymidylate synthesis and hence DNA replication, unless exogenous thy-

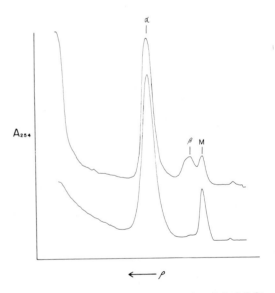

FIG. 12–15. Microdensitometer tracings of analytical CsCl buoyant-density gradients of whole-cell DNA extracted from FdUrd-grown cultures and control cultures of cells. Top tracing: DNA from control culture without FdUrd. Bottom tracing: DNA from culture grown for six doublings in 0.1 mM FdUrd. The marker DNA (M), buoyant density 1.687 g/cc, is from *Tetrahymena pyriformis* (strain HSM); α, buoyant density 1.723 g/cc, and β, buoyant density 1.695 g/cc, are *C. reinhardtii* nDNA and clDNA, respectively. (From Wurtz et al., ref. 79, with permission.)

mine or thymidine is supplied. Since the chloroplast of *C. reinhardtii* takes up thymidine and BrdUrd, phosphorylates these compounds, and incorporates them specifically into clDNA, Wurtz and associates suggested that 5-fluorodeoxyuridine (FdUrd) might also be phosphorylated in the chloroplast and then inhibit the thymidylate synthetase of the chloroplast in the same way that this analog blocks the functioning of this enzyme in *E. coli*. In short, growth of *C. reinhardtii* in FdUrd, if it can be effected, should lead to a reduction in the amount of clDNA per cell, since the chloroplast would be starved for thymidylate.

Wurtz and associates found that FdUrd did behave as a specific inhibitor of clDNA replication as they had predicted (Fig. 12–15). It was possible to reverse the inhibition by addition of exogenous thymidine. They then showed that varying the amount of clDNA in the *mt+* parent had a profound effect on chloroplast marker transmission. An *mt+* strain carrying chloroplast markers for erythromycin (*er-u-37*) and streptomycin (*sr-u-2-60*) resistance was grown in different concentrations of FdUrd. The cells were then induced to form gametes in the presence of FdUrd, and the amount of clDNA was determined. These cells were then mated to untreated *mt−* cells carrying a chloroplast mutation to spectinomycin resistance (*spr-u-1-6-2*). The zygotes were matured and germinated. Maternal, biparental, and paternal zygotes could then be distinguished readily by the resistance markers they carried. Wurtz and associates found that as the amount of clDNA in the maternal parent was reduced, the frequencies of biparental and paternal zygotes increased dramatically (Table 12–5). At the higher FdUrd concentrations most of the zygotes were paternal. The fact that clDNA did not appear to continue to decline at higher FdUrd concentrations, although the shift from maternal to paternal transmission was augmented further, was probably artifactual (Table 12–5). The sensitivity of the radiochemical assay used to estimate clDNA amounts is low at the higher FdUrd concentrations because one is operating very close to background, whereas the sensitivity of the genetic assay remains high.

In another experiment Wurtz and asso-

TABLE 12–5. *Relationship between chloroplast gene transmission patterns and amounts of maternal clDNA as a function of FdUrd concentration[a]*

Experiment	FdUrd concentration	Percentage			
		Maternal zygotes	Biparental zygotes	Paternal zygotes	clDNA
I	0	98	1.5	0.5	6.2
	0.1 mM	66.5	23	10.5	2.0
	0.5 mM	1	11	88	1.8
	1.0 mM	0	3.5	96.5	1.1
II	0	99	1	0	3.5
	1.0 mM	7.5	3	89.5	1.1
III	0.1 mM	54.5	30.5	15	2.7
	1.0 mM	3	2.5	94	1.6

[a] The results of three separate experiments are shown. In each experiment the maternal parent was grown in FdUrd-containing medium for a number of generations (minimum of six), following which gametogenesis was induced in the presence of the appropriate FdUrd concentration, and the treated maternal gametes were mated to untreated paternal gametes. The maternal parent carried chloroplast markers conferring streptomycin (*sr-u-2-60*) and erythromycin (*er-u-37*) resistance, and the paternal parent carried a chloroplast marker conferring spectinomycin resistance (*spr-u-1-6-2*). Maternal, biparental, and paternal zygotes were readily distinguished by the markers they carried.
From Wurtz et al. (79), with permission.

TABLE 12–6. *Chloroplast gene transmission patterns*[a]

FdUrd treatment		Percentage		
mt^+	mt^-	Maternal zygotes	Biparental zygotes	Paternal zygotes
−	−	93	7	0
+	−	40	28	32
−	+	95	5	0
+	+	66	28	6

[a] These patterns were obtained when the mt^+ parent or the mt^- parent or both parents were grown in 0.1 mM FdUrd prior to mating. The parents used in these crosses were the same as those used in the crosses reported in Table 12–5. From Wurtz et al. (79), with permission.

ciates grew both mt^+ and mt^- cells in FdUrd. Gametes derived from the treated cells were mated to each other and to untreated gametes. This experiment showed that FdUrd also affected transmission of paternal chloroplast genes (Table 12–6). This effect was not evident when treated mt^- gametes were mated to untreated mt^+ gametes, probably because the spontaneous frequency of exceptional zygotes was already so low. However, in comparing the frequencies of exceptional zygotes when untreated and treated mt^- gametes were mated to treated mt^+ gametes, the effect was obvious. When untreated mt^- gametes were mated to treated mt^+ gametes, both biparental and paternal zygotes were found in high frequency. When treated mt^- gametes were used, the frequency of paternal zygotes dropped markedly, and these were replaced principally by maternal zygotes. In summary, growth of cells in FdUrd reduces the amount of clDNA per cell. Reduction in the amount of clDNA in either mating type is correlated with a decrease in transmission of the chloroplast genes carried by that mating type. There is no detectable effect on nDNA or on the pattern of segregation of nuclear genes. These results provide strong support for the hypothesis that the maternally inherited genes of *C. reinhardtii* are localized in clDNA. In addition, we should note that FdUrd may prove an invaluable tool in studying chloroplast genetics in *C. rein-*

hardtii because it makes it possible for the first time to vary the input of clDNA molecules in the zygote in a systematic fashion.

CONCLUSIONS

Although there is much indirect evidence that the maternally inherited genes of *C. reinhardtii* are localized in the chloroplast genome, this remains to be proved. In view of the detailed genetic studies carried out with these mutants indicating that they are linked and can be mapped (see Chapter 14), unequivocal demonstration of their location becomes even more important. FdUrd promises to be a particularly useful tool in the future because this analog can be used to vary the input of chloroplast genomes to the zygote systematically. This compound may also prove to be useful in mutagenesis experiments, since by reducing the redundancy of the chloroplast genome it may be possible to uncover hidden recessive mutations and promote the expression of newly induced chloroplast mutations more readily. At the same time, a new nonmendelian genetic system has been discovered in *C. reinhardtii* that may be mitochondrial in location. We may expect this system to be characterized more fully in the next few years.

REFERENCES

1. Sager, R. (1954): Mendelian and non-Mendelian inheritance of streptomycin resistance in *Chlamydomonas reinhardi. Proc. Natl. Acad. Sci. U.S.A.*, 40:356–363.
2. Buffaloe, N. (1968): A comparative cytological study of four species of *Chlamydomonas. Bull. Torrey Bot. Club.*, 85:157.
3. Levine, R. P., and Folsome, C. E. (1959): The nuclear cycle in *Chlamydomonas reinhardi. Z. Vererbungs.*, 90:215–222.
4. Maguire, M. P. (1976): Mitotic and meiotic behavior of the chromosomes of the octet strain of *Chlamydomonas reinhardtii. Genetica*, 46:479–502.
5. McVittie, A., and Davies, D. R. (1971): The location of the Mendelian linkage groups in *Chlamydomonas reinhardii. Mol. Gen. Genet.*, 112:225–228.
6. Levine, R. P., and Goodenough, V. W. (1970): The genetics of photosynthesis and

of the chloroplast in *Chlamydomonas reinhardi*. *Annu. Rev. Genet.*, 4:397–408.

7. Schötz, F. (1972): Dreidimensionale, masstabgetreue Rekonstruktion einer grünen Flagellatenzelle nach Elektronenmikroskopie von Serienschnitten. *Planta*, 102:152–159.

8. Arnold, C. G., Schimmer, O., Schötz, F., and Bathelt, H. (1972): Die Mitochondrien von *Chlamydomonas reinhardii*. *Arch. Microbiol.*, 81:50–67.

9. Schötz, F., Bathelt, H., Arnold, C. G., and Schimmer, O. (1972): Die Architektur und Organisation der *Chlamydomonas*-zelle: Ergelbnisse der Elektronenmikroskopie von Serienschnitten und daraus resultierenden dreidimensionalen Rekonstruktion. *Protoplasma*, 75:229–254.

10. Ris, H., and Plaut, W. (1962): Ultrastructure of DNA-containing areas in the chloroplast of *Chlamydomonas*. *J. Cell Biol.*, 13:383–391.

11. Boynton, J. E., Gillham, N. W., and Chabot, J. F. (1972): Chloroplast ribosome deficient mutants in the green alga *Chlamydomonas reinhardi* and the question of chloroplast ribosome function. *J. Cell Sci.*, 10:267–305.

12. Wells, R., and Sager, R. (1971): Denaturation and the renaturation kinetics of chloroplast DNA from *Chlamydomonas reinhardi*. *J. Mol. Biol.*, 58:611–622.

13. Chiang, K.-S. (1968): Physical conservation of parental cytoplasmic DNA through meiosis in *Chlamydomonas reinhardi*. *Proc. Natl. Acad. Sci. U.S.A.*, 60:194–200.

14. Bastia, D., Chiang, K.-S., and Swift, H. (1971): Studies on the ribosomal RNA cistrons of chloroplast and nucleus in *Chlamydomonas reinhardtii*. In: *Abstracts 11th Annual Meeting American Society of Cell Biology*, p. 25.

15. Ryan, R. S., Grant, D., Chiang, K.-S., and Swift, H. (1973): Isolation of mitochondria and characterization of the mitochondrial DNA of *Chlamydomonas reinhardtii*. *J. Cell Biol.*, 59:297a.

16. Ryan, R. S., Chiang, K.-S., and Swift, H. (1974): Circular DNA from mitochondria of *Chlamydomonas reinhardtii*. *J. Cell Biol.*, 63:293a.

17. Sager, R., and Ishida, M. R. (1963): Chloroplast DNA in *Chlamydomonas*. *Proc. Natl. Acad. Sci. U.S.A.*, 50:725–730.

18. Behn, W., and Herrmann, R. G. (1977): Circular DNA in the satellite DNA of *Chlamydomonas reinhardtii*. *Mol. Gen. Genet.*, 157:25–30.

19. Lambowitz, A. M., Merril, C. R., Wurtz, E. A., Boynton, J. E., and Gillham, N. W. (1976): Restriction enzyme analysis of chloroplast DNA from *Chlamydomonas reinhardtii*. *J. Cell Biol.*, 70:217.

20. Rochaix, J.-D. (1976): Studies with chloroplast DNA-plasmid hybrids from *Chlamydomonas reinhardi*. In: *Genetics and Biogenesis of Chloroplasts and Mitochondria*, edited by T. Bücher, W. Neupert, W. Sebald and S. Werner, pp. 375–378. North Holland, Amsterdam.

21. Howell, S. H., and Walker, L. L. (1976): Informational complexity of the nuclear and chloroplast genomes of *Chlamydomonas reinhardi*. *Biochim. Biophys. Acta*, 418:249–256.

22. Bastia, D., Chiang, K.-S., Swift, H., and Siersma, P. (1971): Heterogeneity, complexity, and repetition of the chloroplast DNA of *Chlamydomonas reinhardtii*. *Proc. Natl. Acad. Sci. U.S.A.*, 68:1157–1161.

23. Kolodner, R., and Tewari, K. K. (1972): Molecular size and conformation of chloroplast deoxyribonucleic acid from pea leaves. *J. Biol. Chem.*, 247:6355–6364.

24. Chiang, K.-S., and Sueoka, N. (1967): Replication of chloroplast DNA in *Chlamydomonas reinhardi* during the vegetative cell cycle: Its mode and regulation. *Proc. Natl. Acad. Sci. U.S.A.*, 57:1506–1513.

25. Sager, R., and Ramanis, Z. (1976): Chloroplast genetics of *Chlamydomonas*. I. Allelic segregation ratios. *Genetics*, 83:303–321.

26. Sager, R., and Schlanger, G. (1976): Chloroplast DNA: Physical and genetic studies. In: *Handbook of Genetics, Vol. 5*, edited by R. C. King, Chapter 12. Plenum Press, New York.

27. Sager, R., and Granick, S. (1953): Nutritional studies with *Chlamydomonas reinhardi*. *Ann. N.Y. Acad. Sci.*, 56:831–838.

28. Sager, R., and Granick, S. (1954): Nutritional control of sexuality in *Chlamydomonas reinhardi*. *J. Gen. Physiol.*, 37:729–742.

29. Eversole, R. A. (1956): Biochemical mutants of *Chlamydomonas reinhardi*. *Am. J. Bot.*, 43:404–407.

30. Ebersold, W. T. (1956): Crossing over in *Chlamydomonas reinhardi*. *Am. J. Bot.*, 43:408–410.

31. Levine, R. P., and Ebersold, W. T. (1958): The relation of calcium and magnesium to crossing over in *Chlamydomonas reinhardi*. *Z. Vererbungs.*, 89:631–635.

32. Hutner, S. H., Provasoli, L., Schatz, A., and Haskins, C. P. (1950): Some approaches to the study of metals in the metabolism of microorganisms. *Proc. Am. Phil. Soc.*, 94:152–170.

33. Sueoka, N. (1960): Mitotic replication of deoxyribonucleic acid in *Chlamydomonas reinhardi*. *Proc. Natl. Acad. Sci. U.S.A.*, 46:83–91.

34. Surzycki, S. (1971): Synchronously grown cultures of *Chlamydomonas reinhardi*. *Methods Enzymol.*, 23:67–73.

35. Droop, M. R. (1974): Heterotrophy of carbon. In: *Algal Physiology and Biochemistry*, edited by W. D. P. Stewart, pp. 530–559. University of California Press, Berkeley.

36. Levine, R. P. (1969): The analysis of photosynthesis using mutant strains of algae and higher plants. *Annu. Rev. Plant Physiol.,* 20:523–540.

37. Levine, R. P. (1974): Mutant studies on photosynthetic electron transport. In: *Algal Physiology and Biochemistry,* edited by W. D. P. Stewart, pp. 424–433. University of California Press, Berkeley.

38. Surzycki, S. J. (1969): Genetic functions of the chloroplast of *Chlamydomonas reinhardi:* Effect of rifampin on chloroplast DNA-dependent RNA polymerase. *Proc. Natl. Acad. Sci. U.S.A.,* 63:1327–1334.

39. Gibbs, M. (1962): Fermentation. In: *Physiology and Biochemistry of Algae,* edited by R. A. Lewin, pp. 91–97. Academic Press, New York.

40. Gibbs, M., Latzko, E., Harvey, M. J., Plaut, Z., and Shain, Y. (1970): Photosynthesis in the algae. *Ann. N.Y. Acad. Sci.,* 175:541–554.

41. Kessler, E. (1974): Hydrogenase, photoreduction and anaerobic growth. In: *Algal Physiology and Biochemistry,* edited by W. D. P. Stewart, pp. 456–473. University of California Press, Berkeley.

42. Walker, D. A. (1974): Chloroplast and cell: The movement of certain key substances, etc. across the chloroplast envelope. *MTP International Review of Science,* 11:1–49.

43. Lewin, R. A. (1954): Utilization of acetate by wild type and mutant *Chlamydomonas dysomos. J. Gen. Microbiol.,* 6:127–134.

44. Wiseman, A., Gillham, N. W., and Boynton, J. E. (1977): Nuclear mutations affecting mitochondrial structure and function in *Chlamydomonas. J. Cell Biol.,* 73:56–77.

45. Kates, J. R., and Jones, R. F. (1964): The control of gametic differentiation in liquid cultures of *Chlamydomonas. J. Cell. Comp. Physiol.,* 63:157–164.

46. Kates, J. R., Chiang, K.-S., and Jones, R. F. (1968): Studies of DNA replication during synchronous vegetative growth and gametic differentiation in *Chlamydomonas reinhardtii. Exp. Cell Res.,* 49:121–135.

47. Chiang, K.-S., Kates, J. R., Jones, R. F., and Sueoka, N. (1970): On the formation of a homogeneous zygotic population in *Chlamydomonas reinhardtii. Dev. Biol.,* 22:655–669.

48. Van Winkle-Swift, K. (1976): The transmission, segregation, and recombination of chloroplast genes in diploid strains of *Chlamydomonas reinhardtii.* Ph.D. thesis, Duke University, Durham, N.C.

49. Van Winkle-Swift, K. (1977): The maturation of algal zygotes: Alternative experimental approaches for *Chlamydomonas reinhardtii. J. Phycology,* 13:225–231.

50. Ebersold, W. T. (1967): *Chlamydomonas reinhardi:* Heterozygous diploid strains. *Science,* 157:447–449.

51. Gillham, N. W. (1969): Uniparental inheritance in *Chlamydomonas reinhardi. Amer. Naturalist,* 103:355–388.

52. Gillham, N. W. (1963): The nature of exceptions to the pattern of uniparental inheritance for high level streptomycin resistance in *Chlamydomonas reinhardi. Genetics,* 48:431–439.

53. Sager, R., and Ramanis, Z. (1967): Biparental inheritance of nonchromosomal genes induced by ultraviolet irradiation. *Proc. Natl. Acad. Sci. U.S.A.,* 58:931–937.

54. Sager, R., and Ramanis, Z. (1976): Chloroplast genetics of *Chlamydomonas.* I. Allelic segregation ratios. *Genetics,* 83:303–321.

55. Sager, R., and Ramanis, Z. (1963): The particulate nature of nonchromosomal genes in *Chlamydomonas. Proc. Natl. Acad. Sci. U.S.A.,* 50:260–268.

56. Gillham, N. W., Boynton, J. E., and Lee, R. W. (1974): Segregation and recombination of non-Mendelian genes in *Chlamydomonas. Genetics,* 78:439–457.

57. Gillham, N. W. (1963): Transmission and segregation of a nonchromosomal factor controlling streptomycin resistance in diploid *Chlamydomonas. Nature,* 200:294.

58. Boynton, J. E., Gillham, N. W., Harris, E. H., Tingle, C. L., Van Winkle-Swift, K., and Adams, G. M. W. (1976): Transmission, segregation and recombination of chloroplast genes in *Chlamydomonas.* In: *Genetics and Biogenesis of Chloroplasts and Mitochondria,* edited by T. Bücher, W. Neupert, W. Sebald, and S. Werner, pp. 313–322. North Holland, Amsterdam.

59. Alexander, N. J., Gillham, N. W., and Boynton, J. E. (1974): The mitochondrial genome of *Chlamydomonas:* Induction of minute colony mutations by acriflavin and their inheritance. *Mol. Gen. Genet.,* 130:275–290.

60. Alexander, N. J. (1977): Acriflavine and ethidium bromide treatment on *Chlamydomonas reinhardtii:* Physiological and genetic effects. Ph.D. thesis, Duke University, Durham, N.C.

61. Flechtner, V. R., and Sager, R. (1973): Ethidium bromide induced selective and reversible loss of chloroplast DNA. *Nature* [New Biol.], 241:277–279.

62. Wiseman, A., Gillham, N. W., and Boynton, J. E. (1977): The mitochondrial genome of *Chlamydomonas.* II. Genetic analysis of non-Mendelian photoautotrophic mutants. *Mol. Gen. Genet.,* 150:109–118.

63. Bastia, D., Chiang, K.-S., and Swift, H. (1969): Chloroplast de-differentiation and re-differentiation during zygote maturation and germination in *Chlamydomonas reinhardi. J. Cell Biol.,* 43:11a.

64. Cavalier-Smith, T. (1970): Electron micro-

scopic evidence for chloroplast fusion in zygotes of *Chlamydomonas reinhardi*. *Nature,* 228:333–335.

65. Chiang, K.-S. (1970): Replication, transmission and recombination of cytoplasmic DNAs in *Chlamydomonas reinhardi*. In: *Autonomy and Biogenesis of Mitochondria and Chloroplasts,* edited by N. K. Boardman, A. W. Linnane, and R. M. Smillie, pp. 235–249. North Holland, Amsterdam.

66. Chiang, K.-S. (1976): On the search for a molecular mechanism of cytoplasmic inheritance: Past controversy, present progress and future outlook. In: *Genetics and Biogenesis of Chloroplasts and Mitochondria,* edited by T. Bücher, W. Neupert, W. Sebald, and S. Werner, pp. 305–312. North Holland, Amsterdam.

67. Sager, R., and Lane, D. (1972): Molecular basis of maternal inheritance. *Proc. Natl. Acad. Sci. U.S.A.,* 69:2410–2413.

68. Sager, R. (1972): *Cytoplasmic Genes and Organelles.* Academic Press, New York.

69. Chiang, K.-S., Eves, E., and Swinton, D. (1975): Variation of thymidine incorporation patterns in the alternating vegetative and sexual life cycles of *Chlamydomonas reinhardtii*. *Dev. Biol.,* 42:53–63.

70. Lee, R. W., and Jones, R. F. (1973): Induction of Mendelian and non-Mendelian streptomycin resistant mutants during the synchronous cell cycle of *Chlamydomonas reinhardtii*. *Mol. Gen. Genet.,* 121:99–108.

71. Chiang, K.-S. (1975): The nuclear and chloroplast DNA replication mechanisms in *Chlamydomonas reinhardtii:* Their regulation, periodicity and interaction. In: *Les cycles cellulaires et leur blocage chez plusiers protistes,* edited by M. Lefort-Tran and R. Valencia, pp. 147–158. Editions du Centre National de la Recherche Scientifique, Paris.

72. Swinton, D., and Hanawalt, P. (1972): *In vivo* specific labelling of *Chlamydomonas* DNA. *J. Cell Biol.,* 54:592–597.

73. Sueoka, N., Chiang, K.-S., and Kates, J. R. (1967): Deoxyribonucleic acid replication in meiosis of *Chlamydomonas reinhardi*. I. Isotopic transfer experiments with a strain producing eight zoospores. *J. Mol. Biol.,* 25: 47–66.

74. Chiang, K.-S., and Sueoka, N. (1967): Replication of chromosomal and cytoplasmic DNA during mitosis and meiosis in the eukaryote *Chlamydomonas reinhardi*. *J. Cell Physiol.,* 70[Suppl. 1]:89–112.

75. Siersma, P. W., and Chiang, K.-S. (1971): Conservation and degradation of cytoplasmic and chloroplast ribosomes in *Chlamydomonas reinhardtii*. *J. Mol. Biol.,* 58:167–185.

76. Sager, R. (1975): personal communication.

77. Schlanger, G., and Sager, R. (1974): Correlation of chloroplast DNA and cytoplasmic inheritance in *Chlamydomonas* zygotes. *J. Cell Biol.,* 63:301a.

78. Gillham, N. W. (1965): Induction of chromosomal and nonchromosomal mutations in *Chlamydomonas reinhardi* with *N*-methyl-*N′*-nitro-*N*-nitrosoguanidine. *Genetics,* 52:529–537.

79. Wurtz, E. A., Boynton, J. E., and Gillham, N. W. (1977): Perturbation of chloroplast DNA amounts and chloroplast gene transmission in *Chlamydomonas reinhardtii* by 5-fluorodeoxyuridine. *Proc. Natl. Acad. Sci. U.S.A.,* 74:4552–4556.

80. Conde, M. F., Boynton, J. E., Gillham, N. W., Harris, E. H., Tingle, C. L., and Wang, W. L. (1975): Chloroplast genes in *Chlamydomonas* affecting organelle ribosomes. Genetic and biochemical analysis of antibiotic-resistant mutants at several gene loci. *Mol. Gen. Genet.,* 140:183–220.

81. Adams, G. M. W., Van Winkle-Swift, K. P., Gillham, N. W., and Boynton, J. E. (1976): Plastid inheritance in *Chlamydomonas reinhardtii*. In: *The Genetics of Algae,* edited by R. A. Lewin, Chapter 5. Blackwell Scientific, Oxford.

82. Sager, R. (1974): Nuclear and cytoplasmic inheritance in green algae. In: *Algal Physiology and Biochemistry,* edited by W. D. P. Stewart, Chapter 11. Blackwell Scientific, Oxford.

83. Sager, R. (1977): Genetic analysis of chloroplast DNA in *Chlamydomonas*. *Adv. Genet.,* 19:287–340.

84. Sager, R., and Ramanis, Z. (1970): A genetic map of non-Mendelian genes in *Chlamydomonas*. *Proc. Natl. Acad. Sci. U.S.A.,* 65:593–600.

85. Sager, R., and Ramanis, Z. (1976): Chloroplast genetics of *Chlamydomonas*. II. Mapping by cosegregation frequency analysis. *Genetics,* 83:323–340.

86. Singer, B., Sager, R., and Ramanis, Z. (1976): Chloroplast genetics of *Chlamydomonas*. III. Closing the circle. *Genetics,* 83: 341–354.

87. Bennoun, P., and Chua, N. H. (1976): Methods for the detection and characterization of photosynthetic mutants in *Chlamydomonas reinhardi*. In: *Genetics and Biogenesis of Chloroplasts and Mitochondria,* edited by T. Bücher, W. Neupert, W. Sebald, and S. Werner, pp. 33–39. North Holland, Amsterdam.

88. Boynton, J. E., Gillham, N. W., Harris, E. H., Tingle, C. L., Van Winkle-Swift, K., and Adams, G. M. W. (1976): Transmission, segregation and recombination of chloroplast genes in *Chlamydomonas*. In: *Genetics and Biogenesis of Chloroplasts and Mitochondria,* edited by T. Bücher, W. Neupert, W. Sebald, and S. Werner, pp. 313–322. North Holland, Amsterdam.

89. Harris, E. H., Boynton, J. E., Gillham, N. W., Tingle, C. L., and Fox, S. B. (1977): Mapping of chloroplast genes involved in

chloroplast ribosome biogenesis in *Chlamydomonas. Mol. Gen. Genet.,* 155:249–265.

90. Surzycki, S. J., and Gillham, N. W. (1971): Organelle mutations and their expression in *Chlamydomonas reinhardi. Proc. Natl. Acad. Sci. U.S.A.,* 68:1301–1306.

91. Chu-Der, O. M. Y., and Chiang, K.-S. (1974): Interaction between Mendelian and non-Mendelian genes. Regulation of the transmission of non-Mendelian genes by a Mendelian gene in *Chlamydomonas reinhardtii. Proc. Natl. Acad. Sci. U.S.A.,* 71: 153–157.

92. Gillham, N. W., Boynton, J. E., and Burkholder, B. (1970): Mutations altering chloroplast ribosome phenotype in *Chlamydomonas.* I. Non-Mendelian mutations. *Proc. Natl. Acad. Sci. U.S.A.,* 67:1026–1033.

93. Schimmer, O., and Arnold, C. G. (1970): Die Suppression der ausserkaryotisch bedingten Streptomycin-abhängigkeit bei *Chlamydomonas reinhardii. Arch. Microbiol.,* 73:195–200.

94. Schimmer, O., and Arnold, C. G. (1970): Über die Zahl der Kopien eines ausserkaryotischen Gens bei *Chlamydomonas reinhardtii. Mol. Gen. Genet.,* 107:366–371.

95. Schimmer, O., and Arnold, C. G. (1970): Hin- und Rücksegregation eines ausserkaryotischen Gens bei *Chlamydomonas reinhardtii. Mol. Gen. Genet.,* 108:33–40.

96. Gillham, N. W., and Levine, R. P. (1962): Studies on the origin of streptomycin resistant mutants in *Chlamydomonas reinhardtii. Genetics,* 47:356–363.

97. Lee, R. W., Gillham, N. W., Van Winkle, K. P., and Boynton, J. E. (1973): Preferential recovery of uniparental streptomycin resistant mutants from diploid *Chlamydomonas reinhardtii. Mol. Gen. Genet.,* 121: 109–116.

98. Boschetti, A., and Bogdanov, S. (1973): Different effects of streptomycin on the ribosomes from sensitive and resistant mutants of *Chlamydomonas reinhardi. Eur. J. Biochem.,* 35:482–488.

99. Chua, N.-H. (1976): A uniparental mutant of *Chlamydomonas reinhardtii* with a variant thylakoid membrane polypeptide. In: *Genetics and Biogenesis of Chloroplasts and Mitochondria,* edited by T. Bücher, W. Neupert, W. Sebald, and S. Werner, pp. 323–330. North Holland, Amsterdam.

100. Bourque, D. P., Boynton, J. E., and Gillham, N. W. (1971): Studies on the structure and cellular location of various ribosome and ribosomal RNA species in the green alga *Chlamydomonas reinhardi. J. Cell Sci.,* 8: 153–183.

101. Surzycki, S. J., and Rochaix, J. D. (1971): Transcriptional mapping of ribosomal RNA genes of the chloroplast and nucleus of *Chlamydomonas reinhardi. J. Mol. Biol.,* 62:89–109.

Chapter 13

Chloroplast Genetics of *Chlamydomonas*. II. Induction and Expression of Chloroplast Mutations

In 1947 von Euler (1) reported that when barley (*Hordeum vulgare*) was germinated in streptomycin solutions, the coleoptile that emerged was thickened and its underpart was pure white; leaves that previously were green could not be made to bleach. These observations were extended by von Euler and his colleagues (2–6) in a series of short articles that dealt with barley and several other seed plants. For example, they found that only leucoplasts were left in leaves of barley plants germinated on streptomycin solutions more concentrated than 2 mg/ml. In 1948 Provasoli et al. (7) reported that light- and dark-grown cells of the flagellate *Euglena gracilis* could be permanently bleached by streptomycin concentrations of 100 μg/ml or less, but they found that the cells continued to grow, even at very high streptomycin concentrations (5,000 μg/ml), as long as they were provided with a fixed carbon source such as glucose. Cells of *Astasia longa,* a colorless analog of *E. gracilis,* were also able to grow at very high streptomycin concentrations.

In 1951 Provasoli et al. (8) published an article in which they compared several different flagellates and algae with respect to streptomycin-induced bleaching and lethality. They showed that there is a profound difference between the responses of *E. gracilis* and *C. reinhardtii* to streptomycin. Streptomycin concentrations as low as 1 to 5 μg/ml were lethal to *C. reinhardtii* and also to its colorless analogs *Polytoma uvella* and *P. obtusum* even in the presence of a fixed carbon source. Resistant stocks of *C. reinhardtii* developed only slowly. Exposure of resistant cells to streptomycin over a long period did yield some white colonies, but

these could not be subcultured. Thus bleaching may well have followed cell death.

The study of streptomycin effects on *C. reinhardtii* was subsequently continued by Sager (9,10), who reported that streptomycin had two effects on this alga. The first effect was nongenetic; it led to blockage of chlorophyll synthesis in the dark. The second effect was genetic, and it appeared to produce these results: induction of chloroplast mutations to streptomycin resistance (*sr-u-2*), induction of auxotrophic chloroplast mutations (mostly to acetate dependence), and mass induction of a class of mutants designated *y* (for yellow in the dark) that were unable to form chlorophyll in the dark, although they could do so in the light.

Study of the phenomenology of streptomycin effects and the effects of other antibiotics that block chloroplast protein synthesis presents a fascinating story in both *Chlamydomonas* and *Euglena*. In this chapter we shall discuss the evidence for streptomycin mutagenesis in *Chlamydomonas,* as well as the use of conventional mutagens and diploids for production and selection of chloroplast mutants in this alga. We shall conclude with consideration of the mechanisms of expression of forward and reverse mutations in the chloroplast.

NONGENETIC EFFECTS OF STREPTOMYCIN ON *Chlamydomonas*

Wild-type *C. reinhardtii,* unlike *Euglena,* forms chlorophyll and a normal chloroplast in darkness as well as in light. Sager and

Tsubo (10) found that sublethal concentrations of streptomycin blocked formation of chlorophyll by wild-type cells of *C. reinhardtii* in the dark but not in the light (Table 13–1). When such cells were transferred to medium lacking streptomycin, they regained the ability to form chlorophyll in the dark. Sager and Tsubo then showed that chloroplast mutants resistant to streptomycin (*sr-u-sm₂*) or dependent on streptomycin (*sd-u-sm₄*) continued to synthesize chlorophyll in the dark even in the presence of very high concentrations of streptomycin (Table 13–1). On the other hand, a mendelian mutation (*sr-1*) resistant to a lower streptomycin concentration (100 μg/ml) became yellow in the dark in the presence of the drug (Table 13–1). This was true even when the *sr-1* mutant was combined with a modifier, *A,* that raised the level of resistance to streptomycin in all mutants to 2,000 μg/ml. Cells carrying *A* alone had a wild-type level of streptomycin resistance, and they were yellow in the dark in the presence of streptomycin.

For the moment, these results are not entirely interpretable. It is known that both *sr-u-sm₂* and *sd-u-sm₄* mutants have chloroplast ribosomes resistant to streptomycin (see Chapter 21). Therefore a straightforward explanation of these results would be that a component required for chlorophyll accumulation in the dark is synthesized on chloroplast ribosomes. Streptomycin blocks this synthesis in wild-type cells, which have sensitive chloroplast ribosomes, but not in the mutants, which have resistant or dependent chloroplast ribosomes. The problem with this explanation is that the *sr-1* mutant tested is yellow in the dark on streptomycin, and the *sr-1* mutants also have antibiotic-resistant chloroplast ribosomes (see Chapter 21). Furthermore, the double mutant *A sr-1* is yellow in the dark, even though the resistance level of the double mutant is much higher than that of *sr-1* by itself. Since *A* does not confer resistance by itself, but only in combination with a mutation to streptomycin resistance or dependence, it seems likely that *A* is a mutation affecting chloroplast ribosomes that acts synergistically with ribosomal mutations to resistance or dependence to raise the level of chloroplast ribosome resistance to streptomycin. Thus in the dark in the presence of streptomycin, *sr-u-sm₂ A* and *sd-u-sm₄ A* cells are green, and *sr-1 A* cells are yellow. Why is this?

TABLE 13–1. *Chlorophyll formation in light and dark at sublethal streptomycin concentrations by various strains of C. reinhardtii*[a]

Genotype	Streptomycin resistance level (μg/ml)	Color of dark-grown cells	
		Without streptomycin	With streptomycin[b]
a	20	green	yellow
A	20	green	yellow
sr-1 a	100	green	yellow
sr-1 A	2,000	green	yellow
sr-u-sm₂ a	500	green	green
sr-u-sm₂ A	2,000	green	green
sd-u-sm₄ a	500	green	green
sd-u-sm₄ A	2,000	green	green

[a] All strains were homozygous for the wild-type allele of the yellow mutant (Yi⁺). A and a are alleles of a nuclear gene that do not affect the level of resistance of wild-type cells. The A allele, but not the a allele, greatly increases the resistance levels of resistant and dependent strains.

[b] The highest sublethal concentration of streptomycin was used in each instance.

Adapted from Sager and Tsubo (10).

One possible explanation is that the chloroplast ribosomes of *sr-1* are more streptomycin-sensitive than those of *sr-u-sm₂* and *sd-u-sm₄,* and although *A* raises the resistance levels of all mutants, it does so to a lesser extent for *sr-1.* Since Sager and Tsubo checked chlorophyll formation in the dark only at the highest sublethal antibiotic concentration, they may have seen what is really a quantitative difference rather than a qualitative difference between genotypes. Careful concentration curves of chlorophyll accumulated in the dark at different antibiotic concentrations remain to be constructed for each genotype. It would also be desirable to examine the effects of other inhibitors, such as spectinomycin and erythromycin, that are known to block chloroplast protein synthesis on chlorophyll accumulation in the dark. In addition, mutants with chloroplast ribosomes resistant to these antibiotics are available (see Chapter 21). It would be of considerable interest to establish definitively whether or not a component required for chlorophyll accumulation in the dark is synthesized on chloroplast ribosomes, as well as to determine the nature of this component and find out why chlorophyll accumulation in the light proceeds normally in the presence of antibiotics blocking chloroplast protein synthesis.

GENETIC EFFECTS OF STREPTOMYCIN ON *Chlamydomonas*

Sager (9–12) reported that streptomycin is a mutagen for chloroplast genes in *Chlamydomonas.* This assertion is based on three distinct sets of observations that may or may not be related. We shall examine each of these and, where appropriate, introduce comparisons with *Euglena.*

The first set of observations concerns the production of mutants that are yellow in the dark (*y* mutants) in the presence of streptomycin, which was studied by Sager and Tsubo (10). These mutants do not form chlorophyll in the dark, nor do they form thylakoid membranes (see Chapter 20). On exposure of *y* mutants to light, chlorophyll accumulation and membrane synthesis begin; within a matter of hours, wild-type chlorophyll levels are attained, and the chloroplast looks perfectly normal. All *y* mutants thus far isolated, regardless of origin, segregate 1:1 in crosses to the wild type, as do normal mendelian mutants; they appear to be allelic, and centromere-linked in a linkage group of their own (10,13). As was shown by Sager and Tsubo (10) and Winig and Gillham (13), *y* mutants arise at a very high frequency spontaneously (approximately 1×10^{-3} mutations/cell), and they also have a very high reversion frequency (13). In addition, among all mutants with the mendelian pattern of inheritance, only these mutants have been reported to be induced by streptomycin (10). The various peculiarities of the *y* mutants have led to speculation that they might be localized in the chloroplast (14); that is, these mutants, unlike all other chloroplast mutations, would show mendelian inheritance rather than uniparental inheritance. Although this hypothesis cannot be ruled out, there is presently no compelling evidence in favor of it.

Sager and Tsubo (10) studied the induction of *y* mutants by streptomycin in an *sr-u-2* mutant stock, since, as was mentioned previously, the mutant remains green in the dark at sublethal streptomycin concentrations, whereas cells of wild type or of an *sr-1* mutant turn yellow in the dark, becoming phenocopies of the *y* mutants. Their experiments revealed that following prolonged exposure of the *sr-u-2* mutant to streptomycin in the dark, most colonies or all colonies clearly became mosaics of yellow and green cells. In controls, on the other hand, yellow colonies and mosaics were present at a frequency of approximately 3×10^{-3}.

The mosaic colonies produced in the presence of streptomycin fell into three classes. Class I consisted of misshapen colonies that were clearly mixed yellow and green from the onset of growth. Class II consisted of green colonies in which yellow papillae appeared suddenly 1 to 2 weeks after the colonies reached full size, following which the yellow cells overgrew the colonies. Class III colonies were light green in color and had no distinct papillae or yellow sectors. These colonies, when replated on streptomycin-free media, produced yellow colonies at the same frequency as controls (approximately 0.15%); but if they were replated on streptomycin-containing media, the frequency of yellow colonies was much higher (approximately 23%). These yellow colonies, when replated, proved to be more than 95% yellow, whether or not streptomycin was present in the medium. Sager and Tsubo argued that colonies of class III contained cells in a metastable state that required further contact with streptomycin to stabilize the mutant phenotype.

Sager and Tsubo proceeded to study the stability of the *y* mutants obtained with streptomycin by replating yellow papillae from class II colonies in the presence of streptomycin and in the absence of streptomycin and by examining the stability of yellow mutants obtained from class III colonies. The papillae produced a variety of colony types, but a principal finding was that few of the papillae were composed of only *y* mutant colonies of normal size. Most of the colonies were mosaics, containing both yellow and green cells, with certain of the cells frequently forming abnormally small colonies. These authors also found that on replating, mosaic colonies tended to produce mosaic colonies. Most mutant strains obtained from class II and class III colonies gave rise to both stable and persistently unstable substrains. In this respect they differed sharply from spontaneously occurring *y* mutants, since these do not produce unstable substrains. Stable *y* mutants derived from the induction experi-

ments segregated 2:2 in crosses with the wild type and appeared to be allelic with the spontaneous y_1 mutant. The mutable strains in crosses to stable *y* mutants segregated 2:2 stable yellow:mutable yellow.

As these authors clearly realized, the major problem with the induction hypothesis that they supported is the possibility that in the presence of streptomycin there is some complex process of selection for spontaneously occurring *y* mutants. After all, the frequency of spontaneous *y* mutants was already greater than 1×10^{-3} mutations/cell, in single-colony isolates. Sager and Tsubo argued that the following points militated against the selection hypothesis: First, although the frequencies of spontaneous *y* mutants were about the same in all of the nine clonal isolates used in the streptomycin induction experiments, they gave rise to three very different classes of mutant colonies in which *y* mutants appeared at very different times. Sager and Tsubo believed that it was difficult to explain the great differences in the times of appearance of *y* mutants in the different clonal isolates solely on a selection hypothesis, since approximately the same spontaneous mutant reservoir was available in each clone. Second, the *y* mutants obtained from the induction experiments differed from spontaneous *y* mutants in their heterogeneity. Most of them produced mosaics and unstable substrains that frequently grew poorly or had altered growth rates on minimal medium. Third, of 16 *y* mutants studied, only five were more resistant to streptomycin than the parental green strain. One might expect the *y* mutants to be more resistant to streptomycin than the parental strain on the basis of a selection hypothesis. Fourth, in liquid culture experiments in the presence of streptomycin, no selective advantage was noted for a streptomycin-resistant *y* mutant obtained in these experiments in competition with the parental green strain.

Although these authors made a reasonable case for the induction hypothesis, their various observations are somewhat disparate. It is not clear that they should all be classed under the same umbrella of induction, nor is it clear that mutant induction in the usual sense is operative. In the first place, y mutants occurred at a high rate, and the selection experiment that the authors described was carried out in liquid medium and did not rule out selection for y mutants on streptomycin plates. After all, as the authors pointed out, and as we have also observed (15), streptomycin-resistant mutants are far more streptomycin-sensitive in liquid than on solid medium. Therefore, much lower concentrations of streptomycin would have to be used in the liquid selection experiments than were used in plates. One would like to know if a reconstruction experiment on solid medium in which yellow and green cells were placed next to each other on streptomycin plates would give the same results. One would also like to know the plating efficiencies of yellow and green cells on streptomycin medium, as well as whether or not the yellow mutants overgrow the green colonies (i.e., in the case of class II mutants), since they are able to continue growing after growth of green cells has ceased. Thus, subtle forms of selection must be ruled out before the streptomycin induction of y mutants can be accepted. The most important point in favor of the induction hypothesis is that most "induced" y mutants are mutable, and in this sense they differ markedly from spontaneous y mutants. Mutability itself segregates 2:2 in crosses with stable y mutants.

The mutability phenomenon itself is complex; it appears to consist of at least two components. The first is the tendency of the induced y mutants to produce mosaics. The second is the tendency of these mutants to produce unstable substrains with altered growth rates, poor growth rates on minimal medium, and streptomycin resistance levels higher or lower than that of the parental strain. We shall consider the first of these

phenomena here, since the second is probably related to the apparent induction of auxotrophic mutants by streptomycin, which will be discussed later. It seems possible that the induced y mutants that produce mosaics are highly mutable strains that are never seen spontaneously because their frequencies in a colony are always low relative to those of stable y mutants because they revert so frequently. In the presence of streptomycin, yellow cells produced by these unstable isolates will be selected continuously.

In summary, the problem appears to be reducible to one of two possibilities. The first is that y mutants are induced by high concentrations of streptomycin in streptomycin-resistant strains, and they are less stable than spontaneous y mutants. The second possibility is that y mutants are all spontaneous and arise at high rates. Although most, if not all, y mutants revert at relatively high rates, there is a spectrum of reversion rates such that those reverting with the highest rates are not normally recovered as spontaneous mutants. During streptomycin selection, both relatively stable and unstable y mutant strains are selected on agar because they have a selective advantage over green cells in the dark. The y mutants pose an interesting and unusual genetic puzzle, and new experiments are needed. What is abundantly clear is that the induction of y mutants in *C. reinhardtii* by streptomycin (if that is what is happening) and the bleaching of *E. gracilis* by streptomycin are entirely different problems. For one thing, untreated *Euglena* cells behave as y mutants of *C. reinhardtii* in the dark, for *Euglena* does not accumulate chlorophyll in the dark. That is, wild-type *Euglena* does not have the ability to form chlorophyll in the dark. Another important difference is that when it is provided with a fixed carbon source, *Euglena* grows on very high streptomycin concentrations and totally loses its chloroplasts. In *C. reinhardtii* it is necessary first to isolate a specific kind of streptomycin-resistant mutant and then isolate *y* mutants in this stock in the presence of streptomycin. These mutants have not lost their chloroplasts, only the ability to form chlorophyll in the dark. We shall return to this comparison in Chapter 16.

The second set of observations used by Sager (11) to support the hypothesis of streptomycin induction of chloroplast mutations involved application of the classic techniques of microbial genetics, the fluctuation test, and the Newcombe respreading experiment (16,17) to determine if streptomycin induces mutations to streptomycin resistance or if these are spontaneous in origin. We discussed the use of the Newcombe respreading experiment in Chapter 5 in connection with the origin of mitochondrial mutations to antibiotic resistance in yeast. Here we shall describe the fluctuation test and consider the results of both kinds of experiments.

In the fluctuation test small numbers of streptomycin-sensitive cells are distributed to a large number of tubes of culture medium lacking streptomycin. After the cells have been allowed to grow for a while, each culture or a portion of each culture is plated onto a separate dish of streptomycin-containing medium, and colonies of resistant cells are allowed to form. If cells resistant to streptomycin arise at random by spontaneous mutation during growth of the liquid cultures, they will divide and form new clones of resistant cells within the cultures. The size of each resistant clone will depend on the generation of growth at which the original mutation arose. Assuming that the probability of spontaneous mutation per generation is constant, small clones of resistant cells will be much more frequent than large clones because at each successive generation the number of cells capable of mutation doubles. Occasionally, by chance alone, a mutation will arise very early in the growth of a clone, and this will produce a very large clone ("jackpot") of

resistant cells. Thus when the clones are plated on streptomycin-containing medium, there will be fluctuation in numbers of resistant cells per clone reflecting the time at which each mutation arose. Some plates will have many more resistant colonies than others.

On the other hand, if streptomycin induces *sr-u-2* mutants, the numbers of resistant colonies per plate should have a random (Poisson) distribution, just as if successive samples had been plated from a single culture. A useful property of the Poisson distribution is that the variance is not significantly different from the mean. By comparing the mean number of resistant colonies per plate with the variance in the number of resistant colonies per plate it can be determined whether or not the distribution of mutants among plates is a Poisson distribution.

Fluctuation tests performed by Sager (11) and Gillham and Levine (18) to determine the origin of streptomycin-resistant mutants in *C. reinhardtii* yielded essentially similar results. It is clear from both sets of experiments that the mendelian *sr-1* mutants are distributed in a clonal fashion that is consistent with a spontaneous origin in the absence of streptomycin (Table 13–2). Hence the variance of the distribution is much greater than the mean, and it is clearly not a Poisson distribution. On the other hand, the distribution of *sr-u-2* mutations is Poisson or nearly so (Table 13–2). In the experiments of Sager (11) the variance of *sr-u-2* mutants was not significantly different from the mean; in the experiments of Gillham and Levine (18) there was a significant difference, although it can be seen by inspection of the data (Table 13–2) that the deviation from a Poisson distribution is not great.

Reconstruction experiments done by Gillham and Levine (18) showed that growth of *sr-u-2* mutants in the absence of streptomycin was not suppressed in mixtures with a great excess of sensitive cells; thus dif-

TABLE 13–2. Colonies of mendelian (sr-1) and chloroplast (sr-u-2) mutations to streptomycin resistance obtained from similar cultures of wild-type cells in fluctuation tests

Experiment Mutant colonies obtained per culture	sr-1 mutants			sr-u-2 mutants		
	1	2	3	1	2	3
0	469	451	425	477	490	443
1	9	17	11	16	4	4
2	5	6	5	3	1	1
3	3	3	3	0	0	0
4	3	2	1	0	0	0
5	1	2	0	0	0	0
6	1	0	0	0	0	0
7	1	1	0	0	0	0
8	0	1	0	0	0	0
9	0	2	0	0	0	0
10	0	0	1	0	0	0
11	1	1	0	0	0	0
12	1	0	0	0	0	0
13	0	3	0	0	0	0
14	1	0	0	0	0	0
17	0	1	0	0	0	0
20	1	0	0	0	0	0
25	0	1	0	0	0	0
31	0	0	1	0	0	0
45	0	1	0	0	0	0
65	1	0	0	0	0	0
70	0	0	1	0	0	0
2,820	0	1	0	0	0	0
3,000	0	1	0	0	0	0
Total cultures	496	495	448	496	495	448
Mean number of resistant colonies obtained per culture	0.35	12.2	0.32	0.044	0.012	0.013
Variance	10.5	3.42×10^4	13.4	0.055	0.016	0.018
Probability (exact method of Fisher)	—	—	—	0.013	0.030	0.033

From Gillham and Levine (18), with permission.

ferential growth rates did not account for the Poisson distribution of sr-u-2 mutants. Gillham and Levine also provided evidence that suppression of growth of sr-u-2 mutants on streptomycin-containing plates in the presence of a great excess of sensitive cells was not the cause of the Poisson distribution either.

Sager (11), in addition to using the fluctuation test, also showed by means of the Newcombe respreading experiment (see Chapter 5) that sr-u-2 mutants did not have a clonal distribution, and by means of reconstruction experiments she proved that the nonclonal distribution of sr-u-2 mutants was not the result of growth suppression in the presence of sensitive cells.

Now we shall consider the interpretation of these experiments. Sager (11) concluded that her results were best explained by the hypothesis that sr-u-2 mutations are induced by streptomycin, a conclusion clearly compatible with her results. Gillham and Levine (18) considered both the induction hypothesis and a second hypothesis of intracellular selection. The intracellular selection hypothesis assumes that all sr-u-2 mutations are spontaneous and that a cell contains a number of discrete particles capable of mutation from sensitivity to resistance. Then the mutated particle is selected against within the cell until cells are plated on streptomycin-containing medium, following which the direction of selection is reversed.

From the vantage point of hindsight the intracellular selection hypothesis now seems unsatisfactory, since the mutated particle is almost certainly a clDNA molecule, and the mutations to resistance confer resistance on chloroplast ribosomes (see Chapter 21). That is, there is no reason to believe that an *sr-u-2* mutation will confer a replicational disadvantage on a clDNA molecule under nonselective conditions. This leaves us with the induction hypothesis, which is not entirely satisfactory either; as Sager (11) showed, streptomycin does not act as a conventional mutagen in that it does not increase the frequency of *sr-u-2* mutations following short-term exposure of sensitive cells in liquid. It may be that the nonclonal distribution of *sr-u-2* mutations in fluctuation tests is a reflection of the way in which spontaneous chloroplast mutations segregate and become expressed in *Chlamydomonas*. If so, chloroplast mutations resistant to any antibiotic should show nonclonal distributions in fluctuation tests, whereas one would not expect this result if streptomycin is a specific mutagen. That is, given the induction hypothesis, chloroplast mutations to erythromycin resistance may be expected to show a clonal distribution unless erythromycin is also a mutagen. To settle this question satisfactorily it will be necessary to examine the distribution of chloroplast mutations resistant to a variety of different antibiotics in fluctuation tests. If the distribution is always nonclonal, the expression hypothesis will be supported, for it is difficult to imagine that all antibiotics are mutagenic, since their structures vary a great deal; but if the distribution of chloroplast mutations to antibiotic resistance is clonal for antibiotics other than streptomycin, the induction hypothesis will be supported strongly.

It should be noted that Birky (19) showed the distribution of mitochondrial mutations resistant to erythromycin and chloramphenicol in *S. cerevisiae* to be nonclonal in Newcombe respreading experiments. However, Birky's results, in contrast to those for *Chlamydomonas,* are readily interpreted in terms of an intracellular selection hypothesis in which resistant mitochondria are selected against in competition with sensitive mitochondria under nonselective conditions. This interpretation would certainly be consistent with the results of mitochondrial competition experiments in *Paramecium* (see Chapter 10).

The third set of observations supporting the hypothesis of streptomycin mutagenesis derives from the finding of Sager and Tsubo (10) that the *y* mutations obtained in the streptomycin induction experiments frequently segregated auxotrophic mutations, many of which were stimulated to grow when acetate was added to the medium.[1] These mutants showed the same pattern of maternal inheritance as other chloroplast mutations (20). Sager has used the streptomycin mutagenesis technique ever since to obtain chloroplast mutations, including those resistant to antibiotics other than streptomycin (12). The advantage of the method appears to be its selectivity for chloroplast mutants. The technique has since been modified so that wild-type cells instead of an *sr-u-2* mutant are plated on sublethal (20 μg/ml) concentrations of streptomycin. As before, the cells are allowed to form colonies over a period of 2 or 3 weeks, after which they are tested for new mutations. Unfortunately, no quantitative data have ever been published to indicate just how effective this regimen is in inducing new chloroplast mutations. This would seem a relatively easy determination, in view of the fact mutations to antibiotic resistance can be used as an assay.

Sager (12) also reported that streptomycin withdrawal is mutagenic for a streptomycin-dependent (*sd-u*) strain. When growth ceases in the absence of streptomycin, a burst of mutants appears, including

---

[1] Sager, in a personal communication, reported that none of these acetate mutants are stringent in the sense that they cannot grow at all on minimal medium. The most stringent (e.g., *ac-2*) form colonies on minimal medium smaller in size than those formed on acetate medium.

apparent revertants to sensitivity, mutations to streptomycin resistance, mutations whose growth is stimulated by acetate, and mutations to temperature sensitivity. Detailed descriptions of the methodology used and the mutants obtained remain to be published.

A fair summary of streptomycin induction of chloroplast mutations in *Chlamydomonas* would probably be that there is a great deal of interesting phenomenology that requires explanation and further experimentation. It is not clear how the three classes of observations used to support the hypothesis of streptomycin mutagenesis are related, if they are related.

INDUCTION OF CHLOROPLAST MUTATIONS WITH MNNG

Thus far MNNG is the only conventional mutagen that has been used extensively to induce chloroplast gene mutations in *Chlamydomonas*. Gillham (21) found that MNNG induced both chloroplast (*sr-u-2*) and mendelian (*sr-1*) gene mutations to streptomycin resistance in *C. reinhardtii*. Since *sr-u-2* mutations were known to be distributed nonclonally in fluctuation tests, which indicated that they might be induced by streptomycin, it was important to determine if MNNG itself induced the *sr-u-2* mutations or if the mutagen simply made sensitive cells more susceptible to streptomycin mutagenesis. Gillham (21) made use of the fluctuation test to examine these possibilities. Cells were treated with MNNG, inoculated into tubes of nonselective medium, and allowed to grow; then the cells were dispensed to plates containing a low concentration of streptomycin so that both *sr-1* and *sr-u-2* mutations would be expressed. These plates were then replicaplated to a high streptomycin concentration to distinguish *sr-1* and *sr-u-2* mutants. It was clear from these experiments that the distributions of both *sr-u-2* and *sr-1* mutants were clonal, which showed that MNNG

actually induced both classes of mutants. In the same article Gillham reported the isolation of several other chloroplast mutations to antibiotic resistance and dependence following MNNG treatment. Until recently MNNG has been the standard mutagen used to obtain chloroplast mutations in the laboratory of Gillham and Boynton. As was mentioned in the preceding chapter, Lee and Jones (22) found that MNNG preferentially mutagenized replicating DNA in *C. reinhardtii*, as it did in *E. coli*; and in synchronous cultures *sr-u-2* mutations were induced preferentially at the time of clDNA replication, whereas *sr-1* mutations appeared at highest frequency at the time of nDNA replication.

USE OF DIPLOIDS FOR SELECTIVE ISOLATION OF CHLOROPLAST MUTATIONS IN *Chlamydomonas*

Gillham (23,24) found that chloroplast mutations, unlike mendelian gene mutations, segregated somatically in diploids during vegetative division (see Chapter 14 for full discussion). This observation served as the basis for development of a technique for selective isolation of chloroplast mutations to streptomycin resistance. Lee et al. (25) mutagenized haploid and diploid cells of *C. reinhardtii* with MNNG and screened for mutations to streptomycin resistance. In the haploid stock, most of the mutants that were recovered (both spontaneous and induced) were resistant to low levels of streptomycin, which indicated that they were mendelian *sr-1* mutations (Fig. 13–1). When the diploid stock was used, it was found that the frequency of mutants recovered was much lower and that most of these mutants were resistant to high levels of streptomycin, which indicated that they were *sr-u-2* chloroplast mutations (Fig. 13–1). It is interesting that a few low-level mutations to resistance were obtained in the diploids, and these also turned out to be chloroplast mutations. The total absence of

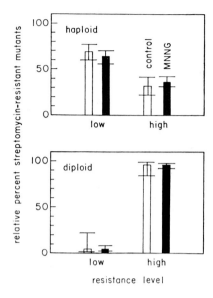

FIG. 13–1. Relative proportions of high- and low-level mutations to streptomycin resistance in haploid and diploid cells of C. *reinhardtii* in control cultures and following treatment of cells with MNNG. High-level resistance mutations are of chloroplast origin. Low-level resistance mutations in haploid cultures are virtually all mendelian in inheritance; those obtained in diploids are nonmendelian. The latter mutants are not normally detected in haploid cell cultures because they are infrequent relative to mendelian low-level resistance mutations. (From Lee et al., ref. 25, with permission.)

sr-1 mutations in the diploid stock was explained by the fact that these mutations were recessive to wild type. Diploid cells in which *sr-1* mutations arose would have been heterozygous for sensitivity and presumably would have died on exposure to streptomycin. Obviously, diploid mutagenesis should be a useful technique for selective recovery of chloroplast mutations, since mendelian mutations, by and large, are recessive.

EXPRESSION OF CHLOROPLAST MUTATIONS

A number of years ago Sager (9) reported that *sr-u-2* mutations had to be isolated on media containing streptomycin concentrations much lower (100 μg/ml) than those to which the mutants would finally be resistant (500 μg/ml). She also found that the yield of *sr-u-2* mutants was 100-fold lower if acetate was omitted from the medium, even at a streptomycin concentration of 100 μg/ml. The yield of *sr-1* mutants, on the other hand, was not influenced by the presence or absence of acetate. Sager's results also suggested that cell division might be an important factor in the expression of *sr-u-2* mutations, since sensitive cells plated on the acetate-containing medium underwent two or three divisions before they died, whereas those plated on minimal medium plus streptomycin divided once at most. Subsequently, Gillham (24) observed that newly isolated *sr-u-2* mutants, unlike *sr-1* mutants, segregated sensitive cells. Cloning of the resistant *sr-u-2* mutants showed that this instability was associated strictly with the newly isolated mutants, since sensitive cells ceased to appear following cloning. In reconstruction experiments it was found that the presence of sensitive cells in the newly isolated mutant clones was not due to accidental contamination of the resistant clones with surviving sensitive cells. The results of Sager and Gillham suggest that when an *sr-u-2* mutation first arises the cell in which it occurs is heterozygous for sensitivity and that resistance is partially dominant in the heterozygous cell. During subsequent cell divisions the cells become homozygous for resistance and attain their final high levels of resistance.

In a detailed study of a chloroplast mutation to streptomycin dependence (*sd-u-3-18*) that reverted to sensitivity, Schimmer and Arnold (26–30) considerably extended the observations on the heterozygous nature of newly arising chloroplast mutations. They found that the frequency of reversion to sensitivity in *sd-u-3-18* was on the order of 4×10^{-7} reversions/dependent cell. They then proceeded to examine the properties of the sensitive revertants, and they found that these segregated streptomycin-dependent cells (secondary *sd* cells) at frequencies much higher than could be accounted for by the mutation rate to streptomycin dependence. Initially the frequency of secondary *sd* cells could be as high as 1×10^{-2} mutations/sensitive cell but the number of dependent cells declined logarithmically as a function of the number of doublings in streptomycin-free medium. One revertant, *sdR_s-u-3-18R20,* was exceptional in this respect in that it maintained the same high level of segregation (0.4 to 4×10^{-2} mutations/sensitive cell) over a long period of time. When this revertant was subcloned, it was found that the subclones behaved very differently. One subclone segregated no secondary *sd* cells; two subclones segregated secondary *sd* cells at a frequency of 1 to 10×10^{-3} mutations/sensitive cell, two clones gave frequencies of 1 to 10×10^{-2} mutations/sensitive cell, and two clones gave frequencies of 1×10^{-2} mutations/sensitive cell. When *sdR_s-u-3-18R20 mt^-* was crossed to wild-type *mt^+*, it was found that virtually all of the progeny were streptomycin-sensitive and did not segregate secondary *sd* cells, despite the fact that the revertant cells used in the cross were still segregating many secondary *sd* cells. In other words, the tendency to segregate secondary *sd* cells was not transmitted by the revertant (the paternal parent), which suggests that this tendency was a property of the chloroplast rather than the nuclear genome.

The secondary *sd* cells were then grown at high concentrations of streptomycin (500 μg/ml) and tested for their ability to segregate sensitive cells (secondary *ss* cells) (29). It was found that the secondary *sd* isolates derived from all the sensitive revertants except one produced secondary *ss* cells at frequencies that were at most two-fold to threefold higher than the frequency of reversion from streptomycin dependence to sensitivity. The one exception was revertant *sdR_s-u-3-18R1,* in which the secondary *sd* isolates produced secondary *ss* isolates at frequencies ranging from 1×10^{-2} revertants/dependent cell to 1×10^{-5} revertants/dependent cell. These authors also

reported that when secondary *sd* cells were isolated on low concentrations of streptomycin (25 μg/ml) and were maintained on this medium they retained the capacity to segregate a high frequency of secondary *ss* cells over a long period of time. Schimmer and Arnold (29) also observed that these secondary *ss* cells segregated streptomycin-dependent cells (tertiary *sd* cells) at frequencies much higher than the spontaneous *sd* mutation frequency. To summarize their results, Schimmer and Arnold (29) presented a flow sheet (Fig. 13–2) in which they suggest that the secondary *sd* cells arise by a process of segregation, but the secondary *ss* cells, at least when the *sd* cells are grown at a high streptomycin concentration, segregate *ss* cells very rapidly, after which *ss* cells arise only by reversion of the streptomycin-dependent cells.

Behn and Arnold (31) extended these studies on reversion of dependent mutants to a neamine-dependent mutant (*nd-u*), and their results were quite different. Neamine-independent revertants arose at a frequency of 7.5×10^{-6} revertants/dependent cell and fell into two classes. About 58% of the revertants were neamine-resistant, and the rest were neamine-sensitive. In the case of reversion from streptomycin dependence, all the revertants had been streptomycin-sensitive. The neamine-independent revertants also differed from the streptomycin-independent revertants in that they failed to segregate secondary neamine-dependent cells (secondary *nd* cells). This was true not only for the neamine-sensitive revertants but also for the resistant revertants, even when the latter were maintained on neamine.

Behn and Arnold (31) interpreted the results as indicating that the *sd-u-3-18* gene was in the mitochondrion and the *nd-u* gene was in the chloroplast. The rationale for this classification was based on Sager's model (see Chapter 15), in which it is assumed that the maternally inherited chloroplast genes we are discussing are present in only two copies per cell. That is, the 80-

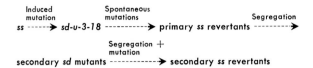

FIG. 13–2. Mutation and segregation mechanisms envisioned by Schimmer and Arnold (29) as being involved in forward mutation to streptomycin dependence (*sd*) and reverse mutation to the wild-type phenotype (*ss*).

fold-redundant fraction does not carry these genes. Schimmer and Arnold (27–29) had previously argued that the propensity of streptomycin-sensitive revertants to segregate secondary *sd* cells at a high frequency was compatible only with a multiple-copy model, and since, according to Sager, there were only two genetic copies of clDNA, the *sd-u-3-18* gene had to be in the mitochondrion. Since the neamine-independent revertants did not segregate secondary *nd* cells, Behn and Arnold (31) supposed that this gene was present in only a few copies per cell and thus was localized in the chloroplast genome.

Behn and Arnold (32) attempted to obtain further proof for their hypothesis by comparing the ultrastructures of wild-type cells grown in the presence or absence of neamine or streptomycin with the ultrastructures of cells of neamine- and streptomycin-dependent mutants grown under conditions of antibiotic deprivation. They hypothesized that in the two dependent mutants both the chloroplasts and the mitochondria had been affected, but one of the two organelles became dependent and the other became resistant. According to this hypothesis, the *sd-u-3-18* mutant had an antibiotic-dependent mitochondrion and an antibiotic-resistant chloroplast, whereas the reverse was true for the *nd-u* mutant. Reversion from streptomycin dependence to sensitivity affected only the mitochondrion, but reversion from neamine dependence to sensitivity altered only the chloroplast.

These ultrastructural studies indicated that treatment of wild-type cells with streptomycin or neamine led to decomposition of thylakoids and damage to the mitochon-

drion, in which atypical cristae were noted after a couple of days. When the neamine-dependent mutant was grown in the absence of neamine, chloroplast defects were seen within 2 days, and mitochondrial anomalies were observed only after 5 to 7 days. On the other hand, mitochondrial defects were seen in the streptomycin-dependent mutant 2 to 3 days after streptomycin withdrawal, but chloroplast defects were noted only after 4 days. When a neamine-resistant revertant was cultured in the presence of neamine, both the chloroplast and the mitochondrion appeared normal; so the authors concluded that both the chloroplast and the mitochondrion were resistant. That is, the mitochondrion was resistant to begin with, and the reversion caused the chloroplast to become resistant.

Sager (12) also discussed the effects of streptomycin withdrawal on an *sd-u* mutant. She reported that growth ceased after four to five doublings in streptomycin-free medium, following which a burst of mutants appeared, including apparent revertants to sensitivity as well as other types of mutations (e.g., streptomycin resistance, loss of photosynthetic ability, temperature sensitivity). Sager also reported that although all of these mutations showed maternal inheritance, two distinct transmission patterns were seen. One class showed the usual 4:0 maternal pattern of transmission, but the other segregated 2:2. For example, if the new mutant was phenotypically *sr mt⁺* and was crossed to a wild-type strain, the progeny segregated 2:2 *sr:sd,* a nonmendelian 2:2 ratio in which both markers come from the *mt⁺* parent and sensitivity is not transmitted by the *mt⁻* parent. Sager reported that the latter mutants represented a special class of heterozygotes called persistent hets. Normally, cells carrying chloroplast genes from each parent segregate these genes rapidly during vegetative cell division (see Chapter 14); however, in persistent hets the parental genotypes segregate during vegetative cell division at a low frequency (1% or less). Sager interpreted the results of Schimmer and Arnold on the persistent het model. She suggested that the segregation of secondary *sd* cells by sensitive revertants did not provide evidence for multiple copies; she believed that Schimmer and Arnold had been dealing with persistent hets that segregated *sd* cells at a low rate mitotically. Since the reversions studied by Schimmer and Arnold were primarily *mt⁻*, they did not see the 2:2 segregation of chloroplast markers in crosses that they would have seen had they used *mt⁺* reversions.

Gillham (33), on the other hand, interpreted the results of Schimmer and Arnold in terms of a multiple-copy model of the chloroplast genome (see Chapter 15). He argued that their results could be explained by assuming that the ratio of *sd* to *ss* alleles varied in different sensitive revertants such that they segregated secondary *sd* mutants at different rates. Similarly, secondary *sd* mutants would segregate secondary *ss* cells at distinctive rates for the same reason. Gillham believed that the Schimmer and Arnold experiments militated against Sager's two-copy model, not only for these reasons but also on the basis of dominance considerations. The *ss* revertants were heterozygous for dependence but phenotypically sensitive, whereas the secondary *sd* mutants were phenotypically dependent but heterozygous for sensitivity. Gillham interpreted this to mean that sensitivity was behaving as if it were dominant to dependence in the first case, whereas the converse situation applied in the second case. Such results would be paradoxical on a two-copy model, but they could easily be explained on a multiple-

copy model, since the phenotype expressed would depend on the ratio of *sd* to *ss* alleles.

It now seems likely that both Sager and Gillham were guilty of some misinterpretation of the Schimmer and Arnold results in terms of their own models. It appears that the revertant *sdRₛ-u-3-18R20* is different from the other revertants in its ability to continue segregating secondary *sd* mutants over a long period of time in the absence of streptomycin. Certainly *sdRₛ-u-3-18R20* would fit Sager's definition of a persistent het, but it is difficult to see how the other more rapidly segregating revertants could be accounted for in the same way. Segregation in these revertants more closely resembles the rapid segregation one normally sees for chloroplast genes among the vegetative progeny of zygotes in which biparental transmission of chloroplast markers has occurred. If we assume that *sdRₛ-u-3-18R20* is a persistent het, then sensitivity is clearly dominant to dependence. This may explain why secondary *sd* mutants generally segregate secondary *ss* cells at very low rates at high streptomycin concentrations. Gillham missed this point in making his dominance arguments. The very low segregation rate of secondary *ss* cells by secondary *sd* mutants at high streptomycin concentrations may, in fact, result because sensitivity is dominant to dependence. Cells in which both sensitivity and dependence become expressed by whatever mechanism may be strongly selected against in the presence of streptomycin. Since at low streptomycin concentrations secondary *sd* cells continue to segregate secondary *ss* cells for much longer, it would appear that sensitivity is not completely dominant to dependence.

The nonsegregation of secondary *nd* mutants by neamine-sensitive and neamine-resistant revertants observed by Behn and Arnold (31,32) is difficult to explain unless one assumes that the dominance of dependence over sensitivity or resistance is complete. If so, reversions in which dependence continues to be expressed in the

absence of antibiotic will be rapidly eliminated, and so the revertants will not appear
to segregate secondary *nd* mutants.

From the foregoing discussion it is obvious that the expression of new chloroplast
mutations, whether assayed in terms of forward or reverse mutation, is a complex
process that is poorly understood and that
has been interpreted in several very different
ways. Nevertheless, the process of expression does seem to be divisible roughly into
two parts. In the first part the mutant cell
usually is heterozygous for the nonmutant
allele, and so the degree of expression of
the new mutation is influenced by its dominance relationships to the nonmutant allele.
During the second part the mutant and nonmutant alleles segregate somatically, so that
cells homozygous for the new mutation begin to emerge. In these cells the phenotype
of the new mutation is fully expressed.

REFERENCES

1. von Euler, H. (1947): Einfluss des Streptomycins auf die chlorophyll Bildung. *Kem. Arb.
II,* 9:1–3.
2. Bracco, M., and von Euler, H. (1947):
Chloroplasten und Chloroplastin in blattzellen
Normaler und mit Streptomycin gekeimter
junger Pflanzen. *Kem. Arb. II,* 10,4pp.
3. Bracco, M., and von Euler, H. (1948): Ein-
wirkung von Streptomycin und anderen Gua-
nidinderiväten auf Nukleotide, Nukleinsäuren,
Nukleoproteide und Nukleoproteidkomplexe.
Kem. Arb. II, 10,4pp.
4. von Euler, H. (1948): Nukleinsäuren als
Wuchstoffe in Gegenwart von Colchicin und
von Streptomycin. *Ark. Kem. Min. Geol.,*
25A:1–9.
5. von Euler, H. (1950): Einfluss von Streptomycin auf samen Normaler und chlorophyll
defekter Gerste. *Z. Naturforsch.* [*B*], 5:448.
6. von Euler, H., Bracco, M., and Heller, L.
(1948): Les actions de la streptomycine sur
les graines en germination des plantes vertes
et sur les polynucléotides. *C. R. Acad. Sci.*
(*Paris*), 227:16–18.
7. Provasoli, L., Hutner, S. H., and Schatz, A.
(1948): Streptomycin-induced chlorophylless
races of *Euglena. Proc. Soc. Exp. Biol. Med.,*
69:279–282.
8. Provasoli, L., Hutner, S. H., and Pintner, I. J.
(1951): Destruction of chloroplasts by streptomycin. *Cold Spring Harbor Symp. Quant.
Biol.,* 16:113–120.
9. Sager, R. (1960): Genetic systems in *Chlamydomonas. Science,* 132:1459–1465.
10. Sager, R., and Tsubo, Y. (1962): Mutagenic
effects of streptomycin in *Chlamydomonas.
Arch. Microbiol.,* 42:159–175.
11. Sager, R. (1962): Streptomycin as a mutagen
for nonchromosomal genes. *Proc. Natl. Acad.
Sci. U.S.A.,* 48:2018–2026.
12. Sager, R. (1972): *Cytoplasmic Genes and
Organelles.* Academic Press, New York.
13. Winig, H., and Gillham, N. W. (1968): unpublished data.
14. Chiang, K.-S. (1971): Replication, transmission and recombination of cytoplasmic DNAs
in *Chlamydomonas reinhardtii.* In: *Autonomy
and Biogenesis of Mitochondria and Chloroplasts,* edited by N. K. Boardman, A. W.
Linnane, and R. M. Smillie, pp. 235–249.
North Holland, Amsterdam.
15. Conde, M. F., Boynton, J. E., Gillham, N. W.,
Harris, E. H., Tingle, C. L., and Wang, W. L.
(1975): Chloroplast genes in *Chlamydomonas* affecting organelle ribosomes. Genetic
and biochemical analysis of antibiotic-resistant mutants at several gene loci. *Mol.
Gen. Genet.,* 140:183–220.
16. Luria, S. E., and Delbrück, M. (1943): Mutations of bacteria from virus sensitivity to
virus resistance. *Genetics,* 28:491–511.
17. Newcombe, H. B. (1949): Origin of bacterial
variants. *Nature,* 164:150–151.
18. Gillham, N. W., and Levine, R. P. (1962):
Studies on the origin of streptomycin resistant mutants in *Chlamydomonas reinhardi.
Genetics,* 47:1463–1474.
19. Birky, C. W., Jr. (1973): On the origin of
mitochondrial mutants: Evidence for intracellular selection of mitochondria in the origin
of antibiotic-resistant cells in yeast. *Genetics,*
74:421–432.
20. Sager, R., and Ramanis, Z. (1963): The
particulate nature of nonchromosomal genes
in *Chlamydomonas. Proc. Natl. Acad. Sci.
U.S.A.,* 50:260–268.
21. Gillham, N. W. (1965): Induction of chromosomal and nonchromosomal mutations in
Chlamydomonas reinhardi with *N*-methyl-*N'*-
nitro-*N*-nitrosoguanidine. *Genetics,* 52:529–
537.
22. Lee, R. W., and Jones, R. F. (1973): Induction of Mendelian and non-Mendelian strep-
mycin resistant mutants during the synchronous cell cycle of *Chlamydomonas reinhardtii.
Mol. Gen. Genet.,* 121:99–108.
23. Gillham, N. W. (1963): Transmission and
segregation of a nonchromosomal factor controlling streptomycin resistance in diploid
Chlamydomonas. Nature, 200:294.
24. Gillham, N. W. (1969): Uniparental inheritance in *Chlamydomonas. Am. Naturalist,*
103:355–388.
25. Lee, R. W., Gillham, N. W., Van Winkle,
K. P., and Boynton, J. E. (1973): Preferential
recovery of uniparental streptomycin resistant

mutants from diploid *Chlamydomonas reinhardtii. Mol. Gen. Genet.,* 121:109–116.

26. Schimmer, O., and Arnold, C. G. (1969): Untersuchungen zur Lokalisation eines ausserkaryotischen Gens bei *Chlamydomonas reinhardi. Arch. Microbiol.,* 66:199–202.

27. Schimmer, O., and Arnold, C. G. (1970): Untersuchungen über Reversions und Segregations verhalten eines ausserkaryotischen Gens von *Chlamydomonas reinhardii* zur Bestimmung des Erbträgers. *Mol. Gen. Genet.,* 107:281–290.

28. Schimmer, O., and Arnold, C. G. (1970): Über die Zahl der Kopien eines ausserkaryotischen Gens bei *Chlamydomonas reinhardi. Mol. Gen. Genet.,* 107:366–371.

29. Schimmer, O., and Arnold, C. G. (1970): Hin- und Rücksegregation eines ausserkaryotischen Genes bei *Chlamydomonas reinhardi. Mol. Gen. Genet.,* 108:33–40.

30. Schimmer, O., and Arnold C. G. (1970): Die Suppression der ausserkaryotisch bedingten Streptomycin-abhängigkeit bei *Chlamydomonas reinhardii. Arch. Microbiol.,* 73:195–200.

31. Behn, W., and Arnold, C. G. (1972): Zur Lokalisation eines nichtmendelnden Gens von *Chlamydomonas reinhardii. Mol. Gen. Genet.,* 114:266–272.

32. Behn, W., and Arnold, C. G. (1973): Localization of extranuclear genes by investigations of the ultrastructure in *Chlamydomonas reinhardii. Arch. Microbiol.,* 92:85–90.

33. Gillham, N. W. (1974): Genetic analysis of the chloroplast and mitochondrial genomes. *Annu. Rev. Genet.,* 8:347–391.

Chapter 14

Chloroplast Genetics of *Chlamydomonas*. III. Segregation, Recombination, and Mapping

Under normal circumstances the chloroplast genes in *Chlamydomonas* are transmitted only by the maternal or mt^+ parent (see Chapter 12). Biparental zygotes in which chloroplast genes from both parents are transmitted are rare. Thus from the beginning in this field there was a methodologic impasse barring the way of those who might wish to analyze the segregation pattern of chloroplast genes in *Chlamydomonas*. In order to break this impasse, several different strategies were developed. Sager and Ramanis (1,2), realizing the importance of pedigree analysis as a tool for gaining an understanding of the segregation pattern of chloroplast genes, sought means for selective recovery of biparental zygotes. They made use of a chloroplast marker selection technique for the isolation of biparental zygotes in which maternal zygotes were selected against, whereas biparental and paternal zygotes were preferentially recovered. The frequency of biparental zygotes was high enough in Gillham's stocks (5 to 10%) to permit tetrad analysis under nonselective conditions, and some information on the pattern of segregation of chloroplast genes was obtained (3,4). Nevertheless, this was a laborious task, since 90% or more of the tetrads analyzed contained only maternal chloroplast genes. Not wishing to make use of marker selection techniques, Gillham (5,6) elected to analyze the progeny of zygote clones after many cell generations of growth under nonselective conditions. Biparental zygote clones could be detected readily by replica plating, and the ratios of different chloroplast genotypes in each clone could then be analyzed. Although this method did not provide information on segregation patterns, it made possible the analysis of large numbers of biparental zygote clones under conditions such that obvious selection for or against a given chloroplast marker was not a complicating factor. In addition, the method worked well for measuring the frequencies of recombination between different chloroplast markers. But, as we shall see, this method was sometimes subject to a different kind of selection problem in which cells of one chloroplast genotype grew faster than those of another within a zygote clone. In 1967 Sager and Ramanis (7) broke the methodologic impasse imposed by the rarity of spontaneous biparental zygotes when they showed that ultraviolet irradiation of the mt^+ parent prior to mating greatly increased the frequency of biparental zygotes.

As a consequence of the early divergences in methods and possible differences between stocks, Sager and Ramanis and Gillham and Boynton developed different methods for mapping chloroplast genes in *C. reinhardtii* and different models to account for their results. We shall begin this chapter with a general account of the methods used to analyze chloroplast gene segregation and recombination in *Chlamydomonas*. The early studies of these processes will be reviewed in an attempt to show how the different methods evolved and why different conclusions were drawn. Next we shall turn to a detailed discussion of the methods presently in use to map chloroplast genes and a comparison of the maps obtained by Sager and by Gillham and Boynton. Finally, the work of Van Winkle-Swift on chloroplast gene segregation and recombination in vegetative diploids will be discussed. The various models set forth to explain the results obtained will be developed only to the extent that is necessary to understand the experimental results. The major hypotheses that seek to explain the inheritance of chloroplast genes in *C. reinhardtii* will be considered and criticized in Chapter 15.

METHODS OF ANALYSIS

Somatic segregation is a characteristic of chloroplast genes in *Chlamydomonas,* as it is of mitochondrial genes in yeast (see Chapters 4 and 5) and nonmendelian genes generally (see Chapter 9). However, in yeast, segregation and recombination have been studied only in vegetative diploid progeny of crosses (see Chapter 7), whereas in *Chlamydomonas* these processes usually have been examined among the haploid meiotic and postmeiotic mitotic progeny of sexual zygotes. Van Winkle-Swift (8,9) recently made an extensive study of segregation and recombination of chloroplast genes among the progeny of vegetative diploids in *Chlamydomonas.* These results, which will be discussed at the end of this chapter, were obtained by examining the genotypes of diploid clones in which chloroplast markers from both parents had been transmitted (*biparental diploids*). This form of zygote clone analysis is methodologically identical to zygote clone analysis in yeast (see Chapter 7). In this section we shall describe each of the methods presently being used for the study of chloroplast gene segregation and recombination in meiotic zygotes of *Chlamydomonas.*

Octospore Daughter Analysis

By using zygotic progeny, Sager and Ramanis (10–13) were able to design the most powerful method of pedigree analysis presently available in organelle genetics. It is the only method other than tetrad analysis that yields direct information about the

segregation pattern of chloroplast genes in *Chlamydomonas,* and tetrad analysis is much more limited in its usefulness.

The octospore daughter method is designed to yield information on the pattern of segregation and recombination of chloroplast genes in pedigrees of individual biparental zygotes from the second meiotic division through the second postmeiotic mitotic division (Fig. 14–1). Germinating zygotes are transferred individually to petri dishes. After one postmeiotic mitotic division, the eight daughter cells are spread over the surface of the plate, and after one further doubling, each pair of octospore daughter cells is locally respread and allowed to form colonies. Each of the resulting 16 colonies is then tested for genotype with respect to both chloroplast and mendelian markers. Three pairs of unlinked mendelian genes are segregating in each cross, and from the mendelian genotype it is usually possible to determine from which meiotic product each pair of octospores was derived. This is true because recombination occurs quite frequently for one or more of the allelic pairs. For example, two of the marker pairs used are actidione resistance (*act-r*) and methionine sulfoximine resistance (*ms-r*). If one parent is actidione-resistant (*act-r ms-s*) and the other is methionine-sulfoximine-resistant (*act-s ms-r*), a tetratype tetrad will result for these marker pairs every time there is a single crossover between either marker pair and its centromere. The nuclear genotypes of the four meiotic products in such a tetratype will be *act-r ms-s, act-r ms-r, act-s ms-r,* and *act-s ms-s,* and the progeny derived from each can easily be distinguished by testing on the appropriate medium (Fig. 14–1). Thus the octospore pair derived from each meiotic product is established by the fact that each carries a similar pair of nuclear markers that differ from the marker combinations carried by the other pairs of octospores. The octospore daughters produced by each octo-

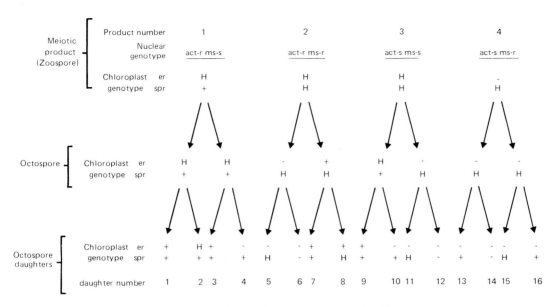

FIG. 14–1. Diagram illustrating octospore daughter analysis among progeny of a biparental zygote. The octospore to which each octospore daughter pair is assigned is readily determined because each cell pair is locally respread. Assignment of each octospore pair to a meiotic product is done by classifying all clones for segregation of a pair of mendelian markers for methionine sulfoximine resistance and sensitivity (ms-r and ms-s) and actidione resistance and sensitivity (act-r and act-s). The tetrad shown is a tetratype for these markers; so assignment of clones to products is unambiguous. If the tetrad had been a parental or nonparental ditype for these markers, other nuclear markers (i.e., mating type) would have been scored in the hope of obtaining a tetratype for one of these. Chloroplast markers for erythromycin resistance and sensitivity (er) and spectinomycin resistance and sensitivity (spr) are also segregated in the pedigree. Key: plus indicates cell homozygous for chloroplast allele from mt⁺ parent; minus indicates cell homozygous for chloroplast allele from mt⁻ parent; H indicates cell heterozygous for both chloroplast alleles. By working backward from the chloroplast genotypes of the octospore daughters, the genotypes of the octospores and meiotic products may be deduced.

spore are known because they have been separated manually. Therefore, for the chloroplast markers in the cross, a complete lineage can be constructed from the second meiotic division through the second postmeiotic mitotic division, and the pattern of segregation and recombination during that period can be established. The method involves working backward from the chloroplast genotypes of the octospore daughters, as shown in Fig. 14–1. Obviously the octospore daughter method of analysis produces a great deal of information when multiply marked stocks are employed and the *mt⁺* gametes are irradiated with ultraviolet light to increase greatly the frequency of biparental zygotes. Nevertheless, the octospore daughter method does have two disadvantages. First, chloroplast marker segregation often does not proceed

very far by the octospore daughter stage, and many cells are still heteroplasmic (*cytohets*) for the chloroplast markers they carry (11). This complicates the analysis of recombination by classic methods involving recombination frequencies. Second, the size of the zygote sample that can be easily handled by the octospore daughter method is considerably smaller than the sample size that can be scored by the zygote clone method.

Zygote Clone Analysis of Haploid Meiotic Progeny

Zygote clone analysis involves examination of chloroplast genotypic ratios in whole biparental zygote clones after many cell generations (9,14,15) (Fig. 14–2). Each parent carries different chloroplast markers

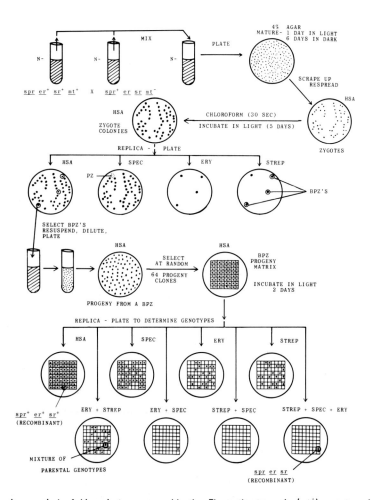

FIG. 14–2. Zygote clone analysis of chloroplast gene recombination. The mating type plus (mt^+) parent receives a low dose of ultraviolet irradiation prior to mating. Both stocks are allowed to undergo gametogenesis and mating in nitrogen-free (N—) medium. The mt^+ parent carries a chloroplast allele for spectinomycin resistance (spr) but is sensitive to erythromycin and streptomycin ($er^+ sr^+$). The mating type minus (mt^-) parent is sensitive to spectinomycin (spr^+) but carries chloroplast markers for erythromycin and streptomycin resistance ($er\ sr$). Media designations are as follows: HSA = nonselective medium; SPEC = spectinomycin-containing medium that selects against spr^+ cells; ERY = erythromycin-containing medium that selects against er^+ cells; STREP = streptomycin-containing medium that selects against sr^+ cells. The protocol was modified after this figure was drawn so that the zygote colonies can be picked into an 8 $\times$ 8 matrix before they are replica-plated to identify which are derived from maternal (MZ), biparental (BPZ), or paternal (PZ) zygotes. (From Conde et al., ref. 14, with permission.)

to antibiotic resistance. In the example given in Fig. 14–2, the mt^+ parent is resistant to spectinomycin (spr), and the mt^- parent carries chloroplast markers for erythromycin (er) and streptomycin (sr) resistance. The zygotes are germinated on nonselective medium and are allowed to form colonies, which are then transferred to a matrix on nonselective medium; following further growth of the zygote clones, this matrix is replica-plated to nonselective medium and to media containing, individually, the antibiotics to which both parents are resistant. Biparental zygote colonies are the only colonies containing cells capable of growth on all of the selective media. Colonies identified as arising from biparental zygotes are picked from the nonselective replica, diluted, plated on fresh nonselective medium, and allowed to form progeny clones. Sixty-four progeny clones from each biparental zygote are transferred as a grid to nonselective medium, incubated for a couple of days, and replicated to single and multiple antibiotic-containing media that permit identification of all genotypes present in the clone. The use of the multiple-antibiotic-containing media is of particular importance, for it permits one to distinguish between clones of antibiotic-resistant recombinants and clones derived from heteroplasmons containing the two parental genotypes. Recombinant clones will grow on multiple-antibiotic-containing media to which the two parents, individually, were resistant, but heteroplasmic clones containing only the two parental genotypes cannot.

Zygote clone analysis is a "black box" method that does not permit us to follow specific patterns of chloroplast marker segregation and recombination, only frequencies of different chloroplast genotypes among the progeny of each biparental zygote after many somatic cell generations. The parameters that can be assessed are the ratios of different pairs of alleles (*allelic ratios*) and the frequencies of different recombinant and parental genotypes among the progeny of biparental zygotes. Subtle "marker" effects in which selection for or against specific genotypes occurs during clonal growth are sometimes a problem. In addition, the method does not provide a check for aberrant nuclear events causing meiotic product abortion unless some tetrads are dissected in each cross.

The advantages of zygote clone analysis are the following: First, biparental zygotes are distinguished by replica plating, and their progeny are then analyzed. In octospore daughter analysis, zygote progeny must be spread out before it can be determined if they were derived from a biparental, a maternal, or a paternal zygote. Therefore, zygote clone analysis is feasible for populations in which biparental zygotes are rare (i.e., for spontaneous biparental zygotes), whereas octospore daughter analysis is not. Second, the frequencies of recombinant genotypes can be established at

a time when most chloroplast markers have segregated and heteroplasmons generally constitute less than 1% of the total. Third, zygote clone analysis is particularly well suited to the measurement of low frequencies of recombination, and it is convenient for allelism tests because large numbers of biparental zygotes can be processed efficiently.

Zygote Clone Analysis of Diploid Vegetative Progeny

Vegetative diploids (8,9) arise at a low frequency in *Chlamydomonas* crosses; they can be selected from the background of sexual zygotes by the use of mendelian auxotrophic mutations that are tightly linked but complementary (see Chapter 12). Biparental diploids occur at high frequencies spontaneously when specific pairs of nuclear markers, notably the arginine-requiring mutations on linkage group I (*arg-1, arg-2,* and *arg-7*), are used for their selection. The same methods of zygote clone analysis used for biparental zygotes are employed for biparental diploids.

Kinetic Studies of Chloroplast Marker Segregation in Liquid Cultures

Sager et al. (11,16,17) devised a method for measuring segregation rates of different chloroplast markers among the progeny of biparental zygotes. In its most recent form (17) this method involves washing the zoospores from a newly germinated zygote population into a liquid growth medium and then making platings every few hours. After several days the resulting colonies are replica-plated and scored for the chloroplast markers they carry. The percentage of hybrid cells for each marker pair is then determined at each generation (Fig. 14–3). Each marker pair has a characteristic segregation rate, and the markers can be ordered by this method (pp. 414–415).

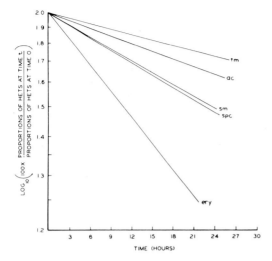

FIG. 14–3. Segregation rates of different chloroplast genes determined by liquid segregation analysis. The slope of each line is established by pooling data from a number of crosses and analyzing the data according to the statistical method described by Singer et al. (17). Key: *tm* = *tm-u-1*; *ac* = *ac-u-1* or *ac-u-2*; *sm* = *sr-u-sm2*; *spc* = *spr-u-sp23*; *ery* = *er-u-11*. (From Singer et al., ref. 17, with permission.)

EARLY STUDIES OF CHLOROPLAST GENE SEGREGATION AND RECOMBINATION

During the period from 1960 to 1970 two principal methods of studying chloroplast gene segregation and recombination in *Chlamydomonas* were evolved. Recombination was shown to be a property of these genes, and certain of them appeared to be linked. Sager and Ramanis found that ultraviolet irradiation of the mt^+ parent greatly increased the yield of biparental zygotes, and so they were able to make the first systematic study of chloroplast gene segregation. Their experiments led them to propose that the chloroplast genome was genetically diploid. One of the key bases on which the diploid model rested was the observation by Sager and Ramanis of a 1:1 allelic segregation ratio for chloroplast genes. Gillham, in his early studies, observed marked deviation from this ratio, which led him to doubt the diploid model. In this section we shall describe these experiments.

Segregation and Recombination of Chloroplast Genes among Progeny of Spontaneous Biparental Zygotes

The first two studies of chloroplast gene segregation by Gillham (3) and Sager and Ramanis (1) were carried out with spontaneous exceptional zygotes. They yielded contrasting results, which ultimately led to divergent views about the mechanism of inheritance of chloroplast genes in *Chlamydomonas*. Gillham (3) studied the segregation of a chloroplast mutation to streptomycin resistance (*sr-u-2-60*) in exceptional tetrads from crosses in which the mt^- parent carried the resistance mutation and the mt^+ parent was streptomycin-sensitive. He dissected 773 tetrads and obtained 25 that were exceptional (Table 14–1). Each meiotic product was classified to determine whether it segregated sensitive cells or resistant cells or a mixture of the two. Among the mixed clones, the proportion of resistant cells to sensitive cells was measured. Segregation of nuclear markers was normal in all exceptional tetrads. In addition, Gillham devised a system of half-tetrad

TABLE 14–1. Segregation of the *sr-u-2-60* mutant (*sr*) and its wild-type allele (+) in exceptional tetrads derived from two crosses of the type sr mt^+ × + mt^{-a}

Tetrad types	Tetrad	
	Number	Frequency
0sr:3+ : 1H	7	0.28
0sr:2+ : 2H	2	0.08
0sr:1+ : 3H	4	0.16
0sr:0+ : 4H	2	0.08
1sr:0+ : 3H	4	0.16
2sr:0+ : 2H	1	0.04
3sr:0+ : 1H	0	0.00
4sr:0+ : 0H	5	0.20
Total	25	

a In each cross a nuclear marker was scored, and in all exceptional tetrads segregation of the marker was normal. The abbreviation H means that the meiotic product was heterozygous (mixed) for resistance and sensitivity. The results of the two crosses are pooled.
Adapted from Gillham (3).

analysis that used undissected tetrads in which half of the meiotic products survived; the progeny of each could usually be distinguished by mating type. This permitted him to expand the same size of meiotic products analyzed for segregation of resistance and sensitivity.

Tetrad analysis (Table 14–1) revealed that most exceptional tetrads segregated resistant and sensitive cells (i.e., they were derived from biparental zygotes). Among these tetrads, meiotic products were found that segregated either resistant cells or sensitive cells or both. The most frequent tetrad class was one in which three meiotic products segregated only sensitive cells and one segregated a mixture of resistant and sensitive cells. Analysis of the total population of meiotic products, including both dissected and undissected tetrads, showed that about half the meiotic products were mixed and the rest were pure for either sensitivity or resistance (35 pure resistant, 55 pure sensitive, 97 mixed). Chi-square analysis of the results showed that the ratio of resistant to sensitive meiotic products was significantly different from 1:1; it fit a ratio of 2:1 in favor of sensitivity ($\chi^2 = 4.44$, $p \cong 0.25$). Analysis of the ratio of resistant cells to sensitive cells in the mixed clones showed that most clones contained great excesses of sensitive cells. A reconstruction experiment in which artificial colonies were synthesized by placing a resistant meiotic product next to a sensitive meiotic product and allowing them to form colonies suggested that the skew seen in mixed clones derived from biparental zygotes could not be accounted for by selection for sensitive cells. From these results Gillham concluded that meiotic segregation was a frequent event and that a skew existed such that pure sensitive cells were segregated more frequntly than resistant cells during both meiotic division and postmeiotic mitotic division.

Sager and Ramanis (1) chose to select biparental zygotes through the use of a chloroplast marker (sd-u-sm4). Reciprocal crosses were made between streptomycin-dependent and streptomycin-sensitive cells. When the mt^+ parent carried the sd marker, full germination occurred only in the presence of streptomycin. In the absence of the drug, about 0.1% of the zygotes germinated, and these were exceptional. In the reciprocal cross, high germination resulted only when streptomycin was absent from the medium. In the presence of streptomycin, about 0.1% of the zygotes germinated, and these were also exceptional. Thus Sager and Ramanis had an effective technique for the recovery of exceptional zygotes, but it had the potential drawback that selection against a specific chloroplast marker might influence the results.

Sager and Ramanis (1) proceeded to study the segregation of two pairs of chloroplast markers in reciprocal crosses using this selective technique. The alleles for streptomycin sensitivity and dependence were the selective pair, and a mutation to acetate dependence (ac^-) and its wild-type allele (ac^+) were the nonselective pair. Three pairs of nuclear markers were included for octospore daughter analysis. With respect to chloroplast markers and mating type, the crosses were as follows: Cross A: $sd\ ac^+\ mt^+ \times ss\ ac^-\ mt^-$. Cross B: $ss\ ac^-\ mt^+ \times sd\ ac^+\ mt^-$. In cross A, exceptional zygotes were selected on acetate medium lacking streptomycin that selected against maternal zygotes carrying the sd allele. No selection was occurring with respect to the acetate alleles. In the reciprocal cross (B), three sets of selective conditions were compared. Exceptional zygotes were selected on medium containing both streptomycin and acetate that selected only for the sd allele from the mt^- parent, on minimal medium that selected only for the ac^+ allele from the mt^- parent, and on medium containing streptomycin but not acetate that selected for both the ac^+ and sd markers from the mt^- parent.

These experiments led to several conclusions. First, the frequencies of exceptional zygotes in cross B were similar under all three sets of selective conditions (0.02–0.1%). This meant that the probability of transmission of the ac^+ and sd markers was not independent. Otherwise, the frequency of exceptional zygotes in which both markers were transmitted would have been the product of the frequencies of exceptional zygotes in which either one or the other marker was transmitted. Second, chloroplast marker segregation occurred during the postmeiotic mitotic divisions after the completion of meiosis, and the two pairs of chloroplast alleles appeared to segregate independently. By the second to fifth postmeiotic mitotic divisions a great deal of segregation had occurred for both pairs of markers. Third, among the segregated progeny of crosses A and B, the ratio $ac^+:ac^-$ was approximately 1:1. It was not possible to establish the segregation ratios of the sd and ss markers because no medium could be devised that supported growth of both genotypes.

In summary, there are two major discrepancies between the observations of Gillham and those of Sager and Ramanis. First, Gillham observed a high rate of meiotic segregation, whereas Sager and Ramanis did not. Second, Gillham observed a 2:1 allelic ratio among meiotic products pure for sensitivity or resistance in favor of the sensitive allele derived from the mt^+ parent. Among meiotic products producing both sensitive and resistant progeny, high proportions of sensitive cells were found in most. Sager and Ramanis found that the ac^+ and ac^- alleles segregated 1:1, and they did not report any evidence of a skew. It was the presence of a skew in Gillham's experiments and its absence in the experiments of Sager and Ramanis that ultimately led to different models for the inheritance of chloroplast genomes in *Chlamydomonas*.

Separate reports that chloroplast genes in *Chlamydomonas* were capable of recombination were published in 1965 by Sager and Ramanis (2) and Gillham (5). Sager and Ramanis selected biparental zygotes from the cross sd-u-$sm4$ ac-u-1 $mt^+ \times sr$-u-$sm2$ ac-u-2 mt^- by plating zygotes on acetate-containing medium that lacked streptomycin. For the sake of simplicity, we shall call this cross sd $ac1$ $mt^+ \times sr$ $ac2$ mt^-, as they did. Octospore daughter analysis was performed on the progeny of the biparental zygotes. Since these progeny included sd clones requiring streptomycin for growth and ss clones killed by the drugs, one set of zygote progeny was analyzed on streptomycin-containing medium and the other on medium lacking streptomycin. Once again Sager and Ramanis reported that chloroplast gene segregation was postmeiotic and that the ratio of $ac1$ to $ac2$ progeny was 1:1. The ratio of sd to sr progeny was not determined because of the selective plating conditions used. The meiotic products gave rise to four clear-cut phenotypic progeny classes: sd $ac1$, sd $ac2$, sr $ac1$, and sr $ac2$. In addition, acetate-independent (ac^+) and streptomycin-sensitive (ss) progeny were obtained. Besides these phenotypic classes, other phenotypes that were less well defined were obtained, including isolates with new levels of acetate requirement or streptomycin dependence and resistance. These results indicated that ac-u-1 and ac-u-2 might be recombining with each other and that the same was true of sd-u-$sm4$ and sr-u-$sm2$. Controls were performed to ensure that the ac^+ and ss progeny could not be explained by reversion of the parental markers, by a sorting-out of wild-type genes present but unexpressed in the parental cells, or by nuclear or cytoplasmic suppression of the mutant phenotype.

If the ac^+ progeny did arise as a result of recombination, it might be possible to identify the reciprocal recombinant (i.e., $ac1$ $ac2$) using appropriate techniques. Sager and Ramanis attempted to do this among cells apparently heterozygous for these markers. It was possible to distinguish heterozygotes by colony morphology on minimal medium from *ac1* and *ac2* homozygotes and by the fact that they segregated *ac⁺* progeny. Sager and Ramanis assumed that the double mutant would have the same colony phenotype as *ac2*, since the latter is more stringent than *ac1* (i.e., *ac2* forms small colonies on minimal medium, whereas *ac1* forms flat colonies). They picked small colonies arising from the presumptive heterozygotes and crossed 10 of these (each *mt⁻*) to an *sd ac⁺ mt⁺* stock; then they selected exceptional zygotes on streptomycin-free medium. Eight of the 10 colonies proved to be pure *ac2*, and they segregated only *ac2* and *ac⁺* progeny; but 2 colonies apparently represented the reciprocal recombinant, since they segregated *ac1* progeny at a frequency of 1 to 5% as well. This cross deserves special emphasis; as can be seen, it is analogous to a test cross for mendelian genes. In theory, the method should be applicable to any pair of chloroplast genes in which it is not clear from the phenotype what the genotype should be, provided that the phenotypes of the two single mutants are distinguishable.

Attempts were also made to identify the reciprocal of the *ss* recombinant, whose genotype would be *sd sr*. Clones in which *ss* recombinants had been detected were examined for *sd sr* double mutants. Since it was not known whether the reciprocal recombinant would be dependent or resistant, clones were selected that had an altered phenotype in that they had a lower level of resistance or dependence than the parental strains. These seemed likely candidates for double mutants. However, when crosses were made to sensitive cells, the low *sr* strains did not yield *sd* progeny, and the low *sd* strain tested did not yield *sr* progeny. Later Sager (11) reported that some *sr* cells segregated *sr* and *sd* cells in test crosses, but this was not true of *sd* cells. Therefore the phenotype of the *sd sr* double mutant appears to be *sr*. The nature of the low *sd* and low *sr* strains was not clarified, although it now appears that at

least the low *sr* strains probably consisted of persistent heterozygotes (pp. 420–421).

The experiments of Sager and Ramanis established that *ac1, ac2, sd,* and *sr* recombined in all possible combinations and that each mutation most likely defined a different gene. On the basis of these data and more recent data, we can rewrite this four-factor cross with respect to chloroplast genes and mating type as *sd-u-sm4 sr-u-sm2⁺ ac-u-1 ac-u-2⁺ mt⁺ × sd-u-sm4⁺ sr-u-sm2 ac-u-1⁺ ac-u-2⁺ mt⁻*. Viewed in this light, it appears that Sager and Ramanis extended their observation that a pair of alleles (*ac⁺* and *ac⁻*) segregates 1:1 among the progeny of biparental zygotes to mutant alleles in two different genes, *ac1* and *ac2*.

Gillham's experiments made use of chloroplast mutations to streptomycin resistance (*sr-u-2-60*) and neamine resistance (*nr-u-2-1*). He sought to determine if reciprocal recombinants could be obtained in coupling and repulsion crosses of these mutants and if the mutants behaved as if they were linked in both sets of crosses. Gillham employed zygote clone analysis of unselected zygotes rather than pedigree analysis of selected zygotes.

In the first cross the chloroplast markers were in repulsion (i.e., *nr-u-2-1 sr-u-2-60⁺ mt⁺ × nr-u-2-1⁺ sr-u-2-60 mt⁻*). Only three zygotes yielded recombinant progeny. Two of these produced *nr-u-2-1 sr-u-2-60* recombinants, and the other produced its wild-type reciprocal. In the coupling crosses (i.e., *nr-u-2-1⁺ sr-u-2-60⁺ mt⁺ × nr-u-2-1 sr-u-2-60 mt⁻*) a much higher proportion of biparental zygote clones contained recombinants, and the two reciprocal types were frequently found among the progeny of the same zygote. However, Gillham concluded that the same recombinational event probably did not produce reciprocal types because so many zygote clones contained only one of the two recombinants and because there was no correlation in the frequencies of reciprocal recombinant types in zygote clones yielding both.

Gillham also measured the frequencies of different genotypes in biparental zygote clones from both coupling and repulsion crosses. His results indicated linkage of the *nr-u-2-1* and *sr-u-2-60* markers, with linkage being much stronger in the repulsion cross. In addition, he found that there was a strong skew toward cells of the maternal parental genotype in the majority of zygote clones from both coupling and repulsion crosses. This appeared to confirm his earlier finding based on tetrad analysis of a one-factor cross (3) that chloroplast alleles did not segregate 1:1. These experiments extended his original observation to alleles at two different loci.

Gillham also did a series of single-factor control crosses to ensure that the recombinants he saw could not be accounted for by new mutations. For example, if *nr-u-2 sr-u-2-60* cells had appeared in the cross *nr-u-2-1⁺ sr-u-2-60⁺ mt⁺ × nr-u-2-1⁺ sr-u-2-60 mt⁻*, this would have indicated that the putative double-mutant recombinants really were not recombinants but that a new mutation to neamine resistance had arisen in an *sr-u-2-60* cell.

Thus by the end of 1965 at least six chloroplast genes had been identified, and the methods of octospore daughter analysis and zygote clone analysis had been established. Recombination was clearly a property of chloroplast genes, and linkage had been reported. However, there was a dispute over whether or not chloroplast genes segregated 1:1.

Gillham and Fifer (6) extended the analysis of chloroplast gene recombination to a three-factor cross that employed a mutation to spectinomycin resistance (*spr-u-1*) in addition to the neamine and streptomycin resistance markers. These experiments showed that the neamine and spectinomycin resistance markers did not recombine with each other, although both recombined with *sr-u-2-60*. Gillham and Fifer then made reciprocal crosses between *nr-u-2-1 spr-u-1⁺ sr-u-2-60* and *nr-u-2-1⁺ spr-u-1 sr-u-2-60⁺*

stocks and replica-plated the resulting zygote colonies to selective media; this permitted them to detect the numbers of exceptional zygote clones containing recombinants resistant to neamine and spectinomycin, resistant to streptomycin and spectinomycin, and resistant to all three antibiotics. Large numbers of exceptional zygotes were screened in this way, and as was to be expected from zygote clone analysis, many contained recombinants resistant to streptomycin and spectinomycin. However, very few clones contained progeny resistant to spectinomycin and neamine. These were checked to make sure that new mendelian mutations to neamine or spectinomycin resistance had not arisen, and in a number of cases new mendelian mutations to neamine resistance were found (18). These experiments indicated extremely tight linkage between the *nr-u-2-1* and *spr-u-1* markers.

Pattern and Rate of Segregation of Chloroplast Genes Following Treatment of *mt⁺* Gametes with Ultraviolet Light

A major breakthrough in the analysis of chloroplast gene segregation and recombination occurred when Sager and Ramanis (7) discovered that ultraviolet irradiation of *mt⁺* gametes prior to mating caused a dramatic increase in the exceptional zygote frequency (Fig. 14–4). The definitive experiments were done with a wild-type *mt⁺* stock that was crossed to an *mt⁻* stock carrying the markers *sr-u-sm2* and *nr-u-2-1*. The *mt⁺* gametes were subjected to various ultraviolet doses and then mated according to each of three protocols. In the first protocol (curve D), the *mt⁺* and *mt⁻* gametes were mated in the dark, zygote formation took place in the dark, and the zygotes were matured in the dark. In the second protocol (curve A), the matings were made in the dark, but the zygotes were plated in the light and kept there for 24 hr so that photoreactivation could occur. In the third proto-

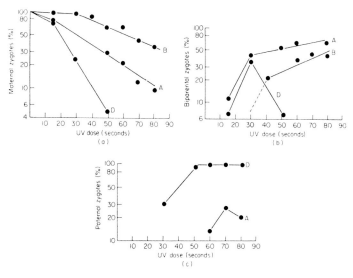

FIG. 14–4. Effectiveness of ultraviolet irradiation in converting maternal zygotes to exceptional zygotes. a: Conversion of maternal to exceptional zygotes as a function of irradiation dose. b: Yield of biparental zygotes as a function of irradiation dose. c: Yield of paternal zygotes as a function of irradiation dose. Maternal (mt^+) gametes were irradiated, mated with unirradiated paternal (mt^-) gametes, and kept in dark until zygote formation was completed. After 2 hr, zygotes were diluted and plated; plates were incubated in dark (D) or in light (A) for photoreactivation. In B, mating gametes were exposed to light during and after zygote formation. After 24 hr in light, A and B were incubated in dark with D series for 1 week; then all plates were exposed to light to induce germination of zygotes. (From Sager and Ramanis, ref. 7, with permission.)

col the ultraviolet-light-treated mt^+ gametes were exposed to light (curve B) during mating, and the zygotes were left in the light for 24 hr. At the end of 24 hr, zygotes from treatments A and B were placed in the dark alongside those from treatment D for maturation.

Sager and Ramanis found that the most rapid rate of conversion of maternal zygotes to exceptional zygotes occurred in the absence of photoreactivation (Fig. 14–4, curve D). The frequencies of both biparental and paternal zygotes increased rapidly (Fig. 14–4, curve D). After 30 sec of ultraviolet irradiation the yield of biparental zygotes declined precipitously, with paternal zygotes replacing them. With either photoreactivation regimen the kinetics of conversion of maternal zygotes to exceptional zygotes as a function of dose proceeded much more slowly (Fig. 14–4, curves A and B), but most of the exceptional zygotes were biparental zygotes rather than paternal zygotes, and the fraction of biparental zygotes remained con-

stant over a wide dosage range (Fig. 14–4, curves A and B).

Sager and Ramanis interpreted their results as indicating the involvement of two distinct processes in the production of biparental zygotes. In process 1 ultraviolet irradiation converted maternal zygotes to biparental zygotes. They argued that this process could be reversed by photoreactivating light more effectively before mating than afterward because the decline in maternal zygotes was less when photoreactivating light was applied during and after mating (Fig. 14–4, curve B) than it was when applied only after mating (Fig. 14–4, curve A). Process 1 was viewed by Sager and Ramanis as a process in which the production of a protein responsible for the elimination of paternal chloroplast genomes is blocked. In process 2 ultraviolet irradiation converted biparental to paternal zygotes. This process was viewed as resulting from the destruction of maternal chloroplast genomes. Both processes were photoreactivable.

The interpretation of ultraviolet effects on the production of exceptional zygotes is an integral part of all models for the inheritance of chloroplast genomes in *Chlamydomonas*. We shall return to this problem later (see Chapter 15); for the moment, our concern is with the use of the ultraviolet method to discern the pattern and rate of segregation of chloroplast genes.

The first definitive study of the pattern and rate of segregation of chloroplast genes in *Chlamydomonas* was published by Sager and Ramanis (19) in 1968. Ultraviolet irradiation of mt^+ gametes coupled with photoreactivation (see protocol A, p. 406) was used to increase the frequency of biparental zygotes. Marker segregation patterns were studied by the technique of octospore daughter analysis, and segregation rates were followed principally by the liquid segregation method (p. 401). The two crosses for which most of the data were obtained employed the *ac-u-1* and *ac-u-2* markers, and the majority of experimental results apply to these.

Segregation rate experiments in liquid revealed that hybrid cells segregated *ac-u-1* and *ac-u-2* progeny in a 1:1 ratio at a rate of 12% per doubling. Octospore daughter analysis demonstrated the existence of three patterns of segregation among the progeny. In type I segregations both daughter cells were hybrid. Type II segregations produced a hybrid daughter and a daughter homozygous for either *ac-u-1* or *ac-u-2*. Type III segregations produced two homozygous daughters, with one being homozygous for *ac-u-1* and the other for *ac-u-2*. The equivalence in frequencies in type II segregations accounted for the 1:1 ratio of cells homozygous for *ac-u-1* and *ac-u-2*. It will be noted that Sager and Ramanis were scoring the segregation patterns of two nonallelic mutants, and some acetate-independent progeny arose as recombinants. Presumably, this was also true of *ac-u-1 ac-u-2* double mutants, which would have been scored as *ac-u-2*. It is unlikely that the frequency of recombinants was great enough to influence the results.

Sager and Ramanis proposed that hybrid zoospores were diploid for those chloroplast genes they were scoring and that these were distributed to daughter cells by an oriented mechanism analogous to mitosis. This postulate was derived frim their observation of a 1:1 allelic ratio of *ac-u-1* to *ac-u-2* and the finding that segregation proceeded at a constant rate per generation. Type I segregations resulted when both daughter cells were diploid and therefore remained heterozygous. Type II segregations resulted when one daughter remained diploid and heterozygous while the other received only one genome and became haploid and thus homozygous for the chloroplast genome it carried. Type III segregation was the result of haploidization in both daughter cells, with one daughter becoming homozygous for one of the parental genomes and the other for the second. One piece of evidence that did not fit the model was that the marker segregation rate was lower than expected, but this problem can be ignored in light of more recent data, because each marker has a characteristic segregation rate that is predicted by the most recent form of the model (p. 409).

It soon became apparent that haploidization was not the explanation for the origin of type II and type III segregations. The problem arises because in multifactor crosses involving genetically linked chloroplast markers, one can observe all three segregation types in the same octospore daughter cell pair (12). A hypothetical example will illustrate the problem. Suppose a zoospore heterozygous for four markers (*a, b, c, d*) and their wild-type alleles produces two daughter cells with a type I segregation pattern for *a* and its wild-type allele, type II segregations for *b* and *c* and their wild-type alleles, and a type III segregation for *d* and its wild-type allele. The haploidization model would necessitate that the two octospore daughters be diploid for *a* and its wild-type allele, that one daughter be haploid for markers *b* and *c* and the other diploid, and that both daughters be haploid with respect to marker *d* and its wild-type allele. The fact that the chloroplast markers are linked would, on the haploidization model, require alternating regions of haploidy and diploidy in the chloroplast genome. As a result, Sager and Ramanis (20) proposed a different model based on diploidy for the chloroplast genome at all times.

CURRENT MODELS OF CHLOROPLAST GENE INHERITANCE

A short exposition of the two models currently being used to explain chloroplast gene segregation and recombination is essential at this point in order to facilitate an understanding of the development of chloroplast gene mapping methods. The two models are completely developed and criti-

cized in Chapter 15. According to Sager's diploid model, recombination takes place at the four-strand stage (Fig. 14–5). Type I segregation for a pair of alleles (e.g., $+/b$) results when there is no recombination for this allelic pair. Since segregation of chloroplast genomes is postulated to occur by a mechanism akin to the segregation of mitotic chromosomes, the two daughter cells will remain heterozygous for the allelic pair (Fig. 14–5). Type II segregations are caused by a nonreciprocal exchange akin to gene conversion. One allele of the pair (e.g., $+$) is converted to the other allele (e.g., b). The result is a daughter cell hybrid for the marker pair (e.g., $+/b$) and a daughter cell homozygous for the converted allele (e.g., b/b) (Fig. 14–5). Type II segregations arise with equal frequencies for alleles of different genes. Originally it was thought that type II events occurred with equal probability for each member of an allelic pair, but deviations from this ratio (which may be allele-specific) were evident

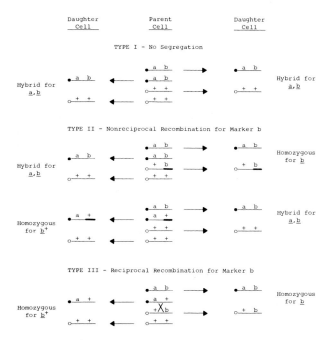

FIG. 14–5. Origin of patterns of chloroplast gene segregation according to Sager's two-copy model. Recombination is depicted as taking place after the two copies have replicated (i.e., at the four-strand stage). The chloroplast genomes are shown as being linear for the sake of clarity, although Sager and Ramanis believed them to be circular. (From Gillham et al., ref. 21, with permission.)

in more recent experiments (see Chapter 15). Type III segregations result from reciprocal recombination and genome segregation by the mitotic mechanism (Fig. 14–5). Genes farthest from a hypothetical attachment point akin to a mitotic centromere have the highest type III segregation frequency, and those close to the attachment point have the lowest. That is, there is polarity of type III segregations with respect to the attachment point (*ap*).

In contrast to the diploid model, Gillham and Boynton (21) proposed a multiple-copy model in which the input of chloroplast genomes to the zygote is variable. Maternal and paternal chloroplast genomes were likened to incompatible bacterial plasmids. It was suggested that the maternal genomes had a preferential affinity for a limited number of attachment sites in the zygote required for replication and segregation of chloroplast genomes according to the oriented mechanism proposed by Sager. The departure from a 1:1 allelic ratio in favor of maternal chloroplast alleles that was consistently seen by Gillham and Boynton among the progeny of spontaneous biparental zygotes was explained in terms of unequal inputs of chloroplast genomes to those zygotes. In other words, if the level of maternal genomes passed below some threshold, paternal genomes could become attached to vacant attachment sites and be replicated. Ultraviolet irradiation of the mt^+ parent inactivated maternal genomes. At low dosages of ultraviolet irradiation the input still favored maternal genomes, but as the dosage of ultraviolet irradiation increased, more and more maternal genomes became inactivated; so a higher fraction of paternal genomes was transmitted, and the allelic ratio approximated 1:1. This model has many similarities to the phage analogy model proposed by Dujon and associates (see Chapter 7) for the transmission of mitochondrial genomes in yeast. An implicit assumption, obviously, is that the output, as measured

by the allelic ratio, reflects the relative input of chloroplast genomes. Coordinate plots of allelic ratios for different pairs of chloroplast alleles showed that the ratios were strongly correlated (22). This suggested that the behavior of whole genomes was being studied. Gillham and Boynton assumed that chloroplast gene segregation took place by the mechanisms proposed by Sager, except that type II segregants arose by reciprocal recombination in cells having multiple copies of the chloroplast genome, so that a skewed allelic ratio could result. Fortunately, the postulated reciprocal recombination is in no way central to the model, as it is almost certainly wrong (see Chapter 15). Nonreciprocal recombination in cells containing multiple copies of the chloroplast genome would work just as well. Type III segregants were assumed to arise in cells with only two copies of the chloroplast genome in which reciprocal recombination had occurred. In any event, for mapping purposes segregation patterns were not taken into account, since they could not be discerned in zygote clones anyway.

MAPPING CHLOROPLAST GENES IN PEDIGREES

Sager and Ramanis (10) published the first map of chloroplast genes in *Chlamydomonas* in 1970. Between 1970 and 1974 this work was reviewed several times by Sager (20,23–25). The mapping methods then in use were in a state of evolution. The original article and the mapping methods used by Sager were subjected to detailed criticism by Adams and associates (22). In 1976 three definitive articles by Sager and Ramanis (12,17,26) appeared in which two of the mapping methods were chosen. One of these methods, cosegregation analysis, is done on pedigrees. The second method, liquid segregation analysis, is done on zygote populations (p. 401); the results obtained with this method will be considered in the next section. The earlier pedigree mapping methods will be outlined first, together with the results they yielded, since these experiments are important conceptually in understanding how the present circular map of Sager and Ramanis was derived. For detailed analysis of these mapping procedures, the reader is referred to Adams et al. (22) and the recent review by Sager (27).

Pedigree analysis allows cells to be classified as homozygous parental (P) or recombinant (R) for a given pair of alleles or as heterozygous (H) for one or both markers. Sager and Ramanis (23) considered two methods of measuring recombination frequencies. In the R_A method, heterozygotes were ignored: $R_A = [R/(P + R)] \times 100$. In the R_B method, heterozygotes were included in the denominator: $R_B = [R/(P + R + H)] \times 100$. The recombination map published by Sager and Ramanis in 1970 (10) employed the R_B method. After two mitotic doublings of each

zoospore, the percentage recombination was calculated (10). The principal problem with this method was the high frequency of heterozygotes still present at the octospore daughter stage (22,23,26). Although this should not affect marker order, it will affect map distances, since the H class is treated as if it contributes preferentially to the P class. In other words, map distances will decline as a function of the frequency of heterozygotes for a given pair of markers in a linear fashion, as was shown by Adams et al. (22).

In Fig. 14–6 the map obtained by Sager and Ramanis using the R_B method is compared for common markers with the map obtained by Gillham and Boynton using the R_A method with zygote clones. Both maps yield the same order. Distances are greater in the Gillham and Boynton map, as would be expected, since there are very few remaining heterozygotes in the zygote clones as compared to the pedigrees. Special note

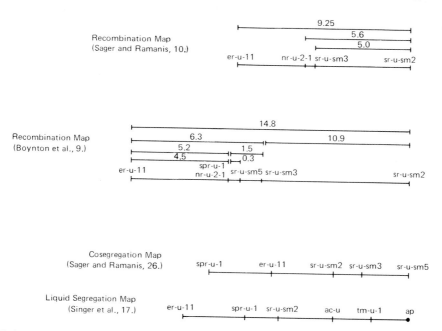

FIG. 14–6. Comparison of recombination and cosegregation maps for chloroplast genes in *Chlamydomonas*. The recombination and cosegregation maps of Sager and Ramanis (10,26) are derived from pedigree data, and the recombination map of Boynton et al. (9) is derived from zygote clone analysis. With the exception of the markers shown in the liquid segregation map, only markers common to at least two of the maps are indicated. Relative spacings of the markers on the cosegregation and liquid segregation maps are the same as in the original articles (ref. 17, p. 348; ref. 26, Fig. 1c), where distances were not given.

TABLE 14–2. *Marker orders obtained in different crosses on basis of polarity of type III segregations*

Crosses giving order[a]	Number of first and second doublings of zoospores	Percentage type III segregations			Reference
		ac	sr	er	
ac sr er	—	10	22.3	30	10
	300	5.7	15.0	20.3	26
	281	6.4	9.6	17.4	26
	277	4.3	4.7	13.7	26
ac er sr	—	4.65	10.7	7.9	10
	262	4.6	8.8	8.0	26
	119	7.6	20.2	18.5	26
sr ac er	191	7.9	4.7	12.6	26
	124	10.5	4.8	12.1	26
er sr ac	71	6.9	5.6	4.2	26
	361	10.1	9.3	7.4	26

[a] ac = ac-u-1 or ac-u-2, which Sager and Ramanis (26) treated as a single gene for type III segregation analysis because of their close linkage; sr = sr-u-sm2; er = er-u-11. Many of the crosses contain markers other than these, but these are the most frequently used.

should be made of the map positions of the *spr-u-1* and *nr-u-2-1* mutations. A map published by Sager (11, Fig. 3–23) using a composite of mapping methods indicated that these mutations were not closely linked, but in zygote clone analysis these mutations behave as if they are extremely tightly linked and probably allelic (p. 418). The *nr-u-2-1* marker is not shown in the more recent maps made by Sager and Ramanis (26).

The second method used by Sager and Ramanis (10) in their original article took advantage of the polarity of type III segregations. Although polarized marker segregation occurred, marker order reversals were apparent in several crosses (22). Recently, more extensive data have been published by Sager and Ramanis (26), and it is apparent that different crosses yield different orders based on type III segregations (Table 14–2). In view of this, the gene order originally published by Sager and Ramanis (10) on the basis of type III segregations should not be taken too seriously, even though it agrees with the map made on the basis of recombination frequencies. The apparent agreement of map distances is

TABLE 14–3. *Cosegregation frequencies of chloroplast genes in* Chlamydomonas *for pairwise gene combinations computed as number of cosegregation events per total first and second doublings*

Gene pair	Number of crosses	Type II segregations only	Type II, type III, and zygote segregations
ac-u sr-u-sm2	8	7.36 ± 0.558	9.45 ± 0.718
ac-u er-u-11	10	4.64 ± 0.510	6.62 ± 0.576
ac-u spr-u-sp23	12	3.55 ± 0.337	5.09 ± 0.401
ac-u tm-u-1	12	5.98 ± 0.678	7.27 ± 0.721
sr-u-sm2 er-u-11	6	16.14 ± 1.648	26.44 ± 3.564
sr-u-sm2 spr-u-sp23	8	9.69 ± 2.058	18.36 ± 3.516
sr-u-sm2 tm-u-1	8	6.42 ± 1.082	8.79 ± 0.921
er-u-11 spr-u-sp23	10	15.11 ± 2.644	28.68 ± 4.183
er-u-11 tm-u-1	10	7.06 ± 1.258	10.91 ± 1.472
spr-u-sp23 tm-u-1	12	12.24 ± 1.179	18.18 ± 1.870

From Sager and Ramais (26), with permission.

probably spurious as well, since two distinct normalizations were applied to the type III segregation data (22). In fact, the order of the *er-u-11* and *sr-u-sm2* mutations has since been reversed on the basis of newer data (17,26).

By definition, type III segregations are the only events that can affect the rate of marker segregation, since type II segregation frequencies are virtually the same for all genes (26). Thus it is the polarity of type III segregations that underlies the liquid segregation method. When the results from a large number of crosses are pooled and the appropriate statistical analysis, is applied, it is possible to establish a characteristic slope for the rate of segregation of each chloroplast marker. In short, there are sampling problems that arise in pedigrees because type III segregants are rare and because it is difficult to obtain satisfactory numbers for good statistics, but these probably do not apply in liquid segregation analysis.

The pedigree method of mapping presently used by Sager and Ramanis (26,27) is called cosegregation analysis. This method was not used in the original 1970 article. As first defined (23), the cosegregation frequency (R_C) was the frequency with which two genes became homozygous at the same doubling. Obviously, two genes can become homozygous in either the recombinant or the parental configuration. More recently, Sager (26,27) defined cosegregation as the frequency with which two or more genes segregate simultaneously in parental configuration at the same doubling. However, cosegregation frequencies are actually computed as the number of cosegregation events per total number of doublings scored. Whichever of these definitions is used, the important point is that genes that cosegregate with high frequencies are assumed to be closely linked, whereas those farther apart are assumed to cosegregate with lower frequencies. The underlying rationale for this assumption is that if type II

events are the result of coconversion, then closely linked markers would be expected to be coconverted at higher frequencies than distantly linked markers. In a formal sense, this rationale is similar to that used for coretention mapping of mitochondrial genes using petite mutants (see Chapter 7).

Sager and Ramanis (26) computed cosegregation frequencies using type II events alone and type II, type III, and zygotic events (Table 14–3). The same linear map was obtained in both cases (Fig. 14–7); however, the number of cosegregation

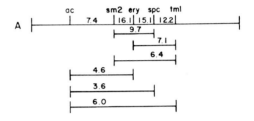

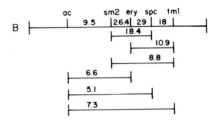

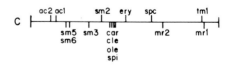

FIG. 14–7. Cosegregation maps of chloroplast markers (see Table 14–3 for pertinent data). A: Arrangement of principal markers employed in most crosses from frequencies of type II cosegregation events. B: Arrangement of principal markers employed in most crosses from frequencies of type II, type III, and zygotic cosegregation events. C: Relative positions of 15 chloroplast genes. Principal markers are above line; "minor" markers present in only one or two crosses are below line. Key: ac = either ac-u-1 or ac-u-2; ac1 = ac-u-1; ac2 = ac-u-2; sm2 = sr-u-sm2; ery = er-u-11; spc = spr-u-sp-23; tm1 = tm-u-1; sm5 = sr-u-sm5; sm6 = sr-u-sm6; sm3 = sr-u-sm3; car = car-u-1; cle = clr-u-1; ole = olr-u; spi = spir-u; mr1 = mr-u-1; mr2 = mr-u-2. (From Sager and Ramanis, ref. 26, with permission.)

events was much larger in the second case, as would be expected, since more events are included in the numerator. For the moment it is not clear how these numbers should be related to map distances measured by recombination, except in a relative sense. The order obtained for cosegregation of gene pairs was generally supported by longer runs of three to five genes.

Although the map order obtained by recombination analysis by Gillham and Boynton is in good agreement with the recombination map of Sager and Ramanis, comparison of these maps with the cosegregation map reveals important differences in marker order (Fig. 14–6). With respect to the antibiotic resistance markers common to both maps, the order is *er-u-11-spr-u-1-sr-u-sm5-sr-u-sm3-sr-u-sm2* in the recombination map of Gillham and Boynton and *spr-u-1-er-u-11-sr-u-sm2-sr-u-sm3-sr-u-sm5* in the cosegregation map of Sager and Ramanis. As we shall see (p. 418), these differences in order bear on the question of whether the map of the chloroplast genome is really linear or circular.

LIQUID SEGREGATION ANALYSIS

Singer et al. (17) examined the segregation of five chloroplast genes by the liquid segregation technique. Several different experiments were done. They then developed a statistical method that allowed them to pool the data for each marker to yield a summary curve characteristic for the segregation of that marker (Fig. 14–3). With respect to the hypothetical attachment point *ap,* this yielded the order *tm-u-1-ac-u-sr-u-sm2-spr-u-sp23-er-u-11,* with *tm-u-1* having the slowest segregation rate and *er-u-11* the highest. In Fig. 14–6 the order of markers in this map is compared to the order seen by recombination analysis and cosegregation analysis. By comparing the maps obtained from cosegregation and liquid segregation analysis, Singer et al. (17) argued that the map must really be circular (Fig. 14–8). The principal points of evidence are as follows: First, *tm-u-1* and *ac-u* appear close together on the basis of liquid segregation analysis; yet they are the markers that are farthest apart on the cosegregation map. If cosegregation does not take place across *ap,* then by circularizing the map these observations can be accounted for. Second, cosegregation analysis yields the map order *spr-u-sp23-er-u-11-sr-u-sm2,* but *spr-u-sp23* and *sr-u-sm2* segregate at slower rates in liquid than *er-u-11,* which has the highest liquid segregation rate of any marker. The two orders can be made coherent by placing

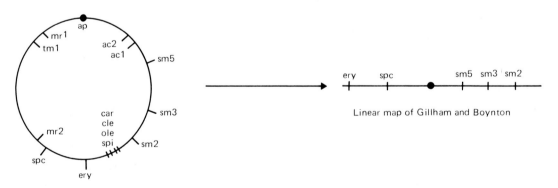

Circular map of Sager and Ramanis

Linear map of Gillham and Boynton

FIG. 14–8. Circular map of chloroplast genome deduced by Sager and Ramanis from cosegregation and liquid segregation mapping compared to linear map of Gillham and Boynton deduced from recombination analysis. Gene symbols are those used by Sager and Ramanis; they are explained in the legend to Fig. 14–7, except for *ap,* which indicates attachment point. Markers common to the two maps are shown on the outside of the circle; those mapped only by Sager and Ramanis are shown on the inside. The five markers that Sager and Ramanis used to establish circularity were *ac, sm2, ery, spc,* and *tm1.* Neither *ac* nor *tm1* was mapped by recombination by Gillham and Boynton.

er-u-11 farthest from *ap* on a circular map and *spr-u-sp23* and *sr-u-sm2* on either side (Fig. 14–8). However, the map obtained by recombination analysis in zygote clones can be made coherent with this map only if the circular map is cut between *er-u-11* and *sr-u-sm2* (Fig. 14–8), as will be discussed later (p. 418).

ANALYSIS OF CHLOROPLAST GENE RECOMBINATION IN ZYGOTE CLONES

Measurement of recombination frequencies in zygote clones (p. 398) has been the sole method used to map chloroplast genes in the laboratory of Gillham and Boynton (9,14,15). Thus far, the crosses carried out have involved only those genes conferring antibiotic resistance on chloroplast genes. The aim of these experiments has been to make a map of these genes based on recombination frequencies and to establish the number of chloroplast genes that can be mutated to a given phenotype (e.g., streptomycin resistance) through allelism tests. As in the case of Sager's experiments, the maternal parent is exposed to a low dose of ultraviolet irradiation prior to mating to increase biparental zygote frequencies. Biparental zygote sample sizes of approximately 100 are generally used in the mapping crosses and in crosses between mutants that appear to be allelic. The data from all biparental zygotes are pooled in each cross, and recombination frequencies are computed by the R_A method, where $R_A = [R/(P + R)] \times 100$. Heterozygotes are ignored, since they form a negligible part ($<1\%$) of the total progeny. Allelism between mutants of similar phenotype is judged solely from the number of antibiotic-sensitive recombinant progeny, since the doubly resistant recombinants cannot be distinguished from either parent. To compute recombination frequencies between such mutants, the formulation used is $R_A = [(2 \times$ antibiotic-sensitive recombinants$)/$ total$] \times 100$. This assumes that the num-

bers of recombinational events yielding resistant and sensitive recombinants are equal. Mutants having phenotypes different from those being tested for allelism are always included in allele tests. These markers serve not only to identify biparental zygotes (p. 400) but also to establish that recombination is proceeding normally. The problem can be illustrated by the following example. Suppose two mutations to streptomycin resistance are subjected to allele testing and no sensitive progeny are found. Two interpretations can be entertained. First, the mutations are alleles. Second, there was failure of chloroplast markers to recombine in the cross. If the other markers included in the cross recombine normally, the latter interpretation can be excluded. Thus far, this has always been the case.

Allelism tests (9,14,15) have established the existence of four loci that can be mutated to streptomycin resistance (*sr-u-sm2*, *sr-u-sm3*, *sr-u-2-60*, and *sr-u-sm3a*). Some of the data are summarized in Table 14–4. These data establish that there are four groups of recombining streptomycin-resistant mutants. Mutants belonging to any one group do not recombine with each other (Table 14–5). Similar allele tests for erythromycin-resistant mutants indicate there are two loci that can be mutated to this phenotype (*er-u-1a* and *er-u-37*). Exhaustive allele tests summarized in great detail by Harris et al. (15) show that spectinomycin-resistant (*spr-u-1*) and neamine-resistant (*nr-u-2-1*) mutants behave as if they are very tightly linked and quite possibly mutations in the same gene. Included in the allele tests are several of the mutants (*spr-u-sp23*, *er-u-11*, *sr-u-sm2*, *sr-u-sm3*, and *sr-u-sm5*) used by Sager in her mapping experiments, as well as her isolate of *nr-u-2-1*, which was supplied originally by Gillham. The significance of this will become apparent later when we consider mapping anomalies between the two laboratories. The erythromycin-resistant mutant (*er-u-1a*) isolated by Mets and Bogorad (28) has also been allele tested and *er-u-11*

TABLE 14–4. *Allele tests of streptomycin-resistant chloroplast mutants yielding evidence for four streptomycin resistance (sr) loci in chloroplast genome of C. reinhardtii[a]*

	sr-u-sm2	sr-u-sm3	sr-u-2-60	sr-u-sm3a
sr-u-sm2	0/1,649	526/3,623	612/8,558	349/3,028
sr-u-sm3		0/486	78/7,646	202/7,890
sr-u-2-60			0/3,182	16/4,250
sr-u-sm-3a				0/2,884

[a] Number of streptomycin-sensitive progeny/total progeny scored is indicated for each cross.

From Harris et al. (15), with permission.

has been found to be an allele of this mutant.

To establish an order for these seven genes, a series of mapping crosses was performed (9,15) in which representatives of each of the four streptomycin resistance loci in the *mt⁺* parent were crossed to a stock of the genotype *er-u-37 nr-u-2-1 spr-u-1-H4* and one of the genotype *er-u-1a nr-u-2-1 spr-u-1-H4*. The results of these crosses are summarized in Table 14–6 and Fig. 14–9. The seven loci can be ordered unambiguously with respect to one another despite the fact that marker effects are evident that lead to differential growth of specific genotypes, notably the reciprocally recombinant wild-type genotype and the quadruply antibiotic-resistant genotype (Table 14–6). The reason is that the crosses were always made in a precisely symmetrical fashion to two different testers; so the bias seen against specific recombinant classes is also symmetrical from cross to cross (Table 14–6). Therefore it seems unlikely that the bias seen between classes of reciprocal recombinants should affect marker ordering, since bias is in the same direction in every cross. Nevertheless, for all crosses marker order has also been established using the rarest-class method. In this method one assumes the correct gene order to be the order that generates the least frequent class of recombinants by two exchange events and the other two more frequent recombinant classes by single exchange events. Using this method, the gene order is always erythromycin (*er*), neamine (*nr*), streptomycin (*sr*), even though mutations to streptomycin resistance at all four loci are being tested (Table 14–7). Finally, all possible reciprocal crosses have been made for a specific set of alleles at the *er-u-37, spr-u-1,* and *sr-u-2-60* loci, and all crosses have yielded the map order *er-spr-sr* (Table

TABLE 14–5. *Allele tests of streptomycin-resistant mutants mapping in the sr-u-sm3a locus[a]*

	sr-u-sm-3a	sr-u-2-23	sr-u-2-NG-1	sr-u-sm5	sr-u-2-60
sr-u-sm-3a	0/2,884	0/3,174	0/3,163	0/6,797	16/4,250
sr-u-2-23		1[b]/2,928	0/3,127	0/1,277	12/3,129
sr-u-2-NG-1			0/3,175	0/636	37/3,437
sr-u-sm5				—	9/6,326[c]
sr-u-2-60					0/3,182

[a] Recombination frequencies of each mutant with *sr-u-2-60* that maps in a different locus are shown for comparison. Number of streptomycin-sensitive progeny/total progeny is indicated for each cross.

[b] Possibly a wild-type contaminant or a back-mutation.

[c] This value is corrected from Table 5 in Boynton et al. (9).

From Harris et al. (15), with permission.

TABLE 14–6. Results of mapping crosses between mutants at seven different chloroplast loci using the zygote clone method[a]

Cross: mt[+b] × mt[−]	Number of zygotes	Number of progeny	Percentage biparental zygotes	Parental		Recombinant									Mixed	Average Paternal allelic ratio[c]	Percentage recombination interval		
				S + + +	+ E N Sp	S E + +	+ + N Sp	+ E + Sp	S + N +	+ E N +	S + + Sp	S E N +	S + N Sp	+ E + +			E-N/Sp	S-N/Sp	S-E
1 sr-u-sm-3a (GB-99) × er-u-37 nr spr (GB-137)	241	15,224	25	9,596	5,078	251	296	1	2	0	0	0	0	0	190	0.35	3.87	0.02	3.59
2 sr-u-2-60 (GB-118) × er-u-37 nr spr (GB-137)	100	6,366	36	5,439	755	64	76	9	0	21	2	2	0	0	33	0.13	2.34	0.50	2.56
3 sr-u-2-23 (GB-181) × er-u-37 nr spr (GB-137)	100	6,359	19	4,546	1,652	54	97	0	0	10	0	0	0	0	14	0.27	2.37	0.16	2.53
4 sr-u-sm3 (GB-266) × er-u-37 nr spr (GB-137)	97	6,057	68	2,891	2,904	40	115	24	9	62	0	12	0	0	148	0.50	3.10	1.77	3.78
5 sr-u-sm2 (GB-270) × er-u-37 nr spr (GB-137)	100	6,357	23	3,257	2,208	72	84	85	1	649	0	1	0	0	40	0.40	3.81	11.6	12.7
6 sr-u-sm-3a (GB-99) × er-u-la nr spr (GB-295)	106	6,384	58	4,063	2,017	88	197	2	3	17	0	1	1	0	79	0.34	4.50	0.30	4.73
7 sr-u-2-60 (GB-118) × er-u-la nr spr (GB-295)	99	5,921	54	4,056	1,505	137	158	14	3	26	1	1	0	0	41	0.28	5.61	0.74	5.78
8 sr-u-2-23 (GB-181) × er-u-la nr spr (GB-295)	100	6,186	60	3,213	2,654	100	217	0	0	2	0	0	0	0	60	0.46	5.12	0.032	5.16
9 sr-u-sm3 (GB-266) × er-u-la nr spr (GB-295)	100	6,237	44	3,707	2,128	36	275	7	3	74	0	7	0	0	98	0.38	5.15	1.46	6.28
10 sr-u-sm2 (GB-270) × er-u-la nr spr (GB-295)	59	3,746	7	2,089	1,058	82	108	46	0	360	0	1	0	1	14	0.35	6.30	10.8	14.7
1 + 3 (pooled data for allelic mutants)	341	21,583		14,142	6,730	305	393	1	2	10	0	0	0	0		0.33	3.25	0.06	3.28

[a] Possible recombinant classes where no progeny were observed are not listed. The designation GB followed a number next to each genotype is the stock number in the collection of Gillham and Boynton.
Key: spr = spr-u-1-H-4, nr = nr-u-2-1, S = streptomycin-resistant, Sp = spectinomycin-resistant, N = neamine-resistant, E = erythromycin-resistant, + = sensitive to a given antibiotic.
[b] Ultraviolet irradiation for 15 sec.
[c] Defined on page 433.
From Harris et al. (15), with permission.

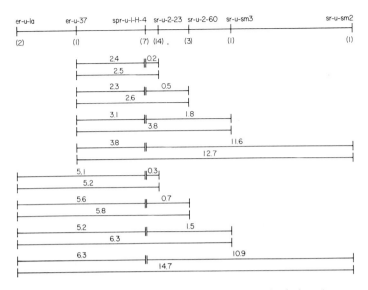

FIG. 14-9. Map of chloroplast genes that mutate to antibiotic resistance in C. *reinhardtii* based on zygote clone analysis of allele tests and mapping crosses. Numbers in parentheses below each gene symbol refer to numbers of alleles presently known at each locus. Map distances determined by the R_A method in mapping crosses are shown below. It should be noted that *nr-u-2-1* behaves as an allele at the *spr-u-1-H4* locus and *sr-u-sm5* behaves as an allele at the *sr-u-2-23* locus. (From Harris et al., ref. 15, with permission.)

14-8). Taking all of these data into consideration, the map shown in Fig. 14-9 seems to fit the zygote clone data best.

Superficially, the linear map obtained by zygote clone analysis and the circular map obtained by a combination of cosegregation and liquid segregation analysis do not agree, but in fact the marker orders are the same, at least for the antibiotic resistance markers, if the circular map (Fig. 14-8) is cut between *sr-u-sm2* and *er-u-11* and remains connected across the attachment point to form a linear map. This mapping anomaly arises because zygote clone analysis yields the order *er-u-11-spr-u-1-sr-u-sm5-sr-u-sm3-sr-u-sm2*. Whether there is an explanatory biologic mechanism to account for these differences in marker order or there is something inherently wrong with one of the mapping methods remains to be seen. A final mapping anomaly deserving mention concerns the *nr-u-2-1* and *spr-u-1* mutations. Zygote clone analysis indicates extremely tight linkage of these mutations, but Sager (11, Fig. 3-23) showed them to be rather distantly linked in an early version of her circular map.

SEGREGATION AND RECOMBINATION OF CHLOROPLAST GENES IN VEGETATIVE DIPLOIDS

A small percentage of the zygotes formed in a mating divide vegetatively as diploids rather than becoming meiotic zygotes. Vegetative diploids can be selected from the population of meiotic zygotes by use of the appropriate complementing nuclear gene mutations (p. 361, Fig. 12-7). In some preliminary studies Gillham (4,29) showed that a chloroplast mutation to streptomycin resistance (*sr-u-2-60*) and its wild-type allele were transmitted in much higher frequencies by the *mt⁻* parent in diploids than in meiotic zygotes. It was found that diploids could be classified (just as meiotic zygotes) as being maternal, biparental, or paternal (Fig. 12-7), but the biparental diploid class was in much higher frequency than the biparental zygote class. As in the case of haploid meiotic progeny carrying chloroplast genes from both parents, somatic segregation occurred rapidly among the progeny of biparental diploids. Gillham (4)

TABLE 14–7. *Ordering sr, er, and nr chloroplast gene mutations by predicted incidence of single- and double-exchange events*[a]

Cross: $mt^{+b} \times mt^-$	Total progeny	Putative order I II	Exchanges in intervals I	II	I + II	Double exchanges Observed	Calcu-lated
1	15,224	nr-er-sr	3	0	547		
		er-nr-sr	547	0	3		
		er-sr-nr[c]	547	3	0[d]		
2	6366	nr-er-sr	9	23	140		
		er-nr-sr[c]	140	23	9	9	1
		er-sr-nr	140	9	23		
3	6359	nr-er-sr	0	10	151		
		er-nr-sr[c]	151	10	0		
		er-sr-nr	151	0	10		
4	6057	nr-er-sr	33	74	155		
		er-nr-sr[c]	155	74	33	33	3
		er-sr-nr	155	33	74		
5	6357	nr-er-sr	86	650	156		
		er-nr-sr[c]	156	650	86	86	16
		er-sr-nr	156	86	650		
6	6384	nr-er-sr	2	17	285		
		er-nr-sr[c]	285	17	2	2	1
		er-sr-nr	285	2	17		
7	5920	nr-er-sr	17	27	315		
		er-nr-sr[c]	315	27	17	17	1
		er-sr-nr	315	17	27		
8	6186	nr-er-sr	0	2	317		
		er-nr-sr[c]	317	2	0		
		er-sr-nr	317	0	2		
9	6237	nr-er-sr	10	81	311		
		er-nr-sr[c]	311	81	10	10	4
		er-sr-nr	311	10	81		
10	3743	nr-er-sr	46	360	190		
		er-nr-sr[c]	190	360	46	46	18
		er-sr-nr	190	46	362		
Pooled crosses 1 + 3	21,583	nr-er-sr	3	10	698		
		er-nr-sr[c]	698	10	3		
		er-sr-nr	698	3	10		

[a] Crosses correspond to those in Table 14–6.

[b] Ultraviolet irradiation for 15 sec.

[c] Indicates presumed correct order.

[d] This cross gives the wrong map order both by the usual recombination equation and by this exchange analysis, but where these data are pooled with the comparable cross incorporating the allelic mutant sr-u-2-23 (cross 3), the correct order is obtained.

From Harris et al. (15).

also presented evidence from a two-factor cross involving the *sr-u-2-60* and *nr-u-2-1* mutations that recombination could occur among the progeny of biparental diploids.

Van Winkle-Swift (8) made an exhaustive study of the transmission, segregation, and recombination of chloroplast genes in diploid *C. reinhardtii;* some of the results are summarized in an article by Boynton et al. (9). Van Winkle-Swift began with an investigation of the factors influencing the frequency of biparental diploids in a cross. She showed that treatments that delayed the first somatic division following mating, e.g., incubation of the mated cells in the dark (Fig. 14–10) or nitrogen deprivation in the light, led to a marked decline in the frequency of biparental diploids and their replacement by maternal diploids. Hence, delay of cell division of a diploid seems to result in systematic loss of paternal chloroplast genes. Van Winkle-Swift, using both

TABLE 14–8. *Effect of marker combination on recombination of chloroplast genes er-u-37 (er), spr-u-1-6-2 (spr), and sr-u-2-60 (sr) analyzed by zygote clone method*[a]

Cross ($mt^+ \times mt^-$)	Percentage recombination		
	er–sr	er–spr	spr–sr
1. er spr$^+$ sr $\times$ er$^+$ spr sr$^+$	4.3	4.0	0.6
2. er$^+$ spr sr$^+$ $\times$ er spr$^+$ sr	5.0	4.8	1.5
3. er$^+$ spr$^+$ sr $\times$ er spr sr$^+$	5.2	4.2	1.5
4. er spr sr$^+$ $\times$ er$^+$ spr$^+$ sr	6.0	5.7	1.3
5. er$^+$ spr sr $\times$ er spr$^+$ sr$^+$	4.1	3.8	2.2
6. er spr$^+$ sr$^+$ $\times$ er$^+$ spr sr	4.8	3.9	2.7
7. er$^+$ spr$^+$ sr$^+$ $\times$ er spr sr	3.7	2.8	2.3
Overall	4.7	4.2	1.7

[a] In each cross, mt^+ gametes were subjected to 15 sec of ultraviolet irradiation prior to mating. Approximately 3,200 progeny from 50 biparental zygotes were analyzed in each cross.
From Boynton et al. (9), with permission.

two-factor (8) and four-factor (9) crosses, obtained the gene order *er-u-37-nr-u-2-1-spr-u-1-sr-u-2-60* by analysis of the frequency of recombination using the R_A method. This is the same order obtained via zygote clone analysis. It should be noted, however, that there was an extreme bias in the four-factor cross against the *mt*⁻ paren-

tal genotype, which carried all four resistance markers (9), and against two recombinant classes carrying three resistance markers. Because of this, the agreement in order obtained with two-factor crosses is very reassuring, since bias is a much less serious problem.

It is not appropriate to discuss the results of Van Winkle-Swift in more detail at this time, since most of them are available only in dissertation form (8). Suffice it to say that her studies of gene transmission, segregation, bias, and recombination are very detailed and have brought our level of understanding of the behavior of chloroplast genes in vegetative diploids almost to the same point as for meiotic zygotes. In yeast, the analysis of mitochondrial gene segregation and recombination is carried out on the vegetative diploid progeny from mitotic zygotes (see Chapters 4–7). Vegetative diploids in *Chlamydomonas* provide a directly comparable system for the study of chloroplast genes.

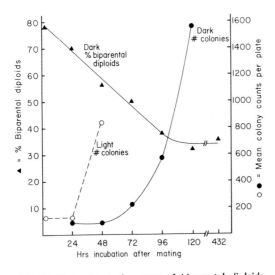

FIG. 14–10. Decline in frequency of biparental diploids (mitotic zygotes carrying chloroplast gene markers from both parents) when the first mitotic division is delayed by incubation in darkness. Respreading experiments with plates incubated in light compared to plates incubated in dark indicate time of the first division. (From Boynton et al., ref. 9, with permission.)

PERSISTENT HETEROZYGOTES

Sager and Ramanis (11,30) described an interesting class of cells that remain heterozygous for the chloroplast genes they carry and are called persistent heterozygotes. Persistent heterozygotes were first detected among progeny of crosses of a streptomycin-dependent mutant (*sd*) that were phenotypically streptomycin-resistant (*sr*). When *sr mt⁺* cells were crossed to wild-type *mt⁻* cells, the zygotes segregated 2:2 *sr:sd* progeny following meiosis. In the reciprocal cross these phenotypes were not transmitted. Thus the *sr mt⁺* parent was carrying both the *sr* and *sd* genes, and it transmitted both genes at meiosis with a segregation process sufficiently regular to produce, in the main (but not exclusively), 2:2 segregation ratios.

To pursue the properties of persistent heterozygotes further, Sager and Ramanis made a cross between strains marked by a series of chloroplast mutations to antibiotic resistance and found a small fraction of progeny carrying resistance markers from both parents that they suspected were not recombinants. During vegetative growth more clones were observed to segregate out the two parental genotypes at a low frequency (1% or less). When persistent heterozygotes were used as the *mt⁺* parent in crosses, they were found to segregate the two parental combinations of chloroplast markers they carried at meiosis with high frequency.

Persistent heterozygotes could provide a powerful tool for organelle genetics, since it might be possible to study complementation of organelle genes in such heterozygotes for the first time. In addition, they could provide important information on the mechanisms governing organelle gene segregation. A full account of the genetics of persistent heterozygotes is eagerly awaited; the two brief reports published thus far are tantalizing.

CONCLUSIONS

Chloroplast genes in *C. reinhardtii* can be mapped and mutants allele tested and detailed maps are now available for those genes that can be mutated to chloro-plast ribosomal antibiotic resistance. Because cosegregation analysis and zygote clone analysis yield different map orders in this region (Fig. 14–8), the form of this map (i.e., whether it is linear or circular) is not yet established. Chloroplast gene mapping can now be done among the progeny of both biparental meiotic zygotes and vegetative diploids. The most pressing need at the moment is for the isolation of new mutants not affecting chloroplast ribosomes that do not map in the ribosomal region. Until now, only the *ac-u-1, ac-u-2,* and *tm-u-1* mutants, which appear to fit this category, have been studied in any detail. The work of Sager and Ramanis (12) indicates that these mutants segregate more nearly 1:1 among the progeny of biparental zygotes and that their rate of segregation in liquid culture analysis (17) is slower than for the ribosomal markers. Thus it is possible that the behavior of the ribosomal genes does not typify that of the chloroplast genome as a whole. The obvious place to search for new chloroplast mutants is among that large class of mutants that require acetate for growth because of photosynthetic defects (see Chapter 12). The mutants should be stringent acetate-requirers (unlike *ac-u-1* and *ac-u-2*) for mapping purposes. Mutants with the appropriate phenotype have been isolated very recently (see Chapter 20) and they should have the same profound effect on chloroplast gene mapping that the *mit⁻* mutants have had in yeast (see Chapter 5).

REFERENCES

1. Sager, R., and Ramanis, Z. (1963): The particulate nature of nonchromosomal genes in *Chlamydomonas. Proc. Natl. Acad. Sci. U.S.A.,* 50:260–268.
2. Sager, R., and Ramanis, Z. (1965): Recombination of nonchromosomal genes in *Chlamydomonas. Proc. Natl. Acad. Sci. U.S.A.,* 53:1053–1061.
3. Gillham, N. W. (1963): The nature of exceptions to the pattern of uniparental inheritance for high level streptomycin resistance

in *Chlamydomonas reinhardi*. *Genetics*, 48: 431–439.

4. Gillham, N. W. (1969): Uniparental inheritance in *Chlamydomonas reinhardi*. *Am. Naturalist*, 103:355–388.

5. Gillham, N. W. (1965): Linkage and recombination between nonchromosomal mutations in *Chlamydomonas reinhardi*. *Proc. Natl. Acad. Sci. U.S.A.*, 54:1560–1567.

6. Gillham, N. W., and Fifer, W. (1968): Recombination of nonchromosomal mutations: A three point cross in the green alga *Chlamydomonas reinhardi*. *Science*, 162:683–684.

7. Sager, R., and Ramanis, Z. (1967): Biparental inheritance of nonchromosomal genes induced by ultraviolet irradiation. *Proc. Natl. Acad. Sci. U.S.A.*, 58:931–937.

8. Van Winkle-Swift, K. (1976): The transmission, segregation, and recombination of chloroplast genes in diploid strains of *Chlamydomonas reinhardtii*. Ph.D. thesis, Duke University, Durham, N.C.

9. Boynton, J. E., Gillham, N. W., Harris, E. H., Tingle, C. L., Van Winkle-Swift, K., and Adams, G. M. W. (1976): Transmission, segregation and recombination of chloroplast genes in *Chlamydomonas*. In: *Genetics and Biogenesis of Chloroplasts and Mitochondria*, edited by T. Bücher, W. Neupert, W. Sebald, and S. Werner, pp. 313–322. North Holland, Amsterdam.

10. Sager, R., and Ramanis, Z. (1970): A genetic map of non-Mendelian genes in *Chlamydomonas*. *Proc. Natl. Acad. Sci. U.S.A.*, 65: 593–600.

11. Sager, R. (1972): *Cytoplasmic Genes and Organelles*. Academic Press, New York.

12. Sager, R., and Ramanis, Z. (1976): Chloroplast genetics of *Chlamydomonas*. I. Allelic segregation ratios. *Genetics*, 83:303–321.

13. Sager, R., and Schlanger, G. (1976): Chloroplast DNA: Physical and genetic studies. In: *Handbook of Genetics, Vol. 5*, edited by R. C. King, Chapter 12. Plenum Press, New York.

14. Conde, M. F., Boynton, J. E., Gillham, N. W., Harris, E. H., Tingle, C. L., and Wang, W. L. (1975): Chloroplast genes in *Chlamydomonas* affecting organelle ribosomes. Genetic and biochemical analysis of antibiotic resistant mutants at several gene loci. *Mol. Gen. Genet.*, 140:183–220.

15. Harris, E. H., Boynton, J. E., Gillham, N. W., Tingle, C. L., and Fox, S. B. (1977): Mapping of chloroplast genes involved in chloroplast ribosome biogenesis in *Chlamydomonas reinhardtii*. *Mol. Gen. Genet.*, 155:249–265.

16. Sager, R., and Ramanis, Z. (1968): The pattern of segregation of cytoplasmic genes in *Chlamydomonas*. *Proc. Natl. Acad. Sci. U.S.A.*, 61:324–331.

17. Singer, B., Sager, R., and Ramanis, Z. (1976): Chloroplast genetics of *Chlamydomonas*. III. Closing the circle. *Genetics*, 83:341–354.

18. Gillham, N. W., and Fifer, W. (1968): Unpublished data.

19. Sager, R., and Ramanis, Z. (1968): The pattern of segregation of cytoplasmic genes in *Chlamydomonas*. *Proc. Natl. Acad. Sci. U.S.A.*, 61:324–331.

20. Sager, R., and Ramanis, Z. (1970): Genetic studies of chloroplast DNA in *Chlamydomonas*. *Symp. Soc. Exp. Biol.*, 24:401–417.

21. Gillham, N. W., Boynton, J. E., and Lee, R. W. (1974): Segregation and recombination of non-Mendelian genes in *Chlamydomonas*. *Genetics*, 78:439–457.

22. Adams, G. M. W., Van Winkle-Swift, K. P., Gillham, N. W., and Boynton, J. E. (1976): Plastid inheritance in *Chlamydomonas reinhardtii*. In: *The Genetics of Algae*, edited by R. A. Lewin, Chapter 5. Blackwell Scientific, Oxford.

23. Sager, R., and Ramanis, Z. (1971): Methods of genetic analysis of chloroplast DNA in *Chlamydomonas*. In: *Autonomy and Biogenesis of Mitochondria and Chloroplasts*, edited by N. K. Boardman, A. W. Linnane, and R. M. Smillie, pp. 250–259. North Holland, Amsterdam.

24. Sager, R., and Ramanis, Z. (1971): Formal genetic analysis of organelle genetic systems. In: *Stadler Symposia*, edited by G. Kimber and G. P. Rédei, *Vols. 1 and 2*, pp. 65–78. University of Missouri, Agricultural Experiment Station, Kansas City, Missouri.

25. Sager, R. (1974): Nuclear and cytoplasmic inheritance in green algae. In: *Algal Physiology and Biochemistry*, edited by W. D. P. Stewart, Chapter 11. Blackwell Scientific, Oxford.

26. Sager, R., and Ramanis, Z. (1976): Chloroplast genetics of *Chlamydomonas*. II. Mapping by cosegregation frequency analysis. *Genetics*, 83:323–340.

27. Sager, R. (1977): Genetic analysis of chloroplast DNA in *Chlamydomonas*. *Adv. Genet.*, 19:287–340.

28. Mets, L. J., and Bogorad, L. (1971): Mendelian and uniparental alterations in erythromycin binding by plastid ribosomes. *Science*, 174:707–709.

29. Gillham, N. W. (1963): Transmission and segregation of a nonchromosomal factor controlling streptomycin resistance in diploid *Chlamydomonas*. *Nature*, 200:294.

30. Sager, R., and Ramanis, Z. (1971): Persistent cytoplasmic heterozygotes in *Chlamydomonas*. *Genetics*, 68:S56.

Chapter 15

Chloroplast Genetics of *Chlamydomonas*. IV. Models to Explain Inheritance of Chloroplast Genes

In the preceding chapter we discussed segregation, recombination, and mapping of chloroplast genes in *Chlamydomonas* and outlined briefly two models that seek to explain the inheritance of these genes. In this chapter we shall begin with a complete elaboration of these models and then turn to the interpretation of the experimental data in terms of the models. One of the most difficult tasks the scientist faces is unbiased discussion of his own hypotheses in relation to conflicting hypotheses of others. It is difficult to resist the temptation to explain away contradictory ideas and apply a double standard in comparing models. The models discussed in this chapter are the model of the author and his colleagues and the model of Sager and Ramanis. In reading this chapter it is important to remember that the author's prejudices probably have crept in, albeit inadvertently.

TWO-COPY MODIFICATION-RESTRICTION MODEL

The two-copy model for the chloroplast genome was proposed by Sager and Ramanis (1) in their initial study of chloroplast gene segregation in pedigrees following ultraviolet irradiation of the maternal parent. The model now assumes that the chloroplast of *Chlamydomonas* contains two species of DNA. The major component is highly reiterated; it has a molecular weight of 1.3×10^8 d (see Chapters 2 and 12). The minor component is present in two copies and contains all the chloroplast genes described to date. In a recent review, Sager (2) suggested that the highly reiterated component might be interspersed with

the single-copy component. In discussing the circular diploid model, we shall refer to this presently undetected clDNA species as the circular diploid chloroplast genome. The circular diploid genome is assumed by Sager and Ramanis to be circular on the basis of genetic mapping experiments described in the preceding chapter. This genetic evidence for circularity should not be confused with the physical evidence discussed earlier (see Chapters 2 and 12) that the reiterated major component is circular.

Before replication there are two copies of the circular diploid genome per chloroplast, each with an attachment point (*ap*) (Fig. 15–1). Each genome replicates to produce two daughter molecules, each with its own attachment point. At this four-strand stage, recombination occurs; it can be either reciprocal or nonreciprocal (Figs. 14–5 and 15–1). Distribution of the four resulting genomes to the two daughter cells is by a regular oriented mechanism whereby the two new attachment points and their associated genomes go to one daughter cell and the old attachment points and their associated genomes go to the other. Thus the number of genomes per cell is conserved. This model ensures that a cell heterozygous for its two chloroplast genomes (i.e., a hybrid zoospore for a biparental zygote) will remain hybrid in the absence of recombination (Figs. 14–5 and 15–1). Nonreciprocal recombination will yield a type II segregation for the marker, and reciprocal recombination will yield a type III segregation, as was described in the preceding chapter (Figs. 14–5 and 15–1).

The mechanism leading to maternal inheritance of chloroplast genomes first came under scrutiny by Sager and Ramanis (3)

TYPE I

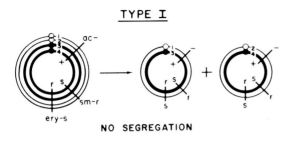

NO SEGREGATION

TYPE II

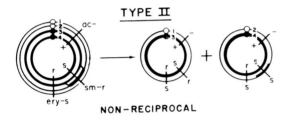

NON-RECIPROCAL

TYPE III

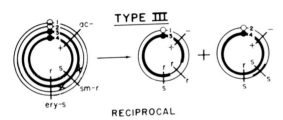

RECIPROCAL

FIG. 15–1. Segregation of chloroplast genes according to circular diploid model of Sager. The chloroplast genome is shown on the left in circular form at a four-strand stage after replication and before cell division; it is shown on the right after cell division. Each line represents a double-stranded DNA molecule. Type I, the most common pattern, resembles mitotic segregation of nuclear chromosomes. All three markers shown remain heterozygous. Type II depicts a gene-conversion-like event at the *sr-u-sm2* (*sm*) locus that produces one daughter heterozygous for that locus and the other two loci and one daughter homozygous for sensitivity (*sm-s*) at that locus only. Type III depicts a reciprocal event giving rise to one pure resistant (*sm-r*) daughter and one pure *sm-s* daughter. (From Sager and Ramanis, ref. 14, with permission.)

when they showed that ultraviolet irradiation of the maternal parent prior to mating led to a large increase in the frequency of exceptional zygotes. They postulated that irradiation had two effects on the inheritance of chloroplast genes in *Chlamydomonas* (see Chapter 14):

$$\text{maternal zygote} \underset{\text{photoreactivation}}{\overset{\text{irradiation}}{\rightleftharpoons}} \text{biparental zygote}$$

1. maternal zygote ⇌ biparental zygote

$$\text{biparental zygote} \underset{\text{photoreactivation}}{\overset{\text{irradiation}}{\rightleftharpoons}} \text{paternal zygote}$$

2. biparental zygote ⇌ paternal zygote

Process 1 could be reversed more effectively by photoreactivation before mating than afterward. This step was viewed as the one that regulated maternal inheritance. It was postulated that ultraviolet irradiation blocked the formation of a gene product (possibly an enzyme) required for the loss of paternal chloroplast genes. In process 2 there was direct interference with the replication of maternal clDNA so that it could not duplicate, and paternal zygotes resulted. This damage was partially repairable by photoreactivating light.

Although process 2 was straightforward enough, the question was how process 1 was effected. Sager and Ramanis (4,5) dealt with the problem in terms of an ingenious model wherein process 1 was governed by a restriction-modification system similar to those in bacteria (6) that led to destruction of paternal clDNA under conditions such that maternal clDNA was protected. They supposed that the modification enzyme protected maternal clDNA (possibly by methylation of specific sites in the DNA) from attack by the restriction enzyme, as in bacteria. However, the restriction enzyme recognized these same sites in paternal clDNA because it was not modified, and so degradation of paternal clDNA ensued. Sager and Ramanis supposed that the chloroplast of the mt^+ gamete contained the modification enzyme (M) in an inactive form, whereas the mt^- chloroplast contained the restriction enzyme (R), also in an inactive form (Fig. 15–2). Activation of M and R was regulated by a substance (G_1) that was produced by mt^+ gametes at the time of mating. In the few hours preceding chloroplast fusion in the zygote, the mt^+ clDNA was modified, and that of the mt^-

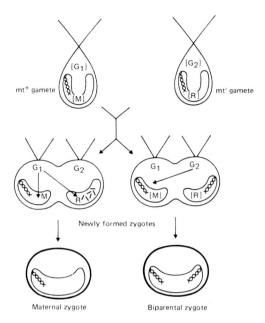

FIG. 15–2. Proposed mechanism of maternal inheritance of chloroplast genes in *Chlamydomonas* modified from Sager and Ramanis (4) according to Sager and Ramanis (5). The female (*mt⁺*) gamete contains inactive modification enzyme [M] in its chloroplast and the regulatory substance G_1 in its cell sap. The male (*mt⁻*) gamete contains inactive restriction enzyme [R] in its chloroplast and the regulatory substance G_2 in its cell sap. In the case of a maternal zygote, G_1 activates both [M] and [R] following gamete fusion, and the maternal clDNA is modified and that from the paternal parent is restricted. By the time of chloroplast fusion, only maternal clDNA remains. In the case of a biparental zygote, G_2 acts to antagonize G_1; so [M] and [R] are not activated, and clDNA from both parents survives in the zygote.

parent was degraded. By the time the chloroplasts had fused, all of the *mt⁺* clDNA was modified; so it could not be restricted. A second regulatory substance (G_2) produced in *mt⁻* gametes at the time of mating acted to inhibit G_1. Thus the activation of M and R depended on the balance between G_1 and G_2. Since G_1 was much more effective than G_2, inheritance of clDNA was usually maternal.

A question that Sager and Ramanis failed to deal with explicitly was the distinction between the two postulated circular diploid genomes and the reiterated fraction of clDNA. If the reiterated copies are interspersed in the same molecule with the single-copy material, as Sager (2) suggested, one would expect both fractions to have the same fate with respect to restriction and modification.

MULTICOPY PLASMID EXCLUSION MODEL

The model proposed by Gillham et al. (7) is based on the principle of competitive exclusion, much as in the case of incompatible bacterial plasmids (8). The plasmid model also contains features found in the model of Dujon et al. (see Chapter 7) that is designed to explain the inheritance of mitochondrial genomes in yeast. That is, the plasmid model of Gillham and Boynton is to some extent a phage analogy model.

According to the plasmid model, the zygote of *C. reinhardtii* contains a fixed number of membrane attachment sites that are required for replication and segregation of chloroplast genomes according to the oriented mechanism proposed by Sager. The sites are occupied preferentially by maternal chloroplast genomes. In a biparental zygote, one or more attachment sites become occupied by paternal chloroplast genomes. In a paternal zygote, attachment sites are occupied by paternal chloroplast genomes because of the loss or destruction of all maternal genomes. Genomes that are not attached are not replicated, and during successive rounds of replication they are diluted out.

Initially Gillham and Boynton assumed a unitary effect of ultraviolet irradiation in which irradiation simply inactivated chloroplast genomes. Irradiation of *mt⁺* gametes led to inactivation of maternal chloroplast genomes. Photoreactivating light converted paternal zygotes to biparental zygotes and biparental zygotes to maternal zygotes at different rates because the former event involved only the conversion of a single nonfunctional genome to a functional state, whereas the latter required conversion of all nonfunctional genomes present to the functional state. In short, the plasmid model predicted that the two processes would show different photoreactivation kinetics because the numbers of maternal chloroplast genomes that had to be photoreactivated were different for the two processes, not because two different irradiation effects were involved. Adams et al. (9), in a slightly more recent version of this model, suggested that ultraviolet irradiation of the maternal parent might also lead to detachment of maternal chloroplast genomes from their attachment sites. It is implicit in the model of Gillham and Boynton that the chloroplast genomes being studied must reside in the reiterated fraction of clDNA. That is, no special undetected diploid chloroplast genome exists.

The multiple-copy model is based on zygote clone analysis experiments, as will be discussed presently. Nevertheless, it is essential to account for the three segregation patterns seen by Sager and Ramanis in pedigrees. Gillham and Boynton proposed that oriented segregation of chloroplast genomes took place by the segregation mechanism proposed by Sager. Type I segregations arose in hybrid cells whose copy number was two or more. Type II segregations occurred by reciprocal recombination in cells whose copy number was greater than two. Type III segregations arose by reciprocal recombination in cells whose copy number was two.

COMPARISON OF THE MODELS

We shall now compare the two models in terms of the assumptions they make concerning the nature of the chloroplast genome and its mechanism of inheritance. This comparison can then be used as a basis for applying the experimental data to the models.

Number and Character of Chloroplast Genomes

The model of Sager and Ramanis assumes genetic diploidy for the chloroplast genome. Diploidy is consistent with much of the genetic data discussed previously (see Chapter 14), as well as data that will be discussed later in this chapter. However, diploidy is not easily reconciled with the existing data, which suggest that the chloroplast genome is present in many copies per chloroplast (see Chapters 2 and 12). This problem was recognized and discussed by Wells and Sager (10). They suggested two possible explanations. First, the major kinetic species is present in tandem repeats in two large DNA molecules in the chloroplast. They pointed out that such an arrangement would be difficult to reconcile with the evidence that genetic recombination in the chloroplast linkage group occurs at each mitotic doubling. Second, there is a slowly renaturing fraction corresponding to the genetic markers being studied. This amounts to 10 to 20% of the clDNA and will not be detected by renaturation kinetics. This explanation is consistent with Sager's recent model (p. 423), which assumes that the highly reiterated species is interspersed with single-copy material.

The model of Gillham and associates assumes that the markers being studied must be localized in the major repetitive species of clDNA, that there is no undetected slowly renaturing clDNA species, and that there are many chloroplast genomes per cell in the genetic sense as well as the physical sense.

Attachment Sites

Each model assumes the existence of membrane attachment sites for chloroplast genomes. These attachment sites were originally invoked by Sager and Ramanis (11) to explain the polarity of type III segregation seen for different chloroplast alleles (see Chapter 14). The model of Gillham and Boynton also adopted the assumption of attachment sites, as well as the notion that segregation of chloroplast genomes in an oriented fashion is governed by the attachment site. However, the models differ in the way they explain the function of the attachment sites in governing the pattern of inheritance of chloroplast genomes. Sager (12) considered attachment sites only in terms of the oriented segregation of the circular diploid chloroplast genomes. Attachment sites were not invoked in explaining the mechanism of maternal inheritance. In contrast, Gillham et al. (7) ascribed an important role to attachment sites in determining the pattern of inheritance of chloroplast genomes. Gillham and Boynton assumed that the zygotic chloroplast contained a fixed number of attachment sites that is less than the total number of input chloroplast genomes. According to the model of Gillham and Boynton, these sites would normally be occupied by maternal chloroplast genomes, and paternal genomes would be excluded from them. Biparental zygotes would arise when one or more attachment sites were occupied by paternal chloroplast genomes. Paternal zygotes would arise when all attachment sites were occupied by paternal chloroplast genomes.

Mechanism of Elimination of Paternal Genomes

The mechanism of elimination of paternal chloroplast genomes was explicitly stated in terms of restriction and modification by Sager and Ramanis (Fig. 15–2). Following gametic fusion, but prior to chloroplast fusion, the modification and restriction enzymes are activated by compound G_1, which is produced in the mt^+ parent. The maternal chloroplast genomes are modified, and the paternal chloroplast genomes are restricted. Compound G_2, produced by mt^- gametes at the time of mating, is a weak inhibitor of G_1. In those rare instances in which G_2 effectively inhibits G_1, a biparental zygote results.

The model of Gillham and Boynton is much more vague about the fate of paternal chloroplast genomes in the zygote. It assumes that a limited number of zygotic attachment sites governs maternal inheritance

and that these are normally occupied by maternal chloroplast genomes. Paternal genomes are either diluted out, since they cannot replicate, or destroyed by a nuclease. The restriction-modification model assumes recognition and destruction of paternal chloroplast genomes by a restriction enzyme. The model of Gillham and Boynton assumes that if a nuclease is involved, it will degrade paternal or maternal genomes at random, so long as they are not bound to an attachment site. The model of Gillham and Boynton also leaves open the possibility that paternal genomes are not physically destroyed, but rather diluted out because they cannot replicate.

Segregation and Recombination

Each of the models makes different assumptions about the mechanism by which chloroplast gene segregation is achieved. Type I segregations produce cells that are heterozygous for a pair of alleles. Sager and Ramanis assumed that such cells are diploid for this pair of alleles; Gillham and Boynton assumed that they are multiploid. Type II segregation produces a cell heterozygous for a given chloroplast allele (e.g., a/a^+) and a cell that is homozygous (e.g., a^+/a^+ or a/a). The model of Sager and Ramanis assumes that type II segregants arise as the result of a process akin to gene conversion, so that one allele (e.g., a) carried by one copy is converted to that carried by the other copy (e.g., a^+). The model of Gillham and Boynton assumes that type II segregants arise by reciprocal recombination in a hybrid cell with more than two chloroplast genomes. Both models agree that type III segregations occur in cells that contain only two copies of the chloroplast genome as the result of reciprocal recombination.

APPLYING EXPERIMENTAL DATA TO THE MODELS

We shall now attempt to illustrate the virtues and shortcomings of each model by relating each to specific observations and facts.

Chloroplast DNA and Its Pattern of Inheritance

The circular diploid model requires the identification of a new slowly renaturing species of clDNA present in the cell in only two copies. Thus far, this DNA has eluded detection. Certainly, such a DNA species would be difficult to detect by any of the available physical methods, given the great predominance of the reiterated clDNA component. No clDNA of large size that renatures slowly has been found in either higher plants or *Euglena,* but clDNA of very large size and high kinetic complexity has been reported in *Acetabularia* (see Chapter 2). Ultimately, acceptance of the circular diploid model will depend on the identification of the predicted species of clDNA. This problem is not relevant to the multicopy model, which assumes that the chloroplast genes thus far identified do reside in the highly reiterated DNA fraction.

The modification-restriction model predicts that paternal clDNA is degraded preferentially in the young zygote and that maternal clDNA is conserved. The density labeling experiments of Sager and Lane (see Chapter 12) are in accord with this interpretation. Furthermore, the density shift reported by these authors for the conserved clDNA could be accounted for by DNA base methylation of about 5% (see Chapter 12), which could be the result of modification of this DNA to prevent its restriction, as in the case of bacteria (6). Sager and Schlanger (13) also briefly mentioned experiments that remain to be published in detail showing that paternal clDNA is recovered in young zygotes following ultraviolet irradiation of the *mt*⁺ parent or when crosses are made that involve the *mat-1* mutant. This mutant, which will be discussed presently, is tightly linked to the *mt*⁻ allele; it causes a large increase in the frequency of exceptional zygotes. That is, the effect of *mat-1* is similar to the effect of ultraviolet irradiation of the maternal parent prior to mating. These observations are accommodated with ease by the restriction-modification model, as will be discussed later. The observations do not violate the multicopy model either, for this model assumes that unattached paternal chloroplast genomes will be destroyed or diluted out and that attached maternal chloroplast genomes will be conserved. Ultraviolet irradiation of the maternal parent will lead to destruction of maternal clDNA and permit transmission of paternal clDNA; so the results of Sager and Schlanger are predicted by this model. The *mat-1* mutant is also readily explained by the multicopy model, as will be seen later. The biochemical basis of the clDNA density shift seen by Sager and Lane (see Chapter 12) is vague enough not to compel any interpretation on the multicopy model at present.

Chiang's experiments, in contrast to those of Sager and her colleagues, indicated extensive degradation of both parental clDNAs in the zygote (see Chapter 12). The destruction of maternal clDNA is not consistent with the restriction-modification model, unless the model is applied specifically to the two postulated circular diploid genomes and it is assumed that only the unmodified paternal copies are destroyed. Thus, the reiterated clDNA molecules from both parents would be degraded, but only the two circular diploid genomes from the paternal parent would be subject to degradation. To explain Chiang's results on the multiple-copy model, it would have to be assumed that the number of zygotic attachment sites is considerably smaller than the number of paternal or maternal clDNA molecules introduced into the zygote. Thus clDNA from both parents would be degraded. Nevertheless, the attachment sites present would be occupied preferentially by maternal chloroplast genomes under normal circumstances.

The foregoing discussion is a tribute to the flexibility of models. The real question is why Sager and Chiang get different results.

Experiments with Irradiation and Inhibitors

It is not surprising that both models have explanations for the effect of ultraviolet irradiation on the inheritance of chloroplast genes, since they were originally designed to do so. In the modification-restriction model, irradiation is assumed to block the release of regulator substance G_1, which activates both the modification and restriction enzymes (process 1) and leads to the conversion of maternal zygotes to biparental zygotes. Process 2, the event that leads to the production of paternal zygotes, results when the maternal chloroplast genomes are inactivated. Both processes are photoreactivable. The mechanism by which the damages induced by process 2 are photoreactivated is clear-cut, since damages introduced into clDNA presumably are repaired by a photoreactivating enzyme. However, the means by which process 1 can be reversed by photoreactivating light is obscure, since repair of damage introduced into DNA, at least clDNA, presumably is not involved. The multicopy model supposes that irradiation causes damage to clDNA exclusively and that biparental zygotes result from damage to a fraction of the maternal chloroplast genomes, permitting transmission of paternal chloroplast genomes. Paternal zygotes arise when all maternal chloroplast genomes have been inactivated. Both processes ought to be photoreactivable.

A key difference in the predictions made by the two models concerns the effect of irradiation dosage on allelic ratios. Sager and Ramanis reported allelic ratios of $1:1$ for chloroplast genes in pedigree analysis; Gillham and Boynton reported a skew in favor of maternal chloroplast alleles (see

Chapter 14). The results of Sager and Ramanis applied to both spontaneous and irradiation-induced biparental zygotes, but the results of Gillham and Boynton applied only to spontaneous biparental zygotes. Gillham and Boynton (7) showed by zygote clone analysis of reciprocal crosses that the skew in favor of maternal chloroplast alleles could be markedly reduced by subjecting the maternal parent to increasing doses of ultraviolet irradiation prior to mating (Fig. 15–3). In a population of biparental zygotes the allelic ratio began to approach $1:1$ with increasing irradiation doses, although the zygotes were not normally distributed around this mean.

On the basis of these results, Gillham and Boynton attempted a quantitative formulation of their model that had its analog in

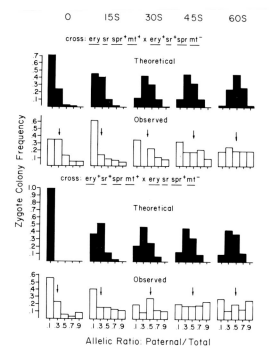

FIG. 15–3. Theoretical and observed allelic ratios among biparental zygotes in reciprocal crosses. The mt^+ parents were treated with various doses (top: 0–60 sec) of ultraviolet irradiation prior to mating. Zygote clones were grouped into classes based on the fraction of progeny in each clone carrying the alleles from the paternal parent. Key: ery = er-u-37, spr = spr-u-1-6-2, sr = sr-u-2-60. (From Gillham et al., ref. 7, with permission.)

phage genetics. They assumed that the different allelic ratios really reflected different genomic inputs into the zygote. On this basis they tried to calculate an "average multiplicity of infection" of zygotes with maternal and paternal chloroplast genomes. If paternal and maternal chloroplast genomes were randomly distributed with respect to zygotes, the number of zygotes not "infected" with maternal chloroplast genomes divided by the total could be used to calculate the mean number of maternal genomes (m_m) per zygote from the P_0 term of the Poisson distribution. That is,

$$\text{paternal zygotes/total zygotes} = e^{-m_m}$$

Similarly, the average multiplicity of infection by paternal genomes (m_p) would be

$$\text{maternal zygotes/total zygotes} = e^{-m_p}$$

These figures were estimated for each irradiation dose; then, using the other terms of the Poisson distribution, it was possible to predict for a given dose of irradiation what the frequency distribution of biparental zygotes should be with respect to allelic ratio.

The model predicted a strong maternal skew, which disappeared with increasing dosage (Fig. 15–3). The observed results fit the model, with the exception that with increasing irradiation dosage the distribution of allelic ratios did not become normal but rather flattened out. These results were interpreted by Gillham and Boynton as the result of "random drift" of the remaining maternal and paternal chloroplast genomes into attachment sites that had been reduced in number through the action of ultraviolet irradiation. Gillham and Boynton emphasized the reliance of their model on parameters derived from phage genetics.

Sager and Ramanis (14) subsequently repeated the ultraviolet irradiation experiments with zygote clones using their stocks and conditions. In their hands there was no effect of increasing irradiation dose on the 1:1 allelic ratio observed for chloroplast alleles in zygote clones. In short, the irradiation dose experiment performed in one laboratory supported one model, but the same experiment repeated in a different laboratory supported the other model.

Sager and Ramanis (4) examined the effects of a number of transcriptional and translational inhibitors on the frequency of exceptional zygotes. Inhibitors were administered to cells of both mating types either during gametogenesis or following gamete formation. Treated gametes were then mated to untreated gametes of opposite mating type. The treated gametes were washed prior to mating in an attempt to eliminate the inhibitor from the mating cell population and resultant zygotes.

Sager and Ramanis found that treatment of paternal gametes as well as maternal gametes with a given inhibitor often increased the frequencies of exceptional zygotes. Further, many compounds (i.e., cycloheximide, erythromycin, spiramycin, rifamycin SV, and butyrl cyclic AMP) were more effective in inducing exceptional zygote formation when paternal gametes were treated. Sager and Ramanis pointed out that these results could not be accounted for by carryover of inhibitor into the zygote by the paternal gamete, since a greater effect of specific inhibitors on these gametes would not be expected if that were the case. After all, gametes of opposite mating type are approximately the same size; they fuse completely during zygote formation, and fusion is accompanied not only by nuclear fusion but also by chloroplast fusion (see Chapter 12). Sager and Ramanis concluded from these results that paternal gametes as well as maternal gametes must play some role in the control of maternal inheritance. Beyond this general conclusion, one can deduce little about the mechanism of maternal inheritance from these experiments.

The experiments of Wurtz et al. (15) with 5-fluorodeoxyuridine (FdUrd) are the most illuminating inhibitor experiments reported to date with respect to the mechanism of chloroplast gene transmission. These experiments were considered at length in Chapter 12. The three major conclusions that can be drawn from them will be stated here and then related to the two models for the inheritance of chloroplast genes. First, growth of mt^+ cells in FdUrd leads to a considerable reduction in the amount of clDNA per cell. When cells grown under these conditions are mated to untreated mt^- cells, the frequency of biparental and paternal zygotes increases greatly. Second, if mt^- cells grown in FdUrd are mated to mt^+ cells grown in the presence of the analog, the normal pattern of maternal inheritance is partially restored. Third, despite the considerable reduction in amount of clDNA caused by FdUrd, growth of cells in the presence of the inhibitor does not lead to cell death or a noticeable decline in zygote germination.

The results of the FdUrd experiments are simply interpreted on the multicopy model. They suggest that the input amount of clDNA from each parent is related to the pattern of chloroplast gene transmission. If the number of maternal chloroplast genomes introduced into the zygote is reduced, the pattern of chloroplast gene transmission becomes predominantly biparental and paternal. If the input number of paternal chloroplast genomes is reduced simultaneously, the inheritance pattern reverts toward maternal.

The results of the FdUrd experiments are difficult to interpret on the modification-restriction model. The experiments show that a treatment that is known to affect only the amount of clDNA present is sufficient to cause major perturbations in the pattern of chloroplast gene transmission. There is no reason why the balance between the postulated G_1 and G_2 compounds should be altered by growth of cells on FdUrd, nor is there any reason why the gene responsible for G_1 production should be inhibited by the analog. In other words, on the modi-

fication-restriction model it is not obvious why FdUrd should alter the pattern of inheritance of chloroplast genes, but it is abundantly clear from a model that assumes that relative input of chloroplast genomes from the two parents determines the inheritance pattern observed for chloroplast genes.

mat Mutants

Sager and Ramanis (5) described two mutants, *mat-1* and *mat-2*, that affected the pattern of inheritance of chloroplast genes in *C. reinhardtii*. The *mat-1* mutation is very tightly linked to mt^-, and *mat-2* has thus far proved inseparable from mt^+. The *mat-1* mutation increased the frequency of exceptional zygotes in Sager's stocks from less than 1% to 20 to 100%, and the *mat-2* mutation causes a marked reduction in the frequency of exceptional zygotes. When a *mat-1* mt^- stock is crossed to an irradiated mt^+ stock, the frequency of exceptional zygotes is even further increased. When a *mat-1* mt^- stock is irradiated and crossed to an unirradiated mt^+ stock, the normal pattern of maternal inheritance is largely restored. If a *mat-2* mt^+ stock is irradiated and mated to a wild-type mt^- stock, the frequency of exceptional zygotes increases, but only biparental zygotes are found.

The linkage of the *mat* mutations to mating type is of considerable interest, and it supports the view that mating type, *sensu lato*, may actually be a region containing at least several genes governing functions related to the mating act and zygote formation. This possibility was suggested originally by Gillham (16), who noted that certain auxotrophic mutations to thiamine dependence, nicotinamide dependence, and acetate dependence were completely linked to each other and to mating type. The same view was elaborated by Sager and Ramanis (5) in discussing the *mat* mutants and more recently by Goodenough et al. (17) in discussing mutations affecting mating functions, one of which is completely linked to *mt*⁺. Goodenough and associates pointed out that mutations affecting at least six different functions are now clustered around the mating type alleles. As suggested by both Gillham and Goodenough and associates, crossover suppression may be operative in this region of the mating type chromosome to keep genes associated with mating functions clustered in a particular way. Association of auxotrophic mutations with this cluster may be fortuitous.

Sager and Ramanis (5) incorporated the behavior of the *mat-1* and *mat-2* mutations into their modification-restriction model. They supposed that *mat-1* increased the effectiveness of G_2 vis-à-vis G_1 and that the *mat-2* mutation made G_1 more effective than its wild-type counterpart. When the *mt*⁺ parent was irradiated and crossed to unirradiated *mat-1 mt*⁻ cells, Sager and Ramanis assumed that the imbalance between the effectiveness of G_1 and G_2 was increased even further, because according to the hypothesis, ultraviolet irradiation interferes with the production of G_1. However, to explain why ultraviolet irradiation of *mat-1 mt*⁻ cells restored the maternal pattern of inheritance of chloroplast genes, Sager and Ramanis had to suppose that irradiation also blocked production of G_2. Thus the modification-restriction model postulates that irradiation specifically blocks production of two regulatory substances, G_1 and G_2, in addition to damaging clDNA. Ultraviolet irradiation of the *mat-2 mt*⁺ parent would be expected to increase the frequency of exceptional zygotes, just as in the case of irradiation of wild-type *mt*⁺ cells, since production of G_1 is interfered with.

The multiple-copy model can also accommodate the behavior of the *mat* mutants. A particularly simple explanation of *mat-1* that is open to direct experimental test is that this mutation does not have a direct effect on clDNA, but rather on gametogenesis. However, the indirect result is an increase in the amount of clDNA in *mat-1*

gametes. When *mat-1 mt*⁻ gametes are crossed to *mat-1*⁺ *mt*⁺ gametes, there is an increase in the frequency of exceptional zygotes because the input of paternal clDNA is perturbed in favor of the paternal parent. This is the same result that obtains in the case of FdUrd (15) as discussed above except that with FdUrd this perturbation is achieved by lowering the amount of maternal clDNA through growth of *mt*⁺ cells in the presence of the analog. Ultraviolet irradiation of the maternal parent prior to mating further increases the bias toward exceptional zygotes, as would be expected. However, the interesting observation is that ultraviolet irradiation of the *mat-1 mt*⁻ parent returns the normal maternal pattern of chloroplast gene transmission. This result is simply explained by the assumption that irradiation inactivation of paternal chloroplast genomes effectively reduces the input amount of paternal clDNA. Support for this interpretation comes from two sources. First, Adams (18) obtained the same result when he compared transmission of paternal chloroplast genes in experiments where the maternal parent alone was irradiated with transmission in experiments in which both parents were irradiated prior to mating. Second, Wurtz and associates (15) found that the increase in exceptional zygotes resulting from FdUrd-induced reduction of clDNA in the maternal parent could be reversed if both parents were grown in FdUrd prior to mating. It is worth noting that an analogous situation may exist in the case of the $\rho 5$ mutation in *S. cerevisiae* (see Chapter 7). It will be recalled that this mutant contains twice as much mtDNA as the wild type and that transmission of mitochondrial markers derived from the $\rho 5$ parent is favored in crosses to the wild type.

It is not clear that perturbation of clDNA inputs can explain the behavior of the *mat-2* mutant. In this case the assumption would be that *mat-2 mt*⁺ gametes contain more clDNA than normal. Therefore, transmission of chloroplast genes by the *mt⁻* parent would be reduced even further below the already very low spontaneous levels in crosses involving *mat-2*. Ultraviolet irradiation of the *mat-2 mt⁺* parent prior to mating would lead to an increase in exceptional zygotes because maternal chloroplast genomes were being inactivated by irradiation. The difficulty is that although irradiation does cause a 100-fold increase in exceptional zygotes (from less than 0.1% to approximately 10%), only biparental zygotes are to be found.

The *mat* mutants can also be accommodated by the multiple-copy model in a less testable form similar to that proposed by Sager with respect to the effects of these mutants on the hypothetical substances G_1 and G_2. For example, the *mat-1* mutant might increase the affinity of paternal chloroplast genomes for attachment sites or the affinity of attachment sites for paternal genomes, or the mutant might cause an increase in the number of attachment sites. The *mat-2* mutant, conversely, would increase the affinity of maternal chloroplast genomes for attachment sites or vice versa, or it might lead to a reduction in the number of attachment sites. The effects of irradiation are easily accommodated by such a model, but in view of its speculative nature, we shall refrain from further elaboration.

The *mat* mutants are interesting because they are linked to the *mt* gene and they affect the transmission pattern of chloroplast genes, but their behavior can be explained in many ways. The isolation of *mat-1* shows definitively that a genetic change in the *mt⁻* parent can affect the pattern of chloroplast gene transmission.

Allelic Ratios in Pedigrees

According to the circular diploid model, type III events result in reciprocal recombination for a pair of alleles and by definition must yield a 1:1 segregation ratio for that pair of alleles. Type II segregations should, in the absence of complicating factors such as preferential gene conversion, also yield 1:1 allelic segregation ratios on this model. Sager and Ramanis consistently reported a 1:1 segregation ratio for the *ac-u* alleles in pedigrees (see Chapter 14). Recently (14) they also published extensive data for allelic ratios of other markers in pedigrees. They tabulated both allelic segregation ratios among meiotic products from biparental zygotes and the ratios of type II events for pairs of alleles during the first and second postmeiotic mitotic divisions. The data are presented in summary form in Table 15–1. The paternal allelic ratio (*P*) is defined in this table as the number of cells that have segregated the paternal allele divided by the number that have segregated both paternal and maternal alleles. Heterozygotes are not included. Thus when $P = 0.5$, the alleles are segregating in a 1:1 ratio.

In the cases of the *ac-u* and *tm-u-1* markers there is no striking bias in favor of the maternal or paternal chloroplast allele at any cell division (Table 15–1). A bias in favor of the maternal allele is seen in the cases of the *sr-u-sm2, er-u-11,* and *spr-u-sp23* markers. This bias is most marked among the meiotic products, but even there it is not great, being at most 2:1 in favor of the maternal marker. Among octospores and octospore daughters the bias is even smaller; Sager and Ramanis presented data indicating that the bias may be allele-specific. That is, the bias was most marked when the maternal parent carried the resistance allele at each locus; it virtually disappeared when the maternal parent carried the sensitive allele. In short, the bias seen in the experiments of Sager and Ramanis is small at best, and their results are in good agreement with the 1:1 allelic segregation ratio predicted by their model. Sager and Ramanis also noted that zygotic gene segregation ratios varied from gene to gene and that this variation was not pre-

TABLE 15–1. *Allelic ratios obtained for different chloroplast markers in pedigrees*[a]

Marker		Total cells scored	Paternal allelic ratio
ac-u	Meiotic products	83	0.47
	Octospores	668	0.43
	Octospore daughters	666	0.53
sr-u-sm2	Meiotic products	257	0.335
	Octospores	693	0.40
	Octospore daughters	555	0.38
er-u-11	Meiotic products	330	0.34
	Octospores	658	0.41
	Octospore daughters	442	0.39
spr-u-sp23	Meiotic products	379	0.26
	Octospores	585	0.43
	Octospore daughters	367	0.44
tm-u-1	Meiotic products	148	0.45
	Octospores	621	0.44
	Octospore daughters	465	0.50

[a] Paternal allelic ratio is the ratio of cells that segregate the paternal allele divided by the total cells that segregate both paternal and maternal alleles. Heterozygous cells are not included, and octospore and octospore daughter data apply only to type II segregants.
Adapted from Sager and Ramanis (14).

dicted by the multicopy model of Gillham et al. (7). However, it would seem that such variation is not explained by the two-copy model, either, without ad hoc hypotheses (2).

In contrast, Boynton et al. (19), in an extensive tetrad analysis of biparental zygote progeny following treatment of the mt^+ parent with a low dose of ultraviolet irradiation, observed a marked skew in allelic ratio for all three markers in the cross that approximated 3:1. They also found that some 50% of the meiotic products were homoplasmic by the end of the second meiotic division. Boynton and associates pointed out that assuming the whole chloroplast genome is transmitted en bloc, including the region delimited by the three markers in the cross, then tetrads in which the meiotic products showed a predominance of maternal genomes comprised 70% of the total. If it was assumed that each meiotic product received a minimum of two copies of the chloroplast genome, then in the most extreme cases one had to assume a minimum ratio of 7:1 maternal:paternal ge-

nomes (21% of observed tetrads) and 1:7 maternal:paternal genomes (2% of observed tetrads) (Table 15–2). Nuclear markers were not included in the cross; thus it is not known if this maternal allelic bias continued during the postmeiotic mitotic divisions. These results support a multicopy model in which most chloroplast genomes are contributed by the maternal parent.

Tetrad data are much more closely akin to pedigree data than are the results of zygote clone analysis; so it is appropriate here to consider several of Sager's criticisms (2) of the multicopy model in light of the tetrad data presented by Boynton and associates (Table 15–2). First, Sager pointed out that in zygote clone analysis one does not know whether all four meiotic products are ordinarily recovered. The inference is that aberrant nuclear events might affect the pattern of chloroplast gene transmission and result in the maternal skew. The data of Boynton et al. (19) were drawn from 253 complete tetrads derived from biparental zygotes; so aberrant segre-

TABLE 15–2. *Inheritance of chloroplast genomes in meiotic products of biparental zygotes*[a]

Chloroplast genotypes of the four meiotic products	Minimum number of chloroplast genomes in zygote after replication	Observed frequency of tetrad type	
		Number	Percentage
Zygotes with maternal bias			
3 M/M 1 M/P —	7 M 1 P	52	21
2 M/M 2 M/P —	6 M 2 P	60	24
3 M/M — 1 P/P	6 M 2 P	0	0
1 M/M 3 M/P —	5 M 3 P	51	20
2 M/M 1 M/P 1 P/P	5 M 3 P	13	5
Zygotes with no bias			
1 M/M 2 M/P 1 P/P	4 M 4 P	27	11
— 4 M/P —	4 M 4 P	15	6
2 M/M — 2 P/P	4 M 4 P	3	1
Zygotes with paternal bias			
— 3 M/P 1 P/P	3 M 5 P	12	5
1 M/M 1 M/P 2 P/P	3 M 5 P	3	1
— 2 M/P 2 P/P	2 M 6 P	11	4
1 M/M — 3 P/P	2 M 6 P	1	0.5
— 1 M/P 3 P/P	1 M 7 P	5	2

[a] Analysis is based on 1,012 meiotic products from 253 complete tetrads formed by biparental zygotes. A biparental zygote is here defined as one in which identifiable chloroplast markers (er-u-37, spr-u-1-6-2, sr-u-2-60) from both parents can still be scored at the end of meiosis. Each meiotic product is presumed to have a minimum of two copies of the chloroplast genome. Cross: M/M mt$^+$ × P/P mt$^-$, where M/M indicates maternal genotype, P/P indicates paternal genotype, and M/P indicates biparental genotype (carrying chloroplast markers from both parents). From Boynton et al. (19), with permission.

gation clearly does not explain the maternal skew they found (Table 15–2). Second, Sager pointed out that the markers used by Boynton and associates were closely linked and might not have been well distributed around the chloroplast genome. In fact, the markers in question cover between a quarter and a third of the cosegregation map (see Fig. 14–7). More important, all of the markers used by Sager, except the closely linked ac-u-1 and ac-u-2 markers and tm-u-1, were resistant to antibiotics and caused chloroplast ribosomal alterations (see Chapter 21). In short, a large part of the argument that a limited portion of the map is covered by the markers used by Boynton and associates hangs on the assumption that the rest of the genome is defined by just two other genes (ac-u-1 and ac-u-2, being so closely linked that they are usually treated as one by Sager and Ramanis). In this respect, cosegregation analysis of the ac-u markers and sr-u-sm5 becomes of par-

ticular interest (20), since this marker maps relatively close to the ac-u genes (see Fig. 14–7). Harris et al. (21) showed that sr-u-sm5 is an allele at the sr-u-2-23 locus that maps between the er and sr-u-sm2 loci. If this map position is correct, it means that the markers used by Boynton et al. (19) very likely span 50% or more of the map. The point is that without new markers, preferably ones not affecting chloroplast ribosomes, it is difficult to know how much of the map is really being studied.

Third, Sager pointed out that Gillham and Boynton did not distinguish between events occurring in zygotes and in zoospores. This is true. The data published by Boynton and associates pertain only to meiotic products. Limited data published for a single marker by Gillham (16,22) suggest that the maternal marker skew is still present in zoospore progeny clones. But these data are much too skimpy; they must be expanded before any decision can be made vis-à-vis chloroplast marker segregation in zoospore clones under the conditions used by Gillham and Boynton.

Frequencies of Spontaneous Exceptional Zygotes

The frequencies of spontaneous exceptional zygotes reported by Sager have consistently been 10- to 100-fold less than that reported by Gillham and Boynton (Table 15–3). These differences raise questions about the natures of the biparental zygotes being sampled in the two laboratories. It may be that Sager and Ramanis simply do not recover that class of biparental zygotes with skewed allelic ratios among the spontaneous class. Perhaps this class is also magnified in the experiments of Gillham and Boynton at low irradiation doses, but not in the experiments of Sager and Ramanis. Sager and Schlanger (13) argued that the strains used by Gillham and Boynton may have contained a molecular mechanism akin to *mat-1* or about 30 sec of ultra-

violet irradiation of the *mt*[+] parent under their conditions. They also suggested the possibility that in the zygotes of Gillham and Boynton there was an imbalance of modification and restriction activities such that partial degradation and marker rescue occurred and gave rise to zygotes carrying some chloroplast genes from the paternal parent, which were thus scored as exceptional, but with an excess of copies per zygote from the maternal parent. Another explanation might be that subtle differences in mating or zygote maturation conditions led to the loss of a large fraction of biparental zygotes in the experiments of Sager and Ramanis. Since germination was high in their experiments, these zygotes presumably would be converted principally to maternal zygotes.

Patterns of Segregation in Pedigrees for Several Markers

Gillham and associates (7) attempted to explain both type II and type III segregations in terms of reciprocal recombination. They assumed that the former exchanges occurred in cells with multiple copies of the chloroplast genome and that the latter occurred in cells with only two copies. They ignored the fact that type II and type III events can occur at the same generation in the same cell for different markers (Table 3 ref. 14). This makes the hypothesis of Gillham and associates untenable, at least if it is assumed that type II and type III events are frequent in the same cell division. It seems much more likely that the type II events do occur by some form of nonreciprocal recombination, as was proposed by Sager (12). Many years ago Gillham (23) showed that there was no correlation between recombinants in zygote clone analysis of the progeny of spontaneous biparental zygotes. Evidently this analysis needs to be extended to biparental zygotes induced by ultraviolet irradiation. In addition, pedigree analysis should now be done at the oc-

TABLE 15–3. *Percentages of spontaneous exceptional zygotes in different crosses.*

Cross	Percentage spontaneous exceptional zygotes	
	Sager and Ramanis	Gillham
1	0.08	1.98
2	0.10	3.28
3	0.03	7.20
4	0.40	2.22
5	0.60	0.98
6	0.10	4.35
7	0.60	4.05
8	0.10	3.40
9	0.10	4.60
10	0.36	13.10
11	0.35	0.96
12	0.06	0.30
13	2.80	3.16
14	0.80	6.33

Data are summarized from Table 10 of Sager and Ramanis (20) and Table 2 of Gillham (16). The original tables should be consulted with reference to the genotypes used and sample sizes.

tospore and octospore daughter stages on the stocks used by Gillham and Boynton. In summary, the simplest explanation of Sager's results is that recombination is occurring between only two copies. How else does one explain type III segregations and the fact that type II and type III events occur in the same cell? The question is whether or not this solution is satisfactory for the results reported by Gillham and Boynton, where there was a strong maternal skew and a high spontaneous biparental zygote frequency.

CONCLUSIONS

Neither of the current models for the inheritance of chloroplast genes in *Chlamydomonas* is really satisfactory. At least three reasons can be cited for the evolution of such divergent models. First, different methods were used by Sager and Ramanis and Gillham and Boynton. Second, distinctly different results were obtained in the two laboratories with respect to allelic ratios, spontaneous exceptional zygote frequencies, etc. Third, both models are probably correct in part; thus the particular method of analysis and the results obtained seem to support the model of choice. It does not help that our knowledge of the inheritance of clDNA in *C. reinhardtii* is also a morass of conflicting results. Despite a nearly irresistable urge to lay out here what he believes to be the ultimate solution, the author will refrain from doing so; prudence and experience dictate that this ultimate solution will in time accumulate difficulties of its own. Suffice it to say that until we are really certain how clDNA is inherited in *C. reinhardtii*, until we know whether or not there is a unique diploid clDNA yet to be discovered, and until we understand the differences in genetic results obtained by Sager and Ramanis and Gillham and Boynton, a really satisfactory model of chloroplast gene inheritance in *C. reinhardtii* is likely to remain an elusive goal.

REFERENCES

1. Sager, R., and Ramanis, Z. (1968): The pattern of segregation of cytoplasmic genes in *Chlamydomonas. Proc. Natl. Acad. Sci. U.S.A.,* 61:324–331.
2. Sager, R. (1977): Genetic analysis of chloroplast DNA in *Chlamydomonas. Adv. Genet.,* 19:287–340.
3. Sager, R., and Ramanis, Z. (1967): Biparental inheritance of nonchromosomal genes induced by ultraviolet irradiation. *Proc. Natl. Acad. Sci. U.S.A.,* 58:931–937.
4. Sager, R., and Ramanis, Z. (1973): The mechanism of maternal inheritance in *Chlamydomonas:* Biochemical and genetic studies. *Theor. Appl. Genet.,* 43:101–108.
5. Sager, R., and Ramanis, Z. (1974): Mutations that alter the transmission of chloroplast genes in *Chlamydomonas. Proc. Natl. Acad. Sci. U.S.A.,* 71:4698–4702.
6. Arber, W. (1974): DNA modification and restriction. *Prog. Nucleic Acid Res. Mol. Biol.,* 14:1–35.
7. Gillham, N. W., Boynton, J. E., and Lee, R. W. (1974): Segregation and recombination of non-Mendelian genes in *Chlamydomonas. Genetics,* 78:439–457.
8. Clowes, R. C. (1972): Molecular structure of bacterial plasmids. *Bacteriol. Rev.,* 36:361–405.
9. Adams, G. M. W., Van Winkle-Swift, K. P., Gillham, N. W., and Boynton, J. E. (1976): Plastid inheritance in *Chlamydomonas reinhardtii.* In: *The Genetics of Algae,* edited by R. A. Lewin, Chapter 5. Blackwell Scientific, Oxford.
10. Wells, R., and Sager, R. (1971): Denaturation and renaturation kinetics of chloroplast DNA from *Chlamydomonas reinhardi. J. Mol. Biol.,* 58:611–622.
11. Sager, R., and Ramanis, Z. (1970): A genetic map of non-Mendelian genes in *Chlamydomonas. Proc. Natl. Acad. Sci. U.S.A.,* 65:593–600.
12. Sager, R. (1972): *Cytoplasmic Genes and Organelles.* Academic Press, New York.
13. Sager, R., and Schlanger, G. (1976): Chloroplast DNA: Physical and genetic studies. In: *Handbook of Genetics,* edited by R. C. King, Chapter 12. Plenum Press, New York.
14. Sager, R., and Ramanis, Z. (1976): Chloroplast genetics of *Chlamydomonas.* I. Allelic segregation ratios. *Genetics,* 83:303–321.
15. Wurtz, E. A., Boynton, J. E., and Gillham, N. W. (1977): Perturbation of chloroplast DNA amounts and chloroplast gene transmission in *Chlamydomonas reinhardtii* by 5-fluorodeoxyuridine. *Proc. Natl. Acad. Sci. U.S.A.,* 74:4552–4556.
16. Gillham, N. W. (1969): Uniparental inheritance in *Chlamydomonas reinhardi. Am. Naturalist,* 103:355–388.

17. Goodenough, U. W., Hwang, C., and Martin, H. (1976): Isolation and genetic analysis of mutant strains of *Chlamydomonas reinhardi* defective in gametic differentiation. *Genetics,* 82:169–186.

18. Adams, G. M. W. (1975): An investigation of the mechanism of inheritance of chloroplast genes in *Chlamydomonas reinhardtii*. Ph.D. thesis, Duke University, Durham, N.C.

19. Boynton, J. E., Gillham, N. W., Harris, E. H., Tingle, C. L., Van Winkle-Swift, K., and Adams, G. M. W. (1976): Transmission, segregation and recombination of chloroplast genes in *Chlamydomonas*. In: *Genetics and Biogenesis of Chloroplasts and Mitochondria,* edited by T. Bücher, W. Neupert, W. Sebald, and S. Werner, pp. 313–322. North Holland, Amsterdam.

20. Sager, R., and Ramanis, Z. (1976): Chloroplast genetics of *Chlamydomonas*. II. Mapping by cosegregation frequency analysis. *Genetics,* 83:323–340.

21. Harris, E. H., Boynton, J. E., Gillham, N. W., Tingle, C. L., and Fox, S. B. (1977): Mapping of chloroplast genes involved in chloroplast ribosome biogenesis in *Chlamydomonas reinhardtii*. *Mol. Gen. Genet.,* 155:249–265.

22. Gillham, N. W. (1963): The nature of exceptions to the pattern of uniparental inheritance for high level streptomycin resistance in *Chlamydomonas reinhardi*. *Genetics,* 48:431–439.

23. Gillham, N. W. (1965): Linkage and recombination between nonchromosomal mutations in *Chlamydomonas reinhardi*. *Proc. Natl. Acad. Sci. U.S.A.,* 54:1560–1567.

Chapter 16
Chloroplast Continuity in *Euglena*

At first glance it might seem odd to discuss *Euglena gracilis* in a book on organelle heredity, for this flagellate has no known mechanism of sexual reproduction, and genetic analysis in the usual sense is not possible. On the other hand, *E. gracilis* has a property that *C. reinhardtii* appears to lack. The organism can be permanently bleached, with complete loss of clDNA, but it will continue to grow normally as long as it is supplied with an appropriate carbon source. In this sense *E. gracilis* has the same advantage for the study of chloroplast heredity that yeast has for the study of mitochondrial heredity. The organelle of interest can be permanently inactivated and its DNA eliminated, but the cells will survive under the appropriate growth conditions.

Spontaneous bleaching in *E. gracilis* was observed as early as 1912 by Ternetz (1), and it was later reported by several other authors. In 1948 Provasoli et al. (2) reported that streptomycin induced mass transformation of *E. gracilis* to the permanently bleached state; thus for the first time there was a method for controlled bleaching of all cells in a population. They also showed that as long as a fixed carbon source was provided, *E. gracilis* could grow in the presence of very high streptomycin concentrations.

The discovery that streptomycin caused bleaching in *Euglena* stimulated considerable interest, and the first phase of the investigations into bleaching came to a close around 1960 with the experiments of DeDeken-Grenson, who attempted to establish the mechanism of loss of plastids (i.e., whether or not they were diluted out) by means of careful kinetic experiments and pedigree studies (3). The second phase began in 1959 when Lyman, Schiff, and Ep-

stein initiated a series of novel studies on the mechanism by which ultraviolet irradiation bleaches *Euglena* cells under conditions such that there is no loss in viability (4–10). These experiments were later extended by Uzzo and Lyman (11,12) in an attempt to explain how heat treatment, which had been shown by Pringsheim and Pringsheim (13) to bleach *Euglena,* acted. Meanwhile, particularly during the second phase, more and more bleaching agents were being discovered. In fact, the hunt for new bleaching agents became something similar to a national pastime for euglenologists. It became apparent that many of these agents could be classified on the basis of whether they affected DNA replication or chloroplast protein synthesis. A major aim of this chapter will be to explore the bleaching experiments in detail. We shall also examine a few of the numerous studies that were made on chloroplast development in *Euglena,* particularly as they relate to the bleaching question, for it is evident that reagents that block clDNA replication do not prevent dark-grown cells from differentiating their proplastids into plastids when placed in the light, whereas this is not true of inhibitors of chloroplast protein synthesis. Finally, we shall make a number of comparisons between the *Chlamydomonas* and *Euglena* systems and consider the properties of *Euglena* mutants affecting chloroplast function, since there is some reason to believe that these may be localized in clDNA.

THE ORGANISM AND ITS GROWTH

The classification of euglenoid flagellates and the biology of *Euglena* species have been reviewed extensively in two volumes to which the interested reader is referred (14,

15). *E. glacilis* (Figs. 16–1 and 16–2) is a unicellular flagellate about 150 μm long (including the whip-like flagellum); as far as is known, it completely lacks a sexual cycle (14,16). The organism possesses numerous mitochondria and 10 to 12 chloroplasts (14,15). The chromosome number has been estimated from cytological examination to be around 45 (14,16). Schiff and his colleagues (9,17), on the basis of target analysis using ultraviolet light and X-rays, claimed that the nucleus was octaploid; but Rawson (18), using DNA renaturation kinetics, estimated that the nucleus was probably diploid. There seems to be no cytological proof of polyploidy, although it is generally assumed that some, and perhaps most, *Euglena* species are polyploid (16).

Three species of DNA can be identified in *E. gracilis* (Table 16–1). The nDNA has a buoyant density of 1.707 g/cc; the two minor species, called S_c and S_m, have buoyant densities of 1.686 and 1.691 g/cc, respectively. The S_c satellite is derived from the chloroplast and the S_m from the mitochondria. The clDNA of *E. gracilis* consists of 44-μm circles (19). These molecules have molecular weights of 9.2×10^7 d. Summation of the molecular weights of fragments produced following digestion by EcoR1 endonuclease digestion gives a molecular weight 5 to 10% lower (20,21). As techniques have been refined, estimates of the number of clDNA molecules have been revised progressively upward, so that it now seems that light-grown cells contain somewhere between 1,400 and 2,900 copies of clDNA per cell, with the number estimated depending on the precise conditions of growth used and on the experimenter (22, 23). This would yield somewhere between 100 and 300 clDNA molecules per chloroplast, which is considerably higher than the number of targets for ultraviolet bleaching

per chloroplast, as we shall see shortly. An early report (24) indicated that the mitochondrial genome of *E. gracilis* was very small, with a molecular weight of 3×10^6 d, but more recently a molecular weight of 4×10^7 to 5×10^7 d has been estimated on the basis of renaturation kinetic data (25).

As with *C. reinhardtii*, the commonly used strains of *E. gracilis*, the *bacillaris* or B strain and the Z strain (14), can be grown phototrophically, mixotrophically, or heterotrophically. Phototrophic conditions are restrictive for photosynthetic function, as they are in *C. reinhardtii*, but the other two growth conditions are permissive. As early as 1900 Zumstein observed that heterotrophic growth of *E. gracilis* caused it to lose its chlorophyll-forming ability (26). More recent studies have revealed that in heterotrophically grown cells the chloroplasts are replaced by structures termed proplastids (Fig. 16–2) that are similar in many respects to the proplastids formed by higher plants grown in the dark (see Chapter 1). The *Euglena* proplastids are 1 to 2 μm in size and roughly spheroidal (Fig. 16–2). They are bounded by a double membrane like that of the chloroplast, but they contain very little chlorophyll or thylakoid membrane material (27). Like the proplastids of higher plants, the *Euglena* proplastids contain prolamellar bodies (27). As Kirk and Tilney-Bassett (28) pointed out, the name proplastid really should be reserved for the undifferentiated plastid-like structures of light-grown cells (see Chapter 1), and the dedifferentiated chloroplast precursors seen in the dark should be termed etioplasts. However, this usage has not gained general acceptance and proplastid is still routinely used instead of etioplast.

When heterotrophically grown cells of

$\longrightarrow$

FIG. 16–1. Cell of *E. gracilis* strain B grown in the light for 72 hr. $\times$12,480. Key: E = endosome; G = Golgi; M = mitochondrion; N = nucleus; P = plastid; Pm = paramylum; V = vacuole. (From Schiff and Zeldin, ref. 79, with permission.)

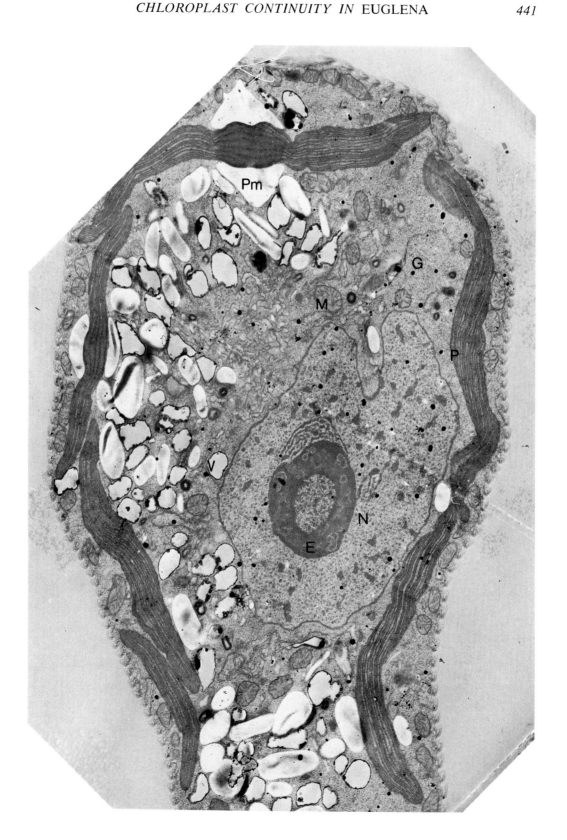

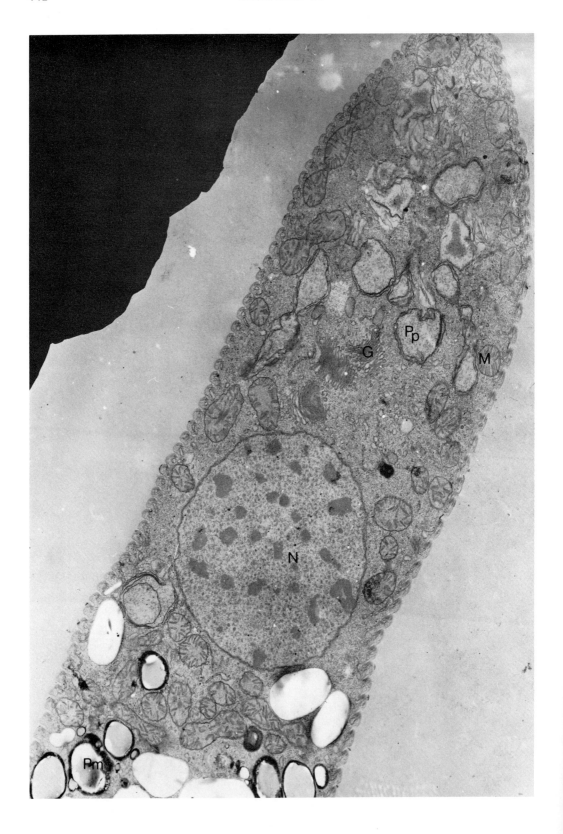

Euglena are exposed to light, chlorophyll and thylakoid membrane accumulation begins (Fig. 16–3). Over a period of hours the proplastids develop into photosynthetically competent chloroplasts. The differentiation process has been studied in great detail, particularly by Schiff and his colleagues (27); it has been used as an experimental system in which to determine the photosynthetic components that are synthesized in chloroplast and cytoplasm, which are determined by the use of antibiotics specific for the chloroplast and cytoplasmic protein-synthesizing systems. In addition, this system can be used to study mutants blocked in chloroplast development in order to determine the point at which the developmental sequence is blocked. We shall have much more to say about these experiments later in the chapter.

A number of different mutants have been described in *E. gracilis,* and their properties have been reviewed by Schiff et al. (17). They include a variety of phenotypes; the most important of these, for our purposes, are the mutants that affect plastid function or that cause cells to become resistant to bleaching by a specific reagent (Table 16–2). The system used to describe mutants in *E. gracilis* consists of a four-part code. The first part describes the phenotype of the mutant (e.g., Y = yellow color, Sm^r = streptomycin resistance); the second part indicates the parental strain in which the mutant was obtained (i.e., B = *bacillaris* strain, Z = Z strain); the third part indicates the mutagen used (e.g., S = spontaneous, X = X-ray, U = ultraviolet light, Ng = nitrosoguanidine); the fourth part indicates whether light- or dark-grown cells were mutagenized (i.e., L = light-grown cells, D = dark-grown cells). All mutants within a given phenotypic class are numbered in order of isolation (or, in the future, in order of appearance in the published literature) by a numerical subscript after the first symbol (e.g., P_1 = the first pale green mutant isolated). For example, P_1BXL is the first pale green mutant isolated, and it was obtained by mutagenizing light-grown cells of strain B with X-rays. The mutant Sm^r_1BNgL is the first streptomycin-resistant mutant isolated, and it was obtained by mutagenizing light-grown cells of strain B with nitrosoguanidine.

BLEACHING EXPERIMENTS WITH ULTRAVIOLET LIGHT

Although the kinetic studies of DeDeken-Grenson (3) on streptomycin-induced bleaching of *E. gracilis* antedate those of Lyman, Schiff, and Epstein, who characterized the ultraviolet irradiation bleaching process, it is easier to understand the whole bleaching phenomenon if we begin with the latter experiments and the model they generated. Lyman et al. (5) first established that 100% of the cells in an irradiated population could be bleached without attendant lethality provided that the doses of ultraviolet irradiation used were low enough. Target analysis of the bleaching curves (see Chapter 4) revealed that there were 30 ultraviolet-sensitive targets per cell, all of which had to be inactivated to produce a permanently bleached cell. This was true whether the cells were grown in the dark (where they contained proplastids) or in the light (where they contained fully differentiated chloroplasts). However, the sensitivity of the proplastids to ultraviolet irradiation was much greater than that of the plastids. The authors considered several explanations for these differences in sensitivity to bleaching; they favored the hypothesis that the ultraviolet-sensitive targets became incorporated into the fully formed

FIG. 16–2. Dark-grown cell of *E. gracilis* strain B. ×10,863. Same abbreviations as in Fig. 16–1, except Pp = proplastid. (Courtesy of Dr. J. A. Schiff.)

TABLE 16–1. DNA of E. gracilis

Source and method	Main band AT	Main band GC[a]	Chloroplast satellite AT	Chloroplast satellite GC	Mitochondrial satellite AT	Mitochondrial satellite GC	References
E. gracilis strain *bacillaris*							
Density	52	48	74	26	69	31	34,68–70
Base composition analysis	47	53	76	24	—	—	71
Thermal denaturation	45	55	70	30	—	—	69
E. gracilis strain Z							
Density	51	49	76	24	67	33	72,73
Base composition analysis	49	51	75	25	—	—	72,73
Thermal denaturation[b]	47–52	48–53	74–79	21–26	—	—	72,73
Density (g/cc)	1.707		1.686		1.691		34,68–73

[a] Includes approximately 2.3% methyl cytosine (71,74). Main band is assumed to represent DNA.
[b] Calculated from data given in references.
Adapted from Schiff (32).

chloroplasts and the chloroplasts created a shielding effect that protected these sites from inactivation by ultraviolet irradiation.

Schiff et al. (6) then showed that ultraviolet irradiation bleaching of *Euglena* could be photoreactivated with 100% efficiency. The action spectrum for photoreactivation of ultraviolet irradiation bleaching in *Euglena* had a broad peak in the blue region (330–460 nm) and was similar to the action spectrum for photoreactivation in phage T2, *E. coli,* and *Neurospora* (29,30). A similar action spectrum is seen for the activity of the photoreactivating enzymes of yeast and *E. coli* (29,30). These enzymes split covalent linkages between adjacent pyrimidines in DNA formed as a result of ultraviolet irradiation. These pyrimidine dimer photoproducts are believed to be largely responsible for the lethal effects of ultraviolet irradiation (29). A comparison between dividing and nondividing *Euglena* cells revealed that only the former cells lost the ability to be photoreactivated, and their number increased at each cell generation.

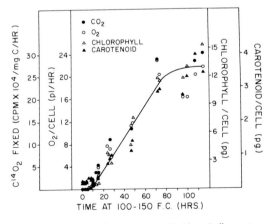

FIG. 16–3. Kinetics of appearance of chlorophyll, carotenoids, photosynthetic oxygen evolution, and photosynthetic carbon dioxide fixation during chloroplast development in *Euglena*. Zero time represents measurements on dark-grown cells; time is measured from inception of light-induced chloroplast development. (From Schiff, ref. 27, with permission.)

TABLE 16–2. Some mutants of E. gracilis

Phenotypic class	Genotypic designations of mutants in each class
Golden color	G_1BU
Olive green color	O_1BS, O_2BX
Pale green color	P_1BXL, P_4ZUL, P_7ZNgL, P_8ZNgL, P_9ZNgL, $P_{10}ZNaIL$, $P_{27}ZNaIL$
Orange color	O_r
White color, permanently bleached	W_1BUL, W_1ZXL, W_2ZUL, W_3BUL, W_4BUL, W_8BHL, $W_{10}BSmL$, $W_{30}BS$, $W_{33}ZUL$
Yellow color	Y_1BXD, Y_2BUL, Y_3BUD, Y_6ZNaIL, Y_9ZNaIL, $Y_{11}P_{27}DL$
Resistant to bleaching by streptomycin	$Sm_1{}^rBNgL$

Adapted from Schiff et al. (17).

Schiff and associates then attempted to calculate the number of photoreactivable entities per cell from the kinetics of loss of photoreactivability in dividing cells so that they could compare that number with the number of irradiation targets. The reasoning behind these calculations was as follows: It is assumed that irradiation treatment prevents replication of the photoreactivable entities because irradiation inactivates the targets whose functioning is required for chloroplast formation. Treatment with photoreactivating light restores the ability

of the targets (and therefore the entities including them) to replicate. In nondividing cells there is no dilution of the irradiation-inactivated entities because there is no cell division; so treatment of a potential irradiation-bleached population of nondividing cells at any time with photoreactivating light can rescue the ability of all the cells in the population to form normal green chloroplasts (Fig. 16–4), presumably through the action of a chloroplast photoreactivating enzyme. In dividing cells, on the other hand, the number of photoreactivable

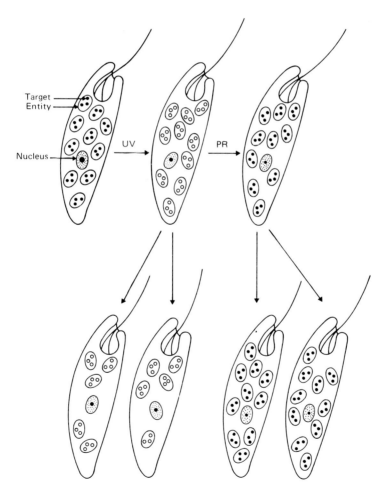

FIG. 16–4. Diagram of model proposed by Schiff and colleagues to explain ultraviolet irradiation bleaching in *Euglena*. Each cell contains 10 entities, presumably chloroplasts, with each entity containing three irradiation-sensitive targets. Following irradiation, all targets are inactivated. In the absence of photoreactivation (PR), cell division results in dilution of the entities. If the targets are photoreactivated the entities are replicated and distributed normally with cell division. Active targets are represented by filled circles and inactive targets by open circles. It should be noted that the number of targets is much lower than the number of clDNA molecules per plastid.

entities will decline at each cell generation because the entities are not dividing, whereas the cells are (Fig. 16–4). If a cell has n photoreactivable entities to begin with, all of which are inactivated by ultraviolet irradiation, then as cell division proceeds, each of the two daughter cells should have the same probability of obtaining a given entity (*Euglena* divides longitudinally into two fairly equal halves). The distribution of entities among cells at the first generation will then be given by the expansion of $(p + q)^n$ or $(0.5 + 0.5)^n$. Since white and green colonies are the only items scored at each generation in the experiments, it is necessary to sum all probability classes containing one or more photoreactivable entities and assume that any cell containing at least one entity can be photoreactivated and will form a green colony. From this calculation the expected percentage of green colonies at each generation can be estimated for different numbers of entities (n).

Theoretical dilution curves were calculated for 4, 8, 10, and 12 entities, and it was found that none of these curves fit the observed data after about three generations, since the percentage of photoreactivable cells was less than predicted (Fig. 16–5). When the assumption was made that a cell containing either 0 or 1 photoreactivable entity would produce a colony scored as colorless, a good fit to the observed data was obtained for 10 photoreactivable entities per cell (Fig. 16–5). The difference between 30 targets for irradiation bleaching and 10 photoreactivable entities from dilution kinetics was reconciled in the following manner. It was hypothesized that the 10 to 12 plastids in each *Euglena* cell contained three targets for irradiation inactivation. All 30 targets had to be inactivated by irradiation to produce a bleached cell. When all 30 targets were inactivated, the 10 to 12 plastids could not replicate and were diluted out according to dilution kinetics, giving a number of 10.

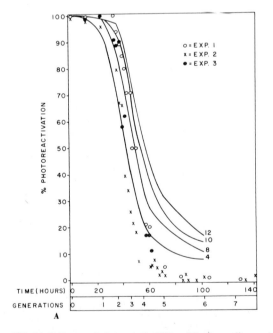 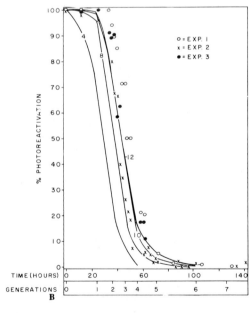

FIG. 16–5. Decay of photoreactivability of *Euglena* cells growing in dark following bleaching dose of irradiation. Lines are theoretical curves based on random assortment of number of photoreactivable entities indicated at end of each curve. A: Uncorrected curves. B: Curves corrected for the possibility that cells with only one photoreactivable entity may yield apparently albino colonies. (From Schiff et al., ref. 6, with permission.)

At first it was thought that the 30 ultraviolet-sensitive targets were equivalent to or were contained within 30 proplastids believed to be present in the dark-grown cell (5). It was believed that on exposure to light the proplastids coalesced in groups of three to form mature chloroplasts. The reason behind this assumption of 30 proplastids per cell was the observation of Epstein and Schiff (31) that there were about 30 red-fluorescing entities per dark-grown cell. It now appears likely that these red-fluorescing entities are prolamellar bodies rather than proplastids, if, as in higher plants, protochlorophyll(ide) and chlorophyll(ide) are confined to the prolamellar bodies and associated thylakoid materials (32). Thus, there will be 10 proplastids per cell, each containing an average of three prolamellar bodies. Each proplastid will give rise to a chloroplast (10 in all, as observed), obviating the need for proplastid aggregation, as previously proposed (6). The significance, if any, of the correlation between the number of irradiation-sensitive targets and the number of prolamellar bodies per proplastid is unknown.

The underlying assumption in all of these experiments was that the targets for irradiation inactivation were required for the replicative continuity of the chloroplast, but it is also possible that the ultraviolet treatment prevented these targets from participating in the formation of a functional chloroplast. Evidence that a functional chloroplast could be formed came from another experiment (7). Red light induced differentiation of proplastids into chloroplasts in *Euglena,* but it did not photoreactivate irradiation damage. If dark-grown *Euglena* cells were treated with ultraviolet irradiation and then exposed to red light, plastid differentiation proceeded normally; but these chloroplasts were diluted out during growth in the dark, and thus irradiation-induced bleaching was still observed (Table 16–3).

Are the irradiation targets clDNA molecules? Lyman et al. (5) showed that the action spectrum of irradiation inactivation of green colony formation had a major peak near 260 nm and a minor peak at 280 nm. Nucleic acids absorb maximally around 260 nm, and the aromatic amino acids present in protein absorb maximally at approximately 280 nm. These results suggested that a nucleic acid was the target and that its inactivation destroyed the replicative continuity of the chloroplast, but replication of this nucleic acid might depend on the proper functioning of some protein or group of proteins. The action spectrum of photo-

TABLE 16–3. *Effect of ultraviolet light on chloroplast development and replication in* E. gracilis[a]

Ultraviolet	Light treatment following ultraviolet irradiation	Chloroplast development[b]	Green colonies (%)	
			Light	Dark
−	White[c]	Normal	100	100
−	Red[d]	Normal	100	100
+	White	Normal	100	100
+	Red	Normal	100	0
+	Dark	—	—	0

[a] After ultraviolet irradiation, cells were exposed to light conditions shown for 72 hr; chloroplast and chlorophyll formation was observed in aliquots by fluorescence microscopy; after 72 hr cells from each set of treatments were plated on duplicate sets of plates; one set of plates was placed in white light for 7 days; the other was placed in darkness for 5 days, followed by 2 days in the light.

[b] Followed over 72 hr by fluorescence microscopy prior to plating.

[c] Photoreactivating light.

[d] Nonphotoreactivating light.

Adapted from Schiff et al. (7).

TABLE 16–4. *Effects on green-colony formation of irradiating nuclei and cytoplasms of single cells of E. gracilis with ultraviolet microbeam*

Target irradiated	Size of ultraviolet microbeam (μ^2)	Total cells irradiated	Percentage bleached or mixed colonies[a]
Nucleus	16	163	0
	32	84	0
Cytoplasm	350	92	52
Total cell	380	98	41

[a] Mixed means that the colony contained both green and white cells.
From Gibor and Granick (33), with permission.

reactivation also suggests a nucleic acid target, since it is similar to the action spectrum of the photoreactivating enzyme that, as mentioned previously, splits irradiation-induced pyrimidine dimers in DNA that are otherwise lethal. The nucleic acid target in question appears to be localized in the cytoplasm, presumably in the chloroplasts, as was shown by the elegant ultraviolet microbeam experiments of Gibor and Granick (33) that were mentioned in Chapter 2. These authors irradiated *Euglena* cells with an ultraviolet microbeam. In some experiments the nucleus was shielded from irradiation, and cytoplasm was irradiated; in other experiments the reverse was obtained. Only when the cytoplasm was irradiated were bleached colonies observed (Table 16–4). Finally, Edelman et al. (34) found that irradiation-bleached mutants lacked the S_c satellite DNA that is known to be clDNA. Although these results are consistent with the idea that clDNA molecules are the targets for ultraviolet irradiation, careful measurements of numbers of clDNA molecules per cell (p. 452) clearly show that there are many more clDNA molecules than targets per cell. This creates a serious problem for the analysis, to which we shall return at the end of this discussion.

BLEACHING EXPERIMENTS WITH NALIDIXIC ACID

Nalidixic acid is a specific inhibitor of DNA replication in certain bacteria (35–37). Lyman (38) found that this inhibitor

also induced bleaching in *Euglena* under conditions such that cell growth and viability were not impaired. Since dark-grown cells regenerated chloroplasts in the presence of nalidixic acid when transferred to the light, it was evident that nalidixic acid was not blocking those functions involved in chloroplast differentiation that were dependent on transcription and translation within the chloroplast. Lyman also found that whereas nalidixic acid effectively bleached light-grown cells, it was ineffective in the case of dark-grown cells.

Subsequently, Lyman et al. (39) followed up these original studies with a more detailed account of the mechanism of nalidixic acid bleaching. They confirmed that nalidixic acid preferentially bleached light-grown cells; the bleaching reaction was greatly stimulated by the presence of a functioning photosynthetic electron transport chain. More important, however, Lyman and associates investigated the effect of nalidixic acid on preexisting clDNA, since it had been shown that DNA degradation occurred in bacteria whose DNA replication had been inhibited by nalidixic acid. They found that within 6 hr after addition of nalidixic acid there was considerable diminution in the amount of clDNA. Ultraviolet irradiation target analysis performed on these cells revealed that the target number had dropped from 30 to 10. It is interesting that cells plated at 6 hr produced only 20% bleached colonies.

The experiments of Lyman et al. established that bleaching of *E. gracilis* by nalidixic acid was accompanied by a decline in clDNA and a decline in irradiation target number. In short, as in some of the experiments described previously (p. 447), they seem to have established a correlation between clDNA and irradiation target number without explaining why there are so few targets in relation to total numbers of clDNA molecules.

HEAT BLEACHING EXPERIMENTS

In 1952 Pringsheim and Pringsheim (13) reported that exposure of *Euglena* cells to elevated temperatures resulted in irreversible bleaching of these cells (heat bleaching). More recently, the heat bleaching process has been examined in detail by Uzzo and Lyman (11,12). These experiments are important because they show that under appropriate conditions more targets than normal can be "visualized" (although this number is still well below the number of clDNA molecules per cell) and because they establish yet more correlations between clDNA and the ultraviolet irradiation targets.

The ability of dividing (but not nondividing) *Euglena* cells to form green colonies is affected by incubation at temperatures between 32°C and 34°C. Light-grown cells begin to lose this ability after three generations and dark-grown cells after two generations (Fig. 16–6). The heat bleaching process in dividing cells is accompanied by a gradual loss, presumably through dilution, of clDNA (Fig. 16–7). Heat probably does not affect clDNA replication per se, for Uzzo and Lyman found that the amount of clDNA doubles when dark-grown cells are transferred to the light under nondividing conditions, and this doubling takes place whether the cells are incubated at 26°C or 32°C. As in the case of ultraviolet irradiation or nalidixic acid inhibition, incubation at elevated temperature does not interfere with differentiation of proplastids into plastids. Thus it can be concluded that heat bleaching does not result because existing clDNA fails to function in the formation of chloroplasts.

Uzzo and Lyman also analyzed the kinetics of heat bleaching in dividing cells and concluded that they were consistent with the hypothesis that light-grown cells contain 20 entities whose replication is sensitive to heat, whereas dark-grown cells contain only 10 (Table 16–5). Since there are 10 proplastids, each of which forms a chloroplast, they argued that these findings were consistent with the hypothesis that each proplastid contains one heat-sensitive entity and each chloroplast contains two.

Despite the fact that light-grown *Euglena* cells appeared to have twice the amount of clDNA of dark-grown cells, the number of

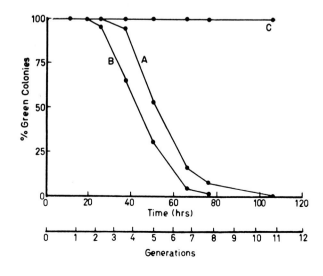

FIG. 16–6. Heat-induced bleaching of *Euglena* cells. Curve A is "bleaching" curve for light-grown cells at 32°C, curve B is curve for dark-grown cells at 32°C; curve C is for light- or dark-grown cells at 26°C. (From Uzzo and Lyman, ref. 12, with permission.)

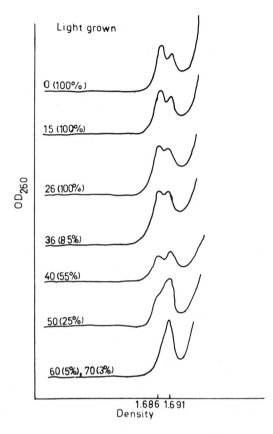

FIG. 16–7. Loss of plastid-associated DNA in *Euglena* during growth of cells in light at 32°C. Tracings were made with photoelectric scanner on DNA samples from whole cells run to equilibrium in CsCl gradients in the Spinco model E ultracentrifuge. Only the two organelle DNA satellites are shown in the figure, with clDNA having a buoyant density of 1.686 g/cc and mtDNA having a buoyant density of 1.691 g/cc. The left-hand number on each trace is the hour the sample was removed during growth; the percentage figure is the percentage green colonies seen when each sample is plated. (From Uzzo and Lyman, ref. 12, with permission.)

irradiation targets in each kind of cell at 26°C was 30. One way of explaining this paradox is that light-grown cells actually contained 60 targets, but only 30 of them could be titrated by irradiation inactivation kinetics. Evidence that this was indeed the case came from experiments with nondividing cells at 32°C. It was found that when cells grown in the light at 26°C were incubated in the light at 32°C under conditions such that they did not divide, the number of irradiation targets doubled over a period of 18 hr to 60, without an increase in the amount of clDNA (Fig. 16–8). Nondividing dark-grown cells incubated in the dark at 32°C showed no such doubling of irradiation targets (Fig. 16–8 and Table 16–5). When dark-grown cells were placed in the light under nondividing conditions at 32°C, it was found that the number of irradiation targets doubled from 30 to 60 (Fig. 16–8) over a period of 72 hr, which is the time necessary for proplastid clDNA to double during chloroplast synthesis. When nondividing cells incubated in the light at 32°C were returned to 26°C, the irradiation target numbers rapidly dropped to the original 30 once again. Uzzo and Lyman repeated most of these experiments with a pale green mutant of *E. gracilis* (P₄ZUL) that has only three chloroplasts (Table 16–5). This mutant has an irradiation target number of 9 at 26°C and 18 at 32°C. Dilution analysis defined three heat-sensitive sites in dark-grown cells of the mutant and six in light-grown cells. It appeared, therefore, that there was a constant relationship among the number of plastids, the number of temperature-sensitive sites, and the number of irradiation-sensitive sites. Uzzo and Lyman also showed that during growth of cells in the light at 32°C, the loss in clDNA (Fig. 16–7) was paralleled by the loss in number of irradiation targets per cell (Fig. 16–9). When the cells were exposed to 32°C, the irradiation target number first increased to 60. Over the first three generations of growth the irradiation target number stayed constant, as did the amount of clDNA. At generation three the irradiation target number and clDNA began precipitous and parallel declines over the next two generations to near zero. The proportion of permanently bleached cells in the population began to increase in parallel about one generation later, at a time when the number of irradiation targets per cell had dropped to somewhat under 30. Thus bleaching began as expected, when the number of irradiation targets dropped below the level normally present in a light-grown cell at 26°C.

Although it is evident from these experi-

TABLE 16–5. *Summary of numbers and types of sites involved in plastid replication in wild-type and mutant* Euglena

Genotype	Number of chloroplasts[a]	Number of proplastids[b]	Number of ultraviolet targets				Number of heat-sensitive sites	
			Light-grown 26°C	Dark-grown 26°C	Light-grown 32°C	Dark-grown 32°C	Light-grown	Dark-grown
Wild type	10	10	30	30	60	30	20	10
Mutant (P$_4$ZUL)	3	3	9	9	18	9	6	3

[a] Light-grown cells only contain chloroplasts.
[b] Dark-grown cells only contain proplastids.
Adapted from Uzzo and Lyman (12).

ments that the number of targets "visualized" by ultraviolet irradiation doubles at 32°C, this number is still much lower than the number of clDNA molecules per cell. At the time, Uzzo and Lyman did not realize this; they assumed that they were titrating all of the chloroplast genomes by target analysis at 32°C, but only half of them at 26°C. Thus they thought that the heat treatment allowed them to "visualize" all copies of clDNA, because they hypothesized that only half of the copies could participate in replication at 26°C but that all of them could do so, at least briefly, at 32°C. It now appears that all estimates of amounts of clDNA per cell of *E. gracilis* made until quite recently were severe underestimates because of the methodology employed. In the past, the amount of clDNA per cell was determined from knowledge of the total amount of DNA per cell or plastid

and the fraction of this total that appeared to band in the clDNA region on CsCl equilibrium gradients. Such gradients do not have a great deal of resolution, particularly with respect to the clDNA and mtDNA peaks (Fig. 16–7). Schiff (27) gave a value of 4.5×10^8 d of clDNA per plastid using this method. If we assume a molecular weight of 9.2×10^7 d for a molecule of clDNA (p. 440), this yields a value of about five clDNA molecules per plastid, which is in excellent agreement with the hypothesis of Uzzo and Lyman, as was pointed out by Schmidt and Lyman (40). More recently, however, Rawson and Boerma (22) and Chelm et al. (23) used renaturation kinetics to estimate the amount of clDNA per cell. In these experiments the renaturation rates of radioactive clDNA to itself and to total cellular DNA were compared, and from a comparison of the

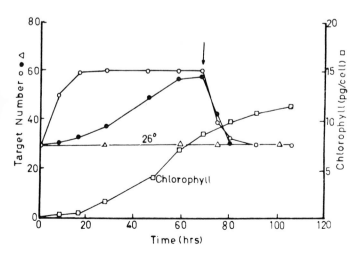

FIG. 16–8. Ultraviolet irradiation target analysis of nondividing *Euglena* cells at 32°C. Nondividing light- and dark-grown cells were incubated at 26°C and 32°C in both the light and darkness. At 70 hr (vertical arrow) the cultures at 32°C were returned to 26°C. Open circles: irradiation target number, light-grown cells in the light at 32°C. Filled circles: irradiation target number, dark-grown cells in the light at 32°C. Triangles: irradiation target number, dark-grown cells in the dark at 32°C. Squares: chlorophyll synthesis in dark-grown cells in the light at 32°C. (From Uzzo and Lyman, ref. 12, with permission.)

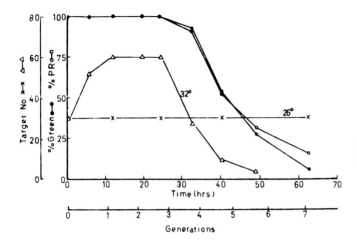

FIG. 16–9. Ultraviolet irradiation target analysis of light-grown *Euglena* cells dividing at 32°C. Light-grown cells were grown at 26°C and 32°C. At intervals, samples were removed for irradiation target analysis. Meaning of each symbol is indicated in figure, except that triangles and crosses represent target numbers of cells growing at 32°C and 26°C, respectively. (From Uzzo and Lyman, ref. 12, with permission.)

rate constants for reassociation, the fraction of clDNA in the cell was estimated. This method yields numbers ranging between 1,000 and 2,900 clDNA molecules per phototrophically grown cell. Assuming the same value as above for the molecular weight of a clDNA molecule, one obtains molecular weights of clDNA per chloroplast between 9.2×10^9 d and 27×10^9 d, or 20 to 60 times more clDNA per plastid than is estimated from buoyant-density centrifugation. Assuming that the values for clDNA measured using reassociation kinetics are correct, the number of irradiation targets per cell is clearly much less than the total number of clDNA molecules; yet the correlation among treatments that vary the amounts of clDNA and target numbers seems indisputable from Lyman's experiments. This paradox can be explained in either of two ways. First, either target analyses are yielding inaccurate estimates of the number of clDNA molecules per cell or renaturation kinetics are giving the wrong amount of clDNA per cell. It seems unlikely that the latter situation applies. Second, both estimates are meaningful, but the genetically important "titratable" clDNA molecules that can be knocked out by ultraviolet irradiation are a small fraction of the total. Perhaps this possibility should not be dismissed summarily in view of the experimental data described previously regarding

target analysis (see Chapters 4 and 5) and the rates of organelle gene segregation (see Chapters 7 and 14); these data suggest that biologic estimates of numbers of organelle genomes always tend to be less than would be expected from a knowledge of the number of organelle DNA molecules per cell.

BLEACHING OF *EUGLENA* BY STREPTOMYCIN

After 1948, when Provasoli et al. (2) discovered that streptomycin induced mass bleaching of *Euglena,* the mechanism of this transformation became the subject of a number of studies notably those of De-Deken-Grenson (41–43). These experiments were reviewed by Grenson (3).

It was soon evident that streptomycin-induced transformation of green cells took place without any loss of cell viability and, under suitable conditions, without even a change in growth rate. DeDeken-Grenson examined a number of physiological parameters related to streptomycin-induced bleaching and found that continuous exposure to the drug was not necessary and that exposure times as short as 30 min to 2 hr at suitable concentrations were sufficient to cause bleaching.

DeDeken-Grenson and Messin (42) made a detailed study of the bleaching process of cells exposed to streptomycin for a short time. They found that treatment of a culture of growing *Euglena* cells with 0.2% streptomycin for 30 min blocked chlorophyll accumulation in the entire culture for several generations without substantially influencing growth rate. At the end of that time, chlorophyll accumulation resumed at an exponential rate. Obviously these results meant that the amount of chlorophyll per cell was declining during the period of cell growth without net chlorophyll accumulation. DeDeken-Grenson and Messin not only showed this to be the case but also found that the number of chloroplasts per cell appeared to decline as well. They concluded that the amount of chlorophyll per cell was a good estimate of the number of plastids per cell; they hypothesized that during the period when chlorophyll was not accumulated, the plastids were diluting out, so that when chlorophyll synthesis resumed, most of the cells were permanently bleached because they lacked plastids, and chlorophyll synthesis resumed in the remainder. Following resumption of chlorophyll synthesis, the normal numbers of chloroplasts were restored rapidly in those cells in which chlorophyll synthesis occurred.

DeDeken-Grenson and Messin attempted to test the dilution hypothesis by examining the kinetics of white-colony formation as a function of the number of generations following exposure of green cells to the drug, assuming that each cell had an average of eight chloroplasts per cell at the beginning of the experiment and that one chloroplast per cell was sufficient to yield a green colony. They also assumed, from their chloroplast counting experiments, that there actually was a one-generation lag following streptomycin treatment before chloroplast dilution began. That is, at generation 1 following treatment, each cell contained eight rather than four chloroplasts, as would be expected if dilution began at generation 0. The experimental results showed that the onset of bleaching was more rapid than

predicted; it was dose-dependent and was even variable at a given concentration.

These authors postulated from their results that the deficiencies in green colonies predicted and expected might result because cells containing a single plastid might frequently produce a bleached colony. For example, if a cell containing one plastid were plated out, it would produce three bleached cells and one green cell two generations later. Since there appeared to be a reasonable amount of mortality among cells within colonies growing on solid medium, such a colony would be scored as bleached 75% of the time and green only 25% of the time, assuming there was no selective mortality. Furthermore, if there were variability in the timing of reinitiation of chloroplast synthesis in cells with one plastid, depending on streptomycin concentration or physiological state of the cells at the time of streptomycin treatment, there would be corresponding increases and decreases in the frequencies of bleached colonies.

An important prediction of the plastid dilution model was that the descendants of a single cell would frequently be a mixture of bleached and green cells. DeDeken-Grenson and Godts (43) investigated this question by examining the progeny of single *Euglena* cells in hanging drop cultures following 2-hr treatment with 1% streptomycin. They found, as expected, that during the first generations after treatment the cells lost chlorophyll and green chloroplasts. After five to six generations, at the time when chlorophyll synthesis resumed in the mass culture experiments of DeDeken-Grenson and Messin (42), a green color reappeared in a number of the cells. However, contrary to the predictions of the dilution hypothesis, all the *Euglena* cells in a given drop were either white or green (Table 16–6). No correlation was noted with the phenotypic appearance of the cells during the first generations following streptomycin treatment.

In contrast, mixed microcultures were

TABLE 16–6. *Clones produced by single cells of E. gracilis following different bleaching treatments*[a]

Treatment	Number of cells isolated	Number of cells producing		
		Pure green clones	Pure white clones	Mixed clones
Streptomycin	60	47	13	0
Culture at 34°C	41	16	23	2
Ultraviolet irradiation	16	7	5	4

[a] Times of isolation depended on the treatment used (see text), but in each case the cell contained green plastids at the moment of isolation.
Adapted from DeDeken-Grenson and Godts (43).

found at a frequency of 25% when ultraviolet irradiation was used to induce bleaching (Table 16–6). These results are certainly consistent with the expectations of the dilution hypothesis, but Lyman et al. (5) pointed out that they may be more fortuitous than real. DeDeken-Grenson and Godts kept their irradiated cells in the dark for only a day to prevent photoreactivation, but Lyman and associates found that 3 days of incubation in the dark were required for complete decay of photoreactivability. Hence, some of the mixed clones may have arisen because of photoreactivation repair of irradiation damage in certain cells. Had there been no photoreactivation, such a clone would have been scored as entirely bleached. Thus 25% is an upper limit for mixed clones following ultraviolet irradiation. Under conditions such that photoreactivation was completely eliminated, Lyman and associates reported that mixed clones, although frequently seen, constituted less than 10% of the total. DeDeken-Grenson and Godts also found that the frequency of mixed clones following heat bleaching was very low (Table 16–6).

DeDeken-Grenson and Godts concluded from their results that the dilution hypothesis was invalid and that all plastids in a given cell underwent the same fate. These results were elaborated in a review article by Grenson. Despite the fact that the results of DeDeken-Grenson and Godts appeared to be at variance with the dilution hypothesis, at least with respect to heat and

streptomycin bleaching, they were largely ignored by *Euglena* workers. An exception is Kirk, who has recognized the significance of the results and dealt with them at length (28). He pointed out that the irradiation bleaching experiments did not pose a problem, as we have discussed previously; then he turned to the results of the heat and streptomycin bleaching experiments. The cells isolated into microcultures for the study of heat bleaching were obtained after 3 to 4 days of culture, and few mixed clones were found. Kirk pointed out that by that time enough dilution might have occurred so that a given cell would be either bleached or green. This interpretation was supported by the results of Uzzo and Lyman (12), who found that after 3 to 4 days heat-treated cells were already well into the bleaching process and the irradiation target number per cell had declined considerably.

In the case of streptomycin, the results were even more difficult to explain, for no mixed clones were found after short-term (2-hr) exposure of cells to the drug. Kirk tried to explain these results by the hypothesis that only a certain fraction of the cells in a population are susceptible to streptomycin bleaching at any given time. As he pointed out, this hypothesis can be tested by using synchronously growing cells and measuring the proportion of bleached clones produced by brief treatment with antibiotic at different points within the division cycle. Clearly the problem is not one of differential cell permeability to streptomycin at different points in the cell cycle, since all cells in a log phase culture stop forming chlorophyll after short-term streptomycin treatment, irrespective of whether or not they become permanently bleached.

Unfortunately, Kirk's hypothesis failed to explain certain other results of DeDeken-Grenson and Godts, as he recognized. DeDeken-Grenson and Godts also examined the effect of plating time following streptomycin treatment on green-colony formation in mass liquid-grown cultures. They found that after a few generations during which the ability of cells to form green colonies declined, the fraction of cells capable of forming green colonies recovered rapidly; thus a population forming only 25% green colonies at five or six generations following streptomycin treatment might later form 100% green colonies. Reconstruction experiments showed that these results could not be attributed to selection for green cells in competition with white cells during liquid culture. Grenson (3) hypothesized that streptomycin-treated cells were in an unstable state following exposure to the drug and that a "decision" putting an end to the period of instability could be reached simply by plating the cells out. If the cells were plated shortly after streptomycin treatment, the decision favored bleached-colony formation. Kirk modified his hypothesis to accommodate these results and suggested that the duration of the unstable state might correspond to the time during which chlorophyll accumulation does not take place in the treated culture. In agreement with Grenson, he supposed that the drastic change in culture conditions engendered by plating cells out led to the formation of bleached colonies. The modified version of the Kirk hypothesis suggested that replication of the cytoplasmic entities responsible for chloroplast maintenance in *Euglena,* presumably clDNA molecules, was blocked irreversibly in about 20% of the cells, whereas the remainder were in the unstable state. When cells are cultured in hanging

drops, about 20% are in the irreversible state and produce the permanently bleached cultures that are found; the rest of the cultures are inoculated with cells in the unstable state, which produce green progeny under these conditions. Many of the cells in the unstable state produce bleached colonies when they are plated out. Clearly the parameters governing the "decision" of a cell to become green or bleached at plating following streptomycin treatment are worthy of more detailed study.

A different approach to the streptomycin bleaching question provided evidence that bleaching is an indirect result of inhibition of chloroplast protein synthesis. Streptomycin acts as a specific inhibitor of chloroplast protein synthesis *in vivo* and *in vitro* in *Euglena* (27,32,44–46), as it does in *Chlamydomonas* (see Chapter 21). Schiff and his colleagues (44,45) reported the isolation and characterization of a streptomycin-resistant mutant of *Euglena* (Sm$_i$BNgL) that was selected by its ability to grow photosynthetically at a concentration of streptomycin (0.05%) that normally causes irreversible loss of plastids. The chloroplast ribosomes, unlike those of the wild type, did not bind radioactively labeled dihydrostreptomycin. This antibiotic occupies the same binding site as streptomycin, as was shown by competition experiments using wild-type chloroplast ribosomes and cold streptomycin. It was also apparent that plastid development occurred normally when dark-grown cells of the mutant were exposed to light in the presence of streptomycin. Streptomycin blocks chlorophyll accumulation and development of photosynthetic capacity in wild type under similar conditions because it blocks chloroplast protein synthesis (44–46). From comparison of the mutant and wild type, it is difficult to escape the conclusion that some product of chloroplast protein synthesis is required for the maintenance of clDNA in *Euglena.* Synthesis of this product is blocked in wild-type cells growing in the presence of streptomycin, but it continues normally in the mutant.

It is well known that streptomycin will bind to negatively charged polymers (such as DNA) in solution (47), and it should be noted that Ebringer et al. (48) proposed a different mechanism of streptomycin bleaching prior to the ribosome binding studies of Schwartzbach and Schiff (44). Ebringer and associates proposed that streptomycin cross-links strands of clDNA and thereby prevents DNA replication. This model was derived from a model suggested earlier by Aronson et al. (47) to explain the action of streptomycin and dihydrostreptomycin. It was found that streptomycin was a more effective cross-linking reagent *in vitro* than dihydrostreptomycin, and it was suggested that the presence of an aldehyde group in the streptose moiety of the molecule might be responsible. According to Aronson et al. (47), this aldehyde group was capable of forming a second cationic attachment site by reacting with the secondary amino group in the *N*-methyl glucosamine moiety of streptomycin, resulting in a cyclic iminium salt. Dihydrostreptomycin, which contains a secondary alcohol group instead of an aldehyde, cannot form a cyclic iminium salt. Ebringer and associates, found that dihydrostreptomycin and other derivatives containing a secondary alcohol group were less effective bleaching reagents than streptomycin. They also found that hydroxylamine reversed the effect of streptomycin, and they suggested that hydroxylamine probably reacted with the aldehyde group on streptomycin, since it did not reverse bleaching by streptomycin analogs in which a secondary alcohol group replaced the aldehyde.

Schwartzbach and Schiff (44) showed that the hydroxylamine effect is easily explained in terms of streptomycin binding to chloroplast ribosomes. Unlike streptomycin, the streptomycin oxime produced by the interaction of streptomycin with hydroxylamine does not compete with radioactive

dihydrostreptomycin in the chloroplast ribosome binding reaction. Therefore the hydroxylamine effect on streptomycin bleaching seen by Ebringer and associates is readily explained in terms of the lack of affinity of the streptomycin oxime for chloroplast ribosomes. At the same time, hydroxylamine has no effect on dihydrostreptomycin binding, which is in keeping with the observation of Ebringer and associates that hydroxylamine does not reduce the effectiveness of dihydrostreptomycin as a bleaching agent. In the absence of direct evidence, prudence dictates that the cross-linking hypothesis of Ebringer and associates, be rejected in favor of the hypothesis that streptomycin bleaching is the result of disruption of chloroplast protein synthesis by the antibiotic.

INDUCTION OF BLEACHING BY OTHER INHIBITORS OF CHLOROPLAST PROTEIN SYNTHESIS AND THE CHLORAMPHENICOL PARADOX

Truly astonishing numbers of reagents have been tested for their ability to bleach *Euglena,* and a high proportion of these have proved effective [for reviews, see Mego (49) and Kirk and Tilney-Bassett (28)]. Bleaching agents include the following: inhibitors of DNA replication, prokaryotic protein synthesis inhibitors, mutagens, amino acid analogs, antihistamines, nitrofurans, magnesium deficiency, and heat. In what is probably the most complete study to date, Ebringer (50) examined a formidable number of different substances for bleaching ability. These were chosen to include inhibitors of DNA replication, transcription, and translation.

Ebringer found that bleaching agents fell in two general categories. In the first category were inhibitors of DNA replication or synthesis; in the second category were inhibitors of protein synthesis. The latter category included a variety of antibiotics known to block organelle protein synthesis in other systems, notably the macrolide and aminoglycoside antibiotics. In general, the antibiotics which blocked organelle protein synthesis also blocked development of chloroplasts from the proplastids of dark-grown cells, as would be expected.

These results support the conclusion that bleaching can result from either direct or indirect blockage of clDNA replication. A blockage in chloroplast protein synthesis is itself most probably an indirect form of blockage of clDNA replication. So far, so good; this umbrella is probably large enough to cover all the bleaching agents tried to date, with the necessary ad hoc assumptions in particular cases. The problem comes when an apparent bona fide exception arises, as in the case of chloramphenicol.

A host of studies have suggested that chloramphenicol blocks chloroplast protein synthesis in *Euglena*. These have included experiments *in vivo* in which dark-grown cells were transferred to the light, where it was found that chloramphenicol inhibited chlorophyll accumulation and synthesis of ribulose bisphosphate carboxylase (27,40, 46,51,52), an enzyme thought to be synthesized in part on chloroplast ribosomes (see Chapter 20), as well as experiments *in vitro* showing that chloramphenicol inhibits protein synthesis in isolated chloroplasts (53,54) and on chloroplast polyribosomes (55). On the other hand, chloramphenicol normally does not induce permanent bleaching.

Richards et al. (56) reported that in cells grown photoautotrophically on 3.3-mM concentrations of chloramphenicol (1 mg/ml), there was no diminution in the amount of clDNA, but the cells were reversibly bleached. That is, during growth in liquid the cells appeared bleached, but when plated they formed normal green colonies. Richards and associates concluded that there were two possibilities: (a) all unstable proteins essential for syn-

thesis of clDNA were made in the cytoplasm; (b) if such proteins were made in the chloroplast, they had extremely long half-lives ($>$ 5.5 generations). Ebringer (50) reported no chloramphenicol-induced bleaching, either reversible or irreversible, in liquid-grown cultures, but it is not clear what drug concentrations he used in these experiments. Under the same experimental conditions, Ebringer showed that macrolides and aminoglycosides were very effective bleaching agents.

In contrast with these reports are two articles demonstrating that chloramphenicol will induce irreversible bleaching under certain special conditions. Miyoshi and Tsubo (57) obtained up to 25% bleaching, provided that treated cells had previously been grown in the dark so that their chloroplasts had been transformed into proplastids. The effective dose was 0.5 mg/ml, and bleaching was seen only if the cells were grown in the dark. Marčenko (58) extended these findings. She plated light-grown cells in the dark at 30°C on chloramphenicol (1 mg/ml) and observed 41% bleaching of wild-type cells and 96% for a yellow mutant incapable of forming chlorophyll in the light. Controls showed that incubation of the cells at 30°C in the absence of chloramphenicol did not cause significant heat bleaching. Marčenko also reported that bleaching did not occur in the presence of chloramphenicol at 25°C and that it was necessary to add ethanol (0.4%) to the medium to cause permanent bleaching. From these experiments, as well as from those of Miyoshi and Tsubo, it is evident that chloramphenicol will cause permanent bleaching under certain circumstances.

It seems likely that these differences in the bleaching actions of chloramphenicol and streptomycin are more apparent than real. Two sorts of explanations can be suggested. First, one gets the impression that streptomycin is a much more effective inhibitor of chloroplast protein synthesis than chloramphenicol in the few experiments in which the two antibiotics have been compared under similar conditions. Ebringer (50) examined the effects of both inhibitors on chlorophyll accumulation in dark-grown cells when transferred to the light. In his experiments there was virtually no inhibition of chlorophyll accumulation by either the D(−)-*threo* or L(+)-*threo* isomer of chloramphenicol at the highest concentrations tested. These were approximately 0.31 mM. With streptomycin, a concentration of 0.017 mM was sufficient to inhibit chlorophyll accumulation 87%. Growth of cells in this concentration of streptomycin produced irreversible bleaching of the culture within 7 days. Schiff and his colleagues (27, 46) also examined the effects of the two inhibitors on chlorophyll accumulation by dark-grown resting (nongrowing) cells of *Euglena* when they were returned to the light. These investigators reported complete inhibition of chlorophyll accumulation with 0.86 mM streptomycin. This streptomycin concentration also caused irreversible bleaching in a high proportion of the cells. In contrast, 6.2-mM chloramphenicol inhibited chlorophyll accumulation only 54%, and permanent bleaching was not reported. Schiff and associates also examined the increase in ribulose bisphosphate carboxylase activity, which is a good indicator of chloroplast protein synthesis *in vivo* (see Chapter 20). If their figures are normalized to give a 100% increase in enzyme activity in control cells after chloroplast differentiation has taken place in the light, then the increase in streptomycin-treated cells is only 0.5%, whereas it is between 14 and 17% in chloramphenicol-treated cells. It is perhaps worth noting that the chloramphenicol concentrations used by Schiff and associates were approaching the aqueous solubility of the drug (2.5 mg/ml, 7.75 mM), whereas the concentration of streptomycin used to induce bleaching and to block chloroplast development was nowhere near the solubility point of this antibiotic (> 20 mg/ml, > 30 mM). In short, one possibility to explain the chloramphenicol paradox may be that irreversible bleaching by a protein synthesis inhibitor requires virtually total inhibition of chloroplast protein synthesis; this is difficult to achieve *in vivo* with chloramphenicol.

Second, it is clear that physiological conditions play an important role in streptomycin bleaching, as was pointed out earlier. The experiments of Marčenko (58) and Miyoshi and Tsubo (57) suggest that the same is true of chloramphenicol. Thus it seems likely that given proper physiological conditions and concentrations, chloramphenicol can be used as an effective bleaching agent in *Euglena*. In reports in the literature there is great variation in the conditions that obtain during measurement of chlorophyll accumulation and bleaching; physiological and concentration parameters obviously play such an important role that one should not take the apparently contradictory results with chloramphenicol too seriously. The important observations are that a mutant of *Euglena* with streptomycin-resistant chloroplast ribosomes is not bleached by streptomycin and that virtually all inhibitors of prokaryotic ribosome function, with the exception of chloramphenicol, do cause irreversible bleaching.

NEGATIVE USE OF PERMANENTLY BLEACHED MUTANTS

The permanently bleached mutants of *Euglena* are analogs of the vegetative petite mutants of yeast with respect to the chloroplast. They have no clDNA, and since chloroplast rRNA in *Euglena* is a transcript of clDNA (59), they have no chloroplast protein-synthesizing system. However, the bleached mutants do have some remnants of chloroplast structure; this point was documented for cells bleached with ultraviolet irradiation in a study by Michaels and Gibor (60) that employed both fluorescence and electron microscopy. These authors treated cells with a bleaching dose of irradiation and subsequently grew them in red light in order to retain differentiated chloroplast structure while preventing photoreactivation. Their results indicated that there was continuity of chloroplast structure in cells for at least five generations after treatment but that structural changes reminiscent of those taking place in senescing higher plant plastids began to occur. These included progressive loss of thylakoids with each generation. Eventually only sac-like structures reminiscent of proplastids were left. By definition, these remaining components must be coded by nuclear genes whose messages are translated in the cytoplasm.

Table 16–7 compares some of the chloroplast-related properties of wild-type cells and bleached mutants. Such mutants, in addition to lacking chlorophyll and being tremendously deficient in total carotenoids, are unable to synthesize ribulose bisphosphate carboxylase. However, they are able to make many of the plastid membrane polypeptides, as was shown in a recent study by Bingham and Schiff (61). These workers used an ingenious trick to purify plastid membranes from a bleached mutant. It will be recalled that plastid membranes contain a unique sulfolipid (see Chapter 1). This sulfolipid is also found in bleached mutants of *Euglena*. Bingham and Schiff labeled cells with radioactive sulfate; then they purified the membrane fraction containing radioactive lipids and compared the polypeptides in that fraction to those found in membranes obtained from green chloroplasts.

Bleached *Euglena* cells also synthesize NADP-linked glyceraldehyde-3-phosphate

TABLE 16–7. *Comparison of various parameters in light-grown wild-type cells of E. gracilis and a bleached mutant (W₃BUL)*

Parameter	W$_3$BUL	Wild type
Chloroplast structure	Largely absent	Present
Chloroplast DNA	Not detectable	Present
Chlorophyll (pg/organism)	0	10.50 ± 0.86
Carotenoids (pg/organism)	0.13	2.62 ± 0.19
$10^2 \times$ photosynthesis (pmol CO_2 fixed/organism/hr)	0	28
$10^{11} \times$ cytochrome 552 (μmol/organism)	0	2.79 ± 0.39
Ribulose bisphosphate carboxylase (μmol CO_2 fixed/mg protein/hr)	0	4.11 ± 0.34
NADP-GPD phosphate dehydrogenase (μmol TPNH oxidized/mg protein/hr)	4.01	27.00 ± 6.83
NAD-linked triose phosphate dehydrogenase (μmol NADH oxidized/mg protein/hr)	40.10	27.00 ± 5.42
Chloroplast isoleucine tRNA	−	+
Chloroplast phenylalanine tRNA	−	+
Chloroplast isoleucyl tRNA synthetase	−	+
Chloroplast phenylalanyl tRNA synthetase	+	+

Adapted from Bovarnick et al. (46), with additional data (62–64).

dehydrogenase (NADP-GPD). This finding is mentioned only because there has been some controversy concerning the site of synthesis of NADP-GPD in the cell. Smillie and associates (51) reported that the synthesis of NADP-GPD was blocked by chloramphenicol, but not cycloheximide, when dark-grown cells were transferred to the light. Schiff (46) found that there was some inhibition of synthesis of this enzyme under similar conditions in the presence of streptomycin, but it was much less than reported by Smillie et al. (51). Schmidt and Lyman (40) found some inhibition of synthesis of NADP-GPD by chloramphenicol under conditions designed to be very similar to those of Smillie and associates, but the inhibition was not great and was not comparable to that reported by Schiff for streptomycin. The NADP-GPD activity found in bleached cells is similar to that of dark-grown cells in amount, and no light-stimulated synthesis of the enzyme occurs (40, 46).

These observations may be interpreted in either of two ways: First, the NADP-GPD activity present in the bleached mutant is distinct from that stimulated by growth of cells in the light. Possibly it is the same enzyme present in dark-grown cells. According to this interpretation, the NADP-GPD enzyme present in bleached cells must be synthesized on cytoplasmic ribosomes, whereas the light-induced enzyme could be synthesized on chloroplast ribosomes. The second possibility is that there is one enzyme, and the results reported by Smillie and associates are incorrect. It should be possible to distinguish these possibilities by use of the bleached mutant. The two-enzyme hypothesis predicts the presence of two enzymes in light-grown cells; the one-enzyme hypothesis predicts a single activity in the bleached mutant, in dark-grown cells, and in light-grown cells. One could perhaps make antibodies to the enzyme synthesized by the bleached mutant and test these against the enzyme(s) present in light- and dark-grown cells. On the one-enzyme hypothesis, these antibodies will cross-react with all the NADP-GPD enzyme present in both light- and dark-grown cells. On the two-enzyme hypothesis, light-grown cells will contain an activity that is not eliminated by the antibodies made against the enzyme from the bleached mutant, in addition to an activity against which these antibodies are effective. Obviously the same approach can be used to make the same determination for other enzyme activities whose syntheses are stimulated by light (40).

COMPARISONS OF THE EFFECTS OF INHIBITORS OF CHLOROPLAST PROTEIN SYNTHESIS IN *CHLAMYDOMONAS* AND *EUGLENA*

As was mentioned in Chapter 13, *C. reinhardtii* and its colorless analogs *Polytoma uvella* and *P. obtusum* are very sensitive to killing by streptomycin, whereas *E. gracilis* and its colorless relative *Astasia longa* are not. The same parallelism exists for other antibiotics believed to inhibit organelle protein synthesis in both *Chlamydomonas* and *Euglena* (Table 16–8). The interesting fact also emerges that bleaching concentrations of aminoglycoside antibiotics in *Euglena* tend to be similar to lethal concentrations for *Chlamydomonas,* but the same relationship does not hold for macrolides (Table 16–8). The lethal effects of both aminoglycoside and macrolide antibiotics in *Chlamydomonas* have been interpreted in terms of two mutually exclusive hypotheses (see Chapter 21). One of these supposes that these antibiotics also block mitochondrial protein synthesis *in vivo* in *Chlamydomonas;* the other assumes that a regulatory substance required for nDNA replication in *Chlamydomonas* must be synthesized on chloroplast ribosomes. Obviously, neither hypothesis is applicable to *Euglena.*

Since the concentrations of antibiotics required to kill wild-type *Chlamydomonas*

TABLE 16–8. *Comparison of the effects of inhibitors of organelle protein synthesis on cell viability in* Chlamydomonas *and* Euglena *and on bleaching in* Euglena

Antibiotic	Lethal concentration ($\mu g/ml$)		Least concentration causing highest percentage bleaching[a]	
	Chlamydomonas[a]	Euglena[a]	Concentration ($\mu g/ml$)	Percentage bleached
Streptomycin	20	2,000	10	100
Spectinomycin	10–30	1,000	50	100
Kanamycin	50	1,000	200	100
Paromomycin	10	15	5	74
Neomycin[b]	>200, >500	400	does not bleach	—
Lincomycin	100	4,000	1,500	100
Erythromycin	10	5,000	800	100
Oleandomycin[c]	<50	5,000	4,000	100
Spiramycin[c]	<100	1,200	600	49
Carbomycin[c]	<50	700	100	100
Chloramphenicol	<50	1,000	does not bleach	—

[a] These figures are approximate, since variations in growth conditions will affect resistance levels, as is clearly evident from the literature.

[b] Figures for *Chlamydomonas* from Gillham (78). Wild-type cells are resistant to >200 $\mu g/ml$ neomycin B and >500 $\mu g/ml$ neomycin C.

[c] These are upper limits of resistance for *Chlamydomonas*, since Sager (77) reported isolation of mutants resistant at these concentrations. Presumably, wild-type cells are sensitive to lower concentrations.

Data for *Euglena* from Ebringer (50); data for *Chlamydomonas* from Surzycki and Gillham (76) and Sager (77).

cells are the same as or lower than those sufficient to bleach *Euglena,* it becomes difficult to determine if these antibiotics can bleach *Chlamydomonas.* However, two classes of *Chlamydomonas* mutants are known whose properties may be pertinent in answering this question. The first class includes two mutants that confer extremely low levels of antibiotic resistance on chloroplast ribosomes (see Chapter 21). One is resistant to streptomycin and the other to spectinomycin. These mutants cannot be grown in the presence of the respective antibiotics under photosynthetic conditions at concentrations greater than those sufficient to block the growth of wild-type cells, but they will grow at very high antibiotic concentrations when supplied with acetate as a carbon source. Under these conditions it appears as if chloroplast protein synthesis *in vivo* is largely blocked, as measured in terms of the cells' ability to synthesize ribulose bisphosphate carboxylase (see Chapter 20), but the cells continue to accumulate normal amounts of chlorophyll. The second class of mutants cannot syn-

thesize normal amounts of chloroplast ribosomes; these mutants show a syndrome of defects that precisely parallels that seen in the aforementioned resistant mutants grown in the presence of antibiotics. These mutants, too, accumulate normal amounts of chlorophyll.

The results obtained with both classes of mutants indicate that inhibition of chloroplast protein synthesis in *Chlamydomonas* does not block chlorophyll accumulation nor cause irreversible bleaching as in the case of *Euglena.* One explanation of these differences may be that the membrane polypeptides synthesized in the chloroplasts of the two organisms are not the same. It is possible that certain polypeptides required for chlorophyll accumulation and clDNA binding or replication are synthesized on chloroplast ribosomes in *Euglena* but not *Chlamydomonas.* At the moment, such an explanation is pure speculation, but it does serve to emphasize what may be important differences in the products of chloroplast protein synthesis in *Chlamydomonas* and *Euglena.*

EUGLENA MUTANTS AND THEIR GENOMIC LOCALIZATION

In addition to bleached mutants, a number of other kinds of mutants have been isolated in *E. gracilis* (17). They include mutations affecting chloroplast function and mutations resistant to antibiotics (Table 16-2). Schiff et al. (17) argued that these mutations arose in the plastid genome. This argument was based on the assumption that *Euglena* had an octaploid nucleus, which would render expression of recessive nuclear mutations unlikely. During vegetative cell division, chromosomal mutants should not segregate, assuming a regular mitosis, whereas organelle mutations should segregate, as they do in diploid strains of *Chlamydomonas* (see Chapter 13) and yeast (see Chapter 5). It was also argued that in *Euglena* the mutations occurring in the plastids were much more likely to be expressed than those that arose in the mitochondria, since mitochondria and mtDNA were far more abundant than plastids and clDNA molecules. As was discussed previously (p. 440), the assumption of octaploidy, which was based on target analysis, now appears to be incorrect, based on renaturation kinetics analysis, which indicates that the nuclear genome is diploid. These data, which are much more reliable, still are not as convincing as cytologic evidence would be. Furthermore, it is now evident that clDNA molecules are much more abundant than was previously thought (p. 440); thus the argument that clDNA molecules are fewer in number than mtDNA molecules probably does not hold either. However, the mutants to be discussed do affect plastids specifically, and so it seems unlikely that they are mitochondrial; the argument that plastid mutations will be selected in preference to nuclear gene mutations applies to diploids as well as octaploids.

If one accepts the argument that most mutations in *Euglena* are in the plastid genome, it becomes of considerable interest to define their phenotypes precisely as a means of elucidating which plastid functions are dependent on clDNA. Of the mutants having photosynthetic defects, several have been studied in some detail (62–67). The pale green (P) mutants were characterized by Russell and Lyman (65,66). Photosynthetic measurements indicate that these mutants are blocked at or near the light reaction in photosystem II, and dark reaction enzyme activities, notably ribulose bisphosphate carboxylase, are normal. These results show that the primary defect in these mutants is not related to chloroplast protein synthesis. Otherwise, ribulose bisphosphate carboxylase would be greatly reduced in activity. A more detailed characterization of the P mutants will be of interest. They might, for example, affect the biosynthesis of the reaction center thylakoid membrane polypeptide(s) of photosystem II, as do similar mutants in *Chlamydomonas* (see Chapter 20).

One mutant of special interest is G_1BU, which is golden in color (see Chapter 21). This mutant has lost the ability to control the synthesis of specific chloroplast tRNA species that are found normally only in light-grown cells (62,63). Goins et al. (64) found that this regulatory mutant, unlike the wild type, continued to synthesize the light-induced chloroplast tRNAs for methionine and isoleucine when grown in the dark. It will be interesting to see if this mutant is constitutive for other light-induced chloroplast functions as well.

If one could cross *Euglena,* it would probably be the best system available for the study of plastid genetics, since the organism can be permanently bleached without loss of viability. In this sense it is comparable to yeast, for the DNA of the organelle of interest is completely dispensable, although deletion mutants equivalent to ρ^- petites, as opposed to ρ^0 petites, have not yet been reported. With the availability of antibiotic-resistant mutants, it may now become possible to look for rare genetic exchanges. For example, one could mix cells resistant to streptomycin with cells resistant to erythromycin and plate under phototrophic conditions on medium containing both antibiotics. If double mutants arise at a frequency greater than is observed when unmixed control cells are plated, a major breakthrough will be achieved. Conditions that maximize the exchange rate could then be sought and mechanisms of exchange worked out later. After all, that is how bacterial genetics began.

REFERENCES

1. Ternetz, C. (1912): Beiträge zur morphologie und physiologie von *Euglena gracilis. Jahrb. Wiss. Botan.,* 51:435.
2. Provasoli, L., Hutner, S. H., and Schatz, A. (1948): Streptomycin-induced chlorophyllless races of *Euglena. Proc. Soc. Exp. Biol. Med.,* 69:279–282.
3. Grenson, M. (1964): Physiology and cytology of chloroplast formation and "loss" in *Euglena. Int. Rev. Cytol.,* 16:37–59.
4. Lyman, H., Epstein, H. T., and Schiff, J. (1959): Ultraviolet inactivation and photoreactivation of chloroplast development in *Euglena* without cell death. *J. Protozool.,* 6:264–265.
5. Lyman, H., Epstein, H. T., and Schiff, J. A. (1961): Studies of chloroplast development in *Euglena.* I. Inactivation of green colony formation by U.V. light. *Biochim. Biophys. Acta,* 50:301–309.
6. Schiff, J. A., Lyman, H., and Epstein, H. T. (1961): Studies of chloroplast development in *Euglena.* II. Photoreversal of the U.V. inhibition of green colony formation. *Biochim. Biophys. Acta,* 50:310–318.
7. Schiff, J. A., Lyman, H., and Epstein, H. T. (1961): Studies of chloroplast development in *Euglena.* III. Experimental separation of chloroplast development and chloroplast replication. *Biochim. Biophys. Acta,* 51:340–346.
8. Schiff, J. A., Epstein, H. T., and Lyman, H. (1961): Ultraviolet inactivation and photoreactivation of chloroplast development in *Euglena.* In: *Proceedings Third International Congress on Photobiology,* edited by M. Avron, pp. 289–292. Elsevier/North Holland, Amsterdam.
9. Hill, H. Z., Schiff, J. A., and Epstein, H. T. (1966): Studies of chloroplast development in *Euglena.* XIII. Variation of chloroplast sensitivity with extent of chloroplast development. *Biophys. J.,* 6:125–133.
10. Hill, H. Z., Epstein, H. T., and Schiff, J. A. (1966): Studies of chloroplast development in *Euglena.* XIV. Sequential interactions of ultraviolet light and photoreactivating light in green colony formation. *Biophys. J.,* 6:135–144.
11. Uzzo, A., and Lyman, H. (1969): Light dependence of temperature-induced bleaching in *Euglena gracilis. Biochim. Biophys. Acta,* 180:573–575.
12. Uzzo, A., and Lyman, H. (1971): The nature of the chloroplast genome of *Euglena gracilis.* In: *Proceedings Second International Congress on Photosynthesis Research,* edited by G. Forti, M. Avron, and A. Melandri, pp. 2585–2599. Junk, The Hague.
13. Pringsheim, E. G., and Pringsheim, O. (1952): Experimental elimination of chromatophores and eye spot in *Euglena gracilis. New Phytol.,* 51:65–76.
14. Leedale, G. F. (1967): *Euglenoid Flagellates.* Prentice-Hall, Englewood Cliffs, N.J.
15. Buetow, D. E. (editor) (1968): *The Biology of Euglena, Vols. I and II.* Academic Press, New York.
16. Leedale, G. F. (1968): The nucleus in *Euglena.* In: *The Biology of Euglena I,* edited by D. E. Buetow, pp. 185–242. Academic Press, New York.
17. Schiff, J. A., Lyman, H., and Russell, G. K. (1971): Isolation of mutants of *Euglena gracilis.* In: *Methods in Enzymology, Vol. 23 (Part A),* edited by A. San Pietro, pp. 143–162. Academic Press, New York.
18. Rawson, J. R. Y. (1975): The characterization of *Euglena gracilis* DNA by its reassociation kinetics. *Biochim. Biophys. Acta,* 402:171–178.
19. Manning, J. E., and Richards, O. C. (1972): Isolation and molecular weight of circular chloroplast DNA from *Euglena gracilis. Biochim. Biophys. Acta,* 259:285–296.
20. Gray, P. W., and Hallick, R. B. (1977): Restriction endonuclease map of *Euglena gracilis* chloroplast DNA. *Biochemistry,* 16:1665–1671.
21. Stutz, E., Crouse, E. J., Graf, L., Jenni, B., and Kopecka, H. (1976): Structural and functional analysis of *Euglena gracilis* chloroplast DNA. In: *Genetics and Biogenesis of Chloroplasts and Mitochondria,* edited by T. Bücher, W. Neupert, W. Sebald, and S. Werner, pp. 339–346. North Holland, Amsterdam.
22. Rawson, J. R. Y., and Boerma, C. (1976): Influence of growth conditions upon the number of chloroplast DNA molecules in *Euglena gracilis. Proc. Natl. Acad. Sci. U.S.A.,* 73:2401–2404.
23. Chelm, B. K., Hoben, P. J., and Hallick, R. B. (1977): Cellular content of chloroplast DNA and chloroplast ribosomal RNA genes in *Euglena gracilis* during chloroplast development. *Biochemistry,* 16:782–786.

24. Shori, L., Ben-Shaul, Y., and Edelman, M. (1970): The size of mitochondrial DNA in *Euglena gracilis*. *Isr. J. Chem.*, 8:117.

25. Crouse, E., Vandrey, J., and Stutz, E. (1974): Comparative analysis of chloroplast and mitochondrial DNAs from *Euglena gracilis*. In: *Proceedings Third International Congress on Photosynthesis, Vol. 3*, edited by M. Avron, pp. 1775–1786. Elsevier/North Holland, Amsterdam.

26. Zumstein, H. (1900): Zur Morphologie und Physiologie der *Euglena gracilis* Klebs. *Jahrb. Wiss. Botan.*, 34:149–198.

27. Schiff, J. A. (1973): The development, inheritance, and origin of the plastid in *Euglena*. *Adv. Morphogenesis*, 10:265–312.

28. Kirk, J. T. O., and Tilney-Bassett, R. A. E. (1967): *The Plastids*. W. H. Freeman, San Francisco.

29. Rupert, C. S. (1975): Enzymatic photoreactivation: Overview. In: *Molecular Mechanisms for Repair of DNA, Part A*, edited by P. C. Hanawalt and R. B. Setlow, Chapter 11. Plenum Press, New York.

30. Jagger, J. (1958): Photoreactivation. *Bacteriol. Rev.*, 22:99–142.

31. Epstein, H. T., and Schiff, J. A. (1961): Studies of chloroplast development in *Euglena*. 4. Electron and fluorescence microscopy of the proplastid and its development into a mature chloroplast. *J. Protozool.*, 8:427–432.

32. Schiff, J. A. (1970): Developmental interactions among cellular compartments in *Euglena*. *Symp. Soc. Exp. Biol.*, 24:277–301.

33. Gibor, A., and Granick, S. (1962): Ultraviolet sensitive factors in the cytoplasm that affect the differentiation of *Euglena* plastids. *J. Cell Biol.*, 15:599–603.

34. Edelman, M., Schiff, J. A., and Epstein, H. T. (1965): Studies of chloroplast development in *Euglena*. XII. Two types of satellite DNA. *J. Mol. Biol.*, 11:769–774.

35. Cook, T. M., Brown, K. G., Boyle, J. V., and Goss, W. A. (1966): Bactericidal action of nalidixic acid on *Bacillus subtilis*. *J. Bacteriol.*, 92:1510–1514.

36. Goss, W. A., Dietz, W. H., and Cook, T. M. (1965): Mechanism of action of nalidixic acid on *Escherichia coli*. II. Inhibition of deoxyribonucleic acid synthesis. *J. Bacteriol.*, 89:1068–1074.

37. Bourguignon, G. J., Levitt, M., and Sternglanz, R. (1973): Studies on the mechanism of action of nalidixic acid. *Antimicrob. Agents Chemother.*, 4:479–486.

38. Lyman, H. (1967): Specific inhibition of chloroplast replication in *Euglena gracilis* by nalidixic acid. *J. Cell Biol.*, 35:726–730.

39. Lyman, H., Jupp, A. S., and Larrinua, I. (1975): Action of nalidixic acid on chloroplast replication in *Euglena gracilis*. *Plant Physiol.*, 55:390–392.

40. Schmidt, G. W., and Lyman, H. (1976): Inheritance and synthesis of chloroplasts and mitochondria of *Euglena gracilis*. In: *The Genetics of Algae*, edited by R. A. Lewin, Chapter 13. Blackwell Scientific, Oxford.

41. DeDeken-Grenson, M. (1959): The mass induction of white strains in *Euglena* as influenced by the physiological conditions. *Exp. Cell Res.*, 18:185–187.

42. DeDeken-Grenson, M., and Messin, S. (1958): La continuité génétique des chloroplastes chez Euglènes I. Mécanisme de l'apparition des lignées blanches dans les cultures traitées par la streptomycine. *Biochim. Biophys. Acta*, 27:145–155.

43. DeDeken-Grenson, M., and Godts, A. (1960): Descendance of *Euglena* cells isolated after various bleaching treatments. *Exp. Cell Res.*, 19:376–382.

44. Schwartzbach, S. D., and Schiff, J. A. (1974): Chloroplast and cytoplasmic ribosomes of *Euglena*: Selective binding of dihydrostreptomycin to chloroplast ribosomes. *J. Bacteriol.*, 120:334–341.

45. Bovarnick, J. G., Chang, S.-W., Schic, J. A., and Schwartzbach, S. D. (1974): Events surrounding the early development of *Euglena* chloroplasts: Experiments with streptomycin in non-dividing cells. *J. Gen. Microbiol.*, 83:51–62.

46. Bovarnick, J. G., Schiff, J. A., Freedman, Z., and Egan, J. M., Jr. (1974): Events surrounding the early development of *Euglena* chloroplasts: Cellular origins of chloroplast enzymes in *Euglena*. *J. Gen. Microbiol.*, 83:63–71.

47. Aronson, J., Meyer, W. L., and Brock, T. C. (1964): A molecular model for chemical and biological differences between streptomycin and dihydrostreptomycin. *Nature*, 202:555–557.

48. Ebringer, L., Mego, J. L., Jurasek, A., and Kada, R. (1969): The action of streptomycins on the chloroplast system of *Euglena gracilis*. *J. Gen. Microbiol.*, 59:203–209.

49. Mego, J. L. (1968): Inhibitors of the chloroplast system in *Euglena*. In: *The Biology of Euglena, Vol. II*, edited by D. E. Buetow, pp. 351–381. Academic Press, New York.

50. Ebringer, L. (1972): Are plastids derived from prokaryotic micro-organisms? Action of antibiotics on chloroplasts of *Euglena gracilis*. *J. Gen. Microbiol.*, 71:35–52.

51. Smillie, R. M., Graham, D., Dwyer, M. R., Grieve, A., and Tobin, N. F. (1967): Evidence for the synthesis *in vivo* of proteins of the Calvin cycle and of the photosynthetic electron-transfer pathway on chloroplast ribosomes. *Biochem. Biophys. Res. Commun.*, 28:604–610.

52. Smillie, R. M., Bishop, D. G., Gibbons, G. C., Graham, D., Grieve, A. M., Raison, J. K., and Reger, B. J. (1971): Determination of

the sites of synthesis of proteins and lipids of the chloroplast using chloramphenicol and cycloheximide. In: *Autonomy and Biogenesis of Mitochondria and Chloroplasts,* edited by N. K. Boardman, A. W. Linnane, and R. M. Smillie, pp. 422–433. North Holland, Amsterdam.

53. Harris, E. H., Preston, J. F., and Eisenstadt, J. M. (1973): Amino acid incorporation and products of protein synthesis in isolated chloroplasts of *Euglena gracilis. Biochemistry,* 12: 1227–1234.

54. Reger, B. J., Smillie, R. M., and Fuller, R. C. (1972): Light-stimulated production of a chloroplast-localized system for protein synthesis in *Euglena gracilis. Plant Physiol.,* 50: 24–27.

55. Avadhani, N. G., and Buetow, D. E. (1972): Isolation of active polyribosomes from the cytoplasm, mitochondria and chloroplasts of *Euglena gracilis. Biochem. J.,* 128:353–365.

56. Richards, O. C., Ryan, R. S., and Manning, J. E. (1971): Effects of cycloheximide and of chloramphenicol on DNA synthesis in *Euglena gracilis. Biochim. Biophys. Acta,* 238: 190–201.

57. Miyoshi, Y., and Tsubo, Y. (1969): Permanent bleaching of *Euglena* by chloramphenicol. *Plant Cell Physiol.,* 10:221–225.

58. Marčenko, E. (1974): On the permanent bleaching of *Euglena* by chloramphenicol and its inhibition by cycloheximide. *Cytobiologie,* 9:280–289.

59. Gruol, D. J., and Haselkorn, R. (1976): Counting the genes for stable RNA in the nucleus and chloroplasts of *Euglena. Biochim. Biophys. Acta,* 447:82–95.

60. Michaels, A., and Gibor, A. (1973): Ultrastructural changes in *Euglena* after ultraviolet irradiation. *J. Cell Sci.,* 13:799–809.

61. Bingham, S., and Schiff, J. A. (1976): Cellular origins of plastid membrane polypeptides in *Euglena.* In: *Genetics and Biogenesis of Chloroplasts and Mitochondria,* edited by T. Bücher, W. Neupert, W. Sebald, and S. Werner, pp. 79–86. North Holland, Amsterdam.

62. Barnett, W. E., Pennington, C. J., Jr., and Fairfield, S. A. (1969): Induction of *Euglena* transfer RNAs by light. *Proc. Natl. Acad. Sci. U.S.A.,* 63:1261–1268.

63. Reger, B. J., Fairfield, S. A., Epler, J. L., and Barnett, W. E. (1970): Identification and origin of some chloroplast aminoacyl-tRNA synthetases and tRNAs. *Proc. Natl. Acad. Sci. U.S.A.,* 67:1207–1213.

64. Goins, D. J., Reynolds, R. J., Schiff, J. A., and Barnett, W. E. (1973): A cytoplasmic regulatory mutant of *Euglena: Constitutivity* for the light-inducible chloroplast transfer RNAs. *Proc. Natl. Acad. Sci. U.S.A.,* 70: 1749–1752.

65. Russell, G. K., and Lyman, H. (1968): Isolation of mutants of *Euglena gracilis. Plant Physiol.,* 43:1284–1290.

66. Russell, G. K., Lyman, H., and Heath, R. L. (1969): Absence of fluorescence quenching in a photosynthetic mutant of *Euglena gracilis. Plant Physiol.,* 44:929–931.

67. Schmidt, G., and Lyman, H. (1974): Photocontrol of chloroplast enzyme synthesis in mutant and wild type *Euglena gracilis.* In: *Proceedings Third International Congress on Photosynthesis,* edited by M. Avron, pp. 1755–1764. Elsevier/North Holland, Amsterdam.

68. Leff, J., Mandel, M., Epstein, H. T., and Schiff, J. A. (1963): DNA satellites from cells of green and aplastidic algae. *Biochem. Biophys. Res. Commun.,* 13:126–130.

69. Edelman, M., Cowan, C. A., Epstein, H. T., and Schiff, J. A. (1964): Studies of chloroplast development in *Euglena.* VIII. Chloroplast-associated DNA. *Proc. Natl. Acad. Sci. U.S.A.,* 52:1214–1219.

70. Edelman, M., Epstein, H. T., and Schiff, J. A. (1966): Isolation and characterization of DNA from the mitochondrial fraction of *Euglena. J. Mol. Biol.,* 17:463–469.

71. Ray, D. S., and Hanawalt, P. C. (1964): Properties of the satellite DNA associated with the chloroplasts of *Euglena gracilis. J. Mol. Biol.,* 9:812–824.

72. Brawerman, G., and Eisenstadt, J. M. (1964): Template and ribosomal ribonucleic acids associated with the chloroplasts and cytoplasm of *Euglena gracilis. J. Mol. Biol.,* 10:403–411.

73. Brawerman, G., and Eisenstadt, J. M. (1964): Deoxyribonucleic acid from the chloroplasts of *Euglena gracilis. Biochim. Biophys. Acta,* 91:477–485.

74. Brawerman, G., Hufnagel, D. A., and Chargaff, E. (1962): On the nucleic acids of green and colorless *Euglena gracilis:* Isolation and composition of deoxyribonucleic acid and transfer ribonucleic acid. *Biochim. Biophys. Acta,* 61:340–345.

75. Diamond, J., and Schiff, J. A. (1974): Isolation and characterization of mutants of *Euglena* resistant to streptomycin. *Plant Sci. Lett.,* 3:289–295.

76. Surzycki, S. J., and Gillham, N. W. (1971): Organelle mutations and their expression in *Chlamydomonas reinhardtii Proc. Natl. Acad. Sci. U.S.A.,* 68:1301–1306.

77. Sager, R. (1972): *Cytoplasmic Genes and Organelles.* Academic Press, New York.

78. Gillham, N. W. (1965): Induction of chromosomal and nonchromosomal mutations in *Chlamydomonas reinhardtii* with *N*-methyl-*N*′-nitro-*N*-nitrosoguanidine. *Genetics,* 52: 529–537.

79. Schiff, J. A., and Zeldin, M. H. (1968): The developmental aspect of chloroplast continuity in *Euglena. J. Cell Physiol.,* 72(Suppl. 1) 103–128.

Chapter 17

Chloroplast Inheritance in Higher Plants. I. Transmission, Segregation, and Mutation of Plastids

The discovery that different plastid phenotypes could be inherited in a nonmendelian fashion may be traced to observations made by Correns and Baur in 1909, just a few years after the rediscovery of Mendel's laws in 1900. Correns (1) reported on the inheritance of green, variegated, and white color patterns in seedlings of the four-o'clock (*Mirabilis jalapa*). He found that the progeny of reciprocal crosses resembled the maternal parent in phenotype (Table 17–1). Thus if the maternal parent was green in color, the progeny seedlings were always green; but if that parent was variegated, the progeny could be variegated, or white. Later, as a result of more extensive work on nonmendelian inheritance of plastid phenotypes in a variety of plant genera, it was shown that variegated plants could yield green progeny as well as white and variegated progeny. Since variegated

plants, by definition, contain green and white sectors, the white plants presumably arise when an egg containing only white plastids is fertilized; the green plants presumably result when an egg pure for green plastids is fertilized, and the variegated plants must originate from eggs containing both plastid phenotypes. This pattern of nonmendelian inheritance has been referred to as uniparental-maternal (2), and it is characteristic of the great majority of flowering plants. The second pattern of nonmendelian inheritance of plastid phenotype was discovered by Baur (3) in the garden geranium (*Pelargonium zonale*). If reciprocal crosses are made between green and variegated plants, green, white, and variegated progeny are obtained in the F_1 (Table 17–1). This pattern of inheritance is called biparental.

There is a vast literature on the inherit-

TABLE 17–1. *Plastid transmission patterns in higher plants*

Pattern of inheritance	Parent ♀　♂	Offspring	Remarks
Uniparental-maternal	G[a] × W	G	
	W × G	W	
	G × V	G	
	W × V	W	
	V × G	± G ± V ± W	From sorting-out of maternal plastids
	V × W	± G ± V ± W	
Biparental	G × W	G ± V ± W	In some species, plants with paternal plastids alone are found in addition to plants that are pure for maternal plastids and variegated plants; in other species, only variegated plants or plants pure for maternal plastids are found
	W × G	W ± V ± G	
	G × V	± G ± V ± W	From transmission of paternal plastids followed by sorting-out
	W × V	± G ± V ± W	
	V × G	± G ± V ± W	From sorting-out of maternal plastids and from paternal plastid transmission
	V × W	± G ± V ± W	

[a] G = green, V = variegated, W = white.
Adapted from Kirk and Tilney-Bassett (2).

467

ance of plastids in higher plants that has been reviewed authoritatively in English by Kirk and Tilney-Bassett (2) in their admirable treatise *The Plastids* and updated in a recent review by Tilney-Bassett (4). It would serve little purpose to cover the same material in detail here. Rather, what appear to be some of the major features of plastid genetics in higher plants will be sketched in outline here, mainly for the purpose of comparison with the organelle genetic systems discussed in preceding chapters. The studies of Wildman and his colleagues using interspecific hybrids of *Nicotiana* as a means for localizing structural genes for specific chlo-

roplast components will be discussed in detail in Chapter 18.

FREQUENCIES OF PATTERNS OF PLASTID INHERITANCE IN NATURE

In the angiosperms, maternal inheritance appears to be the most common pattern of plastid inheritance (Table 17–2). In the dicotyledons, maternal inheritance has been reported in 24 genera, and 14 genera have shown at least some evidence of biparental inheritance. Evidence for biparental inheritance has been obtained in 2 genera of monocotyledons, and maternal inheritance

TABLE 17–2. *Mode of plastid inheritance in different species of higher plants*

Biparental plastid inheritance	Maternal plastid inheritance
Ferns	**Dicotyledons**
Scolopendrium vulgare	*Arabidopsis thaliana*
	Arabis albida
Gymnosperms	*Aubrieta graeca, A. purpurea*
Cryptomeria japonica	*Beta vulgaris*
	Capsicum annuum
Dicotyledons	*Cucurbita maxima*
Acacia decurrens, A. mearnsii	*Epilobium hirsutum*
Antirrhinum majus[a]	*Gossypium hirsutum*
Borrago officinalis	*Lactuca sativa*
Fagopyrum esculentum[a]	*Lycopersicum esculentum*
Geranium bohemicum	*Mesembryanthemum cordifolium*
Hypericum acutum, H. montanum, H. per-	*Mimulus quinquevulnerus*
foratum, H. pulchrum, H. quadrangulum	*Mirabilis jalapa*
Medicago truncatula	*Nicotiana colossea, N. tabacum*
Nepeta cataria	*Petunia hybrida, P. violacea*
Oenothera (Euoenothera) 28 species	*Pharbitis nil*
Oenothera (Raimannia) berteriana	*Pisum sativum*
O. (R.) odorata	*Plantago major*
Pelargonium denticulatum, P. filicifolium,	*Primula sinensis, P. vulgaris*
Pelargonium × *Hortorum (zonale)*	*Stellaria media*
Phaseolus vulgaris	*Trifolium pratense*
Rhododendron 11 species	*Viola tricolor*
Silene pseudotites	
Solanum tuberosum[a]	**Monocotyledons**
	Allium cepa, A. fistulosum
Monocotyledons	*Avena sativa*
Chlorophytum comosum,[a] *C. elatum*[a]	*Hordeum vulgare*
Secale cereale	*Hosta japonica*
	Oryza sativa
	Sorghum vulgare
	Triticum aestivum, T. vulgare
	Zea mays

[a] Plants in which plastid inheritance is predominantly maternal but that show a trace of biparental inheritance.

Adapted from Kirk and Tilney-Bassett (2,4); original references cited therein.

appears to be the rule in 8 genera. Instances of biparental and maternal inheritance have been reported at the familial level (Leguminosae and Gramineae), but at the generic level all species show either maternal or biparental plastid inheritance. In the gymnosperms, extensive breeding data on the inheritance of a plastid mutation are available only for *Cryptomeria japonica,* where plastid inheritance is biparental with a strong paternal bias. Electron microscopic observations on other gymnosperms, as reviewed by Tilney-Bassett (4), have shown that the plastids of the proembryo are mostly, if not entirely, derived from the paternal parent. The one example of a plastid mutation in ferns shows biparental inheritance. Clearly, it would be of great interest to have more extensive data on plastid inheritance in gymnosperms, ferns, and bryophytes for comparative purposes.

ARE PLASTIDS RESPONSIBLE FOR NONMENDELIAN INHERITANCE OF COLOR PATTERNS?

The major evidence that the plastids themselves have mutated comes from studies of somatic segregation of green, white, and variegated tissue within a variegated plant. Kirk and Tilney-Bassett have contrasted the variegation patterns to be expected as a result of nuclear and plastid mutation. A nuclear mutation in a somatic cell will result in a simple cell lineage in which all cells are identical with respect to the new genotype. If the new mutation prevents chlorophyll development and it arises in a green shoot, the new lineage will at once be conspicuous as a single white, yellow, or even pale green sector. In contrast, a mutation in a plastid (or some other extranuclear factor affecting chloroplast development) will produce a complex cell lineage because only a single plastid in the initial cell will have mutated. The progeny of this cell may be green, variegated, or white, depending on the proportions of normal and

mutant plastids each of the progeny cells receives. This "sorting-out variegation" will result in mosaicism or striping of the leaves (Fig. 17–1).

If, in fact, sorting-out variegation reflects the cytoplasmic assortment of mutant plastids (*plastome mutations*) rather than the effects of different cytoplasms on plastid development (*nonplastome mutations*), as postulated by Correns (1,5) many years ago, it should be possible to find mixed cells containing both mutant and normal plastids. The Correns hypothesis predicts that each cell will contain either normal or mutant plastids, but not both, since it is the cytoplasm rather than the plastid that is either normal or diseased. Consequently, mixed cells have been sought assiduously in plants exhibiting sorting-out variegation, and they have been observed in numerous species (2,4) (Figs. 17–1, 17–2, and 17–5).

Until recently, identification of mixed cells was the best evidence that the mutation in question did, in fact, reside in the plastome, but even this evidence was not rigorous. For example, von Wettstein and Erickson (6) observed that a plastome mutant of *Nicotiana* also exhibited mitochondrial defects, and they argued that defects in the mitochondria might have led to the plastid defects that were seen. Although this hypothesis seems unlikely, for reasons that were discussed in detail by Kirk and Tilney-Bassett (2) (e.g., the mixed cells observed were shown to be mixed only with respect to plastids, and mitochondrial mixing was not demonstrated), it cannot be ruled out on the basis of cytological observation. The most direct way of demonstrating that chloroplasts do carry mutant and wild-type specificities would be chloroplast transfer experiments similar to the mitochondrial transfer experiments reported earlier for *Neurospora* (see Chapter 8) and *Paramecium* (see Chapter 10). So far, it appears that successful chloroplast transfer experiment have been reported only for *Nicotiana* (7). And these results are

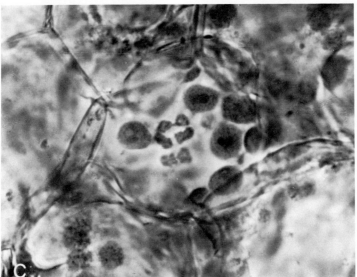

FIG. 17–1. A and B: Plants of *Antirrhinum majus* containing green chloroplasts and mutant white plastids (of the type *en : alba-1*) with typical variegation pattern caused by a sorting-out of two genetically different plastid types (a fine checkered pattern in leaves, as well as pure green and white leaves and shoots). C: Mixed cell of *A. majus* containing normal green chloroplasts and mutant white plastids (*en : alba-1*) side by side in the same cell without any transitions. (A and B courtesy of Dr. R. Hagemann, Halle, GDR. C from Hagemann, ref. 16, with permission.)

themselves suspect (Chapter 18). In Chapter 18 we shall also consider other evidence from protoplast fusion experiments in *Nicotiana* that certain traits are coded by clDNA, as well as some results suggesting that clDNA may be maternally inherited in plants of this genus. We may expect that restriction enzymes will soon be put to use to determine the pattern of inheritance of clDNA in higher plants, showing maternal or biparental inheritance of plastome mutants. Ideally, it would be nice if mtDNA showed a different pattern of inheritance, but there is no reason to expect that this will happen. In the meantime, the best evidence that the pattern of inheritance of plastome mutants really reflects the pattern of inheritance of plastid DNA comes from ultrastructural studies (8–11) showing that the pattern of inheritance of the plastids themselves is the same as the pattern of inheritance of presumptive plastome mutants in any given species. These studies were recently reviewed by Hagemann (12).

In angiosperms the pollen mother cell undergoes meiosis to yield the microspores that are shed as pollen grains. After formation of the microspores, a resting period generally ensues that may last for a few hours or several months, following which the microspore nucleus divides mitotically, and two distinct cells are produced. One of these, the vegetative or tube cell, will form the pollen tube that extends down the style to the micropyle of the ovule. The second or generative cell divides to form the two sperm cells that are enclosed within the pollen tube formed by the vegetative cell. One of these fertilizes the egg cell, and the other fuses with the two polar nuclei of the ovule to form the triploid endosperm. The key question, of course, is whether or not the generative cell contains plastids, since if it does not, plastids will be excluded from the sperm cells by definition.

Hagemann separated the angiosperms into three groups according to plastid content of generative cells. Genera belonging to group I do not contain plastids in the generative cells. These include *Gossypium* and *Mirabilis,* both of which exhibit maternal inheritance of plastome mutants, as well as *Capsella,* for which inheritance of specific plastome mutants does not seem to have been reported (Table 17–2). Plants belonging to group II only infrequently produce generative cells containing single plastids. These include the genera *Beta, Lycopersicum,* and *Antirrhinum.* Plastome mutants in the former two genera are inherited maternally, as they are in group I genera, but *Antirrhinum* rarely transmits a plastome mutant biparentally as well, which presumably reflects the rare inclusion of paternal plastids in the generative cell (Table 17–2). In principle there is no difference between groups I and II, except that plastid exclusion is marginally less efficient in group II plants, as is shown by the *Antirrhinum* example. The normal situation for *Antirrhinum* is shown in Figs. 17–3 and 17–4, where it can be seen that plastids are excluded from the generative cell, although they are abundant in the vegetative cell cytoplasm. Group III angiosperms regularly contain numerous plastids in the generative cell, as is illustrated in the case of *Pelargonium* (Fig. 17–4A). Genera in this group, in addition to *Pelargonium,* are *Oenothera, Castilleia,* and *Lobelia.* The pattern of inheritance of plastome mutants is biparental in both *Oenothera* and *Pelargonium* (Table 17–2), but it does not appear to have been reported for *Lobelia* or *Castilleia.* As Hagemann pointed out, these results strongly suggest that exclusion or inclusion of plastids in the generative cell of the pollen grain determines the pattern of inheritance of plastome mutants in higher plants. These findings should be contrasted with those reported earlier for *Chlamydomonas* (see Chapter 12), in which fusion of maternal and paternal plastids occurs in the zygote, although chloroplast genes are maternally inherited.

Before leaving the subject of plastid

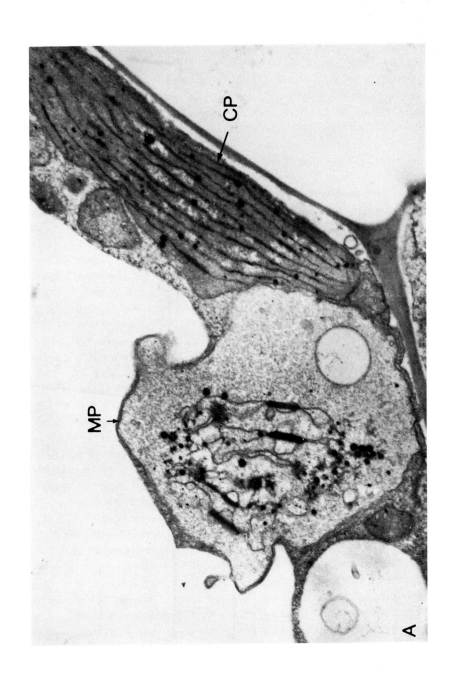

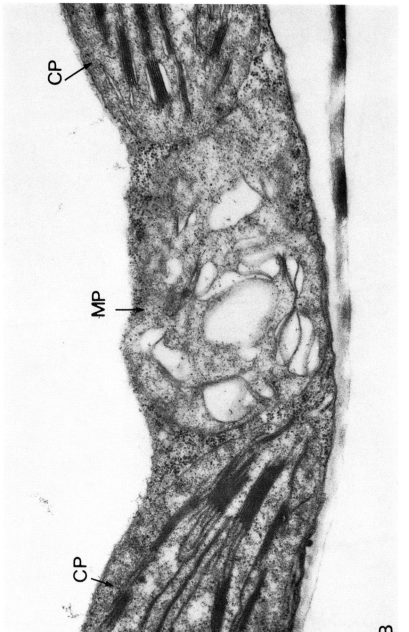

FIG. 17–2. Electron micrographs of mixed cells from A. majus (A) and Helianthus annuus (B) with normal green (CP) and mutant white (MP) plastids in the same cell. Plastid mutations were induced by N-nitroso-N-methylurea. (Courtesy of Dr. R. Knoth and Dr. M. Meyer, Halle, GDR.)

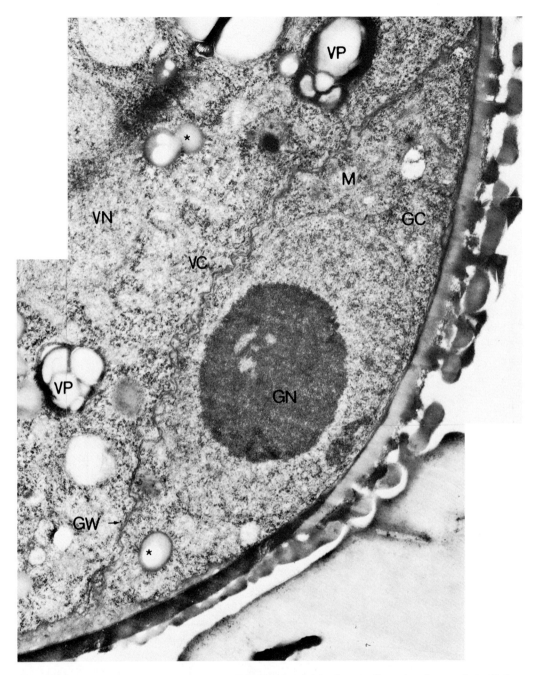

FIG. 17–3. Electron micrograph of a generative cell in a pollen grain from A. *majus* illustrating absence of plastids from such cells. The young generative cell (GC) is seen lying at the pollen grain wall with a large nucleus (GN), mitochondria (M), and osmiophilic granules (*), but no plastids. In contrast, the surrounding cytoplasm of the vegetative cell (VC) contains many plastids (VP) with starch grains. The wall of the generative cell (GW) and the vegetative cell nucleus (VN) are also indicated. (Courtesy of Dr. R. Knoth, Halle, GDR, and Dr. M. Wrischer, Zagreb, Yugoslavia.)

transmission pattern, it should be noted that some experiments of Nilsson-Tillgren and von Wettstein-Knowles (13) clearly showed that pollen grain plastids are functional in *Nicotiana,* where inheritance of plastome mutants is strictly maternal. They used the technique of anther culture (14) to obtain haploid plants from pollen grains derived from wild-type *Nicotiana tabacum* cv Turkish Samsun and variegated plants containing a specific plastome mutant. The anthers from wild-type plants produced green progeny; those from the variegated plants produced green, white, and variegated progeny. Thus the pollen in *Nicotiana* clearly contains viable plastids.

THEORETICAL BASIS OF SORTING-OUT VARIEGATION

Michaelis examined the theoretical properties of sorting-out variegation in several detailed articles whose contents have been reviewed by Michaelis (15), Kirk and Tilney-Bassett (2), and Hagemann (16). Although we cannot do justice to this work here, it is important to consider a few of the implications, since the theoretical analysis devised by Michaelis can be applied to somatic segregation of extranuclear particles of any kind, provided that segregation is random.

Michaelis (15) stated that in variegated plants a random distribution of "cytoplasmic units" is usually found. Therefore the laws of plastid segregation can be derived by comparing observed segregation patterns with model experiments (17) or by making appropriate mathematical calculations (18). Suppose that one starts with cells containing mutant and wild-type particles (e.g., green and white balls in a model experiment) in a mixed cell (a box in a model experiment). It is possible to calculate the rate at which cells will become homoplasmic for mutant or wild-type phenotype as a function of the number of particles per cell and the ratio

of the two particle types per cell. The distribution of unmixed and mixed cells as a function of particle number with equal inputs of wild-type and mutant particles is shown in Table 17–3. The important point is that the number of cell divisions required before unmixed cells are seen increases rapidly as a function of input. In the case of very asymmetric inputs, unmixed cells of the majority type appear rapidly, but cells homoplasmic for the minority genotype appear much more slowly. Thus the asymmetric ratio of particles at the outset is reflected in the preferential segregation of unmixed cells carrying the majority genotype. On completion of segregation, the frequency of unmixed cells of each type will reflect the particle ratio at the outset. Michaelis (18) also calculated "the velocity of sorting-out" as a function of the number of segregating particles per cell at equal inputs. Here the percentage of cells homoplasmic for one of a pair of segregating alleles is plotted as a function of number of cell divisions. If $n = 10$, virtually all of the cells have become homoplasmic for one or the other allele after 50 cell divisions; but if $n = 50$, only a little more than 40% of the cells will be homoplasmic after 50 cell divisions. Michaelis also computed the expected particle distributions for daughter cells derived from mixed cells with different input ratios. Each distribution is normal, and variations in starting ratios do not have a marked effect on the distributions.

The principal conclusions of Michaelis were succinctly summarized by Hagemann (16), and they were translated into English by Kirk and Tilney-Bassett (2). Because of their general interest, we will paraphrase the translation of Kirk and Tilney-Bassett once again. To calculate what might happen to two plastid types during successive cell divisions, Michaelis assumed that only two plastid types exist in the cell. These are randomly mixed before cell division, and they separate randomly at cell division, when they are exactly halved. All plastids

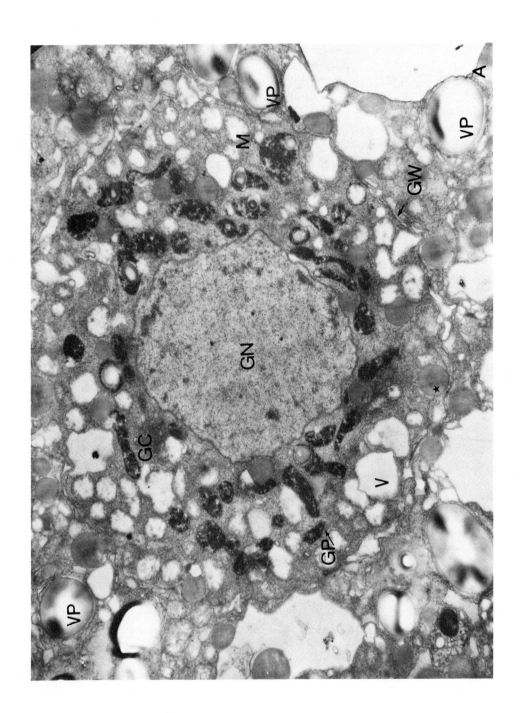

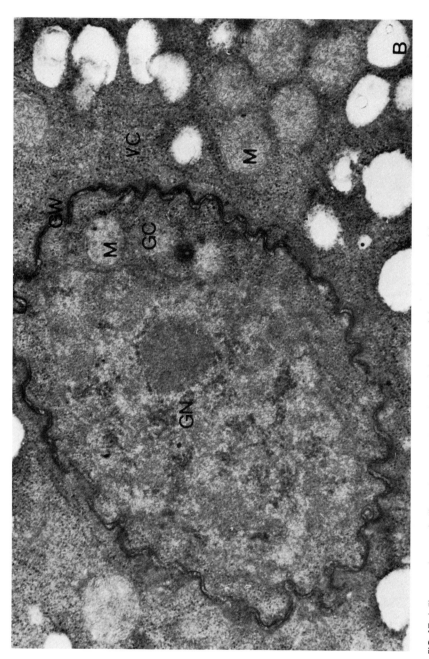

FIG. 17–4. Electron micrographs illustrating generative cells from *Pelargonium* (A) and *Antirrhinum* (B). The large generative cell of *P. zonale* (GC) contains many proplastids (GP) that are unusually enriched with masses of phytoferritin particles and sometimes contain small starch grains; the cytoplasm also contains mitochondria (M), vacuoles (V), and microbody-like structures (*). The generative cell wall (GW) separates the vegetative cell from the generative cell. The plastids of the vegetative cell (VP) are distinctly larger than those in the generative cell and have large starch grains, but phytoferritin was never found. For contrast, a second generative cell of *Antirrhinum* (B) is shown that lacks plastids. The generative cell (GC) with its cell wall (GW) is in a later stage of development than the one shown in Fig. 17–3. The generative cell is surrounded by the vegetative cell (VC). The generative cell contains a small cytoplasmic area surrounding the generative nucleus (GN) and mitochondria (M), but no plastids. (Courtesy of Dr. R. Knoth, Halle, GDR, and Dr. M. Wrischer, Zagreb, Yugoslavia. A from Hagemann, ref. 12, with permission.)

TABLE 17–3. *Decline in number of mixed cells as a function of cell division with random segregation when genetic copy number (n) is variable and half of the copies contain a specific allele (a)*

Division	Copy number (n)	Percentages of cells in which percentage of genomes carrying a is				
		0	1–25	26–74	75–99	100
1	10	1.19	13.10	71.42	13.10	1.19
	20	0	1.15	97.70	1.15	0
	50	0	0.04	99.92	0.04	0
	100	0	0	100.00	0	0
	300	0	0	100.00	0	0
4	10	13.60	15.55	41.70	15.55	13.60
	20	1.63	12.14	72.46	12.14	1.63
	50	0	4.52	90.96	4.52	0
	100	0	0.15	99.80	0.05	0
	300	0	0	100.00	0	0
8	10	27.16	10.04	25.60	10.04	27.16
	20	3.37	15.00	53.26	15.00	8.37
	50	0.27	11.58	76.30	11.58	0.27
	100	0	1.17	98.05	0.78	0
	300	0	0	100.00	0	0
16	10	41.10	3.92	9.96	3.92	41.10
	20	22.33	11.01	33.32	11.01	22.33
	50	3.37	17.78	57.70	17.78	3.37
	100	0.02	7.06	85.72	7.18	0.02
	300	0	0.27	99.42	0.31	0

Adapted from Michaelis (18).

double again following cell division, so that the total number per cell remains constant. Under these conditions, the following mathematical properties are valid:

1. The number of possible combinations of green and white plastids increases with increasing total numbers of plastids per cell; the number of combinations is $n + 1$.

2. The frequency of the appearance of sorting-out decreases with increasing total number of plastids.

3. The number of successive cell divisions that leads to an almost complete sorting-out ($> 99\%$) amounts to approximately $10n$.

4. After the first cell division, the frequency of individual mixture combinations depends on the mixture proportions of the starting cell. With increasing numbers of cell divisions, the frequency with which any combination appears always remains less than the frequency of the starting mixture, even though the percentage frequency of the starting mixture decreases with each successive cell division. The stage when sorting-out is practically complete is solely dependent on the total number of plastids per cell, not on the proportions of the two plastid types per cell.

5. After fully sorting-out, the two kinds of homoplasmic cells (one with only green plastids, the other with only white plastids) appear in the same proportion that the two types of plastids were mixed in the starting cell.

Since the calculations of Michaelis apply to any pair of particle types assorting at random, it is most important to be aware of them. Thus the models of Michaelis can be set up as null hypotheses against which the rates and patterns of segregation of organelles in other systems (e.g., mitochondria in strains of *Paramecium* containing two kinds of genetically marked mitochondria) and organelle genes themselves (e.g., chloroplast genes in *Chlamydomonas* or mitochondrial genes in yeast) can be compared. It is surprising that these detailed and useful calculations have not evoked more widespread interest among workers in the field of organelle genetics.

FORWARD AND REVERSE MUTATIONS OF PLASTIDS

In addition to the numerous examples of spontaneous plastid mutations, recessive nuclear genes that act as plastid mutators have been described in a number of genera of higher plants (Table 17–4). In some cases a single mutant plastid type results, but in other cases various mutant plastid types appear to be induced. Röbbeln (19) reported that the *am* gene in *Arabidopsis* causes a specific plastid mutation in which plastid development is blocked in thylakoid membrane formation at an early stage. On the other hand, the *chm* (chloroplast mutator) mutant of *Arabidopsis,* which was isolated by Redei (20) following ethyl methane sulfonate mutagenesis, causes the induction of a number of structurally distinct plastid mutations. The classic case of

a plastid mutator is, of course, the *iojap* gene of *Zea mays,* whose genetics was worked out in such elegant detail many years ago by Rhoades (21–23). Shumway and Weier (24) showed that the plastid mutations induced by this gene are of a single type; they are aberrant at all developmental stages, and they lack the normal grana system as well as a normal prolamellar body and plastid ribosomes. It would clearly be of interest to know more about the mechanism of action of different plastid mutators. Perhaps some mutators cause deletions in clDNA and others lead to point mutations at different sites. Mutators of the latter type could be particularly valuable in obtaining a wide array of different plastid mutations in the absence of nuclear gene mutations in higher plants in order to study the chloroplast functions governed by the chloroplast genome. Obviously those mutators, such as *chm,* that induce a variety of phenotypically different plastid mutations would be good candidates for study.

If a mutator gene causes point mutations in the plastid genome, one might expect these mutations to revert. Despite the diffi-

TABLE 17–4. *Examples of variegated plants in which repeated plastid mutations (G → W) are induced by a nuclear mutator gene when homozygous recessive*

Plant species	Mutator gene symbol	Forward mutation (G → W)		Reverse mutation (W → G)	Plastid inheritance
		One kind	Many kinds		
Arabidopsis thaliana albomaculatus	*am*	+	−	+	Maternal
Arabidopsis thaliana	*chm*[1]	−	+	+	Maternal
	chm[2]				
Capsicum annuum	—	+	−	−	Maternal
Epilobium hirsutum	*mp*₁	−	+	+	Maternal
	*mp*₂	−	+	−	Not maternal
Hordeum vulgare Okina-mugi and Okina-mugi tricolor	—	+	−	−	Maternal
Hordeum vulgare	*w*	+	−	−	Maternal
Hordeum vulgare albostrians	*as*	+	−	−	Maternal
Hordeum vulgare striata-4	—	+	.	−	Maternal
Nepeta cataria	*m*	+	−	−	Biparental
Oenothera hookeri plastome mutator	*pm*	−	+	−	Biparental
Oryza sativa	—	+	−	−	Maternal
Petunia hybrida	*a⁻*	−	+	+	Maternal
Zea mays iojap	*ij*	+	−	−	Maternal
Zea mays chloroplast mutator	*cm*	+	−	−	Maternal

Adapted from Tilney-Bassett (4); original references cited therein.

culties encountered in distinguishing sorting-out variegation from true reversion, there is evidence that certain plastid mutations can revert. Redei (20) obtained limited evidence that one of the plastid mutations induced by the *chm* mutation could revert. He observed two small islands containing one to five green cells in a white leaf containing an estimated 325,254 cells. He calculated that the reversion rate for this plastid mutant must have been on the order of 10^{-5} per cell. Michaelis (25) made a monumental study of plastid reversion in *Epilobium hirsutum* that was reviewed in detail by Tilney-Bassett (4). Forward mutations were obtained using the nuclear mutator gene mp_1. In order to distinguish reversion from sorting-out variegation, it was necessary to have some estimate of the smallest number of green cells that could be produced in a white leaf by sorting-out variegation. To obtain this estimate, Michaelis calculated the number of green cells that could be produced in a leaf starting with a mixed cell with only one green plastid. It takes about 20 cell divisions to produce an *Epilobium* leaf with up to 1 million cells. Michaelis calculated that with random sorting-out from a mixed cell with only one green plastid, such a leaf would contain 2 to 3% green cells. In two successive leaf whorls, requiring 40 to 55 cell divisions, the frequency would rise to 6 to 9%. Hence, if no visible green flecks appear through three leaf whorls in a white shoot, it can be concluded that sorting-out is complete and that green plastids produced by sorting-out are no longer present. Michaelis kept a white shoot alive for 21 months, during which time it passed through at least 150 cell divisions, and he never found a cluster of green cells larger than four containing up to 16 green plastids between them. It seems very unlikely that these results can be explained by sorting-out; they are best accounted for by plastid reversion. Moreover, the ratios in which reverted plastids appeared in mixed cells corresponded

closely with theoretical predictions made on the assumption that the green plastids in these cells arose by chance reversion of a single plastid. Finally, as we shall discuss, it could be shown in some instances that the reverted plastids actually differed from the original green wild-type plastids.

By means of an ingenious cross between two plastome mutants, Michaelis was able to demonstrate that reversion was a property of the plastome rather than the nucleus. One of the mutants, which was ivory white, did not revert; the other, which was pale gold, contained macroscopically visible flecks of revertant cells. Michaelis found that revertibility was strictly maternally inherited together with the pale gold phenotype. In addition he found that the mp_1 gene not only induced forward plastome mutations, but also reversions. In a cross of a pale yellow ♀ ($Mp_1\ mp_1$) and a white ♂ ($mp_1\ mp_1$), both parent branches were mutant, and the seedlings were colorless, but only the heterozygous seedlings died. The homozygous $mp_1\ mp_1$ seedlings survived because reversions appeared that produced a marbled variegation in which the green tissue enabled the plants to survive. From these plants, entirely green seedlings were obtained. If the mp_1 gene was eliminated from these plants by outcrossing, they remained stably green. These results show that although plastid reversion is a property of the plastome, the reverse mutation rate, like the forward mutation rate, can be influenced by the mp_1 gene.

The availability of revertant plants made it possible to determine if the revertant plastids were identical to the original wild-type plastids. Four distinct clones were obtained. Clone I had thick, dark green leaves with larger plastids than normal, but the plants died during development. Clones II, III, and IV all survived, but with respect to the wild type, they showed inconclusive morphologic differences. Michaelis, therefore, crossed these clones with a series of different test plants. The most important crosses were made to *Epilobium parviflorum*. Normally, *E. hirsutum* Essen ♀ × *E. parviflorum* ♂ hybrids are characterized by developmental disturbances such as stunting, leaf deformities, sterility, etc. When *E. hirsutum* Essen clones II, III, and IV were tested in this combination with *E. parviflorum*, Michaelis found that clones II and IV produced normal fertile hybrids, and some of the developmental disturbances were also removed in crosses with clone III. These results indicated that the reversion process resulted in the production of new plastid types not identical to the original wild-type plastids, since they were compatible with the nuclear genome of the hybrid. These studies are reminiscent of the interspecific mitochondrial compatibility experiments done on *Paramecium* (see Chapter 10) and mammalian tissue culture cells (see Chapter 11). It would appear from the experiments of Michaelis that organelle compatibility relationships may be altered by very subtle changes. Clearly, this is an area that deserves far more attention.

FUNCTIONS IMPAIRED IN PLASTOME MUTANTS

Surprisingly, only limited numbers of plastome mutants have thus far been characterized biochemically (see Chapter 20). In this regard, the most interesting and most informative work was done by Hagemann and his colleagues Herrmann, Börner, and Knoth. Herrmann (26) reported on the pigment-protein complexes of a plastome mutant of *Antirrhinum majus* called *en:alba-1*. This scheme of delineating plastome mutants to distinguish them from nuclear mutants affecting plastid phenotype was devised by Hagemann (27). Thus *en* refers to extranuclear, and *alba* indicates albomaculata. The *en:alba-1* mutant has pale green mutated plastids incapable of photosynthesis. Herrmann compared the chlorophyll-protein complexes I and II (CPI and CPII) from wild-type and mutant plants. It will be

recalled (see Chapter 1) that CPI is enriched for photosystem I activity and CPII for photosystem II activity. The *en:alba-1* mutant totally lacked CPI, but it contained CPII in roughly normal amounts. Gel electrophoresis of total thylakoid membrane polypeptides revealed that the mutant lacked the principal polypeptide associated with CPI. There were other minor differences in the gel patterns as well.

Herrmann (28) also investigated another plastome mutant of *A. majus* called *en:viridis-1*. This mutant is also pale green in color, but it is lacking in photosystem II activity. Both CPI and CPII are present in the mutant, but SDS gel electrophoresis of the thylakoid membrane polypeptides revealed an absence of a major polypeptide (polypeptide 7, approximately 50,000 d), diminution of several other polypeptides (polypeptides 8–12), and the presence of a new polypeptide. These results suggest that photosystem II activity can be lost without loss of CPII. This point, together with further supporting data from *Chlamydomonas*, will be considered again in Chapter 20. It should be emphasized that although these two plastome mutants affect different photosystems, both of them are quite pleiotropic in terms of thylakoid membrane polypeptide phenotype. Such pleiotropy must be expected in nonconditional mutants where the primary defect most probably involves a membrane polypeptide or its insertion. Membrane assembly is a complex process in which the insertion of one component may depend on the presence of a second.

Börner et al. (29) examined a plastome mutant (*en:alba-1*) of *Pelargonium zonale* present in a variety called "Mrs. Parker." This is a white-over-green periclinal chimera. Reciprocal crosses were made between "Mrs. Parker" and a green variety called "Trautlieb." Since plastid transmission in *P. zonale* is biparental, the F_1 consisted of green, variegated, and white seedlings. Leaf material for ultrastructural and biochemical studies was obtained from pure

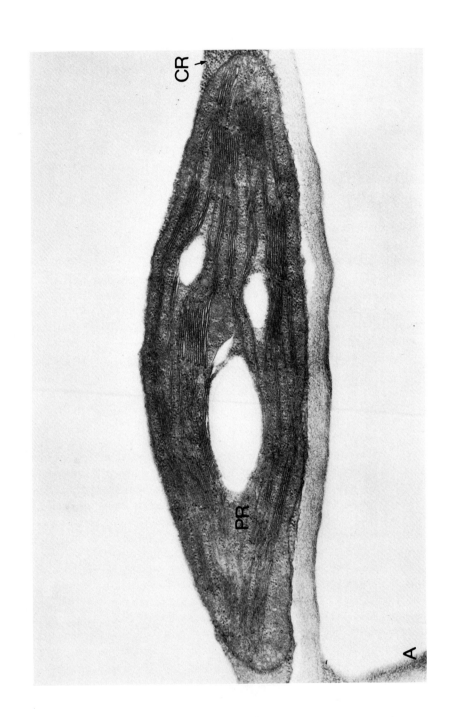

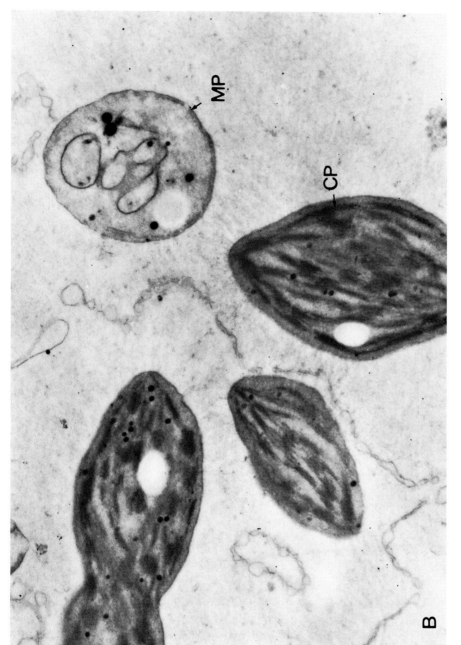

FIG. 17–5. Electron micrographs illustrating normal and mutant plastids of *P. zonale*. A: Normal mesophyll chloroplast of *P. zonale* with well-developed membrane structures, starch grains, and double-layer envelope. Numerous plastid ribosomes (PR) are present that appear smaller than the cytoplasmic ribosomes (CR). B: Mixed cell from embryonic tissue of a hybrid of *P. zonale* 2 weeks after fertilization. The cell contains a mutant white plastid (MP) from the maternal plant (Mrs. Parker, *en : alba-1*) and normal green plastids (CP) transmitted to the zygote by the sperm cell of the pollen tube of the paternal parent (Trautlieb). (Courtesy of Dr. R. Knoth, Halle, GDR.)

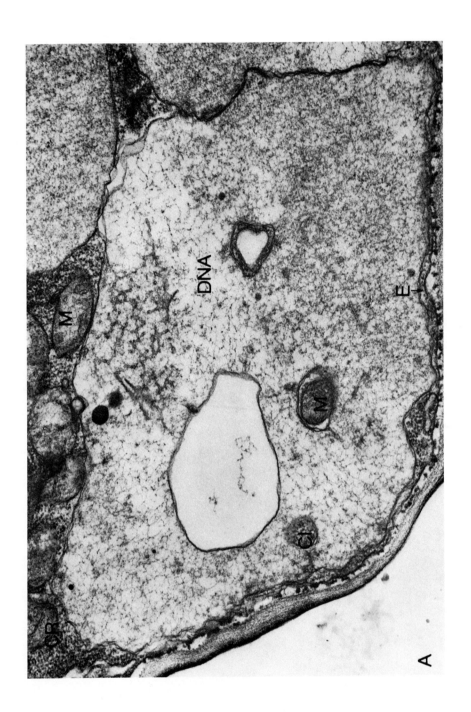

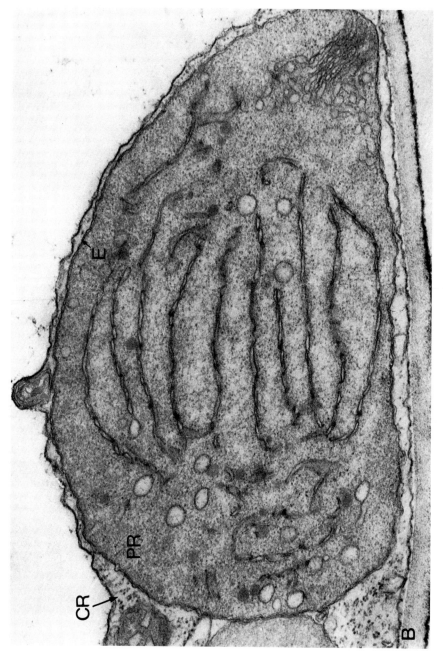

FIG. 17–6. Electron micrographs of plastids from plastome mutants of *P. zonale*. A: White plastid of the plastome mutant *en:alba-1* from Mrs. Parker. Note the absence of plastid ribosomes, but the presence of numerous cytoplasmic ribosomes (CR). The plastid contains a prolamellar-like structure, DNA fibrils (DNA), and a double-layer envelope (E). The plastid stroma seems to have flowed around parts of the cytoplasm (CI) and a mitochondrion (M), which can be seen as inclusions within the plastid. B: Yellow plastid of the plastome mutant *en:gilva-1* from Mrs. Pollock that has a deficiency in photosystem I. The plastid contains normal plastid ribosomes. Plastid differentiation ranges from yellowish green under dim light to yellow under intense light, as in this micrograph. (From Börner et al., ref. 29, with permission; A courtesy of Dr. R. Knoth, Halle, GDR.)

green or white shoots produced by variegated plants. Ultrastructural investigation of the white leaves revealed plastids very different from wild type (Fig. 17–5), with little internal membranous structure and no chloroplast ribosomes (Figs. 17–5B and 17–6A). The absence of chloroplast ribosomes was confirmed by electrophoretic analysis of total rRNA from the white tissue, which showed that only the cytoplasmic rRNAs were present. From these results it seems plausible that the multiple defects seen in the white plastids from "Mrs. Parker" may be partially accounted for in terms of the inability of these cells to carry out chloroplast protein synthesis. However, it is also possible that the absence of chloroplast rRNA in these plastids signals the fact that really gross changes in the clDNA of these plastids has occurred. One might envision changes similar to those seen in the mtDNA of a vegetative petite mutant. Thus the defects seen in white "Mrs. Parker" plastids may also be attributable to other effects of the mutation on clDNA. These results have been extended by Börner et al. (30) to several other chimerical varieties of *P. zonale* (i.e., "Freak of Nature," "Flower of Spring," and probably also "Madame Salleron," "Gnom," and "Griefswald") and the albostrians mutant of barley (31), which as these authors have shown cannot synthesize ribulose bisphosphate carboxylase and several thylakoid membrane polypeptides. Shumway and Weier (24) presented ultrastructural evidence for the absence of chloroplast ribosomes in plastids found in white tissue of the iojap mutant of *Zea mays*. It is interesting that plastid ribosomes are also absent in certain nuclear higher-plant mutations. Thus plastid ribosomes depend on both nuclear and chloroplast genes for their formation, a process we shall discuss in much more detail in Chapter 21.

In summary, we can recognize two general classes of plastome mutants in higher plants. Those in the first class, exemplified by the *Antirrhinum* mutants, probably carry out chloroplast protein synthesis normally and may well be point mutants within the chloroplast genome. The second class, exemplified by the white-green geranium chimeras, consists of mutants that form white plastids with no plastid ribosomes. These could well be deletion mutants. Reversion studies of the kind discussed in the preceding section could provide a crude distinction between the two mutant classes, but much more powerful proof would be demonstration that deletions occur in the clDNA of the mutants lacking chloroplast ribosomes.

PLASTID COMPETITION AND COMPATIBILITY IN *Oenothera*

Perhaps the most detailed studies of competition between genetically different organelles in the same nuclear background and organelle compatibility in different nuclear backgrounds are those done with *Oenothera* hybrids. There is an enormous German literature on these subjects; it was reviewed in detail by Kirk and Tilney-Bassett in *The Plastids* (2) and subsequently was discussed by Sager (32), Kutzelnigg and Stubbe (33), and Tilney-Bassett (4). Our discussion will be brief and will be based largely on these reviews.

Most of the work on the *Oenothera* plastome has been done on species of the subgenus *Euoenothera* by Renner and his students Stubbe and Schötz and by Schwemmle and colleagues on species of the subgenus *Raimannia*. *Oenothera* species contain seven pairs of chromosomes. Cleland (34,35) showed that the chromosomes are linked together in a chain or ring at the first meiotic division as a result of a series of subterminal reciprocal translocations (Fig. 17–7). Crossing-over between these homologous regions serves to stabilize the chromosomes in the ring or chain configuration. The result of this highly symmetrical arrangement is to prevent independent as-

FIG. 17–7. Diagram showing segregation of maternal and paternal chromosome complexes in first meiotic anaphase in *Oenothera*. The seven chromosomes from each parent are lined up opposite each other at metaphase because of reciprocal translocation. The set shown here at anaphase will segregate without recombination, giving rise to gametes of identical genotype as the parental strains. (From Cleland, ref. 35, with permission.)

sortment and to restrict recombination to the chromosome tips. At anaphase I the chromosomes segregate as a unit. In some species all 14 chromosomes are included in a single ring, with one unit of 7 going to one pole and the other unit of 7 to the opposite pole. In other species as few as 4 chromosomes are in the ring formation, with the rest segregating and assorting normally.

Each parental complex of chromosomes has its own distinguishing features; these were given distinctive Latin names by Renner. For example, in *O. lamarckiana* all 14 chromosomes form a ring, and the two 7-chromosome units are called *gaudens* and *velans*. Maintenance of the heterozygous *gaudens·velans* genotype is assured by the presence of recessive lethal factors in each complex. Thus, on selfing a *gaudens·velans* plant, one would expect *gaudens·gaudens*: *gaudens·velans*:*velans·velans* progeny in a 1:2:1 ratio, but in fact only the heterozygous plants develop to the fertile stage. By contrast, *O. hookeri* is a homozygous species that forms no complexes; thus its haploid chromosome set is simply referred to as ⁱʰ*hookeri*.

Plastid compatibility has been studied extensively in *Oenothera* through the use of interspecific hybrids. The first such studies were made many years ago by Renner (36,37). Renner made reciprocal crosses between *O. hookeri* and *O. lamarckiana*. When *O. hookeri* served as the maternal parent the ⁱʰ*hookeri·gaudens* progeny were green, but the ⁱʰ*hookeri·velans* progeny, although predominantly green, included a few variegated plants as well (Table 17–5). In

the reciprocal cross the results were quite different. Although the ⁱʰ*hookeri·gaudens* progeny all were green, the majority of the ⁱʰ*hookeri·velans* progeny were yellow, with a few being variegated (Table 17–5).

These results can be explained in terms of the pattern of plastid transmission in *Oenothera* and the reaction of different plastid types to distinctive nucleocytoplasmic environments. In the *hookeri* ♀ × *lamarckiana* ♂ cross most of the plants receive maternal *hookeri* plastids that remain green and healthy in either the ⁱʰ*hookeri· gaudens* or ⁱʰ*hookeri·velans* background. In the reciprocal cross, *lamarckiana* plastids are transmitted to most of the progeny; these are green in the ⁱʰ*hookeri·gaudens* background, but they are yellow in the ⁱʰ*hookeri·velans* background because they are not compatible with this nucleocytoplasmic environment. However, some variegated plants are found in both crosses in the ⁱʰ*hookeri·velans* background. Plastid transmission in *Oenothera* is not strictly maternal (Table 17–2); when paternal plastids are transmitted, variegated plants result. When *hookeri* is the female parent, variegated plants result when *lamarckiana* plastids are transmitted to F₁ plants having the ⁱʰ*hookeri·velans* genotype, since the *lamarckiana* plastids turn yellow. When *lamarckiana* is the female parent, variegated plants arise when *hookeri* plastids are transmitted, since these are green in the ⁱʰ*hookeri·velans* background. The ⁱʰ*hookeri· gaudens* plants are always green because both plastid types are compatible in this background. Kutzelnigg and Stubbe (33) called the conditional responses of plastids

TABLE 17-5. *Plastid compatibility in interspecific hybrids of O. hookeri and O. lamarckiana*

Cross	Source of maternal plastids	F$_1$ plants	
		Nuclear genotype	Color
O. hookeri × O. lamarckiana	O. hookeri	^{h}hookeri•gaudens	Green
		^{h}hookeri•velans	Green plus a few variegated
O. lamarckiana × O. hookeri	O. lamarckiana	^{h}hookeri•gaudens	Green
		^{h}hookeri•velans	Yellow plus a few variegated

Adapted from Kirk and Tilney-Bassett (2).

such as those of *lamarckiana* to the ^{h}hookeri•velans background "differentiation mutations," whereas nonconditional plastome mutations such as those discussed in the preceding section were called "defect mutations." In differentiation mutations the functions of one or more plastid genes have presumably been modified during the course of evolution of *Oenothera* species in the wild, and this modification can be recognized in terms of plastid compatibility.

The subgenus *Euoenothera* contains 10 North American and 14 European species. Stubbe (38–41) made an exhaustive study of genome-plastome interactions among these species that involved the synthesis of some 400 different combinations. From these experiments Stubbe concluded that, in terms of compatibility, there are five different plastome types (I–V) and three basic haploid genomes that can be combined to form six diploid combinations (Fig. 17–8). The compatibility responses range from normal to lethal. Stubbe considered plastome type IV to be the most primitive, as it is fully compatible with all nuclear genomes except AA, with which it is still partially compatible. It is particularly interesting that Stubbe's three basic haploid genomic types (deduced from genome-plastome interactions in predominantly European species) correspond to the three main populations recognized by Cleland for the North American species on the basis of taxonomic criteria. This is reasonable, for it appears that the European species were only recently derived from the North American species. Stubbe's conclusions regarding genome-

plastome relationships were compared with the actual results obtained by Cleland and other workers from intercrossing North American *Euoenothera* species, and they were found generally to be in good agreement (35,42). Thus there is remarkably good concordance in the classification of *Euoenothera* species, whether the comparison is based on conventional taxonomic grounds or on genome-plastome relationships. So it appears that organelle compatibility may be a reasonable measure of evolutionary relatedness, at least in *Oenothera*.

Plastid competition in *Oenothera* was studied extensively by Schötz (2,4,35, 43–47). His method was to make crosses between species of *Oenothera* that contained green plastids and two chimeras with green and white plastids that produced shoots with white germ layers by sorting-out. Both of the white plastome mutants arose spontaneously. One occurred in a stock of *O. hookeri* that as a result of earlier crosses contained *O. biennis* plastids. The other arose in a stock of *O. blandina* that had *O. lamarckiana* plastids. Among the progeny of these crosses, Schötz determined the number of green, white, and variegated plants. However, among the variegated plants the degree of variegation ranged from weak to intense. To take the amount of variegation into account, Schötz devised a variegation rating. The variegated plants were separated into 12 classes based on the amount of white tissue contained in the cotyledons and, where possible, the first pair of leaves in each seedling. In computing

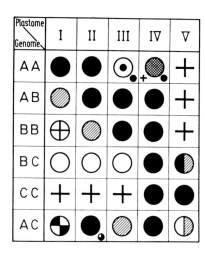

Plastome / Genome	I	II	III	IV	V
A A	●	●	⊙	◉ + ●	+
A B	◍	●	●	●	+
B B	⊕	◍	●	●	+
B C	○	○	○	●	◐
C C	+	+	+	●	●
A C	⊕	●●	◍	●	◑

● normal green
◉ green to grayish green
◍ yellow green (lutescent)
◐ periodically lutescent
◑ yellow green to yellow
○ white or yellow
⊕ white and with inhib. of growth and germin.
+ lethal; but white if occuring as an exception
◓ slightly yellowing
◔ periodically pale (diversivirescent)
⊙   "   "   (virescent)

FIG. 17–8. Compatibility relationships among different genotypes and plastid types in *Oenothera*. The use of more than one sign in some squares depends on slight differences between the A complexes. (Reprinted from Stubbe, ref. 55, with permission.)

the variegation rating, green, white, and variegated seedlings all were used. Green seedlings were counted as zero and white seedlings as one. The numbers obtained from these pure types were added together with the sum of the fractions of white tissue in the variegated seedlings to obtain the variegation rating. Thus the variegation rating is a complex variable; we shall discuss this point in more detail shortly.

Schötz arranged the green (G) × white (W) crosses in a series according to variegation rating (Table 17–6). With the exception of a few inconsistencies, the same species order was obtained whether *biennis* or *lamarckiana* plastids were donated by the male parent. The reciprocal cross was made only with *biennis,* and the variegation ratings tended to be very high, largely because there were many white seedlings (carrying only maternal *biennis* plastids), which have a rating of one in the Schötz system.

In the discussion that follows, it is important that the reader keep in mind that the variegation rating may result from the interaction of several other variables in addition to plastid competition. The first is the probability that paternal plastids will be transmitted at all. The second is the average input of paternal and maternal plastids into the zygote. The third is the probability of destruction of either paternal or maternal plastids in the zygote or shortly thereafter. The fourth is the rate of multiplication of cells containing dissimilar plastids in different crosses. The fifth is the time of onset of replication of specific plastid types, as opposed to their replication rates (12). Of these variables, the first is routinely measured directly, and the second has been shown to have an effect (48). The remaining three variables may or may not be important.

It is evident from the data that in *Oenothera* major control of plastid transmission is exerted by the female parent (Table 17–6). Thus nonvariegated plants comprise a large fraction of the total in all crosses, and these plants always contain plastids derived only from the maternal parent. Even among the variegated plants, the maternal parent would appear to be the major contributor of plastids. In the reciprocal crosses of *biennis* by the other species, white tissue composed a relatively small fraction of the total when white *biennis* plastids were contributed by the male parent, but a large fraction of the total when *biennis* was used as the female parent (Table 17–6). Despite this maternal influence, there was a

TABLE 17–6. *Percentage of variegated offspring, and variegation rating from crosses between a series of Oenothera species[a]*

Source of plastids			Number of offspring			Percentage variegated plants	Variegation rating
			G	V	W		
G × W							
hookeri	×	biennis	797	0	0	0	—
lamarckiana	×	"	555	14	0	2.5	0.14
bauri	×	"	256	26	0	9.3	0.06
rubricaulis	×	"	255	84	0	24.8	0.69
suaveolens	×	"	264	71	0	21.1	1.34
biennis	×	"	118	68	0	36.4	2.85
syrticola	×	"	169	108	0	38.9	4.22
parviflora	×	"	219	158	0	41.9	5.11
rubricuspis	×	"	183	108	0	37.1	7.26
ammophila	×	"	63	51	0	44.6	8.86
atrovirens	×	"	148	344	0	69.9	15.69
hookeri	×	lamarckiana	380	65	0	14.7	0.25
lamarckiana	×	"	376	140	0	27.2	1.31
bauri	×	"	119	91	0	43.3	2.47
rubricaulis	×	"	90	56	0	38.4	3.39
suaveolens	×	"	105	117	0	52.6	7.23
biennis	×	"	77	64	0	45.4	4.53
syrticola	×	"	72	134	0	65.0	9.83
parviflora	×	"	64	83	0	56.4	8.04
rubricuspis	×	"	27	57	0	67.8	14.46
atrovirens	×	"	27	51	0	65.3	22.10
W × G							
biennis × hookeri			0	12	76	13.5	93.9
"	×	lamarckiana	0	124	68	68.2	86.2
"	×	bauri	0	54	159	25.5	95.2
"	×	rubricaulis	0	34	55	38.2	—
"	×	suaveolens	0	8	52	13.3	—
"	×	biennis	0	11	45	20.5	—
"	×	syrticola	0	0	150	—	—
"	×	parviflora	0	1	65	1.5	99.97
"	×	ammophila	0	0	204	—	100
"	×	atrovirens	0	0	207	—	100

[a] In the green (G) ♀ × white (W) ♂ crosses, mutant plants containing either W *biennis* or *lamarckiana* plastids were used as constant sources of paternal plastids, and a series of *Oenothera* species was used to provide G maternal plastids. In the W ♀ × G ♂ crosses, plants with W *biennis* plastids were used as the constant source of maternal plastids, with the same series of *Oenothera* species being used to provide paternal G plastids. This table was adapted by Kirk and Tilney-Bassett (2) (Table 8.3) from original data of Schötz (43).

clear dependence on the paternal parent, as revealed by the fraction of variegated plants, and on "competition," as measured by the variegation rating in different interspecific crosses. This was partly a function of the plastids themselves and partly a function of the nuclear genome (44,45). Schötz (44) extended his interspecific G × W crosses to include 28 species of *Euoenothera* using the same two sources of white plastids for the male side of the cross. He distinguished three groups of green plastids with strong, average, and weak ability to compete with white plastids. These groups correspond closely with the five plastome types recognized by Stubbe. Group 1, with the most competitive plastids, corresponds to plastomes 1 and 2; group 2, of average competitive ability, corresponds to plastome 2; group 3, with weak competitive ability, corresponds to plastomes IV and V (Table 17–7). These competitive abilities can also be modified by nuclear genotype, as defined by Stubbe's classification (Fig. 17–8). For

TABLE 17–7. Summary of Oenothera species[a]

Plastid multiplication rate	Plastome	Basic nuclear genome	Species
Fast	I	A_1A_1	O. bauri, O. cockerelli, O. mollis
		A_2A_2	O. franciscana, O. hookeri
	III	A_2A_2	O. blandina, O. deserens
		A_2B	O. lamarckiana
		BA_1	O. chicaginensis
Medium	II	$A_?A_?$	O. purpurata
		A_1B	O. biennis, O. holscheri, O. nuda, O. rubricaulis, O. suaveolens
		A_2B	O. conferta, O. coronifera
		BB	O. grandiflora
Slow	IV	A_1C	O. ammophila, O. germanica, O. syrticola
		BB	O. pictarubata (hybrid)
		BC	O. atrovirens, O. flexirubata (hybrid), O. parviflora, O. rubricuspis, O. siliesiaca
	V	CC	O. argillicola

[a] Oenothera species classified according to plastid multiplication rates, plastid compatibility relationships (plastomes I–V), and basic nuclear genomes (see text) A (A_1,A_2), B, and C. Rates of plastid multiplication are judged by the variegation rating, and the interpretation may be an oversimplification (see text).

Adapted from Tilney-Bassett (4), Table 4, which is based on the data of Schötz (44).

example, tests of plastomes I–IV showed them to have increasing competitive ability when combined with diploid genomes of the type $BA_2 < A_1A_2 < A_2A_2$ in that succession.

The most important question to be resolved regarding Schötz's studies is the extent to which differential rates of plastid multiplication account for his results. Renner originated the idea that *Oenothera* plastids multiply at different rates, and this notion was popularized by Schötz's experiments. However, as was discussed previously, competition is measured by the variegation rating, which seems to be comprised of several variables. If plastid multiplication rates were the only determining factor, one would expect each zygote to contain both paternal and maternal plastids in a constant ratio. Schötz counted the number of plastids in eggs of several *Oenothera* species and found that they numbered between 26

and 31, with few significant differences between species. Since a constant male was used in each series of crosses, he rejected the hypothesis that variable plastid ratios could account for his results. However, Meyer and Stubbe (48) made an electron microscopic study of egg cells and zygotes in crosses of *O. lamarckiana* by itself. The plastids donated by the sperm cells appeared to be small and lacking in starch grains in comparison to the large, starch-containing plastids of the egg cells. Plastid counts made on the zygotes indicated that plastids contributed by the sperm cell ranged from 8 to 13 and that the egg cell contributed between 25 and 29 plastids. Although these results are in agreement with Schötz's finding that the number of egg cell plastids is relatively constant, they show that paternal plastid input can vary almost twofold in an intraspecific cross. Thus it does not seem that use of a constant male in a series of intraspecific crosses constitutes rigorous grounds for rejecting variable plastid ratios as playing a role in Schötz's experiments.

PLASTID INHERITANCE IN
Pelargonium

Plastid transmission in *Pelargonium* is biparental, as it is in *Oenothera,* with the important difference that plants containing only paternal plastids are found in addition to variegated plants and plants containing only maternal plastids. Tilney-Bassett [for reviews, see (2,4)] made a detailed and extremely careful analysis of plastid transmission in *Pelargonium* that will be outlined here. Tilney-Bassett made reciprocal crosses between chimeric plants of different cultivars of *P. zonale* containing green and white plastids. Green (G), white (W), and variegated (V) progeny resulted from these crosses. In order to obtain some measure of plastid success in different crosses, Tilney-Bassett devised a system of data quantification, much as Schötz had done for *Oenothera*. Tilney-Bassett assumed arbitrarily that each fertilized egg receives at least four plastids. Thus a pure green seedling receives four green plastids, and a pure white seedling receives four white plastids. Variegated plants can then be divided into three classes corresponding to the amounts of green and white tissues they contain. A variegated plant that is mostly green will be classified as 3:1 in terms of the ratio of green to white plastids, whereas a plant with roughly equal amounts of green and white tissues will be classified as 2:2.

Some of the data obtained by Tilney-Bassett are summarized in Table 17–8. For each cross in a pair of reciprocal crosses, the number of seedlings of each phenotype is used to calculate the hypothetical number of green and white plastids transmitted in the cross. The estimated percentage of green plastids transmitted in each cross is then used to calculate the paternal or ma-

ternal advantage. For example, in the reciprocal crosses of Crystal Palace Gem × Dolly Varden, the G × W cross yields an estimated 98.6% green plastids from the female parent, whereas the W × G cross produces 78.2% green plastids from the male parent. The advantage is calculated from the difference between these two figures, and it is maternal. Using this classification system, Tilney-Bassett was able to arrange each pair of reciprocal crosses in an order that proceeds from the greatest maternal advantage to the greatest paternal advantage (Table 17–8). Two other conclusions can be drawn from these experiments. First, the success of green plastids is greater than that of white plastids. Second, in most crosses there is a paternal advantage such that plastids transmitted from the pollen parent tend to be more successful than those transmitted from the egg parent. These results are in clear contrast with those obtained by Schötz for *Oenothera*.

It must be remembered that Tilney-Bassett's estimates of plastid transmission are indirect and are based on the numbers of G, V, and W seedlings. Although he tentatively concluded that green plastids were at an advantage to white plastids, it was evident from his data (Table 17–8) that white seedlings were also much less common than green seedlings. Hence, the results could also be accounted for in terms of the trivial explanation of differential seedling mortality. Tilney-Bassett (49) put this hypothesis to direct experimental test. The young *Pelargonium* seed has a transparent wall, so that wild-type plastids differentiate and become green. Within 10 days of pollination it can be determined if an embryo will give rise to a G, W, or V plant. It is therefore possible to compare G, W, and V embryo and seedling counts directly to determine if there is differential mortality for any class of seedlings. Furthermore, the embryos can be dissected from seeds and scored, irrespective of whether the seeds are healthy or shriveled. It is, of

TABLE 17–8. *Polarity of plastid transmission in reciprocal crosses between different cultivars of P. zonale*[a]

Cross		Number of seedlings			Percentage variegated seedlings	Number of plastids			Plastid distribution (%)		Polarity
		G	V	W		G	W	Total	Female	Male	(green ♀ -green ♂)
CPG × DV	G × W	67	4	0	5.7	280	4	284	98.6	1.4	20.4 (maternal)
	W × G	44	43	0	49.4	272	76	348	21.8	78.2	
CPG × JCM	G × W	208	5	0	2.3	847	5	852	99.4	0.6	15.1 (maternal)
	W × G	83	40	3	31.7	425	79	504	15.7	84.3	
DV × DV	G × W	78	27	4	24.8	370	66	436	84.9	15.1	−2.3 (paternal)
	W × G	36	13	0	26.5	171	25	196	12.8	87.2	
PC × FS	G × W	276	76	28	20.0	1268	252	1520	83.4	16.6	−6.2 (paternal)
	W × G	121	30	2	19.6	548	64	612	10.4	89.6	
PC × MBC	G × W	12	11	11	32.4	70	66	136	51.5	48.5	−9.9 (paternal)
	W × G	11	24	0	68.6	86	54	140	38.6	61.4	
FS × FS	G × W	104	14	73	7.4	451	313	764	59.0	41.0	−30.9 (paternal)
	W × G	115	27	2	18.8	518	58	576	10.1	89.9	
PC × DV	G × W	108	139	83	42.1	716	604	1320	54.2	45.8	−31.8 (paternal)
	W × G	127	50	3	27.7	619	101	720	14.0	86.0	

[a] Method used to calculate plastid numbers is described in the text. Key: CPG = Crystal Palace Gem, DV = Dolly Varden, JCM = J. C. Mapping, PC = Paul Crampel, FS = Flower of Spring, MBC = Miss Burden Coutts.
Adapted from Kirk and Tilney-Bassett (2) (Table VIII.2).

course, possible that W embryos might die at a very early stage when they are too small to be scored accurately. However, such mortality should be reflected in a depression in the overall mean fertility of G × W and W × G crosses with respect to G × G crosses.

Tilney-Bassett (49) performed a careful series of experiments to test the differential mortality or selection hypothesis with the cultivars Dolly Varden and Flower of Spring. There were no significant differences in the segregation ratios of embryos and seedlings except in two cases. These two cases involved G × W crosses with Dolly Varden as the female parent. In these crosses there was a deficit of white seedlings, but since these formed the smallest class from the embryo stage on, the effect of this selection on segregation ratios was mild. Similarly, mean fertility rates showed little difference between crosses.

These experiments by Tilney-Bassett ruled out selection as an important contributing factor determining the frequencies of G, W, and V progeny that he saw in different crosses and also provided him with the tools necessary to detect the effects of selection should it pose a problem. At the same time, Tilney-Bassett (50) also ruled out a second trivial hypothesis that supposed comparisons between crosses to be unreliable because of environmental changes between the times that different crosses were made. With these hypotheses eliminated, Tilney-Bassett could proceed to examine the genetic basis of biparental plastid transmission in *Pelargonium*.

In order to determine the genetic mechanism that was responsible for yielding different proportions of G, W, and V progeny in crosses, Tilney-Bassett (51) made reciprocal G × W crosses between six chimeric cultivars of *P. zonale* containing white and green plastids and green cultivars of Dolly Varden or Flower of Spring. Segregation ratios were measured in the embryos, and mean fertility and embryo survival were

recorded. In G × W crosses employing Dolly Varden as the green female parent, embryos were in the ratio G > V > W (type I), and transmission of white paternal plastids was less than 50%. When Flower of Spring was the green female parent, embryos were in the ratio of G > V < W (type II), and the transmission rate of male plastids was greater than 50%. In W × G crosses with Flower of Spring as the pollen parent, the six crosses could be arranged in a series from high to low green plastid transmission. The same series was found when Dolly Varden was the male parent, except that transmission of green plastids was higher.

Three conclusions could be drawn from these experiments. First, there was clear maternal control of plastid transmission, as shown by elucidation of the type I and type II segregation patterns. Second, there was a lesser and more subtle interaction between plastids from different parents, as shown by the fact that in W × G crosses the six cultivars could be arranged in the same series for both green male cultivars with respect to green plastid transmission. Third, there was generally an excess of green plastids to white plastids in all crosses, except those in which Flower of Spring was the green female parent, which indicated a general advantage of green plastids over white plastids.

The next problem was to elucidate the genetic control of the type I and type II segregation patterns. Tilney-Bassett (52) resolved this problem by inbreeding the two cultivars Flower of Spring and Dolly Varden and making hybrids between them. Genetic analysis revealed that Dolly Varden was homozygous for the type I segregation pattern, since on inbreeding the progeny yielded no segregation pattern other than type I; Flower of Spring behaved as if it was heterozygous, since inbreeding of the progeny gave both type I and type II segregation patterns (Table 17-9). Tilney-Bassett explained these results in terms of

TABLE 17–9. *Summary of segregational patterns of selected progeny produced by inbreeding Flower of Spring (FS) and Dolly Varden (DV) and hybrids between them*[a]

Cross number	Parents	Presumptive parental genotypes		Presumed F_1 genotypes and segregation patterns	
				$Pr_1 Pr_1$ (I)	$Pr_1 Pr_2$ (II)
1	DV $\times$ DV	$Pr_1 Pr_1 \times Pr_1 Pr_1$	expected ratio	1	0
			observed total	17	0
2	FS $\times$ FS	$Pr_1 Pr_2 \times Pr_1 Pr_2$	expected ratio	1	1
			observed total	14	11
3	DV $\times$ FS	$Pr_1 Pr_1 \times Pr_1 Pr_2$	expected ratio	1	1
			observed total	29	18
4	FS $\times$ DV	$Pr_1 Pr_2 \times Pr_1 Pr_1$	expected ratio	1	0
			observed total	24	1
5	F_1 of FS $\times$ DV selfed	$Pr_1 Pr_2 \times Pr_1 Pr_2$	expected ratio	1	1
			observed total	20	18

[a] The $Pr_2 Pr_2$ genotype is not shown because the Pr_2 allele is presumed to be lethal in the egg (see text), and so the homozygous diploid must also be lethal.
Adapted from Tilney-Bassett (4).

a pair of mendelian alleles, Pr_1 and Pr_2, that he supposed affected plastid replication at the time of fertilization. Females whose offspring segregated with a type I pattern were homozygous $Pr_1 Pr_1$, and females whose offspring segregated with a type II pattern were heterozygous $Pr_1 Pr_2$. Although this explanation worked well enough for Dolly Varden, since this cultivar yielded only type I segregations among the progeny and was presumably homozygous $Pr_1 Pr_1$, it did not explain the fact that the segregation ratios obtained on inbreeding Flower of Spring were 1:1. If Flower of Spring were heterozygous $Pr_1 Pr_2$, one would have expected segregation ratios of either 3:1 or 1:2:1 among the progeny.

To explain the results of the Flower of Spring crosses, Tilney-Bassett hypothesized that the Pr_2 allele caused gametophytic lethality on the maternal side but that it had a much smaller effect, if any, on the paternal side. A review of the crosses in Table 17–9 shows that this hypothesis explains the results well. In cross 2, where Flower of Spring is selfed, the Pr_2 ova will be inviable, and only the Pr_1 ova can be fertilized. Since Pr_1 and Pr_2 pollen grains will be produced at equal frequencies, and since

they are equally viable, the resulting progeny plants will be $Pr_1 Pr_1$ and $Pr_1 Pr_2$ in equal frequency, and so type I and type II segregations will occur at equal frequencies. Cross 3 is of the type $Pr_1 Pr_1 \times Pr_1 Pr_2$, and again a 1:1 ratio of type I to type II segregations is observed. Since all of the ova are Pr_1, and since the pollen are Pr_1 and Pr_2 at equal frequencies, the progeny should be $Pr_1 Pr_1$ and $Pr_1 Pr_2$ at equal frequencies, as should type I and type II segregations. It is the reciprocal cross (cross 4), however, that is the key to the hypothesis. Here the maternal parent is $Pr_1 Pr_2$, and the paternal parent is $Pr_1 Pr_1$. In this cross, by definition, almost all of the progeny should be $Pr_1 Pr_1$ and should show type I segregation because the Pr_2 ova will be lethal. This expectation was confirmed.

The biochemical mechanism underlying maternal control of plastid transmission in *Pelargonium* remains to be explored. Tilney-Bassett pointed out that whatever the mechanism, *Pelargonium* and *Oenothera* must be different, for in *Oenothera* maternal control is synonymous with maternal predominance. No plants exclusively carrying paternal plastids are ever found. In *Pelargonium*, in contrast, maternal control does not affect male or female plastid transmission per se, but rather the transmission pattern seen. Sager (53) hypothesized that plastid transmission in *Pelargonium* might be accounted for by the modification-restriction mechanism she proposed for *Chlamydomonas* (see Chapter 15). In a G × W cross, G, V, and W plants would be equivalent to maternal, biparental, and paternal zygotes, respectively. The *Pr* gene would be equivalent to the *mat-1* gene in *Chlamydomonas*. In Pr_1 Pr_1 homozygotes a maternal pattern of inheritance is seen (i.e., type I segregation: G > V > W) because male plastid (presumably male clDNA) transmission is restricted by the restriction enzyme. The Pr_2 allele would be equivalent to the *mat-1⁻* allele and would permit higher transmission of paternal plastid genomes by the mechanism described earlier (see Chapter 15). Hence, Pr_1 Pr_2 heterozygotes would yield the type II segregation pattern (G > V < W), in which paternal plastid transmission is increased.

Plastid competition in *Pelargonium* embryos also appears to play a role in determining plastid inheritance and the resulting variegation that is seen, according to Hagemann (12,54). Hagemann compared the behavior of wild-type plastids obtained from the variety Trautlieb with those of the yellow plastome mutant *en:gilva-1* from Mrs. Pollack and two white plastome mutants *en:alba-1* and *en:alba-2* from Mrs. Parker and Flower of Spring, respectively. Segregation ratios of pure seedlings were studied in the F_1 progeny of reciprocal crosses together with the amount of variegation in embryos, cotyledons, and foliage leaves. Hagemann drew several conclusions from his experiments, including the following: First, the percentage of green seedlings is always high in both directions in reciprocal crosses. In each F_1 the following segregation pattern was found: green > variegated > yellow or white. Second, in F_1 hybrids competition takes place between genetically different plastid types and the

cells containing these plastids, with the level of competition being different in different stages of ontogenetic development. Thus competition is most marked in zygotes and embryos, and it largely vanishes in the foliage leaves. Again, green plastids are favored with respect to yellow plastids, and yellow plastids with respect to white plastids. Hagemann supposed that plastid multiplication rates in the embryo were differential and that the progress of replication of one plastid type somehow inhibited the replication of another. This differential multiplication of the plastids in zygotes and embryos is predetermined by maternal genotype; it disappears after development of the foliage leaves.

CONCLUSIONS

The literature on plastid genetics in higher plants is rich, varied, and historically important. Yet one comes away from a study of this literature with the distinct impression that further work along the same lines is likely to tell us little new about the genetics of chloroplasts. The major problem, even in those species where biparental transmission of plastids provides the opportunity to study recombination and segregation of plastid genomes, is an impoverishment of phenotypic diversity among plastid mutations. In a sense, students of plastid genetics in higher plants are in the same position that yeast mitochondrial geneticists occupied when they had only vegetative petite mutants to study. One cannot really begin a systematic elucidation of such problems as segregation and recombination without a collection of suitable point mutants with clearly defined phenotypes. Neither can one undertake the genetic dissection of an organelle genome without such mutants. Until the appropriate mutants become available, *Chlamydomonas* will remain the model for studies of chloroplast genetics. Hypotheses concerning the rules governing chloroplast genetics will be by extension from

Chlamydomonas to higher plants rather than the other way around. This is not to say that new approaches and new tools are not available. The chloroplast genome in higher plants can now be mapped, at least in part, by hybridization of chloroplast mRNAs known to code for specific chloroplast polypeptides to clDNA restriction fragments (see Chapter 2). Molecular mapping will certainly obviate, in part, the need for classic genetic methods, but this author firmly believes that mutants provide an added dimension. Perhaps the complementary genetic tools will be provided by somatic cell genetics and regeneration of whole plants from callus (see Chapter 18).

REFERENCES

1. Correns, C. (1909): Vererbungsversuche mit blass (gelb) grunen und buntblattrigen Sippen bei *Mirabilis jalapa, Urtica pilulifera* und *Lunaria annua. Z. Vererbungs.,* 1:291–329.
2. Kirk, J. T. O., and Tilney-Bassett, R. A. E. (1967): *The Plastids.* W. H. Freeman, San Francisco.
3. Baur, E. (1909): Das Wesen und die Erblichkeitsverhaltnisse der "varietates albomarginatae hort" von *Pelargonium zonale. Z. Vererbungs.,* 1:330–351.
4. Tilney-Bassett, R. A. E. (1975): Genetics of variegated plants. In: *Genetics and Biogenesis of Mitochondria and Chloroplasts,* edited by C. W. Birky, Jr., P. S. Perlman, and T. J. Byers, Ohio State University Press, Columbus.
5. Correns, C. (1909): Zur Kenntnis der Rolle von Kern und Plasma bei der Vererbung. *Z. Vererbungs.,* 2:331–340.
6. von Wettstein, D., and Erickson, G. (1965): The genetics of chloroplasts. In: *Proceedings Eleventh International Congress of Genetics,* Vol. 3. pp. 591–612, edited by S. J. Geerts. Pergamon Press, Oxford.
7. Carlson, P. S. (1973): The use of protoplasts for genetic research. *Proc. Natl. Acad. Sci. U.S.A.,* 70:598–602.
8. von Wettstein, D. (1967): Chloroplast structure and genetics. In: *Harvesting the Sun,* edited by A. San Pietro, F. A. Greer, and T. J. Army, pp. 153–190. Academic Press, New York.
9. Jensen, W. A., and Fisher, D. B. (1968): Cotton embryogenesis: The sperm. *Protoplasma,* 65:277–286.
10. Jensen, W. A., Ashton, M., and Heckard, L. R. (1974): Ultrastructural studies of the pollen of subtribe Castilleiinae, family Scrophulariaceae. *Bot. Gaz.,* 135:210–218.
11. Lombardo, G., and Gerola, F. M. (1968): Cytoplasmic inheritance and ultrastructure of the male generative cell in higher plants. *Planta,* 82:105–110.
12. Hagemann, R. (1976): Plastid distribution and plastid competition in higher plants and the induction of plastom mutations by nitrosourea-compounds. In: *Genetics and Biogenesis of Chloroplasts and Mitochondria,* edited by T. Bücher, W. Neupert, W. Sebald, and S. Werner, pp. 331–338. North Holland, Amsterdam.
13. Nilsson-Tillgren, T., and von Wettstein-Knowles, P. (1970): When is the male plastome eliminated? *Nature,* 227:1265–1266.
14. Nitsch, J. P. (1972): Haploid plants from pollen. *Z. Pflanzenzuchtg.,* 67:3–18.
15. Michaelis, P. (1959): Cytoplasmic inheritance and the segregation of plasma genes. *Proceedings Tenth International Congress of Genetics,* Vol. 1. pp. 375–385.
16. Hagemann, R. (1964): *Plasmatische Vererbung.* Gustav Fischer Verlag, Jena.
17. Michaelis, P. (1955): Modelversuche zur Plastiden- und Plasmavererbung. *Züchter,* 25:209–221.
18. Michaelis, P. (1955): Über Gesetzmässigkeiten der Plasmon-Umkombination und über eine Methode zur Trennung einer Plastiden-, Chondriosomen- resp. Sphaerosomen-, (Mikrosomen)- und einer Zytoplasmavererbung. *Cytologia,* 20:315–338.
19. Röbbeln, G. (1966): Chloroplasten diffenenzierung nach geinduzierter Plastommutation bei *Arabidopsis thaliana* (L.) Heynh. *Z. Pflanzenphysiol.,* 55:387–403.
20. Redei, G. P. (1973): Extra-chromosomal mutability determined by a nuclear gene locus in *Arabidopsis. Mutat. Res.,* 18:149–162.
21. Rhoades, M. M. (1943): Genic induction of an inherited cytoplasmic difference. *Proc. Natl. Acad. Sci. U.S.A.,* 29:327–329.
22. Rhoades, M. M. (1946): Plastid mutations. *Cold Spring Harbor Symp. Quant. Biol.,* 11:202–207.
23. Rhoades, M. M. (1955): Interaction of genic and non-genic hereditary units and the physiology of nongenic inheritance. In: *Encyclopedia of Plant Physiology,* Vol. 1, edited by W. Ruhland, pp. 19–57. Springer-Verlag, Berlin.
24. Shumway, L. K., and Weier, T. E. (1967): The chloroplast structure of iojap maize. *Am. J. Bot.,* 54:773–780.
25. Michaelis, P. (1969): Über Plastiden-Restitutionen (Rückmutationen). *Cytologia [Suppl.],* 34:1–115.
26. Herrmann, F. (1971): Genetic control of pigment-protein complexes I and Ia of the plastid mutant en:alba-1 of *Antirrhinum majus. FEBS Lett.,* 19:267–269.
27. Hagemann, R. (1971): Struktur und Funktion

der genetischen Information in den Plastiden. I. Die Bedeutung von Plastommutanten und die genetische Nomenklatur extranukleärer Mutationen. *Biol. Zbl.*, 90:409–418.

28. Herrmann, F. (1971): Chloroplast lamellar proteins of the plastid mutant en:viridis-1 of *Antirrhinum majus* having impaired photosystem II. *Exp. Cell Res.*, 70:452–453.

29. Börner, T., Knoth, R., Herrmann, F., and Hagemann, R. (1972): Struktur und Funktion der genetischen Information in den Plastiden V. Das Fehlen von ribosomaler RNA in den Plastiden der Plastommutante 'Mrs. Parker' von *Pelargonium zonale* Ait. *Theor. Appl. Genet.*, 42:3–11.

30. Börner, T., Herrmann, F., and Hagemann, R. (1973): Plastid ribosome deficient mutants of *Pelargonium zonale*. *FEBS Lett.*, 37:117–119.

31. Börner, T., Schumann, B., and Hagemann, R. (1976): Biochemical studies on a plastid ribosome-deficient mutant of *Hordeum vulgare*. In: *Genetics and Biogenesis of Chloroplasts and Mitochondria*, edited by T. Bücher, W. Neupert, W. Sebald, and S. Werner, pp. 41–48. North Holland, Amsterdam.

32. Sager, R. (1972): *Cytoplasmic Genes and Organelles*. Academic Press, New York.

33. Kutzelnigg, H., and Stubbe, W. (1974): Investigations on plastome mutants in *Oenothera*. 1. General considerations. *Subcell. Biochem.*, 3:73–89.

34. Cleland, R. E. (1962): The cytogenetics of *Oenothera*. *Adv. Genet.*, 11:147–237.

35. Cleland, R. E. (1972): *Oenothera*: *Cytogenetics and Evolution*. Academic Press, New York.

36. Renner, O. (1924): Die Scheckung der Oenotherenbastarde. *Biol. Zbl.*, 44:309–336.

37. Renner, O. (1936): Zur Kenntnis der nichtmendelnden Buntheit der Laubblätter. *Flora (Jena) N. F.*, 30:218–290.

38. Stubbe, W. (1959): Genetische Analyse des Zusammenwirkens von Genom und Plastom bei *Oenothera*. *Z. Vererbungs.*, 90:288–298.

39. Stubbe, W. (1960): Untersuchungen zur genetischen Analyse des Plastoms von *Oenothera*. *Z. Bot.*, 48:191–218.

40. Stubbe, W. (1963): Die Rolle des Plastoms in der Evolution der Oenotheren. *Ber. Dtsch. Bot. Ges.*, 76:154–167.

41. Stubbe, W. (1964): The role of the plastome in evolution of the genus *Oenothera*. *Genetica*, 35:28–33.

42. Cleland, R. E. (1962): Plastid behavior in North American *Euoenotheras*. *Planta*, 57:699–712.

43. Schötz, F. (1954): Über Plastidenkonkurrenz bei *Oenothera*. *Planta*, 43:182–240.

44. Schötz, F. (1968): Über Plastidenkonkurrenz bei *Oenothera*. *II*. *Biol. Zbl.*, 87:33–61.

45. Schötz, F. (1974): Untersuchungen über die Plastidenkonkurrenz bei *Oenothera*. *IV*. Der Einfluss des Genoms auf die Durchsetzungsfähigkeit der Plastiden. *Biol. Zbl.*, 93:41–64.

46. Schötz, F. (1958): Beobachtungen zur Plastidenkonkurrenz bei *Oenothera* und Beiträge zum Problem der Plastiden-vererbung. *Planta*, 51:173–185.

47. Schötz, F., and Heiser, F. (1969): Über Plastidenkonkurrenz bei *Oenothera*. *III*. Zahlenverhältnisse in Mischzellen. *Wiss. Z. Pädagogische Hochschule Potsdam*, 13:65–89.

48. Meyer, B., and Stubbe, W. (1974): Das Zahlenverhältnis von mütterlichen und väterlichen Plastiden in den Zygoten von *Oenothera erythrosepala* Borbas (syn. *Oe. lamarckiana*). *Ber. Dtsch. Bot. Ges.*, 87:29–38.

49. Tilney-Bassett, R. A. E. (1970): Genetics and plastid physiology in *Pelargonium* III. Effect of cultivar and plastids on fertilization and embryo survival. *Heredity*, 25:89–103.

50. Tilney-Bassett, R. A. E. (1970): Effects of environment on plastid segregation in young embryos of *Pelargonium* × *Hortorum* Bailey. *Ann. Bot.*, 34:811–816.

51. Tilney-Bassett, R. A. E. (1970): The control of plastid inheritance in *Pelargonium*. *Genet. Res. Camb.*, 16:49–61.

52. Tilney-Bassett, R. A. E. (1973): The control of plastid inheritance in *Pelargonium*. II. *Heredity*, 30:1–13.

53. Sager, R. (1975): Patterns of inheritance of organelle genomes: Molecular basis and evolutionary significance. In: *Genetics and Biogenesis of Mitochondria and Chloroplasts*, edited by C. W. Birky, Jr., P. S. Perlman, and T. J. Byers, Chapter 7. Ohio State University Press, Columbus.

54. von Hagemann, R., and Scholze, M. (1974): Struktur und Funktion der genetischen Information in den Plastiden. VII. Vererbung und Entmischung genetisch unterschiedlicher Plastidensorten bei *Pelargonium zonale* Ait. *Biol. Zbl.*, 93:625–648.

55. Stubbe, W. (1971): Origin and continuity of plastids. In: *Results and Problems in Cell Differentiation. Vol. 2. Origin and Continuity of Cell Organelles*, edited by J. Reinert and H. Ursprung, pp. 65–81. Springer-Verlag, Berlin.

Chapter 18
Chloroplast Inheritance in Higher Plants. II. Localizing Structural Genes for Chloroplast Components in *Nicotiana* by Interspecific Hybridization

Genetic and biochemical characterization of mutations affecting organelle components is one method of determining which structural genes are localized in the nucleus and which are localized in the genome of the organelle itself. A second approach involves the use of hybrids between different strains of a species or crosses between species. The technique uses the pattern of inheritance of specific interstrain or interspecies protein differences as a guide for localization of the structural genes for these proteins. Wildman and his colleagues pioneered the use of hybrids for localizing the structural genes for organelle components. In this chapter we shall discuss their use of interspecific hybrids in *Nicotiana* as a means of localizing the genes for specific chloroplast proteins. The hybridization technique has also been used to localize genes coding for mitochondrial proteins in *Paramecium* (see Chapter 10), and mammalian cells in tissue culture (see Chapter 11).

In his monograph on the genus *Nicotiana,* Goodspeed (1) identified more than 60 species on the basis of careful taxonomic and cytogenetic examination. These species were distributed in nature in the western United States, Central and South America, and Australia, and this distribution led Goodspeed to suppose, with good reason, that the genus *Nicotiana* must have arisen prior to the breakup of Gondwanaland into what are now modern South America, Antarctica, and Australia. Many of the *Nicotiana* species can be hybridized, and several South American and Australian species have been used by Wildman and his colleagues in their work (Table 18–1). The basic genetic method is to examine the pattern of inheritance of a specific chloroplast trait in reciprocal interspecific crosses. If inheritance is biparental, the trait in question is assumed to be determined by a nuclear gene; if inheritance is strictly maternal, it is concluded that clDNA codes for the protein in question. The evidence for maternal inheritance of chloroplast traits in *Nicotiana,* and in particular clDNA, will be discussed presently. Unfortunately,

TABLE 18–1. *Species of Nicotiana used by Wildman et al. in interspecific hybridization experiments*[a]

Subgenus	Species	Geographic origin	Haploid chromosome number
Rustica	*N. glauca*	South America	12
Tabacum	*N. glutinosa*	South America	12
	N. tabacum	South America	24
Petunioides	*N. gossei*	Australia	18
	N. excelsior	Australia	19
	N. suaveolens	Australia	16
	N. langsdorfii	South America	9

[a] Classification, geographic origin, and haploid chromosome number of the species are taken from Goodspeed (1).

TABLE 18–2. *Definitive hybrids used by Wildman et al. to study inheritance of different chloroplast traits in* Nicotiana[a]

Character or protein studied	Inheritance	Definitive hybrids	References
Variegation	Maternal	*N. tabacum* × *N. glauca*	4
Fraction I protein			
Small subunit	Biparental	*N. tabacum* × *N. glauca* *N. tabacum* × *N. glutinosa* *N. excelsior* × *N. gossei*	10,11
Large subunit	Maternal	*N. tabacum* × *N. gossei* *N. tabacum* × *N. glauca* *N. tabacum* × *N. glutinosa* *N. glauca* × *N. langsdorfii*	11,12
Catalytic site	Maternal	*N. gossei* × *N. tabacum* *N. gossei* × *N. excelsior* *N. gossei* × *N. suaveolens*	14
Cold inactivation, heat reactivation	Maternal	*N. gossei* × *N. tabacum* *N. gossei* × *N. excelsior* *N. gossei* × *N. suaveolens*	21
Two proteins of the large chloroplast ribosomal subunit	Biparental	*N. tabacum* × *N. glauca*	19
Photosystem II protein	Biparental	*N. tabacum* × *N. glauca*	18
Ferridoxin	Biparental	*N. glutinosa* ♀ × *N. glauca*	20
Granal size regulation	Maternal	*N. excelsior* × *N. gossei*	22

[a] All hybrids were made in reciprocal directions unless otherwise indicated.

the F₁ hybrid plants are sterile; thus the pattern of inheritance of putative nuclear and chloroplast genes cannot be established with complete satisfaction through further crosses. However, this reservation is not terribly serious, and the technique does appear to be quite powerful.

The chloroplast traits whose inheritance patterns have thus far been investigated by interspecific hybridization in *Nicotiana* are summarized in Table 18–2 together with the interspecific crosses used to establish the inheritance patterns. The most critical work was done on the inheritance of the subunits of fraction I protein, ribulose bisphosphate carboxylase (RuBPCase), which will be discussed at length later in this chapter.

INHERITANCE OF CHLOROPLASTS AND clDNA IN *Nicotiana*

Ten years ago von Wettstein (2) reported that clDNA from an albino plastome mutant of *N. tabacum* had a buoyant density indistinguishable from that of wild type. More recently, Wong-Staal and Wildman (3) reported a very slight difference in the G + C contents of the clDNA of *N. tabacum* cv Turkish Samsun and the clDNA of a variegated plastome mutant of this cultivar. These experiments were important because they provided the principal evidence that clDNA is maternally inherited in *Nicotiana*. This point is crucial to the argument that maternal transmission of specific chloroplast traits in *Nicotiana* means that the genes governing these traits are localized in clDNA.

The *Nicotiana* plastome mutant was initially characterized by Wildman et al. (4). They reported that this was the only mutant of its kind recovered in their own experimental plantings, and they estimated from these data that the frequency of such mutants was less than $\frac{1}{3} \times 10^6$ tobacco plants. Phenotypic characterization of the mutant revealed a marked deficiency in chlorophyll and granal structure in those plastids from white sectors. Mixed cells were also found. RuBPCase and chloroplast ribosomes were present in the white tissue but were reduced about 50%. Since the large subunit of RuBPCase is synthesized on chloroplast ribosomes (see Chapter 20), these results suggest that chloroplast protein synthesis is probably operative in the mutant chloroplasts; however, this point cannot be proved rigorously without purification of the mutant plastids, for it could be argued that the presence of some green plastids in the white tissue accounts for the presence of the enzyme and chloroplast ribosomes. One reason that this seems unlikely is that Wong-Staal and Wildman (3) showed that the number of plastids in mutant leaf tissue is much lower than it is in green tissue. These results would suggest that the white plastids do contain chloroplast ribosomes and RuBPCase and that the reduced levels of these chloroplast components seen in white tissue can be accounted for by the reduced number of plastids. The important point is that the plastome defects seen in the mutant probably do not result because of pleiotropic effects engendered from a deficiency in chloroplast protein synthesis.

Wong-Staal and Wildman (3), in their investigation of the clDNA from mutant and green tissue of variegated plants containing the plastome mutant, found that the buoyant density of clDNA was the same in wild-type and mutant plastids (i.e., 1.700 g/cc). However, the amount of clDNA was reduced in the white tissue, presumably because there were fewer plastids. Wong-Staal and Wildman then observed, in very careful melting experiments, that wild-type and mutant clDNAs obtained from the same plant and melted simultaneously in the same apparatus differed in G+C content by about 1%, with the mutant being higher in G+C content than the wild type. They then sheared clDNA from mutant and wild-type tissue, denatured the DNA, mixed mutant and wild-type DNAs, and looked for heteroduplexes. Of a total of six heteroduplexes

expected as a maximum, on the assumption that one of the 40-odd fragments obtained by shearing would be different between the mutant and the wild type, three heteroduplexes were found. These results supported the observation that mutant and wild-type clDNAs were different in G+C content.

Wildman et al. (4) showed in crosses that the plastome mutant was strictly maternally inherited. In these experiments they were careful to use both pollen and ovaries obtained from white shoots of variegated plants. In reciprocal crosses to wild type, strictly maternal inheritance of the green and white phenotypes was observed. These results, coupled with the finding that the clDNAs of the mutant and wild type differ, support the argument made by Wildman et al. (4) that since the mutant phenotype is transmitted only by the maternal line, maternal inheritance in *Nicotiana* reflects the inheritance of chloroplast genes. Although mtDNA may also be maternally inherited in *Nicotiana,* it is unlikely that Wildman's conclusion will prove invalid, at least for the traits studied by him, since they all involve specific chloroplast components.

LOCALIZATION OF STRUCTURAL GENES FOR SUBUNITS OF RuBPCase

In 1947 Wildman and Bonner (23) reported the discovery of a major protein in green plant leaves that they called fraction I protein. The properties of this protein were reviewed extensively by Wildman and his colleagues (5–7) and by Kung (8) and were considered briefly in Chapter 1. Fraction I protein may be the most abundant protein in the world, as these authors suggested, since it comprises more than 50% of the total soluble protein of green plant leaves. Fraction I protein is similar in structure in all higher plants and green algae. The protein has a sedimentation velocity ($S^{\circ}_{20,w}$) of about 18S and a molecular weight of approximately 560,000 d; it consists of two types of polypeptide subunits having molecular weights of approximately 55,000 and 12,000 d. It is currently believed that each molecule of fraction I protein consists of eight large and eight small subunits (7). The primary function of fraction I protein has long been recognized as the catalysis of the first CO_2 fixation step in photosynthesis, using ribulose-1,5-bisphosphate (RuBP) and CO_2 to produce two molecules of 3-phosphoglyceric acid (6). Recently, however, fraction I protein has also been shown to catalyze the oxygenation of RuBP to form one molecule each of 3-phosphoglyceric acid and phosphoglycollic acid, the first intermediate in the glycolate pathway of photorespiration (8). The catalytic site of the enzyme is thought to be on the large subunits of the enzyme, whereas the small subunits serve to regulate the activity of the enzyme (8).

Kung et al. (9) resolved 55 peptides from the large subunit and 32 from the small subunit of fraction I protein of *N. tabacum* following trypsin digestion and two-dimensional gel electrophoresis. On the basis of amino acid compositions, tryptic

peptide analyses, and immunologic comparisons, Kung (8) concluded that in a given species the large subunit of fraction I protein shared no similarities with the small subunit. However, large subunits from different species were very similar in these properties, whereas the small subunits varied considerably. Wildman and coworkers (5,7), using isoelectric focusing of carboxymethylated fraction I protein, were able to resolve the large subunits of fraction I protein from all *Nicotiana* species into a cluster of three polypeptides, with a total of four clusters representing all 63 species examined. Isoelectric focusing of the small subunit of fraction I protein from these same 63 species resolves from 1 to 4 polypeptides per species, with a total of 13 combinations (Fig. 18–1). Kung (8) raised the obvious question: Do the three large-subunit polypeptides seen on isoelectric focusing actually represent three distinct polypeptides coded by three separate genes, presumably located in the chloroplast or do they result from posttranslational modifications of a single gene product producing three polypeptides differing in charge? On

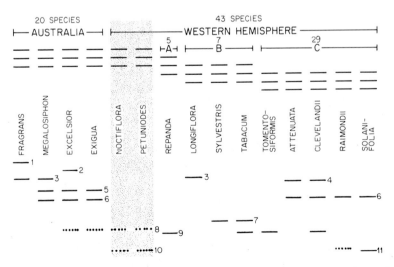

FIG. 18–1. Diagrammatic representation of numbers and positions of large- and small-subunit polypeptides of fraction I protein from 63 species of *Nicotiana* following isoelectric focusing of carboxymethylated material. Small-subunit polypeptide 7′ found in *N. glauca* and 5′ found in *N. glutinosa* are not shown. Small-subunit polypeptides 1 and 2 are found only in Australian species, and polypeptides 4, 7, and 9 are found only in Western Hemisphere species. Polypeptides 8 and 10 are stippled to emphasize the probability that *N. noctiflora* and *N. petunioides* contain proteins closely similar in composition to those existing before continental drift separated the *Nicotiana* species. (From Chen et al., ref. 7, with permission.)

the basis of known arginine-lysine composition, he calculated that if each subunit contained three different polypeptides having the same molecular weight, there should be more than the 55 peptides seen on trypsin digestion. Therefore the nature of the three polypeptides obtained following isoelectric focusing of carboxymethylated material remains to be resolved. In spite of this, the polypeptides seen on isoelectric focusing of the carboxymethylated material are characteristic of individual species, and they serve as reliable genetic markers in interspecific crosses, as we shall shortly see.

Kawashima and Wildman (10) were first to take advantage of interspecific hybrids to study the inheritance of the fraction I polypeptide subunits in *Nicotiana*. These authors prepared tryptic peptide maps for the large and small subunits of fraction I protein from five American species of *Nicotiana*. These experiments revealed several differences that could be employed as the basis for determining the pattern of inheritance of the small subunit in reciprocal interspecific hybrids, but no differences were found in the large subunit. In the small subunit, two sorts of alterations were noted. First, on Sephadex chromatography of the tryptic peptides, it was found that *N. tabacum* yielded four peaks that absorbed at 280 nm, whereas *N. glauca* and *N. glutinosa* produced only three. Peak 3, which was unique to *N. tabacum,* was shown to result because the peptides moving in this region of the column contained tyrosine in *N. tabacum* and therefore absorbed 280-nm ultraviolet light, whereas these peptides in *N. glauca* and *N. glutinosa* contained a different amino acid that did not absorb in this region. Reciprocal crosses between *N. tabacum* and *N. glauca* or *N. glutinosa* showed that the tyrosine-containing peptide was biparentally inherited, which suggested that inheritance of this peptide and therefore inheritance of the small subunit was controlled by a nuclear gene.

Second, paper chromatography of tryptic peptides revealed that *N. glutinosa* and *N. glauca* lacked a peptide (No. 1) present in *N. tabacum;* in addition, in *N. glauca* two peptides (A and B) moved differently on chromatography than they did in *N. tabacum* (Fig. 18–2). In the *N. glutinosa* × *N. tabacum* crosses, peptide 1 (characteristic of *N. tabacum*) was present in both reciprocal hybrids (Fig. 18–2). The same was true in the *N. tabacum* × *N. glauca* crosses; in addition, the position of peptides A and B was similar to that seen for the *N. tabacum* parent in both crosses (Fig. 18–2).

From these results, Kawashima and Wildman drew two conclusions. First, the biparental inheritance of the *N. tabacum* traits indicates that the primary structure of the small subunit of fraction I protein is determined by a nuclear gene. Second, since peptides A and B are positioned in reciprocal hybrids between *N. glauca* and *N. tabacum,* as they are in *N. tabacum* alone, the *N. tabacum* genetic information must somehow suppress the *N. glauca* genetic information for the small subunit. Later, by means of carboxymethylation of small-subunit protein followed by isoelectric focusing, Sakano et al. (11) found that the small subunit of *N. tabacum* could be resolved into two peptides in approximately equal amounts (Fig. 18–3). In *N. glauca* only one of these peptides was found, but at approximately double the amount of the same peptide in *N. tabacum*. The reciprocal interspecific hybrids contained both peptides, which indicated that they contained *N. tabacum* information; but the staining of the peptide that would have corresponded to that derived from the *N. glauca* parent was intense, which indicated that this peptide was present in an amount greater than would have been expected had only the *N. tabacum* gene been expressed in the hybrids. Sakano et al. (11) concluded that the earlier opinion of Kawashima and Wildman (10) was wrong and that *N. glauca* information for the small subunit was, after all, expressed in the hybrids. Although this con-

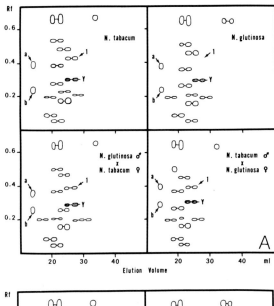

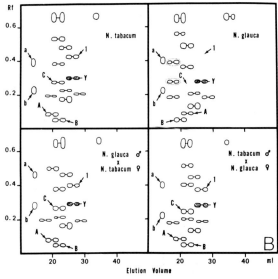

FIG. 18–2. Pattern of inheritance of peptides obtained following trypsin digestion of small-subunit polypeptides of *Nicotiana* fraction I protein. A: Tryptic peptide maps obtained from *N. tabacum, N. glutinosa,* and their F_1 interspecific hybrids. Notice the presence of peptide 1, which is found only in the *N. tabacum* parent in the F_1 progeny of both crosses. B: Tryptic peptide maps obtained for *N. tabacum, N. glauca,* and their reciprocal F_1 hybrids. Notice the presence of peptide 1, which is found only in the *N. tabacum* parent in the F_1 progeny of both crosses. Notice also that peptides A and B are arranged in the pattern characteristic of *N. tabacum* rather than that seen in *N. glauca* in the F_1 progeny of both crosses. (From Kawashima and Wildman, ref. 10, with permission.)

clusion may be valid for the carboxymethylated peptides, it does not really explain the earlier results for the tryptic peptides A and B. One would have expected to see both the *N. tabacum* and *N. glauca* A and B peptides in the hybrids observed by Kawashima and Wildman, but only the *N. tabacum* A and B peptides were in evidence.

Sakano et al. (11), using carboxymethylation and isoelectric focusing, also found that *N. excelsior* small subunits contained four peptides, whereas *N. gossei* contained

three. Inheritance of the extra peptide was biparental, which supported the original conclusion of Kawashima and Wildman that the small subunit was coded by a nuclear gene.

Since Kawashima and Wildman (10) found no tryptic peptide differences among large subunits of fraction I protein obtained from five American *Nicotiana* species, the next step was to examine the Australian species; in an evolutionary sense, these were likely to be far more distant from the

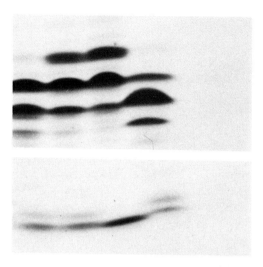

FIG. 18–3. Isoelectric focusing patterns obtained for carboxymethylated large- and small-subunit polypeptides of fraction I protein from *N. tabacum, N. glauca,* and their reciprocal F_1 hybrids. The large subunit yields three species that migrate near the top of the gel. The small subunit of *N. glauca* produces one species and that of *N. tabacum* two, all of which move toward the bottom of the gel (see also Fig. 18–1). Left to right: 1. F_1 hybrid from a cross in which *N. tabacum* was the maternal parent. 2. F_1 hybrid from a cross in which *N. glauca* was the maternal parent. 3. *N. glauca.* 4. *N. tabacum.* (From Sakano et al., ref. 11, with permission.)

American species than the American species were from each other. Chan and Wildman (12) prepared peptide maps of the large subunits of three Australian species (*N. gossei, N. excelsior,* and *N. suaveolens*) and found that they were precisely similar to each other and to the American species, with one exception: the large subunit of all the Australian species contained an extra tryptic peptide. Reciprocal hybrids were prepared between *N. gossei* and *N. tabacum,* and it was found that the extra peptide was maternally inherited. The authors concluded that the large subunit of fraction I protein was a product of the chloroplast genome. Sakano et al. (11) found that the large subunit could be separated into three peptides using carboxymethylation and isoelectric focusing, but the isoelectric points of these peptides showed interspecific differences, as mentioned previously. In reciprocal interspecific hybrids of *N. glauca* × *N. tabacum, N. glutinosa* × *N. tabacum,* and *N. glauca* × *N. langsdorfii,* the isoelectric points of the three peptides were always the same as those of the maternal parents (Fig. 18–3). These results greatly bolstered the original conclusions of Chan and Wildman.

The RuBPCase catalytic site is localized in the large subunit of fraction I protein (13). Singh and Wildman (14) compared the K_m for RuBP of RuBPCase isolated from different *Nicotiana* species and found that *N. gossei* had a lower K_m than the other species studied. In interspecific hybrids employing *N. gossei* as one of the parents, it was found that the K_m for RuBP was maternally inherited. Thus the RuBPCase catalytic site is maternally inherited, which is consistent with its localization in the large subunit of fraction I protein, which has the same inheritance pattern.

The subunits of fraction I protein have also been used as markers for the study of inheritance of chloroplast genomes in interspecific parasexual hybrids of *Nicotiana.* Carlson et al. (15) reported the first such plants following protoplast fusions between *N. glauca* and *N. langsdorfii.* Three plants were obtained. These plants were self-fertile, in contrast to the self-sterile F_1 hybrids obtained from sexual crosses. The reason probably was that the parasexual hybrids (in which the chromosome number was equal to the sum of the diploid chromosome numbers of the two parent species) were behaving as allotetraploids, so that meiosis proceeded normally. In any event, Kung et al. (16) obtained seed from one of the parasexual hybrids; then they grew plants and examined the fraction I protein subunits obtained following isoelectric focusing of the carboxymethylated polypeptides. They found that the hybrid plants contained the three large-subunit polypeptides characteristic of *N. glauca* but showed no trace of the three diagnostic for *N. langsdorfii*. However, the single small-subunit polypeptide characteristic of *N. glauca* was present, together with the two found in *N. langsdorfii*. Thus the nuclear genes coding for small subunits from both species were expressed, whereas the chloroplast gene coding for the large subunit of *N. langsdorfii* fraction I protein was either missing or not expressed. More recently, Chen et al. (17) reported on the properties of fraction I protein obtained from 16 additional parasexual hybrids on *N. glauca* and *N. langsdorfii* produced by protoplast fusions. These plants were also fertile, but in all cases they appeared to contain the diploid chromosome complement from one parent and a tetraploid complement from the other. In each hybrid, small-subunit polypeptides characteristic of both parents were present; but in 15 of the 16 plants only one kind of clDNA coding for the large subunit appeared to be expressed, and it was apparently random as to whether the gene(s) from *N. glauca* or *N. langsdorfii* were expressed. These plants were vigorous in appearance, and a number produced F_2 progeny. In contrast, the single fusion in which both *N. glauca* and *N. langsdorfii* large subunits were in evidence was dwarfed and distorted and unable to produce flowers. Chen and associates theorized that the presence of two different clDNAs in the

same plant resulted in this catastrophe, and they suggested that maternal inheritance of clDNA might have arisen to protect dissimilar chloroplast genomes from recombination. This seems improbable to the present author, in view of the fact that biparental inheritance of plastids does occur in many other plants. In addition, it is simply not clear whether or not plastid genomes recombine in higher plants at all.

Kung et al. (16) also reported on the properties of a second parasexual hybrid plant (thus far the only such plant obtained) that resulted following incubation of chloroplasts from *N. suaveolens* with protoplasts from white tissue of a variegating mutant of *N. tabacum* cv Xanthi NC. Green calluses were obtained at an approximate rate of 2.0×10^{-4}, which is about four orders of magnitude greater than the reversion rate. Attempts to regenerate whole plants from these calluses were unsuccessful, with one exception. The exceptional plant was a variegating albino of abnormal morphology that was sterile. This plant had a chromosome number of 64, whereas the diploid chromosome numbers of the two parent species *N. suaveolens* and *N. tabacum* were 32 and 48, respectively.

Examination of the fraction I polypeptides present in the variegating albino plant revealed that large-subunit polypeptides characteristic of both parent species were present in the hybrid, which indicated that clDNA from each parent was also present. Examination of the small-subunit polypeptides revealed that six were in evidence. Of these, four were characteristic of the protein from *N. suaveolens,* and two were characteristic of *N. tabacum*. These results indicated that nuclear information from *N. suaveolens* was also present in the hybrid. That is, either the chloroplast preparation from *N. suaveolens* contained protoplasts or it contained nuclei. The important point is that the experiment did not rigorously prove that transfer of chloroplasts per se is accompanied by transfer of information specifying the large subunit of fraction I protein.

A particularly gratifying aspect of the fraction I story is that it can also be shown that the large subunit is synthesized on chloroplast ribosomes and that the small subunit is made on cytoplasmic ribosomes (see Chapter 20). Thus the division of labor involved in the coding and synthesis of fraction I protein seems to be strictly along orthodox lines. A nuclear gene codes for the small subunit, which is made on cytoplasmic ribosomes, and a chloroplast gene codes for the large subunit, which is made on chloroplast ribosomes. In view of the abundance of fraction I protein and the relatively clear-cut evidence concerning localization of structural genes for the two polypeptides comprising this protein and their sites of synthesis, fraction I protein would seem to be an excellent model for studying the mechanism(s) by which polypeptides made in the cytoplasmic and chloroplast compartments are joined together in the chloroplast, as well as the regulatory mechanisms that govern the production of the two polypeptides in proper stoichiometry (see Chapter 20).

LOCATION OF GENES CODING FOR OTHER CHLOROPLAST COMPONENTS IN *Nicotiana*

Table 18–2 summarizes the other chloroplast components thus far investigated by the interspecific hybridization technique in *Nicotiana*. Kung et al. (18) isolated chlorophyll-protein complex (CPII) from *N. tabacum* and *N. glauca* and compared tryptic fingerprints of the protein moiety. They found that *N. glauca* and *N. tabacum* each had a unique peptide, designated G and T, respectively. In reciprocal interspecific hybrids the T peptide was found, indicating biparental inheritance and control of the protein by a nuclear gene, but the G peptide was absent. This observation is reminiscent of the finding of Kawashima and Wildman with respect to the A and B tryptic peptides of the small subunit of fraction I protein, where the *N. tabacum* peptides were present in both hybrids, whereas the *N. glauca* peptides were absent. It is surprising in both cases for presumably unrelated proteins that only the *N. tabacum* peptides were in evidence. Bourque and Wildman (19) reported two protein differences between the large chloroplast ribosomal subunits of *N. tabacum* and *N. glauca* on one-dimensional gel electrophoresis of the ribosomal proteins. One protein present in *N. glauca* (*a*) was absent from a corresponding position in the gel of the *N. tabacum* proteins. The second protein (*b*) had greater mobility in *N. tabacum* than in *N. glauca*. In reciprocal interspecific hybrids, protein *a* from the *N. glauca* parent was found in both hybrids, and both hybrids contained a protein with the same electrophoretic mobility as protein *b* of *N. tabacum*. The authors concluded that the two proteins were coded by nuclear genes. However, careful repetition of these experiments using gel techniques with greater resolution (i.e., two-dimensional gels) is desirable before these conclusions are taken for granted.

Isoelectric focusing of carbaminomethylated ferridoxins of different *Nicotiana* species revealed that the *S*-carbaminomethylated ferridoxin of *N. glutinosa* had a slightly more acidic isoelectric point than those of nine other *Nicotiana* species (20). The hybrid *N. glutinosa* ♀ × *N. glauca* ♂ had *S*-carbaminomethylated ferridoxins characteristic of both parents, and it was concluded by Kwanyuen and Wildman (20) that the ferridoxin protein moiety was coded by a nuclear gene. The ferridoxins of *N. tabacum* and *N. glauca* contained methionine, whereas that from *N. glutinosa* did not. The *N. glutinosa* ♀ × *N. glauca* ♂ hybrid contained half the methionine of the *N. glauca* parent, which again suggested the presence of two ferridoxins in the hybrid, one of which contained methionine (as did the *N. glauca* parent) and one of which lacked methionine (as did the *N. glutinosa* parent).

CONCLUSIONS

Until recently, the *Nicotiana* hybrid experiments provided the best evidence concerning localization of the genes coding for the subunits of fraction I protein. Now this evidence has received direct confirmation in the case of *Chlamydomonas*, where it has been shown that the large-subunit mRNA hybridizes to a specific restriction fragment from clDNA (see Chapter 20). We may expect similar experimental evidence to be published soon for higher plants. The *Nicotiana* hybrid system has also proved useful in localization of genes coding for other chloroplast polypeptides, and in the future we may expect more results similar to the ones reported in this chapter. This approach, coupled with experiments on plastome mutants (see Chapter 17) and molecular mapping of chloroplast mRNAs (see Chapter 20), ultimately should yield a better picture of what proteins the higher plant chloroplast genome codes for.

REFERENCES

1. Goodspeed, T. H. (1954): The genus *Nicotiana*. Chronica Botanica, Vol. 16, No. 116. Chronica Botanica, Waltham, Mass.
2. von Wettstein, D. (1967): Chloroplast structure and genetics. In: *Harvesting the Sun*, edited by A. San Pietro, F. A. Greer, and T. J. Army, pp. 153–190. Academic Press, New York.
3. Wong-Staal, F., and Wildman, S. G. (1973): Identification of a mutation in chloroplast DNA correlated with formation of defective chloroplasts in a variegated mutant of *Nicotiana tabacum. Planta*, 113:313–326.
4. Wildman, S. G., Lu-Liao, C., and Wong-Staal, F. (1973): Maternal inheritance, cytology and macromolecular composition of defective chloroplasts in a variegated mutant of *Nicotiana tabacum. Planta*, 113:293–312.
5. Wildman, S. G., Chen, K., Gray, J. C., Kung, S. D., Kwanyuen, P., and Sakano, K. (1975): Evolution of ferridoxin and fraction I protein in *Nicotiana*. In: *Genetics and Biogenesis of Mitochondria and Chloroplasts*, edited by C. W. Birky, Jr., P. S. Perlman, and T. J. Byers, Chapter 9. Ohio State University Press, Columbus.
6. Kawashima, N., and Wildman, S. G. (1970): Fraction I protein. *Annu. Rev. Plant Physiol.*, 21:325–358.
7. Chen, K., Johal, S., and Wildman, S. G. (1976): Role of chloroplast and nuclear DNA genes during evolution of fraction I protein. In: *Genetics and Biogenesis of Chloroplasts and Mitochondria*, edited by T. Bücher, W. Neupert, W. Sebald, and S. Werner, pp. 3–11. North Holland, Amsterdam.
8. Kung, S.-D. (1976): Tobacco fraction I protein: A unique genetic marker. *Science*, 191: 429–434.
9. Kung, S.-D., Sakano, K., and Wildman, S. G. (1974): Multiple peptide composition of the large and small subunits of *Nicotiana tabacum* fraction I protein ascertained by fingerprinting and electrofocusing. *Biochim. Biophys. Acta*, 365:138–147.
10. Kawashima, N., and Wildman, S. G. (1972): Studies on fraction I protein. IV. Mode of inheritance of primary structure in relation to whether chloroplast or nuclear DNA contains the code for a chloroplast protein. *Biochim. Biophys. Acta*, 262:42–49.
11. Sakano, K., Kung, S. D., and Wildman, S. G. (1974): Identification of several chloroplast DNA genes which code for the large subunit of *Nicotiana* fraction I proteins. *Mol. Gen. Genet.*, 130:91–97.
12. Chan, P.-H., and Wildman, S. G. (1972): Chloroplast DNA codes for the primary structure of the large subunit of fraction I protein. *Biochim. Biophys. Acta*, 277:677–680.
13. Sugiyama, T., Matsumoto, C., and Akazawa, T. (1970): Structure and function of chloroplast protein. IX. Dissociation of spinach leaf ribulose-1,5-diphosphate carboxylase by urea. *J. Biochem.*, 68:821–831.
14. Singh, S., and Wildman, S. G. (1973): Chloroplast DNA codes for the ribulose diphosphate carboxylase catalytic site on fraction I proteins of *Nicotiana* species. *Mol. Gen. Genet.*, 124:187–196.
15. Carlson, P. S., Smith, H. H., and Dearing, R. D. (1972): Parasexual interspecific plant hybridization. *Proc. Natl. Acad. Sci. U.S.A.*, 69:2292–2294.
16. Kung, S.-D., Gray, J. C., Wildman, S. G., and Carlson, P. S. (1975): Polypeptide composition of fraction I protein from parasexual hybrid plants in the genus *Nicotiana. Science*, 187:353–355.
17. Chen, K., Johal, S., and Wildman, S. G. (1977): Phenotypic markers for chloroplast DNA genes in higher plants and their use in biochemical genetics. In: *Nucleic Acids and Protein Synthesis in Plants*, J. H. Weil and L. Bogorad, pp. 183–194. Plenum Press, New York.
18. Kung, S.-D., Thornber, J. P., and Wildman,

S. G. (1972): Nuclear DNA codes for the photosystem II chlorophyll-protein of chloroplast membranes. *FEBS Lett.,* 24:185–188.

19. Bourque, D. P., and Wildman, S. G. (1973): Evidence that nuclear genes code for several chloroplast ribosomal proteins. *Biochem. Biophys. Res. Commun.,* 50:532–537.

20. Kwanyuen, P., and Wildman, S. G. (1975): Nuclear DNA codes for *Nicotiana* ferridoxin. *Biochim. Biophys. Acta,* 405:167–174.

21. Singh, S., and Wildman, S. G. (1974): Kinetics of cold inactivation and heat reactivation of the ribulose diphosphate carboxylase activity of crystalline fraction I proteins isolated from different species and hybrids of *Nicotiana. Plant Cell Physiol.,* 15:373–379.

22. Wildman, S., Kawashima, N., Bourque, D., Wong, F., Singh, S., Chan, P., Kwok, S., Sakano, K., Kung, S., and Thornber, J. (1973): Location of DNA's coding for various kinds of chloroplast proteins. In: *The Biochemistry of Gene Expression,* edited by J. K. Pollack and J. W. Lee, pp. 443–456. Australia and New Zealand Book Co., Sydney.

23. Wildman, S. G., and Bonner, J. (1947): The proteins of green beans. I. Isolation, enzymatic properties and auxin content of spinach cytoplasmic proteins. *Arch. Biochem.,* 14:381–413.

Chapter 19
Biogenesis of Mitochondrial Inner Membrane

The key to the riddle of why the mitochondrial genome and protein-synthesizing system have been conserved in all eukaryotes may lie in the mechanism by which the mitochondrial inner membrane is assembled. Most mitochondrial functions are performed normally in vegetative petite mutants of *S. cerevisiae,* even though they lack a mitochondrial protein-synthesizing system (see Chapter 4) and their mtDNA either contains gross deletions or is missing entirely (see Chapter 6). Thus vegetative petites elaborate the enzymes of the mitochondrial matrix (e.g., the citric acid cycle enzymes and ferrochelatase), a morphologically normal outer membrane, and even an inner membrane, together with such components as cytochromes c and c_1 and three of the subunits of cytochrome oxidase (1). However, these mitochondria are deficient in the remaining four cytochrome oxidase polypeptides, in cytochrome b, and in the energy-transfer system associated with the inner membrane of normal mitochondria. In short, the synthesis of some inner-membrane components and the integration of others are disrupted in vegetative petites. In this chapter we shall chronicle the elegant experimental evidence from both yeast and *Neurospora* that has now yielded a clear picture of where most of the polypeptides associated with the respiratory chain and the energy-transfer system of the inner membrane are synthesized. We shall then turn to a discussion of what is known concerning the localization of the structural genes for these polypeptides. Information on the latter point, which was scanty until recently, is now beginning to accumulate rapidly.

Determining the sites of synthesis of a group of organelle proteins that must become assembled into a complex macromolecular structure such as a membrane is a difficult problem. Two general approaches have been taken. The *in vitro* approach asks what polypeptides are made within isolated organelles. Obviously, polypeptides made within an isolated mitochondrion must be synthesized by the mitochondrial protein-synthesizing system. The *in vivo* approach makes use of inhibitors specific for cytoplasmic and mitochondrial protein synthesis. Ideally, proteins made in the mitochondrion will continue to be synthesized in the presence of an inhibitor of cytoplasmic protein synthesis, whereas proteins whose messages are translated on cytoplasmic ribosomes should be made in the presence of inhibitors of mitochondrial protein synthesis. In practice, both methods have serious drawbacks. These were considered at length in the excellent review of the biosynthesis of mitochondrial proteins published by Schatz and Mason (1).

At the time Schatz and Mason wrote their review, relatively little had been learned from *in vitro* protein synthesis experiments with isolated mitochondria regarding the synthesis of specific mitochondrial polypeptides, despite a large literature on the subject. Schatz and Mason cited several reasons for this. First, amino acid incorporation tended to be very slow in isolated mitochondria, and it responded in an unpredictable fashion to variations in the composition of the incubation medium. Second, particularly in early experiments, bacterial and microsomal contamination of mitochondrial fractions posed a problem. Third, the polypeptides made by isolated mitochondria were tightly associated with the inner membrane, and they effectively resisted purification by many of the early techniques. It was also reported that in rat liver the polypeptides synthesized by mitochondria *in vitro* were continuously degraded to acid-soluble products. Finally, the greatest problem at the time of the Schatz and Mason review seemed to be the requirement for partner proteins made in the cytoplasm. It appeared that the mitochondrially synthesized components of cytochrome oxidase and the oligomycin-sensitive ATPase were made in significant amounts only if they were combined with their cytoplasmically synthesized partner proteins. This alone could have explained the slow rate of amino acid incorporation seen in isolated mitochondria and the susceptibility of the uncombined products to proteolytic degradation in the early experiments. In addition, problems such as these coupled with less than satisfactory methods for mitochondrial isolation and electrophoretic separation of proteins probably had much to do with the isolation of a rather nondescript insoluble protein fraction that appeared to be synthesized in mitochondria. This fraction was once believed to contain "miniproteins" (2) or "mitochondrial structural protein" (3,4). However, as was discussed by Schatz and Mason, it subsequently became evident that the miniproteins probably were membrane lipids (5) and the structural protein was a mixture of denatured proteins (6–8).

Despite the foregoing difficulties, Schatz and Mason also pointed out that a number of studies with isolated mitochondria had agreed that at least six polypeptides ranging in apparent molecular weight from 45,000 d to less than 10,000 d were synthesized. Since the same synthetic pattern was obtained when mitochondria were labeled *in vivo* in the presence of cycloheximide, Schatz and Mason concluded that isolated mitochondria probably did synthesize many (perhaps all) of their normal protein products, but at a reduced rate. The recent experiments of Poyton and his colleagues with isolated mitochondria from yeast showed that this conclusion was well justified. These authors showed in unequiv-

ocal fashion that these mitochondria synthesize three of the subunits of cytochrome oxidase, as we shall discuss later in this chapter.

Because of the previously mentioned difficulties inherent in the early *in vitro* experiments with isolated mitochondria, *in vivo* experiments with specific inhibitors of mitochondrial and cytoplasmic protein synthesis have provided us with the most comprehensive picture of the inner-membrane polypeptides synthesized inside and outside the mitochondrion. However, despite its conceptual simplicity, the inhibitor approach is also fraught with pitfalls that must be avoided. For example, the length of time of exposure to an inhibitor is of critical importance. Suppose cells are exposed to an inhibitor of mitochondrial protein synthesis for a relatively long time and either labeled coordinately or pulse-labeled later. Under these conditions the synthesis of polypeptides on mitochondrial ribosomes will surely be blocked, and those polypeptides whose synthesis continues must be made in the cytoplasm. However, certain proteins made on cytoplasmic ribosomes may require the presence of mitochondrial "partner" proteins for their integration into the inner membrane. Pools of these partner proteins will become depleted during long-term incubation. Cytoplasmically synthesized polypeptides that require these partners for their own integration into the membrane will become depleted in the membrane. The end result will be an apparent blockage of synthesis of these cytoplasmically made polypeptides by inhibitors of mitochondrial protein synthesis. Obviously the length of incubation and the pool size will vary in different experimental systems. The pool size of a partner protein may be very small compared to that of the protein that depends on it for integration. In such a case the results of even a short-term pulse labeling experiment can be misinterpreted. A case in point, which will be discussed shortly, concerns cytochrome oxidase in *Neurospora,* where one of the mitochondrially synthesized polypeptides has an extremely small pool size and appears to be the rate-limiting polypeptide in the assembly of the enzyme. Second, it is of prime importance to establish, through the use of concentration curves, that inhibition is as complete as possible. For example, inhibition of mitochondrial protein synthesis in intact cells by chloramphenicol or erythromycin is usually incomplete even at very high concentrations (9). It is also possible that certain proteins are synthesized in the organelle or cytoplasm by mechanisms more resistant to some inhibitors than to others. Third, a given antibiotic may have other cellular side effects. For example, chloramphenicol inhibits mitochondrial electron transport at site 1, and it may uncouple oxidative phosphorylation.

Despite these cautionary remarks, clean results have been obtained using pulse labeling in the presence of inhibitors *in vivo* in yeast and *Neurospora,* and our picture of the sites of synthesis of the inner-membrane polypeptides associated with cytochrome oxidase and cytochromes b, c, and c_1, as well as the oligomycin-sensitive ATPase, is now quite clear. This picture has been confirmed for cytochrome oxidase through the use of complementary *in vitro* experiments.

BIOSYNTHESIS OF CYTOCHROME OXIDASE

The sites of synthesis of the polypeptides comprising the cytochrome oxidase complex were established by Schatz and his colleagues in yeast and by Bücher and his colleagues working with *Neurospora.* The results in both cases were essentially the same. Cytochrome oxidase isolated from each organism was purified, and in each case it contained seven polypeptides (Table 19–1). The apparent molecular weights of

TABLE 19–1. *Inner-membrane polypeptides and their sites of synthesis[a]*

Component	Polypeptide	Molecular weight (d)	Site of synthesis	References
Cytochrome oxidase	I	42,000	Mitochondrion	1,10,22,23
	II	34,500	Mitochondrion	1,10,22,23
	III	23,000	Mitochondrion	1,10,22,23
	IV	14,000	Cytoplasm	1,10,22,23
	V	12,500	Cytoplasm	1,10,22,23
	VI	12,500	Cytoplasm	1,10,22,23
	VII	4,500	Cytoplasm	1,10,22,23
Cytochrome b	1	28,000	Mitochondrion	41
Iso-1-cytochrome c	1	12,667	Cytoplasm	1,50,84
Cytochrome c_1	1	31,000	Cytoplasm	1,10,42,85
ATPase				
F_1	1	58,500	Cytoplasm	31–35
	2	54,000	Cytoplasm	31–35
	3	38,500	Cytoplasm	31–35
	4	31,000	Cytoplasm	31–35
	8a	12,000	Cytoplasm	31–35
OSCP	7	18,500	Cytoplasm	31,36
Membrane factor	5	29,000	Mitochondrion	31,37
	6	22,000	Mitochondrion	31,37
	8b	12,000	Mitochondrion	31,37
	9	8,000	Mitochondrion	31,37

[a] All data are taken from experiments done with *Saccharomyces.* Experiments with *Neurospora* yielded results in good agreement for cytochrome oxidase and cytochrome b, but similar experiments do not seem to have been done with respect to the other components shown.

the polypeptides, as shown by the use of SDS gel electrophoresis, ranged from 42,-000 d (subunit I) to less than 5,000 d (subunit VII). Schatz pointed out on a number of occasions that in the case of a complex enzyme such as cytochrome oxidase, it is difficult to determine where the enzyme ends and the membrane begins. However, he cited the following reasons for believing that all seven polypeptides are involved in oxidation of cytochrome c *in vivo* (1,10,11).

1. Cytochrome oxidase preparations isolated from different organisms by widely different procedures exhibit similar polypeptide compositions (11–15), except the preparation of the enzyme from beef heart, which lacks one of the large polypeptides (12,16).

2. Antiserum to a single polypeptide of the cytochrome oxidase complex of yeast (subunit VI) will precipitate all seven polypeptides in the same stoichiometry as antibody made against the holoenzyme or several polypeptides together (17).

3. Antibodies specific for polypeptides II, VI, or V plus VII inhibit cytochrome c oxidase activity (17).

4. Segregational petite mutants of yeast exist that specifically lack cytochrome c oxidase activity and also lack polypeptide III (18).

5. Polypeptide I is buried inside the enzyme complex, as is shown by the fact that it labels very poorly with surface probes (e.g., iodination of tyrosine with ^{125}I) used against the holoenzyme (19).

These lines of evidence in aggregate argue strongly that all seven polypeptides are integral components of cytochrome oxidase. In an interesting report that deserves mention before we turn to the sites of synthesis of the individual polypeptides, Eytan and Schatz (19) attempted to probe the molecular organization of cytochrome oxidase. To do this they used several specific surface probes to establish the positioning of the cytochrome oxidase subunits in the isolated holoenzyme and within the inner membrane. For example, lactoperoxidase can be used for the enzymatic iodination of tyrosine residues in proteins with [125]I (20). If a polypeptide in a multicomponent complex such as cytochrome oxidase is partially exposed on the surface of the complex, the exposed tyrosine residues will become iodinated. Polypeptides buried within the complex will not. The polypeptides can then be separated from the complex by SDS gel electrophoresis and the level of iodination of each polypeptide determined. Other surface probes, although different in mechanism of action, can be used to the same end. An important reason for comparing probes is, of course, the possibility that some might enter into internal portions of the membrane and not react solely with polypeptides protruding on the external surfaces. Eytan and Schatz considered these limitations carefully in their article. Their results indicated that polypeptides I and II were buried in the isolated holoenzyme and within the enzyme complex when it is embedded in the inner membrane. All the other polypeptides labeled with at least two probes and appeared to be exposed to some degree, with polypeptides VI and VII being the most exposed. It was not clear why polypeptide IV was iodinated in the membrane but not in the isolated enzyme. The explanation may be that its conformation changes on isolation so that the tyrosine residues are no longer exposed. These results, taken together with more recent findings, indicate that polypeptides II, III, VI, and VII are on the external side of the inner membrane, with polypeptides I and V being buried in the inner membrane and polypeptide IV being exposed on the matrix side (Fig. 19–1). The major differences to be noted in the most recent results concern the positions of polypeptides II and V. The former polypeptide is more exposed than the results of Eytan and Schatz indicated, whereas polypeptide V appears to be buried within the inner membrane. These results

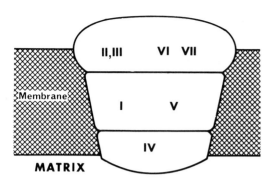

FIG. 19–1. Subunit arrangement of cytochrome c oxidase from bakers' yeast in mitochondrial inner membrane according to Schatz and collaborators. (Courtesy of Dr. G. Schatz.)

with yeast cytochrome oxidase have been confirmed by Eytan et al. (21) for beef heart cytochrome oxidase using surface probes. In short, cytochrome oxidase is a transmembranous enzyme.

Following these remarks about enzyme structure, we shall turn to biosynthesis of the enzyme. The sites of translation of six of the seven cytochrome oxidase polypeptides of yeast were established by Mason and Schatz (22). Cells were grown for a number of generations in the presence of ^{14}C-leucine; then they were washed and labeled for an hour with ^3H-leucine either in the presence of cycloheximide to inhibit cytoplasmic protein synthesis or in the presence of erythromycin to block mitochondrial protein synthesis (Fig. 19–2). The enzyme was then isolated by immunoprecipitation with antibody against the holoenzyme from an extract of submitochondrial particles, and the constituent polypeptides were separated by SDS gel electrophoresis. The results showed clearly that synthesis of the three largest subunits (I–III) was inhibited by erythromycin but not cycloheximide (Fig. 19–2). These, therefore, were made on mitochondrial ribosomes. The converse inhibition pattern was seen for the other three polypeptides (IV–VI) identified in these experiments, which were therefore synthesized on cytoplasmic ribosomes. Later polypeptide VII was identified and

was also shown to be a product of cytoplasmic protein synthesis (1,10). At almost the same time that Mason and Schatz (22) reported their results, Rubin and Tzagoloff (23) published a similar report on the sites of synthesis of the yeast cytochrome oxidase polypeptides; they used methods not greatly different from those of Mason and Schatz (22).

Mason and Schatz reported several other interesting observations on the synthesis of the cytochrome oxidase polypeptides. When anaerobically growing yeast cells were labeled in the presence of cycloheximide, they continued to synthesize polypeptide III, but not polypeptides I and II. Vegetative petites do not make any of these three polypeptides, as expected, since their messages are translated on mitochondrial ribosomes; but they continue to make three of the cytoplasmically synthesized polypeptides (IV–VI), but not the fourth (VII). These observations lead to some interesting questions. How do anaerobically growing yeast cells regulate mitochondrial protein synthesis in such a way that one of the three polypeptides synthesized on mitochondrial ribosomes continues to be made? Why is this regulation important? Why is polypeptide VII, which is made on cytoplasmic ribosomes, lacking in vegetative petites? Since antisera are used to precipitate the polypeptides, and since polypeptides IV–VI, which are made on cytoplasmic ribosomes, are present in vegetative petites, it would appear that faulty assembly cannot account for the absence of polypeptide VII and that its loss is specific. As we shall see shortly, answers to at least the first two questions are likely to come from experiments carried out *in vitro* with isolated mitochondria.

The experiments on the sites of biosynthesis of the *Neurospora* cytochrome oxidase polypeptides were begun by Bücher and his colleagues at about the same time as the yeast experiments (24). The experimental design was similar to that used by

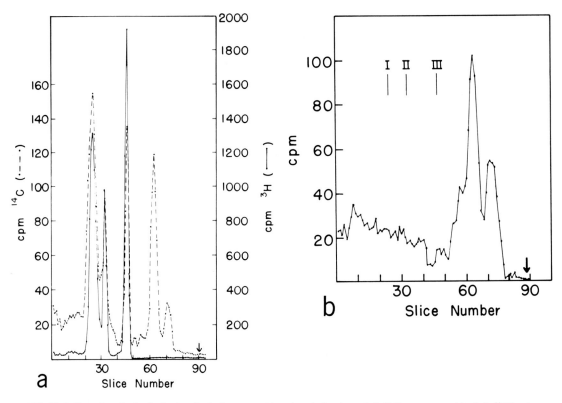

FIG. 19–2. Sites of synthesis of subunits of cytochrome c oxidase from bakers' yeast. A: Cells grown aerobically in ^{14}C-leucine for 8 to 10 generations were pulse-labeled with ^{3}H-leucine for 60 min in the presence of cycloheximide to block cytoplasmic protein synthesis as they entered stationary phase. Cytochrome c oxidase was isolated by immunoprecipitation from mitochondria. All subunits were labeled during continuous incubation of cells in the presence of ^{14}C-leucine in the absence of cycloheximide, but only the three largest subunits became labeled with ^{3}H-leucine in the presence of the inhibitor. B: Cells entering stationary phase were pulse-labeled with ^{3}H-leucine as in A, but in the presence of erythromycin to block mitochondrial protein synthesis. Cytochrome oxidase was isolated by immunoprecipitation of partially purified enzyme. The three largest subunits were not labeled, whereas three of the smaller ones were. Subunit VII was not visualized in either A or B. (From Mason and Schatz, ref. 22, with permission.)

Mason and Schatz, except that chloramphenicol rather than erythromycin was used to inhibit mitochondrial protein synthesis. The conclusions were also the same. The three large cytochrome oxidase subunits were made on mitochondrial ribosomes, and the four small subunits were made on cytoplasmic ribosomes. Despite the similarity of experimental design and results in the yeast experiments and *Neurospora* experiments, there are certain important differences that need to be detailed here. If chloramphenicol and labeled amino acids were added at the same time, no incorporation of radioactivity into any of the polypeptides was seen. However, if the chloram-

phenicol was washed out and the mycelium was allowed to grow for an hour, incorporation took place principally into the four small subunits of the enzyme. If label was added together with cycloheximide, only one of the mitochondrially synthesized subunits (III) became significantly labeled, but if the cells were preincubated in the presence of chloramphenicol and then labeled in the presence of cycloheximide, incorporation into all three of the mitochondrially synthesized subunits took place (Fig. 19–3).

These results were explained by Schwab and associates (25) and Sebald et al. (26) in terms of precursor pool sizes of the cytochrome oxidase polypeptides. To understand why pool problems are more likely to have affected the *Neurospora* experiments than the yeast experiments, it is important to compare the methods used to isolate the cytochrome oxidase polypeptides. In the *Neurospora* experiments the cytochrome oxidase holoenzyme was isolated by standard fractionation procedures rather than immunoprecipitation. This means that if the pool size of a mitochondrially synthesized polypeptide is rate-limiting the incorporation of all of the component polypeptides into newly made holoenzyme will be affected. In a pulse experiment this will result in an apparent reduction in incorporation into all of the constituent polypeptides of the holoenzyme, which reflects exhaustion of the pool of rate-limiting polypeptide and

cessation of holoenzyme synthesis before newly synthesized radioactive polypeptides have had time to replace the unlabeled polypeptides made prior to the pulse in the non-limiting polypeptide pools. If, on the other hand, immunoprecipitation is used, antibody to holoenzyme should precipitate not only holoenzyme molecules but also the labeled and unlabeled polypeptides in each of the pools, provided this complex antibody preparation has components that recognize each of the polypeptides. Therefore the immunoprecipitation experiment should be relatively independent of the rate-limiting effect of a particular polypeptide on assembly, since it should measure radioactive polypeptides both in the holoenzyme and in the pools. Of course, if there is a tight regulatory interlock on synthesis of the

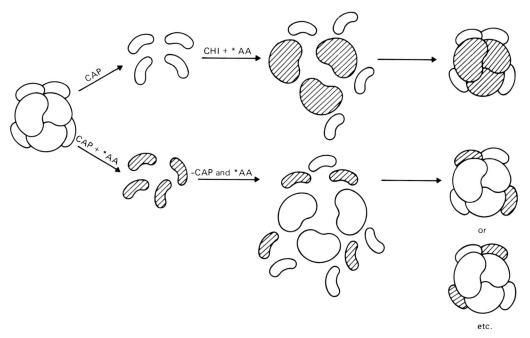

FIG. 19–3. Diagram illustrating how sites of synthesis of cytochrome oxidase subunits were determined in *Neurospora*. The experimental design used to establish the sites of synthesis of the three large mitochondrial subunits is shown at the top. Cells were incubated in the presence of chloramphenicol (CAP) to permit synthesis of unlabeled small cytoplasmic subunits in the absence of synthesis of the three mitochondrial subunits. CAP was then replaced with cycloheximide (CHI) and labeled amino acids (*AA); under these conditions labeled large mitochondrial subunits were made and labeled, but no new small subunits could be synthesized. The lower half of the diagram shows how the site of synthesis of the smaller subunits was determined. Cells were incubated in the presence of CAP and *AA, so that the four small cytoplasmically synthesized subunits became labeled, but the mitochondrially synthesized subunits did not. CAP and *AA were then withdrawn, and unlabeled subunits were made in both cytoplasm and mitochondrion, but the only labeled subunits in the pool were those made in the cytoplasm prior to CAP withdrawal.

component polypeptides, there might also be some sort of rate limitation on polypeptide synthesis that depends on the polypeptide with the smallest pool size.

Schwab et al. (25) showed that in *Neurospora* the pool size of one of the mitochondrially synthesized polypeptides (subunit III) was the smallest of all the polypeptides. This explained why chloramphenicol appeared to block synthesis of all the cytochrome oxidase polypeptides when chloramphenicol and labeled amino acids were added at the same time. When chloramphenicol was washed out following the pulse and incubation continued, the pool size of subunit III could build up again, and the radioactive polypeptides made on cytoplasmic ribosomes during the pulse in the presence of chloramphenicol could be incorporated into holoenzyme.

The cycloheximide inhibition experiments can be explained using similar logic. If cells are pulse-labeled in the presence of cycloheximide, only subunit III becomes labeled. Since this polypeptide has the smallest pool size, isotope will pass into this polypeptide most rapidly, and radioactive subunits will be assembled into intact cytochrome oxidase during the pulse. Since the pool sizes of subunits I and II are larger, radioactive subunits will equilibrate with nonradioactive subunits more slowly. Hence, in the holoenzyme these subunits will label more slowly. If one of the cytoplasmically synthesized polypeptides has the next smallest pool size, it will become the rate-limiting polypeptide for holoenzyme assembly during cycloheximide treatment. The result will be the one observed: only one of the three mitochondrially synthesized subunits will appear to label during the pulse in the presence of cycloheximide because only this polypeptide has a pool smaller than that of the rate-limiting polypeptide made in the cytoplasm. Schwab et al. (25), on the basis of pool size measurements, concluded that subunit VII was probably the rate-limiting cytoplasmic polypeptide.[1] Finally, preincubation with chloramphenicol prior to pulse labeling in the presence of cycloheximide allows all pools of cytoplasmically synthesized cytochrome oxidase polypeptides to build up during the preincubation period, because subunit III, which is made in the mitochondrion, limits the assembly rate. Subsequently, when chloramphenicol has been removed and cycloheximide and labeled amino acids are added together, all three mitochondrially made subunits become labeled because a sufficient pool of the limiting cytoplasmically made polypeptide has been built up during the preincubation period.

We have entered into this lengthy discussion of the experiments of Bücher and his colleagues because they illustrate precisely the kind of trap into which one can fall using inhibitors to assess the sites of synthesis of the polypeptide components of a macromolecular structure such as a membrane or a ribosome. Fortunately, these investigators were fully aware of the effects pools might have on their results; they not only determined the pool size for each polypeptide but also developed methods for correcting for those polypeptides with limiting pool sizes.

Although the *in vivo* studies just described established the sites of synthesis of the cytochrome oxidase polypeptides in yeast and *Neurospora* with reasonable certainty, they did not permit investigation of the mechanism of assembly of these polypeptides nor a direct attack on the mechanism of regulation of their synthesis. Poyton and his colleagues (27–30) succeeded in developing an *in vitro* system using isolated yeast mitochondria that synthesizes subunits I, II, and III of cytochrome oxidase, which previously were shown to be made on mito-

---

[1] This conclusion will have to be reevaluated if it proves that the cytoplasmically synthesized subunits of cytochrome oxidase in *N. crassa* are synthesized as a single large precursor as is the case in yeast *vide infra*.

chondrial ribosomes in the *in vivo* experiments of Mason and Schatz. Groot and Poyton (28) showed that synthesis of these three polypeptides in the *in vitro* system was subject to the same regulation *in vivo* by oxygen that was alluded to earlier by Mason and Schatz. That is, *in vitro* subunits I and II labeled only in the presence of oxygen, whereas subunit III labeled both in the presence and in the absence of oxygen. These results demonstrated directly that oxygen regulation was at the level of the mitochondrion itself. Poyton and Kavanagh (29) subsequently showed that protein synthesis in isolated mitochondria depended on specific cytoplasmic stimulatory factors. These factors proved to be proteins present in a cytoplasmic supernatant from which the polysomes had been removed. The most interesting results of all were obtained with cytochrome oxidase (30). Synthesis of subunits I–III by isolated mitochondria appeared to depend specifically on the presence in the postpolysomal supernatant of the cytoplasmically synthesized cytochrome oxidase polypeptides. If the supernatant was treated with antisera to either subunit IV or subunit VI, incorporation of radioactivity into subunits I, II, and

III was reduced drastically (Table 19–2). Furthermore, the immunoprecipitates from the cytoplasmic supernatants contained a single radioactive polypeptide with a molecular weight of 55,000 d, which is much greater than the molecular weights of polypeptides IV and VI combined (Table 19–1). Examination of the immunologic properties of this polypeptide revealed that it cross-reacted with antisera specific for subunits IV and VI but not with antisera specific for subunits I, II, V, and VII. Despite the fact that cross-reaction was not obtained with polypeptides V and VII, tryptic peptide analysis indicated that the 55,000-d precursor had the same peptide composition as that obtained following trypsin digestion of a mixture of polypeptides IV–VII, with the exception that the precursor contained between four and six additional peptides. Thus the 55,000-d polypeptide is a precursor to all four of the cytoplasmically synthesized cytochrome oxidase polypeptides. Other labeling studies indicated that processing of this precursor takes place after it enters the inner membrane. Entry of the precursor into the membrane also stimulates synthesis of one or all of the mitochondrially made cytochrome

TABLE 19–2. *Removal of stimulatory activity for synthesis of mitochondrially translated subunits of cytochrome oxidase in isolated mitochondria from bakers' yeast by immunotitration of postpolysomal cytoplasmic supernatant with antisera to cytochrome oxidase subunits IV and VI, which are synthesized on cytoplasmic ribosomes*

Cell fraction used to stimulate incorporation of radioactive amino acids by isolated mitochondria	Stimulation (counts per minute per milligram of mitochondrial protein) of incorporation into:	
	Total protein	Cytochrome c oxidase
Postpolysomal supernatant	24,150	2,510
Postpolysomal supernatant titrated with:		
Antiserum against holo-cytochrome c oxidase	20,690	160
Antiserum against subunit IV	21,750	250
Antiserum against subunit VI	19,960	300

From Poyton and Kavanagh (29).

oxidase subunits. Of particular note is the fact that the precursor is found in the post-polysomal supernatant and is able to enter the mitochondrion and the inner membrane directly in the absence of ribosomes bound to the outer mitochondrial membrane. These results show that for cytochrome oxidase, at least, vectorial translation (see Chapter 3) is not the mechanism by which the cytoplasmically translated subunits of the enzyme enter the mitochondrion. Finally, it should be noted that Rabinowitz and his colleagues achieved synthesis of the three cytochrome oxidase subunits synthesized on mitochondrial ribosomes *in vitro* using a cell-free extract from *E. coli* and a poly-A-containing RNA fraction isolated from mitochondria of *S. cerevisiae* (see Chapter 3).

BIOSYNTHESIS OF OLIGOMYCIN-SENSITIVE ATPase

Tzagoloff, in a series of elegant experiments (31), established the sites of synthesis of the numerous polypeptides comprising the oligomycin-sensitive ATPase of yeast. The ATPase of yeast mitochondria consists of three parts (see Chapter 1). The most important of these is the water-soluble ATPase (F_1) that, unlike the entire complex, is not inhibited by oligomycin or rutamycin and is cold-labile. Yeast F_1 has a molecular weight of 340,000 d and is comprised of at least five subunit proteins (Table 19–1). The oligomycin-sensitivity-conferring protein (OSCP) is a single protein that is required for reconstitution of the rutamycin-oligomycin-sensitive ATPase. It may function as a link between F_1 and the other protein components of the complex. The membrane factor contains four highly water-insoluble polypeptides (Table 19–1).

The five polypeptides of the F_1 ATPase are synthesized in cytoplasm. This can be demonstrated in two ways. First, F_1 is present in vegetative petite mutants (32, 33), and these mutants have no mitochondrial protein synthesis (see Chapter 4). Second, Tzagoloff (34,35) showed that synthesis of F_1 was not inhibited by chloramphenicol but was inhibited by cycloheximide. Glucose-repressed cells (see Chapter 4) were inoculated into a low-glucose medium to induce derepression in the presence of chloramphenicol. Under these conditions the ATPase of the mitochondria remained constant, but that of the postribosomal supernatant increased. This unbound ATPase was precipitated by antiserum to F_1. Presumably, it did not become bound to the mitochondrial inner membrane of the derepressing yeast cells because the membrane factor polypeptides were synthesized on mitochondrial ribosomes (*vide infra*). When the chloramphenicol incubation was done in the presence of ^{14}C-leucine, it was found that all five subunits of the F_1 ATPase became labeled. If the same experiment was done in the presence of cycloheximide, no accumulation of F_1 ATPase was seen in the cytoplasm. Furthermore, F_1 did not label in the presence of cycloheximide. The cytoplasmic synthesis of the OSCP was established by Tzagoloff (36) in a similar manner.

The four membrane factor polypeptides appear to be synthesized on mitochondrial ribosomes. To demonstrate this, Tzagoloff (31,37) carried out a sophisticated experiment. If derepressing yeast is first incubated in the presence of chloramphenicol and is then labeled in the presence of cycloheximide, the mitochondrially synthesized polypeptides label more completely than if the chloramphenicol step is omitted. Presumably, the chloramphenicol preincubation prevents pools of mitochondrially synthesized polypeptides from building up, but at the same time it permits synthesis of their cytoplasmic partners. Cells were labeled using this inhibitor protocol, following which mitochondrial particles were extracted from them (Fig. 19–4). One part of the sample was extracted directly with the nonionic detergent Triton, and the ATPase was precipitated with antiserum to the entire rutamycin-sensitive ATPase complex. Four radioactive polypeptides were found following gel electrophoresis; these must have been made on mitochondrial ribosomes. Antiserum to F_1 did not precipitate this portion of the ATPase, which indicated that any F_1 synthesized during the chloramphenicol preincubation had not become attached to the membrane factor polypeptides synthesized during the pulse labeling part of the experiment in the presence of cycloheximide.

The second part of the experiment demonstrated that the four mitochondrially synthesized membrane factor polypeptides were, in fact, part of the rutamycin-sensitive ATPase. In this part of the experiment

the labeled particles were complexed with unlabeled F_1 and OSCP prior to Triton extraction. One part of this sample was purified on a glycerol gradient, and the component polypeptides were subjected to gel electrophoresis. All 10 polypeptides of the complete complex were observed, but only the four previously mentioned membrane factor polypeptides were labeled. This observation confirmed the supposition that the Triton-extracted ATPase particles that were precipitated by antiserum in the first part of the experiment were deficient in F_1 and OSCP. The second portion of the sample was precipitated with antiserum to F_1. Once again the whole complex was precipitated, and the only labeled polypeptides were the same four from the membrane factor. Thus the membrane factor became so tightly complexed with the added

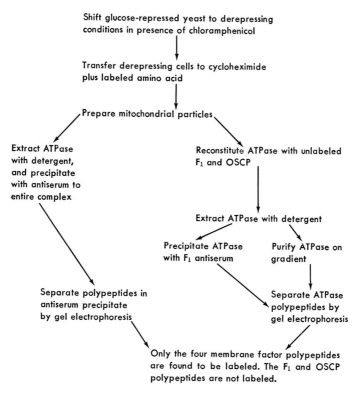

FIG. 19–4. Diagram illustrating experimental design used by Tzagoloff and his colleagues to demonstrate that the four membrane factor polypeptides of the oligomycin-sensitive F_1 ATPase were the only polypeptides of the complex made on mitochondrial ribosomes in bakers' yeast. (Adapted from Tzagoloff et al., ref. 31.)

F_1 and OSCP that it could be precipitated with antiserum to F_1 alone.

These experiments showed conclusively that the membrane factor polypeptides were made on mitochondrial ribosomes and that F_1 and OSCP were tightly bound to the membrane factor. Once again the problem of where the membrane begins and the enzyme ends was dealt with in an ingenious fashion.

BIOSYNTHESIS OF CYTOCHROMES b, c_1, AND c

Weiss (38–40) studied the biosynthesis of cytochrome b in *Neurospora*. This enzyme contains one or possibly two nearly identical polypeptides of approximately 28,000 d molecular weight whose synthesis is inhibited by chloramphenicol but not cycloheximide. By this criterion these polypeptides must be made on mitochondrial ribosomes. Two other polypeptides with molecular weights of 10,000 and 11,000 d were also originally reported by Weiss (38) to be associated with cytochrome b. Both are synthesized on cytoplasmic ribosomes. More recently, Weiss (39) reported that extensive gel filtration of cytochrome b in the presence of bile salts completely removes the 11,000-d polypeptide; the 10,-000-d polypeptide is also partially removed by this treatment. These results suggest that the two polypeptides are probably contaminants. Lin and Beattie (41) published results for yeast substantially in agreement with those of Weiss. They found that cytochrome b contained a single 28,000-d polypeptide that also appeared from inhibitor experiments to be synthesized on mitochondrial ribosomes.

Ross et al. (85) purified cytochrome c_1 from yeast and studied the biosynthesis of the single polypeptide moiety. This polypeptide has a molecular weight of 31,000 d, and the heme is covalently linked to it (Table 19–1). Cytochrome c_1 is extremely sensitive to proteolysis, which very likely accounts for an earlier report that the molecular weight of this polypeptide was 27,000 d (10). Ross and Schatz (42) showed in pulse labeling experiments that cycloheximide (but not acriflavine) blocked labeling of this polypeptide. Since acriflavine acts as a specific inhibitor of mitochondrial protein synthesis in yeast (43), these results indicate that the polypeptide is made on cytoplasmic ribosomes. It was thought for a long time that cytochrome c_1 was absent from vegetative petites, which, of course, lack mitochondrial protein synthesis (see Chapter 4). This led Ross et al. (10) to speculate that it might be necessary for the cytochrome c_1 polypeptide to combine with another polypeptide (x) synthesized on mitochondrial ribosomes before it could accept its heme group. However, Ross and Schatz (42) recently identified cytochrome c_1 in vegetative petites by immunoprecipitation and showed that the protein precipitated contained a covalently linked heme group. In short, it is no longer necessary to invoke a special protein (x) that is synthesized on mitochondrial ribosomes and is responsible for attachment of heme to cytochrome c_1 protein. The polypeptide associated with iso-1-cytochrome c is obviously made on cytoplasmic ribosomes, since vegetative petite mutants contain this cytochrome (see Chapter 4).

Although we have stressed, as have most investigators, the contribution of mitochondrial protein synthesis to inner-membrane formation (see summary in Table 19–1), it is clear from freeze-fracture studies with vegetative petite mutants that these mutants not only have a normal outer mitochondrial membrane, but a distinct inner membrane that does not differ greatly in its morphology from that of the wild type (44). Therefore, many components of the inner membrane must be made on cytoplasmic ribosomes. Now that high-resolution two-dimensional electrophoretic techniques are available for the analysis of membrane proteins (45), it is of interest to compare the inner mitochondrial membranes of vegetative petite and wild-type yeast just to get an idea of how many of the total inner-membrane polypeptides are made on cytoplasmic ribosomes. Ultimately this analysis will have to be extended to pulse labeling experiments with wild-type cells in the presence of specific inhibitors. With these kinds of experiments one could obtain a "global" estimate of the total number of inner-membrane polypeptides and the total numbers made in the cytoplasmic and mitochondrial compartments. These could then be compared with the polypeptides that have already been identified through purification of the specific electron transport and ATPase complexes to yield a complete polypeptide profile of the inner membrane.

GENETICS OF THE INNER MEMBRANE

A concerted effort is now being made to localize the structural genes for the inner-membrane polypeptides in *S. cerevisiae*. To this end, both nuclear and mitochondrial gene mutations are being characterized. Until recently, only nuclear gene mutations, the segregational petites, were known to cause specific inner-membrane defects; all respiratory-deficient mitochondrial mutants were vegetative petites. Now, however, respiratory-deficient point mutants (*mit* mutants) are available (see Chapter 5), and they have been extensively mapped using techniques discussed earlier (see Chapter 7). In addition, both nuclear and mitochondrial mutants resistant to inhibitors of electron transport and oxidative phosphorylation are available (see Chapter 5). With the acquisition of these mutants, prospects are very good that the structural genes for many of the inner-membrane polypeptides will be localized in the near future. The first part of this discussion will consider the nuclear mutants, and the second part will consider the mitochondrial mutants. The approaches taken in the two

sections are different. In discussing the nuclear mutants, most of which are segregational petites, we shall consider all of the mutants as a group without attempting to indicate in more than a general sense how many genes have been found to be involved in the formation of cytochrome oxidase, cytochrome b, etc. The reason for doing this is that the specific gene products affected by segregational petite mutants have not been studied in much detail, with the exception of the mutants deficient in cytochrome c. In addition, little attempt has been made to determine by mapping experiments, fine structure analysis, and complementation tests how many genes can be mutated so that they confer specific defects on inner-membrane components. In contrast, detailed genetic and biochemical characterization of the mitochondrial mutations affecting inner-membrane structure and function is in progress, and it is evident that mutants having distinctly different phenotypes (e.g., *mit⁻* and inhibitor resistance) may map within the same gene. Therefore the second part of the discussion involving the mitochondrial genes is written from the point of view of mitochondrial inner-membrane functions affected in specific groups of mitochondrial mutants.

Nuclear Mutations Affecting Mitochondrial Inner-Membrane Function

Most of the early biochemical and genetic work on segregational petites was done by Sherman (46,47), who examined the properties of a series of segregational petites designated pet_1 through pet_7 as well as ly_6 and ly_8. The latter two mutations, in addition to being respiration-deficient, require lysine and glutamate for growth, and they are deficient in the enzyme aconitate hydratase. Three of the mutants (pet_3, ly_6, and ly_8) produced only petite diploids in the retention test (see Chapter 4), and thus they acquired secondary vegetative petite mutations so rapidly as to be useless for biochemical analysis. Sherman (46) and Roodyn and Wilkie (48) suggested that these mutations might control the retention of ρ. All the remaining *pet* mutants produced some ρ^- cells in the retention test; of these, *pet*$_5$, *pet*$_6$, and *pet*$_7$ produced a predominance of ρ^+ cells and were suitable for biochemical analysis. Experiments with these mutants revealed that all three had pleiotropic effects on respiratory phenotype. All three mutants, like the vegetative petites, lacked cytochrome a+a$_3$, but only two lacked cytochrome b. Cytochrome c was found in excess in all three mutants, as it is in vegetative petites. The *pet*$_7$ and *pet*$_5$ mutants appeared to have morphologically normal mitochondria, although modifications in mitochondrial morphology were evident in *pet*$_6$ (49).

Sherman then turned his attention specifically to mutations deficient in cytochrome c [for review, see Sherman and associates (50)]. He found that such mutants are mendelian in inheritance and fall into at least six distinct loci. Aerobically grown yeast cells contain two forms of cytochrome c designated iso-1-cytochrome c and iso-2-cytochrome c, with the former constituting 95% of the total cellular cytochrome c. Sherman provided unequivocal evidence that the CY_1 gene is the structural gene for iso-l-cytochrome c. Mutations in the other five genes led to reductions in the amounts of both cytochromes. There is some evidence that the CY_2 and CY_3 genes may be directly involved in cytochrome c synthesis and that the CY_4, CY_5, and CY_6 genes may only indirectly influence cytochrome c content. These experiments constitute direct demonstration that a nuclear gene codes for the polypeptide moiety of cytochrome c, the message for which is translated on cytoplasmic ribosomes.

Since Sherman's original studies, a number of other investigators have reported on the properties of different segregational petite mutants (51). By far the most extensive and most detailed of these investigations were those initiated several years ago

by Schatz and more recently by Tzagoloff. Schatz and his colleagues characterized 18 segregational petite mutants (18,52–54). These were obtained by mutagenizing wild-type cells and testing the mutagenized cells for their ability to grow on a nonfermentable substrate. Those mutants that failed to do so were crossed to a ρ^- mutant to see if they would complement. Segregational petites (*pet*) usually complement ρ^- mutants because the ρ^- mutant contains a wild-type (*PET*) nuclear gene and the *pet* mutant normally contains a ρ^+ factor (see Chapter 4). Vegetative petites, on the other hand, cannot complement the ρ^- tester, for they are, by definition, either ρ^- or ρ^0.

Once the *pet* mutants were distinguished from the vegetative petites, it was possible to characterize them biochemically and genetically. It was, of course, important to use only mutants that did not accumulate new ρ^- mutants at high frequency. The 18 mutants studied fulfilled this criterion. When pairwise crosses were made between the mutants, it was found that they fell into seven complementation groups (Table 19–3). It was shown by the use of specific inhibitors that all of the mutants had functional mitochondrial protein-synthesizing systems. Biochemical characterization permitted classification of the mutants into four general classes. In the first class, cytochrome a+a$_3$ and cytochrome oxidase activities were selectively lost (Table 19–3). Mutants belonging to complementation groups II, III, and IV fell into this class. The second class included mutants that had lost cytochrome a+a$_3$ and cytochrome oxidase activities and were also greatly diminished in the activities of succinate cytochrome c reductase and ATPase. Mutants belonging to complementation groups I and V fell into this class. Class III mutants were even more pleiotropic than those of class II, and they lacked cytochromes b and c$_1$ as well. Mutants of complementation group VII belonged to this class. The final class of mutants was similar to class III, except that these mutants were even more

TABLE 19–3. *Mitochondrial defects present in segregational petite mutants retaining mitochondrial protein synthesis*

Phenotypic defect	Ebner and Schatz (52)		Tzagoloff et al. (60)	
	Total mutants	Total complementation groups	Total mutants	Total complementation groups
Cytochrome $a + a_3$	8	3	20	10
Coenzyme Q	—	—	14	7
Coenzyme QH_2–cytochrome c reductase	—	—	18	9
Oligomycin-sensitive ATPase	—	—	3	3
Cytochrome $a + a_3$, ATPase $^{32}P_i$-ATP exchange, succinate cytochrome c reductase	3	2	—	—
Loss of cytochromes $a + a_3$, b, c_1, succinate cytochrome c reductase, $^{32}P_i$-ATP exchange	6	1	—	—
Loss of cytochromes $a + a_3$, b, c_1, $^{32}P_i$-ATP exchange, ATPase	1	1	—	—

deficient in ATPase-related functions. Mutations belonging to complementation group VI fell in this class.

Although Ebner and Schatz (53) examined the ATPase of one of the class VI mutants in detail, showing that it lacked F_1, the principal focus of the biochemical studies was on certain of the cytochrome-oxidase-deficient mutants, in particular a mutant called *pet* 494. Ebner et al. (18) found that the five complementation groups previously defined in which cytochrome oxidase alone or cytochrome oxidase plus succinate cytochrome c reductase were missing corresponded to five unlinked nuclear genes. Three mutants were chosen for further biochemical characterization. The *pet* Ell and *pet* 1030 mutants were reported to be incapable of synthesizing the three mitochondrial subunits of cytochrome oxidase, although the cytoplasmic subunits were made in normal amounts; more recently it has been found that *pet* Ell retains subunit III (54). Neither of these mutants really poses a paradox to the dictum that mitochondrially synthesized polypeptides are coded by the mitochondrial genome, since they are pleiotropic and may simply affect the regulation or assembly of these polypeptides by mechanisms yet to be re-

vealed that are under nuclear gene control. On the other hand, *pet* 494 is another story. This mutant lacks specifically subunit III of cytochrome oxidase (54), which is a product of mitochondrial protein synthesis. It is possible that this mutant could be explained in either of two ways. It might affect a gene, the product of which is synthesized on cytoplasmic ribosomes and is required for the assembly or processing of polypeptide III. Alternatively, it might be the ultimate embarrassment—a mutation in a nuclear gene, the message from which is translated on mitochondrial ribosomes. If so, polypeptide III would be coded in the nucleus.

Ono et al. (55) constructed an elaborate genetic argument that strongly suggested that *pet* 494 was not a mutation in the structural gene for subunit III. The elements of the argument are as follows: Supersuppressor mutants were isolated in yeast some years ago by Hawthorne and Mortimer (56). These mutants, although allele-specific, were not locus-specific. Sherman and his colleagues (57,58) in several definitive studies were able to show through the use of a series of suppressible mutants in the iso-1-cytochrome c structural gene that the yeast supersuppressors fell into two groups and corresponded to the bacterial nonsense suppressors. One group suppressed the UAA termination codon (ochre) and the other the UAG termination codon (amber). In *E. coli* and its bacteriophages, nonsense suppressors cause alterations in specific tRNAs that permit translation of the nonsense codons (59).

The *pet* 494 mutant can either revert or be suppressed. When this mutant becomes suppressed, known amber mutants introduced into the stock become suppressed at the same time. Therefore *pet* 494 is an amber mutant. The amber suppressor genes, like the *pet* 494 gene, are nuclear. The authors argued that if amber suppressors in yeast alter specific tRNAs, as they do in bacteria, then the gene product of the *pet* 494 gene cannot be subunit III. The reason is that the mitochondrion has its own complete set of tRNAs that are quite distinct from their cytoplasmic counterparts, which are coded by nDNA. According to this reasoning, the amber suppressor of *pet* 494 alters a cytoplasmic tRNA; thus the authors argued the product of the *pet* 494 gene was synthesized in the cytoplasm and thus could not be subunit III. Another less palatable possibility is that the particular tRNA in question is present in both cytoplasm and mitochondrion, although it does not hybridize with mtDNA because it is a nuclear gene product. This seems a possible loophole in an ingenious argument.

Tzagoloff et al. (60) reported preliminary characterization of another large group of segregational petite mutants (Table 19–3) using a selection procedure described previously (p. 164). The segregational petites chosen for biochemical characterization produced vegetative petite double mutants at a low rate and fell into three general groups. Thirty-three mutants were found to be deficient in NADH–cytochrome c reductase. These mutants had from 20 to 100% of the normal levels of cytochrome oixdase and the rutamycin-sensitive ATPase and could be divided into two subgroups

(Table 19–3). The first group consisted of 18 mutants in which respiratory chain activity could not be restored by addition of exogenous coenzyme Q. These mutants fell into nine complementation groups. Respiratory chain activity could be restored by addition of coenzyme Q in the second group of mutants. There were 14 of these falling into seven complementation groups. The second biochemically distinct group of mutants was deficient in cytochrome oxidase. There were 20 such mutants belonging to 10 complementation groups (Table 19–3). In the final group there were three mutants belonging in three complementation groups that were deficient in ATPase. Two mutants had no detectable F_1 activity, and F_1 could not be precipitated with antiserum.

Perhaps the most interesting preliminary observations to come out of this study concern the inability of certain mutants to synthesize specific products of mitochondrial protein synthesis. The third ATPase mutant (N9–168) contained soluble active F_1, but it was unable to synthesize one of the membrane factor polypeptides (subunit 9) that is made on mitochondrial ribosomes. Two mutants (N5–26 and N6–70) that were specific for coenzyme QH_2–cytochrome c reductase were missing a major mitochondrial polypeptide that appeared to be the one associated with cytochrome b. Mutants similar to *pet* 494 lacking subunit III of cytochrome oxidase also appeared to have been isolated. In each case a product of mitochondrial protein synthesis was absent in a nuclear gene mutant. It would be of great interest to know why this was so, as it is unlikely that any of these nuclear genes are structural genes for the polypeptides in question. The structural genes most probably lie in the mitochondrial genome, as we shall see in the next section.

In addition to segregational petite mutants, it is possible to isolate nuclear mutations resistant to specific inhibitors of respiration or oxidative phosphorylation.

One particularly interesting mutation deserves mention here. Agsteribbe et al. (61) reported the isolation of nuclear mutations resistant to aurovertin, an antibiotic that binds to F_1 and inhibits its function. These mutants fell into two phenotypic classes. Mutants of class I, of which only two were obtained, had an aurovertin-resistant ATPase; mutants of class II, which were in the vast majority, had an aurovertin-sensitive ATPase, and presumably they altered the accessibility of the ATPase in the cell without altering the ATPase itself. Resistance in the class I mutants could be traced to F_1 itself, although convincing evidence for alteration of a specific F_1 polypeptide remains to be obtained. It seems quite likely that future studies will prove that the class I mutants affect a structural gene for one of the F_1 polypeptides, which would be completely consistent with the synthesis of these polypeptides in the cytoplasm.

Mitochondrial Mutations Affecting Inner-Membrane Structure and Function

Mitochondrial mutations affecting cytochrome oxidase, complex III of the respiratory chain, and the oligomycin-sensitive F_1 ATPase have been isolated and mapped. Certain mutants confer specific deficiencies; others confer inhibitor resistance; still others are pleiotropic. In some instances, as in the case of mutants affecting complex III, all three phenotypes map in what may be the same gene.

Cytochrome oxidase. Thus far, the mutants affecting cytochrome oxidase synthesis specifically have all been of the *mit⁻* phenotype and have fallen in three loci designated *OXI1, OXI2,* and *OXI3* (see Table 5–2). Slonimski and Tzagoloff (62), who isolated these mutants and did much of the preliminary mapping, attempted to ascertain which polypeptides synthesized in the mitochondria were altered in (or were

absent from) mutations in these genes. They labeled cells with $^{35}SO_4$ in the presence of cycloheximide and displayed the labeled polypeptides on SDS polyacrylamide gels. Under these conditions four radioactive polypeptides were resolved and were designated bands 1 through 4, with band 1 having the highest molecular weight and band 4 the lowest. All of these polypeptides labeled in the *OXI1* and *OXI2* mutants, but band 2, which corresponded to the highest-molecular-weight subunit of cytochrome oxidase, was unlabeled in the *OXI3* mutants. Cabral et al. (63), using the same technique, but employing gels having higher resolution, extended these results. In particular, they provided strong evidence that the *OXI1* gene is the structural gene for subunit II of cytochrome oxidase.

The gel technique of Cabral and associates visualizes about eight radioactive bands (Fig. 19–5), of which six have been correlated with specific inner-membrane polypeptides, including all three of the mitochondrially synthesized subunits of cytochrome oxidase. Analysis of the mitochondrial translation products of the *OXI1* mutants revealed that all except two lacked a specific polypeptide of molecular weight 33,500 d that had been identified as subunit II of cytochrome oxidase (Fig. 19–6). However, several of these mutants contained an extra labeled band whose apparent molecular weight was between 28,000 and 31,000 d, and occasionally other extra polypeptides were found with molecular weights between 10,000 and 20,000 d (Fig. 19–6).

Cabral and associates found that the polypeptides of molecular weight 28,000 to 31,000 d that were present in some (but not all) *OXI1* mutants had indeed replaced the wild-type subunit II polypeptide, because these mutant polypeptides cross-reacted with antisera prepared against subunit II. However, no cross-reaction was found with the low-molecular-weight extra polypeptides, which indicated either that

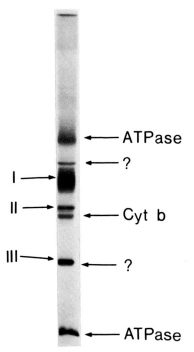

FIG. 19–5. Autoradiogram of SDS polyacrylamide gel showing mitochondrial polypeptides of bakers' yeast labeled
with $^{35}SO_4^{2-}$ in the presence of cycloheximide, which blocks
cytoplasmic protein synthesis. The products of mitochondrial
translation include the apoprotein of cytochrome b and
subunits I, II, and III of cytochrome oxidase. Two of the
membrane factor polypeptides of the ATPase are also
visualized. (Courtesy of Dr. G. Schatz.)

these polypeptides were new species unrelated to subunit II or that they had lost
the antigenic specificity characteristic of
subunit II. Revertants were then obtained
of two mutants that clearly contained a
bona fide variant of subunit II. These revertants, of course, had cytochrome oxidase
activity, but the revertant subunit II species
varied widely in electrophoretic mobility in
different reversions. However, the revertant
polypeptides always migrated more rapidly
than native subunit II, but more slowly than
the subunit II present in the mutants. That
is, they were smaller than wild-type subunit
II, but larger than the subunit characteristic of the mutants. The identity of these
revertant polypeptides with subunit II was
established once again by immunologic
techniques. By means of careful electro-

phoretic studies, Cabral and associates also
showed that there was no evidence for posttranslational modification of the revertant
polypeptides and that limited proteolysis of
revertant and wild-type subunit II polypeptides yielded the same fragment pattern,
with the exception of a single fragment that
was smaller in the revertants by the amount
expected from electrophoresis of the un-
digested subunits.

Cabral and associates considered two
general hypotheses to explain their results.
The first was that the *OXI1* locus coded for
an enzyme that either modified subunit II
specifically or cleaved it proteolytically.
This hypothesis was rejected on the basis of
the studies mentioned previously that dealt
with the mobility of the revertant polypeptides in SDS gels and examination of the
fragments produced as a result of limited
proteolysis. In essence, these studies gave
no indication that the revertant polypeptides
had been modified so that their mobility
had been increased in gels nor that they
were being chewed up in some nonspecific
fashion by a protease. The second hypothesis was that the *OXI1* locus was the structural gene for subunit II; thus mutations at
this locus altered subunit II so that it became more susceptible to proteolysis or
posttranslational modification, or, alternatively, such events were not involved, and
the mutant polypeptides as well as the
revertant polypeptides were simply shorter
than normal. Cabral and associates believed the latter possibility was the most
likely, since posttranslational modification
could be ruled out on the same grounds as
for the first hypothesis, although proteolysis
was more difficult to eliminate as an explanation for the differing sizes of the mutant and revertant polypeptides.

Finally, Cabral and associates mentioned briefly some studies of the polypeptides present in mutants at the *OXI2* and
OXI3 loci. They found, in agreement with
Slonimski and Tzagoloff, that *OXI3* mutants specifically had lost subunit I of cyto-

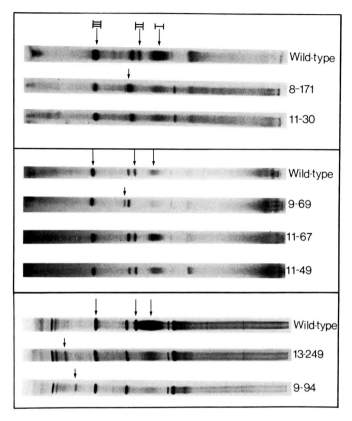

FIG. 19–6. Mutations in OXI1 locus of mitochondrial genome of S. cerevisiae specifically affect cytochrome oxidase subunit II, as shown in this comparison of the mitochondrial translation products of wild-type cells and several OXI1 mutants. The conditions of gel electrophoresis are described in the legend for Fig. 19–5. Roman numerals identify cytochrome oxidase subunits I–III. Notice the absence of subunit II in the mutants and the fact that extra mitochondrial translation products are found in certain of the mutants (small arrows). (From Cabral et al., ref. 63, with permission.)

chrome oxidase. Their results also indicated that *OXI2* mutants were lacking subunit III. These results are summarized in Fig. 19–7, which shows the tentative coding assignments for each of the *OXI* genes with respect to the mitochondrially synthesized subunits of cytochrome oxidase.

Complex III. Complex III of the respiratory chain includes cytochromes b and c_1 (see Chapter 1). This segment of the chain catalyzes the oxidation of ubiquinone (coenzyme Q) by cytochrome c, and its activity is measured by the coenzyme QH_2–cytochrome c reductase assay. Genes in a unique region of the mitochondrial genome appear to control the assembly and function of complex III. Mutations in this region can lead to specific loss of cytochrome b (*COB*)

and, under conditions of catabolite repression, loss of cytochrome b and cytochrome oxidase (*BOX*), as well as resistance to inhibitors such as antimycin (*ANA*), funiculosin (*FUN*), mucidin (*MUC*), and diuron (*DIU*) that act on this part of the respiratory chain (Fig. 19–8). Although mapping experiments show that all of these phenotypes map in this region of the mitochondrial genome, it is not clear how many genes are involved. Studies done on this region illustrate perhaps better than anything else the enormous disadvantage under which the organelle geneticist must labor in the absence of a suitable complementation test for gene function and in the absence of tests for dominance and recessiveness.

Tzagoloff et al. (64,65) reported that

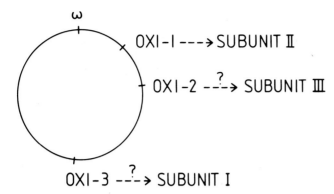

FIG. 19–7. Summary diagram of present evidence concerning which of the three mitochondrial *OX1* genes are structural genes for each of the three mitochondrially translated subunits of cytochrome oxidase in bakers' yeast. (From Cabral et al., ref. 63, with permission.)

mit⁻ mutants deficient in cytochrome b mapped in two loci designated *COB1* and *COB2*. All *COB1* mutants lacked a major mitochondrial translation product that was identical with the apoprotein of cytochrome b (Fig. 19–5). Mutants in the *COB2* locus were of two kinds. Some were similar to *COB1* mutants and lacked the cytochrome b polypeptide; but two, although lacking this polypeptide, had a new translation product that migrated more rapidly than the cytochrome b polypeptide. Tzagoloff and associates speculated from these results that *COB2* might be the structural gene for the cytochrome b apoprotein and that the smaller polypeptide seen in two of the mutants might result because of nonsense mutations in the *COB2* gene that led to

premature chain termination during synthesis of the cytochrome b apoprotein. More recently, this assignment has been disputed, as we shall see shortly.

Kotylak and Slonimski (66) reported a second group of *mit⁻* mutants that mapped in the complex III region of the mitochondrial genome (*BOX*) and seemed to be deficient simultaneously in cytochrome b and cytochrome oxidase. These mutants mapped in four clusters, which they called *BOX1* through *BOX4*. Pajot et al. (67) investigated the effects of catabolite repression and derepression on the *BOX* mutants. They found that whereas mutants in all of the clusters lacked cytochrome b, cytochrome oxidase was present in certain of the mutants under conditions of catabolite

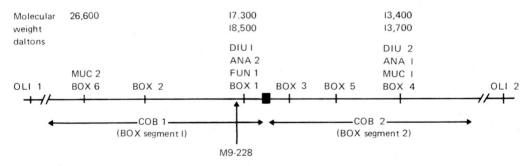

FIG. 19–8. Diagrammatic genetic map of COB-BOX region of mitochondrial genome of *S. cerevisiae*. The six clusters of *BOX* mutants fall into the *COB1* and *COB2* loci (or *BOX1* and *BOX2* segments). In addition, *COB* mutants (not shown except for M9-228) map in these two loci. Mucidin-resistant mutants (*MUC1* and *MUC2*) map in the *BOX4* and *BOX6* clusters; diuron-resistant mutants (*DIU1* and *DIU2*) map in the *BOX1* and *BOX4* clusters; antimycin-A-resistant mutants (*ANA1* and *ANA2*) map in the *BOX4* and *BOX1* clusters; funiculosin-resistant mutants (*FUN1*) map in the *BOX1* cluster. The flanking *OLI1* and *OLI2* genes are also shown. The molecular weights given at the top of the diagram indicate the sizes of putative cytochrome b apoprotein fragments made by specific mutants in the *BOX1*, *BOX4*, and *BOX6* clusters. (Adapted from the work of Slonimski and colleagues, refs. 73–75.)

derepression. This was true for the *BOX1* and *BOX2* mutants and to a slight extent for the single *BOX4* mutant, but catabolite derepression of cytochrome oxidase was not found for the *BOX3* mutants. At the same time, Linnane and his colleagues (68) reported similar observations for certain of their mutants deficient in cytochrome b.

Mitochondrial mutations resistant to antimycin A that mapped in the *COB* region were first reported by Groot Obbink et al. (69) and shortly thereafter by other authors (70,71). Lang et al. (70) published evidence for complex III resistance to antimycin A *in vitro* in the mutants, and Pratje and Michaelis (71) found that their *ANA* mutants were located in two clusters or loci that they designated A_I and A_{II}, with these two loci being separated by about 8% recombination. Pratje and Michaelis also reported that two mutants resistant to funiculosin, another inhibitor that acts on the cytochrome b-c_1 segment of the respiratory chain, mapped in their A_{II} locus. Mitochondrial mutations resistant to mucidin, a third inhibitor acting on this part of the respiratory chain, were reported by Šubik (72) to map in two loci that on the basis of recombination analysis appeared not to be closely linked.

The genetics of the complex III region of the mitochondrial genome of *S. cerevisiae* have been the subject of a series of recent papers by Slonimski and his colleagues (73–76). Their results show that all of the mutants just discussed can be assigned to two adjacent segments, as defined by deletion mapping experiments using a selected series of ρ^- mutants that behave as discriminating testers (see Chapter 7). The first segment is equivalent to the *COB1* locus and the second to the *COB2* locus (Fig. 19–8). Within each segment there are mutants with the *COB* phenotype and the *BOX* phenotype. Recombination analysis shows that the *BOX* mutants fall into six clusters called *BOX1* through *BOX6*, with *BOX1*, *BOX2*, and *BOX6* falling within

segment one and *BOX3, BOX4,* and *BOX5* falling within segment two. One mutant, M9–228, deserves special attention. This was mapped as a *COB2* mutant by Tzagoloff et al. (65), but according to Kotylak and Slonimski (73) it mapped within *COB1*. The significance of this was that M9–228 was one of the two mutants that led Tzagoloff and associates to propose that *COB2* was the structural gene for the cytochrome b apoprotein (p. 530). On the basis of the new map position for M9–228, as well as other results to be discussed below, Kotylak and Slonimski argued that *COB1* was the structural gene for cytochrome b apoprotein, but as we shall see shortly, it is presently not clear whether there are one or three structural genes coding for this polypeptide. In any event, Kotylak and Slonimski did not remap the second mutation on which Tzagoloff and associates based their conclusion; thus, for the moment, one of the two mutants remains assigned to *COB2*.

The inhibitor-resistant mutants are also localized within the two segments of the complex III region of the mitochondrial genome (Fig. 19–8). They fall into three clusters, with the mutants in each cluster being closely linked or allelic with specific *BOX* mutants. The *BOX4* cluster includes mutants that previously were used to define mitochondrial genes for diuron resistance (*DIU2*), antimycin A resistance (*ANA1*), and mucidin resistance (*MUC1*). The *BOX1* cluster includes the *DIU1* and *ANA2* loci as well as the funiculosin resistance locus (*FUN1*). It should be noted that evidence for probable allelism of the *ANA2* and *FUN1* mutants was provided earlier by Pratje and Michaelis (p. 71). The *BOX6* cluster includes the *MUC2* mutants. Thus these two segments of complex III of the mitochondrial genome of *S. cerevisiae* can be mutated either to inhibitor resistance or to the *COB* or *BOX* phenotypes.

The two complex III segments and the clusters of mutants within them do not correspond to genes or cistrons in classic genetic parlance because complementation tests cannot be done between mutants in these clusters. Claisse et al. (75) found that in specific mutants falling in the *BOX1*, *BOX4*, and *BOX6* clusters the cytochrome b apoprotein was replaced by new translation products of lower molecular weight. These have not yet been identified immunologically as fragments of the cytochrome b apoprotein, but this is a reasonable inference. If so, the entire gene coding for this polypeptide might encompass a good part (if not all) of the *COB1* and *COB2* loci. The new fragments found in two *BOX4* mutants have molecular weights of 13,400 and 13,700 d; in two *BOX1* mutants the molecular weights of the new fragments are 17,300 and 18,500 d; in the one *BOX6* mutant the new fragment has a molecular weight of 26,600 d. Claisse and associates pointed out that these results can be explained by assuming that there is a single structural gene for the cytochrome b apoprotein that is read from *BOX4* to *BOX6*. Mutants falling in these clusters that made new fragments would then be nonsense mutations that would cause polypeptide chain termination.

Although this hypothesis seems quite attractive to this author, Claisse and associates preferred another that assumes that there are three structural genes for the cytochrome b apoprotein. Their reasons were as follows: First, the recombination frequency between the *BOX1* and *BOX4* clusters is much higher than would be expected if mutants in these clusters coded for the same polypeptide. However, it is possible that this region is one in which recombination is much higher than normal, for one reason or another. Second, Pajot et al. (76) presented evidence that the *BOX2* and *BOX3* clusters, which map between *BOX4* and *BOX6* (Fig. 19–8), may be involved in assembly processes rather than coding for the cytochrome b apoprotein. Third, the *BOX1*, *BOX4*, and *BOX6* clusters (which are the ones containing mutants that

synthesize modified translation products, presumably fragments of the cytochrome b apoprotein) are also the clusters in which the inhibitor-resistant mutants map (Fig. 19–8). None of these reasons seems particularly persuasive to this author as grounds for rejecting the model of one structural gene in favor of a three-gene model. It also seems to this author that the three-gene model does not explain the fact that certain mutants lack detectable cytochrome b apoprotein, unless some rather questionable ad hoc assumptions are made. It is of interest that *BOX* mutations mapping in segment one seem to have a more moderate effect on cytochrome oxidase than those mapping in segment two.

Oligomycin-sensitive F₁ ATPase. The four polypeptides of the membrane factor of the F_1 ATPase are products of the mitochondrial protein-synthesizing system (Table 19–1); thus it is reasonable to suppose that the structural genes coding for these polypeptides lie in the mitochondrial genome. Several mitochondrial genes that contain mutations that seem to affect the properties of the oligomycin-sensitive F_1 ATPase have been identified. Also identified are four loci with mutations that cause resistance to oligomycin, a specific inhibitor of the ATPase. The *OLI1* and *OLI3* loci are very tightly linked; in fact, they may be a single locus (77,78). These loci are unlinked to the *OLI2* and *OLI4* loci, which themselves are tightly linked (79). Foury and Tzagoloff (80) also reported *mit⁻* mutants that cause specific loss of activity of the oligomycin-sensitive F_1 ATPase. These mutants map in a locus called *PHO1* that is very tightly linked to *OLI2*.

The *OLI1* mutants are much more resistant to oligomycin *in vivo* than the *OLI2* mutants, and this difference is reflected at the level of the ATPase *in vitro* (81,82). The single mutant at the *OLI4* locus is similar to the *OLI2* mutants in this respect (79). Tzagoloff et al. (83) published some suggestive evidence concerning the possible site of action of the *OLI1* mutants. They found that the proteolipid from wild-type mitochondria that contained subunit 9, a polypeptide synthesized on mitochondrial ribosomes (Table 19–1), had a molecular weight of 45,000 or 8,000 d, depending on how the ATPase complex was extracted. When the complex was dissolved directly in SDS, the molecular weight was 45,000 d; if the mitochondria or the ATPase complex itself was pretreated with strong acid or base or with organic solvents, the protein migrated with an apparent molecular weight of 8,000 d. In the case of several *OLI1* mutants tested, the polypeptide was already in the low-molecular-weight form on dissolution of the ATPase complex in SDS. This was not true of a mutant tested that mapped in the *OLI2* locus. Thus there is a clear correlation between the properties of the subunit 9 proteolipid and mutation to oligomycin resistance at the *OLI1* locus.

The *PHO1* mutants, which are tightly linked to the *OLI2* locus, lack significant oligomycin-sensitive F_1 ATPase activity in submitochondrial particles, although F_1-specific ATPase activity can be demonstrated in the high-speed supernatant fraction from which mitochondria and polysomes have been removed. These results suggest that the *PHO1* mutants probably alter the membrane factor such that it does not bind F_1 effectively. It does not seem inconceivable that the *PHO1* and *OLI2* phenotypes are different manifestations of mutations at the same locus.

CONCLUSIONS

The sites of synthesis of the inner-membrane polypeptides belonging to specific complexes of the respiratory chain, as well as the oligomycin-sensitive F_1 ATPase, are now well established (Table 19–1), as the results chronicled in this chapter show. The mitochondrially synthesized inner-membrane polypeptides include three of the seven subunits of cytochrome oxidase, the cytochrome b apoprotein, and the four subunits of the membrane factor of the oligomycin-sensitive F_1 ATPase. Genetic evidence that admittedly is incomplete suggests that the structural genes coding for these polypeptides will also be found in the mitochondrial genome. In the next few years we may expect this evidence to become much more rigorous in nature. At the same time, we may expect considerable progress from *in vitro* studies of inner-membrane assembly using isolated mitochondria. The genetic mechanisms governing the coordination of synthesis of inner-membrane components coded in mitochondrion and nucleus will be explored, and this should provide a fascinating story. In short, we are moving from a descriptive phase in which the products of mitochondrial protein synthesis were identified and catalogued into a phase in which we shall learn how their biogenesis is regulated and the mechanisms by which the respiratory complexes are assembled.

REFERENCES

1. Schatz, G., and Mason, T. L. (1974): The biosynthesis of mitochondrial proteins. *Annu. Rev. Biochem.,* 43:51–87.
2. Dreyer, W. J., Laico, M. T., Ruoslahti, E. I., and Papermaster, D. S. (1970): The fundamental polypeptide subunits of biological membranes: A large peptide or "miniprotein." *Fed. Proc.,* 29 (abstract 2016).
3. Criddle, R. S., Bock, R. M., Green, D. E., and Tisdale, H. (1962): Physical characteristics of proteins of the electron transfer system and interpretation of the structure of the mitochondrion. *Biochemistry,* 1:827–842.
4. Woodward, D. O., and Munkres, K. D. (1966): Alterations of a maternally inherited mitochondrial structural protein in respiratory-deficient strains of *Neurospora. Proc. Natl. Acad. Sci. U.S.A.,* 55:872–880.
5. Dreyer, W. J., Papermaster, D. S., and Kühn, H. (1972): On the absence of ubiquitous structural protein subunits in biological membranes. *Ann. N.Y. Acad. Sci.,* 195:61–74.
6. Schatz, G., and Saltzgaber, J. (1969): Identification of denatured mitochondrial ATPase in "structural protein" from beef heart mitochondria. *Biochim. Biophys. Acta,* 180:186–189.
7. Senior, A. E., and MacLennan, D. H. (1970):

Mitochondrial "structural protein." A reassessment. *J. Biol. Chem.,* 254:5086–5095.

8. Zollinger, W. D., and Woodward, D. O. (1972): Comparison of cysteine and tryptophan content of insoluble proteins derived from wild type and *mi-1* strains of *Neurospora crassa. J. Bacteriol.,* 109:1001–1013.

9. Linnane, A. W. (1968): The nature of mitochondrial RNA and some characteristics of the protein-synthesizing system of mitochondria isolated from antibiotic-sensitive and resistant yeasts. In: *Biochemical Aspects of the Biogenesis of Mitochondria,* edited by E. C. Slater, J. M. Tager, S. Papa, and E. Quagliariello, pp. 333–353. Adriatica Editrice, Bari.

10. Ross, E., Ebner, E., Poyton, R. O., Mason, T. L., Ono, B., and Schatz, G. (1974): The biosynthesis of mitochondrial cytochromes. In: *The Biogenesis of Mitochondria,* edited by A. M. Kroon, and C. Saccone, pp. 477–489. Academic Press, New York.

11. Poyton, R. O., and Schatz, G. (1975): Cytochrome c oxidase from bakers' yeast, III. Physical characterization of isolated subunits and chemical evidence for two different classes of polypeptides. *J. Biol. Chem.,* 250:752–761.

12. Rubin, M. S., and Tzagoloff, A. (1973): Assembly of the mitochondrial membrane system. IX. Purification, characterization, and subunit structure of yeast and beef cytochrome oxidase. *J. Biol. Chem.,* 248:4269–4274.

13. Mason, T. L., Poyton, R. O., Wharton, D. C., and Schatz, G. (1973): Cytochrome c oxidase from bakers' yeast. I. Isolation and properties. *J. Biol. Chem.,* 248:1346–1354.

14. Sebald, W., Weiss, H., and Jackl, G. (1972): Products of mitochondrial protein synthesis in *Neurospora crassa:* Determination of equimolar amounts of three products in cytochrome oxidase on the basis of amino-acid analysis. *Eur. J. Biochem.,* 38:311–324.

15. Weiss, H., Lorenz, B., and Kleinow, W. (1972): Contribution of mitochondrial protein synthesis to the formation of cytochrome oxidase in *Locusta migratoria. FEBS Lett.,* 25:49–51.

16. Capaldi, R. A., and Hayashi, H. (1972): The polypeptide composition of cytochrome oxidase from beef heart mitochondria. *FEBS Lett.,* 26:261–263.

17. Poyton, R. O., and Schatz, G. (1975): Cytochrome c oxidase from bakers' yeast. IV. Immunological evidence for the participation of a mitochondrially synthesized subunit in enzymatic activity. *J. Biol. Chem.,* 250:762–766.

18. Ebner, E., Mason, T. L., and Schatz, G. (1973): Mitochondrial assembly in respiration-deficient mutants of *Saccharomyces cerevisiae.* II. Effect of nuclear and extrachromosomal mutations on the formation of cytochrome c oxidase. *J. Biol. Chem.,* 248:5369–5378.

19. Eytan, G. D., and Schatz, G. (1975): Cytochrome c oxidase from bakers' yeast. V. Arrangement of the subunits in the isolated and membrane-bound enzyme. *J. Biol. Chem.,* 250:767–774.

20. Hubbard, A. L., and Cohn, Z. A. (1972): The enzymatic iodination of the red cell membrane. *J. Cell Biol.,* 55:390–405.

21. Eytan, G., Carroll, R., Schatz, G., and Racker, E. (1975): Arrangement of the subunits in solubilized and membrane-bound cytochrome c oxidase from bovine heart. *J. Biol. Chem.,* 250:8598–8603.

22. Mason, T. L., and Schatz, G. (1973): Cytochrome c oxidase from bakers' yeast. II. Site of translation of the protein components. *J. Biol. Chem.,* 248:1355–1360.

23. Rubin, M. S., and Tzagoloff, A. (1973): Assembly of the mitochondrial membrane system. X. Mitochondrial synthesis of three of the subunit proteins of yeast cytochrome oxidase. *J. Biol. Chem.,* 248:4275–4279.

24. Sebald, W., Machleidt, W., and Otto, J. (1974): Cooperation of mitochondrial and cytoplasmic protein synthesis in the formation of cytochrome c oxidase. In: *The Biogenesis of Mitochondria,* edited by A. M. Kroon and C. Saccone, pp. 453–463. Academic Press, New York.

25. Schwab, A. J., Sebald, W., and Weiss, H. (1972): Different pool sizes of the precursor polypeptides of cytochrome oxidase from *Neurospora crassa. Eur. J. Biochem.,* 30:511–516.

26. Sebald, W., Weiss, H., and Jackl, G. (1972): Inhibition of the assembly of cytochrome oxidase in *Neurospora crassa* by chloramphenicol. *Eur. J. Biochem.,* 30:413–417.

27. Poyton, R., and Groot, G. S. P. (1975): Biosynthesis of polypeptides of cytochrome c oxidase by isolated mitochondria. *Proc. Natl. Acad. Sci. U.S.A.,* 72:172–176.

28. Groot, G. S. P., and Poyton, R. (1975): Oxygen control of cytochrome c oxidase synthesis in isolated mitochondria from *Saccharomyces cerevisiae. Nature,* 255:238–240.

29. Poyton, R., and Kavanagh, J. (1976): Regulation of mitochondrial protein synthesis by cytoplasmic proteins. *Proc. Natl. Acad. Sci. U.S.A.,* 73:3947–3951.

30. Poyton, R., and McKemmie, E. (1976): The assembly of cytochrome c oxidase from *Saccharomyces cerevisiae.* In: *Genetics and Biogenesis of Chloroplasts and Mitochondria,* edited by T. Bücher, W. Neupert, W. Sebald, and S. Werner, pp. 207–214. North Holland, Amsterdam.

31. Tzagoloff, A., Rubin, M. S., and Sierra, M. F. (1973): Biosynthesis of mitochondrial enzymes. *Biochim. Biophys. Acta,* 301:71–104.

32. Kováč, L., and Weissková, K. (1968): Oxidative phosphorylation in yeast. III. ATPase activity of the mitochondrial fraction from a

cytoplasmic respiratory-deficient mutant. *Biochim. Biophys. Acta,* 153:55–59.

33. Schatz, G. (1968): Impaired binding of mitochondrial adenosine triphosphatase in the cytoplasmic "petite" mutant of *Saccharomyces cerevisiae. J. Biol. Chem.,* 243:2192–2199.
34. Tzagoloff, A. (1969): Assembly of the mitochondrial membrane system. II. Synthesis of the mitochondrial adenosine triphosphatase F_1. *J. Biol. Chem.,* 244:5027–5033.
35. Tzagoloff, A., Akai, A., and Sierra, M. F. (1972): Assembly of the mitochondrial membrane system. VII. Synthesis and integration of F_1 subunits into the rutamycin-sensitive adenosine triphosphatase. *J. Biol. Chem.,* 247:6511–6516.
36. Tzagoloff, A. (1970): Assembly of the mitochondrial membrane system. III. Function and synthesis of the oligomycin sensitivity-conferring protein of yeast mitochondria. *J. Biol. Chem.,* 245:1545–1551.
37. Tzagoloff, A., and Meagher, P. (1972): Assembly of the mitochondrial membrane system. VI. Mitochondrial synthesis of subunit proteins of the rutamycin-sensitive adenosine triphosphatase. *J. Biol. Chem.,* 247:594–603.
38. Weiss, H. (1972): Cytochrome b in *Neurospora crassa.* A membrane protein containing subunits of cytoplasmic and mitochondrial origin. *Eur. J. Biochem.,* 30:469–478.
39. Weiss, H., and Ziganke, B. (1974): Biogenesis of cytochrome b in *Neurospora crassa.* In: *The Biogenesis of Mitochondria,* edited by A. M. Kroon and C. Saccone, pp. 491–500. Academic Press, New York.
40. Weiss, H., and Ziganke, B. (1976): Subunit structure and arrangement of mitochondrial cytochrome b. In: *Genetics and Biogenesis of Chloroplasts and Mitochondria,* edited by T. Bücher, W. Neupert, W. Sebald, and S. Werner, pp. 259–266. North Holland, Amsterdam.
41. Lin, L.-F., and Beattie, D. S. (1976): Purification and biogenesis of cytochrome b in bakers' yeast. In: *Genetics and Biogenesis of Chloroplasts and Mitochondria,* edited by T. Bücher, W. Neupert, W. Sebald, and S. Werner, pp. 281–288. North Holland, Amsterdam.
42. Ross, E., and Schatz, G. (1976): Cytochrome c_1 of bakers' yeast. II. Synthesis on cytoplasmic ribosomes and influence of oxygen and heme on accumulation of the apoprotein. *J. Biol. Chem.,* 251:1997–2004.
43. Groot, G. S. P., Rouslin, W., and Schatz, G. (1972): Promitochondria of anaerobically grown yeast. VI. Effect of oxygen on promitochondrial protein synthesis. *J. Biol. Chem.,* 247:1735–1742.
44. Packer, L., Williams, M. A., and Criddle, R. S. (1973): Freeze-fracture studies on mitochondria from wild-type and respiratory-deficient yeasts. *Biochim. Biophys. Acta,* 292:92–104.

45. Ames, G. F.-L., and Nikaido, K. (1976): Two-dimensional gel electrophoresis of membrane proteins. *Biochemistry,* 15:616–623.
46. Sherman, F. (1963): Respiration-deficient mutants of yeast. I. Genetics. *Genetics,* 48:375–385.
47. Sherman, F., and Slonimski, P. (1964): Respiration-deficient mutants of yeast. II. Biochemistry. *Biochim. Biophys. Acta,* 90:1–15.
48. Roodyn, D. B., and Wilkie, D. (1968): *The Biogenesis of Mitochondria.* Methuen, London.
49. Yotsuyanagi, Y. (1962): Études sur le chondriome de la levure. II. Chondriomes des mutants à déficience respiratoire. *J. Ultrastruct. Res.,* 7:141–158.
50. Sherman, F., Stewart, J. W., Parker, J. H., Putterman, G. J., Agrawal, B. B. L., and Margoliash, E. (1970): The relationship of gene structure and protein structure of iso-*l*-cytochrome c from yeast. *Symp. Soc. Exp. Biol.,* 24:85–107.
51. Beck, J. H., Parker, J. H., Balcavage, W. X., and Mattoon, J. R. (1971): Mendelian genes affecting development and function of yeast mitochondria. In: *Autonomy and Biogenesis of Mitochondria and Chloroplasts,* edited by N. K. Boardman, A. W. Linnane, and R. M. Smillie, pp. 194–204. North Holland, Amsterdam.
52. Ebner, E., Mennucci, L., and Schatz, G. (1973): Mitochondrial assembly in respiration-deficient mutants of *Saccharomyces cerevisiae.* I. Effect of nuclear mutations on mitochondrial protein synthesis. *J. Biol. Chem.,* 250:5360–5368.
53. Ebner, E., and Schatz, G. (1973): Mitochondrial assembly in respiration-deficient mutants of *Saccharomyces cerevisiae.* III. A nuclear mutant lacking mitochondrial adenosine triphosphatase. *J. Biol. Chem.,* 248:5379–5384.
54. Cabral, F., Saltzgaber, J., Birchmeier, W., Deters, D., Frey, T., Kohler, C., and Schatz, G. (1976): Structure and biosynthesis of cytochrome c oxidase. In: *Genetics and Biogenesis of Chloroplasts and Mitochondria,* edited by T. Bücher, W. Neupert, W. Sebald, and S. Werner, pp. 215–230. North Holland, Amsterdam.
55. Ono, B.-I., Fink, G., and Schatz, G. (1975): Mitochondrial assembly in respiration-deficient mutants of *Saccharomyces cerevisiae.* IV. Effects of nuclear amber suppressors on the accumulation of a mitochondrially made subunit of cytochrome c oxidase. *J. Biol. Chem.,* 250:775–782.
56. Hawthorne, D. C., and Mortimer, R. K. (1963): Super-suppressors in yeast. *Genetics,* 48:617–620.
57. Gilmore, R. A., Stewart, J. W., and Sherman, F. (1971): Amino acid replacements resulting from super-suppression of nonsense

mutants of iso-*l*-cytochrome c from yeast. *J. Mol. Biol.,* 61:157–173.

58. Sherman, F., Liebman, S. W., Stewart, J. W., and Jackson, M. (1973): Tyrosine substitutions resulting from suppression of amber mutants of iso-1-cytochrome c in yeast. *J. Mol. Biol.,* 78:157–168.

59. Goodman, H. M., Abelson, J., Landy, A., Brenner, S., and Smith, J. D. (1968): Amber suppression: A nucleotide change in the anticodon of a tyrosine transfer RNA. *Nature,* 217:1019–1024.

60. Tzagoloff, A., Akai, A., and Needleman, R. B. (1975): Assembly of the mitochondrial membrane system. Characterization of nuclear mutants of *Saccharomyces cerevisiae* with defects in mitochondrial ATPase and respiratory enzymes. *J. Biol. Chem.,* 250:8228–8235.

61. Agsteribbe, E., Douglas, M., Ebner, E., Koh, T. Y., and Schatz, G. (1976): Mutation in *Saccharomyces cerevisiae* mitochondrial F_1 leading to aurovertin resistance. In: *Genetics and Biogenesis of Chloroplasts and Mitochondria,* edited by T. Bücher, W. Neupert, W. Sebald, and S. Werner, pp. 135–141. North Holland, Amsterdam.

62. Slonimski, P., and Tzagoloff, A. (1976): Localization in yeast mitochondrial DNA of mutations expressed in a deficiency of cytochrome oxidase and/or coenzyme QH_2-reductase. *Eur. J. Biochem.,* 61:27–41.

63. Cabral, F., Solioz, M., Rudin, Y., Schatz, G., Clavilier, L., and Slonimski, P. P. (1977): Identification of the structural gene for yeast cytochrome c oxidase subunit II on mitochondrial DNA. *J. Biol. Chem.,* 253:297–304.

64. Tzagoloff, A., Foury, F., and Akai, A. (1976): Genetic determination of mitochondrial cytochrome b. In: *Genetics and Biogenesis of Chloroplasts and Mitochondria,* edited by T. Bücher, W. Neupert, W. Sebald, and S. Werner, pp. 419–426. North Holland, Amsterdam.

65. Tzagoloff, A., Foury, F., and Akai, A. (1976): Assembly of the mitochondrial membrane system XVIII. Genetic loci on mitochondrial DNA involved in cytochrome b biosynthesis. *Mol. Gen. Genet.,* 149:33–42.

66. Kotylak, Z., and Slonimski, P. P. (1976): Joint control of cytochromes a and b by a unique mitochondrial DNA region comprising four genetic loci. In: *The Genetic Function of Mitochondrial DNA,* edited by C. Saccone and A. M. Kroon, pp. 143–154. North Holland, Amsterdam.

67. Pajot, P., Wambier-Kluppel, M. L., Kotylak, Z., and Slonimski, P. P. (1976): Regulation of cytochrome oxidase formation by mutations in a mitochondrial gene for cytochrome b. In: *Genetics and Biogenesis of Chloroplasts and Mitochondria,* edited by T. Bücher, W. Neupert, W. Sebald, and S. Werner, pp. 443–451. North Holland, Amsterdam.

68. Cobon, G. S., Groot Obbink, D. J., Hall, R. M., Maxwell, R., Murphy, M., Rytka, J., and Linnane, A. W. (1976): Mitochondrial genes determining cytochrome b (complex III) and cytochrome oxidase function. In: *Genetics and Biogenesis of Chloroplasts and Mitochondria,* edited by T. Bücher, W. Neupert, W. Sebald, and S. Werner, pp. 453–460. North Holland, Amsterdam.

69. Groot Obbink, D. J., Hall, R. M., Linnane, A. W., Lukins, H. B., Monk, B. C., Spithill, T. W., and Trembath, M. K. (1976): Mitochondrial genes involved in the determination of mitochondrial membrane proteins. In: *The Genetic Function of Mitochondrial DNA,* edited by C. Saccone and A. M. Kroon, pp. 163–173. North Holland, Amsterdam.

70. Lang, B., Burger, G., Bandlow, W., Kaudewitz, F., and Schweyen, R. J. (1976): Antimycin- and funiculosin-resistant mutants in *Saccharomyces cerevisiae:* New markers on the mitochondrial DNA. In: *Genetics and Biogenesis of Chloroplasts and Mitochondria,* edited by T. Bücher, W. Neupert, W. Sebald, and S. Werner, pp. 461–465. North Holland, Amsterdam.

71. Pratje, E., and Michaelis, G. (1976): Two mitochondrial antimycin a resistance loci in *Saccharomyces cerevisiae.* In: *Genetics and Biogenesis of Chloroplasts and Mitochondria,* edited by T. Bücher, W, Neupert, W. Sebald, and S. Werner, pp. 467–471. North Holland, Amsterdam.

72. Šubik, J. (1976): Mitochondrial inheritance of mucidin resistance in yeast. In: *Genetics and Biogenesis of Chloroplasts and Mitochondria,* edited by T. Bücher, W. Neupert, W. Sebald, and S. Werner, pp. 473–477. North Holland, Amsterdam.

73. Kotylak, Z., and Slonimski, P. P. (1978): Fine structure genetic map of the mitochondrial DNA region controlling coenzyme QH_2-cytochrome c reductase. In: *Mitochondria 1977. Proceedings of the Colloquium on Genetics and Biogenesis of Mitochondria, Schliersee,* edited by W. Bandlow et al. De Gruyter, Berlin (*in press*).

74. Colson, A. M., and Slonimski, P. P. (1978): Mapping of drug-resistant loci in the coenzyme QH_2-cytochrome c reductase region of the mitochondrial DNA in *Saccharomyces cerevisiae.* In: *Mitochondria 1977. Proceedings of the Colloquium on Genetics and Biogenesis of Mitochondria, Schliersee,* edited by W. Bandlow et al. De Gruyter, Berlin (*in press*).

75. Claisse, M. L., Spyridakis, A., and Slonimski, P. P. (1978): Mutations at any one of three unlinked mitochondrial genetic loci *BOX 1, BOX 4* and *BOX 6* modify the structure of cytochrome b polypeptide(s). In: *Mitochondria 1977. Proceedings of the Colloquium on Genetics and Biogenesis of Mitochondria, Schliersee,* edited by W. Bandlow et al., De Gruyter, Berlin (*in press*).

76. Pajot, P., Wambier-Kluppel, M. L., and Slonimski, P. P. (1977): Cytochrome c re-

ductase and cytochrome oxidase formation in mutants and revertants in the "BOX" region of mitochondrial DN*I*. In: *Mitochondria 1977. Proceedings of the Colloquium on Genetics and Biogenesis of Mitochondria, Schliersee,* edited by W. Bandlow et al. De Gruyter, Berlin (*in press*).

77. Lancashire, W. E., and Griffiths, D. E. (1975): Studies on energy-linked reactions: Genetic analysis of venturicidin-resistant mutants. *Eur. J. Biochem.,* 51:403–413.

78. Griffiths, D. E. (1975): Utilization of mutations in the analysis of yeast mitochondrial oxidative phosphorylation. In: *Genetics and Biogenesis of Chloroplasts and Mitochondria,* edited by C. W. Birky, Jr., P. S. Perlman, and T. J. Byers, Chapter 3. Ohio State University Press, Columbus.

79. Clavilier, L. (1976) Mitochondrial genetics. XII. An oligomycin-resistant mutant localized at a new mitochondrial locus in *Saccharomyces cerevisiae. Genetics,* 83:227–243.

80. Foury, F., and Tzagoloff, A. (1976): Localization on mitochondrial DNA of mutations leading to a loss of rutamycin-sensitive adenosine triphosphatase. *Eur. J. Biochem.,* 68:113–119.

81. Somlo, M., Avner, P. R., Cosson, J., Dujon, B., and Krupa, M. (1974): Oligomycin sensitivity of ATPase studied as a function of mitochondrial biogenesis, using mitochondrially determined oligomycin-resistant mutants of *Saccharomyces cerevisiae. Eur. J. Biochem.,* 42:439–445.

82. Somlo, M., and Cosson, J. (1976): Mitochondrially encoded oligomycin-resistant mutants of *S. cerevisiae.* In: *Genetics and Biogenesis of Chloroplasts and Mitochondria,* edited by Γ. Bücher, W. Neupert, W. Sebald, and S. Werner, pp. 143–149. North Holland, Amsterdam.

83. Tzagoloff, A., Akai, A., and Foury, F. (1976): Assembly of the mitochondrial membrane system XVI. Modified form of the ATPase proteolipid in oligomycin-resistant mutants of *Saccharomyces cerevisiae. FEBS Lett.,* 65:391–395.

84. Narita, K., Murakami, H., and Titani, K. (1964): Amino acid composition of bakers' yeast cytochrome c. *J. Biochem.,* 56:216–221.

85. Ross, E., and Schatz, G. (1976): Cytochrome c_1 of bakers' yeast. I. Isolation and properties. *J. Biol. Chem.,* 251:1991–1996.

Chapter 20

Biogenesis of Thylakoid Membrane Polypeptides and Ribulose Bisphosphate Carboxylase

In similarity to the situation with the mitochondrion, nuclear genes whose products are synthesized on cytoplasmic ribosomes are largely responsible for synthesis of the soluble components of the chloroplast, with the notable exception of the enzyme ribulose bisphosphate carboxylase (RuBPCase), or fraction I protein, about which we have already said quite a lot (see Chapter 18). The same holds true for the photosynthetic pigments and for the great majority of proteins present in the chloroplast envelope; see Gillham and associates (1) for discussion and references. On the other hand, the thylakoid membrane polypeptides, like the polypeptides of the mitochondrial inner membrane, fall into two groups. Some are synthesized in the chloroplast and probably are coded by clDNA. Others are made in the cytoplasm, with the genes coding for them being localized in the nucleus.

We shall begin this chapter with a discussion of the sites of synthesis of the thylakoid membrane polypeptides; then we shall turn to what little is known concerning the localization of the genes coding for these proteins. A similar format will be followed in considering RuBPCase. Before beginning, however, it should be pointed out that experiments to determine the sites of synthesis of these proteins, were carried out for the chloroplast much as they were for the mitochondrion (see Chapter 19). That is, the *in vivo* approach employed specific inhibitors of chloroplast and cytoplasmic translation, and the *in vitro* approach made use of the isolated organelles. Needless to say, all cautionary remarks made in the preceding chapter concerning interpretation of such experiments for the mitochondrion also apply for the chloroplast.

SITES OF SYNTHESIS OF CHLOROPLAST MEMBRANE POLYPEPTIDES

The first indication that synthesis of thylakoid membrane polypeptides might depend on both the cytoplasmic and the chloroplast protein-synthesizing systems came from experiments in which the increases in activities of certain membrane-bound enzymes were measured in the presence and in the absence of specific inhibitors. In synchronously growing cultures of *Chlamydomonas,* Armstrong et al. (2) examined the effects of the chloroplast ribosome inhibitors chloramphenicol and spectinomycin, as well as the inhibitor of cytoplasmic ribosome function cycloheximide, on the formation of various chloroplast components that normally increase twofold to threefold during the same period in the growth cycle. Increases in chlorophyll, ferridoxin, and ferridoxin–NADP reductase were blocked only by cycloheximide, which indicates that these components require the activity of cytoplasmic ribosomes. Increases in cytochrome 563 and cytochrome 553 were inhibited by cycloheximide, chloramphenicol, and spectinomycin, which suggests that these components require translational activity of both chloroplast and cytoplasmic ribosomes. A summary of the effects of these inhibitors on similar components associated with thylakoid membranes in *Euglena* can be found in Table 13.6 of the review by Schmidt and Lyman (3). Rather than consider these experiments further, we shall turn to more recent experiments dealing with the sites of synthesis of specific thylakoid membrane polypeptides.

Most recent experiments on the sites of

synthesis of the thylakoid membrane poly-peptides dealt with global polypeptide composition, irrespective of function. In this respect these experiments differed markedly from similar studies on the sites of synthesis of mitochondrial inner-mem-brane polypeptides (see Chapter 19), which dealt almost exclusively with functional complexes (e.g., the oligomycin-sensitive F_1 ATPase and cytochrome oxidase). A nota-ble exception is the chloroplast coupling factor CF1, the thylakoid-bound ATPase that catalyzes the terminal stage of photo-phosphorylation in the chloroplast (4). Biosynthesis of CF1 appears to involve both chloroplast and cytoplasmic protein syn-thesis. The structure and properties of the CF1 membrane-bound enzyme have been reviewed recently (4,5) (see Chapter 1). Briefly, CF1 can be solubilized by treatment of the thylakoid membranes with 1-mM EDTA; the soluble form possesses latent ATPase activity. Results obtained princi-pally by Racker's group have shown that spinach CF1 has a molecular weight of 325,000 d and is made up of five nonidenti-cal subunits: α, β, γ, δ, and ϵ. The α and β subunits are believed to be involved in the catalytic reaction, whereas the ϵ subunit is a potent inhibitor of ATPase activity; the functions of the γ and δ subunits have not yet been determined (4,5). Early inhibitor studies by Horak and Hill (6,7) indicated that the synthesis of CF1 during greening of etiolated leaves of *Phaseolus vulgaris* re-quires the participation of both chloroplast and cytoplasmic ribosomes. In isolated pea chloroplasts, Eaglesham and Ellis (8) found that the labeling pattern of thylakoid membranes is not altered by EDTA wash-ing; accordingly, they concluded that all CF1 subunits are synthesized in the cyto-plasm. However, more careful studies by Mendiola-Morgenthaler et al. (9) revealed that the α, β and ϵ subunits of CF1 are labeled in intact spinach chloroplasts, whereas the remaining two are presumably

synthesized in the cytoplasm. These results are in contrast to those obtained with the mitochondrial coupling factor F_1, in which all the five subunits are made in the cyto-plasm (see Chapter 19).

Since the most complete picture of the biogenesis of individual thylakoid mem-brane polypeptides comes from experiments with *Chlamydomonas reinhardtii,* we shall begin with a description of these experi-ments and then compare the results with those obtained in higher plants. Two ap-proaches have been taken in studying the biogenesis of thylakoid membrane polypep-tides in *Chlamydomonas*. The original stud-ies of Hoober and Ohad and their col-leagues employed a mutant called *y-1*. Unlike wild-type cells of *C. reinhardtii, y-1* does not form chlorophyll in the dark when supplied with acetate as a carbon source; rather, it becomes etiolated, as do higher plants (10–14). The mutant stops making chlorophyll and thylakoid membranes when it is transferred to the dark; it begins to ac-cumulate protochlorophyll(ide). During successive cell divisions, chlorophyll and preexisting lamellar structures are diluted out. If these "yellow" cells are returned to the light, the accumulated protochloro-phyll(ide) is photoconverted to chloro-phyll(ide), and rapid synthesis of chloro-phyll and thylakoid membranes commences (13). Within 6 to 8 hr the regreening proc-ess is complete, and the cells are indis-tinguishable from wild type in their amount of chlorophyll and thylakoid membrane organization.

To study the sites of synthesis of the thylakoid membrane polypeptides, both Hoober and Ohad (14–23) allowed dark-grown cells of *y-1* to regreen in the presence of inhibitors of chloroplast or cytoplasmic protein synthesis and labeled arginine. The polypeptides made in the presence of each inhibitor were radioactively labeled. This approach was similar to that used by Tzago-loff (see Chapter 19) in studying the biosynthesis of the oligomycin-sensitive ATPase and cytochrome oxidase in yeast in that polypeptide synthesis was followed in a derepressing organelle. The differentiation was triggered in the yeast experiments by glucose limitation and in the *Chlamydomonas* experiments with *y-1* by light. Compared to the yeast experiments, the *Chlamydomonas* experiments were generally rather long-term.

A second approach to the study of the sites of synthesis of thylakoid membrane polypeptides was that taken more recently by Chua and Gillham (24). This approach was similar to the one taken by Schatz and Bücher and their colleagues in defining the sites of synthesis of the cytochrome oxidase polypeptides in yeast and *Neurospora* (see Chapter 19). Light-grown wild-type cells of *C. reinhardtii* that contained fully developed chloroplasts were pulse-labeled with radioactive arginine in the presence of inhibitors of cytoplasmic or chloroplast protein synthesis. In the absence of limiting pools of particular polypeptides (p. 516), polypeptides whose synthesis is blocked by the cytoplasmic protein synthesis inhibitor should be made in the presence of the inhibitor of chloroplast protein synthesis and vice versa.

To determine the sites of synthesis of these lamellar polypeptides, Chua and Gillham (24) pulse-labeled the thylakoid membrane polypeptides of exponentially growing cells of *C. reinhardtii* with ^{14}C-acetate. Cytoplasmic protein synthesis was inhibited by anisomycin, and chloroplast protein synthesis was inhibited by chloramphenicol or spectinomycin. The labeling patterns of isolated thylakoid membranes of wild-type cells were compared to those of a mutant (*spr-u-1-6-2*) known from both *in vivo* and *in vitro* experiments to have spectinomycin-resistant chloroplast ribosomes (see Chapter 21). The prediction was that polypeptides made on chloroplast ribosomes would not be pulse-labeled in the presence of either chloramphenicol or spectinomycin in the wild type, but would be labeled in the *spr-u-1-6-2* mutant in the presence of spectinomycin. Thus the results with *spr-u-1-6-2* would serve as positive confirmation for those polypeptides synthesized on chloroplast ribosomes. Following labeling, the thylakoid membranes were isolated; the component polypeptides were subjected to SDS-gradient gel electrophoresis, and autoradiograms were made of the dried gels. Under these conditions approximately 33 polypeptides were visualized; 19 were visualized clearly enough that they could be numbered Fig. 1–18 and Fig. 20–1).

Chua and Gillham (24) found that polypeptides 4.1, 4.2, 5, and 6, as well as two low-molecular-weight polypeptides (LMW-1 and LMW-2), were made on chloroplast ribosomes (Fig. 20–1 and Table 20–1). Polypeptide 2, the only polypeptide associated with chlorophyll-protein complex I (CPI) in *Chlamydomonas* (25), appeared to require functioning chloroplast and cytoplasmic ribosomes for its formation and/or integration into the membrane. Synthesis of this polypeptide was inhibited completely by either chloramphenicol or spectinomycin in wild type and by chloramphenicol alone in the *spr-u-1-6-2* mutant, which indicated that it was made in the chloroplast. However, partial and variable inhibition of incorporation of radioactivity into this polypeptide was also obtained with anisomycin. From these results Chua and Gillham concluded that integration of polypeptide 2 was dependent on one or more cytoplasmically synthesized partner proteins whose pool size tended to be smaller than that of polypeptide 2. Therefore, inhibition of synthesis of this partner polypeptide by anisomycin should block integration of polypeptide 2. If this interpretation is correct, it should be possible to minimize the apparent inhibition of polypeptide 2 synthesis by anisomycin by increasing the pool size of the putative partner protein. In fact,

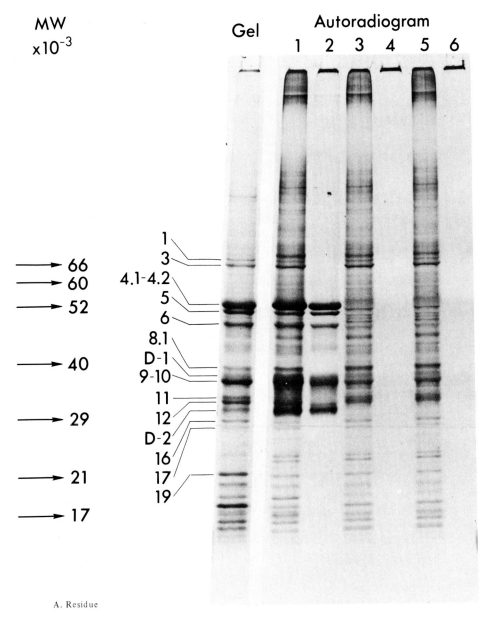

A. Residue

FIG. 20–1. Pulse labeling of thylakoid membrane polypeptides with ^{14}C-acetate in *C. reinhardtii* in presence and absence of inhibitors of cytoplasmic and chloroplast protein synthesis. Chloroplast membranes were isolated, and polypeptides were extracted with chloroform-methanol (C/M) mixture. Certain of the polypeptides were soluble in this mixture (extract), whereas others were not (residue). This prefractionation made it easier to visualize different polypeptides during subsequent electrophoresis on SDS slab gels. A scale of apparent molecular weights is given on the left, and each polypeptide is numbered. Slots on left (labeled *Gel*) show Coomassie blue staining patterns of membrane polypeptides; slots 1–6 show corresponding autoradiograms of gels: 1, control; 2, anisomycin; 3, chloramphenicol; 4, anisomycin plus chloramphenicol; 5, spectinomycin; 6, anisomycin plus spectinomycin. A and B: Labeling patterns for wild-type thylakoid membrane polypeptides. Note that polypeptides 2, 4.1, 4.2, 5, 6, D-1, D-2, LMW-1, and LMW-2 are sensitive to inhibitors of chloroplast (but not cytoplasmic) protein synthesis. C and D: Labeling patterns for thylakoid membrane polypeptides of *spr-u-1-6-2* mutant. This mutant has spectinomycin-resistant chloroplast ribosomes. Note that the polypeptides that are made on chloroplast ribosomes label in the presence of spectinomycin but not chloramphenicol in this mutant, whereas in wild type these polypeptides fail to label in the presence of either inhibitor. (From Chua and Gillham, ref. 24, with permission.)

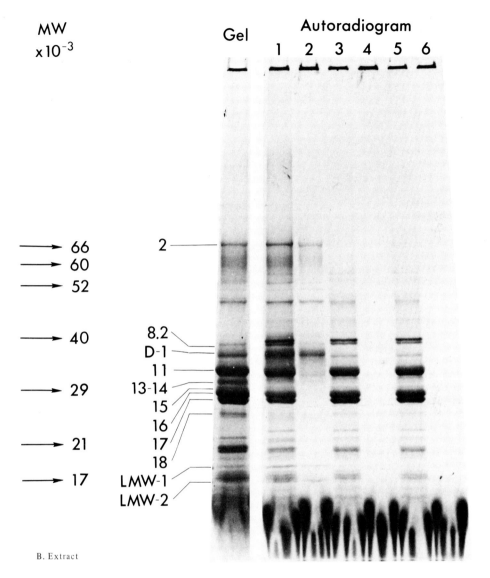

B. Extract

Chua and Gillham found that prior incubation of the cells with chloramphenicol relieved the inhibition of polypeptide 2 synthesis by anisomycin during subsequent treatment with anisomycin.

In addition to these polypeptides, two broad diffuse bands of radioactivity (D-1 and D-2) that were resistant to anisomycin but sensitive to the inhibitors of chloroplast protein synthesis were observed in the gels. These bands of radioactivity corresponded to polypeptides that stained faintly. All remaining polypeptides appeared to be products of cytoplasmic protein synthesis (Table 20–1). Any polypeptide whose synthesis was blocked by chloramphenicol or spectinomycin in wild type was synthesized in the presence of the latter antibiotic in *spr-u-1-6-2*.

Gillham et al. (1) recently interpreted the gel patterns seen by Hoober (15) in terms of the results of Chua and Gillham (24). In brief, three major peaks were seen in Hoober's gels, designated a, b, and c. Peak a seemed to correspond to polypeptides 9, 10, and possibly 8 and D-1; peak

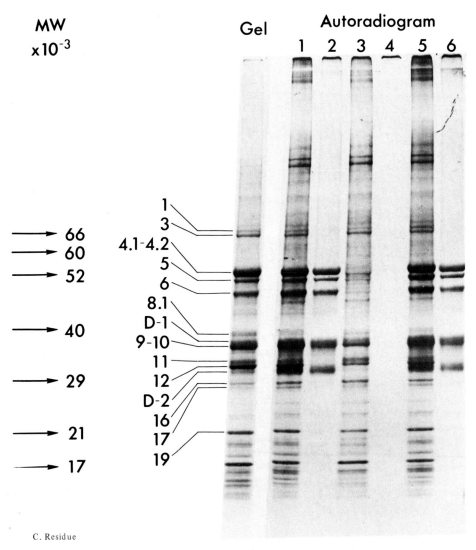

C. Residue

b, to polypeptides 11 and 12; peak c, to polypeptides 13–17 of Chua and Gillham. Polypeptides 2 and 5 seemed to be missing, and polypeptide 6 was greatly reduced in amount. The method of membrane extraction might account for this. Hoober (15) reported both cycloheximide- and chloramphenicol-resistant amino acid incorporation into polypeptide a, which is to be expected if this peak contains polypeptides 8, 9, and 10, which are made on cytoplasmic ribosomes, and D-1, which is synthesized on chloroplast ribosomes. Polypeptide b labeled in the presence of chloramphenicol but not cycloheximide, which is to be expected if it corresponds to polypeptides 11 and 12. Polypeptide c was principally sensitive to cycloheximide, although a small peak of chloramphenicol-sensitive incorporation was seen in this region as well. This, too, is to be expected if this peak includes polypeptides 13–17 and D-2.

Hoober and Stegeman (16–19) also studied the regulation of thylakoid membrane polypeptide synthesis in *y-1*. They reasoned that the conversion of protochlorophyll(ide) might be the key step triggering thylakoid membrane assembly in *Chla-*

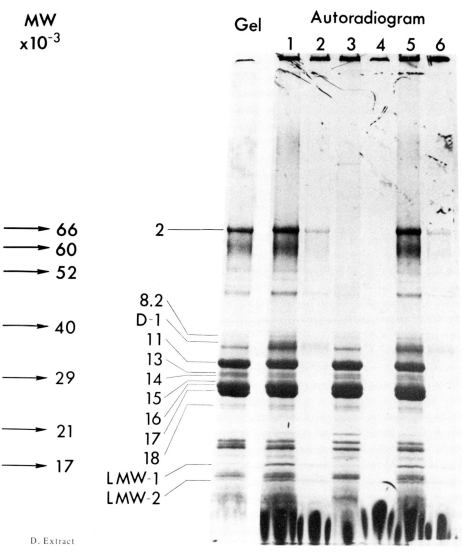

mydomonas, as it seems to be in higher plants (26), since the action spectrum for greening of the *y-1* mutant in *C. reinhardtii* is identical to that in higher plants (27). This proved to be the case, since cells regreened normally if exposed to light above 600 nm, but regreening stopped immediately if the cells were exposed to light above 675 nm (17,18), which is above the 650-nm *in vivo* absorption maximum of the photoconvertible form of protochlorophyll(ide). Synthesis of polypeptides a, b, and c also declined over a period of 2 to 3

hr following transfer of cells to light above 675 nm (17,18). This finding was consistent with the hypothesis that protochlorophyll-chlorophyll conversion and membrane biosynthesis go hand in hand. Hoober (16) had previously reported that the major thylakoid membrane polypeptides, in particular polypeptide c, could be detected in gels of whole-cell protein. When dark-grown cells of *y-1* were transferred to white light in the presence of high concentrations of chloramphenicol, synthesis of chlorophyll and thylakoid membranes was in-

TABLE 20–1. *Sites of synthesis and functions of thylakoid membrane polypeptides in* C. reinhardtii *and the mutants that affect them*

Site of synthesis	Polypeptide	Function	Mutants affecting polypeptides	
			Designation	Probable location
Chloroplast	2	CPI polypeptide essential for PSI activity	F1, F14 C_1	Nucleus Chloroplast
	4.1, 4.2	Polypeptides associated with CF_1 ATPase	F54	Nucleus
	5	Function unknown, but a variant of this polypeptide is uniparentally inherited; so the polypeptide is probably coded by a chloroplast gene	—	—
	6	PSII reaction center	F34	Nucleus
	LMW-1, LMW-2, D-1, D-2	Function unknown	—	—
Cytoplasm	11	Constituent of CPII	—	—
	12, 3, 8–19, and several low-molecular-weight proteins	Function unknown	—	—

Data on sites of synthesis from Chua and Gillham (24); data on polypeptide function and mutants affecting specific polypeptides from Bennoun and Chua (65).

hibited, but polypeptide c continued to be made, appearing in the soluble fraction instead of the membrane pellet. Subsequently, Hoober and Stegeman (17,18) showed that in the presence of chloramphenicol and white light above 675 nm, greening of *y-1* was also inhibited; again polypeptide c continued to be made, but it was not incorporated into thylakoid membranes. When greening cells of *y-1* were transferred to darkness, the same result was obtained with respect to polypeptide c. Similar but less nearly definitive results were reported for synthesis of polypeptide b.

Based on these findings, Hoober and Stegeman (17,18) formulated a model for control of synthesis of polypeptides b and c by protochlorophyll. This model hypothesizes that protochlorophyll controls the synthesis of polypeptides b and c. The model postulates that the mRNAs for these two polypeptides are transcribed from clDNA and exported to the cytoplasm, where they are translated on cytoplasmic ribosomes. Protochlorophyll, in combination with a polypeptide synthesized on chloroplast ribosomes (P_R) that acts as a corepressor, blocks transcription of the mRNAs for polypeptides b and c. Therefore, in *y-1* growing in the dark, the protochlorophyll(ide) accumulated blocks synthesis of these key thylakoid polypeptides, which in turn prevents formation of photosynthetic membranes. When the light is turned on, protochlorophyll is converted to chlorophyll, which cannot combine with P_R, and transcription of the mRNAs for polypeptides b and c occurs. These mRNAs are transported to the cytoplasm, where they are translated, and thylakoid membrane assembly proceeds. If cycloheximide is added to the medium in either the light or dark, translation of the mRNAs for polypeptides b and c is blocked. If regreening cells are exposed to light above 675 nm, transcription of the mRNAs for polypeptides b and c is blocked by the accumulation of protochlorophyll, which combines with P_R. If chloramphenicol is added to the medium, translation of the message for P_R on chloroplast ribosomes is blocked, there is no co-repressor with which protochlorophyll can combine to block transcription of the mRNAs for polypeptides b and c, and synthesis of these polypeptides is not repressed.

This scheme now needs reevaluation in light of the finding of Chua and Gillham (24) that Hoober's polypeptides b and c probably consist of several polypeptides each. The finding that synthesis of polypeptides in this region of the gel continues if regreening y-1 cells are transferred to the dark or to light above 675 nm in the presence of chloramphenicol is interesting. However, polypeptides b and c are identified by gel electrophoresis of radioactively labeled total soluble protein and by comparison of the positions of these radioactive peaks with the protein staining peaks seen in the gels of membrane preparations. Thus at least three interpretations can be made of these results. First, synthesis of some or all of the polypeptides corresponding to peaks b and c is derepressed by chloramphenicol in y-1 under conditions such that regreening is blocked. Second, the radioactive peaks that appear in the soluble fraction in the presence of chloramphenicol do not represent thylakoid membrane polypeptides at all; rather, they represent polypeptides of similar molecular weights whose synthesis is stimulated by an indirect effect of chloramphenicol. Third, during thylakoid membrane biogenesis there is tight coupling between the chloroplast and cytoplasmic protein-synthesizing systems involving insertion of chlorophyll into lamellae assembled from polypeptides made on both chloroplast and cytoplasmic ribosomes. Chloramphenicol uncouples the system by blocking chloroplast protein synthesis, and the cytoplasmic system becomes "free running" so that thylakoid membrane polypeptides continue to be synthesized on cytoplasmic ribosomes. Clearly, the observations made by Hoober and Stegeman (17,18) are of interest, and the system needs to be analyzed further with the use of monospecific antibodies in order to prove that the polypeptides accumulated in the soluble fraction in the presence of chloramphenicol really are thylakoid membrane polypeptides.

Eytan and Ohad (20,21) also used cycloheximide and chloramphenicol to investigate the sites of synthesis of thylakoid membrane polypeptides in the regreening y-1 system of *C. reinhardtii*. However, their acetic acid–urea gel system had low resolution. Only one major polypeptide, as revealed by peaks of staining and radioactivity, appeared to be synthesized during greening; they called it the L protein. They found that L protein synthesis was cycloheximide-sensitive, and they concluded that the L protein was a protein or probably a group of proteins associated with photosynthetic lamellae that was made on cytoplasmic ribosomes. Several other poorly resolved protein peaks appeared to be labeled in the presence of cycloheximide, but not chloramphenicol; thus these would be products of chloroplast protein synthesis. Meaningful comparisons of the gel patterns seen by Eytan and Ohad with those of Chua and Gillham or Hoober are difficult because of the poor resolution of the former gels. However, the L protein of Eytan and Ohad may be commensurate with Hoober's peaks a, b, and c and polypeptides 8–17 in the system of Chua and Gillham.

Based on these results and on more extensive physiological and ultrastructural findings, Eytan and Ohad (21) constructed a model to explain the regreening of y-1 on the assumption that membrane composition can be modulated through the use of inhibitors such as chloramphenicol and cycloheximide. Proteins of cytoplasmic origin (L proteins) assemble together with lipids to form a membrane framework that serves to accept proteins made on chloroplast ribosomes (activation proteins) and chlorophyll. Addition of the activation proteins and chlorophyll results in the formation of active photosynthetic lamellae. This is an interesting model, but it is difficult to test; qualitative comparison of the polypeptide compositions of the few remaining thylakoid membranes present in dark-grown cells of *y-1* and the membranes formed during regreening might give some idea of its validity.

More recently, Bar-Nun and Ohad (22) reexamined the regreening of dark-grown *y-1* cells using SDS gels with somewhat increased resolution; they identified seven stained polypeptide peaks (I, II, IIb, III, IV, Va, and Vb). Peak I polypeptides did not label in the presence of either chloramphenicol (18–19 hr of exposure) or cycloheximide (4 hr of exposure); peak II and IIb polypeptides labeled in the presence of cycloheximide but not chloramphenicol; incorporation into the polypeptides of the remaining four peaks was inhibited by cycloheximide but not chloramphenicol. When CPI and CPII fractions were prepared, the CPI fraction was found to be enriched in peak II polypeptides and the CPII fraction in peak III, IV, Va, and Vb polypeptides. Differences between gel systems make it difficult to relate these findings directly to those of Chua and Gillham, although a recent article by Kretzer et al. (28) made some tentative identifications possible. These authors made use of a temperature-sensitive photosynthetic mutant (T4) that was defective in certain thylakoid membrane polypeptides at the restrictive temperature. These authors divided their gels into five regions designated I (2 polypeptides), II (5 polypeptides), III (2 polypeptides), IV (several polypeptides), and V (2 polypeptides), with polypeptides in region I having the highest molecular weight. Polypeptides II-2 and II-4 were missing in T4 when the mutant was grown at the restrictive temperature. Chua and Bennoun (29) studied the same mutant grown under restrictive conditions; they reported that it is lacking polypeptide 6 and that polypeptides 4.1, 4.2, and 5 also seem greatly reduced in amount. All of these polypeptides run in the same region of the gel as the region II polypeptides of Kretzer and associates. This sug-

gests identity of the two groups of polypeptides in these systems. Further support for this hypothesis comes from the observation that synthesis of these polypeptides is inhibited by chloramphenicol in both systems. If this identification is correct, then at least one of the two region I polypeptides of Kretzer and associates ought to correspond to polypeptide 2 of Chua and Gillham. This hypothesis is supported by the observation that synthesis of these polypeptides is inhibited by both chloramphenicol and cycloheximide. Since the T4 mutant is known to be deficient in the photosystem II reaction center, a fact on which Chua and Bennoun and Kretzer and associates were agreed, these results suggest that the previous identification of peak II polypeptides with CPI by Bar-Nun and Ohad was probably erroneous. Instead, it seems likely that one of the region I polypeptides must have been enriched in the complex obtained by them. Certain of the polypeptides of Bar-Nun and Ohad associated with CPII (regions III and IV) that are synthesized on cytoplasmic ribosomes must be identical to some of the polypeptides (polypeptides 11–17) in the region of 25,000 to 30,000 d of Chua and Gillham.

Biosynthesis of thylakoid membrane polypeptides in higher plants has been studied both *in vivo* using inhibitors of chloroplast and cytoplasmic protein synthesis and *in vitro* in isolated chloroplasts. The *in vivo* studies of Machold and Aurich (30), who exposed excised shoots of *Vicia faba* to radioactive amino acids in the presence of chloramphenicol or cycloheximide for 5 or 25 hr, are the studies most easily compared with those done with *Chlamydomonas*. The chloroplast lamellae were isolated, and polypeptides were subjected to electrophoresis in an SDS gel system that gave high resolution. Machold and Aurich were able to visualize 21 distinct polypeptides; synthesis of 12 of these was inhibited by chloramphenicol. The synthesis of the CPI polypeptide, designated B, was strongly inhibited by chloramphenicol and weakly inhibited by cycloheximide. Chua and Gillham (24) subsequently observed exactly the same inhibition pattern for polypeptide 2 in *C. reinhardtii*. Synthesis of polypeptides E, F, and G, probably the equivalents of polypeptides 4–6 in *C. reinhardtii,* was inhibited by chloramphenicol but not cycloheximide, as would be expected from the *Chlamydomonas* results. Among the polypeptides synthesized on cytoplasmic ribosomes is one associated with CPII that is equivalent to polypeptide 11 in *C. reinhardtii*. Thus the results obtained by Machold and Aurich with *V. faba* parallel those of Chua and Gillham, except that a higher proportion of the total polypeptides (approximately 60%) appears to be inhibited by chloramphenicol. In this regard, it should be noted that the *V. faba* experiments were relatively long-term, and some of the inhibitor effects might be indirect.

Ellis (31,32) analyzed thylakoid membrane polypeptide synthesis *in vivo* in etiolated pea shoots during regreening. Lincomycin and cycloheximide were used to inhibit chloroplast and cytoplasmic protein synthesis, respectively. Membrane polypeptides were labeled with ^{35}S-methionine, and membrane systems were isolated for analysis by SDS gel electrophoresis. Accumulation of a CPI polypeptide, which Ellis called polypeptide 2, was completely inhibited by lincomycin and partially inhibited by cycloheximide (31,32). Thus the inhibition patterns seen for this polypeptide were identical in *Chlamydomonas, Vicia,* and *Pisum*. However, Ellis (31) made the converse hypothesis to that of Chua and Gillham (24); he supposed that polypeptide 2 was synthesized on cytoplasmic ribosomes and required the presence of a partner protein synthesized on chloroplast ribosomes for its integration. Three other cycloheximide-resistant peaks of incorporation were also observed (31). One of these, which was not discussed by Ellis, was in the correct range of molecular weight to correspond to poly-

peptides 4–6 of Chua and Gillham (24) in *C. reinhardtii* and polypeptides E–F of Machold and Aurich (30) in *Vicia*. The other two cycloheximide-resistant peaks were in the region of the gel that probably corresponds to polypeptides 8–17 of Chua and Gillham (24). The major radioactive peak of Ellis, peak D, was in the same position as a stained band. In *C. reinhardtii* this region of the gel comprised polypeptides 8–10, which are made on cytoplasmic ribosomes, and D-1, which is the rapidly labeled component made on chloroplast ribosomes corresponding to a diffuse, poorly stained band. Whether peak D of Ellis corresponds to a stained band will be considered later, for this region of the gel may be as complex as it is in *C. reinhardtii*.

Both the gel positions and the sites of synthesis of the CPII polypeptides in Ellis's experiments are the same as in *V. faba* and *C. reinhardtii*. The third cycloheximide-resistant peak of incorporation observed in Ellis's experiments is also in this part of the gel, and it may be equivalent to D-2 in *C. reinhardtii,* which also corresponds to a faintly stained polypeptide.

Most other studies on the biogenesis of thylakoid membrane polypeptides in higher plants employed isolated chloroplasts. Eaglesham and Ellis (8) examined light- or ATP-driven incorporation of ^{35}S-methionine into thylakoid membranes of isolated pea plastids; they observed five radioactive peaks, all of which were pronase-digestible. However, only peak D (32,000 d) appeared to correspond to a stained polypeptide. Siddell and Ellis (33) later performed a similar experiment with etioplasts isolated in various stages of the regreening process. Incorporation occurred first into the large subunit of RuBPCase and later into peak D. No other incorporation of any significance was observed in the thylakoid membranes. Bottomley et al. (34) examined incorporation of ^{35}S-methionine into thylakoid membrane polypeptides of isolated spinach chloroplasts and reported incorporation of label into at least nine discrete products in the membrane fraction. One major peak of labeling corresponded to a stained band of molecular weight 72,000 d. One of the highly radioactive peaks that did not correspond to a stained band was equivalent to a polypeptide with a molecular weight of 36,000 d. Morgenthaler and Mendiola-Morgenthaler (35), also using isolated spinach chloroplasts, observed five radioactive peaks in the thylakoid membrane fraction. The major peak of radioactivity corresponded to a stained band and had a molecular weight similar to that of peak D of Ellis and that of the nonstaining radioactive peak observed by Bottomley and associates. The experiments of Bottomley and associates and Morgenthaler and Mendiola-Morgenthaler differ only in the method of preparation of the spinach thylakoid membrane polypeptides for electrophoresis. Bottomley and associates acetone-extracted their membranes prior to electrophoresis, whereas Morgenthaler and Mendiola-Morgenthaler did not. Chua and Gillham (24) found in *Chlamydomonas* that acetone extraction led to selective loss of specific polypeptides from the gels. It is conceivable that peak D is not a homogeneous protein, but a group of two or more proteins of similar molecular weights. The staining protein or proteins would not run in the SDS gels following acetone extraction of the membranes, whereas the rapidly labeling components would. The true nature of peak D will not be revealed until gels with better molecular weight resolution are employed and more sophisticated methods of membrane polypeptide fractionation (such as differential solubility in chloroform-methanol) are used. The stained polypeptides in the peak D region that do not label in isolated chloroplasts may well prove to be equivalent to polypeptides 8–10 in *Chlamydomonas,* which are made on cytoplasmic ribosomes; the highly radioactive component may prove to be equivalent to peak D-1 in *Chlamydomonas,* which labels rap-

idly, does not stain well, and is a product of chloroplast protein synthesis.

As in the case of the mitochondrial inner membrane (see Chapter 19), the most satisfactory picture of the sites of synthesis of individual polypeptides in the thylakoid membranes has come from *in vivo* experiments with specific inhibitors carried out over a short term rather than from *in vitro* experiments with isolated organelles. In order to determine by rigorous means if the results obtained by Chua and Gillham (24) for *Chlamydomonas* can be generalized satisfactorily to higher plants, one must establish the identity of specific polypeptides through immunologic techniques. Until this is done, the comparisons made here between *Chlamydomonas* and the higher plants must be regarded as suppositional.

Recently, Joy and Ellis (36) showed that 2 of some 25 or more chloroplast envelope proteins become labeled in isolated pea chloroplasts. This labeling is inhibited by chloramphenicol but not cycloheximide. Morgenthaler and Mendiola-Morgenthaler (35) observed that envelope membranes from spinach chloroplasts labeled *in vitro* contained one major radioactive component and two minor components.

LOCATIONS OF GENES CODING FOR THYLAKOID MEMBRANE POLYPEPTIDES

Our understanding of the number and location of genes coding for specific thylakoid membrane polypeptides is at a primitive stage. However, these matters are under study, and three methods are currently being employed. The first makes use of strains of a single species or different species that are interfertile to study the inheritance of molecular weight or other differences that can be found with respect to a specific polypeptide. This approach was discussed in Chapter 18 for interspecies hybrids in *Nicotiana;* in this chapter we shall consider similar experiments that pertain to the inheritance of a variant thylakoid membrane polypeptide in *Chlamydomonas*. The second method makes use of mutants having photosynthetic defects. The pattern of inheritance of these mutants is examined, and at the same time, specific thylakoid membrane defects are sought. Third, messages for membrane polypeptides can be isolated and hybridized to clDNA restriction fragments. This method is only now coming into use; thus far, it has been applied only to the large subunit of RuBPCase, as we shall discuss presently. In the near future we may expect to see increasing numbers of reports in which messages for specific thylakoid membrane polypeptides are localized to restriction fragments of clDNA. Ultimately, this molecular mapping method will likely prove the most powerful method for determining chloroplast gene function in higher plants. In *Chlamydomonas* the method can be used in conjunction with chloroplast gene mapping.

Both nuclear and chloroplast mutants have been described in *Chlamydomonas* that specifically block the synthesis of polypeptide 2 and therefore CPI. Chua et al. (25) characterized two mendelian mutants of *C. reinhardtii,* F1 and F14, that lacked polypeptide 2 and the ability to form CPI. Very recently, Bennoun and associates (37) described a chloroplast gene mutant of *C. reinhardtii,* C_1, that contains only 10 to 15% of the normal amount of CPI in its thylakoid membranes. This deficiency is caused by a drastic reduction in the amount of polypeptide 2 in the membranes. As Bennoun and associates pointed out, although polypeptide 2 is synthesized on chloroplast ribosomes, its insertion into thylakoid membranes seems also to depend on cytoplasmic protein synthesis. Therefore it is not surprising that the presence or absence of this polypeptide in the thylakoid membranes is under the control of both nuclear and chloroplast genes. Mutant C_1 is the first chloroplast mutant of *Chlamydomonas* to which

a specific photosynthetic defect can be assigned. In this sense it is analogous to the *mit⁻* mutants of *S. cerevisiae* that have specific defects in inner-membrane-related functions (see Chapter 19). More such mutants have been reported by Shepherd et al. (38), and they seem to map in several chloroplast genes. Thus we may now expect this line of attack on chloroplast gene function to open up rapidly. In *Antirrhinum* Herrmann (39) described a plastome mutant (*en:alba-1*) (see Chapter 17) that lacks both CPI and the polypeptide associated with this complex that presumably is the equivalent of polypeptide 2 in *Chlamydomonas*.

Chua and Bennoun (29) investigated two mendelian mutants in *Chlamydomonas* that are deranged in the reaction center complex of PSII. On the basis of fluorescence induction kinetics, the primary electron acceptor of PSII, Q, is either missing or inactive in both mutants. One of the mutants, *F34,* is nonconditional, but it is suppressed by another nuclear mutation, *SU-1*. The second mutant, *T4,* is temperature-sensitive and has a wild-type phenotype at 25°C but a mutant phenotype at 35°C. Analysis of the membrane polypeptides shows that *F34* lacks polypeptide 6 and that polypeptide 5 is reduced by 50% as compared to the wild type. When *F34* is coupled with *SU-1*, the activity of PS II is partially restored to about 60%; this is paralleled by partial recovery of polypeptide 6 to about 50%. The stoichiometric recovery of polypeptide 6 in *F34 SU-1* strongly suggests that this polypeptide is a component of the PSII reaction center complex. Growth of *T4* at the restrictive temperature causes the mutant to behave generally like *F34*, except that polypeptides 4.1 and 4.2 become reduced in addition to 5 and 6. These polypeptides are all products of chloroplast protein synthesis (Table 20–1). If polypeptides translated in the chloroplast are coded in the chloroplast, then the products of the *F34* and *T4* genes must be involved indirectly in the assembly or processing of polypeptides 5 and 6 and possibly 4.1 and 4.2. The fact that each mutant is to some extent pleiotropic is consistent with such a hypothesis.

Chua (40) characterized a chloroplast gene mutant (*thm-u-1*) that synthesizes a variant polypeptide in place of lamellar polypeptide 5 (apparent molecular weight 50,000 d). The function of this polypeptide is unknown. The variant, designated 5′, is larger than the wild-type counterpart by about 1,000 d. Replacement of polypeptide 5 by 5′ does not have any apparent effect on the photosynthetic capacities of the membranes. The *thm-u-1* mutant is transmitted to all meiotic progeny when carried by the maternal parent, but it is rarely transmitted by the paternal parent. When biparental transmission occurs, *thm-u-1* segregates somatically during mitosis of haploid zoospores and vegetative diploids with respect to its wild-type allele. A small percentage of the diploids remain heterozygous and continue to segregate the 5 and 5′ polypeptides for many cell generations. Since these biparental diploids synthesize both polypeptides 5 and 5′, these results strongly suggest that *thm-u-1* codes for polypeptide 5. Inhibitor studies have shown that polypeptide 5′, like polypeptide 5 (Table 20–1), is made on chloroplast ribosomes (41).

The membrane polypeptides of *C. reinhardtii* mutants deficient in PSII were also studied by Levine et al. (42–44) using the gel system devised by Hoober (15). Polypeptides a, b, and c of Hoober were designated IIa, IIb, and IIc by Levine and associates (42) because submembrane fragments enriched in PSII activity were also enriched in these polypeptides. Using membrane preparations that had been extracted with acetone, Levine et al. (42) examined two mendelian photosystem II mutants, *ac-141* and *lfd-27;* they found that polypeptide IIa was reduced in *ac-141* but that all three PSII polypeptides were present in normal amounts in *lfd-27*. Since in wild-

type *Chlamydomonas* polypeptide 5 is absent from gels of acetone-extracted material and polypeptide 6 is reduced in amount (24), the possibility that these two polypeptides are deficient in the *ac-141* mutant cannot be ruled out.

In a later study with a pigment-deficient mendelian mutant, *ac-5,* Levine and Duram (43) reported that polypeptides IIa, IIb, and IIc were present in reasonably normal amounts in cells grown photosynthetically but were very deficient in cells grown with acetate as a carbon source. Nevertheless, the Hill reaction rates (a measure of PSII activity) were similar in both groups of cells since the Hill reaction rate on a chlorophyll basis in *ac-5* is much higher than it is in the wild type, but is the same whether cells are grown with CO_2 or acetate as carbon source (45), despite the great fluctuations in amounts of polypeptides IIa, IIb, and IIc under the two sets of conditions; one can only conclude that these polypeptides do not play a direct role in the PSII reaction. In light of these findings the experiments with *ac-141* cannot be regarded as establishing that polypeptide IIa is important in PSII activity, especially when acetone-extracted materials instead of unextracted membranes are analyzed. In any event, polypeptide IIa of Levine et al. (42) is equivalent to at least three polypeptides (polypeptides 8–10) in the system of Chua et al. (24,25,29).

Anderson and Levine (44) examined the thylakoid membrane polypeptides of a chlorophyll-deficient pea mutant and a mendelian mutant of barley lacking chlorophyll b whose chlorophyll-protein complexes were also studied by Thornber and Highkin (46). The barley and pea mutants both contained diminished amounts of polypeptides IIb and IIc, but polypeptide IIa was present in normal amounts. By analogy with *Chlamydomonas,* the deficiency in polypeptide IIb was to be expected since polypeptide IIb is a constituent of the light harvesting chlorophyll-protein complex (44). Herrmann (47) (see Chapter 17) described a second plastome mutant of *Antirrhinum, en:viridis-1,* that is pale green and contains 37% of the chlorophyll of wild type. This mutant has an almost normal ratio of chlorophyll a to chlorophyll b, but it cannot perform the Hill reaction. In the mutant the two polypeptides, 2 and 13, associated with the two pigment-protein complexes are present in almost normal amounts; but a major polypeptide, 7, is missing, and polypeptides 8–12 are decreased in amount. Polypeptide 7 in *Antirrhinum* is very likely equivalent to polypeptide 6 in *C. reinhardtii,* which could explain why the mutant shows no Hill reaction. Since this polypeptide is a product of chloroplast protein synthesis in *C. reinhardtii,* it would not be unreasonable if it were also coded by the chloroplast genome.

The foregoing discussion should indicate that at the moment we have only a faint idea which thylakoid membrane polypeptides are coded by nuclear genes and which by chloroplast genes (Table 20–1). With the discovery of chloroplast mutants in *Chlamydomonas* having specific photosynthetic defects, it now becomes possible to attack the question of chloroplast gene function directly. We may hope that in the next few years this approach will prove as fruitful in *Chlamydomonas* as analysis of the *mit⁻* mutants of *S. cerevisiae* (see Chapter 19) has been.

BIOSYNTHESIS OF FRACTION I PROTEIN

The properties of the enzyme RuBPCase were discussed in Chapter 18. Briefly, the enzyme catalyzes the first CO_2 fixation step in photosynthesis and also the first oxygenation step in photorespiration. It accounts for a large fraction of the soluble stromal protein of most green plants, and it has the same structure in eukaryotic algae and higher plants. RuBPCase is an aggregate of two polypeptide subunits having molecular weights of approximately 55,000 d (large subunit) and 12,000 d (small subunit).

The total molecular weight of the aggregate is approximately 560,000 d, and each molecule of RuBPCase is believed to contain eight large and eight small subunits. Evidence from *Nicotiana* (see Chapter 18) indicates that the large subunit is coded by clDNA and the small subunit by nDNA.

The large subunit of RuBPCase is synthesized in the chloroplast and the small subunit in the cytoplasm. Involvement of both protein-synthesizing systems was suggested by Criddle et al. (48) on the basis of a double-labeling experiment in barley. They found that chloramphenicol affected formation of the large subunit and that cycloheximide blocked synthesis of the small subunit. Subsequent inhibitor studies in other plants yielded similar results (2,49,50). Blair and Ellis (51) then showed that isolated pea chloroplasts could synthesize the large subunit but not the small subunit of RuBPCase and that this was the only soluble chloroplast polypeptide made in this system. The *in-vitro*-labeled product was shown to be identical with the native large subunit in terms of tryptic peptide maps. Gray and Kekwick (52) described results complementary to those of Blair and Ellis; they used an *in vitro* protein-synthesizing system in which 80S polysomes from bean incorporated ¹⁴C-labeled amino acids into a protein precipitated by antisera to the small subunit of RuBPCase. Gooding et al. (53) made antibodies to the large and small subunits of RuBPCase from wheat. The 70S chloroplast ribosomes and the 80S cytoplasmic ribosomes were then separated by density-gradient centrifugation; nascent peptides were released with ³H-puromycin, and the puromycin-labeled peptides precipitated with antisera to the two subunits. 70S ribosomes were found to have only large-subunit peptides attached to them, but the 80S ribosomes had peptides derived from both subunits associated with them. Gooding and associates suggested that this association probably reflected the complexing of completed large subunits with nascent chains of small subunits being made on 80S ribosomes and did not mean that large subunits were being made in both chloroplast and cytoplasm. Roy et al. (54) recently established that the small subunit of RuBPCase is the product of a small proportion of cytoplasmic polyribosomes during greening of etiolated wheat seedlings.

As discussed earlier (see Chapter 3) Hartley and associates (55) were the first to obtain *in vitro* translation of mRNA for the large subunit of fraction I protein. They translated total chloroplast RNA from spinach using a cell free extract from *E. coli* and found that two products, one of 52,000 d and the other of 35,000 d, were made. The 52,000 d product was slightly smaller (ca. 1,500 d) than the native large subunit of RuBPCase made *in vitro* by isolated pea chloroplasts, and contained seven of the nine chymotryptic peptides of these native large subunits. Wheeler and Hartley (66) subsequently showed that the mRNA programming large subunit synthesis was not polyadenylated nor was the message coding for the 35,000 d polypeptide. Sagher and associates (67) confirmed that the large subunit mRNA from *Euglena* was not polyadenylated either. Furthermore, they succeeded in isolating a product identical with native large subunit using the wheat germ *in vitro* translation system. They also showed that this message was absent from the nonpolyadenylated RNA fraction of the aplastidic mutant W_3BUL. Howell and associates (64) also report that large subunit mRNA from *Chlamydomonas* is not polyadenylated. These experiments are discussed in more detail below. Suffice it to say here that Howell and associates found as had Hartley and associates that the product formed following *in vitro* translation of large subunit message in the *E. coli* system was incomplete large subunit.

Because of its abundance and ease of isolation, RuBPCase should be an ideal model for investigating how polypeptides (in this case the small subunit) are trans-

ported into the chloroplast and how the synthesis of a protein (part of which is coded in the nucleus and part of which is coded by clDNA) is regulated. Chua and his colleagues (56,57) initiated such an investigation in *Chlamydomonas*. Iwanij et al. (56) first demonstrated a very tight control of synthesis of the two subunits of RuBPCase *in vivo* using synchronous cultures of *C. reinhardtii*. Both subunits were made in synchrony during the light part of the light–dark cycle. Dobberstein and associates (57) then succeeded in demonstrating that the small subunit was made as a precursor *in vitro*. This precursor was then clipped at a specific point by an endoprotease present in a postpolysomal supernatant of *C. reinhardtii*. Since the small subunit was made on cytoplasmic ribosomes, and since there was little evidence that 80S ribosomes were attached to the outer envelope of the chloroplast, Dobberstein and associates speculated that the extra sequence present on the precursor was not the functional equivalent of the signal sequence found at the amino terminal end of membrane-bound polysomes (see Chapter 3). Instead, they supposed that the fragment cleaved from the precursor might be required for recognition of a receptor on the chloroplast envelope, perhaps in regions where the inner and outer chloroplast envelopes come into contact. This fragment would mediate transfer of the small subunit into the chloroplast, at which time it would be cleaved off so that only the authentic small subunit would traverse the chloroplast membranes and enter the stroma.

LOCATIONS OF GENES CODING FOR RuBPCase LARGE AND SMALL SUBUNITS

Until recently, experiments with interspecific hybrids of *Nicotiana* had provided the strongest evidence that the large subunit of RuBPCase was coded by clDNA (see Chapter 18). Now, strong supporting evidence that this assignment is correct has come from *in vitro* hybridization experiments between large-subunit mRNA and clDNA from *Chlamydomonas* (p. 556). The small subunit appears to be coded by DNA, and the best evidence for this is still the results of the *Nicotiana* experiments. To date, no mutants have been reported that specifically affect the structure or catalytic activity of the large subunit of RuBPCase. In contrast, two nuclear gene mutations have been identified that alter the carboxylase/oxygenase activity ratio of RuBPCase. Kung and Marsho (58) described a dominant mendelian mutant in tobacco (*Su*) that produces yellow plants when in the heterozygous (*Su su*) condition, but is lethal when homozygous. The mutant shows a twofold to threefold higher rate of photorespiration than wild type, as well as reduced carboxylase and oxygenase activities. Although no differences in the isoelectric focusing patterns of either the large or small subunits could be detected between mutant and wild type, Kung and Marsho were able to demonstrate that the ratio of oxygenase activity to carboxylase activity was lower in purified fraction I protein from the mutant, as compared to the wild type. Kung and Marsho concluded that since the *su* mutant specifically affects the activity of fraction I protein, particularly its carboxylase/oxygenase activity ratio, and since the mutant was inherited in a mendelian fashion, it must modify the small subunit of this enzyme in some manner that in turn affects the catalytic activity of the large subunit. Hence they suggested that the small subunit plays a regulatory role in the function of this enzyme aggregate. They also questioned the ability of the oxygenase activity of this enzyme to account totally for photorespiration, since the low oxygenase activity of purified fraction I protein from the mutant could not account for the observation that its photorespiration rate was twofold to threefold higher than that of the wild type.

The mendelian acetate-requiring mutant *ac-i72* of *C. reinhardtii* isolated by Nelson and Surzycki (59,60) was also reported to alter the carboxylase/oxygenase activity of fraction I protein. This mutant is similar to the yellow mutant of tobacco studied by Kung and Marsho (58) in having subnormal levels of CO_2 fixation and RuBPCase activity together with altered oxygenase activity and high glycolate production. The *ac-i72* mutant is believed to die under normal phototrophic growth conditions because of accumulated glycolate, but it will survive under an amber light regime, where total light intensity is presumably reduced, or in a low-oxygen environment. Purified RuBPCase from *ac-i72* shows reduced specific activity, lower V_{max}, increased requirements for Mg^{2+} for maximal activity, greater sensitivity to inhibition by Cl^-, and a slightly altered isoelectric point, as compared to the wild type, when isolated in Hepes buffer, but not when isolated in Tris buffer of the same molarity and pH. However, the RuBPCase from the *ac-i72* mutant and that from the wild type are identical in amount, pH requirement, temperature sensitivity, molecular weight, subunit structure, and sedimentation coefficient, as well as in K_m for the CO_2 and ribulose-1,5-bisphosphate substrates. Clearly, this nuclear gene mutant appears to affect the functioning of fraction I protein; but this alteration, like that in the yellow tobacco mutant characterized by Kung and Marsho (58), has not yet been shown to lead to alteration of a specific peptide in either subunit of the enzyme nor to affect directly either subunit in any physical way. Until point mutants with better-defined effects on the RuBPCase molecule are isolated and characterized, we cannot use this direct genetic approach to determine where the genes coding for the peptides of the two subunits of fraction I protein are localized or how many genes are involved.

Howell and his colleagues (61–64) took a direct molecular approach to the localization of the genes coding for the large subunit of RuBPCase. These investigators made antibodies to the large subunit and then precipitated chloroplast polysomes synthesizing this subunit using the antibody. They found that the large subunit was made on very small chloroplast polysomes ($n = 2$–5), which was surprising in view of the high molecular weight of this polypeptide (approximately 55,000 d). These small polysomes could be isolated on affinity columns to which anti-large-subunit antibodies were attached. Large-subunit mRNA was isolated from these polysomes; it was found to have a molecular weight of 4×10^5 d, and it was translated *in vitro* to yield peptides precipitable by antibodies to the large subunit. Radioactive large-subunit mRNA isolated by this method hybridized to a specific EcoR1 restriction fragment of clDNA having a molecular weight of 3.2×10^6 d. This same fragment also contained portions of at least one set of 16S and 23S chloroplast rRNA cistrons. However, these rRNAs did not compete with large-subunit mRNA in hybridization competition experiments; thus the mRNA must hybridize to a sequence distinct from that hybridizing the rRNA, as would be expected.

If we assume that only one strand of the fragment of clDNA of molecular weight 3.2×10^6 d codes for large-subunit mRNA, as well as 16S and 23S rRNA, then all three cistrons cannot be present intact on the same fragment. The molecular weight of the three RNAs summed is 2.03×10^6 d, but the molecular weight of the complementary clDNA strand is only 1.6×10^6 d.

From their hybridization experiments, Howell and his colleagues calculated that there was one large-subunit gene per clDNA molecule, or about 75 per chloroplast. They also argued that a high rate of transcription of these genes would be required to maintain the high level of RuBPCase molecules present in the chloroplast and that the juxtaposition of the 23S, 16S, and large-subunit genes might relate to their mutually high rates of transcription. If this argument is correct, it is perhaps not surprising that large-subunit mutants residing in clDNA have not been isolated. After all, if the 75 clDNA copies of the large-subunit gene are all active, a mutation will rarely be expressed. The next logical question is why mutations in other chloroplast genes can be isolated (see Chapters 14 and 21). Are only a few copies of a given chloroplast gene normally expressed? Are the genes for which mutations have been isolated in nonredundant fractions of the clDNA (see Chapter 15) with the RuBPCase genes being localized in the redundant fraction? Is there some other parameter of chloroplast gene segregation that permits mutation expression even when the mutation arises in a highly redundant fraction of the DNA? We do not know the answers to these questions. We can only say that thus far there is no evidence for a nonredundant fraction of clDNA in *Chlamydomonas*. Obviously, if a mutant can be isolated in the structural gene for the large subunit, and if it shows the same pattern of inheritance as other chloroplast genes as well as linkage to them, it will be necessary to conclude that the highly redundant chloroplast genes behave genetically as a small number of segregating units.

CONCLUSIONS

Evidence presented in this chapter and in Chapter 18 indicates that chloroplast and nuclear genes code for the large and small subunits, respectively, of RuBPCase. This dichotomy in the locations of the structural genes for the two subunits is reflected in their sites of translation as well, with the large subunit being made in the chloroplast and the small subunit in the cytoplasm. This, then, is the orthodox model in which protein transport into the organelle, rather than RNA transport, is involved (see Chapter 3 for discussion). In fact, the recent studies of Chua and his colleagues (57) have identified a small-subunit precursor with a fragment that may be important in the transport process. It is also evident that some of the 33 or more thylakoid membrane polypeptides of the chloroplast are synthesized on chloroplast ribosomes, although it is not yet certain that this is the same group in both *Chlamydomonas* and higher plants. These include polypeptides important for the functioning of photosystems I and II and the CF_1 ATPase. The remainder of these polypeptides are, of course, made on cytoplasmic ribosomes. At the moment, no structural gene, either nuclear or chloroplast, that codes for any thylakoid membrane polypeptide has been identified. However, it is apparent that specific polypeptides, at least with respect to their integration into the thylakoid membranes, can be affected by both chloroplast and nuclear genes. In the next few years we can expect to see much more relevant information on the structural genes coding for thylakoid membrane polypeptides, particularly in the case of the chloroplast genome. One approach will involve careful characterization of newly discovered chloroplast mutants, particularly in *Chlamydomonas*. A second approach will involve molecular hybridization of specific mRNAs to restriction fragments of clDNA. The latter approach should prove particularly fruitful in deciphering the function of the chloroplast genome in higher plants, which thus far have proved rather intractable to formal genetic analysis.

REFERENCES

1. Gillham, N. W., Boynton, J. E., and Chua, N. -H. (1978): Genetic control of choloroplast proteins. *Curr. Top. Bioenergetics, 7,* in press.
2. Armstrong, J. J., Surzycki, S. J., Moll, B., and Levine, R. P. (1971): Genetic transcription and translation specifying chloroplast components in *Chlamydomonas reinhardi. Biochemistry,* 10:692–701.
3. Schmidt, G. W., and Lyman, H. (1976): Inheritance and synthesis of chloroplasts and mitochondria of *Euglena gracilis.* In: *The Genetics of Algae,* edited by R. A. Lewin, Chapter 13. Blackwell Scientific, Oxford.
4. Penefsky, H. S. (1974): Mitochondrial and chloroplast ATPases. In: *The Enzymes, Vol. 10,* edited by P. D. Boyer, pp. 375–394. Academic Press, New York.
5. Panet, R., and Sanadi, D. R. (1976): Soluble and membrane ATPases of mitochondria, chloroplasts and bacteria: Molecular structure, enzymatic properties and functions. *Curr. Top. Membranes Transport,* 8:99–160.
6. Horak, A., and Hill, R. D. (1971): Coupling factor for photophosphorylation in bean etioplasts and chloroplasts. *Can. J. Biochem.,* 49: 207–209.
7. Horak, A., and Hill, R. D. (1972): Adenosine triphosphatase of bean plastids. Its properties and site of formation. *Plant Physiol.,* 49:365–370.
8. Eaglesham, A. R. J., and Ellis, R. J. (1974): Protein synthesis in chloroplasts. II. Light-driven synthesis of membrane proteins by isolated pea chloroplasts. *Biochim. Biophys. Acta,* 335:396–407.
9. Mendiola-Morgenthaler, L. R., Morgenthaler, T. T., and Price, C. A. (1976): Synthesis of coupling factor CF_1 protein by isolated spinach chloroplasts. *FEBS Lett.,* 62:96–100.
10. Hudock, G. A., and Levine, R. P. (1964): Regulation of photosynthesis in *Chlamydomonas reinhardi. Plant Physiol.,* 39:889–897.
11. Hudock, G. A., McLeod, G. C., Moravkova-Kiely, J., and Levine, R. P. (1964): The relation of oxygen evolution to chlorophyll and protein synthesis in a mutant strain of *Chlamydomonas reinhardi. Plant Physiol.,* 39: 898–906.
12. Ohad, I., Siekevitz, P., and Palade, G. E. (1967): Biogenesis of chloroplast membranes. I. Plastid dedifferentiation in a dark-grown algal mutant (*Chlamydomonas reinhardi*). *J. Cell Biol.,* 35:521–552.
13. Ohad, I., Siekevitz, P., and Palade, G. E. (1967): Biogenesis of chloroplast membranes. II. Plastid differentiation during greening of a dark-grown algal mutant (*Chlamydomonas reinhardi*). *J. Cell Biol.,* 35:553–584.
14. Hoober, J. K., Siekevitz, P., and Palade, G. E. (1969): Formation of chloroplast membranes in *Chlamydomonas reinhardi y-1. J. Biol. Chem.,* 244:2621–2631.
15. Hoober, J. K. (1970): Sites of synthesis of chloroplast membrane polypeptides in *Chlamydomonas reinhardi y-1. J. Biol. Chem.,* 245: 4327–4334.
16. Hoober, J. K. (1972): A major polypeptide of chloroplast membranes of *Chlamydomonas reinhardi.* Evidence for synthesis in the cytoplasm as a soluble component. *J. Cell Biol.,* 52:84–96.
17. Hoober, J. K., and Stegeman, W. J. (1973):

Control of the synthesis of a major polypeptide of chloroplast membranes in *Chlamydomonas reinhardi. J. Cell Biol.,* 56:1–12.

18. Hoober, J. K., and Stegeman, W. J. (1975): Regulation of chloroplast membrane biosynthesis. In: *Genetics and Biogenesis of Mitochondria and Chloroplasts,* edited by C. W. Birky, Jr., P. S. Perlman, and T. J. Byers, Chapter 6. Ohio State University Press, Columbus.

19. Hoober, J. K. (1976): Synthesis of the major thylakoid membrane polypeptides during greening of *Chlamydomonas reinhardtii* Y-1. In: *Genetics and Biogenesis of Chloroplasts and Mitochondria,* edited by T. Bücher, W. Neupert, W. Sebald, and S. Werner, pp. 87–94. North Holland, Amsterdam.

20. Eytan, G., and Ohad, I. (1972): Biogenesis of chloroplast membranes. VII. The preservation of membrane homogeneity during development of the photosynthetic lamellar system in an algal mutant (*Chlamydomonas reinhardi y-1*). *J. Biol. Chem.,* 247:112–121.

21. Eytan, G., and Ohad, I. (1972): Biogenesis of chloroplast membranes. VIII. Modulation of chloroplast lamellae composition and function induced by discontinuous illumination and inhibition of ribonucleic acid and protein synthesis during greening of *Chlamydomonas reinhardi y-1* mutant cells. *J. Biol. Chem.,* 247:122–129.

22. Bar-Nun, S., and Ohad, I. (1974): Cytoplasmic and chloroplastic origin of chloroplast membrane proteins associated with PSII and PSI active centers in *Chlamydomonas reinhardi y-1. Proceedings Third International Congress on Photosynthesis,* edited by M. Avron, Elsevier/North Holland, Amsterdam, pp. 1627–1638.

23. Ohad, I. (1975): Biogenesis of chloroplast membranes. In: *Membrane Biogenesis. Mitochondria, Chloroplasts and Bacteria,* edited by A. Tzagoloff, pp. 279–347. Plenum Press, New York.

24. Chua, N.-H., and Gillham, N. W. (1977): The sites of synthesis of the principal thylakoid membrane polypeptides in *Chlamydomonas reinhardtii. J. Cell Biol.,* 74:441–452.

25. Chua, N.-H., Matlin, K., and Bennoun, P. (1975): A chlorophyll-protein complex lacking in photosystem I mutants of *Chlamydomonas reinhardtii. J. Cell Biol.,* 67:361–377.

26. Henningsen, K. W., and Boynton, J. E. (1974): Macromolecular physiology of plastids. IX. Development of plastid membranes during greening of dark-grown barley seedlings. *J. Cell Sci.,* 15:31–55.

27. McLeod, G. C., Hudock, G. A., and Levine, R. P. (1963): The relation between pigment concentration and photosynthetic capacity in a mutant of *Chlamydomonas reinhardi.* In: *Photosynthetic Mechanisms of Green Plants,* pp. 400–408. Publication 1145, National Academy of Science, National Research Council, Washington, D.C.

28. Kretzer, F., Ohad, I., and Bennoun, P. (1976): Ontogeny, insertion, and activation of two thylakoid peptides required for photosystem II activity in the nuclear temperature sensitive T4 mutant of *Chlamydomonas reinhardi.* In: *Genetics and Biogenesis of Chloroplasts and Mitochondria,* edited by T. Bücher, W. Neupert, W. Sebald, and S. Werner, pp. 25–32. North Holland, Amsterdam.

29. Chua, N.-H., and Bennoun, P. (1975): Thylakoid membrane polypeptides of *Chlamydomonas reinhardtii:* Wildtype and mutant strains deficient in photosystem II reaction center. *Proc. Natl. Acad. Sci. U.S.A.,* 72: 2175–2179.

30. Machold, O., and Aurich, O. (1972): Sites of synthesis of chloroplast lamellar proteins in *Vicia faba. Biochim. Biophys. Acta,* 281: 103–112.

31. Ellis, R. J. (1975): Inhibition of chloroplast protein synthesis by lincomycin and 2-(4-methyl-2,6-dinitroanilino)-*N*-methylpropionamide. *Phytochemistry,* 14:89–93.

32. Ellis, R. J. (1975): The synthesis of chloroplast membranes in *Pisum sativum.* In: *Membrane Biogenesis. Mitochondria, Chloroplasts and Bacteria,* edited by A. Tzagoloff, pp. 247–276. Plenum Press, New York.

33. Siddell, S. G., and Ellis, R. J. (1975): Protein synthesis in chloroplasts. Characteristics and products of protein synthesis *in vitro* in etioplasts and developing chloroplasts from pea leaves. *Biochem. J.,* 146:675–685.

34. Bottomley, W., Spencer, D., and Whitfeld, P. R. (1974): Protein synthesis in isolated spinach chloroplasts: Comparison of light-driven and ATP-driven synthesis. *Arch. Biochem. Biophys.,* 164:106–117.

35. Morgenthaler, J. J., and Mendiola-Morgenthaler, L. (1976): Synthesis of soluble, thylakoid and envelope membrane proteins by spinach chloroplasts purified from gradients. *Arch. Biochem. Biophys.,* 172:51–58.

36. Joy, K. W., and Ellis, R. J. (1975): Protein synthesis in chloroplasts. IV. Polypeptides of the chloroplast envelope. *Biochim. Biophys. Acta,* 378:143–151.

37. Bennoun, P., Girard, J., and Chua, N.-H. (1977): A uniparental mutant of *Chlamydomonas reinhardtii* deficient in the chlorophyll-protein complex CPI. *Mol. Gen. Genet.,* 153: 343–348.

38. Shepherd, H. S., Boynton, J. E., and Gillham, N. W. (1977): Mutations in chloroplast DNA affecting photosynthetic function in *Chlamydomonas reinhardtii. J. Cell Biol.,* 75:308a.

39. Herrmann, F. (1971): Genetic control of pigment-protein complexes I and Ia of the plastid mutant *en:alba-1* of *Antirrhinum majus. FEBS Lett.,* 19:267–269.

40. Chua, N.-H. (1976): A uniparental mutant

of *Chlamydomonas reinhardtii* with a variant thylakoid membrane polypeptide. In: *The Genetics and Biogenesis of Chloroplasts and Mitochondria,* edited by T. Bücher, W. Neupert, W. Sebald, and S. Werner, pp. 323–330. North Holland, Amsterdam.

41. Chua, N.-H. (1977): unpublished data.

42. Levine, R. P., Burton, W. G., and Duram, H. A. (1972): Membrane polypeptides associated with photochemical systems. *Nature* [*New Biol.*], 237:176–177.

43. Levine, R. P., and Duram, H. A. (1973): The polypeptides of stacked and unstacked *Chlamydomonas reinhardi* chloroplast membranes and their relation to photosystem II activity. *Biochim. Biophys. Acta,* 325:565–572.

44. Anderson, J. M., and Levine, R. P. (1974): Membrane polypeptides of some higher plant chloroplasts. *Biochim. Biophys. Acta,* 333: 378–387.

45. Goodenough, U. W., and Staehelin, L. A. (1971): Structural differentiation of stacked and unstacked chloroplast membranes. *J. Cell Biol.,* 48:594–619.

46. Thornber, J. P., and Highkin, H. R. (1974): Composition of the photosynthetic apparatus of normal barley leaves and a mutant lacking chlorophyll b. *Eur. J. Biochem.,* 41:109–116.

47. Herrmann, F. (1971): Chloroplast lamellar proteins of the plastid mutant *en:viridis-1* of *Antirrhinum majus* having impaired photosystem II. *Exp. Cell Res.,* 70:452–453.

48. Criddle, R. S., Dau, B., Kleinkopf, G. E., and Huffaker, R. C. (1970): Differential synthesis of the ribulose diphosphate carboxylase subunits. *Biochem. Biophys. Res. Commun.,* 41: 621–627.

49. Givan, A. L., and Criddle, R. S. (1972): Ribulose diphosphate carboxylase from *Chlamydomonas reinhardi:* Purification, properties and its mode of synthesis in the cell. *Arch. Biochem. Biophys.,* 149:153–163.

50. Margulies, M. (1971): Concerning the sites of synthesis of proteins of chloroplast ribosomes and fraction I protein (Ribulose-1,5-diphosphate carboxylase). *Biochem. Biophys. Res. Commun.,* 44:539–545.

51. Blair, G. E., and Ellis, R. J. (1973): Protein synthesis in chloroplasts. I. Light-driven synthesis of the large subunit of fraction I protein by isolated pea chloroplasts. *Biochim. Biophys. Acta,* 319:223–234.

52. Gray, J. C., and Kekwick, R. G. O. (1974): The synthesis of the small subunit of ribulose-1,5-bisphosphate carboxylase in the French bean *Phaseolus vulgaris. Eur. J. Biochem.,* 44:491–500.

53. Gooding, L. R., Roy, H., and Jagendorf, A. T. (1973): Immunological identification of nascent chains of wheat ribulose diphosphate carboxylase on ribosomes of both chloroplast and cytoplasmic origin. *Arch. Biochem. Biophys.,* 159:324–335.

54. Roy, H., Patterson, R., and Jagendorf, A. T. (1976): Identification of the small subunit of ribulose-1,5-bisphosphate carboxylase as a product of wheat leaf cytoplasmic ribosomes. *Arch. Biochem. Biophys.,* 172:64–73.

55. Hartley, M. R., Wheeler, A., and Ellis, R. J. (1975): Protein synthesis in chloroplasts. V. Translation of messenger RNA for the large subunit of fraction I protein in a heterologous cell-free system. *J. Mol. Biol.,* 91:67–77.

56. Iwanij, V., Chua, N.-H., and Siekevitz, P. (1975): Synthesis and turnover of ribulose bisphosphate carboxylase and of its subunits during the cell cycle of *Chlamydomonas reinhardtii. J. Cell Biol.,* 64:572–585.

57. Dobberstein, B., Blobel, G., and Chua, N.-H. (1977): *In vitro* synthesis and processing of a putative precursor for the subunit of ribulose-1,5-bisphosphate carboxylase of *Chlamydomonas reinhardtii. Proc. Natl. Acad. Sci. U.S.A.,* 74:1082–1085.

58. Kung, S.-D., and Marsho, T. V. (1976): Regulation of RuDP carboxylase/oxygenase activity and its relationship to plant physiology. *Nature,* 259:325–326.

59. Nelson, P. E., and Surzycki, S. J. (1976): A mutant strain of *Chlamydomonas reinhardi* exhibiting altered ribulose bisphosphate carboxylase. *Eur. J. Biochem.,* 61:465–474.

60. Nelson, P. E., and Surzycki, S. J. (1976): Characterization of the oxygenase activity in a mutant of *Chlamydomonas reinhardi* exhibiting altered ribulose bisphosphate carboxylase. *Eur. J. Biochem.,* 61:475–480.

61. Howell, S., Heizmann, P., and Gelvin, S. (1976): Localization of the gene coding for the large subunit of ribulose bisphosphate carboxylase on the chloroplast genome of *Chlamydomonas reinhardi.* In: *Genetics and Biogenesis of Chloroplasts and Mitochondria,* edited by T. Bücher, W. Neupert, W. Sebald, and S. Werner, pp. 625–628. North Holland, Amsterdam.

62. Howell, S. H., Heizmann, P., Gelvin, S., and Walker, L. L. (1977): Identification and properties of the messenger RNA activity in *Chlamydomonas reinhardi* coding for the large subunit of *d*-ribulose-1,5-bisphosphate carboxylase. *Plant Physiol.,* 59:464–470.

63. Gelvin, S., and Howell, S. H. (1977): Identification and precipitation of the polyribosomes in *Chlamydomonas reinhardi* involved in the synthesis of the large subunit of *d*-ribulose-1,5-bisphosphate carboxylase. *Plant Physiol.,* 59:471–477.

64. Gelvin, S., Heizmann, P., and Howell, S. H. (1977): Identification and cloning of the chloroplast gene coding for the large subunit of ribulose-1,5-bisphosphate carboxylase from *Chlamydomonas reinhardi. Proc. Natl. Acad. Sci. U.S.A.,* 74:3193–3197.

65. Bennoun, P., and Chua, N.-H. (1976): Methods for the detection of photosynthetic mu-

tants in *Chlamydomonas reinhardi.* In: *Genetics and Biogenesis of Chloroplasts and Mitochondria,* edited by T. Bücher, W. Neupert, W. Sebald, and S. Werner, pp. 33–39. North Holland, Amsterdam.

66. Wheeler, A. M., and M. R. Hartley. (1975): Major mRNA species from spinach chloroplasts do not contain poly(A). *Nature,* 257: 66–67.

67. Sagher, D., Grosfeld, H., and Edelman, M. (1976): Large subunit ribulosebisphosphate carboxylase messenger RNA from *Euglena* chloroplasts. *Proc. Natl. Acad. Sci. U.S.A.,* 73:722–726.

Chapter 21
Biogenesis of Organelle Protein-Synthesizing Systems

In many ways, organelle protein-synthesizing systems are like random assemblages of parts gathered from various places throughout the cell. A priori, it would seem that an orderly mind would have designed the organelle to be responsible for the entire blueprint, manufacture and assembly, of its protein-synthesizing system. Certainly, such a straightforward approach would have made life simpler for the student of organelle heredity. However, the cell did not evolve by such a simple route; instead, some components are made inside the organelle and some outside. On the assumption that there is always a good reason for what happens in a cell, some investigators have attempted to unravel this complexity, believing that hidden somewhere is a grand design. So far the main efforts have focused on organelle tRNAs, their synthetases, and the RNA and protein components of organelle ribosomes. The biogenesis and genetics of these elements are the focus of this chapter.

BIOGENESIS OF ORGANELLE tRNAs and AMINOACYL tRNA SYNTHETASES

Barnett and his colleagues (1–4) were the first workers to make a systematic study of organelle tRNAs and their synthetases. Their experiments with *Neurospora* revealed unique mitochondrial tRNAs for most of the amino acids. They also reported the existence of unique mitochondrial tRNA synthetases for aspartic acid, leucine, and phenylalanine in this fungus. Since then, abundant evidence has been obtained, particularly for *S. cerevisiae* and HeLa cells, that the mitochondrial genome may code for a complete set of tRNAs (see Chapter 3). Although these may be sufficient for mitochondrial protein synthesis in these systems, Suyama and associates (5) reported that this was not the case in *Tetrahymena* mitochondria, where they found "imported" species of tRNA in addition to those coded by mtDNA. These authors suggested that the aminoacyl tRNA synthetases for the imported tRNAs serve as transport proteins, in addition to their function of catalyzing aminoacylation, thus allowing the polar tRNA molecules to gain entry to the mitochondria via their protein carriers. This interesting hypothesis serves to highlight the fact that the fragmentary information currently available to us indicates that the mitochondrial tRNA synthetases are coded by nuclear genes whose messages are translated in the cytoplasm. In this respect, we can distinguish three sorts of situations regarding these enzymes. First, the mitochondrion contains an enzyme clearly distinct from its cytoplasmic counterpart. Second, the mitochondrion contains a modified form of the cytoplasmic enzyme. Third, the cytoplasmic and mitochondrial enzymes are identical.

The first situation can be illustrated by studies of the mitochondrial leucyl tRNA synthetases of *Tetrahymena* by Chiu and Suyama (6,7) and studies of phenylalanyl tRNA synthetases of *S. cerevisiae* by Schneller et al. (8). The *Tetrahymena* experiments showed that the mitochondrial and cytoplasmic enzymes were immunologically and physically distinct from one another, and the yeast experiments provided similar immunologic criteria for distinguishing the two phenylalanyl tRNA synthetases. In another report, Schneller et al. (9) showed that the mitochondrial phenylalanyl tRNA synthetases, as well as the mitochondrial leucyl and methionyl tRNA synthetases, were present in a ρ^0 mutant, which indicated that they must all be coded by nuclear genes.

Until recently, the results of Gross et al. (10,11) served as the classic example of the second situation, in which a mitochondrial tRNA synthetase, in this case the leucyl tRNA synthetase, appears to be a modified form of the cytoplasmic enzyme. However, it now appears that the two enzymes are distinct from one another and probably are coded by two adjacent genes (12). Originally Gross et al. (10) reported that a temperature sensitive leucine-requiring nuclear mutant of *N. crassa* (*leu-5^{ts}*) that produced a cytoplasmic leucyl tRNA synthetase with a twofold to threefold higher Michaelis constant (K_m) for leucine than the wild type simultaneously lacked the mitochondrial enzyme. Weeks and Gross (11) then obtained a "reversion" of this mutant that appeared to map at the same locus (*leu-5*) as the original mutant. The mitochondrial enzyme was restored in the revertant, although its properties were somewhat altered. The cytoplasmic enzyme in the revertant had a K_m for leucine intermediate between that of the mutant and that of the wild type. Thus, their revertant was not a true revertant, but rather a second-site mutation within the *leu-5* locus or a revertant at the original site in which an amino acid other than that present in the wild type was substituted at the same position. Since both cytoplasmic and mitochondrial synthetases were altered simultaneously in the mutant and the revertant, Weeks and Gross (11) suggested that either the two enzymes shared a common component or one enzyme was a modified version of the other.

Beauchamp et al. (12) tested the foregoing hypothesis directly using antisera prepared against the cytoplasmic and mitochondrial enzymes. The antisera made against the mitochondrial enzyme were specific for that enzyme and did not affect the activity of the cytoplasmic enzyme. The converse was true of the cytoplasmic enzyme. Furthermore, the enzymes were found to differ in size and other physical properties, and so they were clearly different proteins. The next important event was the discovery of an *N. crassa* strain called Mauriceville-lcA that synthesized a variant of the mitochondrial enzyme designated *leu-5^{M}*. Mapping experiments established that *leu-5^{M}* was located very close to *leu-5^{ts}* in the *leu-5* region. Finally, new revertants of *leu-5^{ts}* were isolated that caused reversion of only the temperature-sensitive part of the phenotype and did not restore normal amounts of the mitochondrial enzyme. On the basis of these results, Beauchamp and associates proposed a model that supposed that the structural genes for the cytoplasmic and mitochondrial leucyl tRNA synthetases were arranged in tandem, with transcription proceeding from the cytoplasmic structural gene to the mitochondrial structural gene. Not only did the gene coding for the cytoplasmic enzyme contain its own promoter for transcription, but the promoter for transcription of the mitochondrial gene ($P_m{}^*$) was located within the structural gene for the cytoplasmic enzyme at its distal end. The *leu-5^{ts}* mutation was imagined as falling within $P_m{}^*$, with its effect being to alter the structure of the cytoplasmic enzyme, and hence its K_m, while largely abolishing transcription of the gene coding for the mitochondrial enzyme at the same time. The reversion described by Weeks and Gross (11) would then map within this region too. The reversions affecting temperature sensitivity but not restoring the mitochondrial enzyme were then postulated to be intragenic suppressors lying to the left of $P_m{}^*$ within the gene coding for the cytoplasmic enzyme. The *leu-5^{M}* allele, on the other hand, mapped to the right of $P_m{}^*$ in the structural gene coding for the mitochondrial enzyme. In another article Beauchamp and Gross (13) reported that growth of mycelia in chloramphenicol or ethidium bromide led to an increase in the amount of mitochondrial leucyl tRNA synthetase and a decrease in the amount of the cytoplasmic enzyme. A similar increase in mitochondrial phenylalanyl tRNA synthetase was also seen under these conditions. Beauchamp and Gross suggested that repressor molecules of the kind proposed by Barath and Küntzel (see Chapter 8) were involved. These molecules were presumed to be coded by mtDNA with their messages being translated on mitochondrial ribosomes. Inhibition of synthesis of these regulatory molecules would then lead to derepression of synthesis of mitochondrial enzymes coded by nuclear genes, including, of course, the mitochondrial tRNA synthetases. To accommodate these results, Beauchamp et al. (12) theorized that the gene coding for the mitochondrial leucyl tRNA synthetase also contained a site that recognized the repressor.

Whether or not the precise model of Beauchamp and associates is correct, these experiments established four important points. First, the mitochondrial and cytoplasmic leucyl tRNA synthetases are distinct enzymes. Second, they seem to be coded by two tightly linked genes mapping in the *leu-5* region. Third, mutations mapping in this region can affect one of the enzymes separately or both simultaneously. Fourth, inhibition of mitochondrial protein synthesis or transcription seems to uncouple regulation of synthesis of the two enzymes.

In the third situation, the cytoplasmic and mitochondrial tRNA synthetases for a given amino acid appear to be very similar if not identical. Suyama and Hamada (5) gave some examples for *Tetrahymena* in the cases of the valyl, arginyl, isoleucyl, and lysyl tRNA synthetases, all of which are quite similar in their abilities to charge both mitochondrial and cytoplasmic tRNAs, as well as in their kinetic properties. Whether or not the enzymes are immunologically identical remains to be determined. It is interesting that these enzymes charge

tRNAs that appear to be imported into the *Tetrahymena* mitochondrion. Suyama and Hamada reported that those enzymes that charge native tRNAs (i.e., those coded by mtDNA) are clearly functionally and structurally distinct from the comparable cytoplasmic enzymes.

Chloroplasts, like mitochondria, contain their own tRNAs and aminoacyl tRNA synthetases, although we have far less information, particularly concerning the tRNAs, than we do for the mitochondrion (see Chapter 3). Tewari and Wildman (14) isolated a tRNA fraction from tobacco chloroplasts and showed that these tRNAs could be charged by a synthetase fraction from chloroplasts. The tRNAs thus charged could be hybridized to clDNA; they included leucyl, valyl, and phenylalanyl tRNAs. Hybridization experiments with total 4S RNA indicated that the tobacco chloroplast genome contained information sufficient to code for 20 to 30 tRNAs of molecular weight 25,000 d. More recently, Schwartzbach et al. (15) showed by hybridization of total iodinated chloroplast tRNA that the chloroplast genome of *Euglena* contains 26 tRNA cistrons. They also showed that these included the cistrons for the tRNAs for aspartic acid and phenylalanine.

The most extensive studies of chloroplast tRNAs and their synthetases were carried out by Barnett et al. (16) and Weil et al. (17) on *Euglena* and French bean (*Phaseolus vulgaris*), respectively. We shall review the results of both groups of workers here because they illustrate two points of significance. First, production of chloroplast tRNAs and their synthetases can show a remarkable degree of coordination. Second, complex relationships may exist in green plants between tRNAs and their synthetases in chloroplast, mitochondrion, and cytoplasm. It will be evident from the discussion to follow that this interesting area of organelle biogenesis is in need of much more study.

Barnett et al. (18) found that when dark-grown cells of *Euglena* that contained pro-plastids (see Chapter 16) were transferred to the light, new chromatographic species of isoleucyl, phenylalanyl, and glutamyl tRNAs were formed during the course of chloroplast development. None of these tRNA species was found in light-grown cells of the permanently bleached mutant W_3BUL, which lacks clDNA; this is consistent with the previously mentioned results of Schwartzbach et al. (15) indicating that clDNA from *Euglena* probably codes for a complete set of tRNAs. Reger et al. (19) studied formation of aminoacyl tRNA synthetases for these tRNAs during greening and found that there were two isoleucyl tRNA synthetases. One was present constitutively in light- and dark-grown cells, and it aminoacylated only the cytoplasmic species of isoleucyl tRNA. The second enzyme was light-inducible and was localized in the chloroplast; it acylated only the light-inducible chloroplast isoleucyl tRNA. In contrast, all three of the phenylalanyl tRNA synthetases were constitutive in dark-grown *Euglena* cells. However, one of these (synthetase I) could aminoacylate the light-inducible phenylalanyl tRNA; it was found in the chloroplast. Hecker et al. (20) demonstrated that synthetase I was really a mixture of mitochondrial phenylalanyl tRNA synthetase and chloroplast phenylalanyl tRNA synthetase. Although the chloroplast enzyme was present constitutively in dark-grown cells, it clearly increased in amount when those cells were transferred to the light. Chloroplast phenylalanyl tRNA synthetase and also a valyl tRNA synthetase specific for the chloroplast were found to be present in the bleached mutant W_3BUL, which lacks clDNA, but the isoleucyl tRNA synthetase could not be found in the mutant. This suggests that the former two tRNA synthetases are coded by nuclear genes and are translated in the cytoplasm. Further evidence that the phenylalanyl and

valyl tRNA synthetases are synthesized on cytoplasmic ribosomes was provided by Hecker and associates who showed that synthesis of these enzymes was inhibited by cycloheximide but not by streptomycin, which is a chloroplast protein synthesis inhibitor in *Euglena*. Similar results were reported for the leucyl, phenylalanyl, and lysyl tRNA synthetases of the chloroplast (16).

Parthier (21) studied the sites of synthesis of chloroplast aminoacyl tRNA synthetases of *Euglena* using a somewhat different approach. He employed the regreening *Euglena* system to detect light-inducible synthetases, but instead of assaying these synthetases with chloroplast tRNA from *Euglena,* he used tRNA extracted from the blue-green alga *Anacystis nidulans*. The obvious assumption is that prokaryotic tRNA obtained from this alga has greater similarity to chloroplast tRNA than does *Euglena* cytoplasmic tRNA. To assay for increases in cytoplasmic synthetases during regreening, Parthier used tRNA extracted from permanently bleached mutants lacking clDNA. Low concentrations of cycloheximide were found to stimulate both chlorophyll accumulation and synthesis of the light-inducible leucyl tRNA synthetase, but higher cycloheximide concentrations were found to inhibit both. The levels of plastid leucyl tRNA synthetases were similar in cells grown in the presence and absence of chloramphenicol, but addition of an inhibitory concentration of cycloheximide in either case blocked the enzyme increase. Parthier obtained the same results with the other light-induced synthetases (i.e., for (aspartyl, lysyl, phenylalanyl, and valyl tRNAs), as well as with the synthetases not induced by light (i.e., for arginyl, isoleucyl, seryl, and threonyl tRNAs). Thus all of the aforementioned experiments point to most light-stimulated aminoacyl tRNA synthetases of *Euglena* being localized in the chloroplast and being made on cytoplasmic ribosomes. In at least two cases these enzymes are also known to be coded by nuclear genes. Thus during chloroplast development in *Euglena,* synthesis of chloroplast-specific tRNAs coded by the chloroplast and synthesis of their aminoacyl synthetases coded by the nuclear genome occur coordinately. Translation of the mRNAs for these synthetases occurs on cytoplasmic ribosomes, and the synthetases are then imported into the chloroplast.

Nothing is presently known about how this light-triggered regulation of chloroplast tRNAs and their synthetases is achieved in *Euglena.* Goins et al. (22) observed that in a golden mutant of *Euglena* (G_1BU) with reduced levels of chlorophyll, light-inducible isoleucyl and methionyl tRNAs of the plastid were formed constitutively in the dark. In some direct or indirect way the G_1BU mutation appears to block normal regulation of these two light-induced tRNAs in the dark, but there is no indication whether or not this is true of their synthetases as well.

Weil and his colleagues studied leucyl, methionyl, and phenylalanyl tRNAs and their synthetases in *P. vulgaris.* Fractionation of leucyl tRNA synthetases yielded three peaks of activity (23). Peak I was a chloroplast-specific leucyl tRNA synthetase; peaks II and III were found not only in chloroplasts but also in the cytoplasm and mitochondria. When leucyl tRNAs were fractionated, two species were found in the cytoplasm, four in the mitochondrion, and six in the chloroplast. Synthetases II and III charged the mitochondrial and cytoplasmic tRNAs. The chloroplast leucyl tRNAs were separable into three groups. The first included two tRNAs indistinguishable from the cytoplasmic leucyl tRNAs that were aminoacylated by synthetases II and III. The second class included two very small peaks that looked as if they could be the result of contamination by mitochondrial tRNAs. These tRNAs were

aminoacylated by synthetases II and III. The third class consisted of three leucyl tRNAs found only in the chloroplast that were aminoacylated only by the chloroplast enzyme (synthetase I). Weil et al. (17) recently reported that these three leucyl tRNAs hybridized with clDNA. Even more interesting is the fact that the hybridization studies showed these tRNAs to be coded by the same chloroplast gene(s). Similar results were obtained for the two phenylalanyl tRNAs specific for the chloroplast. In sum, these results indicate that each set of isoaccepting tRNAs is coded by a single chloroplast gene, but posttranscriptional modifications of the tRNAs occur to yield three leucyl and two phenylalanyl tRNAs. In the case of the leucyl tRNAs, there seems to be a single chloroplast gene that codes for all three forms of this tRNA, all of which are charged by the same enzyme (synthetase I). On the other hand, the cytoplasmic and mitochondrial leucyl tRNAs differ from each other and from the three chloroplast leucyl tRNAs, and they are charged by two other enzymes (synthetases II and III). The presence of significant amounts of two tRNAs in the chloroplast apparently similar in all respects to two of the cytoplasmic tRNAs remains to be explained. Are they contaminants? Or are they imported in some fashion into the chloroplast?

The story of the methionyl tRNAs in *P. vulgaris* is, if anything, more intricate. Guillemaut et al. (24) demonstrated the presence of *N*-formylmethionyl tRNA in mitochondria and etioplasts, as well as the presence of an active transformylase in both organelles. In a later study of the methionyl tRNAs of *P. vulgaris* (25) it was shown that the two formylmethionyl tRNAs could be distinguished from each other by several criteria. Weil and his colleagues (26) then made an even more exhaustive analysis of the methionyl tRNAs. Three methionyl tRNA synthetases have been distinguished.

Two of these are cytoplasmic enzymes, and one is the chloroplast enzyme. The latter enzyme is not distinguishable from mitochondrial methionyl tRNA synthetase. Two cytoplasmic methionyl tRNAs can be differentiated. Both are aminoacylated by the cytoplasmic enzyme, but only one by the chloroplast enzyme. Mitochondrial preparations contain, in addition to the two cytoplasmic tRNAs, two specific peaks which can be charged with the methionyl tRNA synthetase from chloroplasts, mitochondria, or *E. coli,* but not with the cytoplasmic enzyme. One of these tRNAs can be formylated using the mitochondrial or bacterial transformylase. Therefore, *Phaseolus* mitochondria contain unique nonformylatable and formylatable tRNAs, both of which are charged only with the chloroplast or mitochondrial enzyme. The chloroplast contains the two cytoplasmic tRNAs plus three chloroplast-specific methionyl tRNAs. The three chloroplast-specific tRNAs can be aminoacylated by the chloroplast, mitochondrial, or *E. coli* methionyl tRNA synthetases. One tRNA (number 2) is formylatable, and another (number 3) can be charged by the cytoplasmic enzyme. In summary, the chloroplast contains three unique methionyl tRNAs, one of which is formylatable. Only one of these can be charged by the cytoplasmic enzyme, but all three can be charged by enzymes from the chloroplast, mitochondrion, or *E. coli.*

Weil and his colleagues (26) summarized their own work on the leucyl and methionyl tRNAs of *P. vulgaris* as follows:

Our comparative studies on the tRNAs^{Leu} and Leu-tRNA synthetases in the three cell-compartments of *P. vulgaris* had suggested the existence of two groups: one consisting of cytoplasmic and mitochondrial tRNAs and enzymes, the other of chloroplastic and *E. coli* tRNAs and enzymes; cross-aminoacylation reactions were possible within the group but not between the two groups [(23)]. Our studies on the tRNAs^{Met} and the Met-tRNA synthetase in the three compartments show one group consisting of chloroplastic, mitochondrial and *E. coli* tRNAs and enzymes; cross-aminoacylation reactions are possible within this group, whose components clearly differ from cytoplasmic tRNAs^{Met} and Met-tRNA synthetase. These results suggest a certain degree of similarity between bacterial and organellar tRNAs and aminoacyl-tRNA synthetases, [and they show] that tRNAs and enzymes in the organelles often differ from their cytoplasmic counterparts.

It should be evident from the foregoing discussion that we still have a lot to learn about organelle aminoacyl tRNA synthetases and their cytoplasmic counterparts. What little evidence there is uniformly suggests that these enzymes are coded by nuclear genes whose messages are translated in the cytoplasm. It is also clear that chloroplasts and mitochondria do contain certain synthetases that are quite distinct from the comparable cytoplasmic enzymes. In at least one case (the leucyl tRNA synthetases of *Neurospora*) the cytoplasmic and mitochondrial enzymes seem to be coded by tightly linked genes. At the moment we do not know what fraction of the total complement of organelle tRNA synthetases is unique, and there is good reason to believe that not all of them are distinct from the cytoplasmic enzymes. Finally, there is the problem of determining if all the tRNAs necessary for translation via organelle protein-synthesizing systems are really coded by organelle DNA. The results obtained by Suyama and Hamada (5) with *Tetrahymena* suggest that the mitochondria of this ciliate contain imported tRNAs not coded by mtDNA, in addition to native tRNAs coded by mtDNA. These authors postulated that cytoplasmic tRNA synthetases act as carriers for these tRNAs and permit their entry. In this context it goes without saying that we have no knowledge of the mechanisms and processing reactions involved in the entry of these water-soluble enzymes into the matrix of the mitochondrion or the chloroplast stroma.

ASSEMBLY OF MITOCHONDRIAL RIBOSOMES IN *Neurospora* AND *poky* MUTANT REVISITED

Mitochondrial ribosome assembly has been studied principally in *Neurospora* and to a lesser extent in yeast. The experiments of Luck and his associates concerning the assembly of mitochondrial ribosomes in *Neurospora* are by far the most complete and most rigorous carried out in any organelle system, and we shall discuss them in detail in this section. Following that, we shall continue the story of the search for the primary defect in the *poky* mutant of *N. crassa,* whose complex history was largely related in Chapter 8. As we shall see, even now the final episode remains to be written.

As was described in Chapter 3, mitochondrial rRNA in *N. crassa* is synthesized as a 32S precursor that is then cleaved to form precursors of the large and small mitochondrial rRNAs. These precursors are trimmed to form the mature rRNAs of the *Neurospora* mitochondrial ribosome. The major topics we shall address now are the sites of synthesis of the *Neurospora* mitochondrial ribosomal proteins and the assembly of the mitochondrial ribosomal precursor particle.

In the absence of antibodies to all the ribosomal proteins, studying the sites of synthesis of organelle ribosomal proteins can be a tricky business indeed. The problems encountered are precisely analogous to those that obtain for a membrane (see Chapters 19 and 20) or any other multicomponent system. Principal among these is the pool problem. If inhibitor experiments are carried out *in vivo,* inhibition of synthesis of a single protein, if its pool size is limiting and it is at an early step in assembly, can drastically affect the experimental results. For example, suppose that only one mitochondrial ribosomal protein is made on cytoplasmic ribosomes; suppose that its pool size is small and that it becomes inserted into the assembling ribosome early in the sequence. In a pulse labeling experiment carried out in the presence of an inhibitor of cytoplasmic protein synthesis, it will appear as if synthesis of all mitochondrial ribosomal proteins has been blocked. Those proteins made on mitochondrial ribosomes cannot be assembled into mature ribosomes because of the bottleneck created by blockage of synthesis of the limiting protein. Fortunately, the existence of the problem will probably become apparent, for synthesis of the same set of proteins will seem to be blocked by inhibitors of both cytoplasmic and mitochondrial protein synthesis.

The problem of the sites of synthesis of mitochondrial ribosomal proteins in *Neurospora* was first studied by Küntzel (27) and Neupert et al. (28). Küntzel isolated mitochondrial ribosomal proteins from the two subunits and compared them to their cytoplasmic counterparts by cochromatography on carboxymethyl cellulose columns. He then went on to study the sites of synthesis of mitochondrial ribosomal protein, having shown that chloramphenicol, but not cycloheximide, blocked protein synthesis in isolated *Neurospora* mitochondria. Following pulse labeling of cells in the presence of cycloheximide or chloramphenicol, Küntzel determined the amount of incorporation of radioactivity into whole mitochondrial ribosomes. For some reason he did not fractionate the mitochondrial ribosomal proteins in these experiments. Cycloheximide inhibited incorporation of labeled lysine into mitochondrial ribosomes 97%, whereas inhibition by chloramphenicol was only 2%. Küntzel concluded that the mitochondrial ribosomal proteins were synthesized in the cytoplasm. Neupert et al. (28), in a similar kind of experiment, got the same results and reached the same conclusion.

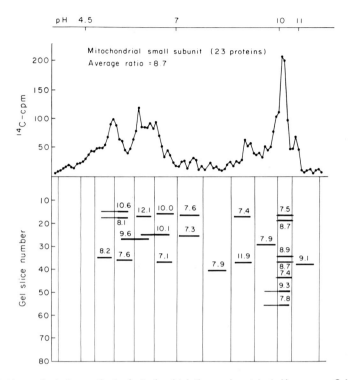

FIG. 21–1. Effect of chloramphenicol on synthesis of mitochondrial ribosomal proteins in *Neurospora*. Cells were labeled continuously with ^{14}C-leucine, following which chloramphenicol was added for 45 min and ^{3}H-leucine for 30 min. After the pulse, cells were allowed to recover for 2 hr in the absence of chloramphenicol to permit incorporation of ^{3}H-labeled proteins made during the chloramphenicol incubation into mature ribosomes. Mycelia were then harvested, and mitochondrial ribosomes and ribosomal proteins were isolated. Proteins from each subunit were then separated by a two-dimensional method involving isoelectric focusing fractionation in the first dimension followed by polyacrylamide gel electrophoresis in the second. Each gel was sliced and counted, and the ^{14}C mass label profile was plotted. The peaks are plotted diagrammatically as horizontal lines in each gel; the number above each line is the ^{3}H:^{14}C ratio. The elution profile obtained following isoelectric focusing is shown at the top of each diagram in terms of ^{14}C counts per minute. (From Lizardi and Luck, ref. 29, with permission.)

Neither set of experiments dealt adequately with the problems of limiting polypeptide pools alluded to earlier.

Lizardi and Luck (29) examined the sites of synthesis of the individual mitochondrial ribosomal proteins in a series of classic experiments in which the pool problem was clearly recognized and dealt with. They developed a two-dimensional gel technique that allowed them to visualize virtually all of the *Neurospora* mitochondrial proteins. The first dimension employed an isoelectric focusing column. As expected, most of the ribosomal proteins were quite basic and eluted at high pH (Fig. 21–1). Fractions were collected from the column, and the proteins were then separated on

SDS-urea gels in the second dimension in terms of molecular weight (Fig. 21–1). In this way, 23 proteins were distinguished in the small subunit and 30 in the large.

Lizardi and Luck (29) made use of both *in vitro* and *in vivo* approaches, carefully pointing out the advantages and disadvantages of each. In the *in vitro* experiments, mitochondria were incubated with a ^{3}H-labeled amino acid and extracted *in toto,* together with mitochondrial ribosomes that had been labeled *in vivo* with a ^{14}C-labeled amino acid. The proteins were then separated on the isoelectric focusing column, and it was evident that the products of *in vitro* protein synthesis did not include mitochondrial ribosomal proteins, or at least not

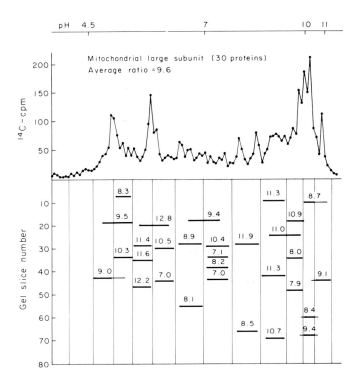

very many of them. Only three of the *in vitro* ³H-labeled protein peaks showed any coincidence with the ¹⁴C-labeled marker proteins. The significance of those three cases of coincidence was uncertain. It should be remarked that the reason Lizardi and Luck extracted whole mitochondria labeled *in vitro* rather than mitochondrial ribosomes had to do with their observation that synthesis of mitochondrial rRNA did not appear to take place in the isolated mitochondria. Thus new ribosomes were not being made, and any ribosomal proteins synthesized could be expected to be in the soluble phase of the mitochondrion.

The *in vivo* experiments employed chloramphenicol as the inhibitor of mitochondrial protein synthesis and anisomycin to block cytoplasmic protein synthesis. In these experiments ¹⁴C-leucine was added to the cultures for 12 hr. This was followed by addition of the inhibitor; then after a short period (7–15 min) ³H-leucine was added as a pulse (approximately 30 min). Following the ³H-leucine pulse, cold amino

acids were added, and the cells were allowed to grow in the absence of the inhibitor for 2 hr longer. The purpose of this pulse-chase-recovery design was to permit ³H-labeled proteins made in the presence of the inhibitor to become incorporated into mature ribosomes; it is essential to allow recovery in order to minimize the effects of problems caused by limiting pools of specific proteins. As was mentioned previously, blockage of synthesis of just one protein with a very small pool size by an inhibitor will prevent incorporation of all ribosomal proteins into mature ribosomes. Thus the inhibitor will exhibit a "pleiotropic" effect. Its primary effect will be to block synthesis of just one protein or a few proteins, but its secondary effect will be to prevent incorporation of all other newly synthesized proteins into mature ribosomes. During the recovery period in the absence of inhibitor, the pool of the limiting protein(s) is again built up, allowing those radioactive proteins made during the pulse in the presence of inhibitor to be incorporated into mature ribosomes. Under these conditions the $^3H : ^{14}C$ ratio in the individual proteins should reflect directly the effect of the inhibitor during the pulse. It will be recalled that Bücher and his colleagues had to incorporate a somewhat similar recovery scheme in their studies of the sites of synthesis of cytochrome oxidase polypeptides in *Neurospora* because of the "pleiotropic" effect of one polypeptide that was made in the mitochondrion and had a limiting pool size (see Chapter 19).

The experimental design devised by Lizardi and Luck for these *in vivo* inhibitor experiments worked quite successfully for chloramphenicol. It was apparent from the isotope ratios that synthesis of none of these proteins was sensitive to chloramphenicol. The relatively small variations in isotope ratios seen could be explained entirely by pool size differences. However, the pulse-chase-recovery design was not effective for the anisomycin part of the experiment; this inhibitor reduced total protein synthesis so much that little ^3H-leucine was incorporated into anything during the pulse, and most of the isotope was incorporated at random during the chase-recovery period. Therefore, a pulse-chase design was employed, with anisomycin present throughout, so that there was no recovery period. The validity of this design was verified by demonstration that mitochondrial rRNA labeled under these conditions became incorporated into mature ribosomes. Therefore, ribosomal protein pool sizes during the radioactive pulse and subsequent chase were sufficient to permit ribosome assembly to proceed normally. These experiments showed that only one of the small- or large-subunit proteins was labeled in the presence of anisomycin. This protein was recovered some of the time with small subunits, and it appeared at the time to be either a contaminant or a ribosomal protein with special properties. In summary, in the Lizardi and Luck experiments all mitochondrial ribosomal proteins in *N. crassa,* with the possible exception of one, appeared to be made on cytoplasmic ribosomes.

Lambowitz et al. (30,31) pursued the mechanism of assembly of mitochondrial ribosomes in *Neurospora* further. They showed that the 32S mitochondrial rRNA precursor was associated with several ribosomal proteins derived from both the small and large subunits. The precursor particle sedimented in approximately the same position as the small mitochondrial ribosomal subunit. This presented difficulties for the identification of the precursor particle, since the precursor fractions were always contaminated with mature small subunits. The precursor particles were identified initially in pulse labeling experiments in which it could be demonstrated that the newly synthesized rRNA and ribosomal proteins sedimented slightly more slowly than mature small subunits. Isolation of the pulse-labeled RNA from these particles showed that it was the 32S precursor and not the 19S rRNA of the mature small subunit. The precursor particle also differed from the mature small subunit in terms of the salt stability of the proteins associated with it. Most of the precursor particle protein became dissociated from the particles in high concentrations of salt, but this was not true of the protein associated with mature small subunits. Pulse-chase experiments substantiated these inferences; following the chase, it could be shown that most of the radioactive protein sedimented with the small and large subunits and was not dissociated by high concentrations of salt.

The protein composition of the precursor particle was analyzed using a pulse of ^{35}S followed by SDS gel electrophoresis of the proteins in a system with high resolution followed by autoradiography. The labeled proteins were then compared to proteins of the small and large subunits that had also been separated on SDS gels and then stained. This comparison revealed that the precursor particles contained both large- and small-subunit proteins.

These experiments employed a ^{3}H mass label as well as the ^{35}S pulse; thus isotope ratios could be calculated and used to estimate free pool sizes of the different ribosomal proteins, provided mature subunits were used for protein isolation. The proteins with small free pool sizes were expected to have high ^{35}S:^{3}H ratios, and those with larger pool size were expected to have lower ratios. Mature subunits were separated from the precursor particles by use of high-salt sucrose gradients that stripped the proteins off the precursor particles and left the mature subunits intact. By means of the isotope ratio, it was possible to identify six small-subunit proteins present in both precursor particles and mature small subunits that had exceptionally small free pool sizes. This was also true of three other proteins found only in mature small subunits. We shall have more to say of one of these proteins, S4a, shortly. It is interesting that large-subunit protein pools appeared to be relatively extensive.

Lambowitz et al. (30) reexamined the sites of synthesis of the mitochondrial ribosomal proteins using a high-resolution SDS gel system, principally to determine if the one small-subunit protein whose synthesis was reported to be resistant to anisomycin by Lizardi and Luck could be identified with certainty as a product of mitochondrial protein synthesis. They were able to show that a single protein of the small subunit, S4a, was synthesized in the presence of anisomycin but not chloramphenicol. No other protein of either subunit was synthesized in the presence of anisomycin, but all were made in the presence of chloramphenicol. Thus it appears that S4a is the only mitochondrial ribosomal protein synthesized in the mitochondrion. As was mentioned previously, this protein is present in mature small subunits but not in precursor particles, and it has a small free pool size.

Lambowitz et al. (30) then turned their attention to the *poky* mutant in the hope of determining the nature of the primary defect

in this mutant that causes it to form subnormal amounts of small mitochondrial ribosomal subunits. They had previously shown that the mutant probably did not have a defect in its processing of the rRNA of the small ribosomal subunit, as was originally supposed, and that fingerprints of this rRNA revealed no differences with respect to wild type (see Chapter 8). Therefore it seemed likely that the primary defect in *poky* might involve a mitochondrial ribosomal protein, specifically S4a. Lambowitz et al. (30) found that although *poky* precursor particles appeared generally similar to those from wild type, the mature small subunits were deficient in S3, S4a, and S23, as shown by a paucity of label in these proteins in pulse labeling experiments of the type already described. Since S4a has a small pool size and is a product of mitochondrial protein synthesis, it should be an excellent candidate for the primary defect in *poky*. One can imagine that the *poky* mutation occurs in the mitochondrial gene coding for this protein and alters its properties such that its rate of synthesis is altered or it is not assembled into mitochondrial ribosomes efficiently. According to this hypothesis, the deficiency in S3 and S23 would be a secondary consequence of the deficiency in S4a. If, in fact, tryptic peptide analysis does prove that S4a is altered in *poky,* the primary defect in this mutant will at last be established with reasonable certainty. This will mark the end of the search for the primary defect in a highly pleiotropic mutant that began over 20 years ago.

Groot (32) carried out experiments on the sites of biosynthesis of mitochondrial ribosomal proteins in yeast; although they were less complete than the experiments of Luck and his associates, they pointed in the same direction. Mitochondrial ribosomal proteins in yeast appear to be made on cytoplasmic ribosomes. However, there is one protein associated with the small subunit whose synthesis is cycloheximide-resistant and partially resistant to erythromycin. For various reasons, Groot discounted this protein as a probable contaminant that sticks to mitochondrial ribosomes, in particular the small subunit; but one cannot help wondering if it might not be the equivalent of S4a. It is difficult to account for the four mitochondrial ribosomal loci thus far identified in terms of antibiotic resistance mutations (see Chapters 5 and 7) solely on the basis of mitochondrial rRNA. Perhaps at least one of these codes for a mitochondrial ribosomal protein.

Mitochondrial ribosome assembly in yeast and *Neurospora* should also be affected by nuclear gene mutations, since it is likely that most of the structural genes coding for specific mitochondrial ribosomal proteins, processing enzymes, etc., are located in the nucleus. Kientsch and Werner (33) recently showed that the *cni-1* mutant of *N. crassa* (which was characterized originally as a mutant having a high level of cyanide-insensitive respiration and a deficiency of cytochromes $a+a_3$ and b) is just such a mutant. Like a number of similar mutants in *E. coli, cni-1* behaves as a cold-sensitive ribosome assembly mutant. At low temperatures small mitochondrial ribosomal subunits are not assembled, although large subunits are. As a result, mitochondrial translation products are not synthesized at low temperatures. Therefore, at low temperatures the *cni-1* mutant behaves as a phenocopy of the *poky* mutant in early log phase, where the small-subunit deficiency is most exacerbated, or as wild-type *N. crassa* grown in the presence of high concentrations of chloramphenicol, where mitochondrial translation is blocked (34–36). There is induction of the cyanide-insensitive alternate oxidase accompanying the decline in cytochrome-mediated respiration that results because essential mitochondrial translation products for the latter pathway are not made. In short, mitochondrial protein synthesis mutants in *Neurospora* are characterized by certain specific pleiotropic effects, including deficiencies of cyto-

chromes $a+a_3$ and b and an increased level of cyanide-insensitive respiration. In the case of assembly-defective chloroplast ribosome mutants in *Chlamydomonas,* a similar syndrome of photosynthetic defects is associated with the inability of these mutants to carry out normal levels of chloroplast protein synthesis, as we shall see presently. In both situations these characteristic pleiotropic effects can be used to identify potential organelle protein synthesis mutants.

ASSEMBLY AND GENETICS OF CHLOROPLAST RIBOSOMES

Nothing is known about the sites of synthesis of individual chloroplast ribosomal proteins, but experiments with *Chlamydomonas* are beginning to reveal a great deal about the genetic control of chloroplast ribosome formation. Honeycutt and Margulies (37) obtained a general impression of where the bulk of chloroplast ribosomal proteins are made in *Chlamydomonas* in experiments analogous to those described for *Neurospora* mitochondrial ribosomes by Küntzel and Neupert and associates. They examined the effects of chloramphenicol and cycloheximide on incorporation of labeled arginine into chloroplast and cytoplasmic ribosomes of exponentially growing cells of the *arg-1* mutant of *C. reinhardtii.* The basic protocol was to incubate the cells with either chloramphenicol or cycloheximide for 1 hr before adding labeled arginine. Under these conditions many of the chloroplast ribosomes became attached to the thylakoid membranes by nascent polypeptide chains, for chloramphenicol prevents translocation of chloroplast ribosomes along the message (see Chapter 3). Since Honeycutt and Margulies isolated their ribosomes from supernatants of whole-cell extracts in which the membrane fragments had been pelleted out by centrifugation, the supernatants from the chloramphenicol-treated cells were depleted of chloroplast ribosomes. To overcome this problem, Honeycutt and

Margulies included a 4-hr posttreatment incubation in their experiments in which the cells labeled in the presence of chloramphenicol were subsequently incubated in the absence of chloramphenicol and labeled arginine. During this period chloroplast ribosomes "frozen" onto the thylakoid membranes by chloramphenicol were slowly released into the supernatant.

Honeycutt and Margulies detected no changes in relative specific activities of chloroplast ribosomes labeled in the presence of chloramphenicol. On the other hand, cycloheximide appeared to block incorporation into chloroplast ribosomes almost completely. Honeycutt and Margulies concluded from their results that most chloroplast ribosomal proteins are made on cytoplasmic ribosomes. Although this conclusion is probably correct, it should be mentioned that these experiments in no way rule out the possibility that some chloroplast ribosomal proteins are made in the chloroplast, as Honeycutt and Margulies were careful to point out.

Ellis and Hartley (38) made a brief comment on the sites of synthesis of chloroplast ribosomal proteins. They studied the accumulation of chloroplast ribosomes in greening pea apices in the presence of lincomycin. They found that chloroplast rRNA continued to be made in the presence of lincomycin but that accumulation of chloroplast ribosomes was blocked under these conditions. They therefore suggested that some chloroplast ribosomal proteins were made on chloroplast ribosomes and that the newly synthesized chloroplast rRNA was degraded because it could not be assembled into mature chloroplast ribosomes due to the absence of specific ribosomal proteins.

Genetic control of chloroplast ribosome biogenesis is under active investigation in *C. reinhardtii,* and some 20 genes involved in this process have now been identified through the use of appropriate assembly-defective and antibiotic-resistant or antibiotic-dependent mutants (Table 21–1).

Since chloroplasts are difficult to isolate intact in *Chlamydomonas,* chloroplast and cytoplasmic ribosomes are generally separated from whole-cell supernatants on sucrose gradients, with the assumption being that mitochondrial ribosomes represent an insignificant fraction of the total. This assumption seems justified on the basis of electron microscopic studies of median sections of whole cells indicating that mitochondrial ribosomes are extremely rare in comparison to cytoplasmic and chloroplast ribosomes (39). Five principal classes of ribosomes have been recognized by Bourque et al. (39) in whole-cell extracts of *C. reinhardtii* cells in low-salt buffers; these have sedimentation velocities of 83S, 70S, 66S, 54S, and 41S as determined by linear extrapolation against the 83S monomer peak (Fig. 21–2). The 70S ribosomes are derived from the chloroplast and contain large (54S) and small (41S) subunits. In certain mutants 66S ribosomes virtually replace the 70S chloroplast ribosomes. The large chloroplast ribosomal subunit contains 5S and 23S rRNA, and the small subunit contains 16S rRNA (44). The 16S and 23S rRNAs are coded by clDNA (40–43), and transcriptional mapping studies have suggested that each clDNA molecule of *C. reinhardtii* contains two to three copies of each rRNA cistron arranged in tandem repeats of a transcriptional unit coding for 16S and 23S rRNA (41). Electrophoretic and chromatographic analyses have indicated that the large subunit contains 26 to 34 proteins and the small subunit contains 19 to 25 proteins; for reviews, see Harris et al. (44) and Gillham et al. (45).

Harris et al. (46,47) identified seven or possibly eight nuclear genes involved in chloroplast ribosome biogenesis by means of isolation of assembly-defective mutants (Table 21–1). They are designated *cr-* with the exception of *ac-20,* the first mutant described as belonging to this class (48). The mutants thus far isolated fall into two

TABLE 21–1. *Summary of nuclear and chloroplast genes in C. reinhardtii known to affect the function or assembly of chloroplast ribosomes*

				Effect of mutant on chloroplast ribosomes		
			Number	Level of antibiotic resistance		
Type of alteration	Location of gene	Locus designation	of alleles	*In vitro*	*In vivo*	Other phenotypic modifications
Antibiotic resistance						
Streptomycin	N[a]	sr-1	3	low		
	C	sr-u-sm-3a	13	low to high (S)[b]		
	C	sr-u-sm3	1	intermediate (S)		
	C	sr-u-2-60	2	high (S)	high	sr-u-2-60 affects assembly of the small subunit
	C	sd-u-3-18[c]	1	high		sd-u-3-18 affects assembly of the small subunit
	C	sr-u-sm2	2	very high (S)		sr-u-sm2 alters a protein of the small subunit
	C	sr-u-35[c]	1		high	sr-u-35 reported to alter a protein of the large subunit[e]
Spectinomycin, Kanamycin, or Neamine	C	spr-u-1-27-3[d]	4	low to high (S)	low to high	spr-u-1-27-3 alters a protein of the small subunit[e]
Erythromycin	N	ery-M1	4			All 4 mutant alleles alter the same protein of the large subunit.
	N	ery-M2	4			ery-M-2d alters a protein of the large subunit[e]
	C	er-u-1a	2	high (L)	high	er-u-1a alters a protein of the large subunit[e]
	C	er-u-37	2	high (L)	high	
Carbomycin	C	car[c]	1	high (L)		
Cleocin	C	cle[c]	1	high (L)		
Assembly-defective						
Small subunit	N	cr-1	1			Accumulates large subunits, deficient in 70S monomers
	N	cr-2	1			Accumulates large subunits, deficient in 70S monomers
	N	cr-3	1			Accumulates large subunits, deficient in 70S monomers
	N	cr-5	1			Accumulates large subunits, deficient in 70S monomers
	N	cr-7	1			Accumulates large subunits, deficient in 70S monomers
Both subunits	N	ac-20	1			Deficient in 70S monomers
	N	cr-4	1			Deficient in 70S monomers
	N	cr-6	1			Deficient in 70S monomers

[a] N = gene located in nucleus; C = gene located in chloroplast.
[b] S = *in vitro* resistance localized to small ribosomal subunit; L = *in vitro* resistance localized to large ribosomal subunit; (?) = data not entirely convincing to the reviewers.
[c] Allelic relationship to other chloroplast genes not determined.
[d] Neamine- and spectinomycin-resistant chloroplast mutants appear to be allelic, based on recombination analysis. From Gillham et al. (73), with permission with additional subunit specificity data from Fox et al. (56).
[e] Data not entirely convincing.

classes on the basis of chloroplast ribosome phenotype (Fig. 21–2). Mutants in three genes cannot synthesize normal amounts of either ribosomal subunit; they are believed to be blocked in ribosomal assembly at a very early point (e.g., at the level of rRNA processing or ribosomal precursor (formation). Mutants in five other loci accumulate large subunits and are deficient to varying degrees in small subunits. These genes are

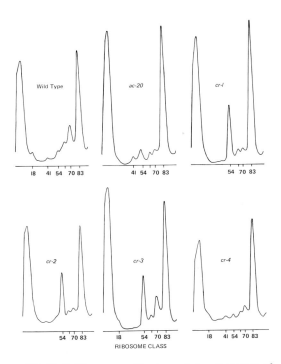

FIG. 21–2. Ribosome profiles from whole-cell extracts of wild-type C. *reinhardtii* and five mutants deficient in chloroplast ribosomes following sucrose-gradient centrifugation of the extracts. Each profile shows the absorbance tracing of a gradient at 254 nm. At the left end of each gradient is a meniscus fraction that includes chlorophyll, most of the proteins present in the supernatant, and nuclei acids. To the right of this is a peak at 18S that represents RuBPCase. This protein, because of its high molecular weight and abundance, is easily separated and visualized with respect to the rest of the cellular proteins. Notice the conspicuous absence or reduction in size of this peak in the ribosome-deficient mutants. The peaks labeled 41, 54, 70, and 83S represent the small and large subunits of the chloroplast ribosome, the intact chloroplast ribosome, and the intact cytoplasmic ribosome fractions, respectively. Notice that ac-20 and cr-4 cannot make normal amounts of either chloroplast ribosomal subunit, but cr-1, cr-2, and cr-3 accumulate normal amounts of large subunits and are deficient only in small subunits (From Harris et al., ref. 46, with permission.)

presumed to be involved in small-subunit assembly.

The assembly-defective mutants behave as photosynthetic mutants because they are unable to carry out chloroplast protein synthesis at normal rates. As a consequence they develop a characteristic syndrome of defects that is readily predicted from the experiments on the products of chloroplast protein synthesis described in the pre-

ceding chapter. Thus the assembly defective mutants are also deficient in RuBPCase and photosystem I and II activity because the large subunit of RuBPCase and specific polypeptides associated with both of the photosystems are made on chloroplast ribosomes. However, these mutants do make normal amounts of chlorophyll, presumably because the constituent enzymes in chlorophyll biosynthesis are made on cytoplasmic ribosomes; they also make normal amounts of chloroplast thylakoids, although these do not stack properly (49). The pleiotropic photosynthetic defects characteristic of the assembly-defective mutants have proved useful in distinguishing them from other kinds of photosynthetic mutants in mutagenesis experiments.

With the possible exceptions of cr-7 and ac-20, all the nuclear assembly-defective mutants are nonallelic, as shown by recombination analysis; and all of them, including ac-20 and cr-7, appear to complement in diploids (46,47). By comparing the photosynthetic capacities and chloroplast ribosome profiles of the individual mutants and of their diploids (with each other and with wild type), a working model can be formulated for the assembly of chloroplast ribosomes that is analogous to the formal schemes used to deduce enzymatic pathways from the behavior of auxotrophic mutants. Relationships of dominance and epistasis among this group of mutants can be used to rank them in a probable sequential order of function (Fig. 21–3). Thus ac-20 and cr-4 appear to affect the formation of components common to both the small and large ribosomal subunits, whereas cr-1, cr-2, and cr-3 (all of which accumulate 54S particles) are probably impaired in a specific pathway leading to the formation of the 41S subunit. Both ac-20 and cr-4 are epistatic to cr-1, cr-2, and cr-3, which suggests that these mutants affect steps leading to formation of both subunits that occur prior to a branch point early in ribosome assembly. In diploids cr-1 is partially dominant to the wild type,

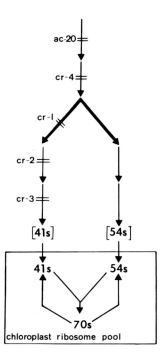

FIG. 21–3. Working model for chloroplast ribosome assembly in *Chlamydomonas* based on genetic and phenotypic analysis of five nonallelic mutants deficient in chloroplast ribosomes; ac-20 and cr-4 appear to be blocked early in ribosome assembly, since they are deficient in both ribosomal subunits, whereas cr-1, cr-2, and cr-3 all accumulate large chloroplast ribosomal subunits and are deficient only in small subunits. This would place them after the first two mutants and in the small-subunit pathway. (From Harris et al., ref. 46, with permission.)

whereas the other four mutants are recessive. This suggests that *cr-1* blocks synthesis of a ribosomal component needed in stoichiometric amounts (i.e., a structural component); thus diploids containing one *cr-1* allele and one *cr-1⁺* (wild-type) allele make fewer functional ribosomes per cell than do wild-type/wild-type diploids or wild-type/mutant diploids, in which the mutant allele affects only a catalytic function.

This model for ribosome assembly indicates possible biochemical roles for each of the individual genes in chloroplast ribosome formation, and it suggests experimental approaches to verify the molecular defects in each of the mutants.

Although cells of *C. reinhardtii* are sensitive to antibacterial antibiotics known to

block protein synthesis in both chloroplasts and mitochondria, one-step mutations resistant to these antibiotics are readily isolated. Mutations at three nuclear and at least seven to nine chloroplast gene loci confer antibiotic resistance or dependence on chloroplast ribosomes (Table 21–1). In *E. coli,* mutations that confer resistance to these antibiotics at the ribosomal level do so by altering specific ribosomal proteins by amino acid substitution or deletion (50). Davidson et al. (51) studied four mutant alleles conferring erythromycin resistance on chloroplast ribosomes at the *ery-M-1* locus, which is a nuclear gene. Three of these alleles altered the charge of a specific protein (LC6) of the large chloroplast ribosomal subunit, and the fourth *ery-M-1b* led to deletion of 30% of this protein (Fig. 21–4). This is the most rigorous evidence to date that any gene, organelle or nuclear, is the structural gene for a specific organelle ribosomal protein. Bogorad and associates (74,75) have shown that chloroplast ribosomes from vegetative diploids between the *ery-M-1b* mutant and wild type contain both the mutant and wild type forms of the LC 6 protein. These results further strengthen the hypothesis that the *ery-M-1* gene codes for the LC 6 protein. If the *ery-M-1* gene were instead responsible for enzymatically modifying the structure of the LC 6 protein, one would expect mutants in this gene to be recessive to wild type rather than being co-dominant as observed. In addition Davidson and Bogorad (76) have mutagenized the *ery-M-1b* mutant and isolated four mutants more sensitive to erythromycin than *ery-M-1b*. All four mutants carried the original *ery-M-1b* mutation plus a second mutation (*es*) which partially suppressed resistance. The suppressor mutations were mapped in at least three different mendelian genes. Although some of the *es* mutations also suppressed resistance of mutations at the *ery-M-2* locus, which is also nuclear, none of them had any effect on the chloroplast gene mutation to erythromycin

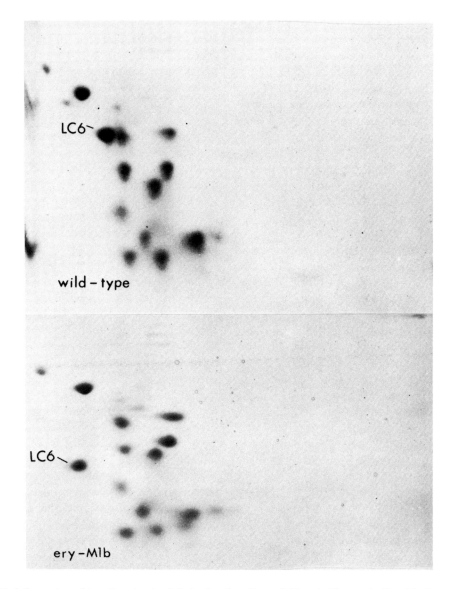

FIG. 21–4. Comparison of two-dimensional gel electrophoretic patterns of chloroplast large-subunit proteins from wild-type *C. reinhardtii* and *ery-M1b*. What is shown here is the second dimension SDS pattern in which proteins are separated on the basis of molecular weight. Notice that the LC6 protein, which appears to be the product of the *ery-M1* locus, moves closer to the front in the *ery-M1b* mutant than in the wild type, indicating that it has a lower molecular weight in the mutant than in the wild type. (From Davidson et al., ref. 51, with permission.)

resistance *ery-u-1a*. Binding experiments employing ^{14}C erythromycin and chloroplast ribosomes from the *es ery-M-1b* double mutants and the parent *ery-M-1b* mutant by itself revealed that in every case the chloroplast ribosomes from the double mutants bound more antibiotic than those from

ery-M-1b. These results suggest that the *es* mutations probably modify chloroplast ribosome structure such that the *ery-M-1b* mutation no longer confers resistance on these ribosomes. The *ery-M-2* locus may specify a different protein of the large chloroplast ribosomal subunit, but here the evidence is not so compelling (52). Nuclear mutations to streptomycin resistance (*sr-1*) are all allelic and confer resistance on chloroplast ribosomes *in vitro* (45,55).

Chloroplast gene mutations confer resistance directly on chloroplast ribosomes to the antibiotics streptomycin, spectinomycin, neamine, erythromycin, carbomycin, and cleocin (53–56). In addition, mutants dependent on streptomycin and neamine are known (57,58), and one streptomycin-dependent mutant (55) has been shown to confer antibiotic resistance on chloroplast ribosomes *in vitro* (Table 21–1).

By mapping and allele testing, four chloroplast loci for streptomycin resistance, two for erythromycin resistance, and one for spectinomycin and neamine resistance have been identified (see Chapter 14, Fig. 14–9). Allele tests of the cleocin- and carbomycin-resistant mutations with respect to each other and to other chloroplast mutations to antibiotic resistance remain to be done. The same is true of the streptomycin- and neamine-dependent mutants. Present evidence suggests that all these antibiotic-resistant and antibiotic-dependent mutants map in a single linkage group in the chloroplast genome (see Chapter 14).

Perhaps the simplest and most direct method of demonstrating that these antibiotic-resistant mutations directly affect chloroplast ribosomes is an *in vitro* assay in which isolated ribosomes are assayed for their ability to incorporate a particular radioactive amino acid in response to a given synthetic polynucleotide in the presence of the antibiotic. This method was first adapted from the *E. coli* system by Schlanger and Sager (53) as a polyuridylic acid/phenylalanine assay for streptomycin, spectinomycin, neamine, carbomycin, and cleocin resistance of chloroplast ribosomes. Subsequently the method was modified to include the polyuridylic acid/isoleucine misreading reaction for streptomycin resistance and a polycytidylic acid/proline assay for erythromycin resistance (54,55) (Fig.

21–5). Schlanger and Sager (53) also performed subunit exchange experiments between mutant and wild-type chloroplast ribosomes and then analyzed these "hybrid" ribosomes in the polyuridylic acid/phenylalanine system to localize the site of antibiotic resistance to either the small or large ribosomal subunit. This method was recently adapted by Fox et al. (56) to determine which chloroplast ribosomal subunit is resistant for each of the seven chloroplast genes (Figs. 14–9 and 21–6) in which mutations conferring antibiotic resistance on chloroplast ribosomes have been identified (Table 21–1).

From the results of the *in vitro* protein synthesis experiments, several important conclusions can be drawn. First, many, if not all, of the antibiotic-resistant mutants isolated in *Chlamydomonas* confer resistance directly on chloroplast ribosomes. Second, each mutant allele has a unique phenotype, and in cases where several different chloroplast gene loci confer resistance to the same antibiotic (e.g., to streptomycin), the general level of resistance tends to be locus-specific as well. Third, subunit exchange experiments have now been done with mutants at many of the nuclear and chloroplast resistance loci, and resistance to a given antibiotic appears to be a property of the same ribosomal subunit in the chloroplast of *Chlamydomonas* and in *E. coli*. Fourth, the chloroplast genes conferring antibiotic resistance in *Chlamydomonas* show certain similarities in their map order to similar genes in *E. coli* and *Bacillus subtilis* (Fig. 21–6), although more loci appear to confer streptomycin resistance on the chloroplast ribosomes of *Chlamydomonas* than on the 70S ribosomes of either bacterial species. Where the levels of resistance of chloroplast protein synthesis in individual mutants have been assessed *in vivo,* they have been found to mirror the resistance levels measured *in vitro* in every case examined (54) (Fig. 21–5). For example, a comparison of two allelic muta-

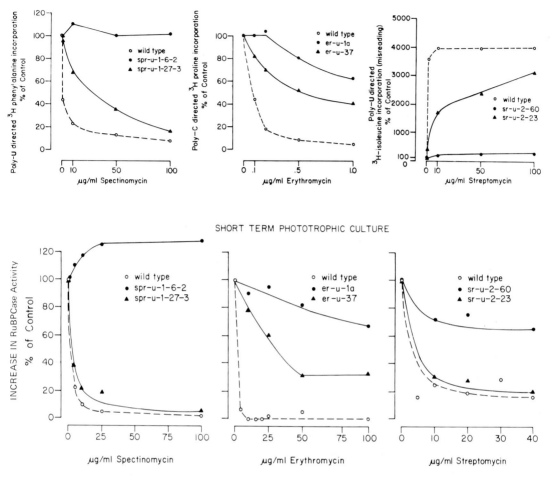

FIG. 21–5. Measurement of chloroplast protein synthesis *in vitro* and *in vivo* in wild-type *C. reinhardtii* and three groups of mutants having chloroplast ribosomes resistant to different antibiotics. The top three panels present the results of *in vitro* protein synthesis experiments employing isolated chloroplast ribosomes, synthetic polynucleotides, and radioactive amino acids. The bottom three panels show the corresponding *in vivo* experiments in which synthesis of RuBPCase over a short time was used as a measure of chloroplast protein synthesis. The two panels on the left present the results for the spectinomycin-resistant mutants *spr-u-1-6-2* and *spr-u-1-27*, where polyuridylic-acid-directed phenylalanine incorporation is used as the assay. Note that both the *in vitro* and *in vivo* assays show that *spr-u-1-27* chloroplast ribosomes are almost as sensitive to spectinomycin as those from the wild type, whereas those from *spr-u-1-6-2* are completely resistant to the antibiotic. The middle panels present the results for the erythromycin-resistant mutants er-u-1a and er-u-37. The *in vitro* assay employs polycytidylic acid and proline, since wild-type chloroplast ribosomes behave as if they are erythromycin-resistant in the polyuridylic acid phenylalanine assay. Note that both mutants are quite resistant to erythromycin, but they can be differentiated from one another both *in vivo* and *in vitro*. The two panels on the right present the results for the streptomycin-resistant mutants sr-u-2-60 and sr-u-2-23. In this case streptomycin-stimulated misreading of isoleucine for phenylalanine in the polyuridylic-acid-directed system proves to be the most sensitive assay for resistance. Note that the sr-u-2-23 mutant is quite sensitive to misreading stimulation, but the sr-u-2-60 mutant is not. These differences are reflected *in vivo*, where RuBPCase synthesis proves to be very sensitive to streptomycin in sr-u-2-23, but not in the sr-u-2-60 mutant. (From Conde et al., ref. 54, with permission.)

tions to spectinomycin resistance, *spr-u-1-6-2* and *spr-u-1-27,* revealed that they had different levels of resistance both *in vitro* and *in vivo*. In the polyuridylic acid/ phenylalanine assay, chloroplast ribosomes

from *spr-u-1-6-2* proved to be completely resistant to the antibiotic, but those from *spr-u-1-27* were almost as sensitive to the inhibitor as wild-type chloroplast ribosomes (Fig. 21–5). These differences were re-

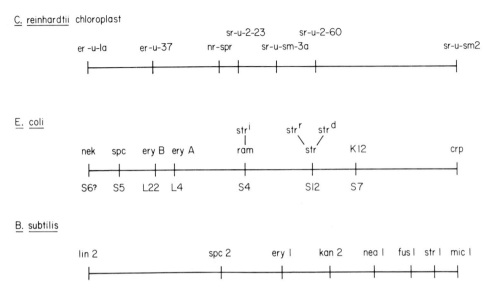

FIG. 21–6. Comparison of chloroplast gene map for C. *reinhardtii* with that for E. *coli* and B. *subtilis* in region where antibiotic resistant mutations altering 70S ribosomes fall. Key: erythromycin resistance = er or ery; spectinomycin resistance = spr o spc; streptomycin resistance = sr or str; neamine resistance = nr or nea; neomycin — kanamycin resistance = nek; kana mycin resistance = kan; fusidic acid resistance = fus; lincomycin resistance = lin; microccin = mic; ribosomal ambiguity = ram. The specific ribosomal proteins affected by the mutations in E. *coli* are shown with the prefix S for small subunit and L for large subunit. (From Gillham et al., ref. 73, with permission.)

flected *in vivo,* with the synthesis of RuBPCase being almost as sensitive to spectinomycin inhibition in *spr-u-1-27* as that of the wild-type enzyme, whereas synthesis of this enzyme in *spr-u-1-6-2* was completely resistant to the antibiotic (Fig. 21–5). In several cases (52,59–61) specific chloroplast gene mutants have been reported to alter particular proteins of a given chloroplast ribosomal subunit, although each of these cases needs to be substantiated further. To prove unequivocally that a specific gene codes for a given ribosomal protein, a minimal requirement is the demonstration that a series of allelic mutants at the same gene locus all affect that same protein. Thus far this has been done only by Davidson et al. (51) for the mendelian mutants at the *ery-M1* locus.

In conclusion, it can certainly be said that a number of chloroplast and nuclear genes involved in the biogenesis of chloroplast ribosomes have now been identified in *Chlamydomonas,* and their functions are beginning to be understood. Chloroplast

rRNA is known to be a chloroplast gene product (p. 573). Although some of the nuclear and chloroplast genes identified by antibiotic resistance mutations could conceivably act directly on the rRNA or its processing, their phenotypes are strikingly similar to bacterial mutations known to alter ribosomal proteins. In each case examined the same ribosomal subunit is involved in conferring resistance to a given antibiotic in both the *Chlamydomonas* chloroplast and in E. *coli* and B. *subtilis.* Assuming, therefore, that chloroplast ribosomal proteins may be products of both nuclear and chloroplast genes, there would appear to be no correlation between the location of the gene and either the ribosomal subunit affected or the type of antibiotic resistance conferred. To account for the ability of mutations at any one of four gene loci (two nuclear and two chloroplast) to confer erythromycin resistance on chloroplast ribosomes, and mutations at any one of five gene loci (one nuclear and four chloroplast) to confer streptomycin resistance, ribosomal "neighborhoods" have been postulated to exist on the large and small subunits, respectively, where these two drugs bind and block protein synthesis (54). Assuming that these ribosomal neighborhoods are composed of a number of proteins, each of which is coded by a separate gene, an appropriate alteration of any one of these proteins by gene mutation must be sufficient to change the conformation of the binding site and thus confer antibiotic resistance on the chloroplast ribosome. Alternatively, certain of the antibiotic resistance genes could code for enzymes that process ribosomal proteins coded by another gene. In this case, resistance would result from improper processing of a given ribosomal protein rather than from a direct change in its primary structure.

ORIGIN OF ONE-STEP MUTATIONS TO ANTIBIOTIC RESISTANCE IN *Chlamydomonas*

As was mentioned previously, mitochondrial protein synthesis in many organisms (see Chapters 3, 5, 10, and 11) is susceptible *in vivo* to at least some of the antibiotics (e.g., erythromycin) that block chloroplast protein synthesis in *Chlamydomonas*. Thus the question arises how one-step mutations that confer antibiotic resistance and dependence on chloroplast ribosomes originate in *C. reinhardtii*. Two hypotheses have been suggested to explain this phenomenon. The first supposes that the mitochondrial and chloroplast protein-synthesizing systems in *Chlamydomonas* are sensitive to the same set of antibiotics. However, chloroplast protein synthesis is dispensable, provided that cells are grown under conditions where mitochondrial protein synthesis and respiratory chain function are unimpaired on a carbon source (i.e., acetate) that can be respired. According to this hypothesis, one-step mutations may confer resistance either on mitochondrial ribosomes alone or on both chloroplast and mitochondrial ribo-

somes simultaneously. The second hypothesis ignores the mitochondrion entirely and assumes that chloroplast protein synthesis is indispensable for cell survival. This indispensability results because a regulatory protein required for nDNA replication is synthesized on chloroplast ribosomes. One-step mutations, by conferring resistance on chloroplast ribosomes, make it possible for cells to continue production of the regulatory compound when grown in the presence of the antibiotic to which they are resistant.

The notion that chloroplast protein synthesis might be dispensable seems to have been formulated first by Surzycki (62,63), who found that rifampin blocked the synthesis of chloroplast rRNA and therefore chloroplast ribosomes. He also found that rifampin blocked the phototrophic growth of cells where CO_2 and light were the sole sources of carbon and energy (see Chapter 12), but *C. reinhardtii* was able to grow heterotrophically in the dark on acetate in the presence of rifampin. Cells grown in the latter fashion for five generations were observed to be greatly depleted in chloroplast rRNA. From these experiments Surzycki concluded that chloroplast ribosomes and therefore chloroplast protein synthesis in *C. reinhardtii* were not vital for growth and division of the organism so long as acetate was provided as a carbon source, since this compound could be respired by the mitochondrion. At about the same time Goodenough et al. (48,64) demonstrated that the nuclear acetate-requiring mutant *ac-20* was defective in chloroplast ribosome assembly and as a consequence exhibited the pleiotropic syndrome of photosynthetic defects discussed earlier in this chapter. Despite the fact that this mutant was leaky (i.e., capable of slow growth under phototrophic conditions), its behavior was consistent with the notion that chloroplast protein synthesis was at least partially dispensable in *C. reinhardtii*.

If chloroplast and mitochondrial protein synthesis were sensitive to the same set of

antibiotics, then one-step mutations to resistance, assuming they were not permeability mutations, would have to confer resistance on the protein-synthesizing systems
of one or both organelles. This idea was
formulated by Surzycki and Gillham (65),
who also set forth a three-part system for
classifying antibiotic-resistant mutants in *C.
reinhardtii*. The first part is really the only
part relevant to this discussion, as it made
use of growth conditions as guides to the
sites of phenotypic expression of the resistance mutations within the cell. Surzycki
and Gillham distinguished three mutant
classes on this basis. Mutants that did not
grow heterotrophically in the presence of
the antibiotic to which they were resistant,
but did grow phototrophically or mixotrophically under these conditions, were assumed to have antibiotic-resistant chloroplasts and antibiotic-sensitive mitochondria.
Mutants that did not grow phototrophically
in the presence of antibiotic, but did grow
when provided with acetate, were assumed
to have resistant mitochondria and sensitive
chloroplasts. Mutants that grew under all
three conditions in the presence of antibiotic were assumed either to be permeability
mutations or to be mutants in which both
organelle protein-synthesizing systems were
antibiotic-resistant.

Surzycki and Gillham found that most of
the mutants they studied were resistant to a
given antibiotic, whether the cells were
grown photosynthetically or with acetate as
a carbon source. According to their classification, this meant that both chloroplast and
mitochondrial protein synthesis was resistant in these mutants (assuming they
were not permeability mutants). These included the spectinomycin-resistant mutant
spr-u-1-27 that, as we shall soon see, was
misclassified. Two other mutants also resistant to spectinomycin (*spA-1* and *spA-2*)
exhibited resistance only when grown on
spectinomycin in the presence of acetate,
which suggested that they had a sensitive
chloroplast protein-synthesizing system, but

a resistant mitochondrial protein-synthesizing system. No mutants that were resistant
when grown photosynthetically, but sensitive when grown on acetate in the dark,
were found.

Further study of the *spr-u-1-27* mutant
by Boynton et al. (59) revealed that it
could not grow photosynthetically in the
presence of spectinomycin as Surzycki and
Gillham had thought. Surzycki and Gillham
had made their classification on solid medium, and on such medium the differential
effects of spectinomycin on growth of the
mutant were not observed, even at relatively
high antibiotic concentrations. However,
Boynton and associates observed that in
liquid medium a clear distinction could be
made, and cells were much more sensitive
to spectinomycin when grown phototrophically than they were when grown in the
presence of acetate. This sensitivity to spectinomycin under phototrophic conditions
could be accounted for, as we have seen
previously, by the fact that chloroplast protein synthesis in the *spr-u-1-27* mutant is
very sensitive to spectinomycin *in vivo* and
in vitro (Fig. 21–5). Boynton and associates also found that cells of this mutant
could be grown for long periods in high
concentrations of spectinomycin, provided
that acetate was present in the medium.
Under these conditions the mutant became
an extreme phenocopy of the assembly-
defective mutants described in the preceding
section in terms of functions dependent on
chloroplast protein synthesis. RuBPCase
and Hill reaction activities, two good measures of chloroplast protein synthesis *in vivo*,
were virtually obliterated, but cell growth
continued. Thus it appeared that chloroplast
protein synthesis in the mutant *in vivo* was
completely sensitive to spectinomycin.

To explain these rather bizarre findings,
Boynton and associates proposed that the
gene product altered by the *spr-u-1-27* mutant was a ribosomal protein common to
the small subunits of both chloroplast and
mitochondrial ribosomes. *In vivo* this pro-

tein conferred a high level of resistance on mitochondrial ribosomes but not chloroplast ribosomes. Thus the mutant could grow in the presence of acetate and spectinomycin, but it would not grow photosynthetically if spectinomycin was present in the medium. Since the *spr-u-1-27* mutant mapped in a chloroplast gene (55), and since chloroplast protein synthesis was presumed to be spectinomycin-sensitive, the hypothesis also assumed that the messenger for this protein must be translated on cytoplasmic ribosomes. That is, this mRNA had to be exported from the organelle.

Before leaving the *spr-u-1-27* mutant it is important to note that *in vitro* spectinomycin binding studies gave vastly different results from the *in vitro* protein synthesis experiments (59). The former experiments suggested that chloroplast ribosomes from the mutant were highly resistant to spectinomycin. These findings are at variance with the *in vivo* studies as well. The reason for the discrepancy seen for the antibiotic binding studies is really not understood, although Boynton and associates suggested the possibility that the protein responsible for spectinomycin binding may be lost in this mutant during preparation of the ribosomes for binding assays, as opposed to *in vitro* protein synthesis experiments.

In another article Conde et al. (54) showed that a streptomycin-resistant mutant (*sr-u-2-23*) behaved in many respects as *spr-u-1-27,* but several other mutants had highly resistant chloroplast ribosomes by both *in vitro* and *in vivo* criteria (Fig. 21–5). They also attempted to measure mitochondrial protein synthesis indirectly using cytochrome oxidase and cyanide-sensitive respiration as indicators. The assumption was that some of the cytochrome oxidase polypeptides are made on mitochondrial ribosomes in *Chlamydomonas,* as they are in yeast and *Neurospora* (see Chapter 19). All mutants, including *spr-u-1-27* and *sr-u-2-23,* had normal levels of cytochrome oxidase during long-term

growth in the presence of antibiotics. Conde and associates concluded that in each mutant proteins shared between chloroplast and mitochondrial ribosomes were altered, and these were coded by chloroplast genes, since each mutant was uniparentally inherited. In some cases both classes of organelle ribosomes became highly resistant to antibiotics; in other cases only the mitochondrial ribosomes did so.

Behn and Arnold (66–69) considered both the chloroplast and the mitochondrion in discussing the mechanism of antibiotic dependence in *C. reinhardtii.* They argued that a uniparentally inherited mutation to streptomycin dependence (*sd-u-3-18*) was located in the mitochondrion, whereas a second mutation dependent on neamine (*nd-u*) was located in the chloroplast. Initially these arguments were based on the pattern of reversion seen in the two mutants (see Chapter 13). In any event, Behn and Arnold attempted to obtain further proof for their hypothesis through ultrastructural studies. The idea was to compare chloroplast and mitochondrial ultrastructure in wild-type cells and cells of the two mutants in the presence and absence of the two antibiotics. Behn and Arnold argued that the neamine-dependent mutant would have a dependent chloroplast, since their genetic data led them to localize the mutant in the chloroplast. But the mitochondria would be either dependent or resistant; they could not be neamine-sensitive, for then cell lethality would result. They postulated that the mitochondria must be resistant rather than dependent, since neamine-sensitive revertants were obtained. This would not be expected if both organelles were antibiotic-dependent. The opposite result was predicted for the *sd* mutant and its revertants. That is, the mitochondrial *sd* mutant reverted to sensitivity, but the chloroplast ribosomes remained antibiotic-resistant. In summary, the chloroplasts (but not the mitochondria) of the *nd* mutant and its sensitive revertants should respond differently to the presence and absence of neamine. In the *sd* mutant and its revertants, only the mitochondria should vary in their response to streptomycin.

Behn and Arnold reported that 48-hr treatment of wild-type cells with either neamine or streptomycin affected the ultrastructures of both chloroplasts and mitochondria. When neamine was withdrawn from a culture of the *nd* mutant, chloroplast damage appeared within 2 days, but the mitochondria looked normal. The reverse was true of the *sd* mutant. Finally, Behn and Arnold examined a neamine-resistant revertant derived from a dependent strain. They found no ultrastructural anomalies, and they concluded that the mitochondria remained antibiotic-resistant but that the neamine-dependent chloroplast mutated to resistance as they had predicted. The Behn and Arnold hypothesis is susceptible to one major criticism with respect to the *sd* mutant. Gillham et al. (70) reported that the *sd-u-3-18* mutant has subnormal amounts of chloroplast ribosomes and accumulates large subunits, as does an assembly-defective mutant. Streptomycin withdrawal accentuates the problem. These observations are consistent with the notion that the chloroplast of this mutant contains streptomycin-dependent ribosomes and that assembly of these ribosomes does not take place properly in the absence of streptomycin. In addition, it can be argued that the reversion data for the *sd-u-3-18* mutant are also consistent with a chloroplast location (see Chapter 13); Sager's *sd* mutant, which has been mapped (see Chapter 14), seems to be in the same linkage group as other chloroplast genes. It is also not clear why the two dependent mutants should be double mutants, as it would appear they must be if one organelle is antibiotic-dependent and the other is antibiotic-resistant, whereas both organelles are antibiotic-sensitive in wild-type cells.

So much for the hypothesis that one-step mutations to antibiotic resistance in *C. rein-hardtii* confer resistance on the chloroplast and mitochondrion or on the latter organelle alone. The other major hypothesis is that of Blamire et al. (71). These authors examined the effects of low concentrations of antibiotics such as neamine, streptomycin, and spectinomycin on incorporation of labeled adenine into clDNA and nDNA of wild-type cells of *C. reinhardtii*. They found that adenine incorporation into nDNA (but not clDNA) was inhibited by the antibiotics. They also showed by means of a density transfer experiment that nDNA replication, rather than repair, was being inhibited. Inhibition was examined over a wide concentration range. For all antibiotics except chloramphenicol, inhibition occurred rapidly and at low concentrations similar to those lethal to wild-type cells. It is interesting that wild-type cells can be grown at rather high chloramphenicol concentrations even though chloroplast protein synthesis appears to be inhibited by the available *in vivo* criteria (72).

Blamire et al. (71) postulated that a chloroplast gene product synthesized on chloroplast ribosomes was required for nDNA replication in *C. reinhardtii*. Antibiotics such as streptomycin, spectinomycin, etc., blocked synthesis of this product and thus blocked synthesis of nDNA. Obviously, one-step mutations to antibiotic resistance would be able to make the regulatory product in the presence of antibiotic, and dependent mutants would not be able to do so in the absence of the antibiotic, which they require as a cofactor. This attractively simple hypothesis assumes that chloroplasts and not mitochondria are primary targets for the action of antibacterial antibiotics, as seems to be the case in *Euglena* (see Chapter 16). The important difference between the two organisms would then be that the chloroplast of *Euglena* does not synthesize a product that regulates nDNA synthesis. If it did, bleached cells would never be found. The hypothesis of Blamire and associates predicts that a bleached mutant of

C. reinhardtii that lacks clDNA can never be found because it will be lethal. So far, this has been the case. The hypothesis also predicts that chloroplast protein synthesis is indispensable under all growth conditions. Therefore, nonleaky chloroplast ribosome assembly mutants should never be found. So far, all assembly-defective mutants reported are, in fact, leaky. However, the hypothesis of Blamire and associates does not account satisfactorily for the experiments of Surzycki with rifampicin or the behavior of the *spr-u-1-27* mutant. In both cases cells can be grown on acetate under conditions such that chloroplast protein synthesis appears to be blocked effectively *in vivo*. However, if the regulatory product is required in very small amounts and continues to be synthesized preferentially under conditions such that chloroplast protein synthesis is severely stressed, these results can be accounted for.

For the moment it is not possible to choose between the two hypotheses on the basis of experiment. The shared-component hypothesis is cumbersome and has at least two unattractive features. The first is that mRNA export from the chloroplast is required to explain how mutants such as *spr-u-1-27* and *sr-u-2-23* are able to grow on acetate in the presence of antibiotic with antibiotic-sensitive chloroplast ribosomes. By definition, the shared proteins cannot be made on these ribosomes under such growth conditions; thus the mRNA for the shared components must be exported from the chloroplast and translated elsewhere. Second, *C. reinhardtii* is sensitive to a wide variety of aminoglycoside antibiotics (i.e., neamine, kanamycin, spectinomycin, and streptomycin) that have not been reported to be toxic to growth or to act as mitochondrial protein synthesis inhibitors *in vivo* in yeast, *Neurospora, Euglena,* or *Paramecium.* Although the latter question needs more detailed study, the results thus far suggest, on the shared-component hypothesis, that the spectrum of mitochondrial antibiotic

sensitivity *in vivo* would have to be much broader in *C. reinhardtii* than has proved to be the case in other organisms. The hypothesis that a regulatory product required for nDNA replication is made on chloroplast ribosomes is simple and direct. However, it does not explain all the previously mentioned results supporting the hypothesis that chloroplast protein synthesis is largely if not entirely dispensable in *C. reinhardtii*. A key question to which there is presently no direct answer at all is whether mitochondrial protein synthesis in *C. reinhardtii* is sensitive *in vivo* to antibacterial antibiotics.

CONCLUSIONS

Our knowledge of where the components of organelle protein synthesis are coded is still limited largely to organelle rRNA and tRNA. These RNAs are partially, if not entirely, coded by the chloroplast and mitochondrial genomes. We know much less about the proteins. Existing evidence indicates that most, but not all, are likely to be nuclear gene products whose messages are translated in the cytoplasm. Two interesting questions arise from these observations: Which proteins of the organelle protein-synthesizing system are coded by organelle DNA, and why? How is the organelle protein-synthesizing system assembled into a functioning unit when it requires the participation of nuclear and organelle genomes and cytoplasmic and organelle protein-synthesizing systems? There must be an intricate regulatory apparatus that we can presently only begin to guess at.

REFERENCES

1. Barnett, W. E., and Brown, D. H. (1967): Mitochondrial transfer ribonucleic acids. *Proc. Natl. Acad. Sci. U.S.A.,* 57:452–458.
2. Barnett, W. E., Brown, D. H., and Epler, J. L. (1967): Mitochondrial-specific aminoacyl-RNA synthetases. *Proc. Natl. Acad. Sci. U.S.A.,* 57:1775–1781.
3. Epler, J. L. (1969): The mitochondrial and cytoplasmic transfer nucleic acids of *Neurospora crassa. Biochemistry,* 8:2285–2290.

4. Epler, J. L., Shugart, L. R., and Barnett, W. E. (1970): *N*-formylmethionyl transfer ribonucleic acid in mitochondria from *Neurospora. Biochemistry,* 9:3575–3579.
5. Suyama, Y., and Hamada, J. (1976): Imported tRNA: Its synthetase as a probable transport protein. In: *Genetics and Biogenesis of Chloroplasts and Mitochondria,* edited by T. Bücher, W. Neupert, W. Sebald, and S. Werner, pp. 763–770. North Holland, Amsterdam.
6. Chiu, A. O. S., and Suyama, Y. (1973): Immunologic studies on intracellular isoenzymes; the mitochondrial leucyl-tRNA synthetases. *Biochim. Biophys. Acta,* 299:557–563.
7. Chiu, A. O. S., and Suyama, Y. (1975): The absence of structural relationship between mitochondrial and cytoplasmic leucyl-tRNA synthetases from *Tetrahymena pyriformis. Arch. Biochem. Biophys.,* 171:43–54.
8. Schneller, J. M., Schneller, C., and Stahl, A. J. C. (1976): Immunological study of yeast mitochondrial phenylalanyl-tRNA synthetase. In: *Genetics and Biogenesis of Chloroplasts and Mitochondria,* edited by T. Bücher, W. Neupert, W. Sebald, and S. Werner, pp. 775–778. North Holland, Amsterdam.
9. Schneller, J. M., Schneller, C., Martin, R., and Stahl, A. J. C. (1976): Nuclear origin of specific yeast mitochondrial aminoacyl-tRNA synthetases. *Nucleic Acids Res.,* 3:1151–1165.
10. Gross, S. R., McCoy, M. T., and Gilmore, E. B. (1968): Evidence for the involvement of a nuclear gene in the production of the mitochondrial leucyl-tRNA synthetase of *Neurospora. Proc. Natl. Acad. Sci. U.S.A.,* 61:253–260.
11. Weeks, C. O., and Gross, S. R. (1971): Mutation and "reversion" at the *leu-5* locus of *Neurospora* and its effect on the cytoplasmic and mitochondrial leucyl-tRNA synthetases. *Biochem. Genet.,* 5:505–516.
12. Beauchamp, P. M., Horn, E. W., and Gross, S. R. (1977): Proposed involvement of an internal promoter in regulation and synthesis of mitochondrial and cytoplasmic leucyl-tRNA synthetases of *Neurospora. Proc. Natl. Acad. Sci. U.S.A.,* 74:1172–1176.
13. Beauchamp, P. M., and Gross, S. R. (1976): Increased mitochondrial leucyl- and phenylalanyl-tRNA synthetase activity as a result of inhibition of mitochondrial protein synthesis. *Nature,* 261:338–340.
14. Tewari, K. K., and Wildman, S. G. (1970): Information content in the chloroplast DNA. *Symp. Soc. Exp. Biol.,* 24:147–179.
15. Schwartzbach, S. D., Hecker, L. I., and Barnett, W. E. (1976): Transcriptional origin of *Euglena* chloroplast tRNAs. *Proc. Natl. Acad. Sci. U.S.A.,* 73:1984–1988.
16. Barnett, W. E., Schwartzbach, S. D., and Hecker, L. I. (1976): The tRNAs and aminoacyl-tRNA synthetases of *Euglena* chloroplasts. In: *Genetics and Biochemistry of*

Chloroplasts and Mitochondria, edited by T. Bücher, W. Neupert, W. Sebald, and S. Werner, pp. 661–666. North Holland, Amsterdam.
17. Weil, J. H., Burkard, G., Guillemaut, P., Jeannin, G., Martin, R., and Steinmetz, A. (1976): tRNAs and aminoacyl-tRNA synthetases in plant organelles. In: *Genetics and Biogenesis of Chloroplasts and Mitochondria,* edited by T. Bücher, W. Neupert, W. Sebald, and S. Werner, pp. 667–675. North Holland, Amsterdam.
18. Barnett, W. E., Pennington, C. J., Jr., and Fairfield, S. A. (1969): Induction of *Euglena* transfer RNAs by light. *Proc. Natl. Acad. Sci. U.S.A.,* 63:1261–1268.
19. Reger, B. J., Fairfield, S. A., Epler, J. L., and Barnett, W. E. (1970): Identification and origin of some chloroplast aminoacyl-tRNA synthetases and tRNAs. *Proc. Natl. Acad. Sci. U.S.A.,* 67:1207–1213.
20. Hecker, L. I., Egan, J., Reynolds, R. J., Nix, C. E., Schiff, J. A., and Barnett, W. E. (1974): The sites of transcription and translation for *Euglena* chloroplastic aminoacyl-tRNA synthetases. *Proc. Natl. Acad. Sci. U.S.A.,* 71:1910–1914.
21. Parthier, B. (1973): Cytoplasmic site of synthesis of chloroplast aminoacyl-tRNA synthetases in *Euglena gracilis. FEBS Lett.,* 38: 70–74.
22. Goins, D. J., Reynolds, R. J., Schiff, J. A., and Barnett, W. E. (1973): A cytoplasmic regulatory mutant of *Euglena:* Constitutivity for the light-inducible chloroplast transfer RNAs. *Proc. Natl. Acad. Sci. U.S.A.,* 70: 1749–1752.
23. Guillemaut, P., Steinmetz, A., Burkard, G., and Weil, J. H. (1975): Aminoacylation of tRNA^{leu} species from *Escherichia coli* and from cytoplasm, chloroplasts and mitochondria of *Phaseolus vulgaris* by homologous and heterologous enzymes. *Biochim. Biophys. Acta,* 378:64–72.
24. Guillemaut, P., Burkard, G., and Weil, J. H. (1972): Characterization of *N*-formyl-methionyl-tRNA in bean mitochondria and etioplasts. *Biochemistry,* 11:2217–2219.
25. Guillemaut, P., Burkard, G., Steinmetz, A., and Weil, J. H. (1973): Comparative studies on the tRNAs^{met} from the cytoplasm, chloroplasts and mitochondria of *Phaseolus vulgaris. Plant Sci. Lett.,* 1:141–149.
26. Guillemaut, P., and Weil, J. H. (1975): Aminoacylation of *Phaseolus vulgaris* cytoplasmic, chloroplastic and mitochondrial tRNAs^{met} and of *Escherichia coli* tRNAs^{met} by homologous and heterologous enzymes. *Biochim. Biophys. Acta,* 407:240–248.
27. Küntzel, H. (1969): Proteins of mitochondrial and cytoplasmic ribosomes from *Neurospora crassa. Nature,* 222:142–146.
28. Neupert, W., Sebald, W., Schwab, A. J., Massinger, P., and Bücher, T. (1969): In-

corporation *in vivo* of ^{14}C-labelled amino acids into the proteins of mitochondrial ribosomes from *Neurospora crassa* sensitive to cycloheximide and insensitive to chloramphenicol. *Eur. J. Biochem.,* 10:589–591.

29. Lizardi, P. M., and Luck, D. J. L. (1972): The intracellular site of synthesis of mitochondrial ribosomal proteins in *Neurospora crassa. J. Cell Biol.,* 54:56–74.
30. Lambowitz, A. M., Chua, N.-H., and Luck, D. J. L. (1976): Mitochondrial ribosome assembly in *Neurospora.* Preparation of mitochondrial ribosomal precursor particles, site of synthesis of mitochondrial ribosomal proteins and studies of the *poky* mutant. *J. Mol. Biol.,* 107:223–253.
31. Lambowitz, A. M. (1976): The *poky* mutant of *Neurospora crassa.* In: *Genetics and Biogenesis of Chloroplasts and Mitochondria,* edited by T. Bücher, W. Neupert, W. Sebald, and S. Werner, pp. 713–720. North Holland, Amsterdam.
32. Groot, G. S. P. (1974): The biosynthesis of mitochondrial ribosomes in *Saccharomyces cerevisiae.* In: *The Biogenesis of Mitochondria,* edited by A. M. Kroon and C. Saccone, pp. 443–452. Academic Press, New York.
33. Kientsch, R., and Werner, S. (1976): Cold sensitivity of mitochondrial biogenesis in a nuclear mutant of *Neurospora crassa.* In: *Genetics and Biogenesis of Chloroplasts and Mitochondria,* edited by T. Bücher, W. Neupert, W. Sebald, and S. Werner, pp. 247–252. North Holland, Amsterdam.
34. Lambowitz, A. M., and Slayman, C. W. (1971): Cyanide-resistant respiration in *Neurospora crassa. J. Bacteriol.,* 108:1087–1096.
35. Lambowitz, A. M., Smith, E. W., and Slayman, C. W. (1972): Oxidative phosphorylation in *Neurospora* mitochondria: Studies on wild type, *poky* and chloramphenicol-induced wild type. *J. Biol. Chem.,* 247:4859–4865.
36. Edwards, D. L., Rosenberg, E., and Maroney, P. A. (1974): Induction of cyanide-insensitive respiration in *Neurospora crassa. J. Biol. Chem.,* 249:3551–3556.
37. Honeycutt, R. C., and Margulies, M. M. (1973): Protein synthesis in *Chlamydomonas reinhardi:* Evidence for synthesis of chloroplastic ribosomes on cytoplasmic ribosomes. *J. Biol. Chem.,* 248:6145–6153.
38. Ellis, R. J., and Hartley, M. R. (1971): Sites of synthesis of chloroplast proteins. *Nature* [*New Biol.*], 233:193–196.
39. Bourque, D. P., Boynton, J. E., and Gillham, N. W. (1971): Studies on the structure and cellular location of various ribosome and ribosomal RNA species in the green alga *Chlamydomonas reinhardi. J. Cell Sci.,* 8:153–183.
40. Bastia, D., Chiang, K.-S., and Swift, H. (1971): Studies on the ribosomal RNA cistrons of chloroplast and nucleus in *Chlamy-*

domonas reinhardi. In: *Abstracts Eleventh Annual Meeting of Cell Biology,* p. 25.
41. Surzycki, S. J., and Rochaix, J. D. (1971): Transcriptional mapping of ribosomal RNA genes of the chloroplast and nucleus of *Chlamydomonas reinhardi. J. Mol. Biol.,* 62: 89–109.
42. Lambowitz, A. M., Merril, C. R., Wurtz, E. A., Boynton, J. E., and Gillham, N. W. (1976): Restriction enzyme analysis of chloroplast DNA from *Chlamydomonas reinhardtii. J. Cell Biol.,* 70:217a.
43. Howell, S., Heizmann, P., and Gelvin, S. (1976): Localization of the gene coding for the large subunit of ribulose bisphosphate carboxylase on the chloroplast genome of *Chlamydomonas reinhardi.* In: *Genetics and Biogenesis of Chloroplasts and Mitochondria,* edited by T. Bücher, W. Neupert, W. Sebald, and S. Werner, pp. 625–628. North Holland, Amsterdam.
44. Harris, E. W., Boynton, J. E., and Gillham, N. W. (1976): Genetics of chloroplast ribosome biogenesis in *Chlamydomonas reinhardtii.* In: *The Genetics of Algae,* edited by R. A. Lewin, Chapter 6. Blackwell Scientific, Oxford.
45. Gillham, N. W., Boynton, J. E., Harris, E. H., Fox, S. B., and Bolen, P. L. (1976): Genetic control of chloroplast ribosome biogenesis in *Chlamydomonas.* In: *Genetics and Biogenesis of Chloroplasts and Mitochondria,* edited by T. Bücher, W. Neupert, W. Sebald, and S. Werner, pp. 69–76. North Holland, Amsterdam.
46. Harris, E. H., Boynton, J. E., and Gillham, N. W. (1974): Chloroplast ribosome biogenesis in *Chlamydomonas:* Selection and characterization of mutants blocked in ribosome formation. *J. Cell Biol.,* 63:160–179.
47. Harris, E. H., Boynton, J. E., and Gillham, N. W. (1977): unpublished data.
48. Goodenough, U. W., and Levine, R. P. (1970): Chloroplast structure and function in *ac-20,* a mutant strain of *Chlamydomonas reinhardi* lacking components of the photosynthetic apparatus. *Plant Physiol.,* 44:990–1000.
49. Boynton, J. E., Gillham, N. W., and Chabot, J. F. (1972): Chloroplast ribosome deficient mutants in the green alga *Chlamydomonas reinhardi* and the question of chloroplast ribosome function. *J. Cell Sci.,* 10:267–305.
50. Jaskunas, S. R., Nomura, M., and Davies, J. (1974): Genetics of bacterial ribosomes. In: *Ribosomes,* edited by M. Nomura, A. Tissieres, and P. Lengyel, pp. 333–368. Cold Spring Harbor Laboratory, Cold Spring Harbor, N.Y.
51. Davidson, J. N., Hanson, M. R., and Bogorad, L. (1974): An altered chloroplast ribosomal protein in *ery-M1* mutants of *Chlamydomonas reinhardi. Mol. Gen Genet.,* 132: 119–129.

52. Mets, L., and Bogorad, L. (1972): Altered chloroplast ribosomal proteins associated with erythromycin-resistant mutants in two genetic systems of *Chlamydomonas reinhardi. Proc. Natl. Acad. Sci. U.S.A.,* 69:3779–3783.

53. Schlanger, G., and Sager, R. (1974): Localization of five antibiotic resistances at the subunit level in chloroplast ribosomes of *Chlamydomonas. Proc. Natl. Acad. Sci. U.S.A.* 71:1715–1719.

54. Conde, M. F., Boynton, J. E., Gillham, N. W., Harris, E. H., Tingle, C. L., and Wang, W. L. (1975): Chloroplast genes in *Chlamydomonas* affecting organelle ribosomes: Genetic and biochemical analysis of antibiotic-resistant mutants at several gene loci. *Mol. Gen. Genet.,* 140:183–220.

55. Harris, E. H., Boynton, J. E., Gillham, N. W., Tingle, C. L., and Fox, S. B. (1977): Mapping of chloroplast genes involved in chloroplast ribosome biogenesis in *Chlamydomonas reinhardtii. Mol. Gen. Genet.,* 155:249–265.

56. Fox, S. B., Grabowy, C. T., Harris, E. H., Gillham, N. W., and Boynton, J. E. (1977): Genetic and biochemical analysis of the ribosomal region of the chloroplast genome of *Chlamydomonas reinhardtii. J. Cell Biol.,* 75:307a.

57. Sager, R. (1972): *Cytoplasmic Genes and Organelles.* Academic Press, New York.

58. Adams, G. M. W., Van Winkle-Swift, K. P., Gillham, N. W., and Boynton, J. E. (1976): Plastid inheritance in *Chlamydomonas reinhardtii.* In: *The Genetics of Algae,* edited by R. A. Lewin, Chapter 6. Blackwell Scientific, Oxford.

59. Boynton, J. E., Burton, W. G., Gillham, N. W., and Harris, E. H. (1973): Can a non-Mendelian mutation affect both chloroplast and mitochondrial ribosomes? *Proc. Natl. Acad. Sci. U.S.A.,* 70:3463–3467.

60. Ohta, N., Sager, R., and Inouye, M. (1975): Identification of a chloroplast ribosomal protein altered by a chloroplast mutation in *Chlamydomonas. J. Biol. Chem.,* 250:3655–3659.

61. Brügger, M., and Boschetti, A. (1975): Two-dimensional gel electrophoresis of ribosomal proteins from streptomycin-sensitive and streptomycin-resistant mutants of *Chlamydomonas reinhardi. Eur. J. Biochem.,* 58:603–610.

62. Surzycki, S. J. (1969): Genetic functions of the chloroplast of *Chlamydomonas reinhardi:* Effect of rifampin on chloroplast DNA-dependent RNA polymerase. *Proc. Natl. Acad. Sci. U.S.A.,* 63:1327–1334.

63. Surzycki, S. J., Goodenough, U. W., Levine, R. P., and Armstrong, J. J. (1970): Nuclear and chloroplast control of chloroplast structure and function in *Chlamydomonas reinhardi. Symp. Soc. Exp. Biol.,* 24:13–37.

64. Goodenough, U. W., Togasaki, R. K., Paszew-ski, A., and Levine, R. P. (1971): Inhibition of chloroplast ribosome formation by gene mutation in *Chlamydomonas reinhardi.* In: *Autonomy and Biogenesis of Mitochondria and Chloroplasts,* edited by N. K. Boardman, A. W. Linnane, and R. M. Smillie, pp. 224–234. North Holland, Amsterdam.

65. Surzycki, S. J., and Gillham, N. W. (1971): Organelle mutations and their expression in *Chlamydomonas reinhardi. Proc. Natl. Acad. Sci. U.S.A.,* 68:1301–1306.

66. Behn, W., and Arnold, C. G. (1972): Zur Lokalisation eines nichtmendelnden Gens von *Chlamydomonas reinhardii. Mol. Gen. Genet.,* 114:266–272.

67. Behn, W., and Arnold, C. G. (1973): Localization of extranuclear genes by investigations of the ultrastructure of *Chlamydomonas reinhardi. Arch. Microbiol.,* 92:85–90.

68. Behn, W., and Arnold, C. G. (1974): Die Wirkung von Streptomycin und Neamin auf die chloroplasten- und mitochondrien Struktur von *Chlamydomonas reinhardii. Protoplasma,* 82:77–89.

69. Behn, W., and Arnold, C. G. (1974): Unterscheidliche genetische Konstitutionen von Chloroplast und Mitochondrien bei antibiotikaabhängigen Mutanten von *Chlamydomonas reinhardii. Protoplasma,* 82:91–101.

70. Gillham, N. W., Boynton, J. E., and Burkholder, B. (1970): Mutations altering chloroplast ribosome phenotype in *Chlamydomonas,* I. Non-Mendelian mutations. *Proc. Natl. Acad. Sci. U.S.A.,* 67:1026–1033.

71. Blamire, J., Flechtner, V. R., and Sager, R. (1974): Regulation of nuclear DNA replication by the chloroplast in *Chlamydomonas. Proc. Natl. Acad. Sci. U.S.A.,* 71:2867–2871.

72. Chua, N.-H., and Gillham, N. W. (1977): The sites of synthesis of the principal thylakoid membrane polypeptides in *Chlamydomonas reinhardtii. J. Cell Biol.,* 74:441–452.

73. Gillham, N. W., Boynton, J. E., and Chua, N.-H. (1978): Genetic control of chloroplast proteins. *Curr. Top. Bioenergetics,* 7, in press.

74. Bogorad, L., Davidson, J. N., and Hanson, M. R. (1976): Genes affecting erythromycin resistance and sensitivity in *Chlamydomonas reinhardi* chloroplast ribosomes. In: *Genetics and Biogenesis of Chloroplast and Mitochondria,* edited by T. Bücher, W. Neupert, W. Sebald, and S. Werner, pp. 61–67. North Holland, Amsterdam.

75. Hanson, M. R., and Bogorad, L. (1977): Complementation analysis at the *ery*-M1 locus in *Chlamydomonas reinhardi Mol. gen. Genet.,* 153:271–277.

76. Davidson, J. N., and Bogorad, L. (1977): Suppression of erythromycin resistance in *ery*-M1 mutants of *Chlamydomonas reinhardi. Mol. gen. Genet.,* 157:39–46.

Questions for the Future

The structure and function of both chloroplasts and mitochondria depend in large part on nuclear genes whose products are made in the cytoplasm. Even in those cases where the organelle genome and protein-synthesizing system are involved in organelle biogenesis, the effort is collaborative in nature. Thus in every sense the genetic apparatus of the organelle acts as the junior partner in construction of the organelle. The root question of organelle heredity is why this genetic division of labor between organelle and nucleus has been persisted during evolution. In order to answer this question, we must first understand what organelle genomes do and why organelle protein-synthesizing systems have been conserved. In the last 10 years there has been rapid growth in research in this area. This research has yielded a "taxonomy" of proteins whose messages are translated in chloroplasts and mitochondria, and it is beginning to yield a similar taxonomy of structural genes for organelle polypeptides that lie within the organelle genome. As we begin to understand better what organelle genomes do, it is hoped that we will be able to clarify matters about which we currently have only vague ideas. It seems fitting to conclude this book with a statement of those questions that currently appear to be most pressing: Are the structural proteins that are coded by the genomes of the chloroplasts and mitochondrion endowed with special physical properties that make it imperative that they be synthesized within the organelle? How do proteins get in and out of organelles? (It appears that a multiplicity of mechanisms may exist.) How do organelle and nuclear genomes signal to each other? What sorts of regulatory mechanisms are involved? Why is it that organelle genomes appear to be highly redundant in a physical sense but not a genetic sense? These are some of the questions we must address from a broadened base of knowledge in the decade to come.

Subject Index

Date Due